87 Springer Series in Solid-State Sciences

Edited by Klaus von Klitzing

Springer Series in Solid-State Sciences

Editors: M. Cardona P. Fulde K. von Klitzing H.-J. Queisser

Managing Editor: H. K. V. Lotsch

Volumes 1–49 are listed on the back inside cover

50 **Multiple Diffraction of X-Rays in Crystals**
By Shih-Lin Chang

51 **Phonon Scattering in Condensed Matter**
Editors: W. Eisenmenger, K. Laßmann, and S. Döttinger

52 **Superconductivity in Magnetic and Exotic Materials**
Editors: T. Matsubara and A. Kotani

53 **Two-Dimensional Systems, Heterostructures, and Superlattices**
Editors: G. Bauer, F. Kuchar, and H. Heinrich

54 **Magnetic Excitations and Fluctuations**
Editors: S. Lovesey, U. Balucani, F. Borsa, and V. Tognetti

55 **The Theory of Magnetism II**
Thermodynamics and Statistical Mechanics
By D. C. Mattis

56 **Spin Fluctuations in Itinerant Electron Magnetism** By T. Moriya

57 **Polycrystalline Semiconductors,**
Physical Properties and Applications
Editor: G. Harbeke

58 **The Recursion Method and Its Applications**
Editors: D. Pettifor and D. Weaire

59 **Dynamical Processes and Ordering on Solid Surfaces**
Editors: A. Yoshimori and M. Tsukada

60 **Excitonic Processes in Solids**
By M. Ueta, H. Kanzaki, K. Kobayashi, Y. Toyozawa, and E. Hanamura

61 **Localization, Interaction, and Transport Phenomena**
Editors: B. Kramer, G. Bergmann, and Y. Bruynseraede

62 **Theory of Heavy Fermions and Valence Fluctuations**
Editors: T. Kasuya and T. Saso

63 **Electronic Properties of Polymers and Related Compounds**
Editors: H. Kuzmany, M. Mehring, and S. Roth

64 **Symmetries in Physics** Group Theory Applied to Physical Problems
By W. Ludwig and C. Falter

65 **Phonons: Theory and Experiments II**
Experiments and Interpretation of Experimental Results By P. Brüesch

66 **Phonons: Theory and Experiments III**
Phenomena Related to Phonons
By P. Brüesch

67 **Two-Dimensional Systems: Physics and New Devices**
Editors: G. Bauer, F. Kuchar, and H. Heinrich

68 **Phonon Scattering in Condensed Matter V**
Editors: A. C. Anderson and J. P. Wolfe

69 **Nonlinearity in Condensed Matter**
Editors: A. R. Bishop, D. K. Campbell, P. Kumar, and S. E. Trullinger

70 **From Hamiltonians to Phase Diagrams**
The Electronic and Statistical-Mechanical Theory of sp-Bonded Metals and Alloys
By J. Hafner

71 **High Magnetic Fields in Semiconductor Physics**
Editor: G. Landwehr

72 **One-Dimensional Conductors**
By S. Kagoshima, H. Nagasawa, and T. Sambongi

73 **Quantum Solid-State Physics**
Editors: S. V. Vonsovsky and M. I. Katsnelson

74 **Quantum Monte Carlo Methods** in Equilibrium and Nonequilibrium Systems
Editor: M. Suzuki

75 **Electronic Structure and Optical Properties of Semiconductors**
By M. L. Cohen and J. R. Chelikowsky

76 **Electronic Properties of Conjugated Polymers**
Editors: H. Kuzmany, M. Mehring, and S. Roth

77 **Fermi Surface Effects**
Editors: J. Kondo and A. Yoshimori

78 **Group Theory and Its Applications in Physics**
By T. Inui, Y. Tanabe, and Y. Onodera

79 **Elementary Excitations in Quantum Fluids**
Editors: K. Ohbayashi and M. Watabe

80 **Monte Carlo Simulation in Statistical Physics**
An Introduction
By K. Binder and D. W. Heermann

81 **Core-Level Spectroscopy in Condensed Systems**
Editors: J. Kanamori and A. Kotani

82 **Introduction to Photoemission Spectroscopy**
By S. Hüfner

83 **Physics and Technology of Submicron Structures**
Editors: H. Heinrich, G. Bauer, and F. Kuchar

84 **Beyond the Crystalline State**
An Emerging Perspective
By G. Venkataraman, D. Sahoo, and V. Balakrishnan

85 **The Fractional Quantum Hall Effect**
Properties of an Incompressible Quantum Fluid
By T. Chakraborty and P. Pietiläinen

86 **The Quantum Statistics of Dynamic Processes**
By E. Fick and G. Sauermann

87 **High Magnetic Fields in Semiconductor Physics II**
Transport and Optics Editor: G. Landwehr

High Magnetic Fields in Semiconductor Physics II

Transport and Optics

Proceedings of the International Conference,
Würzburg, Fed. Rep. of Germany,
August 22–26, 1988

Editor: G. Landwehr

With 441 Figures

Springer-Verlag Berlin Heidelberg New York
London Paris Tokyo Hong Kong

Professor Dr. Gottfried Landwehr

Physikalisches Institut, Universität Würzburg, Am Hubland
D-8700 Würzburg, Fed. Rep. of Germany

Series Editors:

Professor Dr., Dres. h. c. Manuel Cardona
Professor Dr., Dr. h. c. Peter Fulde
Professor Dr., Dr. h. c. Klaus von Klitzing
Professor Dr. Hans-Joachim Queisser

Max-Planck-Institut für Festkörperforschung, Heisenbergstrasse 1
D-7000 Stuttgart 80, Fed. Rep. of Germany

Managing Editor:

Dr. Helmut K. V. Lotsch

Springer-Verlag, Tiergartenstrasse 17
D-6900 Heidelberg, Fed. Rep. of Germany

ISBN-13:978-3-642-83812-5 e-ISBN-13:978-3-642-83810-1
DOI: 10.1007/978-3-642-83810-1

2154/3150-543210 – Printed on acid-free paper

Preface

This volume contains contributions presented at the International Conference "The Application of High Magnetic Fields in Semiconductor Physics", which was held at the University of Würzburg from August 22 to 26, 1988.

In the tradition of previous Würzburg meetings on the subject – the first conference was held in 1972 – only invited papers were presented orally. All 42 lecturers were asked to review their subject to some extent so that this book gives a good overview of the present state of the respective topic. A look at the contents shows that the subjects which have been treated at previous conferences have not lost their relevance. On the contrary, the application of high magnetic fields to semiconductors has grown substantially during the recent past. For the elucidation of the electronic band structure of semiconductors high magnetic fields are still an indispensable tool. The investigation of two-dimensional electronic systems especially is frequently connected with the use of high magnetic fields. The reason for this is that a high B-field adds angular momentum quantization to the boundary quantization present in heterostructures and superlattices. A glance at the contributions shows that the majority deal with 2D properties. Special emphasis was on the integral and fractional quantum Hall effect. Very recent results related to the observation of a fraction with an even denominator were presented. It became obvious that the polarization of the different fractional Landau levels is more complicated than originally anticipated.

The volume contains 56 contributions which were presented as posters. Altogether, 79 posters were shown; unfortunately it was not possible to include all of them in the book. Because the deadlines were rather close to the date of the conference, many contributions contain very new results.

I am convinced that the present book is not only of interest to scientists who are active in the field, it should also provide a good introduction to a rapidly developing area of research for newcomers.

The organizing committee consisted of G. Landwehr (Chairman), J. Hajdu, K. von Klitzing and W. Ossau.

The financial support of the conference by the following sponsors is gratefully acknowledged:

Deutsche Forschungsgemeinschaft

Bayerisches Staatsministerium für Wissenschaft und Kunst

Regionalverband Bayern der Deutschen Physikalischen Gesellschaft

Industrial sponsors: Siemens AG, München; IBM Stuttgart; Bruker Analytische Meßtechnische Meßtechnik, Karlsruhe; Oxford Instruments, Wiesbaden; Cryophysics/Odelga Physics, Nürtingen.

Würzburg, November 1988 *G. Landwehr*

Contents

Part III Heterostructures and Superlattices: Transport and Electronic Structure

Part IV Heterostructures and Superlattices: Optics

Part VIII **Semimagnetic Semiconductors, 2D and 3D**

Part IX Electron-Phonon Interaction, Magneto-phonon Effect

Part XII Reports from High Magnetic Field Laboratories

Part I

Integral Quantum Hall Effect, Electronic States in High Magnetic Fields

Universality and Scaling of Electronic Transport in the Integral Quantum Hall Effect

H.P. Wei[1], *D.C. Tsui*[1], *M.A. Paalanen*[2], *and A.M.M. Pruisken*[3]

[1]Department of Electrical Engineering, Princeton University, Princeton, NJ 08544, USA
[2]AT&T Bell Laboratory, Murray Hill, NJ 07974, USA
[3]Pupin Physics Laboratory, Columbia University, New York, NY 10027, USA

Quantitative experimental results are obtained on the electronic transport in the integral quantum Hall effect in InGaAs/InP heterojunctions. Both the maximum of $d\rho_{xy}/dB$ and the inverse of the half width for ρ_{xx} diverge like $\sim T^{-\kappa}$. We obtain $\kappa = 0.42 \pm 0.04$, independent of the Landau level index. These results confirm the prediction of the scaling theory that the characteristic power law behavior in the transport coefficients is a universal feature of delocalization in the integral quantum Hall effect.

Low temperature (T) experiments on the integral quantum Hall effect (IQHE) show large regions in the magnetic field (B) where the Hall resistance ρ_{xy} is accurately quantized according to h/ie^2 (where i is an integer, h the Planck constant and e the electron charge). The existence of very sharp steps connecting the successive quantum Hall plateaus is an indication that there is a very narrow range in energy in each Landau level, where the electronic states are different from those at other energies [1]. More detailed experimental studies of this narrow transition region have been reported recently in references [2,3,4]. It was found that there is a characteristic temperature, T_{sc}, above which the longitudinal conductance σ_{xx} increases and below which σ_{xx} decreases with decreasing T. This decrease of σ_{xx} with decreasing T has been attributed to genuine scaling in the electronic transport [2], which is intimately related to the properties of delocalization of the electronic wave functions near the Fermi energy $E_{\rm F}$.

The present theoretical understanding of IQHE assumes a collection of two-dimensional non-interacting electrons in a random potential. The important result based on this assumption is that under high B, the electronic states in a Landau level are localized except for those at a singular energy $E_{\rm c}$ at the center of each level. The concept of delocalization (at $T = 0$) can be described by saying that the localization length ξ (i.e. the spatial extent of the electronic wave functions) diverges when $E_{\rm F} \to E_{\rm c}$ according to a power law $\xi \sim (E - E_{\rm c})^{-\nu}$ with a universal critical exponent ν [5]. Such a power law is generic in the scaling theory and it is intuitively clear that this singular behavior should be in a one-to-one correspondence with the sharpness of the Hall steps mea-

Springer Series in Solid-State Sciences, Vol. 87
High Magnetic Fields in Semiconductor Physics II Editor: G. Landwehr

sured at finite T. Experimentally, by sweeping B between two quantum plateaus, E_F is passing through the center E_c of a Landau level. Therefore, the electronic states at E_F undergo a transition from localized to delocalized states. It is thus obvious that the singular behavior of ξ near E_c should be reflected in some physical quantities when B is swept from one plateau to another. The physical quantities we studied are $d\rho_{xy}/dB$, the half width of ρ_{xx} and $d^2\rho_{xx}/dB^2$.

In Fig.1a and Fig.1b, we plot the ρ_{xy} and ρ_{xx} vs. B data taken from an InGaAs/InP heterostructure at $T = 4.2\,\mathrm{K}$, $1.3\,\mathrm{K}$ and $0.35\,\mathrm{K}$. The 2DEG have a density $n_{2D} = 3.3 \times 10^{11}\,\mathrm{cm}^{-2}$ and a mobility $\mu = 34,000\,\mathrm{cm}^2/\mathrm{Vs}$ at $T = 0.8\,\mathrm{K}$. The corresponding $d\rho_{xy}/dB$ vs. B is shown in Fig.1c. We notice that there is a qualitative similarity between $d\rho_{xy}/dB$ and ρ_{xx} at each T, (where ρ_{xx} has a minimum or maximum, $d\rho_{xy}/dB$ also has a minimum or maximum), as was pointed out by Chang and Tsui [6]. However, the T dependence of $(d\rho_{xy}/dB)^{max}$ in each

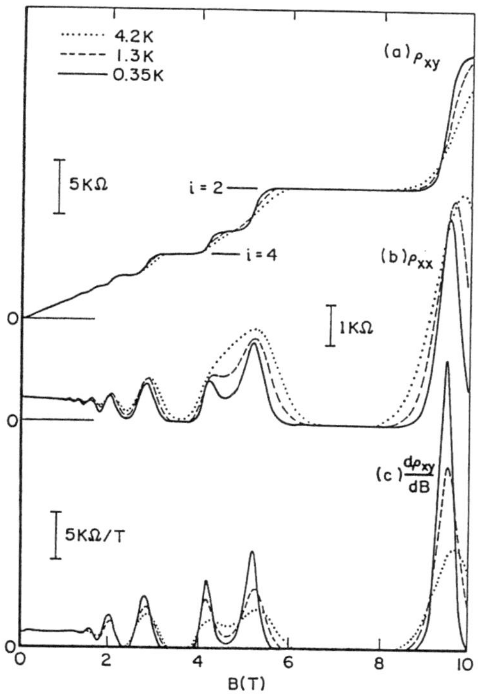

Figure 1: Quantum transport coefficients ρ_{xy} in (a) and ρ_{xx} in (b) as a function of B at three temperatures, $T = 4.2\,\mathrm{K}$, $1.3\,\mathrm{K}$ and $0.35\,\mathrm{K}$. (c) shows the corresponding $d\rho_{xy}/dB$. The sample is an $In_{0.57}Ga_{0.43}As/InP$ heterostructure with a two dimensional electron density $n_{2D} = 3.3 \times 10^{11}\,\mathrm{cm}^{-2}$ and mobility $\mu = 34,000\,\mathrm{cm}^2/\mathrm{Vs}$ at $T = 0.8\,\mathrm{K}$.

Landau level is opposite to that of the peak value of ρ_{xx}. While the ρ_{xx} peak decreases with decreasing T, $(d\rho_{xy}/dB)^{max}$ is clearly increasing with decreasing T albeit the position of $(d\rho_{xy}/dB)^{max}$ in B shifts to the lower B side by about 3% for T decreasing from 4.2 K to 0.35 K. Furthermore, we plot the corresponding $d\rho_{xx}/dB$ vs. B in Fig.2, and we define the width in ρ_{xx} to be $\Delta B \equiv B_2 - B_1$, where B_1 and B_2 are the positions in B that $d\rho_{xx}/dB$ has extrema. We notice that in each transition region between two plateaus the postion B_1 remains constant to within 1% in the whole T range. However, B_2 decreases monotonically with decreasing T, so does the width of ρ_{xx}.

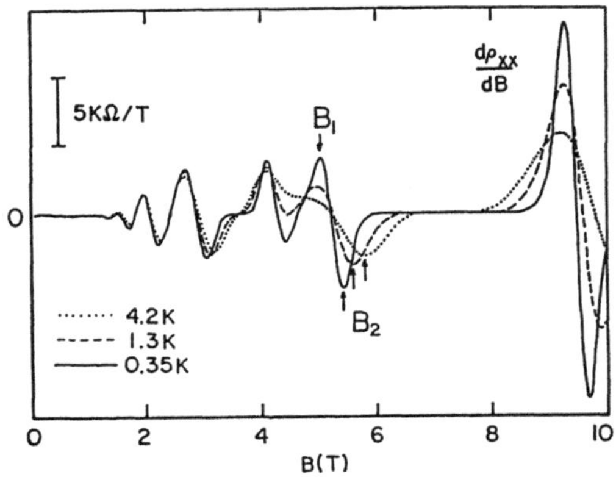

Figure 2: $d\rho_{xx}/dB$ vs. B at three different T's in Fig.1.

We plot the T dependence of $(d\rho_{xy}/dB)^{max}$ and $1/\Delta B$ in Fig.3 for Landau levels $N = 0\downarrow, 1\uparrow$ and $1\downarrow$ on a log-log scale in the T range from 4.2 K to 100 mK. The points drawn are data taken in a ^{3}He system and a dilution refrigerator. We could not obtain $1/\Delta B$ for the $N = 0\downarrow$ level because the value of B_2 falls outside the range of B accessible in the dilution refrigerator. Furthermore, ΔB for the $N = 1\downarrow$ level is not well defined for $T > 1$ K. Nevertheless, the data show strikingly the same linear relation. This indicates that $(d\rho_{xy}/dB)^{max}$ and ΔB have a power law dependence on T, i.e.

$$\left(\frac{d\rho_{xx}}{dB}\right)^{max} \sim T^{-\kappa} \tag{1}$$

and

$$(\Delta B) \sim T^{\kappa} \quad . \tag{2}$$

We obtain $\kappa = 0.42 \pm 0.04$ independent of the Landau level index.

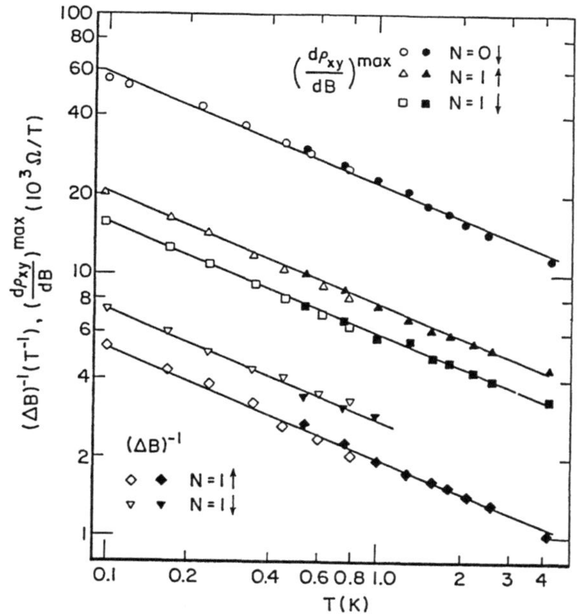

Figure 3: The upper portion shows the T dependence of $(d\rho_{xy}/dB)^{max}$ for Landau levels $N = 0\downarrow, 1\uparrow$ and $1\downarrow$; the lower portion shows the T dependence of $1/\Delta B$ for the $1\uparrow$ and $1\downarrow$ Landau levels. The slope of the straight lines is 0.42 ± 0.04.

The relation between the macroscopic exponent κ for the transport coefficients and the microscopic exponent ν for the localization length ξ has been derived theoretically [7] to be $\kappa = p/2\nu$, where p is the exponent for the inelastic scattering time ($\tau_{in} \sim T^{-p}$). However, it can also be deduced intuitively from the following simple argument. In Fig.4, we plot ξ as a function of E illustrating that the divergence of ξ at E_c follows $(E - E_c)^{-\nu}$. We can define the electronic states to be extended once ξ is larger than the sample size. Furthermore, we assume that at finite T the effective sample size is determined by the inelastic scattering length l_{in} [8], then the region of extended states at a given T is ΔE in Fig.4. Therefore, $\Delta E \sim (l_{in})^{-1/\nu} \sim T^{p/2\nu}$. Since our experimentally measured $\Delta B \sim \Delta E$, we then deduce that $\kappa = p/2\nu$. The absolute values for p and ν are presently unknown. Stressing the analogy with the theory of ordinary metallic conductors in two dimensions, we would expect $p \sim 1$ [8,9,10] and hence $\nu = 1.2 \pm 0.1$, considerably larger than the estimates ($\nu = 0.5$) of Refs [11]. In any case, our results do not agree with the numerical calculations by Aoki and Ando [12], who predict that ν is dependent upon Landau level index (ν=2 and

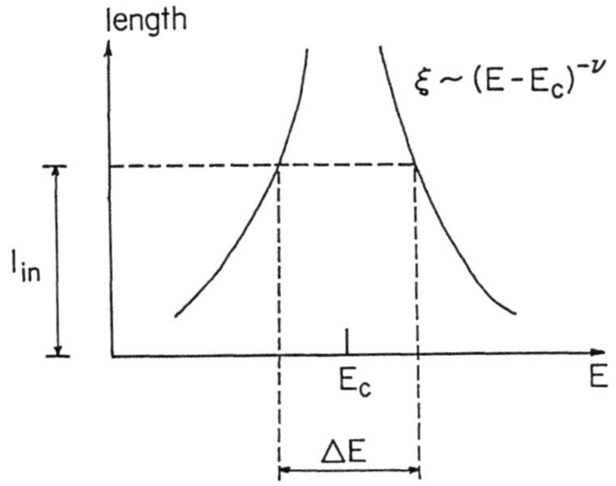

Figure 4: A schematic illustration of localization length ξ as a function of energy. E_c is at the center of a Landau level and l_{in} is the inelastic scattering length at a finite T.

4 for N=0 and 1 respectively). The self-consistent approach by Ono [13], on the other hand, gives a divergent localization length according to $\xi = \exp(a/(E-E_c)^2)$, which effectively leads to a band of extended states and is inconsistent with the observed power law dependence in the transport parameters. It should also be emphasized that the universal scaling behavior observed in both ρ_{xx} and ρ_{xy} is manifestation of a quantum interference phenomenon, due to impurity scattering, and cannot be obtained from purely semiclassical considerations [14,15]. It differs fundamentally from the predictions of Ref.[14]. The scaling predictions from the percolation picture, strictly speaking, apply only in the limit of infinitely long-ranged potential fluctuations. On the other hand, the identical power law dependence on T for both ρ_{xx} and ρ_{xy} confirms the prediction of the scaling theroy [5,7] that the electronic transport shows scaling and κ is a characteristic critical exponent uniquely describing the localization to delocalization trasnsition of 2DEG in high B.

The scaling theory of Ref.[7] goes one step further to predict that all the B derivatives of the magneto-resistance diverges as a power law in T with integer multiples of κ in the exponent. Recently, we have studied $d^2\rho_{xx}/dB^2$ in the same sample and we find its minimum value, denoted by $(d^2\rho_{xx}/dB^2)^{min}$, has a power law dependence on T independent of Landau level index with an exponent twice as large as that in Eq.1 and 2. i.e.

$$\left(\frac{d^2\rho_{xx}}{dB^2}\right)^{min} \sim T^{-2\kappa} \quad . \tag{3}$$

This result further confirms the predictions in the scaling theory that the electronic trasport shows scaling, and that when $E_F \rightarrow E_c$ the transport coefficients are a function of $(T^{-\kappa} \times (B - B^*))$ where B^* is the B at which $E_F = E_c$. It is the latter property that gives rise to our experimentally observed critical behavior.

Our new results naturally lead to the question whether the exponent κ is indeed a universal number, i.e. independent of the sample. We have measured other InGaAs/InP samples with $n_{2D} = 3.8 \times 10^{11}\,\mathrm{cm}^{-2}$ and $\mu = 33,000\,\mathrm{cm}^2/\mathrm{Vs}$ in the range $0.35\,\mathrm{K} < T < 4.2\,\mathrm{K}$. We find the same exponent κ for $N = 0 \downarrow$, $1 \uparrow$ and $1 \downarrow$ levels. The same exponent κ is also obtained for $N = 0 \downarrow$ level in both an InGaAs/InP sample with $n_{2D} = 2.2 \times 10^{11}\,\mathrm{cm}^{-2}$, $\mu = 17,000\,\mathrm{cm}^2/\mathrm{Vs}$ and an InGaAsP/InP sample with $n_{2D} = 3.5 \times 10^{11}\,\mathrm{cm}^{-2}$, $\mu = 21,000\,\mathrm{cm}^2/\mathrm{Vs}$ in the range $0.35\,\mathrm{K} < T < 2.6\,\mathrm{K}$. Although more data in the larger T range are still needed, these preliminary data strongly suggest that the exponent κ is indeed a characteristic number uniquely describing the localization to delocalization transition of 2DEG at high B. However, in the latter two samples, the $i = 3$ plateaus is not observable for T down to 0.35 K. Our data indicate that $(d\rho_{xy}/dB)^{max}$ and $1/\Delta B$ between $i = 4$ and $i = 2$ plateaus show power law dependence on T for $0.35\,\mathrm{K} < T < 4.2\,\mathrm{K}$ with an exponent half of that in Eq.1 and 2 [16]. The precise meaning of this result is not clear at present. Lower T experiments are in progress in order to investigate the formation of the $i = 3$ plateau.

A few comments should be made regarding the AlGaAs/GaAs heterostructures. First, in a sample with $n_{2D} = 2.2 \times 10^{11}\,\mathrm{cm}^{-2}$, $\mu = 60,000\,\mathrm{cm}^2/\mathrm{Vs}$, our data suggest that $T_{sc} = 0.3\,\mathrm{K}$ and in the range $60\,\mathrm{mK} < T < 0.3\,\mathrm{K}$, $d\rho_{xy}/dB \sim T^{-0.4}$ for $N = 0 \downarrow$ level. However, it needs much lower T in order to establish the power law behavior over a sufficiently large region. The fact that T_{sc} in this case is considerably lower than that for the InGaAs/InP sample is attributed to the existence of longer range potential fluctuations in AlGaAs/GaAs [17]. On the other hand, the results for the $N = 1 \uparrow$ and $1 \downarrow$ levels are not conclusive in this T range. This may have something to do with the fact that the peak values of σ_{xx} in the two levels differ by a factor as large as three. In contrast, in the InGaAs/InP samples we studied, they are the same to within 5% in the whole T range. The origin for the asymmetry in σ_{xx} in AlGaAs/GaAs sample is not resolved [18,19] at present.

Finally, we have extended this work to the fractional quantum Hall effect (FQHE) [20]. Phenomelogically, in analogy to the discrete Landau levels in the IQHE, one can also assume that there is discrete energy spectrum due to electron-electron interaction in the FQHE. These discrete collective excitation states [21] are separated by a gap with localized electronic states. Therefore, it is speculated that there may be similar scaling behavior in the FQHE [22,23]. In a sample with $n_{2D} = 8.5 \times 10^{10}\,\mathrm{cm}^{-2}$, $\mu = 400,000\,\mathrm{cm}^2/\mathrm{Vs}$ and in the transition region from 2/5 to 1/3 plateaus, we find that there is a characteristic $T(\sim 0.6\,\mathrm{K})$, anologous to T_{sc}, where $(d\rho_{xy}/dB)^{max}$ changes its T dependent behavior from $T^{-0.8}$ to $T^{-0.4}$ with decreasing T. The latter is observed in $80\,\mathrm{mK} < T < 0.6\,\mathrm{K}$. In another sample with $n_{2D} = 2.5 \times 10^{11}\,\mathrm{cm}^{-2}$ and $\mu = 450,000\,\mathrm{cm}^2/\mathrm{Vs}$, we observe both $(d\rho_{xy}/dB)^{max}$ and $1/\Delta B \sim T^{-0.36}$ in the transition region from 2 to 5/3 plateaus [24] for $50\,\mathrm{mK} < T < 0.5\,\mathrm{K}$. On the other hand, the same power law behavior was not observed for the transition from 4/3 to 1 plateaus. Although our data in the FQHE are not conclusive at present, they suggest the possibility that the electronic transport in the IQHE and the FQHE show the same scaling behavior.

In summary, we have observed a characteristic T dependent power law behavior in the transport coefficients in the IQHE in InGaAs/InP heterostructures. The critical exponent κ we obtained is a number uniquely describing the electron localization to delocalization transition of 2DEG under high B. Clearly, more experiments are needed, especially in AlGaAs/GaAs samples, in which the existence of long range potential fluctuations and the existence of electron-electron interactions make the electron transport more involved than that in InGaAs/InP samples. Moreover, data in other materials (e.g. Si) are definitely interesting to explore.

We gratefully acknowledge the important contribution from Dr M. Razeghi. The work is supported in part by the Office of Naval Research under Contract No. N00014-82-K-0450 and the National Science Foundation through grant No. DMR-8212167 and grant No DMR-8600009. AMMP has been supported in part by a grant from NSF and by the Alfred P. Sloan Foundation.

References

1. M. A. Paalanen, D. C. Tsui and A. C. Gassard, *Phys. Rev.* **B 23**, 4566 (1982).

2. H. P. Wei, D. C. Tsui and A. M. M. Pruisken, *Phys. Rev.* **B 33**, 1488 (1986).

3. H. P. Wei, D. C. Tsui, M. A. Paalanen and A. M. M. Pruisken, "Scaling of the integral quantum Hall effect", In G. Landwehr, editor, *High magnetic fields in semiconductor Physics*, page 11, Springer Verlag, (1987).

4. H. Z. Zheng, H. P. Wei and D. C. Tsui, *Bull. Am. Phys. Soc.* **32**, 892 (1987).

5. A. M. M. Pruisken, "Field theory, scaling and the localization problem", In R. Prange and S. Girvin, editors, *The Quantum Hall Effect*, page 117, Springer Verlag, (1986).
6. A. M. Chang and D. C. Tsui, *Solid State Comm.* **56**, 153 (1985).
7. A. M. M. Pruisken, "Universal singularities in the integral quantum Hall effect", Columbia University preprint CU-TP-387 (submitted to *Phys. Rev. Lett.*).
8. D. J. Thouless, *Phys. Rev. Lett.* **39**, 1167 (1977).
9. E. Abrahams, P. W. Anderson, P. A. Lee and T. V. Ramakrishnan, *Phys. Rev.* **B 24**, 6783 (1981).
10. B. L. Altshuler, A. G. Aronov and D. E. Khmelnitsky, *J. Phys.* **C 15**, 7367 (1982).
11. I. Affleck. *Nucl. Phys.* **B 257**, [FS14], 397 (1985).
12. H. Aoki and T. Ando, *Phys. Rev. Lett.* **54**, 831 (1985).
13. Y. Ono, *J. Phys. Soc. Jpn.* **51**, 3544 (1982); **52**, 2492 (1983); **53**, 2342 (1984).
14. S. A. Trugman, *Phys. Rev.* **B 27**, 7539 (1983).
15. B. Shapiro, *Phys. Rev.* **B 33**, 8447 (1986).
16. S. W. Hwang, H. P. Wei and D. C. Tsui, (unpublished).
17. H. P. Wei, S. Y. Lin, D. C. Tsui and K. Alavi, *Bull. Am. Phys. Soc.* **33**, 665 (1988).
18. R. J. Haug, K. v. Klitzing and K. Ploog, *Phys. Rev.* **B 35**, 5933 (1987).
19. B. E. Kane, D. C. Tsui and G. Weimann, *Surf. Sci.* **196**, 183 (1988).
20. D. C. Tsui, H. L. Störmer and A. C. Gossard, *Phys. Rev. Lett.* **48**, 1559 (1982).
21. R. B. Laughlin, *Phys. Rev. Lett.* **50**, 1395 (1983).
22. R. B. Laughlin, M. L. Cohen, J. M. Kosterlitz, H. Levine, S. B. Libby and A. M. M. Pruisken, *Phys. Rev.* **B 32**, 1311 (1985).
23. R. G. Clark, J. R. Mallett, A. Usher, A. M. Suckling, R. J. Nicholas, S. R. Haynes, Y. Journaux, J. J. Harris and C. T. Foxon, *Surf. Sci.* **196**, 219 (1988).
24. H. P. Wei, D. C. Tsui and R. A. Webb, (unpublished).

Scaling and the Integer Quantum Hall Effect

A. MacKinnon

Blackett Laboratory, Imperial College, London SW7 2BZ, UK

1 Introduction

The scaling description of the localisation of electrons in disordered systems[1], which has been so successful in describing the transport properties of quasi 2-dimensional systems in the absence of a magnetic field, remains controversial when it is applied to the quantum Hall effect[2].

The usual approach is that due to Pruisken et al.[3] in which a σ_{xx}-σ_{xy} renormalisation group flow diagram is derived from a non-linear σ model by including an extra, topological Θ term, related to σ_{xy}. The Pruisken flow diagram is periodic in σ_{xy} and contains a single fixed point for each Landau level located at half integer σ_{xy} and at σ_{xx} of order unity (in units of e^2/h). This saddle point directs the flow towards σ_{xx} = integer and $\sigma_{xx} = 0$; except at half integer σ_{xy}, where the flow is towards the fixed point. This diagram seems to be confirmed by experimental results deriving a temperature driven flow diagram [4,5].

On the other hand numerous numerical results, especially those by Ando and Aoki[6], but also by Chalker and Coddington[7] and by the present author[8] are difficult to reconcile with the Pruisken theory. Perhaps surprisingly, although these numerical approaches have very different starting points, they agree quantitatively with one another. This suggests that the behaviour they predict is model independent. Aoki claims that the numerical results are inconsistent with the 2 parameter scaling theory, It is not clear to me whether he is referring to 2 parameter scaling theories per se, or to the Pruisken theory in particular. Indeed Aoki and Ando[6] have calculated their own σ_{xx}-σ_{xy} flow diagram which contains no crossing flow lines, and is therefore consistent with the basic principle of 2 parameter scaling.

More recently Clark et al.[9] have measured a temperature driven flow diagram for the fractional effect which seems to be consistent with a very speculative theoretical diagram published some years ago[10]. In this diagram the fractionally charged quasiparticles[11,12,13] behave as the electrons do in the integer effect, but with the σ_{xx}-σ_{xy} flow diagram rescaled to reflect the fractional charges. In order to derive a consistent diagram the authors of [10] were forced to include an extra unstable fixed point between 2 stable fixed points on each half integer (or equivalent) line. There was no other physical justification for the inclusion of this feature.

In this paper I shall present some numerical results and shall attempt to analyse them in terms of a 2 parameter scaling theory. I shall show that the results are consistent with a flow diagram containing 2 fixed points rather than one on each half integer line. Hence the extra fixed point in the generalised diagram can be attributed to the behaviour of non-interacting electrons in a magnetic field rather than to the effects of the interactions.

Springer Series in Solid-State Sciences, Vol. 87
High Magnetic Fields in Semiconductor Physics II Editor: G. Landwehr

2. Numerical Calculations

The calculations were carried out using a tight binding model with the magnetic field represented by Peierls factors, thus

$$H = \Sigma_{mn}\varepsilon_{mn}|mn\rangle\langle mn| + \Sigma_{mnm'n'}V_{mnm'n'}|mn\rangle\langle m'n'| \quad , \tag{1}$$

where

$$V_{mnm'n'} \begin{cases} = 1 & m = m' \;\&\; n = n' \pm 1 \\ = \exp(\pm i2\pi n/L_B) & m = m' \pm 1 \;\&\; n = n' \\ = 0 & \text{otherwise.} \end{cases} \tag{2}$$

The (m,n) represents points on a square lattice and ε is a random number chosen from a rectangular distribution of width W ($-\frac{1}{2}W < \varepsilon \le \frac{1}{2}W$).

Using the transfer matrix method[14] the smallest Lyapunov exponent, α, was calculated for long strips with periodic boundary conditions and width M, where $4 \le M \le 64$. This exponent can be interpreted as the inverse localisation length of states on an infinitely long strip. Only a small sample of the results are presented here, with $L_B = 8$ and W = 0.5 and 1.0. The energies are chosen to scan the lowest Landau level which is located at about E = -3.29 in the absence of disorder.

Experience from the simpler localisation problem without a magnetic field[14] has shown that the appropriate quantity to study is the renormalised localisation length $\Lambda = 1/\alpha M$, which displays 1 parameter scaling behaviour. In other words the way Λ changes with increasing M can be expressed as a function of Λ alone. The results are summarised in Fig.1 where Λ is plotted

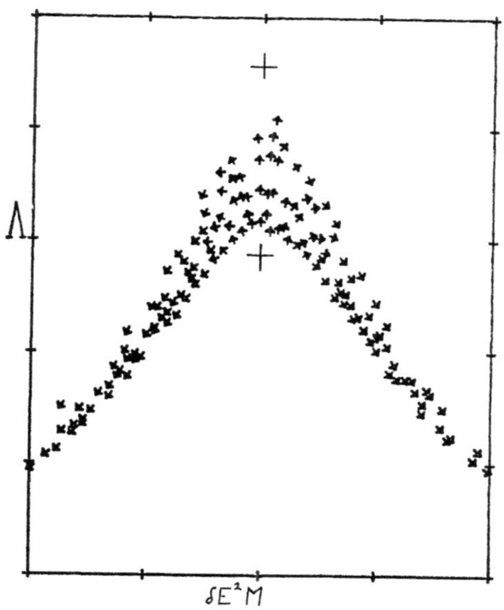

Fig.1 Renormalised Localisation Length Λ vs. $\delta E^2 M$ for W=0.5 & W=1.0. The arrows show the direction of the flow.

against $\delta E^2 M$ in the form of a flow diagram for increasing M. The points are marked by arrows whose directions show the direction of flow. These have been calculated using a cubic spline, joining the different M corresponding to a single energy. The arguments leading to this unconventional way of representing the data are discussed in the next section. Here it suffices to note that the flow lines rarely cross. (The lines are not shown as the diagram would be too cluttered). Those crossings which are present are at small system sizes and can be accounted for by the statistical errors in the data (relative error 1%).

Figure 1 has all the characteristics of a renormalisation group flow diagram. Indeed it looks rather similar to the σ_{xx}-σ_{xy} diagram of Aoki and Ando[6]. There is clearly an unstable fixed point at about $\Lambda \approx 0.8$ and possibly a saddle point at the top of the diagram. The diagram is consistent with 2 parameter scaling but not with the Pruisken theory.

3. Data Analysis

Let x and y be the deviations from the fixed point in the direction of Λ and in the direction of the unknown second parameter respectively. Since the data in Fig.1 are symmetric around $\delta E = 0$, the general equation for the flow around the fixed point reduces to

$$x' = \alpha x + \tau x^2 + \varepsilon y^2 \ , \qquad y' = \beta y + \delta xy \ , \tag{3}$$

where the ' refers to differentiation with respect to lnM. The unknown quantity y can be eliminated to give a second order differential equation which can then be linearised in x to give

$$x'' - (\alpha + 2\beta)x' + 2\alpha\beta x = 0 \ , \tag{4}$$

which has the general solution

$$x = AM^{\alpha} + BM^{2\beta}. \tag{5}$$

Since Λ must be an analytical and symmetric function of δE for all finite M its behaviour near the fixed point can be described by

$$\Lambda = \Lambda^{*} + (a + b\delta E^2)M^{\alpha} + (c + d\delta E^2)M^{2\beta} \ . \tag{6}$$

This general form can be fitted to the raw numerical data to find the 7 unknowns. There is more than enough data for this to be a valid procedure. We can distinguish between the otherwise indistinguishable terms in (6) by considering the behaviour at $\delta E = 0$. Here the α term must dominate, since $y = 0$ in (3). Thus we expect to find $c = 0$.

In practice $2\beta = 1.0\pm0.05$, $c \approx 0.01d$ when the units of δE are chosen such that δE is typically of order unity, and $\Lambda^{*} = 0.83\pm0.01$. Since the β term in (6) can also be written in terms of M/μ, where μ is the localisation length, $2\beta = 1$ implies $\mu \sim 1/\delta E^2$ in agreement with other results[6,7]. The other parameters have rather large error bars, in particular α, which seems to be very small. This may reflect the presence of another fixed point just above the range of the present data.

The above analysis provides the justification for plotting Λ against $\delta E^2 M$ in Fig.1 in order to obtain a 2-dimensional flow diagram to describe the behaviour of Λ.

4 Conclusions

So far the discussion has been in terms of the renormalised localisation length Λ and the unknown second scaling variable y. Although there is no simple mapping between these and σ_{xx}-σ_{xy} such a mapping probably exists in principle. The flow diagram for σ_{xx}-σ_{xy} may differ somewhat from that for Λ-y but the basic structure should be the same. In particular the unstable fixed point and probably a saddle point should be important features of both.

The Pruisken theory involves an expansion in $1/\sigma_{xx}$ which may break down at small values of σ_{xx}. The presence of an unstable fixed point at smaller σ_{xx} than Pruisken's saddle point is thus not necessarily inconsistent with his results. In fact, as mentioned in the introduction, such a fixed point is a necessary part of the generalisation of Pruisken's flow diagram to the fractional quantum Hall effect. The results presented here together with those of Aoki and Ando[6] and of Chalker and Coddington[7] are consistent with such a picture.

A full understanding of the quantum Hall effect requires an understanding of the role of both disorder and interactions. I have shown that the behaviour of non-interacting electrons in a magnetic field is described by a pair of fixed points, in the renormalisation group description. This unit is repeated on a different scale to describe also the behaviour of the fractionally charged quasiparticles in the presence of disorder.

References

1. E.Abrahams, P.W.Anderson, D.C.Licciardelo, T.V.Ramakrishnan: Phys.Rev.Lett. 42 673 (1979)
2. K.von Klitzing, G.Dorda, M.Pepper: Phys.Rev.Lett. 45 494 (1980)
3. A.M.M.Pruisken: In The Quantum Hall Effect, ed. by R.E.Prange & S.M.Girvan (Springer, New York, 1987) and references therein.
4. H.P.Wei, D.C.Tsui, A.M.M.Pruisken: Phys.Rev. B33 1488 (1985)
5. S.Kawaji, J.Wakabayashi: J.Phys.Soc.Japan 56 21 (1987)
6. H.Aoki: Rep.Prog.Phys. 50 655 (1987) and references therein
7. J.T.Chalker, P.D.Coddington: J.Phys. C21 2665 (1988)
8. L.Schweitzer, B.Kramer, A.MacKinnon: J.Phys. C17 4111 (1984)
9. R.G.Clark, J.R.Mallett, A.Usher, A.M.Suckling, R.J.Nicholas, S.R.Haynes, Y.Journaux, J.J.Harris, C.T.Foxon: Proceedings of the 17th International Conference on the Electronic Properties of 2 Dimensional Systems, Santa Fe, New Mexico. (1987)
10. R.B.Laughlin, M.L.Cohen, J.M.Kosterlitz, H.Levine, S.B.Libby, A.M.M.Pruisken: Phys.Rev. B32, 1311 (1985)
11. R.B.Laughlin: Phys.Rev.Lett. 50 1395 (1983)
12. B.I.Halperin: Phys.Rev.Lett. 52 1583 (1984)
13. F.D.M.Haldane: Phys.Rev.Lett. 51 605 (1983)
14. A.MacKinnon, B.Kramer: Z.Phys. B53 1 (1983)

Density of States and Coulomb Interactions in the Integer Quantum Hall Effect

V. Gudmundsson and R.R. Gerhardts*

Max-Planck-Institut für Festkörperforschung, Heisenbergstr. 1,
D-7000 Stuttgart 80, Fed. Rep. of Germany
*Present address: Science Institute, University of Iceland,
Dunhaga 3, IS-107 Reykjavik, Iceland

We review recent Hartree calculations on the screening properties of a two-dimensional electron gas in a quantizing magnetic field in connection with a phenomenological statistical model invented to describe the density of states of an inhomogeneous system in the quantum Hall regime. A consistent picture of screening, inhomogeneous electron density and effective density of states at the Fermi level results which emphasizes the importance of Coulomb interactions even in the integer quantum Hall effect.

1. INTRODUCTION

Coulomb interactions are generally believed to play a crucial role in the fractional quantum Hall effect (FQHE), which is attributed to a highly correlated liquid state of the two-dimensional electron gas (2DEG), as was first proposed by Laughlin [1,2]. The integer quantum Hall effect (IQHE), on the other hand, is usually discussed within a strict single-particle picture without any reference to Coulomb interactions [1]. The observed plateaus of the Hall resistance are explained by localized states of an electron in a random potential. Two extreme cases with different localization mechanisms have been considered, random potentials with short-range fluctuations which are believed to produce Anderson type localization [1] but also long-range potentials which vary slowly on the scale of the magnetic length $l = (\hbar c/eB)^{1/2}$, and which lead to a quasi-classical localization of electrons moving along closed equipotential lines [3]. Although a satisfying theory of Anderson localization in a strong magnetic field B is still missing, the density of states (DOS) within the single-particle picture is well understood [4]-[7]. For material parameters consistent with the high mobility of typical GaAs-(AlGa)As- heterostructures, one obtains in the quantum Hall regime (typical B$\geq$ 4 T) well separated Landau levels (LLs) with an exponentially small DOS between the LLs. This prediction of the single-particle theory is, however, in sharp disagreement with many experiments (see, e.g. [8,9]) which revealed an unexpectedly large DOS between the LLs.

In order to resolve this apparent discrepancy, it is important to note that all these experiments "measure" only the DOS at the Fermi level. If in a macroscopic sample the electron density is slightly inhomogeneous, complete filling of a certain LL will not occur simultaneously everywhere in the sample and, on the average, the DOS at the Fermi level will always appear to be finite. The reason for such inhomogeneities is the screening of long-range fluctuations in the donor distribution which must exist, at least for statistical reasons, on a mezzoscopic scale of, say, 10^{-4} cm in epitaxially grown samples. The aim of the present paper is to draw a consistent picture of the related topics of spatial inhomogeneities, of screening properties, and of the effective DOS of

Springer Series in Solid-State Sciences, Vol. 87
High Magnetic Fields in Semiconductor Physics II Editor: G. Landwehr

the 2DEG in the quantum Hall regime, thereby reviewing briefly our recent work in this field.

In Sect. 2 we consider the screening properties of the 2DEG in a smoothly varying electrostatic potential. For simplicity, we treat the Coulomb interactions among the electrons within the Hartree approximation and thus neglect the exchange and correlation effects which may lead to the FQHE. We obtain, however, strongly non-linear screening effects, notably a pinning of the effective single-particle energy spectrum to the Fermi level, which provides a qualitative explanation of the observed DOS in the regime of the IQHE. In Sect. 3 we discuss a "statistical model of inhomogeneities" which includes both the screening of long-range potential fluctuations and the line-broadening effects of short-range potential fluctuations in a phenomenological manner. The line-broadening effects are included by a model DOS and the most relevant results of the microscopic Hartree theory of screening are simulated by statistical model assumptions. With reasonable model parameters we are able to explain the experimental results on the DOS at the Fermi level quantitatively. A summary of the paper is given in Sect. 4.

2. MICROSCOPIC HARTREE THEORY OF SCREENING

The screening properties of a 2DEG in a heterostructure can of course be calculated from a realistic three-dimensional model. Qualitatively the same result can, however, be obtained within a strictly two-dimensional model of the electronic motion, as is shown elsewhere [10]. Here we consider electrons in a strip $|x| < \frac{1}{2}L_x$ of the x-y-plane, take $A = (0, Bx, 0)$ as vector potential and assume translational invariance in the y-direction. Then, in the Hartree approximation used, the single-electron eigenfunctions are of the form

$$\phi_\alpha(x,y) = L_y^{-\frac{1}{2}} e^{iyk} \phi_{nx_0}(x), x_0 = -l^2 k, kL_y/2\pi = 0, \pm 1, \dots, \tag{1}$$

where the normalization length L_y eventually approaches infinity. The energy eigenvalues $\epsilon_\alpha = \epsilon_{nx_0}$ follow from the reduced Schrödinger equation

$$[-\frac{\hbar^2}{2m}\frac{d^2}{dx^2} + \frac{1}{2}\omega_c^2(x-x_0)^2 + V(x) - \epsilon_{nx_0}]\phi_{nx_0}(x) = 0 \tag{2}$$

with the boundary conditions $\phi_{nx_0}(\pm\frac{1}{2}L_x) = 0$, where m is the effective mass [$m = 0.067\ m_0$ for GaAs] and $\omega_c = eB/mc$ is the cyclotron frequency. The Hartree potential $V(x)$ may contain an external contribution $V_{ext}(x)$ and it contains the electrostatic part originating from the electron density

$$n_s(x) = \frac{2}{2\pi l^2}\sum_n \int_{-\infty}^{\infty} dx_0 |\phi_{nx_0}(x)|^2 f(\epsilon_{nx_0} - \mu) \tag{3}$$

with $f(E) = \{1 + exp(E/k_BT)\}^{-1}$ the Fermi function, and from the immobile backgrounds charges. We assume a neutralizing positive background charge of area density $n_b(x)$ in the stripe occupied by the 2DEG and obtain the solution of Poisson's equation in the form [10]

$$V(x) = V_{ext}(x) - \int_{-\frac{1}{2}L_x}^{\frac{1}{2}L_x} dx' G(x,x')[n_s(x') - n_b(x')] \tag{4}$$

with

$$G(x,x') = \frac{2e^2}{\kappa} ln|\frac{x-x'}{L_x}| \quad , \tag{5}$$

if we assume that the potential created by a point charge vanishes at infinity ($\kappa = 12.4$ is the dielectric constant of GaAs). For the self-consistent solution of eqs. (1)-(5) a diagonalization procedure using the exact wavefunctions of the non-interacting 2DEG (i.e. $V(x) \equiv 0$) [10,11] has been employed, which allows the use of only three LLs to achieve the desired accuracy if the filling factor $\nu = 2\pi l^2 \overline{n}_s$ of the spin-degenerate LLs is less than 2.5 ($\overline{n}_s$ is the average electron density in the stripe). Figure 1 shows calculated energy spectra for a sinusoidal modulation of the positive background according to

$$n_b(x) = \overline{n}_s\{1 + \alpha cos[\frac{\pi x}{L_x}(2p+1)]\}/[1 + \frac{2\alpha(-1)^p}{\pi(2p+1)}] \quad , \tag{6}$$

with $V_{ext}(x) = 0$. For values of the "center coordinate" x_0 near the edges (at $|x| = 120$ nm) and outside the sample the energy bands bend rapidly upwards. In the interior region, the Landau bands closely reflect the spatial variation of the self-consistent potential $V(x)$. Indeed, if this varies slowly, so that $l|dV/dx| \ll \hbar\omega_c$, the energy eigenvalues for small Landau quantum numbers n can be well approximated by [10,12]

$$\epsilon_{nx_0} \approx \hbar\omega_c(n + \frac{1}{2}) + V(x_0) \quad . \tag{7}$$

For partly filled LLs, $\nu = 1.6$ and $\nu = 2.4$ in Fig. 1, screening of the external potential created by the background modulation is good and the energy spectrum shows narrow bands in the interior region. For complete filling, $\nu = 2.0$, screening is considerably poorer and the width of the Landau bands (neglecting the edge effects) becomes much larger.

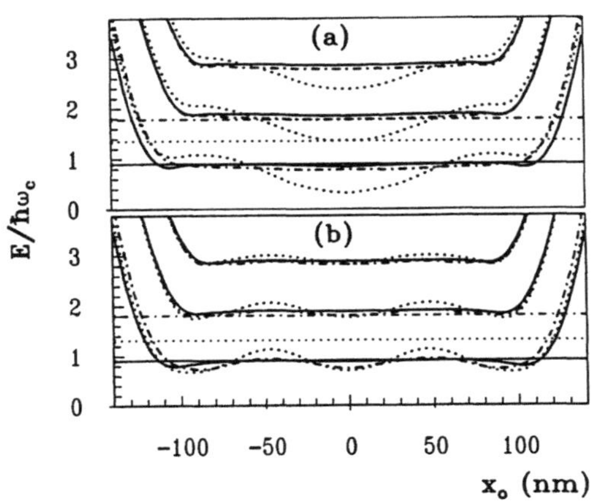

Fig.1 Energy spectrum and Fermi energy with 20% modulation of the background density. The modulation parameters are $\alpha = 0.2$ and (a) $p = 1$, (b) $p = 2$ for the periods. Results are shown for the filling factors $\nu = 1.6$ (solid), 2.0 (dotted), and 2.4 (dash-dotted). $L_x = 240nm$, $n_S = 2.25 \cdot 10^{11} cm^{-2}$, $\kappa = 12.4$, $m = 0.067m_0$ and $T = 1.0K$.

We want to put special emphasis on a nonlinear screening effect which is easily understood from eq. (7): the pinning of the energy spectrum to the Fermi level. If the $n = 0$ LL is more than half filled and the variation of the external modulating potential is not so large that stripes with vanishing electron density are created, then, according to eq. (3), the maximum values of ϵ_{0x_0} cannot exceed μ by more than k_BT. If, for larger ν, the $n = 1$ LL starts to be occupied, the minimum of ϵ_{1x_0} is pinned to the Fermi level. With increasing ν this minimum becomes broader, the whole band becomes narrower ($V(x)$ becomes flatter), and so a piling-up of electron density in a narrow strip is avoided. Close to half filling of one of the lowest LL's, the total variation of the screened potential $V(x)$ is only of the order of k_BT. Near filling factor ν=2, on the other hand, both the maximum of the $n = 0$ band and the minimum of the $n = 1$ band touch the Fermi level, if the total variation of the external potential is larger than $\hbar\omega_c$ [10]. Then the total variation of the screened potential $V(x)$ is just $\hbar\omega_c$ (if we neglect terms of the order $k_BT \ll \hbar\omega_c$ and edge effects) and the DOS at the Fermi energy is considerably large, even for $\nu = 2$. Only if the total variation of the unscreened modulation potential is less than $\hbar\omega_c$, the Fermi level for ν=2 can be pinned by a few edge states, as shown in Fig. 1b, and the bulk DOS drops to zero between the LLs.

Figure 2 shows calculated electron densities corresponding to Fig. 1a. Apart from edge effects arising from the boundary conditions $\phi_{nx_0}(\pm\frac{1}{2}L_x) = 0$, the electron density follows closely the background density for ν=1.6. For ν=2.0, the 2DEG is more rigid and cannot screen the background so well (the structure near $x = 0$ comes from the beginning population of the $n = 1$ LL). Thus the variance of the electron density, defined by

$$\Delta n_s = \{\frac{1}{L_x}\int_{-L_x/2}^{L_x/2} dx[n_s(x) - \overline{n}_s]^2\}^{1/2} \quad , \tag{8}$$

has a minimum at complete filling, as is shown for several modulations in Fig. 3.

The screening calculations presented here and elsewhere [10,12] can be summarized as follows. If the total variation of the slowly varying external potential is sufficiently large ($\geq \hbar\omega_c$) but not too large (no areas without electrons), then the total variation of the screened potential varies drastically as a function of the filling factor ν, being of the order $k_BT \ll \hbar\omega_c$ at half filling and of the order of $\hbar\omega_c$ at complete filling of LLs. The

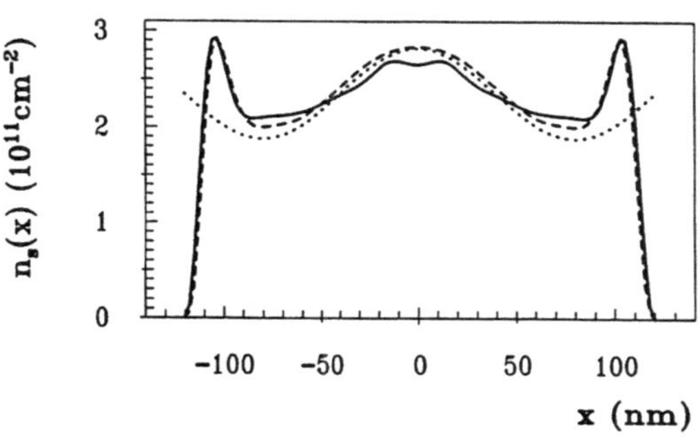

Fig.2 Electron density with a modulated background density ($p = 1$, $\alpha = 0.2$, indicated by the dotted curve), for filling factors $\nu = 2.0$ (solid) and 1.6 (dashed). Parameters same as in Fig.1.

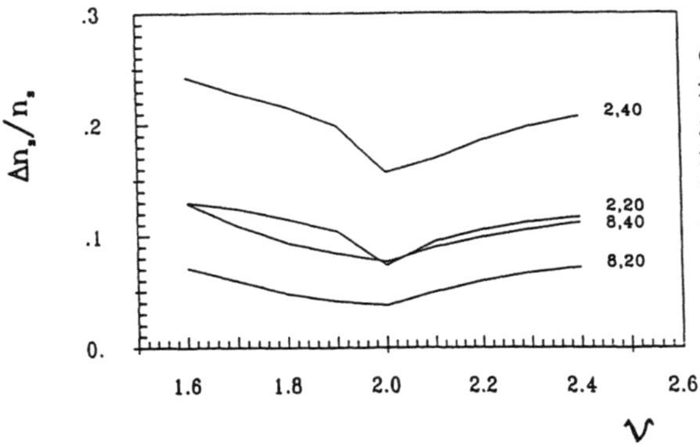

Fig.3 Variance of the electron density vs. filling factor for different modulations of the background density: $p = 2$ and $p = 8$ for 20% and for 40% modulation. The pairs of numbers labelling the curves indicate p and 100α.

variance of the electron density which produces this screening changes only little with ν, being minimum at complete filling. Due to the pinning effect, the DOS at the Fermi level is always considerably large, even in the complete filling situation.

3. PHENOMENOLOGICAL MODEL OF INHOMOGENEITIES

In a typical high-mobility heterojunction the dominating scattering mechanism at low temperatures is the interaction of electrons with the charged donors. Let us assume that the area density of these donors fluctuates on the mezzoscopic length scale of say 10^{-4}-10^{-3}cm by amounts of about 1 %. For resulting electron densities of the order of several $10^{11}cm^{-2}$, such fluctuations must be expected for general statistical reasons [13], and clustering effects during the epitaxial growth of the samples will tend to enhance these fluctuations. Then, the donors give rise to short-range potential fluctuations (typical scale ≤ 10 nm) superimposed on long-range fluctuations with the total variation of the order of several tens of meV. Since there is no microscopic theory which allows us to treat the short-range and the long-range fluctuations on the same footing, we use the following phenomenological approach [14,15].

For $B = 0$ and also at half filled LLs in large magnetic fields, the long-range potential fluctuations are effectively screened by the 2DEG and the mobility for B=0 is determined by the short-range fluctuations. We take these into account by a level broadening $\Gamma \sim B^{1/2}$ and a Gaussian model DOS [5]

$$D(E) = \frac{1}{(2\pi)^{1/2}\Gamma}\frac{2}{2\pi l^2}\sum_{n=0}^{\infty} exp\{-\frac{1}{2\Gamma^2}[E - \hbar\omega_c(n+\frac{1}{2})]^2\} \quad . \tag{9}$$

The treatment of the long-range potential fluctuations is motivated by the following considerations. Their screening leads to a spatial variation of the electron density n_s $(\vec{r})$ and to a screened potential $V(\vec{r})$, which in high magnetic fields ($\hbar\omega_c \sim 10\ meV$) changes drastically with the average filling factor of the LLs. Spatial variation of $V(\vec{r})$ means for the 3D heterostructure that the electrostatic confinement potential and thus the energy eigenvalue $E_0(\vec{r})$ of the lowest, occupied, electric subband varies on a semi-macroscopic scale along the interface. Thus, the long-range fluctuations of $V(\vec{r})$ lead to corresponding fluctuations of the local chemical potential $\mu(\vec{r}) = E_F - E_0(\vec{r})$ determining the local electron density $n_s(\vec{r})$ according to

$$n_s = \int dE D(E) f(E-\mu) \quad . \tag{10}$$

Now we replace the spatially inhomogeneous sample by a statistical ensemble of samples which are homogeneous in the x-y directions and have a probability distribution $P(n_s)$ of densities. Accordingly, we replace spatial averages by ensemble averages. Since the members of the ensemble are meant to simulate homogeneous subregions of the spatially inhomogeneous sample, we choose the distribution P(n_s) such that typical aspects of the magnetic field dependent screening are incorporated. Since the electron number is an additive random variable, it is reasonable to choose a Gaussian distribution, as required by the central limit theorem in the limit of non-interacting subregions. Furthermore, in view of the results of Sect. 2, we take the mean value $\overline{n}_s$ and the variance Δn_s to be independent of the magnetic field, thereby defining the "n_s-Gaussian model" [14,15]. To compare with measurable quantities such as the thermodynamic DOS $\partial n_s/\partial\mu$ or the heat capacity C_v, one calculates these as function of the chemical potential μ, considers μ as a random variable connected with the random variable n_s by eq. (10), and calculates the average values of the desired quantities with the statistical distribution $\tilde{P}(\mu) = P(n_s(\mu))dn_s/d\mu$. Due to this procedure, the distribution of μ becomes strongly B-dependent and the variance $\Delta\mu$ shows qualitatively the behaviour obtained in Sect. 2 for the variation of the screened potential $V(x)$.

For very reasonable values of the linewidth Γ and the variance $\Delta n_s/n_s (\leq 3\%)$ this n_s-Gaussian model nicely explains many experiments revealing an apparently enhanced DOS. For details we refer to the original work [14,15]. Here we want to discuss only the thermal activation of the longitudinal resistivity $\rho_{xx} \sim exp(-E_a/k_B T)$ in the plateau regime of the IQHE [8,16]. The idea with the experiment was that extended states which can contribute to the macroscopic current exist only in a narrow energy region in the center of the LLs. Thus, if the Fermi energy E_F is in the high-energy tail of the $n = 0$ LL, only activation processes from the center of that LL to the Fermi level are relevant for the current. Measuring the activation energy $E_a = E_F - \frac{1}{2}\hbar\omega_c$ then yields E_F as a function of B, and from this the DOS at E_F was evaluated [8,16]. In Fig. 4 the experimental data on E_a vs. B are compared with the statistical model of inhomogeneities [17]. Here we argue that long-range potential fluctuations (which are assumed to be on the average symmetric with respect to the mean potential) lead to fluctuations of the Landau bands around the bare energies $E^n_{ext} = \langle E_0(\vec{r}) \rangle + \hbar\omega_c(n+\frac{1}{2})$, where we expect extended states, while states at other energies are quasi-classically localized on closed equipotential lines [3]. Since only activation from extended states at E^0_{ext} to the Fermi level are relevant for the macroscopic current, we obtain (cf. Fig. 5) $E_a = E_F - E^0_{ext} = \langle \mu \rangle - \frac{1}{2}\hbar\omega_c$. Evaluation of $\langle \mu \rangle$ with the n_s-Gaussian model (cf. solid line in Fig. 4) yields very good agreement with the experiment. We have also considered a "μ-Gaussian model", where we assumed the distribution of μ to be Gaussian for all values of B, taking again $\overline{n}_s$ and Δn_s independent of B. This model also creates an apparent DOS between LLs but it cannot be justified by a central-limit-theorem argument and it does not explain the experiments satisfactorily (dashed curve in Fig. 4). Also shown in Fig. 4 is a best fit of $E_F - \frac{1}{2}\hbar\omega_c$ within a strict single-particle

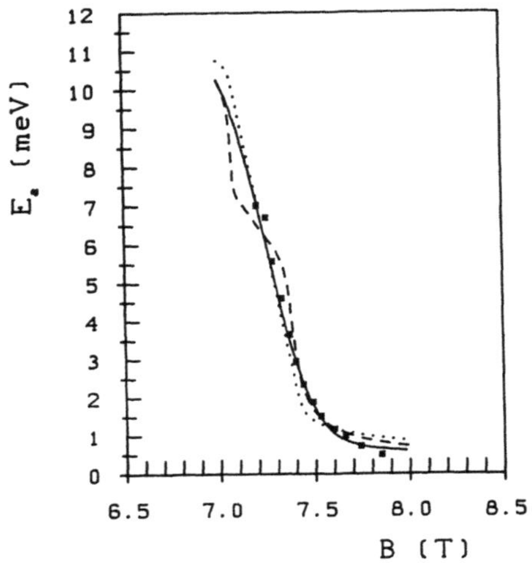

Fig.4 Activation energy vs. magnetic field,(a) for sample 1 in [16] (squares) with $n_s = 3.5 \cdot 10^{11} cm^{-2}$ and mobility $\mu = 550{,}000 cm^2/Vs$, (b) (dotted) for the simple background model with linewidth $\Gamma = 0.2\sqrt{B[T]}\ meV$ and background DOS= $2 \cdot 10^9 cm^{-2} meV^{-1}$, (c) (solid) for the n_s-Gaussian model with $\Gamma = 0.15\sqrt{B[T]}\ meV$ and $\Delta n_s/\bar{n}_s = 0.026$, (d) (dashed) for the μ-Gaussian model with $\Gamma = 0.15\sqrt{B[T]}\ meV$ and $\Delta n_s/\bar{n}_s = 0.12$.

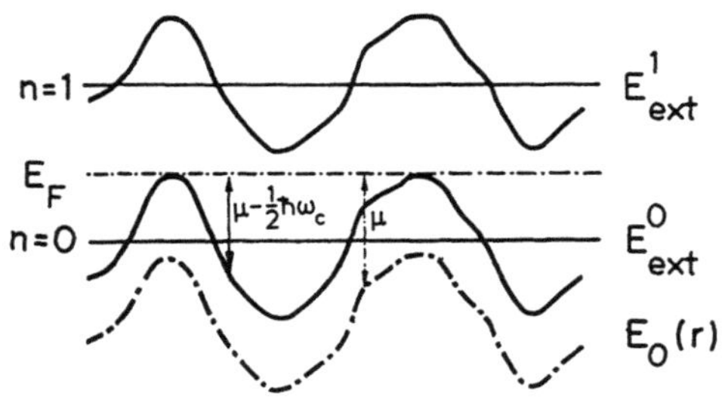

Fig.5 Schematic plot showing the spatial variation of the lowest Landau bands (solid) around the energies E^n_{ext} of non-localized motion, and the electric subband energy (dash-dotted). The Fermi energy E_F, the local chemical potential μ, and a non-relevant activation energy are also indicated.

model assuming a broad Gaussian DOS superimposed on a constant background DOS (dotted line). For this ad hoc background model [8,9,16] no justification is known.

In the n_s-Gaussian and the μ-Gaussian models the mean electron density of the 2DEG is kept constant, independent of the magnetic field. In order to explain experiments in gated heterostructures which explore the B-dependence of the capacitance [8,18], a truly 3D model of a heterostructure was considered. Assuming a homogeneous 2DEG and the validity of eqs. (9) and (10) the electron density n_s was calculated self-consistently as a function of the gate voltage V_g and the magnetic field B. Expressing V_g in terms of the electron density n_{so} for $B = 0$, a relation $n_s = F(n_{so}, B)$ was obtained and used to define the statistical "n_{so}-Gaussian model", which considers n_{so} as a Gaussian distributed random variable with fixed variance Δn_{so} [15]. The distributions of the random variables n_s and μ can then be calculated, and reflect characteristic features of the B-dependent screening. The variance Δn_s of the electron density as a function of the filling factor shows dips with minima at complete filling of LLs [cf. Fig. 3], whereas μ simulates the behaviour of the screened potential with maximum variance at complete filling of LLs. Typical results are shown in Fig. 6.

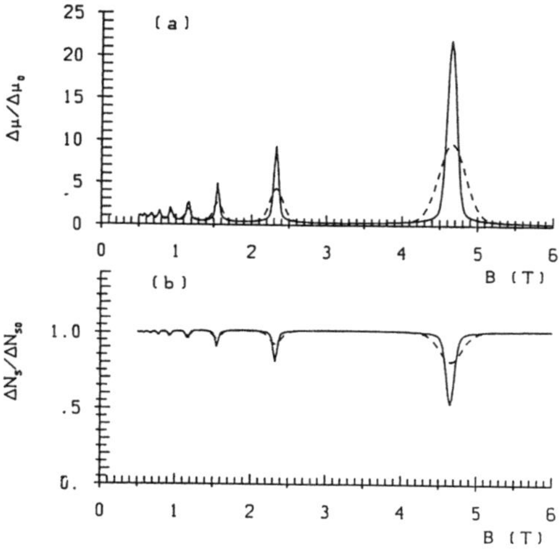

Fig.6 (a) Standard deviation of (a) the chemical potential $\Delta\mu$ and (b) the electron density δn_s as a function of B for the n_{s0}-Gaussian model with $\Delta n_{s0}/\bar{n}_{s0} = 0.01$ (solid) and 0.03 (dashed). $\bar{n}_{s0} = 2.25 \cdot 10^{11} cm^{-2}$, $N_A d = 1.44 \cdot 10^{11} cm^{-2}$, and $\Gamma = 0.3\sqrt{B[T]}\ meV$, $\Delta\mu_0$ is the deviation of μ at $B = 0.0T$.

Calculations within this n_{s0}-Gaussian model nicely explain results on capacitance, floating gate voltage and gate current measured on gated heterostructures while sweeping the magnetic field under different experimental conditions [18].

4. SUMMARY

A slowly varying electrostatic potential is effectively screened by the 2DEG in a strong magnetic field. The spatial variation of the (screened) potential lifts the degeneracy of LLs and leads to Landau bands. The total variation of the screened potential and thus the width of the Landau bands varies strongly with the filling of the LLs, being only of the order k_BT for partial filling. If the total variation of the bare potential is larger than the cyclotron energy $\hbar\omega_c$, the width of the Landau bands at complete filling is just $\hbar\omega_c$. The effective single-particle energy spectrum is pinned to the Fermi energy. As a result, the density of states at the Fermi level is always relatively high, even for complete filling of LLs.

The very reasonable assumption of mezzoscopic fluctuations of the donor density by an amount of only a few per cent in typical heterostructures leads to such long-range electrostatic potentials with a total variation of several tens of meV. The screening properties of the 2DEG in such external potentials explain qualitatively the experimental results concerning a drastically enhanced DOS between LLs, which cannot be understood within a strict single-particle picture neglecting the Coulomb interactions between the electrons.

For a quantitative understanding of the experiments it is important to take line broadening effects by short-range scattering processes into account. This is done in the statistical model of inhomogeneities which also includes the most important aspects of screening of long-range potentials in a phenomenological way. All the experiments on the DOS in the regime of the IQHE can be consistently understood within this model, even the activation energy of ρ_{xx} although no transport theory was attempted. We argued, however, with quasi-classical localization of electrons moving along closed equipotential lines [3]. We believe that these concepts are adequate in the poor screening situation of nearly complete filling of LLs, but not within a strict single-particle picture, since for half-filled LLs, e.g., the long-range potentials producing this localization will be absent owing to screening. Nevertheless, the breakdown of screening of long-range potential fluctuations in situations with nearly complete filling of LLs is a possible localization mechanism, and perhaps the most effective one in the IQHE.

Finally, we want to emphasize that the present discussion of the very effective screening ability of the 2DEG in a strong magnetic field was strictly restricted to thermal equilibrium situations. There is strong evidence that the dynamical screening properties are completely different, and that even a quasistatic electric field is essentially not screened if it is switched on adiabatically [19].

References

[1] For a recent review see: R.E. Prange and S.M. Girvin (eds.): *The Quantum Hall Effect* (Springer, New York, Berlin, Heidelberg 1987)

[2] T. Chakraborty and P. Pietiläinen, *The Fractional Quantum Hall Effect: Properties of an incompressible quantum fluid*, K. v. Klitzing (ed.), Vol. **85** Springer Series in Solid State Sciences (Berlin 1988)

[3] S.M. Apenko and Yu.E. Lozovik, *J. Phys. C: Solid State Phys.*, **18**, 1197 (1985); *Eksp. Teor. Fiz.* **89**, 573 (1985) [*Sov. Phys. JETP***62**, 328 (1985)]

[4] T. Ando, *J. Phys. Soc. Jpn.* **37**, 622 (1974)

[5] R.R. Gerhardts, *Z. Physik* **B21**, 285 (1975)

[6] F. Wegner, *Z. Physik* **B51**, 279 (1983)

[7] E. Brézin, D.J. Gross, and C. Itzykon, *Nucl. Phys.* **B235** [FS 11] 24 (1984)

[8] D. Weiss and K. v. Klitzing, in: *High Magnetic Fields in Semiconductor Physics*, G. Landwehr (ed.), Vol. **71** Springer Series in Solid State Sciences (Berlin 1987), p. 57

[9] E. Gornik, W. Seidenbusch, and G. Strasser, in: *High Magnetic Fields in Semiconductor Physics*, G. Landwehr (ed.), Vol. **71** Springer Series in Solid State Sciences (Berlin 1987), p. 193

[10] U. Wulf, V. Gudmundsson, and R.R. Gerhardts, *Phys. Rev. B*, in press

[11] V. Gudmundsson, R.R. Gerhardts, R. Johnston, and L. Schweitzer, *Z. Physik* **B70**, 453 (1988)

[12] U. Wulf and R.R. Gerhardts, in: *Physics and Technology of Submicron Structures*, H. Heinrich, G. Bauer, and F. Kuchar (eds.), Vol. **xx**, Springer Series in Solid State Sciences (Berlin 1988), in press

[13] B.I. Shklovskii and A.L. Éfros, *Pis'ma Zh. Eksp. Teor.Fiz.* **44**, 520 (1986) [*JETP Lett.* **44**, 669 (1987)]

[14] R.R. Gerhardts and V. Gudmundsson, *Phys. Rev.* **B34**, 2999 (1986)

[15] V. Gudmundsson and R.R. Gerhardts, *Phys. Rev.* **B35**, 8005 (1987); see also *High Magnetic Fields in Semiconductor Physics*, G. Landwehr (ed.), Vol. **71** Springer Series in Solid State Sciences (Berlin 1987), p. 67

[16] E. Stahl, D. Weiss, G. Weimann, K. v. Klitzing, and K. Ploog, *J. Phys. C* **18**, L783 (1985)

[17] V. Gudmundsson and R.R. Gerhardts, *Phys. Rev.* **B37**, 10361 (1988)

[18] D. Weiss, V. Mosser, V. Gudmundsson, R.R. Gerhardts, and K. v. Klitzing, *Solid State Commun.* **62**, 89 (1987)

[19] R.R. Gerhardts and V. Gudmundsson, *Solid State Commun.*, in press

Electronic States in Two-Dimensional Random Systems in the Presence of a Strong Magnetic Field

B. Kramer [1], *Y. Ono* [2], *and T. Ohtsuki* [3]

[1] Physikalisch-Technische Bundesanstalt, Bundesallee 100, D-3000 Braunschweig, Fed. Rep. of Germany

[2] Toho University, 2-1 Miyama 2 chome, Funabashi-shi, Chiba, 274 Japan

[3] University of Tokyo, 7-3-1 Hongo, Bunkyo-ku, Tokyo 113, Japan

We discuss the status of the theory of localisation in two dimensions in the presence of a strong perpendicular magnetic field. The model potentials which have been used are described. The importance of the various length scales in the problem is pointed out. A summary of the currently available definitions of localisation is given. The results for the localisation length, the inverse participation number of the states, and their shape, obtained so far are summarised.

1. INTRODUCTION

For the transport properties of inversion layers in a strong perpendicular magnetic field it is most important to understand the basic features of the electronic quantum states in some detail. In the simplest case, neglecting interactions, one has to solve the Schrödinger equation for an electron (mass m, charge e) which moves in the x-y plane under the influence of a homogeneous magnetic field **B** parallel to the z-axis (represented by a suitable vector potential **A**), and a random potential V(**r**).

$$\left\{\frac{1}{2m}(\mathbf{p}-e\mathbf{A})^2 + V(\mathbf{r})\right\}\psi(\mathbf{r}) = E\,\psi(\mathbf{r}) \ . \tag{1}$$

In general, the potential must be specified by some distribution function P([V]). Some of the currently used models are discussed in the following section.

Since the evaluation of the transport properties involves only integrals of the eigenstates most of the information which would be obtained by solving eq. (1) is superfluous. Only certain (average) properties are of importance. With respect to the quantised Hall effect [1] the question whether or not the states can contribute to a current has to be answered [2]. This can be investigated by considering their asymptotic behavior since that determines the convergence properties of the matrix elements in the Kubo formula [3]. A survey of the relevant definitions is given in the third section [3, 4].

Two quantities can be defined, the localisation length λ, and the participation number P. They depend on the parameters of the random potential ("disorder"), on the energy and on the magnetic field. If they are finite, the states at E do not contribute to transport. They are localised. Presently, the common belief is that in the quantum Hall regime there are non-localised states only near the centres of the (disorder-broadened) Landau bands. At all other energies the states are localised, in a system without edges [2]. The available theories which lead to these statements are critically reviewed in the fourth section.

Springer Series in Solid-State Sciences, Vol. 87
High Magnetic Fields in Semiconductor Physics II Editor: G. Landwehr

2. RANDOM POTENTIALS

Having in mind impurities as the origin of the disorder we write [5]

$$V(\mathbf{r}) = \sum_{j=1}^{N} v_j(\mathbf{r} - \mathbf{R}_j) \tag{2}$$

where v_j is the potential of an impurity at the site R_j. The simplest choice is to take $v_j(\mathbf{r}) = v_j\delta(\mathbf{r})$ ($|v_j| = v_0$, and $\Sigma v_j = 0$ for convenience). Taking $\mathbf{R}_j$ to be independent and randomly distributed within the area $A = L_xL_y$ of the system the potential is spatially uncorrelated (white noise). The parameter which characterises the amount of the disorder is the variance of V, $W = v_0^2N/A$. Spatially correlated potentials are obtained either by taking the positions of the impurities as correlated or to take the potential of the single impurity to be of finite range, a, i. e. $v_j(\mathbf{r}) = v_j/(\pi a^2)\exp(-r^2/a^2)$. More general forms of the potential can be generated by taking non-identical impurities in the superposition eq. (2). Model potentials of this kind have been used together with the Landau model in many of the approximative analytical, and numerical works concerning the electronic structure [5], and the localisation problem [6].

Instead of considering the continuous space we can also take the Hamiltonian to be defined on a discrete space, for instance on a square lattice in 2d. The simplest model is

$$H = \sum_{lm,l'm} V_{lm\,l'm'}\,|lm\rangle\langle l'm'| + \sum_{lm} \varepsilon_{lm}\,|lm\rangle\langle lm| \ , \tag{3}$$

where

$$V_{lm\,l'm'} = \exp\{\pm 2\pi i l\alpha\}\,\delta_{l,l'}\,\delta_{m',m\pm1} + \delta_{l',l\pm1}\,\delta_{m,m'} \ . \tag{4}$$

$\alpha = B/(h/e)$ is the number of the magnetic flux quanta in the unit cell (area "1"). The first term in eq.(3) corresponds to the kinetic energy. The magnetic field is introduced here via the so-called Peierls substitution [7]. The second term corresponds to $V(\mathbf{r})$ in eq. (1). The model has been used for the study of the electronic properties in the presence of a magnetic field and a periodic potential [8, 9] (no randomness) as well as for the localisation problem [10-13]. In this model there is a finite width of the magnetic subbands even without randomness.(Harper broadening [14]). This might influence the localisation properties.

Since the macroscopic quantities should not depend on the microscopic details of the potential it is reasonable, instead of specifying the explicit form of the Hamiltonian, to start from the statistical distribution. A convenient choice is the generalized Gaussian,

$$P([V]) = C\exp\{-\int V(\mathbf{r})K(\mathbf{r},\mathbf{r}')V(\mathbf{r}')\,d\mathbf{r}\,d\mathbf{r}'\} \ , \tag{5}$$

where the normalisation constant C is determined by the functional integral $\int D[V]P([V]) = 1$, and the function K is the inverse of the correlation function of the potential [15].

$$\int d\mathbf{r}''\,K(\mathbf{r},\mathbf{r}'')\,\langle V(\mathbf{r}'')V(\mathbf{r}')\rangle = \delta(\mathbf{r}-\mathbf{r}') \ . \tag{6}$$

In connection with the Landau model a useful choice is

$$\langle V(\mathbf{r})V(\mathbf{r}')\rangle = W \exp\{-|\mathbf{r}-\mathbf{r}'|^2/a^2\} \ . \tag{7}$$

Here $\langle \ldots \rangle$ denotes the configurational average defined by the functional integral $\int D[V]\ldots$. The average of the potential vanishes, its variance is given by W, and the correlation length is a. The generalisation of eqs. (5) to (7) to the lattice model of eqs. (3) and (4) is obvious.

The Gaussian white noise potential (a = 0) as well as the more general spatially correlated potential ($a \neq 0$) was used in perturbation theories, see, e. g. [16, 17], respectively, and in the recently developed two-parameter scaling theories [18-20]. The asymptotic limit of an infinite correlation length describes a slowly varying potential. Here, the electronic problem is equivalent to a percolation problem [21-27] as we shall discuss below. It should be most interesting to study the influence of quantum mechanical effects (tunneling, interference) on this model. One possibility is to put in these effects by hand, as has been done recently by *Chalker and Coddington* [28]. It would certainly be more satisfactory to have a model which is able to cover both of the limits $a\rightarrow 0$ (white noise limit) and $a\rightarrow\infty$ (percolation limit) on an equal footing.

Besides the size of the system which eventually is taken to be infinite, the physical situation described by the Hamiltonian of eq. (1) can be characterised by the magnetic length $l = (h/(eB))^{1/2}$, and the spatial correlation length, a. In the lattice model, eq. (3), there is still another length scale, namely the lattice distance (which is usually taken as unity). *A priori*, it is not obvious, whether or not these additional length scales are important for the localisation problem. Without magnetic field one can set up a one-parameter scaling theory of localisation near the mobility edge [29-31]. This has been confirmed numerically [32, 33].It implies that the spatial correlation of the potential energy is no longer of importance when the critical regime is approached [34, 35]. One can start without loss of generality from the Gaussian white noise potential. For non-vanishing magnetic field it is not yet clear whether this is allowed. The fact that the presently available numerical data indicate a breakdown of the one-parameter scaling theory, according to the conclusions of the authors,e. g. [2, 6, 12, 36], might indicate that it is necessary to take the spatial correlation length finite.

3. DEFINITION OF LOCALISATION

The localisation length λ describes the average asymptotic behavior of the states near an energy E. It may be defined as the exponential decay length of an eigenstate ψ at infinity, i. e.

$$\psi(\mathbf{r}) = f(\mathbf{r}) \exp\{-|\mathbf{r}|/\lambda\} \tag{8}$$

in the limit of $|\mathbf{r}| \rightarrow \infty$. f is a randomly varying, complex function of the order unity. Practically, this definition is not very useful, since its application would require the calculation of the eigenstates.

It is more convenient to define λ by employing the relation to the transport properties [37, 38]

$$\lambda^{-1} = -\lim_{\eta\rightarrow 0} \lim_{|\mathbf{r}-\mathbf{r}'|\rightarrow\infty} \frac{1}{2|\mathbf{r}-\mathbf{r}'|} \langle \log|G(E^+; \mathbf{r}, \mathbf{r}')|^2\rangle \ . \tag{9}$$

G denotes the one-electron Green's function of the Hamiltonian of the system. If at the energy E the eigenstates are of the form eq. (8) then this definition yields the same result for λ. $|G(E, \mathbf{r}, \mathbf{r}')|^2$ measures the probability for an electron to go from **r** to **r'**. Therefore, this definition connects also with the transmission properties of the system.

For numerical purposes it is necessary to have a definition of localisation which yields information about the states already for a finite system size.

There are essentially two possibilities. The first is obtained by applying the above criterion, eq. (9) to a system with the shape of a "bar" which is taken to be very long in one direction, and to calculate λ from the matrix elements of G(E) between its ends. For the lattice model, eqs. (3, 4), we can define [32, 33, 38]

$$\lambda^{-1} = -\lim_{L\to\infty} \frac{1}{2L} \log \mathrm{Tr}\, |G_{1L}(E^+)|^2 \,, \tag{10}$$

where the trace is taken within the subspaces which correspond to the 1st and the Lth layer, and $G_{1L}(E^+)$ denotes the respective matrix which can be calculated recursively [38-41]. λ depends on the the cross-sectional area, say M^{d-1}, and is a finite number since, for $L\to\infty$, the system is quasi-1d. The d-dimensional limit can be obtained in a systematic way by investigating the scaling properties of λ with respect to M, E, and W. The resulting scaling parameter can be shown to be identical with the localisation length in the infinite system in the localised region, and with the inverse of the dc-conductivity σ in the conducting regime [32, 33]. In order to obtain a meaningful d-dimensional limit it is imperative to establish the scaling properties

Alternatively, one can investigate the shift of the energy eigenvalues of a finite system due to small changes in the boundary conditions, as was proposed by *Edwards and Thouless* [37, 42]. The average energy shift, δE, in second order perturbation theory, is related to the conductivity by

$$\sigma = \frac{e^2}{h} f \,\frac{\delta E}{\Delta E} = \frac{e^2}{h} f\, g(L) \,, \tag{11}$$

where ΔE is the average energy spacing of the eigenvalues, and f a numerical factor that depends on the details of the model used (square or triangular lattice, for instance). g(L) is called the Thouless number. For localised states the mean energy shift becomes very small for large sytem sizes such that g(L) vanishes exponentially.In the metallic regime the boundary conditions will always influence the energy levels, even for infinite systems. Although the assumptions made in deriving eq. (11) which concern the statistical behavior of the energy differences in the system, and the corresponding momentum matrix elements, do not seem to be extremely well satisfied [43], the Thouless number turns out to be very useful in numerical calculations, e. g. [44], besides being conceptually important [29, 30].

A special case of this criterion has been introduced for the study of localisation in the disordered Landau model [45]. Introducing a gauge transformation $\mathbf{A} \to \mathbf{A}+\Delta\mathbf{A}$, $\Delta\mathbf{A}=\Delta\Phi\mathbf{e}_y/L$, is equivalent to changing the flux through a looped ribbon (periodic boundary conditions in the y-direction) by an amount $\Delta\Phi$. When an electric field is applied the energy levels that correspond to extended states will change with $\Delta\Phi$, whereas those of the localised states do not, in the thermodynamic limit. All of these criteria are related to the *asymptotic behavior* of the states. We wish also to have insight into their *shape* . This is provided by the inverse participation participation number, P_ψ^{-1}, which is given by the integrated fourth power of the amplitude of a wavefunction [46].

$$P_\psi^{-1} = \int_\Omega d^d r \; |\psi(\mathbf{r})|^4 . \tag{12}$$

Here, $\|\psi\| = 1$, and Ω is the volume of the d-dimensional system. The generalisation to the lattice model is again obvious. For a free electron state $P_\psi = \Omega$. For a state which is confined to a finite volume Ω_0, $P_\psi = \Omega_0$. For an exponentially localised state (eq. (8)) with $|f(\mathbf{r})| = 1$, $P_\psi = \pi\lambda^3$ in 3-d. ψ is extended if P_ψ^{-1} tends to zero in the thermodynamic limit, and localised if P_ψ stays finite. In general, P_ψ is not related to the localisation length.

For disordered systems one considers the configurational average

$$P^{-1} = \left\langle \sum_\psi P_\psi^{-1}\, \delta(E - E_\psi) \Big/ \sum_\psi \delta(E - E_\psi) \right\rangle . \tag{13}$$

It may be related to the transport properties via [47]

$$P(E)^{-1} = 2\pi \rho(E)^{-1} \lim_{\eta \to 0} \eta A(\eta, E) , \tag{14}$$

where $\rho(E)$ is the density of states, and

$$A(\eta, E) = \frac{1}{\pi^2} \left\langle (\mathrm{Im}\, G(E+i\eta; \mathbf{r}, \mathbf{r}))^2 \right\rangle . \tag{15}$$

The quantity $\eta A(\eta, E)$ gives, in the limit of.$\eta \to 0$, and integrated over E, the return probability to the site $\mathbf{r}$ when $t \to \infty$,

$$\lim_{t \to \infty} \left\langle |G(t; \mathbf{r}, \mathbf{r})|^2 \right\rangle = \int_{-\infty}^{\infty} dE\, \rho(E)\, P^{-1}(E) . \tag{16}$$

If $\eta A(\eta, E)$ has a finite value the particle at energy E cannot diffuse away.

Whether or not the states near E are localised can also be decided from the "fractal dimensionality", d*, which may be defined as

$$d^* = \lim_{L \to \infty} \frac{d \log P}{d \log L} . \tag{17}$$

Localised states have $d^* = 0$ whereas extended states have $0 < d^* \le d$, depending on their shape, i. e. whether or not they are "space filling"[4].

4. THE STATUS OF THE THEORY OF LOCALISATION IN STRONG MAGNETIC FIELDS

The main results for the problem of localisation in a strong magnetic field are obtained from perturbational [16, 17, 48-52], numerical [4, 6, 10-13, 45, 53-60], and field theoretical [18-20] treatments of the conductivity, the localisation length, the Thouless number, and the participation number. In the limit of extremely high magnetic fields percolation arguments have been used [21-27]. The main results are summarised in Table 1.

Table 1: The status of the theory of localisation in a strong magnetic field. Explanation of methods and symbols see text

	method	ν	π	d^*
	perturbation expansion [16]	∞	–	–
	Thouless number study [53]	<2	–	–
	numerical scaling [10-12,57]	≈2	–	–
asymptotic behavior	Padé fit [6, 36]	2 (4)	–	–
	field theory [63]	1/2 (?)	–	–
	classical percolation [4, 25]	0, 4/3	–	–
	percolation, quantum correction [28]	≈2.5	–	–
	perturbation expansion [50,51]	1.9±0.2	3.8±0.4	–
shape of the states	numerical diagonalisation [58, 67]	–	–	< 1.7
	percolation [4]	91/72	91/36	1.75

4.1 Perturbation Theory

Most elaborately, the perturbational approach has been carried out by Ono [16, 48]. The starting point is the density-density relaxation function

$$\chi(\mathbf{r}, \mathbf{r}'; t) = i\theta(t)\langle[\rho(\mathbf{r}, t), \rho(\mathbf{r}', t)]\rangle \quad , \tag{18}$$

which can be evaluated systematically as a perturbation series for weak randomness by using diagrammatic techniques. The localisation length is obtained from the static density response function $\chi(\mathbf{q}, 0)$ (Fourier transformed density correlation function) according to

$$\chi(\mathbf{q}, 0) = n(E_F)q^2/(q^2 + \lambda^{-2}) \quad . \tag{19}$$

The explicit result, obtained for the case of *δ-function impurity potentials* is

$$\lambda(E) = \text{const} \cdot \exp\{\alpha/(E-E_N)^2\} \quad , \tag{20}$$

where α is of the order of the square of the disorder-induced bandwidth, and E_N is the energy of the unperturbed Landau level. There is an essential singularity of λ in the band centre. Although there is a number of assumptions in this approach which are not easily justifiable, this was the first hint obtained from theory that extended states might exist only at a single energy in each of the Landau bands in the quantum mechanical high-magnetic field limit.

4.2 Thouless Number

The earliest numerical study of localisation has been performed by Ando [53] by applying the Thouless number criterion to the random Landau model, eq. (1). Calculations have been done for δ-function potentials, as well as for finite range potentials for as many as three Landau bands. An inverse localisation length α(E) was defined via the exponential behavior of g(L).

$$g(E; L) = g(0)\exp\{-\alpha(E)/L\} \ . \tag{21}$$

In order to be consistent with the definition in the previous section we have to identify α with 2/λ. The results for short-range scatterers were consistent with a divergence of the localisation length only at the centres of the Landau bands, but $\alpha(E) \propto |E-E_N|^{\nu}$ with ν being less than 2 in contrast to the essential singularity obtained from the renormalised perturbation expansion. For finite range potentials (a =o(l)) no conclusive result about the critical exponent could be obtained due to too large error bars in the numerical data.

4.3 Numerical Scaling

The scaling properties of the exponential decay length of the modulus of the Green's function, eq. (9), with respect to energy, disorder, and the width of the system have been studied for both, the lattice model, eqs. (3, 4, 10) [10-12, 57], and the random Landau model, eq. (1). In the latter δ-function potentials as well as Gaussian potentials have been considered [2, 6, 36]. Numerically, these have been the most elaborate studies of the localisation problem. We briefly summarize the essential ideas in the calculation [38, 40, 41], and describe the main results.

The inverse localisation length as defined by eq. (10) is calculated recursively as a function of the length of the system, L, for a finite width, say M [32, 33]. The result, $\lambda_M(E, W, B; L)$ is a finite number with relative uncertainty, $\Delta\lambda/\lambda$, which vanishes asymptotically proportional to $L^{-1/2}$, because $\log|G|$ is self-averaging [41]. In the calculations L is increased as much as is necessary to reach a given $\Delta\lambda/\lambda$. The localisation length in the infinite 2d system is obtained from the limit of λ_M when $M \to \infty$.

In order to be able to do the extrapolation a scaling function f(x) is established,

$$\frac{\lambda_M}{M} = f\left(\frac{\xi}{M}\right) \ . \tag{22}$$

f(x) depends on the dimensionality, and on the "universality class" of the Hamiltonian. The scaling parameter ξ depends on E, W, B,.... but not on M. Eq. (22) is a one-parameter scaling relation. In general, especially in the presence of a magnetic field, it might be necessary to include more than one scaling parameter [60]. Having established f(x) the extrapolation with $M \to \infty$ can be trivially done. In the localised regime, ξ is the localisation length.

If a scaling relation of the type eq. (22) exists, and this is the most essential point of the procedure, it is possible to determine e. g. the critical exponent of the localisation length which is defined by

$$\xi(E, W, B\ldots) = |E - E_C(W, B, \ldots)|^{-\nu} \tag{23}$$

directly from the raw data, without referring to the scaling paramater which contains already some (although controllable) fitting procedure [33, 60, 61].

An early attempt to apply this method to the 2d localisation problem in the presence of a magnetic field was based on the lattice model [10-12, 57]. It was tried to find a one-parameter scaling relation of the type eq. (22). This turned out to be successful to a certain degree. However, due to the small system sizes, localised states could only be established in the outermost parts of the magnetic subbands. In the centres of the bands finite energy regions were found where the states are larger than the diameter of the systems considered. No serious attempt was made at that time to extract the critical exponent although the data seemed to be consistent with results obtained from the Thouless number study. There are theoretical reasons for the belief that the one-parameter scaling relation, eq. (22), has to be replaced by a two-parameter scaling law in the presence of a magnetic field [18-20]. Attempts have been made to establish such a relation on the basis of more and refined numerial data for the lattice model [60]. The data are consistent with a singularity of the localisation length near theband centres, and $\nu \cong 2$. There are, however, still inconsistencies which have to be removed by future work [60].

The scaling method has been applied also to the random Landau model [6], although here the situation is more complicated due to the large number of coupling matrix elements between the Landau states. Again, both, δ-function impurity potentials as well as Gaussian potentials were used. No attempt was made to establish either a one-parameter scaling relation in these studies, or a two-parameter one. Instead, a Borel-Padé fit of the data for $\lambda_M(0)/\lambda_M(E)$ as function of M/l,was performed from which the localisation length as a function of the energy could be determined. The critical exponent was extracted. In the lowest Landau band $\nu \cong 2$, and $\nu \cong 4$ in the first Landau band. Also, the critical exponents seem to be independent to the range of the impurity potentials according to the results. It must be noted that the error bars on the localisation lengths near the centres of the bands have been rather large, and increasing towards E_N. Therefore, we feel that these studies must be repeated, and supported by systematically establishing a scaling relation.

4.4 Non-Linear σ-Model

Field theoretical methods have been extremely successful in treating the localisation problem [31], at least for B = 0. For B ≠ 0 these methods have been used for the white noise limit [18, 19], and for the correlated potential [20]. Without going into details, and repeating the standard notation like "generating functions", "replica method", "effective Lagrangian", "Grassmann variables", "non-linear σ-model", "topological term", "dilute instanton gas", etc, which can anyway only be understood by thoroughly reevaluating the original papers, we state that these treatments yield results that are qualitatively consistent with those mentioned above. The essential point is that an "effective Lagragian" is derived by evaluating the configurational average over the randomness exactly (this is possible due to the assumption of a Gaussian distribution), and subsequently performing the remaining integration over the Fermion fields approximately by using the saddle point method after a transformation to boson coordinates. The effective Lagrangian contains two terms, one being related to the longitudinal conductivity σ_{xx}, the other to the Hall conductivity σ_{xy}. The second of these is due to "topological properties" of the model, and, according to the authors, not accessible to perturbational treatments. The two-parameter scaling picture [62] which can be deduced by considering the two conductivity components as scaling variables eventually leads to the singularities in the centres of the Landau bands which are indentified with a non-vanishing σ_{xx}, and divergencies of the localisation length. An estimate of the critical exponent has been given by Affleck, $\nu = 1/2$ [63].

4.5 The Percolation Limit

In the limit $B \rightarrow \infty$ the quantum mechanical problem can be replaced by a classical percolation problem, as has been repetitively argued. The basis of the percolation arguement has been formulated by Tsukada [21]. It can be shown that the Schrödinger equation in the single band approximation, which is justified in the high field limit, can be written as

$$V(X, Y)\, C(X) = (E - E_0)\, C(X) \ , \tag{24}$$

where X, Y are the centre coordinates of the cyclotron motion defined by $X = kl^2$, and $Y = i^{-1}\partial/\partial k$, respectively. k is the wave number of the Landau state, C(X) is the expansion coefficient of the eigenstate with respect to the Landau basis. Eq. (24) holds in the limit of $l \ll a$. It is seen that all Landau states that correspond to the equipotential line $V(X, Y) = (E - E_0)$ contribute with an equal amplitude to the eigenstate at energy E.

The question whether or not an eigenstate is localised may be decided by investigating the percolation problem for the equipotential lines [25] (cf. Fig. 1). An intuitive insight is obtained by considering the analogy with a hilly landscape (potential energy) that is gradually filled with water, the water level corresponding to the energy E. For very low water levels the water runs into the deep valleys. All shore lines (equipotential lines) are closed path, i. e. the states are localised. The same is true for very high water levels. Only a few mountains are high enough to elevate above the water level. The shore lines are again closed paths. It is clear that in between there must be one water level at which it is possible to travel either by boat or by foot from one side of the system to the other. This corresponds to an extended state. Percolation theory says that there is exactly one percolating path. The length t of a cluster (equipotential line) is given by $t = |E - E_0|^{-1/\sigma}$ with $\sigma = 36/91$. The diameter ξ of the area covered is $\xi = |E - E_0|^{-\nu}$ with $\nu = 4/3$.

This may be used to estimate the critical behavior of the states. Of course, there is some ambiguity. If we identify the localisation length with ξ then the corresponding critical exponent is ν. Another possibility is obtained by identifying the area covered by the cluster with $t \cdot l$. This gives the possibility of deriving the critical behavior of the participation number (see below). However, adopting the asymptotic definition of the localisation length it will be zero(!) and does not show any critical behavior at all.

4.6 The Participation Number

The (inverse) participation number (ratio) has been studied for the random Landau model in different geometries using a variety of methods [4, 50, 51,58, 65-67].

The shape of the wave functions have been studied numerically [58]. Fractal behavior has been concluded near the critical energy. The fractal dimension d* of the states near the centres of the Landau bands has been estimated from the behavior of the amplitudes of the wave function, d* = 1.57. From the behavior of the inverse participation number [67] when changing the size of the system a fractal dimension of the same order of magnitude, and even smaller, was obtained. It turned out in the latter studies that it was rather complicated to draw definite conclusions because of the limitations in the sizes of the systems considered. In any case, the fractal dimension of the states near the band centre seems to be much smaller than d = 2. The extended states, if they exist, must therefore be of the "spaghetti type", i. e. they are not space filling. This conclusion is supported by percolation theoretical arguments [4]. In this case one can derive the fractal dimension exactly, d* = 1.75.

The critical behavior of the inverse participation number has been studied systematically for the lowest Landau level by Hikami using large order perturbation theory [50,51]. Taking $P^{-1} = A|E - E_0|^{\pi}$ he obtained $\pi = 3.8 \pm 0.4$. Assuming P to be given by the localisation length, i. e. $P \approx \lambda^2$, this can be used to estimate $\nu = 1.9 \pm 0.2$. However, as mentioned above, this relation between the participation number and the localisation length is not necessarily fulfilled.

From the percolation limit one obtains the critical behavior of the participation number from the critical behavior of the length of the percolating equipotential line [4]. This gives $\pi \approx \sigma^{-1} \approx 2.52$, and, correspondingly, we would have $\nu \approx 1.26$. This is considerably smaller than the above perturbational estimate.

5. CONCLUSION

A few critical remarks concerning the status of the theory of localisation in the magnetic field seem to be in order.

All of the numerical treatments yield results that are more or less in agreement with each other, at least qualitatively. However, it is not yet clear whether or not this agreement is merely an artifact of the limited capacity of the available computers. In this connection it might be of interest to remember the situation concerning the minimum metallic conductivity before 1979, i. e. the advent of the scaling theory of localisation.

A very promising approach seems to me the field theoretical two-parameter scaling theory. However, the approximations involved are only partly justifiable and sometimes rather arbitrary. In addition, the results obtained so far could not yet be fully verified, either experimentally, or theoretically by numerical analyses [60, 64]. Concerning experimental verification, see the contribution by Wei in this volume [68, 69].

The results for the percolation limit seem to agree qualitatively with the quantum mechanical theories. There is only one energy at which extended states exists. Quantitatively, they are inconsistent. Further clarification is needed, especially concerning quantum mechanical corrections [28].

Although the results about the shape of the states near the crtitical energy seem to agree quite well qualitatively, there are serious quantitative differences which have to be sorted out in the future.

Most of the theoretical results have been obtained by using rather specialised microscopic models. On the other hand, localisation is believed to be a general phenomenon. The transition from localisation to non-localisation is supposed to be universal, and should not depend on the microscopic details of the system. So far, the only theoretical approach that takes this fact into account is the field theory. For numerical calculations, a model which is simple enough in order to be accessible to some rigorous mathematical treatment but nevertheless contains the essential physics, is still not at hand.

ACKNOWLEDGEMENTS

BK gratefully acknowledges useful discussions with Walter Apel, Wolfgang Weller and Hajo Leschke. This work was begun when BK visited Toho University, Funabashi, Japan, as a guest scientist. The warm hospitality of the members of the faculty of science has been a great experience, and should be especially mentioned here.

REFERENCES

1. K. von Klitzing, G. Dorda and M. Pepper, Phys. Rev. Letters **45** 494 (1980)
2. H. Aoki, Rep. Progr. Phys. **50** 655 (1987)
3. B. Kramer, preprint, to be published in Procs. 5th Int. Winterschool on New Dev. Sol. St.Phys., Mauterndorf Austria, (1988)
4. B. Kramer, Y. Ono and T. Ohtsuki, Surface. Sci. **196** 127 (1988)
5. T. Ando, A. B. Fowler and F. Stern, Rev. Mod. Phys. **54** 437 (1982)
6. T. Ando and H. Aoki, Phys. Rev. Letters **54** 832 (1985); J. Phys. Soc Japan **54** 2238 (1985)
7. J. M. Luttinger, Phys. Rev. **84** 814 (1951)
8. D. R. Hofstadter, Phys. Rev. **B14** 2239 (1976)
9. G. H. Wannier, G. M. Obermair, and R. Ray, phys. stat. sol. (b) **93** 337 (1979)
10. L. Schweitzer, B. Kramer, and A. MacKinnon, J. Phys. **C17** 4111 (1984)
11. A. MacKinnon, B. Kramer, and L. Schweitzer, Proc. LITPIM Suppl. (PTB-PG-1) 314 (PTB Braunschweig 1984)
12. A. MacKinnon, B. Kramer, Lecture Notes in Physics **177** 74 (1983)
13. H. Aoki, J. Phys. **C18** L67 (1985); Procs. LITPIM Suppl. (PTB-PG-1) 302 (PTB Braunschweig 1984)
14. P. G. Harper, Proc. Phys. Soc. London **A68** 874 (1955)
15. A. Papoulis, Probability, Random Variables, and Stochastic Processes (McGraw Hill New York) 1984)
16. Y. Ono, J. Phys. Soc. Japan **51** 2055, 3544 (1982); **52** 2492 (1983); Suppl. 247; **53** 2342 (1984)
17. S. Hikami and E. Brezin, J. Physique **46** 2021 (1985)
18. H. Levine, S. B. Libby and A. M. M. Pruisken, Nucl. Phys. **B240**[FS12] 30, 40, and 57 (1984)
19. A. M. M Pruisken, Nucl. Phys. **B235** [FS11] 277 (1984)
20. H. A. Weidenmüller, Nucl. Phys. **B290** [FS20] 87 (1987)
21. Masaru Tsukada, J. Phys. Soc. Japan **41** 1466 (1976)
22. S. V. Iordansky, Sol. St. Commun. **43** 1 (1982)
23. R. F. Kazarinov and Serge Luryi, Phys. Rev. **B25** 7626 (1982)
24. Y. Ono, J. Phys. Soc. Japan **51**, 237 (1982)

25. S. A. Trugman, Phys. Rev. **B27** 7539 (1983)
26. S. Luryi and R. F. Kazarinov, Phys. Rev. **B27** 1386 (1983)
27. S. M. Apenko and Yu. E. Lozovik, preprint 196, Acad. Sci (Moscow 1984)
28. J. T. Chalker, P. D. Coddington, J. Phys. **C21** 2665 (1988)
29. E. Abrahams, P. W. Anderson, D. C. Licciardello, T. V. Ramakrishnan, Phys. Rev. Letters **42** 673 (1979)
30. F. J. Wegner, Z. Phys. **25** 327 (1976)
31. F. J. Wegner,Springer Ser. Sol. St. Sci. **61** 99 (1985)
32. A. MacKinnon and B. Kramer, Phys. Rev. Letters **47** 1546 (1981)
33. A. MacKinnon and B. Kramer, Z. Phys. **B53** 1 (1983)
34. J. Sajeev, M. J. Stephen, Phys. Rev. **B28** 6358 (198
35. R. Johnston and B. Kramer, Z. Phys. **63** 273 (1986)
36. T. Ando, Prog. Theor. Phys. Suppl. 84 69 (1985)
37. D. J. Thouless, Phys. Rep. **13** 93 (1974)
38. B. Kramer M. Schreiber, Lecture Notes in Physics **268** 351 (1987)
39. P. Dean, Rev. Mod. Phys. **44** 127 (1972)
40. A. MacKinnon, Z. Phys. **B59** 385 (1985)
41. B. Kramer, M. Schreiber, to be published
42. J. T. Edwards, D. J. Thouless, J. Phys. **C5** 807 (1972)
43. H. Kühl, PHD Thesis, unpublished (Universität zu Köln 1980)
44. D. C. Licciardello, D. J. Thouless, J. Phys. **C8** 4157 (1975); J. Phys. **C11** 925 (1978)
45. H. Aoki, J. Phys. **C15** L1227 (1982); Lecture Notes in Physics **177** 11 (1983); J. Phys. **C16** 1893 (1983); **C18** 1885 (1984)
46. J. Bell, P. Dean, Disc. Faraday Soc. **50** 55 (1970)
47. F. Wegner, Z. Phys. **B36** 209 (1980)
48. Y. Ono, Progr. Theor. Phys. (Suppl.) **84** 138 (1985)
49. S. Hikami, Phys. Rev. **B29** 3726 (1984); Progr. Theor. Phys. **72** 722 (1984)
50. S. Hikami, Progr. Theor. Phys. **76** 1210 (1986)
51. S. Hikami, E. Brezin, Surf. Sci **170** 262 (1986)
52. R. R. P. Singh, S. Chakravarty, Nucl. Phys. **B265** [FS15] 265 (1986)
53. T. Ando, J. Phys. Soc. Japan **52** 1740 (1983); **53** 3101, 3126 (1984)
54. T. Ando,Springer Ser. Sol. St. Sci. **39** 176 (1982) ;**71** 2 (1987); Surf. Sci **196** 120 (1988); Springer Procs. in Phys. **28** 252 (1988)
55. H. Aoki, Surf. Sci. **196** 107 (1988)
56. H. Aoki, T. Ando, Springer Ser. Sol. St. Sci. **71** 45 (1987)
57. B. Kramer, A. MacKinnon, Teubner Texte zur Physik **3** 157 (Leipzig 1984)
58. H. Aoki, J. Phys. **C16** 1205 (1983); J. Phys. **C17** 1875 (1984); Phys. Rev. **B33** 7310 (1986)
59. G. C. Aers, A. H. MacDonald, J. Phys. **C17** 5491 (1984)
60. A. MacKinnon, this volume
61. M. Schreiber, B. Kramer, A. MacKinnon, Phys. Scripta, to be published (1988)
62. D. E. Khmel'nitskii, JETP Lett. **38** 552 (1983)
63. I. Affleck, Nucl. Phys. **B265** [FS15] 409 (1986)
64. H. Aoki, T. Ando, Surf. Sci. **170** 249 (1986)
65. H. Kunz, and B. Soulliard, J. Physique Lett. **43** L39 (1982)
66. M. R. Zirnbauer, Nucl. Phys. **B235** [FS15] 375 (1986)
67. T. Ohtsuki, Y. Ono, to be published
68. H.P. Wei, this volume
69. H.P. Wei, D.C. Tsui, A.M.M. Pruisken, Phys. Rev. **B33** 1488 (1986)

Conductance Fluctuations on the Quantum Hall Plateaus in GaAs/AlGaAs

R.G. Mani and J.R. Anderson

Joint Program for Advanced Electronic Materials, Department of Physics, University of Maryland, and
Laboratory for Physical Sciences, College Park, MD 20742, USA

1 Introduction

The gauge invariance arguments of Laughlin imply perfect quantization of the quantum Hall plateaus /1/ and Hall voltage measurements using digital time-averaging, null detection techniques at fixed values of the magnetic field have shown that the Hall plateaus are flat to ~ 0.01 parts in 10^6 (i.e. 0.01 *ppm*) /2/ even though the current path within the sample switches continuously with the magnetic field over the field range defined by the plateau region /3/. Although the Hall steps are flat at the lowest temperatures, ρ_{xy} can deviate from $h/\nu e^2$ by up to several *ppm*, following the relation $\Delta\rho_{xy} = \rho_{xy}(T) - \rho_{xy}(0) \sim s\rho_{xx}^{min}$ with s a sample dependent parameter of order 1 /2/. This high degree of quantization has possibly set new resistance standards while motivating considerable theoretical work on the role of disorder in high field 2D transport.

We have investigated the quantum Hall plateaus and associated zero-resistance state using field modulation techniques which are sensitive to magnetic field dependent fluctuations of ρ_{xx} and ρ_{xy}. Our transport study reveals aperiodic fluctuations in $d^2\rho_{xx}/dH^2$ and $d^2\rho_{xy}/dH^2$ only over the field range defined by the Hall plateaus and zero-resistance state. Although previous works have observed breakdown related fluctuations of ρ_{xx} in large samples /4/ and possible quantum interference effects in narrow, doubly connected geometries fabricated from GaAs/AlGaAs heterostructures /5/, Hall voltage fluctuations have not been observed in the quantum Hall regime to our knowledge.

2 Data and Discussion

Our transport measurements were made on a four terminal van der Pauw pattern which was photo-lithographically inscribed on MBE GaAs/AlGaAs heterostructure. The sample consisted of 1μm of intrinsic GaAs on Cr-doped GaAs substrate followed by 250A of intrinsic $Al_{0.3}Ga_{0.7}As$, 500A of Si-doped ($8x10^{17}$ cm^{-3}) $Al_{0.3}Ga_{0.7}As$ and capped by 200A of Si-doped ($2x10^{18}$ cm^{-3})GaAs. Ohmic contacts were made by alloying In in a hydrogen atmosphere at 400C. The sample was immersed in pumped liquid helium at T=1.55K for all transport measurements reported here.

In Fig. 1, we show experimental traces of the diagonal and off-diagonal resistivities, i.e. ρ_{xx} and ρ_{xy} respectively, vs the magnetic field, H. Here, the resistivity ρ_{xx} was obtained from $V_{14,32}$ (see Fig. 1) while ρ_{xy} was obtained from $V_{13,42}$. We define $V_{ij,kl}$ as a voltage measurement across contacts i and j with a constant current flowing from contact l to contact k. The figure shows that ρ_{xx} develops Shubnikov-de Haas oscillations with

Springer Series in Solid-State Sciences, Vol. 87
High Magnetic Fields in Semiconductor Physics II Editor: G. Landwehr

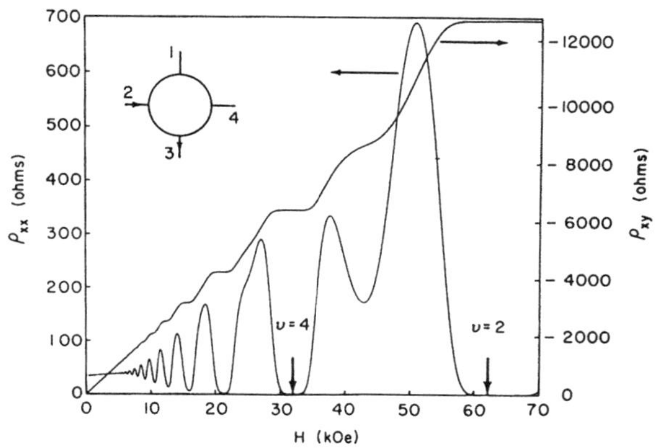

Fig. 1

Figure illustrates the magnetic field dependence of ρ_{xx} and ρ_{xy} in GaAs/AlGaAs heterostructure at 1.55K. Arrows indicate the $\nu=2$ and $\nu=4$ zero-resistance states.

increasing H and $\rho_{xx} \to 0$ as ρ_{xy} is quantized to values $\rho_{xy} = h/\nu e^2$ with ν an integer. Thus, the transport behavior of our sample is typical of GaAs/AlGaAs heterostructures. Our dc measurements of ρ_{xx} and ρ_{xy} at T=1.55K indicated that the carrier density, N_s, and the Hall mobility, μ, varied over the range 3×10^{11} cm^{-2} < N_s < 4×10^{11} cm^{-2} and 120,000cm^2/V s < μ < 350,000cm^2/V s depending upon the sample-state which could be controlled by illumination with a red LED or thermal cycling to 200K.

We have also measured the field derivatives of ρ_{xx} and ρ_{xy} by superimposing upon the static field a small amplitude (80 Oe), ac (μ=100hz) magnetic field and detecting the resulting signal at harmonics of the modulation frequency. For small amplitude modulation, the signal measured at the m'th harmonic of the modulation frequency corresponds to the m'th field derivative of the resistivity, $d^m\rho/dH^m$; we have successfully checked this interpretation for m=2 by comparing the numerically integrated second derivative signal with the measured first derivative signal. Thus, in Fig. 2, we show $d^2\rho_{xx}/dH^2$ (Fig. 2a), $d^2\rho_{xy}/dH^2$ (Fig. 2b) and ρ_{xy} (Fig. 2c) vs the magnetic field over the range of fields defined by the $\nu=2$ Hall plateau. Here, $\nu=hN_s/eH$ is the filling factor. The figure indicates correlated, aperiodic fluctuations in $d^2\rho_{xx}/dH^2$ and $d^2\rho_{xy}/dH^2$ with similar amplitudes over the range of fields where ρ_{xx} vanishes and ρ_{xy} is quantized (refer to Fig.1) even though the current paths differ for the two sets of measurements. Also shown in the figure are concurrent dc measurements of ρ_{xy} which show quantization to better than 4 parts in 10^5. We point out that the step-like structure seen in Fig. 2c reflects the resolution of our 5 1/2 digit digital-voltmeter while the deviation of ρ_{xy} from $h/2e^2$ by 1 part in 10^3 mirrors the precision of the current supply used in these experiments. From the resolution of our Hall voltage measurements, we estimate that the largest amplitude fluctuations in $d^2\rho_{xy}/dH^2$ correspond to changes in ρ_{xy} of less than 50*ppm*. Also, the sensitivity of this measurement technique is better than 1 *ppm* for the modulation amplitudes used here.

Although the fluctuations of ρ_{xx} and ρ_{xy} are similar at each current, the "magneto - fingerprint" develops more structure with increasing current. We have plotted the magnetic field dependent critical current, I_c, and the "magneto - fingerprint" (i.e. $d^2\rho_{xx}/dH^2$) for I = 0.1 I_c^{peak} in Fig.3 to show that these fluctuations are observed at currents well below

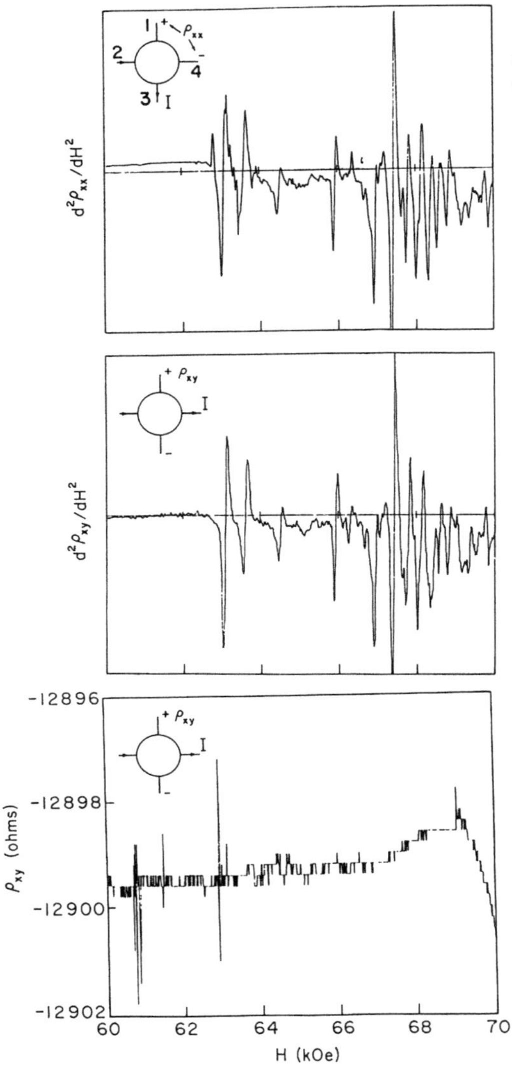

Fig. 2

Conductance fluctuations of $d^2\rho_{xx}/dH^2$(top) and $d^2\rho_{xy}/dH^2$(middle) observed on the $\nu=2$ Hall plateau. A dc measurement of ρ_{xy} is shown on the bottom.

the peak critical current, I_c^{peak}, for the breakdown of the dissipationless state. Here, I_c measures the current necessary for the onset of dissipation in the quantum Hall regime.

We briefly summarize other general features associated with these conductance fluctuations: First, the "magneto-fingerprint" is reproducible at each current without any sweep-direction dependent hysteresis as long as the sample is maintained at low temperatures, T<4.2K. Thermal cycling or illumination produces a new "magneto-fingerprint" which remains in effect until subsequent modification of the sample-state. We have investigated the

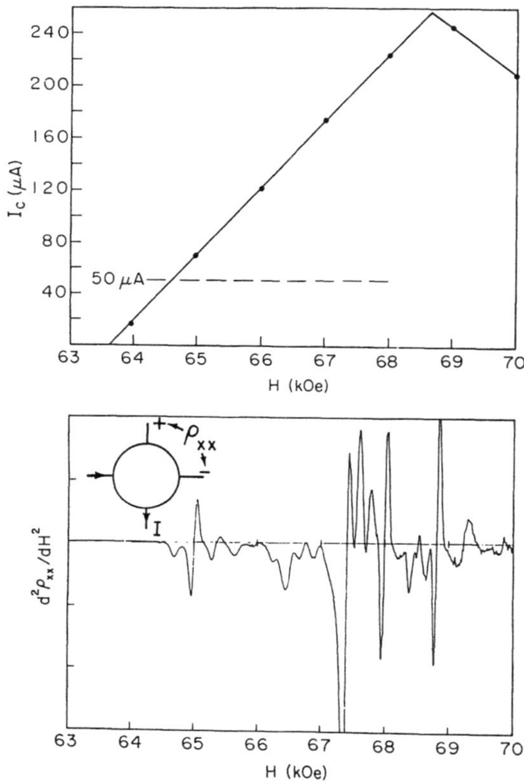

<u>Fig. 3</u>

Above, we show the critical current, I_c, and $d^2\rho_{xx}/dH^2$ vs H for $I=50\mu A$ on the $\nu=2$, $\rho_{xx} \rightarrow 0$ state. Typically, fluctuations are observed for $I < I_c^{peak}$. The magneto - fingerprint shown here differs from the results of Fig. 2 due to a change in the impurity potential resulting from thermal cycling of the sample.

effect of illumination or thermal cycling on I_c and the scattering lifetimes. Typically, thermal cycling enhances small angle scattering while reducing N_s and I_c; illumination reverses these effects. These results indicate that metastable, charged DX centers in the AlGaAs layer enhance charged impurity scattering /6/ and possibly produce spatial inhomogeneities in the carrier distribution with accumulated thermal cycling. Second, correlated conductance fluctuations of ρ_{xx} and ρ_{xy} have also been observed for filling factor $\nu=4$. Typically, at a fixed current, there is less structure for $\nu=4$ than for $\nu=2$. Higher filling factor quantum Hall plateaus have not been investigated since they are not as well developed as those corresponding to $\nu=2$ and $\nu=4$. Third, the "magneto-fingerprint" changes upon magnetic field-reversal but the ρ_{xx} and ρ_{xy} patterns are still similar and do not show a simple even/odd field reversal symmetry like ρ_{xx} and ρ_{xy}. Fourth, current reversal also produces a new "magneto-fingerprint". Finally, temperature-dependence studies indicate that the "magneto-fingerprint" grows stronger with decreasing temperature without any appreciable change in the overall structure; the pattern disappears entirely at high temperatures, $T > 4.2K$.

Although a time varying magnetic field perpendicular to the 2-D EG induces an azimuthal electric field, possibly producing radial charge flow under quantum Hall effect conditions /7/, we rule out eddy current effects since they would be insensitive to the current and vary with the modulation amplitude in contradiction to experimental results which indicate a current dependence and no variation in the "magneto - fingerprint" for a factor - of - two variation in the modulation amplitude. Here, we also point out that the overall ρ_{xx} or ρ_{xy} vs H lineshapes obtained from dc measurements are unaffected by field modulation.

Possible mechanisms which may be relevant for describing these results include current induced localized-breakdown of the quantum Hall effect and variable range hopping through localized states at the Fermi energy/8/. The latter model is especially attractive since i) it would be possible to attribute the "magneto-fingerprint" to a fluctuating hopping probability resulting from the movement of the Fermi energy through localized states and ii) the variation of the "magneto-fingerprint" with thermal cycling or illumination can be attributed to a change in the impurity potential at the 2-D EG resulting from the metastability of the DX centers in the AlGaAs layer. However, this picture fails to describe the temperature dependence since we would expect hopping to be suppressed with decreasing temperature while the "magneto-fingerprint" is enhanced with decreasing temperatures.

Our samples were grown by Mr. R. Wilson at Martin Marietta Corp. This research was partly supported by the U.S. Defense Advanced Research Projects Agency under grant No. DAAG29-85-K-0052.

3 Literature

1. For a review see The Quantum Hall Effect, ed. by R. E. Prange and S. M. Girvin, Graduate Texts in Contemp. Phys., (Springer, New York, 1987).
2. M. Cage, B. Field, R. Dziuba, S. Girvin, A. Gossard, and D. Tsui, Phys. Rev. B30, 2286 (1984).
3. G. Ebert, K. von Klitzing, and G. Weimann, J. Phys. C: Sol. St. Phys. 18, L257 (1985); H. Zheng, D. Tsui, and A. Chang, Phys. Rev. B32, 5506 (1985)
4. M. Cage, R. Dziuba, B. Field, E. Williams, S. Girvin, A. Gossard, D. C. Tsui, and R. Wagner, Phys. Rev. Lett. 51, 1374 (1983).
5. G. Timp, A. M. Chang, J. E. Cunningham, T. Y. Chang, P. Mankiewich, R. Behringer, and R. E. Howard, Phys. Rev. Lett. 58, 2814 (1987).
6. R. G. Mani and J. R. Anderson, Phys. Rev. B37, 4299 (1988).
7. D. Syphers, K. Martin, and R. J. Higgins, Surf. Sci. 170, 222 (1986).
8. G. Ebert, K. von Klitzing, C. Probst, E. Schuberth, K. Ploog, and G. Weimann, Sol. St. Comm., 45, 625 (1983).

Quantum Hall Effect and Related Magneto-transport in Silicon (001) MOSFETs Under Uniaxial Stress

J. Lutz [1], F. Kuchar [1], and G. Dorda [2]

[1] Institut für Festkörperphysik der Universität and Ludwig Boltzmann Institut für Festkörperphysik, Kopernikusgasse 15, A-1060 Wien, Austria

[2] Siemens Forschungslaboratorien, D-8000 München, Fed. Rep. of Germany

The phase diagram for the population of the lowest subbands in stressed Si (001) MOSFETs is deduced from Quantum Hall effect and Shubnikov-de-Haas data. Very good qualitative agreement is found with results of the random-phase approximation by Takada and Ando.

1. INTRODUCTION

The many-valley structure of the conduction band of silicon is responsible for a variety of peculiar electronic effects in this semiconductor. The application of uniaxial stress provides a useful tool to study these effects since the six lowest conduction band minima can be shifted in energy relative to each other. In a silicon MOSFET the relevant band structure is obtained by a projection of the three-dimensional structure on the plane of the two-dimensional electron gas (2DEG) [1]. The energy of the electronic subbands is proportional to $(m_\perp)^{-1/3}$ where $m_\perp$ is the effective mass component perpendicular to the 2D plane. For a (001) silicon surface the situation is shown in Fig. 1a. The isotropic [001] subband is the lowest in energy and two-fold degenerate. The dependence of the subband energies on stress is shown in Fig. 1b. This picture represents a simplified case where the interaction of the electrons in the subbands is neglected. Takada and Ando [2] took these interactions into account within the Hartree, Hartree-Fock, and random-phase approximations. One essential result was that the anisotropic [100] subbands are shifted upwards in energy relative to the isotropic [001] subbands with increasing electron density.
In this contribution we report on experimental results from Shubnikov-de-Haas (SdH) and Hall effect measurements under uniaxial compressive stress which allow one to determine a phase diagram for the occupation of the [100] and [001] subbands in a (001) MOSFET. Previous investigations on lower mobility samples [3] relied on magnetoresistance and SdH data only. In the work presented here

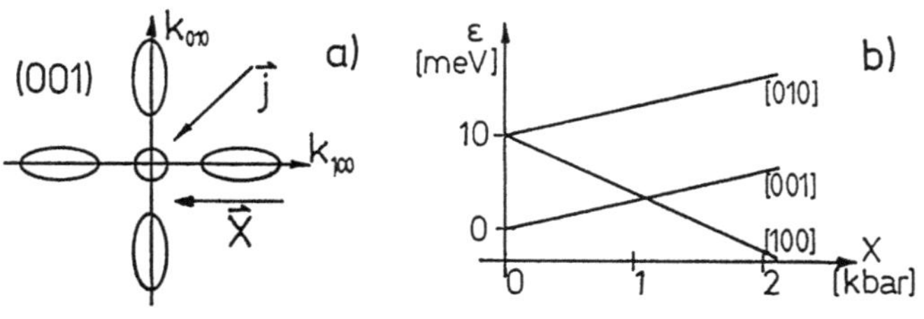

Fig. 1 (a) Constant energy surfaces projected on the (001) plane of the 2 DEG. (b) Stress dependence of 100 subbands.

a conclusive interpretation of the data was made possible by the additional information on the Landau level filling obtained from the observation of the quantized Hall resistance in the high-mobility samples used.

The experiments were performed with uniaxial stress X||[100], current j||[110], and magnetic field B||[001] on silicon (001) MOSFETs at T = 2.1 K. Maximum values of the stress and the magnetic field were 2.5 kbar and 11 T, respectively. The MOSFETs had oxides with a thickness of 680 nm. The mobility at V_G=10V was $2x10^4$ cm^2/Vs. The threshold voltage for n-inversion ($V_t \approx 2V$) was observed to be independent of the uniaxial stress.

2. EXPERIMENTAL RESULTS

2.1 Population of Subbands

Figure 2 shows recordings of ρ_{xx} as a function of gate voltage V_G at B=10.7T for different stress values. Above about V_G=50 V the SdH minima clearly shift to higher V_G values with increasing stress above X ≈ 0.6 kbar. At low V_G values the structure of the ρ_{xx} curves is more complex and just a few selected curves are shown in Fig. 2b. In the low V_G regime the i=2 minimum in ρ_{xx} (i...integer filling factor of Landau levels) disappears at relatively small stress values and appears again at higher values. The i=1 minimum is observable at small stress values only. These observations show that an electron transfer occurs from the isotropic [001] to the anisotropic [100] subbands with increasing stress. Under stress with increasing carrier density electrons are quickly transferred back to the [001] subband causing the complex behavior of the SdH oscillations at intermediate V_G values. At high V_G values the combined effect of electron-electron interaction and uniaxial stress is much easier to analyse since the ratio of the [001] and the [100] subband populations varies only slowly. There, at zero and low stress the electrons mainly occupy the isotropic [001] subband. The shift of the ρ_{xx} minima with increasing stress together with the corresponding values of the quantizied Hall resistance are a quantitative measure for the occupation of the two types of subbands involved (V_t remains

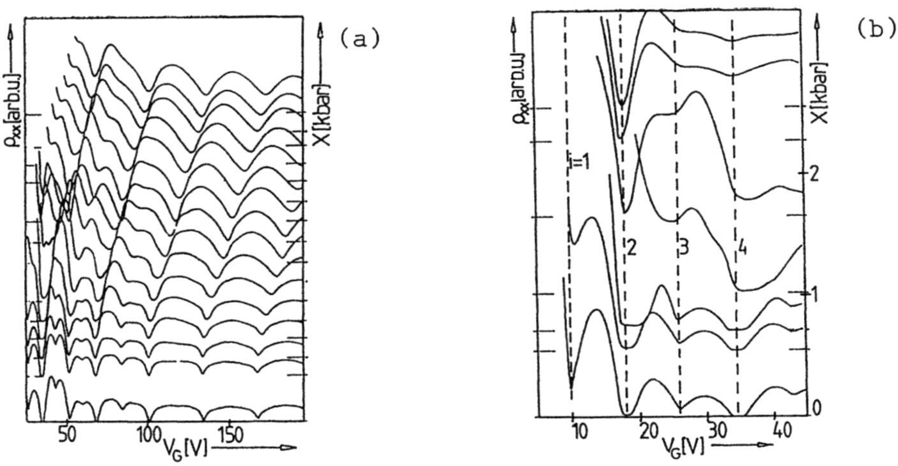

Fig. 2 (a) ρ_{xx} as a function of gate voltage V_G with the stress X as the parameter at B=10.7 T. The zero-offset of the X ≠ 0 curves is proportional to the stress and indicated by horizontal lines. (b) Same as (a) but for the range of lower gate voltages. The maximum stress is 2.3 kbar in both diagrams. Filling factors i=1...4 are indicated.

unchanged). We also observe a 10 % increase of the period of the SdH oscillations with stress at high V_G. The data show that at high V_G about 20 % of the electrons are transferred from the [001] to the [100] subband at the highest stress. From this analysis we can construct a phase diagram for the occupation of the subbands as a function of carrier density n_s and uniaxial stress (Fig. 3a). The curves with constant n_1/n_s are accurate within about 5 %. At lower V_G values the transition from $n_1/n_s=0$ to $n_1/n_s=1$ occurs within a narrow V_G range particularly at stress values around 0.6 kbar. The comparison with the results of the random-phase approximation [2] shows very good qualitative agreement(Fig. 3b). This approximation takes electron-electron interaction into account including exchange and correlation effects. In the Hartree-Fock approximation (no correlation effects) the sharp transition from $n_1/n_s=0$ to $n_1/n_s=1$ extends over a much wider range of n_s. In the Hartree approximation (Coulomb interaction only) the transition is very broad also at low n_s values.

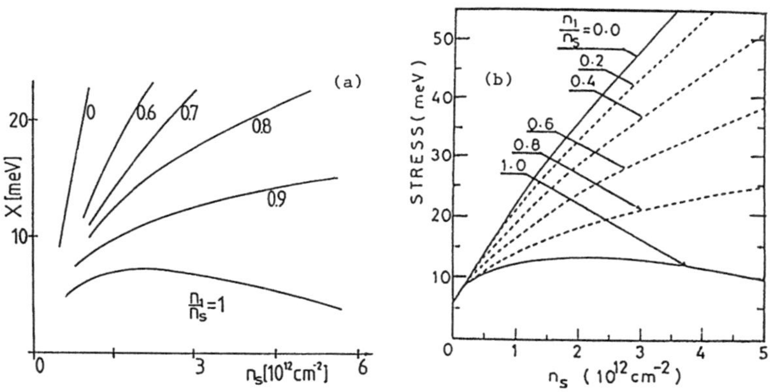

Fig. 3 Phase diagram for the occupation of the two subbands. (a) experimental, n_1 is the electron density in the [001] subband (b) theoretical, random-phase approximation (from [2]).

2.2 Integral Quantum Hall Effect

As an example for the development of the Quantum Hall effect (QHE) with uniaxial stress data are shown for the i=4 plateau in Fig. 4a. The plateau in ρ_{xy} and the minimum in ρ_{xx} are well pronounced at zero stress and positioned at high enough V_G values where the stress induced electron transfer between the subbands is slow. At first sight one recognizes the surprising result that the plateau width increases with stress. This can be interpreted within the widely accepted model of the QHE where localization in the tails of the Landau levels plays an important role. Additionally one has to consider the many-valley structure of silicon (Fig. 4b). The i=4 plateau occurs when the Fermi level ε_F lies within the uppermost range of localized states in the N=0 and the lowest range in the N=1 Landau level of the isotropic [001] subband. With increasing stress the [100] subband is lowered relative to the [001] subband and localized states in these two subbands can overlap as shown schematically in Fig. 4b. In this overlap regime more electrons ($\hat{=}$ higher V_G) are necessary to fill all the localized states. This also explains the asymmetry of the broadening of the Hall plateau. Since the [100] subband is pushed upwards in energy with increasing n_s it is filled very slowly with electrons and the corres-

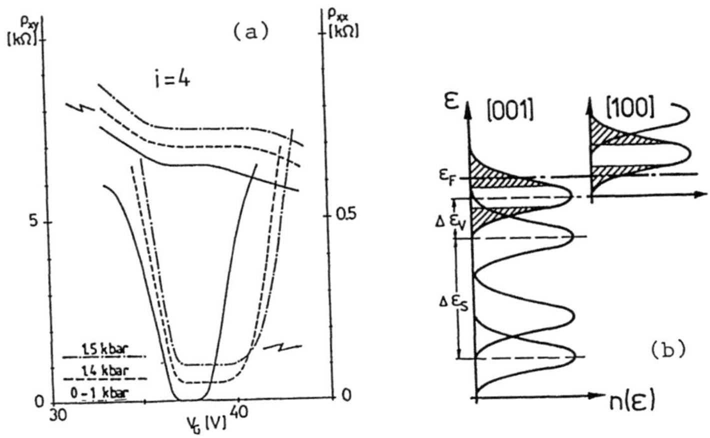

Fig. 4 (a) ρ_{xy} and ρ_{xx} at the i=4 QHE plateau. No variation with stress was observed between 0 and 1 kbar. (b) Schematic level scheme in the case of the broadened i=4 plateau. $\Delta\varepsilon_s$...spin splitting, $\Delta\varepsilon_v$...valley splitting. Shaded areas...loclized states.

ponding SdH oscillations cannot be separated within the applicable V_G range. An indication of their contribution is probably the varying depth of the ρ_{xx} minima with n_s. These minima are primarily caused by the complete filling of a sublevel in the [001] subband. If in addition a sublevel of the [100] subband is completely filled simultanously, the ρ_{xx} minimum is best developed.

3. CONCLUSIONS

The results presented in this paper confirm the assumption of two separate sets of Landau levels in the two subbands also when both are populated. This observation is in agreement with cyclotron resonance experiments [5] where two different values of effective masses were observed when two subbands are occupied. It is not necessary to assume a mixed state which was introduced in a theory by Kelly and Falicov [4].

This work was partly supported by "Fonds zur Förderung der wissenschaftlichen Forschung", Austria

Literature

1. T. Ando, A. B. Fowler, F. Stern: Rev. Mod. Phys. 54, 437 (1982)
2. Y. Takada, T. Ando: I. Phys. Soc. Japan 44, 905 (1978)
3. I. Eisele, H. Gesch, G. Dorda: Solid State Commun. 1E, 743 (1976)
4. M. J. Kelly, L. M. Falicov: Solid State Commun. 22, 447 (1977)
5. H. Gesch, G. Dorda, P. Stallhofer, J. P. Kotthaus: Solid State Commun. 32, 543 (1979)

Effect of Additional Irradiation Induced Scattering Centres on the Quantum Hall Plateau Widths in GaAs-$Al_xGa_{1-x}As$ Heterostructures

W. Möhle[1], *H. Adrian*[1], *L. Bliek*[2], *G. Weimann*[3,*], *and W. Schlapp*[3]

[1]Physikalisches Institut, Friedrich-Alexander-Universität, Erwin-Rommel-Str. 1, D-8520 Erlangen, Fed. Rep. of Germany
[2]Physikalisch-Technische Bundesanstalt, Bundesallee 100, D-3300 Braunschweig, Fed. Rep. of Germany
[3]FTZ, Deutsche Bundespost, Am Kavalleriesand 3, D-6100 Darmstadt, Fed. Rep. of Germany
*Present address: Walter-Schottky-Institut, TU München, Garching, Fed. Rep. of Germany

1. Introduction and Experiment

The presence of random scattering centres in a two-dimensional electron gas (2DEG) is essential for the occurrence and the widths of the quantum Hall effect (QHE) plateaus within the localization model [1]. Scaling theory indicates that within a infinite two-dimensional system all states are localized by quantum coherence effects for any degree of disorder [2]. Concerning the QHE the localization model describes the effect of disorder on the localization of Landau states. Within this model the QHE plateau width should be proportional to the number of electrons in localized states above the upper mobility edge of the highest filled Landau band plus those below the lower mobility edge in the next higher band in energy. In a recent paper we studied the effect of low temperature proton irradiation on the resistivity ρ, mobility μ, carrier concentration n and QHE structures of GaAs-$Al_xGa_{1-x}As$ heterostructures [3]. As a continuation of this work we present a systematic investigation of these quantities on irradiation induced disorder with additional emphasis on the comparison of different samples. To this purpose GaAs-$Al_xGa_{1-x}As$ heterostructures were successively irradiated with 6 MeV protons and the longitudinal and transversal resistivities ρ_{xx} and ρ_{xy} were measured in magnetic fields after each dose.

The samples were prepared from modulation doped GaAs-$Al_xGa_{1-x}As$ heterostructures grown by MBE. An undoped $Al_xGa_{1-x}As$-layer, followed by a Si-doped $Al_xGa_{1-x}As$-layer and an undoped 20 nm cap layer of GaAs, was deposited on top of about 3 μm nominally undoped GaAs. Several samples from three different wafers were used for the experiments. A survey of the sample parameters before irradiation is given in Table 1.

The irradiation experiments were carried out with 6 MeV protons at low temperature ($T \leq 20K$). Details of the experiment are given in Ref. 3. The effect of the irradiation is the displacement of atoms due to the transfer of kinetic energy from the projectiles via Coulomb interaction without implantation in the area of the 2DEG. Using minimal displacement energies of 10 eV we estimate for a fluence $\Phi = 10^{12}$ protons/cm^2 a concentration of displaced atoms $c \approx 10^{-7}$.

Table 1: Sample parameters for the three types of samples used

	W1440	W1786	W1787
d (Si-doped $Al_xGa_{1-x}As$) [nm]	42	50	50
Si-concentration in $Al_xGa_{1-x}As$ [cm^{-3}]	$2 \cdot 10^{18}$	$2.6 \cdot 10^{18}$	$2.6 \cdot 10^{18}$
d (undoped $Al_xGa_{1-x}As$) [nm]	25	12	12
x	0.46–0.48	0.38	0.38
n (4.2K) [cm^{-2}]	$1.8 \cdot 10^{11}$	$2.8 \cdot 10^{11}$	$3.6 \cdot 10^{11}$
μ (4.2K) [cm^2/Vs]	$3 \cdot 10^5$	$9 \cdot 10^4$	$3 \cdot 10^5$

2. Results and Discussion

The irradiation experiments show, that the passage of charged fast particles causes the excitation of a large concentration of additional carriers similar to the persistent photoconductivity [4]. Because of the long relaxation times of the additional carriers the samples were warmed up to 270 K after each dose, which presumably leads to the annealing of part of the irradiation induced lattice defects. The influence of the stable defects on ρ_{xx}(B) and ρ_{xy}(B) due to the equilibrium concentration of charge carriers is shown exemplarily for two samples in Fig. 1. The Hall plateaus and the regions with $\rho_{xx} = 0$ are broadened and shifted to lower magnetic fields with increasing defect concentration. After high fluences, $\Phi = 8.8 \cdot 10^{12}$ cm^{-2} for W1440, $\Phi = 1.06 \cdot 10^{13}$ cm^{-2} for W1786 and $\Phi = 2.56 \cdot 10^{13}$ cm^{-2} for W1787, the measurements show that complete quantization into Landau bands breaks down, because the condition $\mu B \gg 1$ is no longer fulfilled. A survey of the changes of ρ_{xx}, μ and n in the 2DEG with fluence Φ is given in Fig. 2. ρ_{xx}(B=0) increases and μ decreases strongly with Φ. Furthermore, n decreases nearly linearly according to $n(\Phi) = n_o - \alpha \cdot \Phi$ with $\alpha \approx 10^{-2}$ per proton up to fluences of about 10^{13} cm^{-2}. From high-energy (1 – 2 MeV) electron irradiation experiments with GaAs the formation of a number of electrons traps is known [5], which are believed to be essential for the observed decrease of n after proton irradiation. Furthermore, a shift of the position and an increase of the widths of cyclotron resonance lines due to electron beam induced lattice defects have been observed with GaAs-$Al_xGa_{1-x}As$ heterostructures [6]. As can be seen from Fig. 2 the variation of the initial properties of the investigated samples lead to significant differences in the irradiation behavior. Nevertheless, the relative width $\Delta B/B$ (using as criterion a 10^{-3} deviation from the quantized plateau values) of the QHE-plateaus exhibits a universal behavior when it is plotted versus B/μ_c as depicted in Fig. 3. Here μ_c is the mobility as deduced from ρ(1.4K,B=0) and n after correcting localization effects evident in ρ(T,B=0). The physical significance of the uniform behavior shown in Fig. 3 is not clear. However it may be useful to point out that B/μ_c can be taken to be approximately proportional to the number of localized states, since the width of the Landau bands due to life-time effects is proportional to μ_c^{-1} and the degeneracy is proportional to B.

It is evident that the observed irradiation induced broadening of the quantum Hall plateaus is in good qualitative agreement with the localization model. An increasing density of scattering centres leads to a broadening of the structures in the density of states of the 2DEG in magnetic fields due to a reduction of the elastic scattering time.

Additionally the localization length, which increases rapidly for states near the centre of a Landau band, is reduced. Assuming that the cut-off for localization due to inelastic processes remains essentially unchanged, enhanced disorder leads to a shift of the effective mobility edges towards the centres of the Landau bands. Hence the number of electrons in localized states and therefore the Hall plateau widths increase.

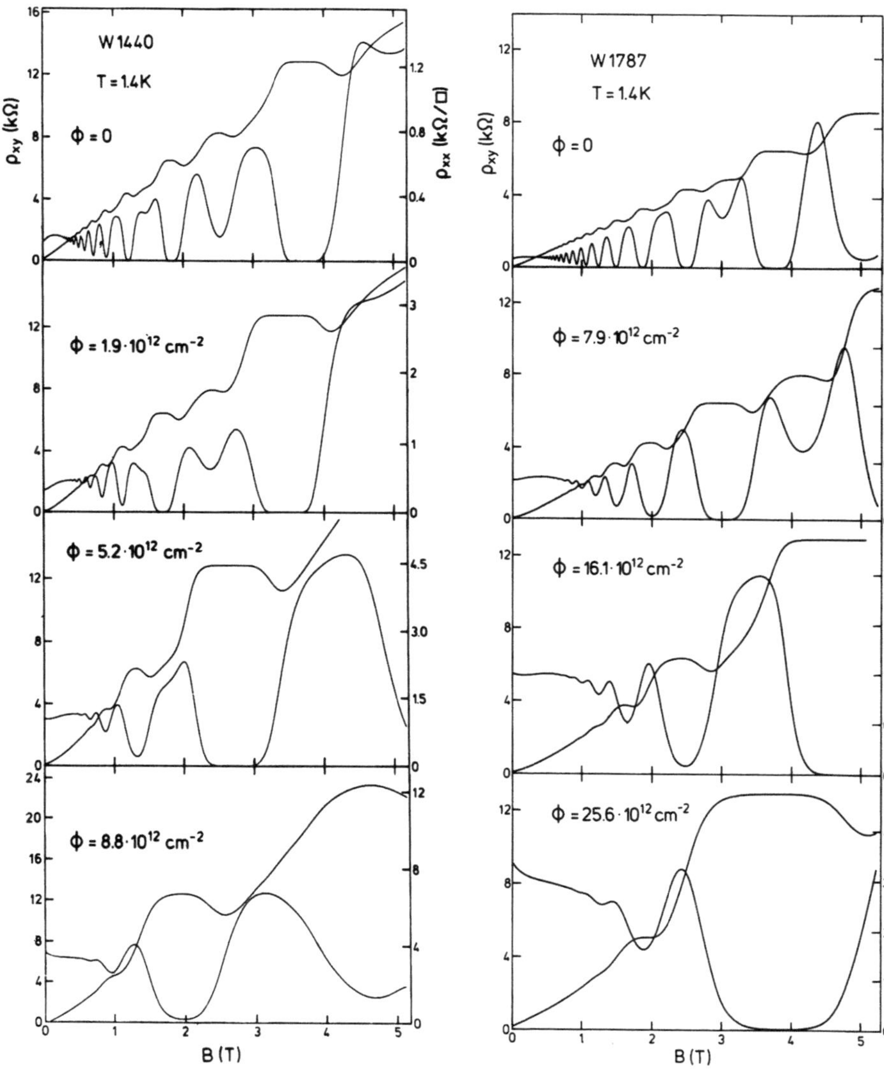

Fig. 1: Resistivities ρ_{xx} and ρ_{xy} versus magnetic field B for the samples W1440 and W1787 for various fluences Φ

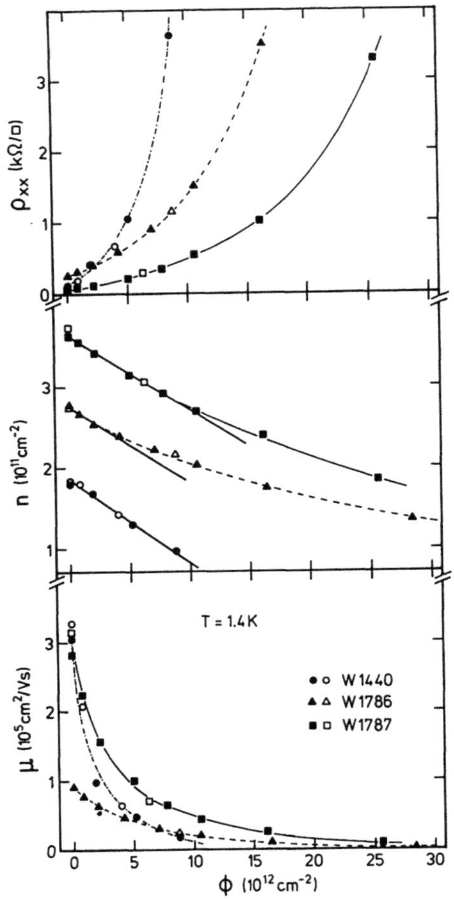

Fig. 2: Mobility $\mu(B=0)$, carrier concentration n and resistivity $\rho_{xx}(B=0)$ as functions of fluence Φ for the three types of samples

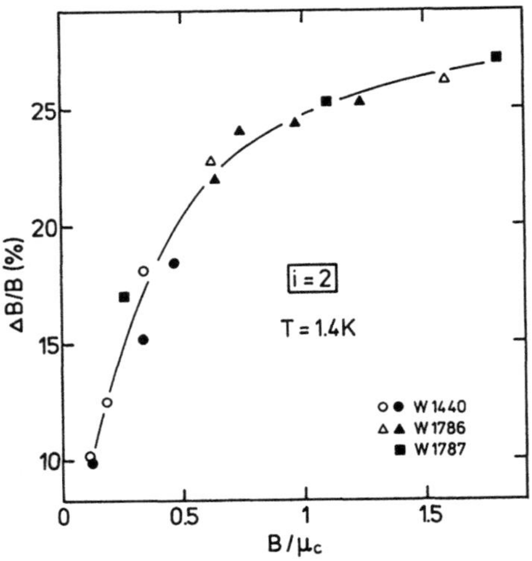

Fig. 3: Relative Hall plateau widths $\Delta B/B$ for i=2 versus B/μ_c. μ_c is explained in the text

Alternative to the localization model, the reservoir model has been proposed [7], in which the Hall plateau widths are proportional to the number of carriers which can be either mobile or immobile in defect states. our observation for the Hall effect and the magnetoresistance, though being in good agreement with the localization model do not decisively rule out the reservoir model.

Acknowledgement

The authors thank the members of the low temperature physics group in Erlangen for experimental assistance during the irradiation experiments and acknowledge financial support by the Bundesminister für Forschung und Technologie.

References

1. H. Aoki, and T. Ando, Solid State Commun. **38**, 1079 (1981)
2. E. Abrahams, P.W. Anderson, D.C. Licciardello, and T.V. Ramakrishnan, Phys. Rev. Lett. **42**, 673 (1979)
3. H. Adrian, W. Möhle, L. Bliek, G. Weimann, and W. Schlapp, submitted for publication
4. R.J. Nelson, Appl. Phys. Lett. **31**, 351 (1977)
5. For a review see: D. Pons, and J.C. Bourgoin, J. Phys. **C18**, 3839 (1985)
6. H. Sigg, D. Weiss, and K. v. Klitzing, Surf. Sci., **196**, 293 (1988)
7. G.A. Baraff, and D.C. Tsui, Phys. Rev. **B24**, 2274 (1981)

On the Effect of the Coulomb Interaction in the Quantum Hall Regime

H. Nielsen

Physics Laboratory, University of Copenhagen, Universitetsparken 5, DK-2100 Copenhagen Ø, Denmark

A characteristic feature of the quantum Hall effect (QHE) is that in the plateau regimes of the Hall voltage the parallel current vanishes and thus no dissipation occurs. Thus one may try to treat the system as being in thermal equilibrium under the constraint of a constant electric current.

It can be shown /1/, however, that for an ideal system (no phonons and no impurities) with no confinement in the direction perpendicular to the current the Hall conductivity is (for any value of the magnetic field B and the average electron concentration $n = N/Ar$) the same as for free electrons, $\sigma_H = en/B$ in spite of the interaction between the electrons.

To avoid this we consider a confined system where the electron density vanishes at the boundaries. It should be pointed out in passing that for a confined system the use of the constant current (or shifted) distribution is essential. This can be demonstrated most easily for free electrons where the usual (unshifted) equilibrium distribution leads to vanishing Hall current /2/. Indeed, the contribution of a single Landau level to the total current is in the limit of vanishing temperature

$$I_n = -\frac{e}{L_y}\sum_X \frac{1}{eB}\frac{\partial\varepsilon}{\partial X} f(\varepsilon(X)-\mu) = -\frac{e^2}{h}\int_{X_1}^{X_2}\frac{\partial\varepsilon}{\partial X}\,dX =$$

$$= -\frac{e^2}{h}[\varepsilon(X_2)-\varepsilon(X_2)] = 0\ , \qquad (1)$$

the boundary values X_1 and X_2 being defined by $\varepsilon(X_i) = \mu$. In (1) f is the Fermi distribution function, μ is the chemical potential and the quantity X is related to the wave number k in the y-direction by $X = \ell^2 k$ where $\ell^2 = \hbar/eB$; in the bulk X is the center coordinate of the cyclotron motion, i.e. of the oscillator wave function. (The gauge $\vec{A} = (0,Bx,0)$ is used). Obviously, (1) holds also for an interacting electron system in any self-consistent single electron approximation and also at finite temperatures.

In the following we consider a confined system of interacting electrons and describe the state of this system by a constant current ensemble. In self-consistent field approximation the resulting single electron distribution function is a shifted Fermi function equivalent to the one applied by Heinonen and Taylor /3/. The use of such a distribution function may physically be justified by the following argument, the constant current is due to the conservation of the total momentum in the y-direction. Each constant momentum state has in the presence of phonons a finite but very long relaxation time. This is due to the very strong selection rules for

Springer Series in Solid-State Sciences, Vol. 87
High Magnetic Fields in Semiconductor Physics II Editor: G. Landwehr

the electron-phonon interaction in the presence of a strong (quantizing) magnetic field /4/. Obviously also the electron electron interaction cannot destroy the total momentum.

In the Hartree approximation the self-consistent single electron potential (which is assumed to be independent of y) is determined by the following set of equations

$$V(x,z) = \frac{1}{4\pi\varepsilon_b} \int \frac{\rho(x',z')dx'dy'dz'}{\sqrt{(x-x')^2 + (y-y')^2 + (z-z')^2}} + V_b \,, \tag{2}$$

$$\rho(x,z) = \frac{-e}{L_y} \sum_a |\psi_a(x,z)|^2 f(\varepsilon_a' - \mu) \,, \tag{2.1}$$

$$(H + V(x,z))\,\psi_a = \varepsilon_a \psi_a \,, \tag{2.2}$$

$$H = \frac{p_z^2}{2m^*} + \frac{p_x^2}{2m^*} + \frac{m^*\omega_c^2}{2}(x-X')^2 + V_c(x) \,, \tag{2.3}$$

where ε_b is the background dielectric constant (for GaAs $\varepsilon_b \simeq 12\,\varepsilon_0$), V_b is the background potential, $\varepsilon_a' = \varepsilon_a - e\lambda\omega_c X$, $X' = X - \lambda/B$, and V_c is the confinement potential. One of the quantum numbers contained in a is the wave number in the y-direction $k = -X/\ell^2$. The potential V(x,z) depends on the parameter λ and can be determined by numerical solution of the self-consistent equation. In the following, however, we restrict ourselves to the bulk of the system in which the potential is slowly varying on the scale of the magnetic length ℓ). Approximations similar to those made by Stern and Howard /5/ and by MacDonald et al. /6/ yield the Poisson equation

$$\frac{\partial^2 V}{\partial x^2} + \frac{\partial^2 V}{\partial z^2} + \frac{g(z)e^2}{\varepsilon_b \hbar\omega_c}\left(i - 1 + f\left(\frac{\varepsilon_i - eV(x) - e\lambda\omega_c x - \mu}{k_B T}\right)\right)\frac{\partial^2 V}{\partial x^2}$$

$$= \frac{e}{\varepsilon_b}\frac{g(z)}{2\pi\ell^2}\left(f\left(\frac{\varepsilon_i - eV(x) - e\lambda\omega_c x - \mu}{k_B T}\right) - \upsilon\right) \,. \tag{3}$$

Since $|\Psi_a(x,z)|^2 \simeq g(z)|\Psi_\alpha(x)|^2$, $g(z) \cong \delta(z)$, $\alpha = (n,k)$ in the argument of the Fermi distribution function V(x,z) has been replaced by V(x) $\equiv V(x,z = 0)$. Furthermore, we have assumed that i-1 Landau levels $\varepsilon_n = \hbar\omega_c(n + \frac{1}{2})$ are occupied and the i'th level is partly filled, $i-1 < \upsilon = 2\pi\ell^2 n < i$. For integer filling the polarization term contained in (3) corresponds to the one obtained by MacDonald et al. /6/. We emphasize that the shifted distribution also used by Heinonen and Taylor /3/ is necessary in order to obtain a current in a confined system.

To get an idea of a physically reasonable solution we may first consider the screening problem in two dimensions in Thomas-Fermi approximation. In the absence of magnetic field (and a total current $\lambda = 0$) Stern and Howard /5/ found that the potential of a point charge placed at the origin falls off with the distance r as ℓ_{sc}^2/r^3 (for $r \gg \ell_{sc}$) where $\ell_{sc} = \pi\varepsilon_b\hbar^2/e^2m^*$ is the Thomas-Fermi screening length which is independent both of the temperature and the carrier concentration (for GaAs $\ell_{sc} \cong 10$ Å). In contrast to this, in a strong magnetic field the degeneracy of the energy levels gives rise to a screening length which depends both on the temperature and the occupation,

$$\frac{1}{\ell_{sc}(B)} = \frac{e^2}{\varepsilon_B 2\pi\ell^2} \frac{\upsilon(1-\upsilon)}{k_B T} \tag{4}$$

$(0 < \upsilon < 1)$. This result holds as long as υ is not too close to unity.

Taking the ratio between the screening length we get

$$V_B(r) = \left(\frac{mk_B}{e\hbar}\right)^2 \left(\frac{2}{\upsilon(1-\upsilon)}\right)^2 \left(\frac{m^*}{m}\right)^2 \left(\frac{T}{B}\right)^2 V_{B=0}(r) , \tag{5}$$

where the numerical value of the first factor is close to 1 in units $K^{-2}T^2$. Therefore, the screening may be very important at low temperatures. Thus, we may guess that (3) has a solution such that in some regions the potential is constant and $f(x) = \upsilon$, and in some other (current carrying) regions $f \cong 1$ (for $T \to 0$). A simple argument supporting this guess is to integrate (3) along the z-direction from a point behind the positive donors to an equivalent point on the other side of the electron system. Then $\int \partial E_z/\partial_z dz = 0$, and since $\int E_x(x,z)dz$ is probably slowly varying only the terms containing $g(z)$ will be important. We then get an equation which in the low temperature limit reduced to

$$\frac{\partial^2 V}{\partial x^2} = \frac{\omega_c}{i} (B_i - B) , \tag{6}$$

where B_i is determined by $2\pi\ell^2 n = i$. For a filled region the current is always proportional to the potential drop ΔV across the region, $I = i(e^2/h)\Delta V$, since the local velocity $v_y(x)$ is proportional to $\partial V/\partial x$.

Therefore, (6) allows us to determine the width of the current carrying region

$$w = \sqrt{\frac{2hI}{e^2\omega_c |B-B_i|}} \tag{7}$$

The limit $w \to \infty$ for $B \to B_i$ means that for integer filling the current flows across the whole sample.

There may be given some experimental support to (6) and (7). Like Eaves et al. /7/ we assume that the breakdown of the QHE occurs when the slope $\partial V/\partial x$ reaches a critical value then (6) yields

$$I_{crit} = \frac{ie^2}{h} V_{crit} = -\tfrac{1}{2} \frac{e^2}{h} \omega_c \, w^2(B-B_i) + I_0 \quad . \tag{8}$$

Measurements by Bliek et al. /8/ showed a linear relation between I_{crit} and $B-B_i$ for a narrow channel with a width of the order 1μ. The observed slope can be obtained from (8) by using w = 0,4μ which has the right order of magnitude.

Finally we should mention that recently Kane et al. /9/ found that current regions at the edges have a width which is large compared with the magnetic length. If the ideas presented in this work are applied to magnetic fields slightly below a plateau value it seems to be possible to explain the reason for these broad current carrying edge regions.

Literature

1. J. Hajdu, Private communication
2. J. Hajdu, U. Gummich, Sol. Stat. Comm. 52, 985 (1984)
3. O. Heinonen, P. L. Taylor, Phys. Rev. B. 28, 6119 (1983)
4. O. Heinonen, P. L. Taylor, S. M. Girvin, Phys. Rev. B. 30, 3016 (1984)
5. F. Stern, W. E. Howard, Phys. Rev. 163, 816 (1967)
6. A. H. MacDonald, T. M. Rice, W. F. Brinkman, Phys. Rev. B. 28, 3648 (1983)
7. L. Eaves, F. W. Sheard, Semicond. Sci. Techn. 1, 346 (1986)
8. L. Bliek, G. Hein, E. Braun, V. Kose, J. Niemeyer, G. Weimann, W. Schlapp, Semicond. Sci. Techn. 1, 110 (1986)
9. B. E. Kane, D. C. Tsui, G. Weimann, Phys. Rev. Lett. 59, 1353 (1987)

Structures in the Breakdown Curves of the Quantum Hall Effect in Narrow Channel GaAs/AlGaAs Heterostructures

A. Sachrajda, M. D'Iorio, D. Landheer, P. Coleridge, and T. Moore

National Research Council of Canada, Ottawa, Ontario, Canada K1A OR6

Measurements of the breakdown of the quantum Hall effect (Q.H.E.) have been made on single heterostructures with a narrow constriction (~2 μm). Structures corresponding to sharp increases in dissipation were observed in both the source-drain current sweeps at constant magnetic field and the magnetic field sweeps at constant source-drain current.

1. Introduction

Several mechanisms for the breakdown of the Q.H.E. have been proposed [1,2,3,4] and recent experimental data [1,5] provide strong evidence for an inter-Landau level Zener mechanism. This mechanism may be expected to be sensitive to the ratio of the Hall energy to the cyclotron energy, $z = eV_H/h\omega_c$. This ratio can only be changed by a few percent during a field sweep at constant current, in contrast to the large variation that can be effected during a current sweep at constant magnetic field (typically 0-20).

Measurements of the breakdown of the Q.H.E. were made at 1.2 K and 30 mK in the magnetic field and source-drain current sweep modes. The single GaAs/AlGaAs heterostructure had a mobility of 97000 cm^2/Vs and a carrier concentration of $3.73*10^{11} cm^{-2}$ at 4.2 K. The design of the Hall bar is similar to that of BLIEK et al. [5] and is shown in the inset of Fig. 1. A pair of contacts which excluded the constriction was used to check that the longitudinal resistance was negligibly small in the wide part of the sample. The dimensions of the constriction were defined by the mesa to be 4 μm in width and 15 μm in length while the "electrical" width defined by the zero field resistivity was 2.4 μm.

2. Results and Discussion

After cooling the sample in the dark, the critical current density was measured to be 1.9 A/m corresponding to a critical current of ~4.5 μA. Instantaneous carrier density losses involving a small percentage of electrons and occurring in the constriction lowered the critical current density by a half typically, whereas illumination with a red light emitting diode raised it by up to a factor of four. Similarly the direction of the current flow and that of the magnetic field affected the data. These measurements will be reported elsewhere.

Springer Series in Solid-State Sciences, Vol. 87
High Magnetic Fields in Semiconductor Physics II Editor: G. Landwehr

Figure 1 shows a typical current sweep in which step-like structures are observed. The position of these structures in both magnetic field and current sweeps is taken to be the centre of the steeply rising part of the structures. No significant temperature dependence of the features of the breakdown curves was observed between 30 mK and 1.2 K.

Figure 2 displays the position of the step-like features as a function of $eV_H/\hbar\omega_c$ for current sweeps at constant magnetic field. The corresponding plot for magnetic field sweeps at constant current is shown in Figure 3. These measurements were made on the same cooldown cycle of the sample. The step-like structures in Figure 2 fall on roughly parallel lines spaced ~0.8 $eV_H/\hbar\omega_c$. No hysteresis was observed in either type of measurements. Within the uncertainty of the data, the slightly negative slope of the lines in Figure 2 is attributed to the oscillatory behaviour of the g-factor as a function of magnetic field for a two-dimensional electron system.

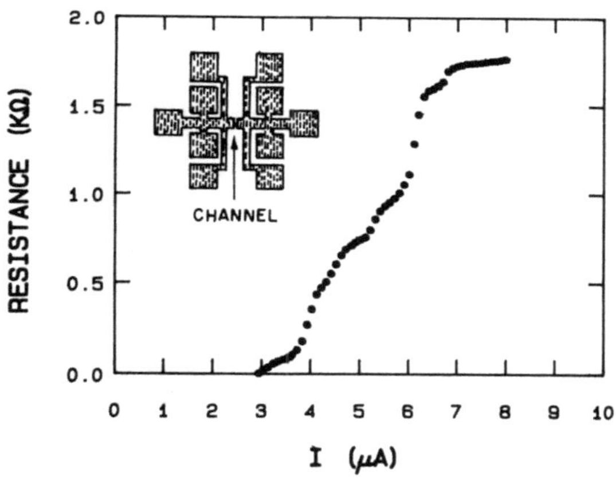

Figure 1:
Current sweep at 1.2 K and 7.9 T. The sample geometry is shown in the inset

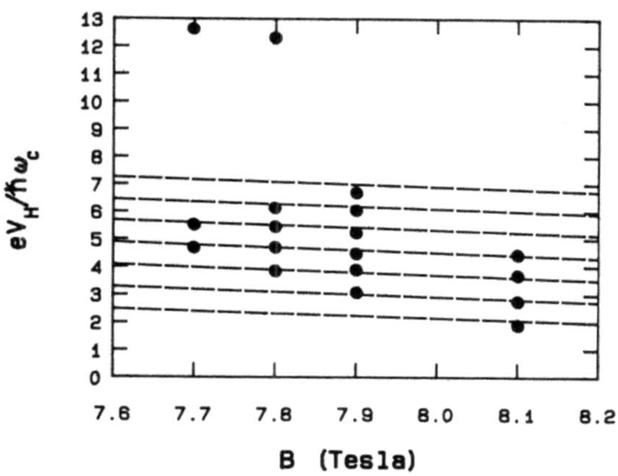

Figure 2:
The position of the structures obtained from current sweeps at 1.2 K (i=2 plateau centered at 7.7 T). The lines are intended as a guide to the eye

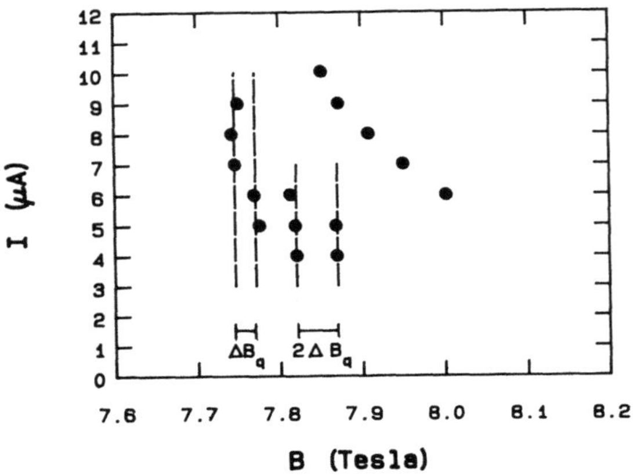

Figure 3:
The position of the structures obtained from magnetic field sweeps at 1.2 K. The lines are intended as a guide to the eye

The quasi-elastic inter-Landau level scattering model (QUILLS) with the assumption of a uniform Hall field [2] has successfully accounted for the critical current observed by BLIEK et al. [5]. This model would predict a critical current of 94 μA for our sample whereas the model of DE VEGVAR et al. [3], in which the critical current is limited by the number of current carrying states, predicts only 0.48 μA. In the model of KIRTLEY et al. [1], the current flows at the edges of individual current paths through the constriction; the critical current can be taken to be a rough measure of the number of such paths since the Hall voltage is divided amongst them. Our observed critical current of 4.5 μA is consistent with about five current paths through the constriction. The breakdown mechanism in this case is an inter Landau level Zener effect at the edges of the conduction paths.

In this multiple conduction path model, it is suggested [1] that the number of conduction paths can be decreased as a function of magnetic field because of the lowering of the Fermi energy. This provides an explanation of the shift to a lower current value of the first step-like structure with increasing magnetic field. The sequence of roughly equally spaced structures at a particular field is attributed to a variation of the number of conduction paths along the length of the constriction due to the percolating nature of the current flow. The breakdown process is thus believed to be spatially varying.

The magnetic field spacing of the structures observed by BLIEK et al. [5] have been interpreted as an increase in dissipation occurring when the sum of the widths of the two wavefunctions involved in an inter-Landau level transition is an integer multiple of the spacing between the centers of the eigenfunctions [2]. The field spacing corresponding to a change of one eigenfunction centre ΔBq is shown in Figure 3. Despite the good agreement the critical currents in our sample are not consistent with a uniform Hall field and the calculated spacing between eigenfunction centers is only 0.4 Å. These structures may be a signature of the loss of conduction paths during a magnetic field sweep.

Measurements of the breakdown of the quantum Hall effect in a single GaAs/AlGaAs heterostructure with a narrow constriction have revealed step-like structures in the longitudinal resistance for both current and magnetic field sweeps. The relevance of the existing models for the breakdown has been studied with respect to our experimental data. The role of inhomogeneities and conduction path pinch-off is currently being investigated.

Acknowledgements

The authors wish to thank R. Boulet, D. Brown, M. Buchanan and P. Marshall for their technical expertise in the sample preparation process and P. Hawrylak for useful discussions.

References

1. J.R. Kirtley, Z. Schlesinger, T.N. Theis, F.P. Milliken, I. Wright and L.F. Palmateer, Phys. REv. B34, 1384 (1986).
2. L. Eaves and F.W. Sheard, Semicond. Sci. Technol. 1, 346 (1986).
3. P.G.N. de Vegvar, A.M. Chang, G. Timp. P.M. Mankiewich, J.E. Cunningham, R. Behringer and R.E. Howard, Phys. Rev. B36, 9366 Cunningham, R. Behringer and R.E. Howard, Phys. Rev. B36, 9366 (1987).
4. O. Heinonen and P.L. Taylor, Phys. Rev. B32, 633 (1983).
5. L. Bliek, G. Hein, V. Kose, J. Niemeyer, G. Weimann and W. Schlapp: In Proceedings of the Inernational Conference on the Applications of High Magnetic Fields in Semiconductor Physics, Springer Ser. Solid-State Sci., Vol. 71 (Springer, Berlin, Heidelberg 1987).

Quantum Hall Effect in Wide Parabolic $GaAs/Al_xGa_{1-x}As$ Wells

E.G. Gwinn [1], *P.F. Hopkins* [1], *A.J. Rimberg* [1], *R.M. Westervelt* [1], *M. Sundaram* [2], *and A.C. Gossard* [2]

[1]Division of Applied Sciences and Department of Physics, Harvard University, Cambridge, MA 02138, USA

[2]Department of Electrical and Computer Engineering and Materials Department, University of California, Santa Barbara, CA 93106, USA

1. INTRODUCTION

The ability to produce narrow two-dimensional (2D) layers of high mobility electron gas in semiconductor structures has permitted the study of the 2D electron gas in strong magnetic fields under nearly ideal conditions, and led to the discovery of the integer and fractional quantum Hall effects [1]. Despite many interesting experiments, comparable progress has not been made for the three dimensional (3D) electron gas in semiconductors, due in part to the strong influence of the potential from dopant atoms. Remotely-doped parabolic $GaAs/Al_xGa_{1-x}As$ wells can be used to produce relatively thick (>1000 Å) layers of high mobility electron gas [2,3] which are spatially uniform due to the action of the parabolic potential. We report here high magnetic field measurements of both the 2D integer and fractional quantum Hall effects for a partially-filled wide parabolic $GaAs/Al_xGa_{1-x}As$ well [2] with very high low temperature mobility $\mu = 2.5\times10^5\ cm^2/Vsec$. These measurements demonstrate both the feasibility of the parabolic well technique and the high quality of the structures.

2. DENSE ELECTRON GAS IN WIDE PARABOLIC WELLS

Figure 1 is a schematic illustration of the conduction band edge vs. position for empty and partially-filled parabolic wells. For the empty well in Fig. 1a, the parabolic potential of depth Δ_1 is confined by a barrier of height Δ_2. Remotely located dopant layers in the barrier are used to fill the well with electrons. The essential idea needed to understand the filling of the well is that the parabolic potential is identical to that produced by a spatially-uniform positively charged background of 3D density $n_+ = 2\varepsilon\Delta_1/\pi e^2 w^2$. The electrons act to screen this fictitious positive charge as they enter the well. When the well is full, the electron density n is constant and equal to the design density n_+ over the entire width of the well, and the total potential is simply that of a square well of width w and depth Δ_2. Even when the well is partially full, as in Fig. 1b, the electrons act to screen the parabolic potential and create a thinner slab of comparable density $n \cong n_+$ over part of the width w_e as indicated. For this case the total potential is approximately that of a square well of width w_e. Thus this process is self-correcting and does not require precise control

Springer Series in Solid-State Sciences, Vol. 87
High Magnetic Fields in Semiconductor Physics II Editor: G. Landwehr

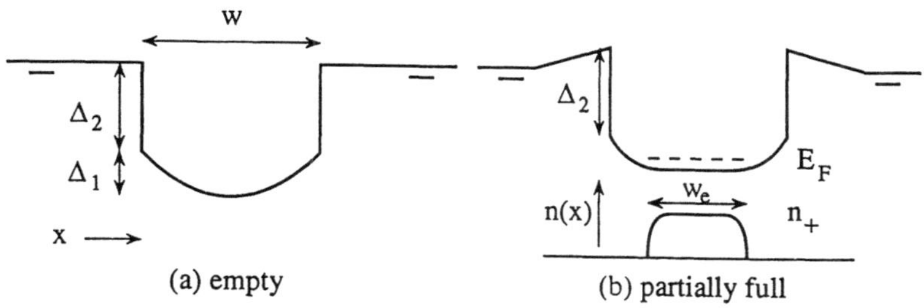

Fig. 1 Illustration of the spatial profile of the conduction band edge for (a) empty and (b) partially full parabolic wells

of the doping to generate spatially uniform electron density profiles. These concepts are completely general, and can be used to create density profiles other than those in Fig. 1.

Parabolic well heterostructures were grown [3] using molecular beam epitaxy on semi-insulating GaAs substrates. The parabolic potential was created by changing the relative thickness of alternating GaAs and $Al_{0.4}Ga_{0.6}As$ layers in a fine superlattice of period 20 Å, with a $Al_{0.4}Ga_{0.6}As$ barrier. Electrons were deposited in the well from Si-doped layers in the barrier symmetrically set back from both sides of the well. Five structures have been grown and characterized to date. For the structure discussed here the well parameters are: width w = 2000 Å, parabolic well depth Δ_1 = 155 meV (for which $n_+ = 2.25\text{x}10^{16}$ cm^{-3}), and barrier height Δ_2 = 143 meV. Low field Hall effect measurements were used to determine the mobility $\mu = 2.5\text{x}10^5$ cm^2/Vsec and the areal electron density $n_{2D} = 1.9\text{x}10^{11}$ cm^{-2} at T = 1.5 K. These measurements indicate that the well is partially full with $w_e \sim n_{2D}/n_+ \cong 800$ Å.

3. QUANTUM HALL EFFECT

Parabolic well samples were prepared in a Hall bar geometry defined using optical lithography. Electrical contacts were made by applying small amounts of In to the contact pads and baking at 400 C for ~ 5 min in a reducing atmosphere. The samples were mounted stress-free and immersed in liquid ^{3}He in a pumped ^{3}He cryostat. Strong magnetic fields to H ~ 20 Tesla were applied using a Bitter magnet, and the Hall resistance and transverse magnetoresistance were simultaneously measured and recorded using separate PAR 113 preamplifiers and a digital lock-in technique with a microcomputer. Typical values of the applied current were ~ 1 µA.

Figure 2 presents Hall effect and magnetoresistance data which clearly show both the integer and fractional quantum Hall effects. Here the Hall resistance and magnetoresistance measured at the temperature T = 0.5 K are plotted vs. magnetic field H over the ranges 0 to 8 Tesla and 0 to 20 Tesla. Because the degeneracy of the lowest Landau level of the lowest sub-band of the well is sufficient to contain the measured areal density

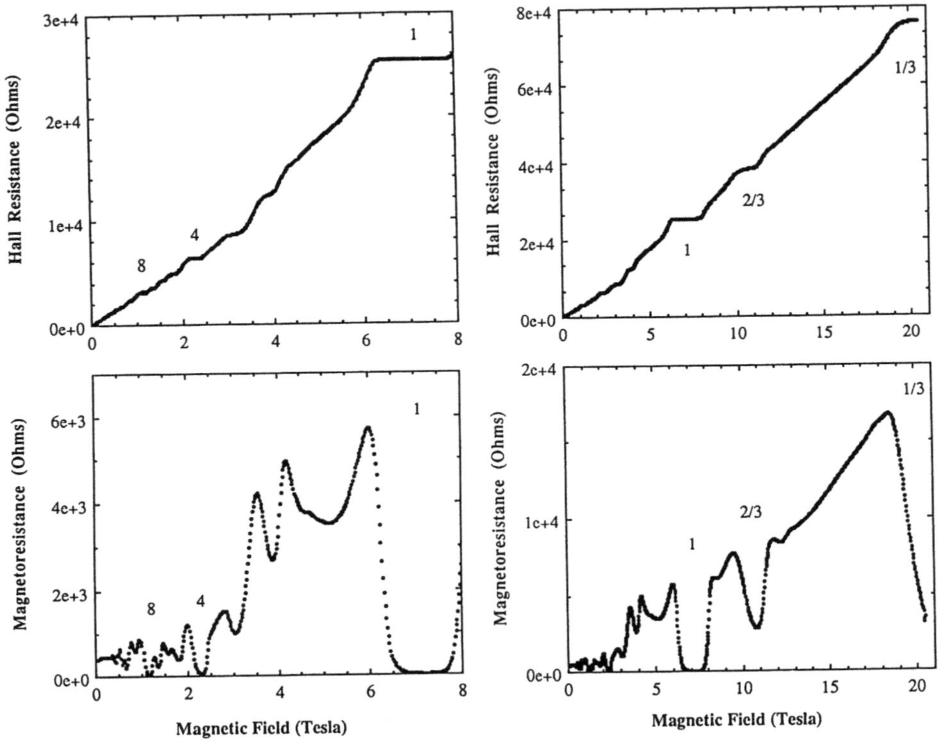

Fig. 2 Integer and fractional quantum Hall effect at T = 0.5 K for a partially-filled parabolic well in the two-dimensional high field limit

n_{2D} for H > 4 Tesla, the electron gas is expected to be two-dimensional at moderately high fields. Well developed Hall plateaus for the integer quantum Hall effect are shown at filling factors $\nu = 1$, 4 and 8 as indicated in Fig. 2, and indications of developing plateaus in both the Hall effect and magnetoresistance occur for $\nu = 2$, 3, 5, 6, and 7. The Hall resistances and fields associated with these steps are in agreement with 2D theory. However, the Shubnikov-de Haas oscillations in the magnetoresistance at fields H < 1 Tesla are not simply periodic, suggesting that more than one sub-band may be occupied at the lowest fields. Plateaus are seen for the fractional quantum Hall effect at $\nu = 2/3$ and $\nu = 1/3$ (cut off by limit of the field sweep) and suggested at $\nu = 3/5$ and 4/5, in agreement with 2D theory. The integer quantum Hall effect has also been observed at a higher temperature T = 1.5 K in samples from 3 of the 4 other parabolic well structures studied.

Figure 3 is a plot of the longitudinal magnetoresistance of the same sample at T = 0.5 K with the magnetic field parallel to the current direction. The presence of a Shubnikov-de Haas oscillation peak in Fig. 3 suggests that at least two sub-bands

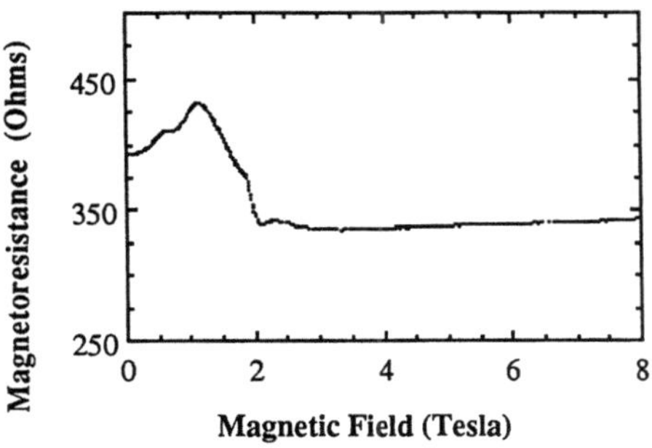

Fig. 3 Longitudinal magnetoresistance at T = 0.5 K

of the well are occupied below H ~ 1 Tesla. This result is consistent with the filling of a square well of width w_e by the measured areal density n_{2D}. The spectrum for this case is $E_n = n^2 E_o$ with $E_o \sim 0.8$ meV and the Fermi energy from n_{2D} is $E_F - E_o \sim 5$ meV, which predicts that two sub-bands are occupied.

4. ACKNOWLEDGEMENTS

We thank B.I. Halperin for many helpful discussions, and J.H. English for help in sample preparation. Part of this work was performed at the Francis Bitter National Laboratory supported at M.I.T. by the National Science Foundation. This research was supported in part by NSF grants DMR-85-08733 and DMR-86-14003.

REFERENCES

1. See for example R.E. Prange and S.M. Girvin eds.: The Quantum Hall Effect (Springer Verlag, New York, 1987)

2. M. Sundaram, A.C. Gossard, J.H. English, and R.M. Westervelt: Superlattices and Microstructures, in press

3. B.I. Halperin, Japan J. of Appl. Phys. 26, Suppl. 26-3, 1913 (1987)

Correlation Between Magnetotransport and Photoluminescence in the Quantum Hall Effect Regime

R. Stepniewski [1;*], *W. Knap* [2;*], *A. Raymond* [2], *G. Martinez* [1], *T. Rötger* [3], *J.C. Maan* [3], *and J.P. André* [4]

[1]SNCI/CNRS, BP 166X, F-38042 Grenoble Cedex, France
[2]GES-USTL, Place Eugene Bataillon, F-34060 Montpellier Cedex, France
[3]MPI-Hochfeld-Magnetlabor, BP 166X, F-38042 Grenoble Cedex, France
[4]LEP, 3, Avenue Descartes, F-94450 Limeil-Brevannes, France
*Permanent address: Institute of Experimental Physics, Warsaw University, Hoza 69, PL-00-681 Warsaw, Poland

The energy level structure of the two-dimensional electron gas in the Quantum Hall Effect regime is studied with simultaneous measurements of magnetotransport and luminescence of a modulation doped single quantum well. Non-linearities in the field dependence of the luminescence maxima energies are observed which are correlated with plateaus of the Quantum Hall Effect. The results are explained by the change in the level structure associated with the occupation of the subbands, on the basis of a single particle self consistent calculation.

Simultaneous measurements of magnetotransport and luminescence at low temperatures and high magnetic fields allow a detailed analysis of the energy level structure of quasi two-dimensional systems in the Quantum Hall Effect (QHE) regime [1]. With such experiments the total density of states (DOS) can be studied optically, whereas the transport properties are governed by the delocalized part of the DOS.

The investigated sample was a Hall bar shaped modulation doped 300 Å $Al_{.28}Ga_{.72}As/GaAs/Al_{.28}Ga_{.72}As$ single quantum well grown by MOCVD technique. Experiments were performed at 1.5K and at magnetic fields up to 12.5 T. The photocarriers were excited using 514.5 nm Ar^{+} laser radiation, with an excitation power density $I_0 < 10 W/cm^2$, since it was found that in that range ρ_{xx} and ρ_{xy} did not depend on the illumination power. ρ_{xx} and ρ_{xy} as a function of B were measured by sweeping the magnetic field, while the photoluminescence spectra were taken at fixed magnetic field, controlling at the same time that the value of ρ_{xy} did not change.

In Fig.1 the energies of the photoluminescence lines are plotted versus magnetic field, as well as the magnetic field dependence of ρ_{xy}. It can be seen that plateaus in ρ_{xy} coincide with the disappearance of the highest luminescence lines. At these magnetic fields a flattening of the field dependence of the lower energy luminescence lines can be seen. A similar behaviour has also been observed previously by other authors [2, 3, 4, 5].

Springer Series in Solid-State Sciences, Vol. 87
High Magnetic Fields in Semiconductor Physics II Editor: G. Landwehr

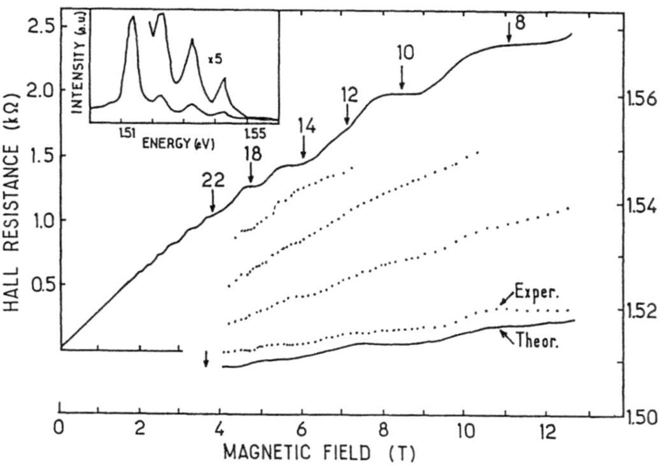

Fig.1. Experimentally obtained energies of the photoluminescence lines (points) and ρ_{xy} (upper solid line), as a function of the magnetic field. The lower solid line shows the result of the fitting procedure for the lowest luminescence line (0→0). For clarity the theoretical line has been shifted as indicated by an arrow. The inset represents a spectrum of photoluminescence at 6T.

Several reasons for the observed step-like behaviour of the luminescence energies with magnetic field can be invoked. First, with increasing fields the Fermi energy jumps between different Landau levels which may be originating from different electric subbands. Since these have different spatial extent of their wavefunctions, different occupation implies different charge distributions. This gives rise to potential deformation causing a change of subband energies[4]. Secondly, a change of the total number of carriers in the QHE regime due to charge transfer from (to) an external reservoir (doped layer) [6] would obviously change the subband energies [2,3,7,5]. Finally, many body effects due to different screening for different filling factors can be invoked [3,5,8,9].

Here we want to investigate to what extent the first two mechanisms, based on a single electron picture, can explain the experimental data. For this purpose we solved self-consistently, for each value of the magnetic field, the Schrödinger and the Poisson equations, for the ideal structure shown in Fig.2. The charge density we used, takes into account the ionized donors in the doped layers of GaAlAs and the free electrons in the well. In the calculations of the energy of Landau levels of electrons and holes we used the single band parabolic approximation with the following parameters: E_d=-14meV [10], m^*_e=0.068 m_o , m^*_h=0.45 m_o , ϵ/ϵ_o=13.1 , E_G=1.519 eV , ΔE_G=0.352 eV , ΔE_c=0.67ΔE_G.

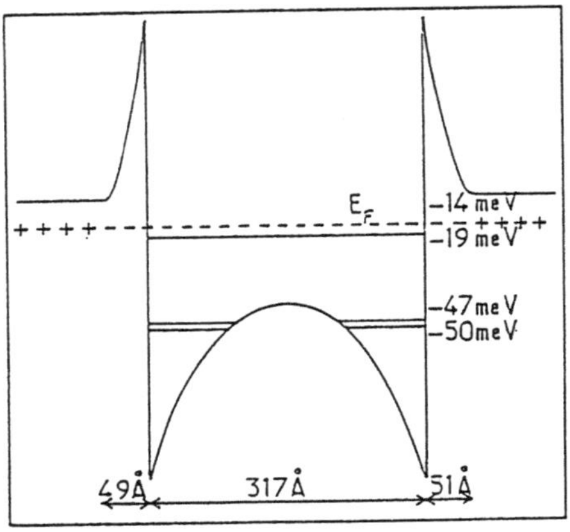

Fig.2. : Energy scheme of the quantum well obtained from the fitting procedure.

We assumed that the optical transitions observed in the experiment obey the $\Delta n=0$ selection rule and correspond to the first electric subband for electrons and holes. The spacer thicknesses ds_1 and ds_2, the width of the well d, (see Fig. 2), the broadening Γ of the Landau levels and the donor concentration N_d in the doped layers have been used as fitting parameters to obtain a good agreement with the experimental energy of the (0→0) transition, in the magnetic field range 5 - 12.5T.

The experimentally observed luminescence energies are quite well reproduced provided we assume that the band gap is 4meV smaller than that of bulk GaAs. This band gap renormalization is probably due to many-body interactions (neglected in our calculation) [11].

The best agreement (see Fig.1.) has been obtained for ds_1=51Å, ds_2=49Å, d=317Å, Γ=2.7meV, $N_d=2\cdot10^{18}$ cm^{-3}. The obtained values are slightly different from those expected from the growth process. This difference is probably associated with the diffusion of the Si impurities during growth.

We find that the two-dimensional electron gas (2 DEG) density, calculated for B=0T, $N_s=2.07\cdot10^{12}cm^{-2}$ corresponds to the Hall voltage in low magnetic fields ($N_{s\ Hall}=2.08\cdot10^{12}cm^{-2}$). The calculated energy diagram is shown in Fig.2. It can be seen that two slightly separated electric subbands for electrons are populated and about 6% of the free electrons occupy the third one at zero magnetic field.

In Fig.1. the positions of the integer filling factors ($\nu=N_sh/eB$) obtained by the calculations are marked. For a magnetic field lower than 6T, the difference between the filling factors associated with the following plateaus is 4. This is caused by the simultaneous depopulation of the Landau levels originated from the two lowest electric subbands, each twofold spin degenerate.

From our calculations it follows that two mechanisms are responsible for the observed step-like behaviour of the luminescence energy: i) oscillations of the 2 DEG density due to transfer of charge to the donors, ii) oscillations of the population of the third electric subband. Both these mechanisms induce a potential deformation inside the quantum well causing a change of the subband energies. The effect of both mechanisms on the energy is about equally strong. However, it should be noted, that the redistribution of charge between different subbands, for a constant electron density in the well, can already lead to steps in the luminescence structure. To separate the influence of many body effects [8, 9] from these "trivial" effects one should use samples where the above-mentioned mechanisms are ineffective.

We would like to stress that, in our case, the transfer of charge from the parent donors to the well, of the order of few percents of the total charge, has a notable influence on the energy of the levels as observed in luminescence, however the effect is by far too weak to explain the Quantum Hall plateaus in ρ_{xy} as has been proposed recently [7].

1. I.V. Kukushkin and V.B. Timofeev, Surf. Science 196, 196-208 (1988),
2. M.C. Smith, A. Petrou, C.H. Perry, J.M. Worlock, R.L. Aggarwal, Proc. of 17th Int. Conf. on the Phys. of Semicond, p.547 (ed. J.D. Chadi and W.A. Harrison, Springer, New York), (1984)
3. J. Orgonasi, J.A. Brum, C. Delalande, B. Bastard, T. Rötger, J.C. Maan, G. Weimann and W. Schlapp, J. de Physique 48, C5-407 (1987),
4. T. Rötger, J.C. Maan, P.Wyder, F. Messeguer and K. Ploog, J. de Physique 48, C5-389 (1987),
5. M.S. Skolnick, K.J. Nash, S.J. Bass, P.E. Simmonds and M.J. Kane, Solid State Commun. 67, 637 (1988),
6. G.A. Baraf and D.C. Tsui Phys. Rev.B 24, 2274 (1981)
7. A. Raymond and K. Karrai, J. de Physique 48, C5-491 (1987),
8. T. Ando, this conference,
9. L.J. Sham, this conference,
10. J.B. Beerens, G. Gregoris, J.C. Portal, F. Alexandre and M.J. Aubin, Proc. 18th Int. Conf. Phys. Semiconductors, Stockholm 1986, p.457,
11. G.E.W. Bauer and T. Ando, J. Phys. C: Solid State Phys.19, 1537 (1986)

The Influence of Contacts on the Quantized Hall Effect

R. Woltjer[1], *M.J.M. de Blank*[1], *J.J. Harris*[2], *C.T. Foxon*[2], *and J.P. André*[3]

[1]Philips Research Laboratories, NL-5600 JA Eindhoven, The Netherlands
[2]Philips Research Laboratories, Redhill, Surrey RH1 5HA, UK
[3]Lab. d'Electronique et de Physique Appliquée,
F-94450 Limeil-Brevannes, France

The influence of electrical contacts on the magnetotransport in the quantized Hall regime is measured on various geometries in GaAs - $Al_xGa_{1-x}As$ heterostructures. The observed effects are interpreted in terms of a local resistivity tensor without taking into account the possible existence of macroscopic quantum states or localization. In special geometries with large contacts we measure, and explain, a two-terminal resistance that is smaller than the Hall resistivity. Furthermore the effects of large Hall contacts in a normal Hall bar geometry are calculated and we show why the measured Hall resistance can be smaller than the true Hall resistance, with their difference proportional to the magnetoresistance, leading to a dip at the strong-field side of Hall plateaus. This extends our interpretation in terms of a (inhomogeneous) local resistivity tensor to real samples with metal contacts.

1. Introduction

The influence of contacts and other geometrical effects in the quantized Hall regime has been studied experimentally by many authors, e.g. FANG and STILES [1], SYPHERS and STILES [2] and VAN DER WEL et al. [3]. The results of these studies have been compared with theories by, e.g. AL'TSHULER and TRUNOV [4], NIU and THOULESS [5], NEUDECKER and HOFFMANN [6] and RIKKEN et al. [7] mainly at the magnetic fields where the Hall plateaus occur. Our model for the quantized Hall effect in inhomogeneous samples [8,9], assuming the existence of a local resistivity tensor, does offer the opportunity to calculate the magnetoresistance and the Hall resistance. The model has so far been developed for a simple Hall bar geometry, and calculations have been restricted to the still simpler case of an infinitely long Hall bar geometry without contacts and with inhomogeneity over the width only.

In this paper, we will extend our local description to more complicated situations. We perform calculations for special geometries and compare them with experimental results. The ultimate test for our description of the magnetotransport in the quantized Hall regime in terms of a local resistivity tensor would require the solution of Kirchhoff's equations in two dimensions for an inhomogeneous resistivity tensor, but this leads to serious numerical difficulties. Instead, we will extend our local resistivity description such as to take into account the Hall voltages that are built up when a current flows between a metal (contact) and a semiconductor. Furthermore, we will study the influence of the finite size of a contact on measurable voltages. The approach chosen in this paper is to divide a sample with a complex geometry into quasi-one-dimensional sections that are connected according to Kirchhoff's laws to calculate the current distribution over the entire sample. The magnetotransport within these sections is described by a Hall resistance and a (homogeneous) magnetoresistance per square. These resistances are obtained either from our measurements or from our model for the quantized Hall effect [9]. Comparison of calculated results with exper-

Springer Series in Solid-State Sciences, Vol. 87
High Magnetic Fields in Semiconductor Physics II Editor: G. Landwehr

imental data will give information on whether this simplified local description is meaningful. The same approach was used earlier to calculate the voltage distribution on a ring-shaped geometry [8]. The use of a local resistivity tensor is worth mentioning, because some theories describe the quantized Hall effect in terms of macroscopic quantum states that extend over a sample. Other theories are based on localized states that exclude certain regions in a sample from participation in the current transport. Our results will show that, in spite of these nonlocal effects, a local description of the resistivity can be useful.

For all the calculations in this paper it is important to realize that in our local description the current distribution within the two-dimensional electron system is (in a good approximation) determined by the local magnetoresistivity only. For small currents, the distribution of the Hall voltage over the sample has no influence on the current distribution within the semiconductor when the Hall resistivity shows no appreciable variations. When a current flows between a semiconductor (where $\rho_{xx} < \rho_{xy}$) and a metal contact (where $\rho_{xy} \approx 0$), the Hall voltage corresponding to this current is built up in one corner of the contact due to the short-circuiting of the Hall resistance by the metal. Comparison of our calculations with experiments in special geometries can only lead to meaningful results in samples of excellent quality. The material has to be fairly homogeneous and the electrical contacts must be ohmic with low resistances. It is worth noting here that the same experiments in worse materials can possibly lead to quite different results. All our experiments were done in high quality samples with good contacts and expected symmetries were checked by changing the direction of the magnetic field or the current to see whether the observed effects are due to the intended geometrical effects, inhomogeneity or imperfections of the samples.

In section 2, we describe the current transport through two parallel Hall bars that are multiply connected by metal wires between both their current contacts and their Hall contacts. The current concentration in a point-like corner of a contact is an approximation that disregards penetration of part of the current in the contact, when current flows along a metal-semiconductor interface. In section 3 we discuss this additional effect. In section 4 we compare calculated and experimental results for GaAs - $Al_xGa_{1-x}As$ heterostructures in special geometries with large contacts. Especially, we will discuss the influence of large Hall contacts on standard magnetotransport measurements in a Hall bar geometry.

2. Two parallel Hall bars with short-circuited contacts

We extend the description of a Hall bar in terms of an infinite one-dimensional stripe without contacts [9] to a more realistic one. A standard Hall bar geometry with two pairs of Hall contacts is given on the left side of figure 1. We consider the two-dimensional electron system as a proper connection of seven one-dimensional parts, as schematically given on the right side of the same figure, in which we describe the electrical transport by their Hall resistance R_H and their magnetoresistance per square $R_\square$. Where sections are connected we obey Kirchhoff's laws by requiring current conservation and equal potentials. Note that for small currents only ρ_{xx} determines the current distribution, making the Hall effect unimportant at the interface between two semiconductor sections which have equal ρ_{xy}. The Hall voltages that are generated at the boundary between a metal contact and the two-dimensional electron system (in the corners where the arrows denote the currents) are taken into account, neglecting the effects of the finite sizes of the contacts. By the symbols $\square$ with their superscripts, we denote the number of squares of a section of the sample, with the superscript referring to the section. The number of squares is the ratio of the length (parallel to the current)

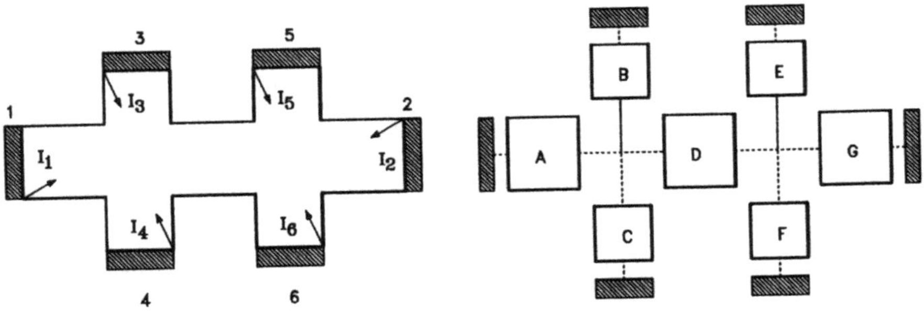

Figure 1. Standard Hall bar geometry with two pairs of Hall contacts. The corners where (most of) the current enters the metal contacts, due to the magnetic field, are denoted by the numbered arrows in the Hall bar. To simplify the calculations we split this geometry in seven semiconductor sections and six metal contacts which are connected (dotted lines) according to Kirchhoff's laws, as shown in the right part of the figure.

over the width (perpendicular to the current) of a rectangular section. As an example, we consider how $V_{3,2}$ is obtained. Between contact 2 and contact 3 we have a Hall voltage $-I_2R_H$ generated by the current between contact 2 and the semiconductor and a Hall voltage $-I_5R_H$ generated by the current passing through contact 5. The magnetoresistance causes a voltage drop $-I_2\square^G R_\square$ in section G, $(I_1+I_3+I_4)\square^D R_\square$ in section D and $I_3\square^B R_\square$ in section B.

$$V_{3,2} = -(I_2+I_5)R_H - I_2\square^G R_\square + (I_1+I_3+I_4)\square^D R_\square + I_3\square^B R_\square, \tag{1}$$

Comparable reasoning leads to all five voltages as a function of five independent currents (using $I_1+I_2+I_3+I_4+I_5+I_6=0$). We will use this six-pole description of a Hall bar to calculate the voltages on the contacts of two Hall bars electrically connected in parallel and compare our results with experimental data on this geometry. The experimental situation is schematically given on the right side of figure 2. Both Hall bar geometries are made on standard GaAs - $Al_xGa_{1-x}As$ heterostructures. The mean electron densities for sample A and sample B are 4.17×10^{15} m^{-2} and 4.84×10^{15} m^{-2}, respectively. The mobilities are 13.1 m^2/Vs and 9.63 m^2/Vs, respectively. The temperature is 2.2 K and the current I_0 is 1 μA.

We calculate the voltages on the contacts by using the six-pole equations. The magnetoresistance per square $R_\square$ and the Hall resistance R_H. are calculated with the inhomogeneity model [9], using the experimentally determined parameters for the samples and including an inhomogeneity of 2.6%. The number of squares of each section is determined from the geometry of the samples used in the experiments. We have $\square^A=\square^G=2.3$, $\square^D=2.7$ and $\square^B=\square^C=\square^E=\square^F=4$ in both Hall bars. Figures 2(a) and 2(b) give the results of experiments and calculations, respectively. The connections on one side of the Hall bars lead to a peculiar oscillatory behaviour of the voltage on that side with a minimum amplitude around 3 T. This feature as well as the different Hall voltages are clearly reproduced in our calculated results. The result shows that our description of the magnetotransport by a resistivity tensor, taking into account the Hall voltages that are built up when current enters metal contacts, leads to reasonable results for small contacts. In addition to the Hall voltages caused by a current which enters a (small) contact (for $\rho_{xx}<\rho_{xy}$ in the semiconductor), we can have an important additional effect when (for $\rho_{xx}\neq0$ in the semiconductor) a current flows along a large contact. This will be discussed in the next section.

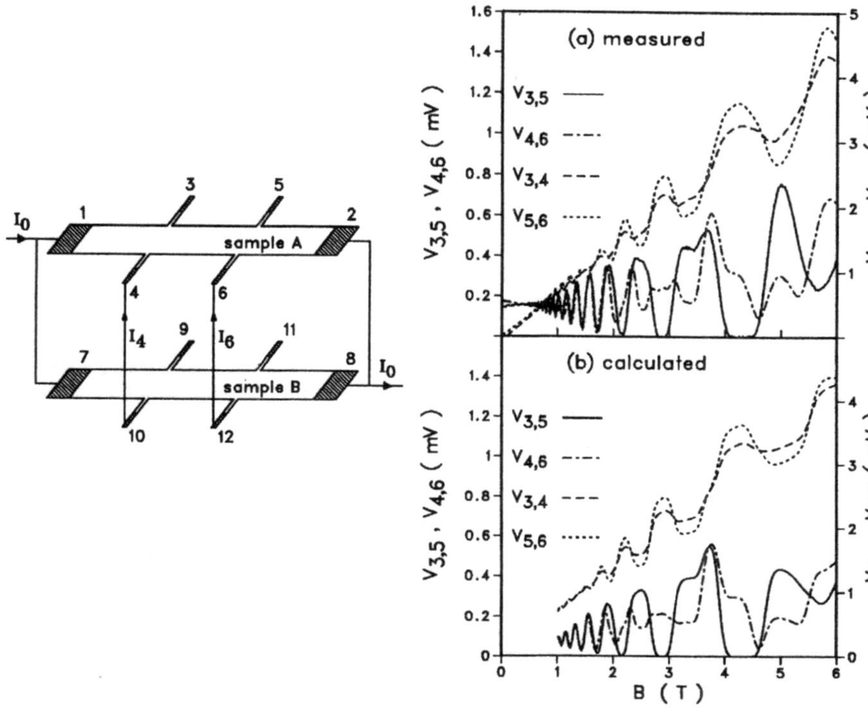

Figure 2. Two Hall bars connected electrically in parallel by metal wires are shown on the left side, with their metal contacts shaded. The current contacts (1, 2, 7 and 8) and two pairs of Hall contacts (4, 10 and 6, 12) are connected by metal wires. On the right we show (a) the measured and (b) the calculated voltages.

3. Model for the current distribution near contacts

In section 4 we will investigate the influence of large contacts in experiments, but first we analyse what we would expect when the magnetotransport in the vicinity of a metal contact is described in terms of a local resistivity tensor. In figure 3 we have sketched a two-dimensional electron system between a metal contact ($y < -w$) and an insulator ($y > 0$). We denote physical quantities in the metal by a superscript m to distinguish them from quantities in the semiconductor without superscripts. Inside the metal we take $\rho^{m}_{xx} = 0$ and $\rho^{m}_{xy} = 0$. In this ideal metal no potential differences can exist, leading to $E_x = 0$ at the metal-semiconductor interface. The Hall angle $\theta_H =$ arctan (ρ_{xy}/ρ_{xx}) is the angle between the electric field and the current. For $\rho_{xx} \neq 0$ the current crosses equipotential lines. Thus part of the current flowing along the metal contact leaves the two-dimensional electron system and penetrates into the metal (for one direction of the magnetic field). We describe this penetration for the simple case of a homogeneous sample with $\underline{\rho}$ independent of the spatial coordinates. Of course, $J_y(x,0) = 0$ at the upper boundary $y = 0$. A stationary current distribution for $x \rightarrow \infty$ leads, in combination with the translation symmetry of the geometry in the x direction, to $J_y(\infty,y) = 0$. Using Kirchhoff's laws for this situation results in

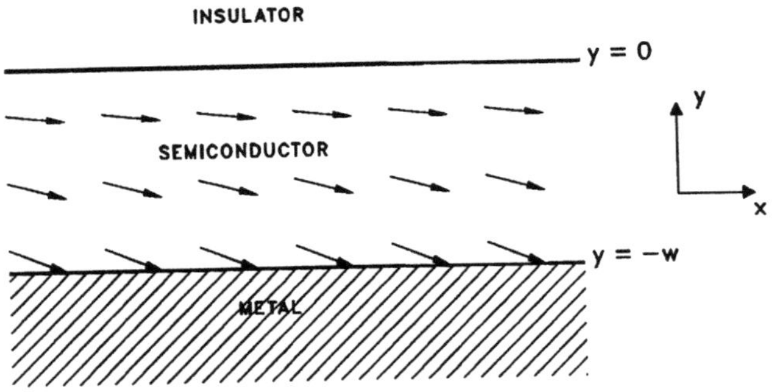

Figure 3. Detail of a metal-semiconductor interface. The x and y directions are given on the right. The metal-semiconductor interface is an equipotential line. A nonzero ρ_{xx} inside the semiconductor makes the current cross the equipotential lines and flow into the metal. The arrows indicate the directions of the current flow for the magnetic field in the positive z direction.

$$J_x(x,y) = J_0\, e^{-\frac{x}{\alpha}} \cos\frac{y}{\alpha} + J_\infty \quad \text{and} \quad J_y(x,y) = J_0\, e^{-\frac{x}{\alpha}} \sin\frac{y}{\alpha}\,. \tag{2}$$

The continuity of $E_x(x,-w) = \rho_{xx}J_x + \rho_{xy}J_y = 0$ at the metal-semiconductor interface determines the lengthscale α. Integration of J_x over the width w of the semiconductor gives the current I flowing in the positive x direction through the semiconductor. For the homogeneous case we can replace the components of the local resistivity tensor by the measurable resistances: $\rho_{xx} \to R_\square$ and $\rho_{xy} \to R_H$. We can thus relate the current in the semiconductor to measurable resistances. We introduce a reduced unit of length $\square = x/w$ (the number of squares) and we denote $I(0) = I_0$, leading to

$$I(\square) = I_0\, e^{-\square \arctan(R_\square/R_H)} \approx I_0\, e^{-\square R_\square/R_H} \approx I_0(1 - \square R_\square/R_H). \tag{3}$$

A current flowing in a homogeneous two-dimensional electron system along a metal contact will indeed partially flow into the contact when the magnetoresistivity in the semiconductor is finite; the current in the two-dimensional electron system will decrease exponentially with distance, when passing the contact. Our treatment so far contains two simplifying assumptions. We have dealt with an ideal metal and a homogeneous semiconductor. The influence of these assumptions is discussed in detail elsewhere [9]. We will consider the magnetoresistance per square to be an appropriate averaging over the local (inhomogeneous) resistivity tensor. For a non-ideal metal (3) must be replaced by

$$\begin{aligned} I(\square) &= \frac{R_\square I_0}{R_\square + R_\square^m} \exp\left(-\square \arctan\frac{R_\square + R_\square^m}{R_H - R_H^m} \right) + \frac{R_\square^m I_0}{R_\square + R_\square^m} \\ &\approx I_0(1 - \square R_\square/\{R_H - R_H^m\}). \end{aligned} \tag{4}$$

Comparison of (3) and (4) clearly shows that the influence on the current flowing along a good metal is negligible already at moderate magnetic field strengths. Even at integer filling factor the resistance of the metal is not important for a current flowing along a contact. The effects of a current leaving a two-dimensional electron system in the corner of a (current carrying) metal contact were studied for the two-terminal resistance near integer filling factor by RIKKEN et al. [7] and found to be at the ppm level. Thus the assumption of an ideal metal will not severely disturb our results.

4. Interpretation of experiments with large contacts

We have performed experiments on GaAs - $Al_xGa_{1-x}As$ heterostructures with special geometries to test the model described in section 3. We will describe our geometries as a series connection of rectangles without contacts and rectangles with contacts, described according to section 3. We describe the magnetotransport in these rectangles by an effective resistivity tensor that is a proper average over the inhomogeneous local resistivity tensor when a sample contains no large-scale inhomogeneities like gradients or large regions with different mean electron densities. These simplifications make it easier to get insight into the physics of two-dimensional current transport in the neighbourhood of a metal contact.

The results are described first for the 'open square' geometry as sketched at the left side of figure 4. The magnetic field is perpendicular to the sample and the current $I_0 = 1\ \mu A$ is flowing from contact 1 to contact 3. The voltages measured between contacts 4 and 2 as well as between 2 and 7 are shown on the right for one direction of

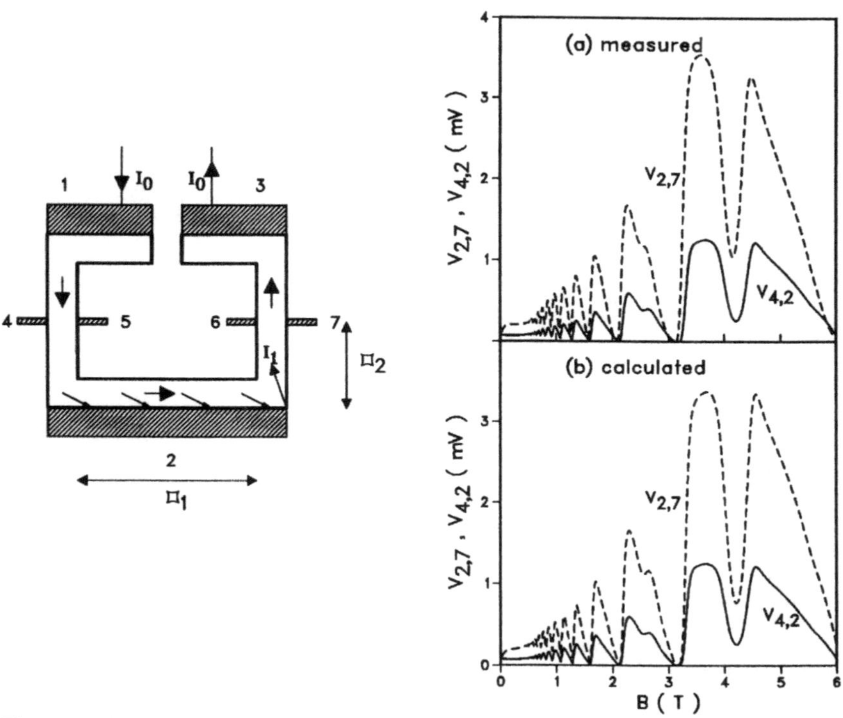

Figure 4. The 'open square' geometry used in experiments to measure the influence of metal contacts on the potential distribution. The current $I_0 = 1\ \mu A$ is flowing from contact 1 to 3. The direction of the current flow, for one direction of the magnetic field, is indicated by the arrows. The current I_1 enters contact 2 over its length and leaves it in its right corner. On the right side we show (a) the measured and (b) the calculated voltages between two contacts. The calculations use measured R_H and $R_\square$.

the magnetic field. Both curves show the expected behaviour for a magnetoresistance measurement. The main asymmetry is in the amplitude of the two signals. The voltage measured between 7 and 2 is nearly three times larger above 0.3 T. It is clear that in a symmetric geometry this can only occur due to the presence of the magnetic field. Reversion of the direction of the magnetic field leads to the expected interchange in behaviour of the two voltages, $V_{4,2}$ and $V_{2,7}$. Reversion of the direction of the current through the sample only changes the signs of the voltages, showing that our observations are not due to unintentional asymmetries in the geometry of the sample.

The sample is described by a series connection of three sections; two sections of semiconductor (lengths $2\Box_2$) connected by one section of semiconductor parallel to a metal (length $\Box_1$). We obtain the following equations,

$$V_{4,2}/I_0 = \Box_2 R_\Box, \tag{5}$$

$$V_{2,7}/I_0 = R_H[1 - \exp(-\Box_1 R_\Box / R_H)] + \Box_2 R_\Box \approx (\Box_1 + \Box_2) R_\Box. \tag{6}$$

The first contribution in the right hand side of (6) is due to the current $I_1 = I_0 - I(\Box_1)$ that enters contact 2 over its full length and leaves that contact in the right corner. For $R_H > \Box_1 R_\Box$, we see that $V_{4,2}$ and $V_{2,7}$ both exhibit a magnetoresistance type of behaviour. From the geometry used in the experiments we determined $\Box_1 \approx 6$ and $\Box_2 \approx 3$ and find a factor $(\Box_1 + \Box_2)/\Box_2 \approx 3$ between both voltages, in accordance with our observations. The resistivity tensor, used in our model for the quantized Hall effect, is only applicable for stronger magnetic fields, making detailed comparison with experiments in weaker magnetic fields ($B < 1$T), where the exponential dependence is important, impossible. For comparison of our calculations with the experiments we use the measured magnetoresistance per square $R_\Box = V_{4,2}/(3I_{1,3})$ and Hall resistance $R_H = V_{5,4}/I_{1,3}$, respectively. The results of these calculations are shown in figure 4 and show convincing agreement with the experimental results. In the calculations it is obvious that the voltages $V_{2,7}$ and $V_{4,2}$ interchange in behaviour for the other polarity of the magnetic field. In weaker magnetic fields we expect to see the influence of the exponential factor in (6), resulting in a lower value for $V_{2,7}$. In figure 4 the expected decrease of $V_{2,7}$ is clearly visible below 0.3 T, whereas $V_{4,2}$ is nearly constant. The physical picture for this smaller resistance in weaker magnetic fields is that all current will enter contact 2 (low resistance) at the left side and leave it at the right side, thus avoiding the magnetoresistance in the semiconductor parallel to the metal contact.

Another geometry that gives peculiar results is the 'open Corbino' geometry shown in the inset of figure 5. This is essentially a normal Corbino geometry with a small sector of the electron system cut out to prevent circulating currents. By conformal mapping it can be shown that this geometry is comparable with a short Hall bar geometry as described by BEER [10]. The two-terminal resistance between the inner and the outer contacts is given in the same figure. At the magnetic field strengths where the magnetoresistance is zero we observe the quantized Hall effect. For comparison we give the Hall resistance measured in a Hall bar geometry of the same material. Between the plateaus the two-terminal resistance is smaller than the Hall resistance and their difference is proportional to the magnetoresistance. This behaviour is very different from the observations in a long Hall bar geometry, where the two-terminal resistance is the sum of Hall resistance and magnetoresistance [1]. The important difference between the long Hall bar and the 'open Corbino' (or the short Hall bar) is that in the last case the current flows along a metal contact over a large distance, where part of the current leaves the semiconductor when $\rho_{xx} \neq 0$. This leads to a negative contribution to the two-terminal resistance R_{io} because the Hall voltage that is built up in the corner of the contacts is generated by a smaller current according to (3),

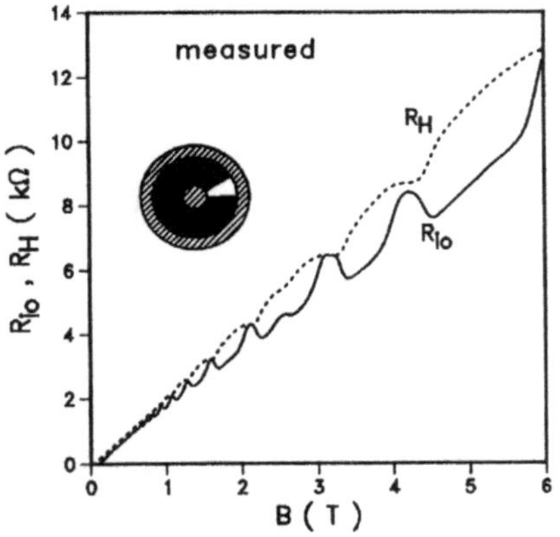

Figure 5. Two-terminal resistance R_{io} between the two contacts (shaded) of the 'open Corbino' geometry sketched in the inset. For comparison we give the Hall resistance R_H as measured in a Hall bar geometry made of the same material. The two-terminal resistance is smaller than the Hall resistance, due to the large contacts.

$$R_{io} = R_H \exp(-\square_1 R_\square / R_H) + \square_2 R_\square \approx R_H + (\square_2 - \square_1) R_\square. \tag{7}$$

Here $\square_1$ is the length over which the current flows parallel to a contact and $\square_2$ is the remaining length of the current path.

We now calculate the effects of large Hall contacts on the standard magnetotransport measurements, using the geometry given in figure 6. This is a standard Hall bar geometry with two large Hall contacts. The direction of the current flow for one direction of the magnetic field is indicated by arrows in the Hall bar to show what contributions we can expect. The current entering contact 3 in the left corner is called I_1 and that leaving contact 7 in the right corner is called I_2. The resistances $R_{3,7}$, $R_{2,3}$ and $R_{3,4}$ are given by

$$R_{3,7} = R_H(1 - I_1/I_0) = R_H \exp(-\square_2 R_\square / R_H) \approx R_H - \square_2 R_\square, \tag{8}$$

$$R_{2,3} = R_H I_2/I_0 + \square_1 R_\square = R_H[1 - \exp(-\square_2 R_\square / R_H)] + \square_1 R_\square \approx (\square_1 + \square_2) R_\square, \tag{9}$$

$$R_{3,4} = \square_1 R_\square. \tag{10}$$

From these equations we learn that large contacts will reduce the measured Hall resistance $R_{3,7}$. The resistance between two contacts $R_{2,3}$ divided by the number of squares between the contacts can be larger (by a factor of $(1 + \square_2/\square_1)$ in this case) than the true magnetoresistance per square. We have not performed experiments to test these calculations, but we have calculated the resistances starting from measured $R_\square$ and R_H to show the effects. The results of these calculations are presented in figure 6 for $\square_1 = 1$, $\square_2 = \frac{1}{2}$. The calculated Hall resistance is lower than the true Hall resistance, proportional to the magnetoresistance. This is most clearly visible by the dip at the strong-magnetic-field side of the Hall plateaus. This is independent of the magnetic field as observed by e.g. Cage et al. [11]. Comparable deviations that do change sign under magnetic field reversal can be caused by a component of the current flowing in the direction of the Hall contacts, as suggested earlier [8]. An enlargement of the Hall resistance compared to the zero-temperature value follows from our inhomogeneity

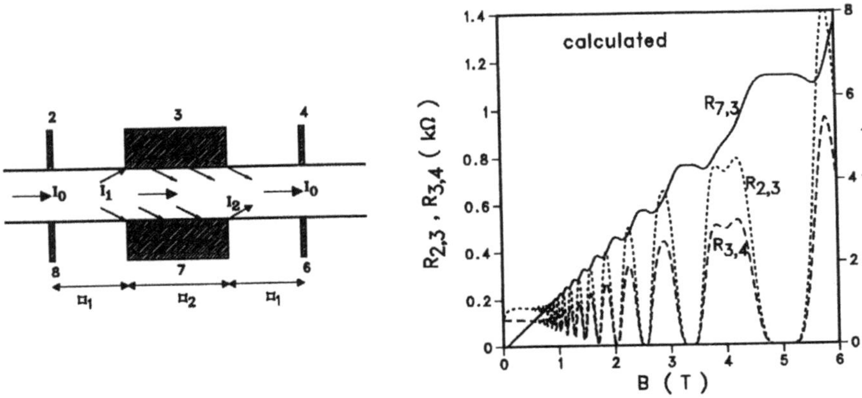

Figure 6. Hall bar geometry with two large Hall contacts in the middle. The direction of the current flow for one direction of the magnetic field is indicated by arrows. The total current flowing from left to right is I_0. The current passing through contact 3 is I_1 and that passing through contact 7 is I_2. On the right we show the calculated Hall and magnetoresistance for this geometry. The Hall voltage between 3 and 7 is smaller than the true Hall voltage between 2 and 8 owing to the effects of the large contacts. The calculations start with measured R_H and $R_\Box$.

model for the quantized Hall effect [8,9]. These last two contributions are (nearly) independent of the size of the Hall contacts. From the presented calculations we can conclude that a Hall bar geometry which is intended to be used for high precision measurements on the quantized Hall effect should have small Hall contacts, or it must have long and narrow side arms where hardly any current flows, as sketched for our standard Hall bar geometries in figure 2.

5. Concluding remarks

In this paper we have studied the influence of the sample geometry and the metal contacts on the measured voltages in the quantized Hall regime. We have not concentrated on the exact potential distribution within one sample, but on the voltages that can be measured on the contacts at the boundaries of the (sometimes multiply connected) samples. To interpret our experiments we have performed calculations using a homogeneous description that can be thought to originate from an adequate averaging over the inhomogeneous local resistivity tensor. This is meaningful when the sample has no large-scale inhomogeneities like gradients, or large regions with different mean electron densities. For this reason we used GaAs - $Al_xGa_{1-x}As$ heterostructures of good homogeneity and used symmetry considerations to verify that the measured effects are not due to deviations from the intended geometries. Our homogeneous description leads to questionable results for the geometries where the current distribution over parts of the sample is determined by the magnetoresistivity only. Experiments and calculations for a ring-shaped geometry, taking into account the measured inhomogeneity of the sample, are described elsewhere [12]. The influence of large contacts on the measured voltages, even in the normal Hall bar geometry, can be very important for high precision experiments at nonzero temperature. The measured Hall voltage can be smaller than the true Hall voltage and a dip in Hall resistance at the high-magnetic-field side of a Hall plateau can be formed. The measured two-terminal resistance can be much smaller (or larger) than the Hall resistivity in geometries with large current contacts.

Our experimental results on special geometry samples can be explained by simply describing the sample by appropriately connected homogeneous rectangular semiconductor sections, some of them with metal contacts. The current is distributed over these sections in accordance with Kirchhoff's laws and Hall voltages are built up where the current crosses a metal-semiconductor interface. The good agreement with experiments lends a further support to the usefulness of a description of the quantized Hall effect as an effect that originates from a local resistivity tensor [9]. We do not take into account the existence of macroscopic quantum states or localization in this local description. The only quantum-mechanics we use is the quantization of the density of states in a magnetic field, leading to vanishing magnetoresistivity in regions with integer filling factor. In inhomogeneous samples this resistivity has a spatial dependence that determines the distribution of the current over the sample. This spatial distribution (in real samples in combination with localization) causes the Hall plateaus to be quantized at the well-known values.

References

1 F.F. Fang and P.J. Stiles: Phys.Rev. B 29 , 3749 (1984).

2 D.A. Syphers and P.J. Stiles: Phys.Rev. B 32 , 6620 (1985).

3 W. van der Wel, J.E. Mooij and C.J.P.M. Harmans: J.Phys. C 21 , L171 (1988)

4 E.L. Al'tshuler and N.N. Trunov: Meas.Tech. 29 , 796 (1986)

5 Q. Niu and D.J. Thouless: Phys.Rev. B 35 , 2188 (1987).

6 B. Neudecker and K.H. Hoffmann: Solid State Comm. 62 , 135 (1987).

7 G.L.J.A. Rikken, J.A.M.M. van Haaren, W. van der Wel, A.P. van Gelder, H. van Kempen, P. Wyder, J.P. André , K. Ploog and G. Weimann: Phys.Rev. B 37 , 6181 (1988)

8 R. Woltjer, R. Eppenga and M.F.H. Schuurmans: In High magnetic fields in semiconductor physics , ed. G. Landwehr, (Springer, Berlin 1987) p.104

9 R. Woltjer: PhD thesis University of Utrecht, The Netherlands (1988)

10 A.C. Beer: Galvanomagnetic effects in Semiconductors , Sol.St.Phys.Suppl. 44 , eds. H. Ehrenreich, F. Seitz and D. Turnbull, (Academic, New York 1963)

11 M.E. Cage, B.F. Field, R.F. Dziuba, S.M. Girvin, A.C. Gossard and D.C. Tsui: Phys.Rev. B 30 , 2286 (1984)

12 R. Woltjer, R. Eppenga, J. Mooren, C.E. Timmering and J.P. André : Europhys. Lett. 2 , 149 (1986)

On the Consistency of Approximations to the Landau-Level Broadening by Random Potentials with Large Correlation Length

K. Broderix, N. Heldt, H. Leschke

Institut für Theoretische Physik, Universität Erlangen-Nürnberg, D-8520 Erlangen, Fed. Rep. of Germany

We discuss the uniqueness and positivity of previously proposed approximations to the density of states for an electron in two dimensions subject to a perpendicular uniform magnetic field and a Gaussian random potential with large correlation length.

The model can be described by the Hamiltonian

$$H := H_0 + V \quad , \quad H_0 := \frac{1}{2}\left(\frac{\partial}{i\partial x_1}\right)^2 + \frac{1}{2}\left(\frac{\partial}{i\partial x_2} + x_1\right)^2 \tag{1}$$

for one (spinless) electron in the infinite (x_1, x_2)-plane under the influence of a perpendicular uniform magnetic field and a random potential V. We have chosen the Landau gauge and units such that Planck's constant, the elementary charge, the (effective) electron mass and the strength of the magnetic field equal one.

In the sequel we are mainly concerned with a Gaussian random potential characterised by

$$\overline{V(x)} = 0 \quad , \quad \overline{V(x)V(x')} = \sigma^2 \mathrm{e}^{-(x-x')^2/L^2} \quad , \quad x := (x_1, x_2) \ . \tag{2}$$

Here the bar denotes the average with respect to the probability distribution, σ^2 is the single-site variance and L the correlation length of the fluctuations of the potential.

The two-sided Laplace transform of the averaged density of states (per area)

$$\rho(E) := \langle x | \ \overline{\delta(E-H)} \ | x \rangle \tag{3}$$

is the averaged partition function (per area)

$$Z(T) := \langle x | \ \overline{\mathrm{e}^{-TH}} \ | x \rangle = \int_{-\infty}^{\infty} \mathrm{d}E \ \rho(E) \ \mathrm{e}^{-TE} \quad , \quad T > 0 \ . \tag{4}$$

The latter can be represented by a Wiener-type path integral

$$Z(T) = \int_{q(0)=0} \mathrm{D}^2 q \ \delta(q(T)) \ \exp\left[-S_0(T|q) + \frac{\sigma^2}{2}\int_0^T \mathrm{d}t \int_0^T \mathrm{d}t' \ \mathrm{e}^{-(q(t)-q(t'))^2/L^2}\right] \ . \tag{5}$$

Here

$$S_0(T|q) := \frac{1}{2}\int_0^T \mathrm{d}t \left(\dot{q}_1^2(t) + \dot{q}_2^2(t) + 2iq_1(t)\dot{q}_2(t)\right) \tag{6}$$

Springer Series in Solid-State Sciences, Vol. 87
High Magnetic Fields in Semiconductor Physics II Editor: G. Landwehr

is the Euclidean action functional corresponding to the unperturbed Hamiltonian H_0. It is understood that only those paths $q(t) := (q_1(t), q_2(t))$ in the plane contribute to the integration (5) which are subject to the boundary conditions $q(0) = q(T) = 0$.

Applying the Golden-Thompson inequality (e.g. [1]) one finds [2]

$$Z(T) \leq \frac{e^{\sigma^2 T^2/2}}{4\pi \operatorname{sh}\frac{T}{2}} \quad . \tag{7}$$

Therefore, for all L the partition function is bounded by its value for $L = \infty$.

Since the integral (5) is not known exactly for $L < \infty$, it has been argued [3] that whenever the correlation length L is large compared with the magnetic length, i.e. $L \gg 1$, an approximation $\tilde{\rho}$ to the density of states may be obtained from the following path integral

$$\tilde{Z}(T) := \int_{q(0)=0} D^2 q \; \delta(q(T)) \; \exp\left[-S_0(T|q) + \frac{\sigma^2}{2} \int_0^T dt \int_0^T dt' \; \left(1 - (q(t) - q(t'))^2/L^2\right)\right] \; . \tag{8}$$

The calculation of the integral [4] leads to

$$\tilde{Z}(T) = \frac{\sigma^2 T^2 \; e^{\sigma^2 T^2/2}}{2\pi L^2 \left[\operatorname{ch}\left(\frac{T}{2}\sqrt{1 + \frac{8\sigma^2 T}{L^2}}\right) - \operatorname{ch}\left(\frac{T}{2}\right)\right]} \quad . \tag{9}$$

One should expect a reasonable approximation $\tilde{\rho}$ to the density of states to fulfil the commutative diagram shown in Fig.1 for a wide class of functions f. Otherwise there would be no unique way to calculate approximations to physical quantities $\langle x | \, \overline{f(H) \exp(-TH)} \, | x \rangle$ like the filling factor or the specific heat.

Concerning the function (9) we assert that the well defined Fourier transform

$$\tilde{\rho}(E) := \lim_{\varepsilon \to 0} \int_{-\infty}^{\infty} \frac{dt}{2\pi} \; e^{itE} \; \tilde{Z}(\varepsilon + it) \tag{10}$$

does not fulfil the diagram even for $f = 1$.

A proof starts by analysing the pole structure of $\tilde{Z}$. Since the algebraic expressions for the poles of function (9) are unwieldy, we show in Fig.2 a plot of the poles' positions in the complex T-plane. The poles are of first order and symmetrically distributed with respect to the real axis. A theorem for the Laplace transformation of analytic functions [5, Chap. 11] only ensures that in every open vertical strip lying in the right half-plane and between the poles a unique inverse Laplace transform of $\tilde{Z}$,

$$\frac{1}{2\pi i} \int_{\tau - i\infty}^{\tau + i\infty} dT \; e^{TE} \; \tilde{Z}(T) \quad , \tag{11}$$

exists, where τ is an arbitrary point in a given strip. The functions (11) corresponding to different strips do not coincide, because the residues of these poles generally take nonvanishing imaginary parts.

By the way, a numerical computation shows that the Fourier transform (10) is negative for some values of E. Lessening the denominator in equation (9) to $4\pi\sigma^2 T^2 \left(1 - \frac{2\sigma^2 T}{L^2}\right) \operatorname{sh}\left(\frac{T}{2}\right)$ results in a positive Fourier transform [3], but does not remove the difficulty of non-uniqueness.

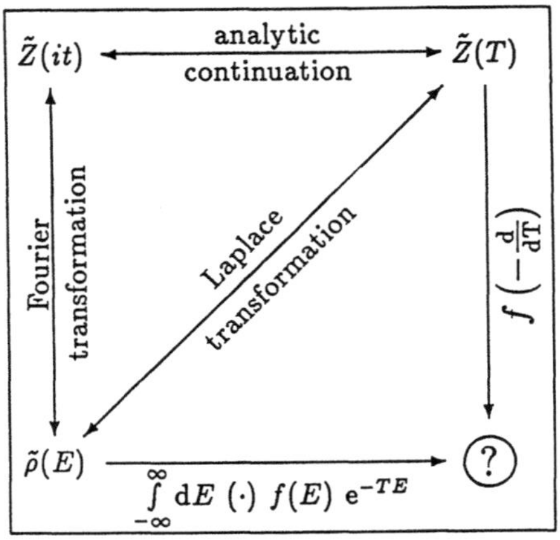

Figure 1: Commutative diagram defining a unique approximation

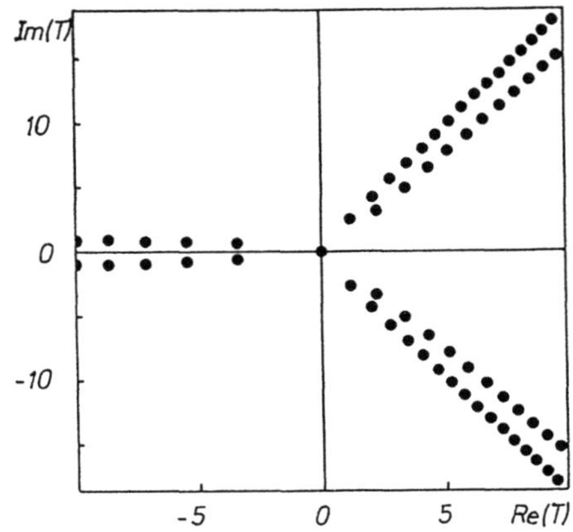

Figure 2: Pole structure of the approximate partition function (9) for $\sigma^2/L^2 = 0.5$. For $\sigma^2/L^2 = 0$ all poles are located on the imaginary axis.

To avoid this problem one may perform a perturbative expansion of the partition function (5) in the inverse correlation length L^{-1}, i.e.

$$Z(T) = e^{\sigma^2 T^2/2} \int_{q(0)=0} D^2q\ \delta(q(T))\ e^{-S_0(T|q)} \left[1 - \frac{\sigma^2}{2L^2} \int_0^T dt \int_0^T dt'\ (q(t)-q(t'))^2 + O(\tfrac{1}{L^4})\right] . \quad (12)$$

The result of the integrations (12)

$$Z(T) = \frac{e^{\sigma^2 T^2/2}}{4\pi \mathrm{sh}\frac{T}{2}} \left[1 - \frac{\sigma^2 T^2}{L^2} \left(\mathrm{cth}\frac{T}{2} - \frac{2}{T}\right) + O(\frac{1}{L^4})\right] \quad (13)$$

can be inferred either from a formal expansion of equation (9) or from a direct calculation, see, e.g., Appendix A of [7]. In contrast to expression (9), this one has a unique inverse Laplace transform,

$$\rho(E) = \sum_{n=0}^{\infty} \frac{e^{-(n+\frac{1}{2}-E)^2/2\sigma^2}}{2\pi\sqrt{2\pi\sigma^2}} \left[1 + \tfrac{2}{L^2}\left(2n+1-E-\tfrac{2n+1}{2\sigma^2}(n+\tfrac{1}{2}-E)^2\right) + O(\tfrac{1}{L^4})\right] , \quad (14)$$

because it takes no poles in the right half-plane. The L^{-1}-expansion for ρ has been calculated earlier within a Green's function approach [6]. As an aside we mention that the L^{-2}-term in (14) is four times larger than found in [6]. Analysing equation (14) and neglecting corrections of order L^{-4} shows that this approximation is negative for $L > \sqrt{2}$ and $0 < \sigma^2 < (2L^2+4)^{-1}$ at all $E = 1, 2, \ldots$.

The problem of negative values of approximate expressions for the density of states may also occur for non-Gaussian correlation functions. For example, the choice

$$\overline{V(x)V(x')} = \sigma^2 e^{-|x-x'|/L} \quad (15)$$

in [8] leads to negative values of the approximate partition function for large L similar to (13).

The problems of uniqueness and positivity for the density of states do not exist in Gerhardt's lowest-order cumulant approximation [9] for

$$Z_n(T) := \langle x|\; P_n \overline{e^{-TH}} P_n \;|x\rangle \tag{16}$$

which is not restricted to large L. Here P_n is the projector on the subspace of the n-th Landau level. While uniqueness is due to the absence of singularities, positivity follows from criterions to be found, e.g., in [10].

The complete partition function can be calculated according to

$$Z(T) = \sum_{n=0}^{\infty} Z_n(T) \tag{17}$$

because of [11,9,2]

$$P_n \overline{e^{-TH}} = \overline{e^{-TH}} P_n \quad . \tag{18}$$

However, in the lowest-order cumulant approximation (without an additional long-time approximation) the complete partition function, with the choice (2), diverges for $T \geq \ln(1 + L^2/2)$. By a formal expansion of this partition function up to the order L^{-2} one recovers the right-hand side of (13) and faces again the associated problem.

In conclusion, we remark that all mentioned approximations are exact for the trivial case $L = \infty$ (e.g. [12]). Surely, the inverse Laplace transform of the exact partition function (4,5) is unique and positive for all L. To maintain uniqueness and positivity in approximations it is sufficient to ensure that $\tilde{Z}(it)$, t real, can be interpreted as an average over the random potential of a positive definite quantity. Following these lines we have recently proposed [2] a consistent approximation for the complete density of states based on a lower bound to $Z(T)$.

[1] *H. Leschke*: In: "Path Summation: Achievements and Goals". S. Lundqvist, A. Ranfagni, V. Sa-yakanit, L. S. Schulman (eds.), Singapore: World Scientific 1988

[2] *K. Broderix, N. Heldt, H. Leschke*: In preparation

[3] *M. Nithisoontorn, R. Lassnig, E. Gornik*: Phys. Rev. B **36**, 6225 (1987)

[4] *F.M. Peeters, J.T. Devreese*: Phys. Rev. B **25**, 7281 (1982)

[5] *G. Doetsch*: "Handbuch der Laplace-Transformation", Bd.1. Basel: Birkhäuser 1950

[6] *K.A. Benedict, J.T. Chalker*: J. Phys. C **18**, 3981 (1985)

[7] *K. Broderix, N. Heldt, H. Leschke*: Z. Phys. B **66**, 507 (1987)

[8] *Zh.S. Gevorkyan, Yu.E. Lozovik*: Fiz. Tverd. Tela **26**, 2852 (1984) Engl. Transl.: Sov. Phys. Solid State **26**, 1725 (1984)

[9] *R.R. Gerhardts*: Z. Phys. B **21**, 275 (1975) ; 285 (1975)

[10] *R.A. Horn*: Z. Wahrscheinlichkeitstheorie verw. Geb. **8**, 219 (1967)

[11] *H. Scher, T. Holstein*: Phys. Rev. **148**, 598 (1966), App. IV
H. Keiter: Z. Phys. **198**, 215 (1967)

[12] *K. Broderix, N. Heldt, H. Leschke*: Z. Phys. B **68**, 19 (1987)

Collective Excitations of Two-Dimensional Electron Solids and Correlated Quantum Liquids in High Magnetic Fields

G. Meissner and U. Brockstieger

Theoretische Physik, Universität des Saarlandes,
D-6600 Saarbrücken, Fed. Rep. of Germany

A simple but rigorous approach will be presented for studying the nature of intra-Landau level collective excitations of interacting electrons fully polarized in a perpendicular magnetic field with partial filling of the lowest Landau level. A finite gap, due to correlations in the density fluctuations of the centers of cyclotron orbits, or no gap, due to crystalline-like correlations because of broken magnetic translational invariance are shown to exist rigorously in the long-wavelength limit of low-lying collective excitations in two condensed phases, i.e., an incompressible quantum liquid and a two-dimensional electron quantum solid, respectively.

1. INTRODUCTION

Interacting electrons of charge -e in a plane with a uniform charge-compensating background perpendicular to a magnetic field are a particularly interesting quantum many-body system, if the magnitude B of the magnetic field is so large that the number ν^{-1} of magnetic flux quanta ch/e per electron exceeds one [1], [2]. For a given area A this then corresponds to a fractional filling factor $\nu = 2\pi r_L^2 n_e < 1$, i.e., a fractional ratio of the mean areal density of electrons $N_e/A = n_e$ and of flux quanta $N_s/A = (2\pi r_L^2)^{-1}$, respectively, where $r_L = (c\hbar/eB)^{1/2}$ denotes the Larmor radius.

Many-body correlations due to Coulomb interaction are considered to be of significance in such two-dimensional electron systems. The many-body ground state is expected to be a correlated incompressible quantum liquid (IQL) giving rise to a quantized Hall conductivity $\sigma_{12} = \nu\, e^2/h$ for certain fractional filling factors $\nu = p/q < 1$, with p and q denoting mutual primes, until, below a critical filling factor ν_c, another solid-like ground state with nonvanishing shear modulus μ is anticipated to exist. In the limit $\nu \to 0$ the triangular electron lattice formed by a classical two-dimensional (2D) Wigner crystal [3] may be identified with the ground state of this sort of a 2D electron quantum solid to be called quantum Hall crystal (QHC), since it results from broken magnetic translational invariance at zero temperature [4].

Early efforts to find experimentally the 2D Wigner crystal in high magnetic fields did not turn out to have been successful. Unexpectedly, however, they led to the discovery of the fractional quantum Hall effect (FQHE) [5]. Only very recently experimental evidence for a gapless collective mode of a magnetically induced Wigner solid in such 2D correlated electron systems was claimed to have been established [6]. Therefore, it is of great interest to develop an approach which provides an analysis of basic differences in the collective excitations of the two phases not resting on too restrictive approximations and assumptions.

Springer Series in Solid-State Sciences, Vol. 87
High Magnetic Fields in Semiconductor Physics II Editor: G. Landwehr

2. COLLECTIVE EXCITATIONS

The dispersion relation between frequencies ω and conserved wave vectors $\underset{\sim}{k}$ of the collective excitation spectrum in our approach is being determined from zeros of the real part of the inverse of a density response function

$$\mathrm{Re}\ \chi^{-1}_{\Delta\Delta}(\underset{\sim}{k},\omega(\underset{\sim}{k})) \equiv 0 , \qquad (1)$$

where the operators $\underset{\sim}{X}(\ell)$ in the density-fluctuation operators

$$\Delta(\underset{\sim}{k}) = \sum_{\ell} \exp(-i\underset{\sim}{k}\cdot\underset{\sim}{X}(\ell)) \qquad (2)$$

denote the centers of the cyclotron orbits, i.e., the so-called guiding center coordinates. The noncommutivity of these projected density fluctuations

$$[\Delta(\underset{\sim}{k}),\Delta(\underset{\sim}{k}')] = -\ 2i\ \sin(k_\alpha\varepsilon_{\alpha\beta}k'_\beta r_L{}^2/2)\ \Delta(\underset{\sim}{k} + \underset{\sim}{k}') \qquad (3)$$

results from the noncommutivity of guiding center coordinates

$$[X_\alpha(\ell),X_\beta(\ell')] = i r_L{}^2 \delta_{\ell\ell'}\varepsilon_{\alpha\beta} \ . \qquad (4)$$

By $\varepsilon_{\alpha\beta}$ in (3) and (4) we have denoted components of the antisymmetric tensor in two dimensions, with the elements $\varepsilon_{11} = \varepsilon_{22} = 0$ and $\varepsilon_{12} = -\ \varepsilon_{21} = 1$.

2.1 Incompressible quantum liquid (IQL)

Applying sum rule techniques we have obtained the following exact spectral representation for the inverse of our density response function

$$\overline{\omega}\ \chi^{-1}_{\Delta\Delta}(z = \omega + i\eta) = -\ z^2 + \overline{\omega}\ \chi^{-1}_{\Delta\Delta}(0) - z\ \overline{\omega} \int \frac{d\omega'}{\pi} \frac{\Gamma_{\Delta\Delta}(\omega')}{\omega' - z} , \qquad (5)$$

where all $\underset{\sim}{k}$-variables have been omitted. The spectral width function $\Gamma_{\Delta\Delta}(\underset{\sim}{k},\omega)$ in (5) has the same symmetry properties as the spectral function $\chi''_{\Delta\Delta}(\underset{\sim}{k},\omega)$. For the static (z = 0) limit, $\chi^{-1}_{\Delta\Delta}(\underset{\sim}{k},0)$, we obtained the rigorous sum rule

$$\overline{\omega}(\underset{\sim}{k})\ \chi^{-1}_{\Delta\Delta}(\underset{\sim}{k},0) = \frac{\overline{\omega^3}(\underset{\sim}{k})}{\overline{\omega}(\underset{\sim}{k})} - \overline{\omega}(\underset{\sim}{k}) \int \frac{d\omega}{\pi}\ \Gamma_{\Delta\Delta}(\underset{\sim}{k},\omega) \ . \qquad (6)$$

We have evaluated the first and third frequency moment explicitly

$$\overline{\omega^n}(\underset{\sim}{k}) = \int \omega^n\ \chi''_{\Delta\Delta}(\underset{\sim}{k},\omega)\ d\omega\ /\ \pi = A_n\ k^4 + \ldots \qquad (n = 1,3)$$

and found both of them to behave like k^4 in the $k \to 0$ limit. The expression for A_1 turns out to depend on two-particle density correlations only, whereas the rather lengthy one for A_3 also contains three-particle correlations.

Hence, at k=0, a gap in the collective excitation spectrum $\omega^2(\underset{\sim}{k})$ of the IQL is determined by the ratio A_3/A_1, as long as the ω-integral over $\tilde{\Gamma}_{\Delta\Delta}$ diverges slower than k^{-4} for $k \to 0$. Thus, $\chi_{\Delta\Delta}(\underset{\sim}{k},0) \sim k^4$ vanishes rapidly for small k, i.e., long-wavelength density fluctuations are strongly suppressed in our incompressible system [7]. Rapidly varying parts, which may appear in $\Gamma_{\Delta\Delta}$, could cause the gap to vanish, if they led to a faster divergence of the ω-integral.

2.2 Quantum Hall crystal (QHC)

For a general investigation of the collective excitations in the QHC-phase it is important to notice that there is a continuous symmetry in our correlated electron system, since the corresponding many-body Hamiltonian is invariant under infinitesimal magnetic translations [4]. Spontaneous symmetry breaking can thus give rise to a new ground state with crystalline-like correlations.

The transition from the IQL to the QHC may occur via softening of the magneto-roton gap [7] of the collective excitations $\omega^2(\underset{\sim}{k})$ of the IQL with decreasing ν, until the gap collapses at some critical ν_c at a certain finite wave vector $\underset{\sim}{k} = \underset{\sim}{G} \neq 0$, becoming a reciprocal lattice vector of the QHC-phase. Due to lattice periodicity, the resulting gapless GOLDSTONE mode, restoring the broken magnetic translational invariance in the QHC, will also develop at $\underset{\sim}{k}=0$.

Since expectation values of guiding center coordinates, $\langle X_\alpha(\ell)\rangle \equiv R_\alpha(\ell)$, do not vanish in the QHC, the dispersion relation between frequencies ω and wave vectors k can directly be determined from zeros of the real part of the inverse response function for guiding center displacement fluctuations

$$u_\alpha(\underset{\sim}{k}) = N_e^{-1/2} \sum_\ell [X_\alpha(\ell) - R_\alpha(\ell)]\exp(i\underset{\sim}{k}\cdot\underset{\sim}{R}(\ell)) \ . \tag{7}$$

The exact spectral representation of the inverse of the displacement response function replacing (5) now takes the form

$$\langle\omega^0_{\alpha\beta}\rangle\, \chi^{-1}_{\beta\alpha'}(z) = -z\delta_{\alpha\alpha'} + \langle\omega^0_{\alpha\beta}\rangle\, \chi^{-1}_{\beta\alpha'}(0) - z\langle\omega^0_{\alpha\beta}\rangle \int \frac{d\omega}{\pi} \frac{\gamma_{\beta\alpha'}(\omega)}{\omega - z} \ . \tag{8}$$

A rigorous sum rule relating components $\chi^{-1}_{\alpha\beta}(\underset{\sim}{k},0)$ of the inverse of the static susceptibility and the ω-integral over components of the spectral width function $\gamma_{\beta\alpha}(\underset{\sim}{k},\omega)$ is now determined by the zeroth and first frequency moment

$$\langle\omega^n_{\alpha\beta}(\underset{\sim}{k})\rangle = \int \omega^n \chi''_{\alpha\beta}(\underset{\sim}{k},\omega)\, d\omega / \pi \qquad (n = 0,1)$$

of the spectral function $\chi''_{\alpha\beta}(\underset{\sim}{k},\omega)$ of displacement fluctuations.

From an expansion into powers of k we have arranged the long-wavelength limit of the collective excitations for a triangular lattice in the form

$$\omega^2(\underset{\sim}{k}) = \frac{2\pi e^2}{m^2\omega_c^2\varepsilon}\,\mu k^3 + O(k^4) \ . \tag{9}$$

In (9), $\omega_c = \hbar/(r_L^2 m)$ denotes the cyclotron frequency with the effective electron mass m and ε is a background dielectric constant. The gapless $k^{3/2}$ magneto-phonon dispersion of $\omega(\underset{\sim}{k})$ holding quite generally, is precisely the same as that found in harmonic approximation (HA) [8], and agrees also with that of the lower hybrid mode of the 2D solid [9]. The isothermal shear modulus μ in (9), however, is rigorously related to the second derivative of the free energy with respect to displacement deformations according to the generalized elastic sum rule [10]. In HA ist may be reduced to

$$\mu_0 = \frac{n_e e^2}{16\varepsilon r_L} \sum_{G\neq 0} \left(G^2 \frac{\partial^2}{\partial G^2} + 3G \frac{\partial}{\partial G}\right) \frac{\nu}{G r_L} \left\{ e^{-G^2 r_L^2/2} - e^{-G^2 r_L^2/2\nu^2} \right\} \tag{10}$$

which remains positive for all filling factors $\nu < 1$. Our approach, however, allows one to be beyond this HA in various respects: we may readily include particle-hole symmetry, certain aspects of Fermi statistics, and anharmonic corrections to one-loop order.

Although we cannot discuss these effects in detail in this short communication, it is worth noting, that their inclusion reduces the range of filling factors, where a stable QHC with $\mu > 0$ exists, already to values of $\nu \lesssim 0.4$ (Fig. 1). A second feature to notice in Fig. 1 is the minimum in the shear modulus $\bar{\mu}_1 = \bar{\mu}_0 + \Delta\bar{\mu}_0$ within a particle-hole symmetric approach including anharmonic contributions $\Delta\bar{\mu}_0$ in one-loop approximation. This already indicates clearly the importance of such effects also for a solid-to-liquid transition. Quantitative estimates of the critical filling factor separating the two phases, however, cannot be drawn yet from the presented results, e.g., since anharmonic effects are not likely to dictate primarily this 2D melting transition. The predicted reduction of the classical value 0.245 $e^2 n_e^{3/2}/\varepsilon$ of the shear modulus for all filling factors where a 2D Wigner solid exists, however, should hold rather generally, although seemingly not in agreement with experimental findings [6].

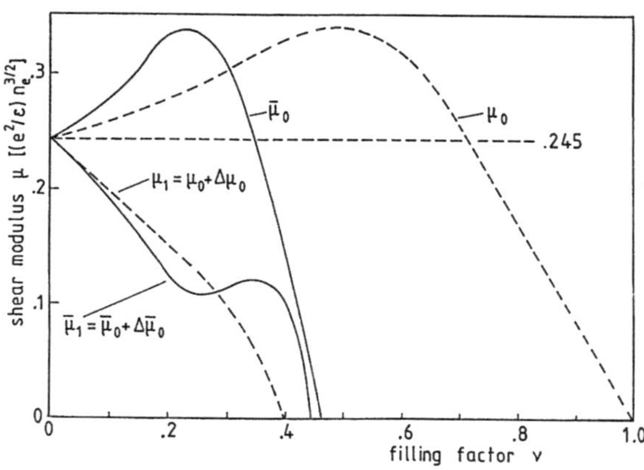

Fig. 1: Shear modulus μ versus filling factor ν of 2D triangular electron lattice in units of $(e^2/\varepsilon)n_e^{3/2}$ with a value 0.245 for the classical Wigner crystal [3]; μ_0 and $\bar{\mu}_0$: HA according to Eq. (10) and modified HA including particle-hole symmetry, respectively; $\mu_1 = \mu_0 + \Delta\mu_0$ and $\bar{\mu}_1 = \bar{\mu}_0 + \Delta\bar{\mu}_0$: including anharmonicities to one-loop order without and with particle-hole symmetry, respectively.

One of the authors (G.M.) would like to thank Dr. V.J. Emery for the hospitality afforded at Brookhaven National Laboratory and for useful discussions.

3. REFERENCES

1. R.B. Laughlin, Phys. Rev. Lett. 50, 1395 (1983).
2. D. Arovas, J.R. Schrieffer, and F. Wilzcek, Phys. Rev. Lett. 53, 722 (1984).
3. G. Meissner, H. Namaizawa, and M. Voss, Phys. Rev. B13, 1370 (1976); L. Bonsall and A.A. Maradudin, Phys. Rev. B15, 1959 (1977).
4. G. Meissner: In Recent Developments in Mathematical Physics, Eds. H. Mitter and L. Pittner (Springer, Berlin, Heidelberg, 1987), p.275.
5. D.C. Tsui, H.L. Störmer, and A.C. Gossard, Phys. Rev. Lett. 48, 1599 (1982).
6. E.Y. Andrei, G. Deville, D.C. Glattli, F.B. Williams, E. Paris, and B. Etienne, Phys. Rev. Lett. 60, 2765 (1988).
7. The k^4 behaviour of $\chi_{\Delta\Delta}(\underset{\sim}{k},0)$ for $k \to 0$ has been conjectured previously within the single-mode approximation by S.M. Girvin, A.M. MacDonald, and P.M. Platzman, Phys. Rev. Lett. 54, 581 (1985).
8. G. Meissner and U. Brockstieger: In Series in Solid State Sciences, Ed. G. Landwehr, 71 (Springer, Berlin, Heidelberg, 1987), p.85.
9. G. Meissner, Z. Physik, B23, 173 (1976).
10. G. Meissner, Phys. Rev. B1, 1822 (1970).

Many-Valley 2D Electron Systems in Strong Magnetic Fields

Yu. Bychkov and S. Iordansky
Landau Institute for Theoretical Physics, Chernogolovka, USSR

For two-dimensional systems in strong magnetic fields the electron-electron interaction is of crucial importance. It defines the structure of the ground state and phase transitions depending on various physical parameters.

The present paper deals with a series of questions related to the formation of magnetic excitons in 2D-systems. Attention to this problem has been attracted by LERNER and LOZOVIK (see [1] and references therein,[2]). Their model corresponds to one electron and one hole band with the same parabolic dispersion law which results in no interaction between excitons in the ground state. On the other hand, BYCHKOV and RASHBA [3] have considered nonsymmetrical models in which not all matrix elements of e-e,h-h,e-h interactions coincide. In these models the exact compensation of interactions, which, in the symmetric case, makes the exciton gas ideal, has disappeared. The difference between the models [1-3] can be completely understood on the basis of symmetry analysis suggested in [4]. The aim of the present work is to analyze different exciton systems on the symmetry basis and to study corresponding phase transitions and collective excitations. We do not take into account impurities and all effects considered below are only due to the electron-electron interaction.

1.PHYSICAL MODEL.

Our model is based upon results obtained in [5,6]. Consider two Landau levels in a magnetic field. If the vector-potential is taken in the form A=(0,Hx,0) the electron states are

Springer Series in Solid-State Sciences, Vol. 87
High Magnetic Fields in Semiconductor Physics II Editor: G. Landwehr

characterized by a set of quantum numbers (n, k_y, τ), where n is the Landau level number and k_y is the wave number. The meaning of an additional parameter τ will be explained by two additional examples.

A. Consider a Si-MOSFET with surface orientation (110). In this case electrons have, due to time inversion, two degenerate valleys [7] with energy minima at the points $\pm\mathbf{Q}/2$ (partial removal of the degeneracy taken into account). The index $\tau=\pm1$ characterizes a valley.

B. For the case of a heterojunction in the absence of inversion spin-orbit interaction removes the two-fold degeneracy [7]. The index τ characterizes a spectrum branch.

We consider the case of strong magnetic fields, when the interaction occurs only between two partially occupied levels $(n_1, -1)$ and $(n_2, +1)$. In particular, for the model A $n_1=n_2$. The interaction does not transfer electrons from one Landau level to another assuming the characteristic energy of interaction $e^2/\varepsilon l_h$ much less than $\hbar\omega_c$, where ω_c is the cyclotron frequency, ε is the dielectric constant, $l_h = (c\hbar/eH)^{1/2}$ is the magnetic length.

If operators of creation and annihilation are introduced

$$a_k^+ = a^+_{n,k,-1}\ ,\ a_k = a_{n,k,-1}\ ,\ b_k^+ = a^+_{n,k,+1}\ ,\ b_k = a_{n,k,+1}\ ,$$

$$A_k = \begin{pmatrix} a_k \\ b_k \end{pmatrix} \qquad A_k^+ = \begin{pmatrix} a_k^+ \\ b_k^+ \end{pmatrix}$$

the Hamiltonian of electron-electron interaction takes the form [5,6] (all lengths in magnetic length units):

$$H_{int} = \frac{1}{2}\sum \int d^2q \left\{ V_1(\mathbf{q})\, a^+_{k_1} a^+_{k_2} a_{k_2+q_y} a_{k_1-q_y} + \right.$$
$$\left. + V_2(\mathbf{q}) b^+_{k_1} b_{k_2} b_{k_2+q_y} b_{k_1-q_y} + 2V_{12}(\mathbf{q})\, a^+_{k_1} b^+_{k_2} b_{k_2+q_y} a_{k_2-q_y} \right\} . \quad (1.1)$$

The operator (1.1) conserves the number of particles with a given τ. As shown in [5] for the model A it is valid to an accuracy of exponentially small terms of the order of

$\exp(-l_h^2/a^2)$, whereas for model B such a form is exact in the case of strong magnetic fields. The concrete form of potentials $V_{1,2}$, V_{12} depends on the model.

Consider the problem of a single exciton. Let the level with $\tau = -1$ be fully occupied and the level with $\tau = +1$ free. The exciton is formed due to transition of an electron from the level with $\tau = -1$ to the level with $\tau = +1$. Using the results of [5] we can write down the operator of exciton creation

$$B_q = \sum_k \exp i q_x (k+q_y)\, b_k^+ \, a_{k+q_y} \qquad (1.2)$$

and the energy of exciton creation is given by

$$E_{ex} = E_1(0) - E_{12}(\mathbf{q}) - \frac{V_1(0)}{2\pi} + \frac{V_{12}(0)}{2\pi}$$

$$E_i = \int \frac{d^2p}{(2\pi)^2} \{\exp i[\mathbf{qp}]\mathbf{n}\} V_i(\mathbf{p}),$$

$$E_{12} = \int \frac{d^2p}{(2\pi)^2} \{\exp i[\mathbf{qp}]\mathbf{n}\} V_{12}(\mathbf{p}) \qquad (1.3)$$

with the vector $\mathbf{n} = \mathbf{H}/H$.

For the model A the difference between $V_1 = V_2$ and V_{12} is related to small terms, arising due to a non-symmetrical form of dispersion law near the valley bottom, giving a positive contribution to the exciton energy of order $(a/l_h)^2$ and to intervalley scattering giving a negative contribution to the exciton energy of order $(a/l_h)^3$ [5], but the unknown numerical factors can easily give any sign to $E_{ex}(0)$ and it is not possible to make any predictions based on existing data.

If $V_1 = V_2 = V_{12}$ the exciton energy $E_{ex}(0) = 0$. This means that $[H_{int}, \sum_k b_k^+ a_k] = 0$. Since H_{int} does not change the number of particles with given τ in any case we have

$$[H_{int}, \sum_k a_k^+ a_k] = [H_{int}, \sum_k b_k^+ b_k] = 0.$$

Let us introduce isospin operators following [4]

$$S_i = \frac{1}{2} \sum_k A_k^+ \sigma_i A_k \qquad (i = x, y, z) \qquad (1.4)$$

where σ_i are Pauli matrices, which satisfy the usual commutation relations for operators of infinitesimal rotations. For an isotropic model ($V_1=V_2=V_{12}$) all S_i commute with H_{int}, whereas for arbitrary $V_{1,2}$, V_{12} only S_z commutes with H_{int}.

For the isotropic model the ground state is degenerate because the wave functions [4]

$$\Phi = \prod_k (\cos\theta\, a_k^+ + e^{i\Psi}\sin\theta\, b_k^+) |0\rangle \qquad (1.5)$$

($|0\rangle$ is the vacuum state), at arbitrary θ, Ψ have the same energy at the complete filling of one Landau level ($\nu = 1$).

For low temperature thermodynamics only soft modes of the Hamiltonian (1.1) are essential, i.e. weakly nonuniform rotations of the operators S_i are important. To describe them we introduce operators

$$S_i(\mathbf{q}) = \frac{1}{2} \sum_k \exp i q_x(k + \frac{1}{2} q_y)\, A_k^+ \sigma_i A_{k+q_y}. \qquad (1.6)$$

It is easy to obtain the commutation relations

$$[S_n(\mathbf{q}_1),S_m(\mathbf{q}_2)] =$$
$$\frac{1}{2} e_{nmk} S_k(\mathbf{q}_1+\mathbf{q}_2) \{ \exp \frac{i}{2} [\mathbf{q}_1\mathbf{q}_2]\mathbf{n} +\exp(\frac{i}{2})[\mathbf{q}_2\mathbf{q}_1]\mathbf{n} +$$
$$+ \frac{1}{2} \hat{J}\, \delta_{nm}(\mathbf{q}_1+\mathbf{q}_2)\{\exp\frac{i}{2} [\mathbf{q}_1\mathbf{q}_2]\mathbf{n} - \exp\frac{i}{2}[\mathbf{q}_2\mathbf{q}_1]\mathbf{n} \}, \qquad (1.7)$$

where

$$\hat{J}(\mathbf{q})= \frac{1}{2} \sum_k A_k^+ \hat{I} A_{k+q_y} \exp i q_x(k+ \frac{q_y}{2}),$$

e_{nmk} is a completely antisymmetric tensor of third rank, $\hat{I}$ is the unit isospin matrix.

The Hamiltonian (1.1) can be easily expressed through the operators $S_i(\mathbf{q})$ however the complex commutation relations (1.7) make its diagonalization rather difficult. At small $\mathbf{q}$ the values $S_i(\mathbf{q})$ become macroscopic, if there is a spontaneous isospin moment with $\mathbf{q}=0$ and exponents in the commutation relations can be replaced by 1 and $S_k(\mathbf{q}_1+\mathbf{q}_1)$ by $S_k(0)\delta(\mathbf{q}_1+\mathbf{q}_2)$ (see e.g. [9]).

Retaining all terms in the Hamiltonian (1.2) containing $S_i(\mathbf{q})$ with small $\mathbf{q}$ we obtain the effective Hamiltonian, neglecting density fluctuations

$$E_{int}=\int\frac{d^2q}{(2\pi)^2}\{-[E_1(\mathbf{q})-E_{12}(\mathbf{q})-\frac{V_1(\mathbf{q})}{2\pi}+\frac{V_{12}(\mathbf{q})}{2\pi}]|S_z(\mathbf{q})|^2- \\ - E_{12}(\mathbf{q})|S(\mathbf{q})|^2\} , \qquad (1.8)$$

where we have for simplicity assumed $V_1= V_2$, i.e. restricted ourselves to the model A. At $\mathbf{q}\to 0$ the coefficient at $|S_z(\mathbf{q})|^2$ is $-E_{ex}(0)$ (see (1.3)).

The effective Hamiltonian (1.8) corresponds to a three-component planar ferromagnet with weak anisotropy $\lambda=-E_{ex}(0)$, so that if $E_{ex}(0)>0$, there is an easy magnetization axis, and if $E_{ex}(0)<0$ there is an easy magnetization plane.

Restricting ourselves to small $\mathbf{q}$ we assume

$$E_{12}(\mathbf{q})\simeq E_1(\mathbf{q})\simeq E_1(0)-\frac{1}{2}Jq^2 \qquad (1.9)$$

so that the effective Hamiltonian acquires a form usual in the theory of magnetism :

$$H_{int}=\int d^2\mathbf{r}\,\left(\frac{\lambda}{2}S_z^2+\frac{1}{2}\frac{\partial S_i}{\partial x_k}\frac{\partial S_i}{\partial x_k}\right) . \qquad (1.10)$$

The phase transition into the ordered phase in the symmetric case ($\lambda= 0$, $V_1=V_2=V_{12}$) occurs at zero temperature and for weak anisotropy λ (see for example [10]) at a critical temperature $T_c=4\pi J/\ln J/|\lambda|$. In our model this expression for T_c is valid only near the full filling of the Landau level $\nu= 1$. Note that for the model A ordering corresponds to spontaneous symmetry breaking in the valley occupation ($\lambda<0$) and to a charge density wave with wave vector $\mathbf{Q}$ for $\lambda > 0$. In the latter case there are gapless collective excitations, corresponding to spin waves of the effective Hamiltonian (1.8) which can be obtained from the equation of motion according to simplified commutation rules (1.7) at small $\mathbf{q}$ (see also [9]). The corresponding dispersion law has an acoustic-like form

$$\omega(\mathbf{k}) = S_0\sqrt{J\lambda}\;k = ck \quad (\lambda> 0) . \qquad (1.11)$$

For the model A (Si-MOSFET) $c\approx e^2/\varepsilon\hbar(Ql_h)^{3/2}$ yielding $c\approx 10^4\cdot H^{3/4}$ where the magnetic field H is in tesla. The velocity c can be compared with the sound velocity.

At $\lambda<0$ excitations have a small gap

$$\omega(k)= S_0(|\lambda| + Jk^2) \quad (1.12)$$

The value S_0 entering equations (1.11), (1.12) is the average value of "magnetic moment". It equals to 1/2 at T= 0, $\nu=1$ and decreases with increased temperature and decreased ν (S_0 coincides with the parameter of anomalous pairing introduced in [5]).

In deriving equations (1.11), (1.12) the temperature has been supposed to be low. Using these expressions and commutation relations it is easy to obtain the low temperature thermodynamics.

2. MOSFETS ON THE (111) SILICON SURFACE.

Though the present calculations for Si-MOSFETs are of mere theoretical interest since for magnetic fields presently achieved the condition $\hbar\omega_c \gg e^2/\varepsilon l_h$ does not hold, we discuss briefly the situation arising for (111) silicon surfaces when there is six-fold valley degeneracy. The valleys are situated as shown in fig.1.

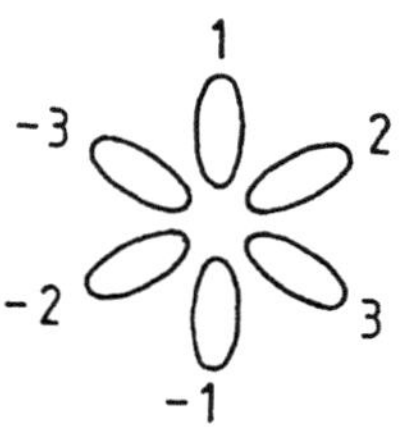

Fig.1

In the quadratic approximation the dispersion law of τ-valley for momenta counted off the minimum has the form

$$\varepsilon_\tau(\mathbf{p})=\frac{1}{2m_x}(p_x\cos\alpha_\tau+p_y\sin\alpha_\tau)^2+\frac{1}{2m_y}(p_x\cos\alpha_\tau-p_y\sin\alpha_\tau)^2. \quad (2.1)$$

In the gauge A=(0,Hx,0) the wave functions for dispersion law (2.1) are

$$\Psi_{n,k,\tau} = \left(\frac{\beta_\tau}{L_y}\right)^{1/2}\exp(iky+i\gamma(\alpha_\tau)(x+k)^2)\Phi_n[\beta_\tau(x+k)], \quad (2.2)$$

where L_y is the dimension of the system along the y-axis, k is the wave vector, n is the Landau level number, $\Phi_n(z)$ are oscillator functions (the magnetic length chosen as unity), $\beta_\tau = (m_x^*/m_y^*)^{1/4}$ and

$$\gamma(\alpha_\tau) = \frac{1}{2}\frac{(m_x - m_y)\sin\alpha_\tau \cos\alpha_\tau}{m_y\cos^2\alpha_\tau + m_y\sin^2\alpha_\tau}$$

$$m_x^* = \frac{m_x m_y}{m_y\cos^2\alpha_\tau + m_x\sin^2\alpha_\tau}, \quad m_y^* = m_y\cos^2\alpha_\tau + m_x\sin^2\alpha_\tau \quad . (2.3)$$

Consider now the interaction Hamiltonian. In the leading approximation we retain only the terms which contain at any point (see (1.1)) operators belonging to the same valley. Thus we get

$$H_{int} = \frac{1}{2}\int\frac{d^2q}{(2\pi)^2}\sum_{\tau,k_1,k_2} V_\tau(\mathbf{q})\, a^+_{\tau,k_1} a^+_{\tau,k_2} a_{\tau,k_2+q_y} a_{\tau,k_1-q_y} +$$
$$+ \sum_{\substack{\tau\neq\tau' \\ k_1k_2}} V_{\tau\tau'}(\mathbf{q})\, a^+_{\tau,k_1} a^+_{\tau',k_2} a_{\tau',k_2+q_y} a_{\tau,k_1-q_y} \quad . \qquad (2.4).$$

The coefficients in (2.4) are connected with the Fourier transform of the interaction potential $V(\mathbf{q})$ by the relations

$$V_\tau(\mathbf{q}) = V(\mathbf{q})\, W_\tau^2(\mathbf{q}'), \qquad V_{\tau\tau'}(\mathbf{q}) = V(\mathbf{q})\, W_\tau(\mathbf{q}')W_{\tau'}(\mathbf{q}') \;, \qquad (2.5)$$

where $\quad W_\tau(\mathbf{q}') = \int d\xi(\exp i\xi q'_x)\,\Phi_n(\xi + q_y/2)\,\Phi_n(\xi - q_y/2)$

$$q'_x = (q_x + 2\gamma(\alpha_\tau)\, q_y)\,\beta_\tau^{-1} \quad , \quad q'_y = q_y\beta_\tau \quad . \qquad (2.6)$$

For the spectrum (2.1) [3]

$$W_\tau(\mathbf{q}') = \exp\left(-\frac{(q')^2}{4}\right) L_n(q'^2/2) \;,$$

where $q'^2 = q'^2_x + q'^2_y$, $L_n(z)$ are Laguerre polynomials.

To single out the most important terms in Hamiltonian (2.4) we find energies of different excitons. Note that there are three classes of excitons. The first class corresponds to the transition between the nearest valleys (e.g. 1-2 etc.), the second to transitions over a valley (1-3 etc.) and the third to transitions between equivalent valleys $\tau,-\tau$ connected by the

Kramers degeneracy analogous to the degeneracy of the valleys on the surface (110).

The exciton energy with zero momentum is [3] $E_{ex}(0)= E_\tau - E_{\tau'}$, where

$$E_\tau = \int \frac{d^2q}{(2\pi)^2} V_\tau(\mathbf{q}) \ , \qquad E_{\tau\tau'} = \int \frac{d^2q}{(2\pi)^2} V_{\tau\tau'}(\mathbf{q}) \ .$$

It is evident that the values $E_\tau = E$ do not depend on the valley number and the values $E_{\tau\tau'}$ are determined only by the exciton class and $E_{\tau,-\tau}=E$. To find the exciton energy we calculate it for a pair of valleys with the opposite sign of the angle ($\alpha_\tau = -\alpha_{\tau'}$). This can be always achieved by some rotation. In this case $\gamma_\tau = -\gamma_{\tau'}$ and we obtain $E_\tau - E_{\tau\tau'} > 0$ for $\tau \neq -\tau'$. This effect is brought by the dispersion anisotropy and is of zero order in the small parameter a/l_h. The excitons belonging to various pairs of valleys are independent, since the Hamiltonian conserves the number of particles in any valley.

Thus we have shown that the formation of excitons due to transition of electrons between nonequivalent valleys ($\tau \neq -\tau'$) is not energetically favorable. Only equivalent valleys can compete. This means that the situation on (111) silicon surfaces is in fact analogous to that on the surfaces (110) analyzed above.

Note in conclusion that the developed symmetry considerations allow us to understand the absence of interaction between excitons in the symmetric model of LERNER and LOZOVIK. Symmetry violation due to intervalley scattering leads to formation of a charge density wave in the ground state of the MOSFET inversion layer for integer filling. Symmetry violation due to the asymmetry of spectrum near the bottom of equivalent valleys (if it prevails) leads to a phase transition with the filling of only one valley. The number of equivalent valleys is insignificant, since only transitions between valleys connected by Kramers degeneracy are important. Transitions to other valleys require energies of order $e^2/\varepsilon l_h$ due to the anisotropy of the effective mass, so that at low temperatures $T \ll e^2/\varepsilon l_h$ only two opposite valleys will be filled with spontaneous breaking of the symmetry with respect to rotation (e.g. a six-fold axis is changed into a two-fold axis).

The problem of the valley degeneracy in MOSFET inversion layers on (111) Si surfaces is rather old. It has been shown experimentally [7] that in strong magnetic fields the degeneracy is two fold. On the other hand direct measurements without magnetic field have shown [11] six-fold valley degeneracy. Our theoretical considerations explain this difference and connect it with electron-electron interaction.

References.

[1] I.Lerner, Yu.Lozovik ZhETPh 1982, v.82,p.1188

[2] Yu.Bychkov, S.Iordansky, G.Eliashberg Poverchnost, 1983, 3, p.3

[3]Yu.Bychkov, E.Rashba ZhETPh 1983, v.85, p.1826

[4] M.Rasolt,B.Halperin,D Vanderbilt Phys.Rev.Lett. 1986, v.57, p126

[5] Yu.Bychkov, S.Iordansky ZhETPh 1987, v. 93, p

[6] Yu.Bychkov, S.Iordansky Physica Tv.Tela 1987, v.29, p.2442

[7] T.Ando, A.Fowler, F.Stern Rev.Mod.Phys.,1982 v.54 N 2

[8] Yu.Bychkov, E.Rashba Pisma ZhETP, 1984, v.39, p.66

[9] E.Lifshits, L.Pitaevsky in Statistical Physics part 2, Moscow 1978,p.447

[10] A.Patashinsky, V.Pokrovsky in Fluctuation theory of phase transitions, Moscow, 1982, p.381

[11] G.M.Gusev,N.V.Zavaritsky, Z.D.Kwon, A.A.Yurgens Pisma v ZhETPh 1984 v.40, p.275

Interaction of Surface Acoustic Waves with Inversion Electrons on GaAs in Quantizing Magnetic Fields

A. Wixforth and J.P. Kotthaus

Institut für Angewandte Physik, Universität Hamburg, Jungiusstr. 11, D-2000 Hamburg, Fed. Rep. of Germany

Investigations of the interaction between surface acoustic waves (SAW) and a two-dimensional electron system in GaAs/AlGaAs heterojunctions at low temperatures and in high magnetic fields are reviewed. Experimental results demonstrating the dependence on quantizing magnetic fields and on SAW power and frequency are presented and discussed. It is shown that the interaction is strongest in the Quantum Hall regime and thus can serve as an additional tool to study quantum phenomena in two-dimensional electron systems.

1. INTRODUCTION

The interaction of acoustic lattice vibrations with a two-dimensional electron system (2DES) has been investigated in various ways both experimentally and theoretically [1]. Direct studies of phonon absorption [2] and emission [3] by a 2DES have been carried out with ballistic phonons of THz frequencies. There the deformation potential interaction dominates and phonon wavelenghts are comparable to intrinsic lengths in the 2DES such as the Fermi-wavelength. Here we want to summarize recent investigations of a 2DES with coherent surface acoustic waves (SAW) via the piezoelectric interaction at frequencies in the 100 MHz regime [4,5,6]. There SAW wavelengths are still larger than intrinsic lengths in the 2DES. Generally, SAWs are ideally suited to study acoustic interactions with electrons near surfaces since their acoustic energy is concentrated in a thin sheet of a few wavelenghts depth beyond the surface. This has been used to investigate such interactions on semiconductor [7] and metal [8,9] surfaces and films. An additional important aspect of SAWs is their application in RF signal processing based on SAW propagation on piezoelectric insulators [10]. The possibility of integrating such SAW devices with semiconductor devices on a common substrate also makes a detailed understanding of the interaction between a SAW and a 2D inversion channel in piezoelectric semiconductors highly desirable.

2. SURFACE WAVES ON PIEZOELECTRICS

Acoustic waves travelling in a piezoelectric medium are electromechanical waves and cause both a mechanical stress T and an electrical displacement D. In a simplified one-dimensional form these may be expressed in terms of the mechanical strain S and the electric field E by:

$$T = cS - eE \qquad \text{and} \qquad D = eS + \varepsilon E \quad . \tag{1}$$

Here, c is the stiffness constant, e the piezoelectric constant, and ε the

Springer Series in Solid-State Sciences, Vol. 87
High Magnetic Fields in Semiconductor Physics II Editor: G. Landwehr

electric permittivity. More generally, they are tensor components. The piezoelectric effect thus yields both an additional stress term eE as well as an additional piezoelectric polarization eS and couples mechanical and electronic motion. In comparison to the nonpiezoelectric limit the stiffness constant of a piezoelectric medium is increased and may be written as

$$c_{pe} = c\ (1+K^2_{eff}) \qquad \text{with} \qquad K^2_{eff} = e^2/c\varepsilon \ . \tag{2}$$

Here, K^2_{eff} is the so-called electromechanical coupling coefficient and is a measure for the strength of the piezoelectric effect for the appropriate crystal cut and propagation direction of the wave. The difference between bulk and surface acoustic waves in piezoelectric media is not visible in the simplified one-dimensional equations above. SAWs are generally elliptically polarized and thus have both transverse and longitudinal components, both of which decay essentially exponentially with distance from the surface. The influence of mobile bulk carriers on the propagation of a SAW on a piezoelectric medium has been considered by Hutson and White [11]. They have calculated the additional damping and the change in SAW velocity caused by the presence of mobile carriers. The more special case of mobile carriers confined to an infinitely thin conducting layer at or very near the surface of a piezoelectric medium has been considered by several authors [8,12,13]. For such a 2D conducting layer the interaction arises from the coupling of the longitudinal electric field components of the SAW in the 2D plane with the mobile carriers. The damping Γ and the sound velocity change $\Delta v/v$ caused by a 2DES of finite areal conductivity σ then may be written as

$$\Gamma = k \cdot \frac{K^2_{eff}}{2} \frac{(\sigma/\sigma_M)}{1+(\sigma/\sigma_M)^2} \qquad \text{and} \qquad \frac{\Delta v}{v} = \frac{K^2_{eff}}{2} \frac{1}{1+(\sigma/\sigma_M)^2} \ . \tag{3}$$

Here $k=2\pi/\lambda$ is the wave vector of the SAW and $\sigma_M=v(\varepsilon_0+\varepsilon_s)$ a characteristic conductivity where maximum attenuation occurs with v the SAW velocity and ε_0 and ε_s the dielectric constants of the media above and beneath the interface, respectively. Fig.1 illustrates this dependence of Γ and $\Delta v/v$ on σ.

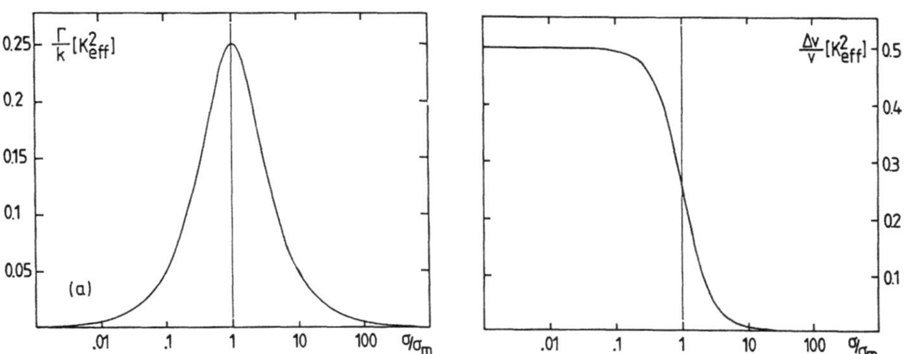

(a) (b)

Fig.1: (a) Attenuation Γ per unit lenght and (b) change in phase velocity $\Delta v/v$ in units of K^2_{eff} as a function of the sheet conductivity σ in units of σ_M as defined in the text. Γ/k shows a maximum at $\sigma = \sigma_M$, whereas $\Delta v/v$ drops in this region.

If the conductivity σ is high, most of the electric field of the SAW is screened and the surface of the crystal behaves as if it was not piezoelectric. At vanishing conductivity one deals with a piezoelectric insulator with stiffened elastic constants. In the intermediate range currents are induced and power is absorbed from the SAW.

Usually SAWs are excited by means of interdigital transducers (IDT) [14]. Such IDTs consist in the simplest case of two comblike metallized structures on top of the piezoelectric substrate. Application of an RF voltage of frequency f=v/2d where 2d is the periodicity of one comblike structure leads to the emission of a SAW with λ=2d. By special choice of the IDT design it is possible to also excite higher harmonics, a feature we also have used in our experiments.

3. EXPERIMENTAL TECHNIQUES

To study the interaction of SAW with a 2DES we prepare GaAs/AlGaAs heterostructures as depicted in Fig.2. On standard heterojunctions [15] we etch away the MBE grown layers containing the 2DES. Two IDTs prepared directly on the GaAs substrate define the SAW delay line. Short RF pulses (0.25 .. 1 μsec) are fed into one of the transducers and generate a SAW packet that is detected after typical delay times of 2 μsec at the other transducer. This signal is analyzed in intensity and phase by use of homodyne detection and standard boxcar integration techniques. The sample is located in a superconducting solenoid providing magnetid fields $B \leq 12$ Tesla at $T \geq 2$ K. The experimental setup is sketched in Fig.3.

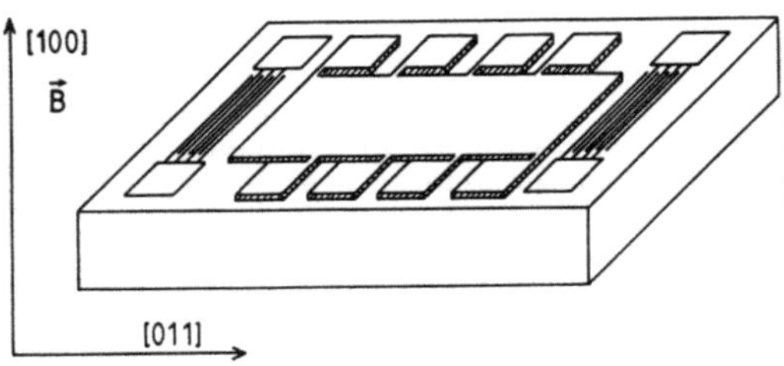

Fig.2.: Sketch of a typical sample for use in SAW experiments. In the propagation path of the SAW a mesa etched Hall bar with DC contacts contains the 2DES. The magnetic field B is oriented normal to the sample surface [4].

4. EXPERIMENTAL RESULTS AND DISCUSSION

At low temperatures we use high magnetic fields to vary the conductivity σ_{xx} in the 2DES by orders of magnitude. We thus cover the range of conductivities where the SAW amplitude and phase are most affected. The magnetic field dependence of the transmitted SAW intensity and phase at low SAW amplitudes are shown in Figs. 4(a) and 4(b), respectively. Fig.4(c) displays the magnetoconductivity $\sigma_{xx}(B)$ as calculated from measurements of the resistivity components $\rho_{xx}(B)$ and $\rho_{xy}(B)$ on a reference Hall bar sample fabricated on the same wafer [4].

At low magnetic fields ($B \leq 1$T) the SAW intensity and phase are nearly constant, whereas $\sigma_{xx}(B)$ drops dramatically. In Fig. 4(c) this region has been omitted for clarity. For magnetic fields $B \geq 1$T both the SAW intensity I/I_0 and phase exhibit quantum oscillations which coincide in position with the Shubnikov-de Haas oscillations of $\sigma_{xx}(B)$. The phase signal, measured as a change in the SAW frequency $\Delta f/f$ via the phase-locked loop of our set-

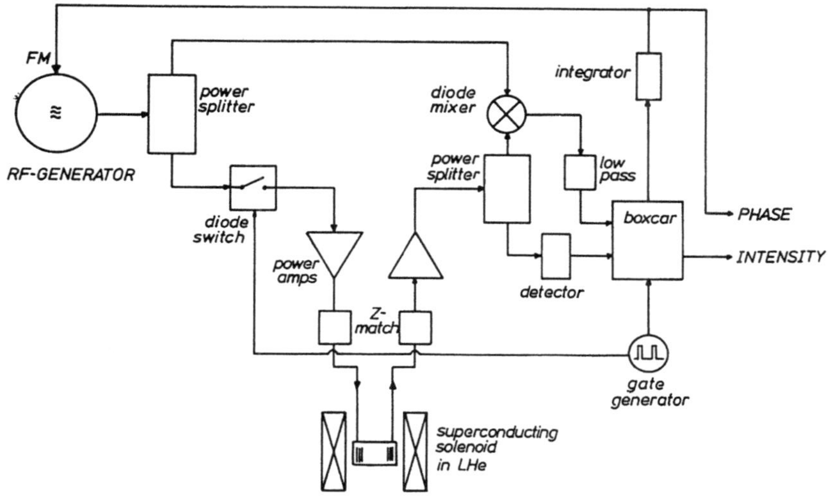

Fig. 3: Block diagram of the setup used in SAW experiments.

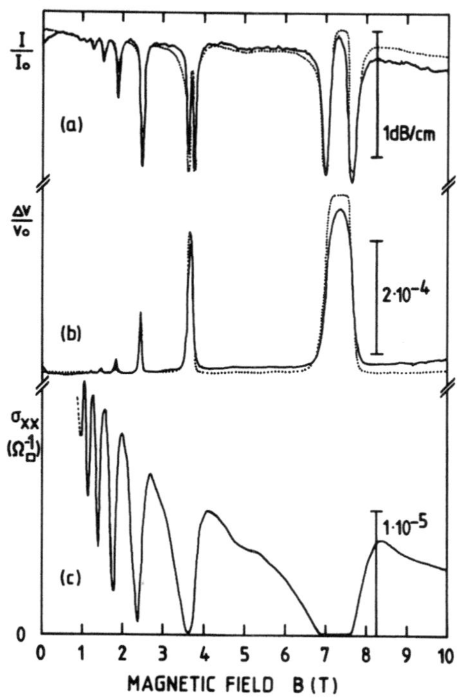

Fig.4: Magnetic field dependence of (a) the SAW intensity and (b) the change in velocity after transmission through the 2DES confined in a single layer AlGaAs-GaAs heterojunction with electron density $n_s = 3.5\cdot10^{11}$ cm^{-2} and mobility $\mu = 5\cdot10^5$ cm^2 V^{-1}sec^{-1} at T= 4.2 K. The SAW frequency is 70.5 MHz. In (c) σ_{xx} as extracted from measurements on a reference sample is shown. The dotted lines in (a) and (b) indicate the calculated quantities according to (3) using σ_{xx} from (c), $\sigma_M=4\cdot10^{-7}(\Omega/\square)^{-1}$, and $K^2{}_{eff}=6.4\cdot10^{-4}$ [4].

up, is calibrated as a relative change of phase velocity $\Delta v/v$. A characteristic feature of the magnetoabsorption of a SAW by a 2DES is the splitting of the quantum oscillations in I/I_0 in higher magnetic fields, i.e., for low filling factors $\nu=(h/eB)\cdot n_s$, where n_s is the areal electron density in the 2DES. This splitting is not observed in $\sigma_{xx}(B)$ nor in $\Delta v(B)/v$ and occurs in I/I_0 if $\sigma_{xx}(B)$ drops to values below $\sigma_M \approx 3.3\cdot10^{-7}\Omega_\square^{-1}$ where the attentuation shows a maximum. Thus a deep minimum in $\sigma_{xx}(B)<\sigma_M$ gives rise to two minima in $I(B)/I_0$. The piezoelectric stiffening of the surface due to the vanishing conductivity is clearly observed in $\Delta v/v$. The dotted lines in Figs. 4(a) and 4(b) are the computed quantitites using (3) and experimental values for $\sigma=\sigma_{xx}(B)$ taken from Fig.4(c). The parameters used are $K^2=6.4\cdot10^{-4}$ and $\sigma_M=4\cdot10^{-7}\Omega_\square^{-1}$, in good agreement with expected values. The difference between the measured and computed traces in Figs. 4(a) and 4(b) are small. Not predicted is the observation that for a given minimum in σ_{xx} the double mimima in I/I_0 do not have the same value I_{min}/I_0. This phenomenon will be discussed below. As expected, the interaction between a SAW and a 2DES becomes strongest in magnetic field regions, where σ_{xx} drops to very low values, i.e., in the region of the Quantum Hall effect (QHE).

In the regime of the QHE we have investigated the dependence of the SAW attenuation from the incident power. With increasing SAW intensity a given quantum oscillation in I/I_0 disappears as depicted in Fig. 5(a). With increasing SAW power the center maximum in $I(B)$, which corresponds to a minimum in $\sigma_{xx}(B)$, decreases whereas the double minima in $I(B)$ decrease their separation on the B-scale until they merge at the highest SAW power in Fig.5(a). The experiment is performed with short SAW packets (~1μsec) at low repetition rates (≈100 Hz) to avoid thermal heating. To demonstrate that this effect is instantaneous within the time resolution of our experiments, we have fabricated a sample with two pairs of transducers operating at

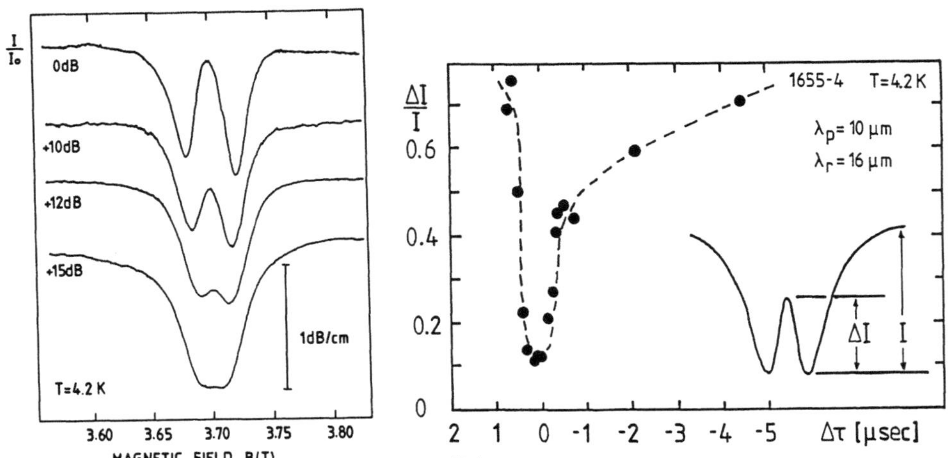

Fig.5: (a) Dependence of the normalized transmitted SAW intensity on incident power for the split oscillation at $\nu=4$. With increasing power the splitting of the structure disappears as the two minima merge. (b) Time dependence of the effect shown in (a). A pumping pulse of wavelength $\lambda_p=10\mu m$ and a reading pulse with $\lambda_r=16\mu m$ are used. A time delay of $\Delta\tau=0$ corresponds to complete overlap of the pumping pulse and the reading pulse.

different frequencies. We then use one of the SAW packets at frequency f_p as 'pumping' pulse to activate the minima in $\sigma_{xx}(B)$ and thus to decrease the center maximum in the split oscillations of $I(B)/I_0$. The other SAW packet at frequency f_r is used to 'read' the effect of the pumping pulse. By changing the time delay between the two packets propagating in the same direction, we can estimate the upper limit of the time that is needed to activate the minima in $\sigma_{xx}(B)$.

Results of this experiment are displayed in Fig.5(b), where the amplitude $\Delta I/I$ defined in the figure is plotted versus the time delay between the two waves. The time resolution of the experiment is given by the length of the SAW pulse and is $\tau \approx 500$nsec. Within this resolution the activation is instantaneous. We thus interpret the disappearance of the splitting of a quantum oscillation in $I(B)/I_0$ with increasing SAW intensity as thermal activation of $\sigma_{xx}(B)$, i.e., by heating of the 2DES in the electric field of the SAW. For quantitative comparison with electron heating in DC source-drain fields we have to include screening of the SAW-field by the mobile carriers to calculate a self-consistent SAW electric field given by

$$E = E_0 \frac{1}{\sqrt{1+(\sigma_{xx}/\sigma_M)^2}} , \tag{4}$$

where E_0 is the electric field corresponding to the bare potential at $\sigma_{xx}=0$. Comparison with data obtained by varying the lattice temperature of the sample yields an increase of the electron temperature of about 10K to 15K for the highest SAW power in Fig.5(a). Heating experiments in DC-fields [16] at liquid helium temperatures show that an increase of the electron temperature of about 10 K is related to an average power absorption per electron of about 10^{-14}Watts. Such power absorption via SAW at $\sigma=\sigma_M$ requires a mechanical amplitude of the SAW of 0.6Å which is in good agreement with the optically measured SAW amplitudes at power levels comparable to the highest ones in Fig. 5(a).

To avoid preparation of IDTs directly onto the precious heterostructure samples as well as to increase the available power in the SAW, we take advantage of a proximity coupling technique [5,17]. The sample geometry is shown in Fig.6(a). Here, the SAW is excited on a YZ-$LiNbO_3$ single crystal. This has one of the highest electromechanical coupling coefficients ($K^2_{eff}=0.048$). A heterostructure is mechanically pressed face down on top of this delay line. The gap between the heterostructure and the piezoelectric substrate is kept small compared to λ and the electrical field of the SAW penetrates into the heterojunction and interacts with the 2DES. We take no special care to achieve a well-defined spacing s between the $LiNbO_3$ substrate and the heterostructure and thus do not expect to be able to reliably describe the absolute damping. Nevertheless, (3) may be modified by a prefactor $\alpha(s)$ as well as by $\sigma_M(s)$ to take the air gap into account [17]. A typical trace of the SAW intensity versus B for such a sandwich structure is shown in Fig.6(b). Comparison to Fig.4(a) demonstrates that the sandwich technique serves as a useful alternative for the investigation of the interaction between a SAW and a 2DES. Since it is contactless and non-destructive, it makes possible a fast characterization of the samples under investigation.

The asymmetry of the intensity minima in the split oscillations in I/I_0 as in Figs. 4(a) and 6(b) has been shown to arise from spatial inhomogeneity of the 2D electron density on the sample area [6]. Spatial inhomogeneities

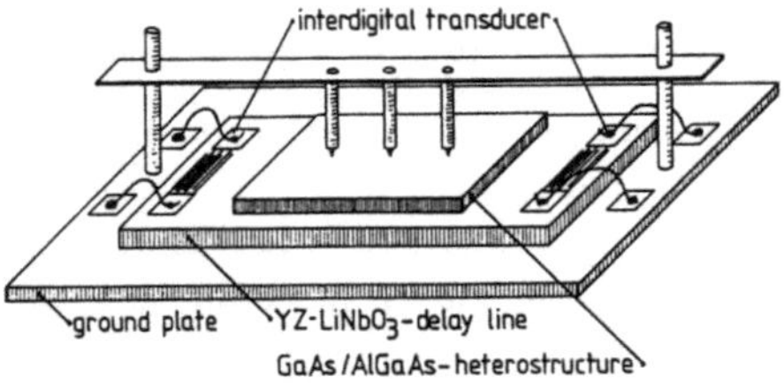

(a)

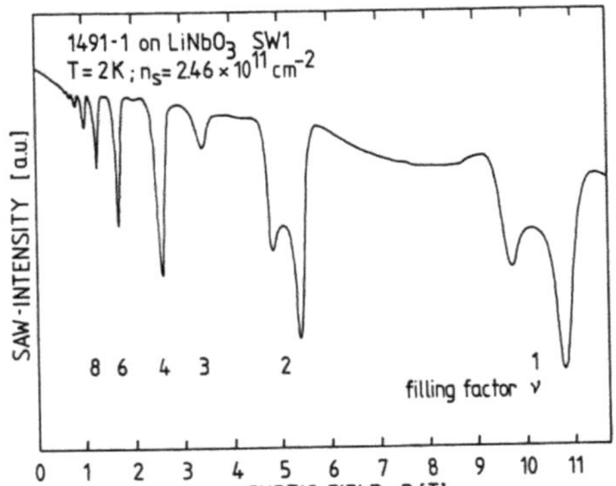

(b)

Fig.6.: (a) Sample arrangement for the sandwich technique in which a SAW is excited on a YZ-$LiNbO_3$ delay line and interacts by proximity coupling with the 2DES in the heterojunction. (b) Typical result for the B-dependence of the SAW intensity in such a sandwich structure [5].

in the density n_s of a 2DES significantly influence the apparent density of states (DOS) between Landau levels in quantizing magnetic fields [18-20]. Experiments that yield insight into spatial inhomogeneities of n_s therefore could contribute to a better understanding of many quantities that depend on the DOS such as capacitance [21] or specific heat [22]. In the small signal approximation we may write the transmitted SAW intensity as

$$I = I_0 \cdot \exp(-\Gamma \cdot \ell) \approx I_0(1-\Gamma \cdot \ell) \quad . \tag{5}$$

A homogeneous $\sigma_{xx}(B)$ over the whole interaction area would lead to all double minima in I(B) to be equal and $I_{min} \approx I_0(1-kK^2L/4)$. If one assumes an inhomogeneous carrier distribution and hence an inhomogeneous distribution in $\sigma_{xx}(B)$ in the interaction area, this is no longer true. For simplicity we consider several independently homogeneous domains of area F_i, each of which has a different $\Gamma_i(B)$. The total transmitted SAW intensity then is

$$I_{tot} \approx I_0(1-\Sigma\Gamma_i \ell_i) \text{ with } \ell_i=F_i/W \text{ and } \Sigma F_i=F \quad . \tag{6}$$

The different $\Gamma_i(B)$ may be caused by slightly different densities $N_{s,i}$ in the domains, which for a given magnetic field correspond to different filling factors ν_i. To confirm our assumption we have intentionally introduced a domain of higher carrier density on a sample by illuminating a small part ($\approx$15%) of the sample area. Via the persistent photoeffect thus n_s in this region is increased gradually by short light pulses. Fig.7 shows experimental results in comparison to a simulation of the detected SAW intensity which only assumes two domains of slightly different carrier densities N_{si}. Here, the magnetoconductivity $\sigma_{xx}(B)$ is simply modeled by a parabola. An inhomogeneity of only 5% on an area of about 15% clearly introduces an asymmetry in the double minimum of I(B) similar as is observed.

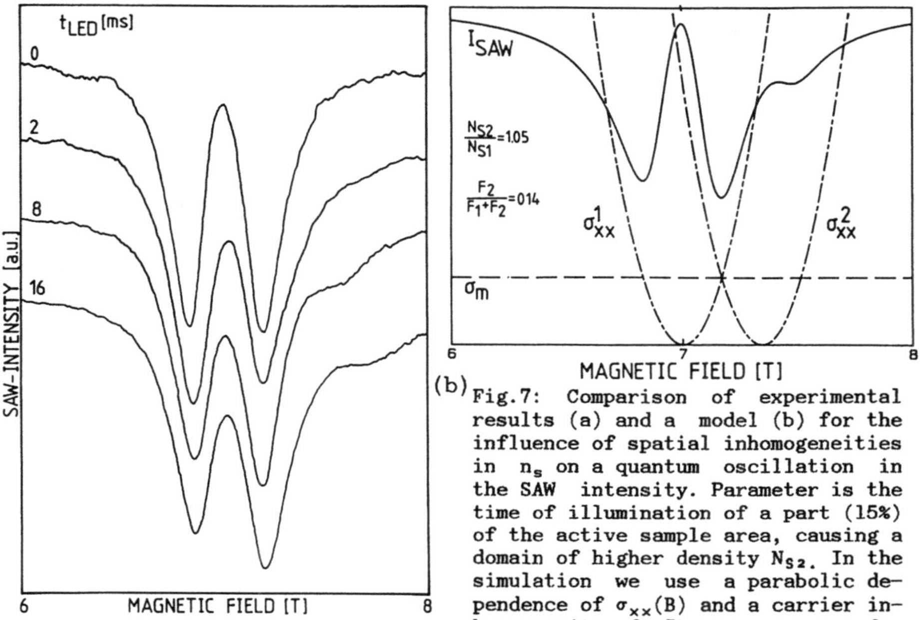

(a) (b)

Fig.7: Comparison of experimental results (a) and a model (b) for the influence of spatial inhomogeneities in n_s on a quantum oscillation in the SAW intensity. Parameter is the time of illumination of a part (15%) of the active sample area, causing a domain of higher density N_{S2}. In the simulation we use a parabolic dependence of $\sigma_{xx}(B)$ and a carrier inhomogeneity of 5% on an area of 14% of the active sample area [6].

The higher density area in both cases is reflected in a shoulder on the high field side. The decrease of the local maximum at integer ν in the experimental I(B) traces with light indicates that the illumination induces a slight parallel conductivity in the AlGaAs layer. It should be noted, that such small inhomogeneities in the carrier distribution are difficult to detect in $\rho_{xx}(B)$ measurements, though the Hall plateaus in $\rho_{xy}(B)$ are found to be broadened by the additional inhomogeneity. On several samples we found that inhomogeneities in the carrier distribution can also be introduced by high drift pulses either with DC or with SAW excitation. The study of these inhomogeneities and their dynamics is easily done by use of SAW. Fig.8 demonstrates the effect of an increase in the carrier density by the application of a DC drift pulse. With increasing pulsed source-drain field E_p a second, well defined density domain appears on part of the sample and grows in area with pulse height and the number of pulses. Here a low power SAW only serves to test the amount of inhomogeneity in the sample. After the drift pulses are switched off, the higher density domain remains persistent. Again, in DC-transport measurements this is difficult to observe though the Hall-plateaus get broadened.

The frequency dependence of SAW attenuation by a 2DES in quantizing magnetic fields has been studied on various GaAs/AlGaAs samples with different IDT geometries up to 550 MHz. For frequencies up to 300 MHz no deviation from the behavior predicted by (3) is observed within experimental error as shown in Fig.9. There the maximum attenuation is extracted from the minima in a double oscillation in I(B)/Io. Since the measurements for the high frequencies are performed using harmonics of the fundamental wave

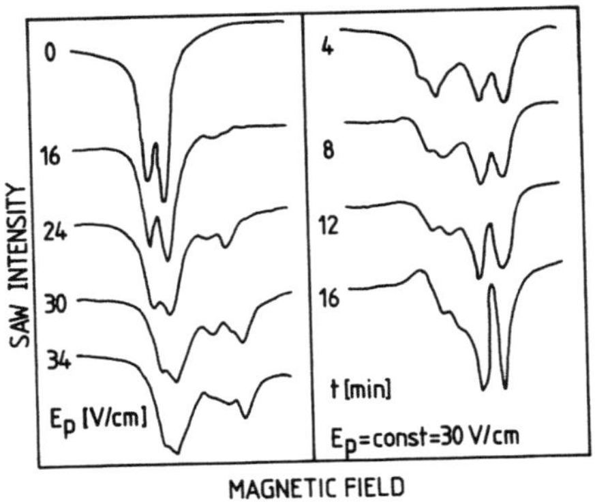

Fig.8: Drift pulse induced inhomogeneity in n_s as observed in the SAW signal. Below a certain pulse height (E_p<20 V/cm) the 2DES is only heated by the pulse. For higher electric fields a second electron density appears as a structure on the high field side. This second density domain grows in area with increasing pulse height and, for a given electric field with the number of pulses as counted by the total pulse time . The second density remains persistent over days, indicating a mechanism comparable to the persistent photoeffect.

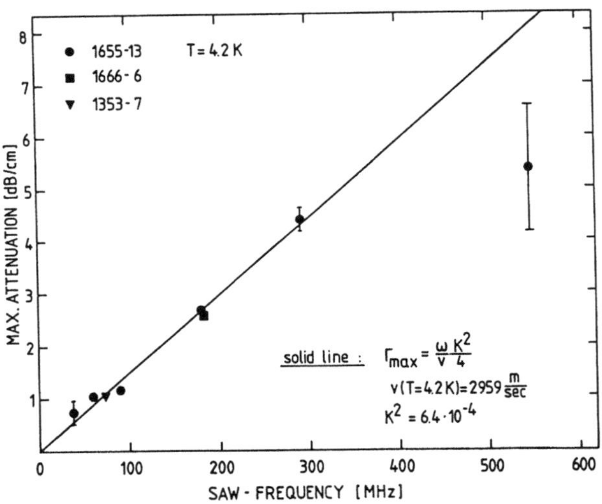

Fig.9: Frequency dependence of the maximum attenuation Γ_{max} of a SAW by a 2DES in a magnetic field. For frequencies up to 300 MHz, i.e. wavelenghts down to 10 μm, no deviation from the expected linear dependence indicated by the solid line is observed.

the signal to noise ratio decreases, leading to quite a large experimental error. Thus preparation of high quality IDT for use at high frequencies, i.e., short SAW-wavelenghts is desirable. Possibly, the SAW wavelengths then become comparable to intrinsic lengths of the 2DES and allow observation of non-local effects.

5. CONCLUSION

In summary, we have demonstrated that the investigation of the interaction of surface acoustic waves with a 2DES can yield valuable information of the properties of a 2DES in quantizing magnetic fields. As the interaction is strongest in the quantum Hall regime SAW seem to be a useful tool to study

QHE phenomena. We have shown that the interaction of a SAW with a 2DES in quantizing magnetic fields can be described classically up to frequencies of 300 MHz, corresponding to SAW-wavelengths of 10μm. The influence of small spatial inhomogeneities on the lineshape of the magnetoabsorption of SAW has been demonstrated both experimentally and in a simple model. The study of the dynamics of such inhomogeneities in the quantum Hall regime may yield information about the apparent DOS between Landau levels. Finally, a simple contactless and non-destructing sandwich-technique for the characterization of heterostructure samples in quantizing magnetic fields has been presented.

We wish to thank J.Scriba and M.Wassermeier for fruitful collaboration in part of the experiments summarized here. The heterojunctions used have been kindly provided by G.Weimann and W.Schlapp (Deutsche Bundespost), $LiNbO_3$-substrates by C.Grabmaier (Siemens). We also thank A.V.Chaplik for stimulating discussions and acknowledge the financial support of the Deutsche Forschungsgemeinschaft.

[1] for a recent review see, e.g., L. J. Challis, G.A.Toombs, and F.W.Sheard: In Physics of Phonons, ed. by T.Paskiewicz, Springer Lect. Notes in Phys., Vol 285, (Springer, Berlin 1987), 348
[2] J. C. Hensel, R. C. Dynes and D. C. Tsui: Phys. Rev. B 28, 1124 (1983)
[3] M. Rothenfusser, L. Köster, and W. Dietsche: Phys. Rev. B 34, 5518 (1986)
[4] A. Wixforth, J. P. Kotthaus and G. Weimann: Phys. Rev. Lett. 56, 2104 (1986)
[5] A. Wixforth, J. Scriba, M. Wassermeier, J. P. Kotthaus, G. Weimann, and W. Schlapp: J. Appl. Phys., in press
[6] A. Wixforth, J. Scriba, M. Wassermeier, J. P. Kotthaus, G. Weimann, and W. Schlapp: Proc. of the 19th ICPS (Warsaw 1988), in press
[7] P. Das, M. E. Motamedi, and R. T. Webster: Appl. Phys. Lett. 27, 120 (1975)
[8] P. Bierbaum: Appl. Phys. Lett. 21 595 (1972)
[9] J. Heil, I. Konroudis, B. Lüthi, and P. Thalmeier: J. Phys. C 17, 2433 (1984)
[10] see, e.g., "Acoustic Surface Waves" (Topics of Appl. Physics), Vol.24 ed. A. A. Oliner, (Springer, Berlin 1978)
[11] A. R. Hutson and D. L. White: J. Appl. Phys. 33, 40 (1969)
[12] R. Adler: IEEE Trans. on Sonics and Ultrasonics SU-18, 115 (1971)
[13] B. K. Ridley: Semicond. Sci. Techn. 3, 542 (1988)
[14] R. M. White and T. W. Voltmer: Appl. Phys. Lett. 7, 314 (1965)
[15] G. Weimann: In Festkörperprobleme (Advances in Solid State Physics) Vol.26, ed. P.Grosse (Vieweg, Braunschweig 1986), 231
[16] H. Sakaki, K. Hirakawa,J. Yoshino, S. P. Svensson, Y. Sekiguchi, T. Hotta, S.Nishii, and N.Miura: Surf. Sci. 142, 306 (1984)
[17] A. Schenstroem, Y. J. Quian, M. F. Xu, H. P. Baum, H. Levy, and B. K. Sarma: Solid State Commun. 65, 739 (1988)
[18] T. P. Smith, B. B. Goldberg, P. J. Stiles, and M. Heiblum: Phys. Rev. B 32, 2696 (1985)
[19] R. T. Zeller, F. F. Fang, B. B. Goldberg, S. L. Wright, and P. J. Stiles: Phys. Rev. B 33, 1529 (1986)
[20] V. Gudmundsson and R. R. Gerhardts: Phys. Rev. B 35, 8005 (1987)
[21] D. Weiss, K. v. Klitzing, and V. Mosser: In "Two Dimensional Systems: Physics and New Devices", eds. G.Bauer, F.Kuchar, and H.Heinrich (Springer, Berlin 1986), 204
[22] R. Lassnig and E. Gornik: In "Two Dimensional Systems: Physics and New Devices", eds. G.Bauer, F.Kuchar, and H.Heinrich (Springer, Berlin 1986), 218

Part II

Fractional Quantum Hall Effect

The Fractional Quantum Hall Effect at Even Denominators

J.P. Eisenstein

AT&T Bell Laboratories, Murray Hill, NJ07974, USA

The recent discovery of the fractional quantum Hall effect (FQHE) at the *even – denominator* filling fraction $\nu=5/2$ represents yet another surprise in the area of two-dimensional electron physics. Previous to the work of WILLETT, et al.[1] the FQHE was strongly believed to occur solely at odd-denominator rational fractional fillings ($\nu=p/q$, q-odd) of the Landau levels. This belief was generated by both a wealth of experimental data and strong theoretical arguments[2] going as deep as the Pauli exclusion principle. How then does this new observation fit in with our overall understanding of the FQHE and 2D correlation effects in general? Here we present data[3] strongly suggesting that the underlying many-body ground state for the $\nu=5/2$ FQHE is not spin polarized, in contrast to the canonical $\nu=1/3$ and $2/3$ incompressible quantum liquid states. Further, the apparent absence of even-denominator states in the lowest Landau level $\nu<2$ points to important differences between electron correlations in the first and second Landau levels. The purpose of the present paper is to review the existing experimental data[1,3] on the $\nu=5/2$ FQHE and show how this startling new result can, in principle at least, be accommodated by a generalized picture of the FQHE.

The $\nu=5/2$ FQHE has been observed in high quality GaAs/AlGaAs heterostructures grown by molecular beam epitaxy. The high mobility ($\sim 1.3x10^6 cm^2/V/s$) and high carrier concentration ($3x10^{11} cm^{-2}$) result from low-temperature illumination by a red LED. Hall and diagonal resistivity measurements have been performed at temperatures down to 20mK and at magnetic fields up to 30T using two different dilution refrigerator/magnet systems. Achieving such low electron temperatures requires low measurement currents, careful thermal grounding of all electrical leads, and shielding from extraneous RF radiation.

Figure 1 presents an overall view of the low-temperature magneto-transport parameters ρ_{xx} and ρ_{xy}. The field range includes both the extreme quantum limit $\nu<1$ and the regime $\nu>2$ in which the fermi level has passed into the second and higher Landau levels. Throughout the range $\nu<2$ copious evidence of the odd-denominator FQHE is present in both the complex oscillations of ρ_{xx} and the large number of plateaus in ρ_{xy}. Fractional features with odd denominators as high as 13 are apparent in the high-field oscillations of ρ_{xx} in this lowest (N=0) Landau level regime.

Figure 2 contains data from the restricted field range corresponding to $3>\nu>2$ for which the fermi level lies in the lower spin subband of the second (N=1) Landau

Springer Series in Solid-State Sciences, Vol. 87
High Magnetic Fields in Semiconductor Physics II Editor: G. Landwehr

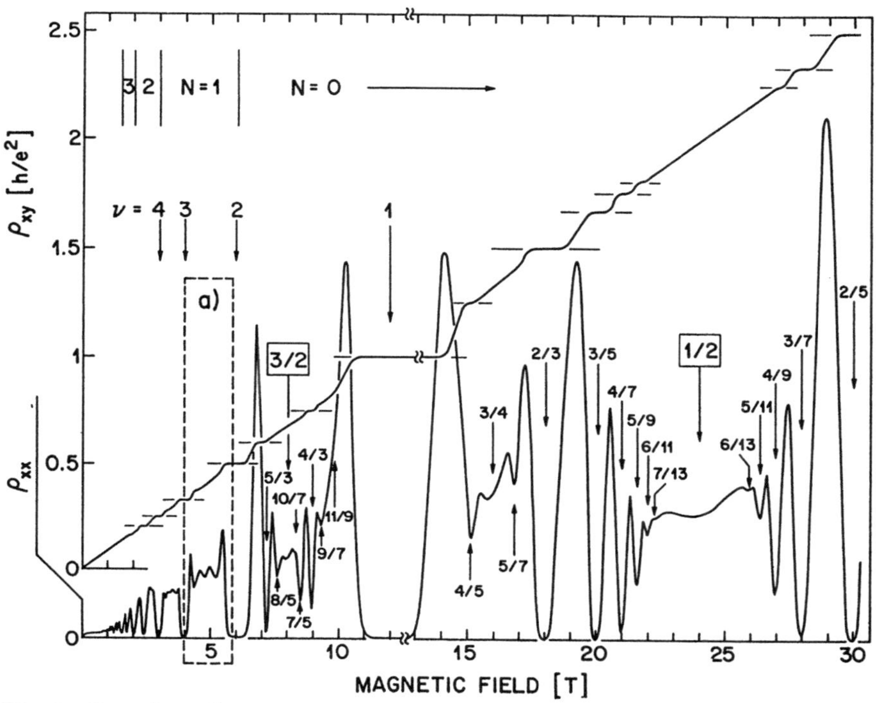

Fig. 1. Overview of magnetotransport parameters ρ_{xx} and ρ_{xy} for sample described in text. Figure is a composite of low and high field data (break at 12T). Temperature 150mK except 85mK for high field Hall trace. High field ρ_{xx} data reduced by a factor of 2.5 for clarity.

level. The ρ_{xy} data, taken at 25mK, clearly reveal a plateau forming at the even-denominator filling factor $\nu=2+1/2=5/2$ and straddling the classical Hall line determined by the 2D density. We have found this plateau to be centered at $\rho_{xy}=(h/e^2)/(5/2)$ to within 0.5%. Also shown in the figure are ρ_{xx} data showing the development of a deep minimum at $\nu=5/2$ below 100mK. These data represent the first observation[1] of the FQHE at an even-denominator filling factor. CLARK, et al.[4] have observed a weak minimum in ρ_{xx} at $\nu=5/2$ and conjectured the existence of a FQHE at this filling factors. Without quantization in ρ_{xy} though, their data remained speculative. It is the observation[1] of the $\nu=5/2$ plateau that has led to a reevaluation of our understanding of the FQHE.

Given the overwhelming experimental evidence in favor of solely odd denominators in the FQHE it can be suggested that the $\nu=5/2$ plateau results in fact from the spurious conjunction of high-order odd-denominator plateaus. To address this we have indicated the positions of the fractions 32/13 and 33/13 in the figure. They were chosen since 13 is the largest denominator at which vestigial features have been observed in the lowest Landau level (see Fig. 1) at the high magnetic fields where the FQHE is greatly enhanced. These fractions lie well clear of the $\nu=5/2$

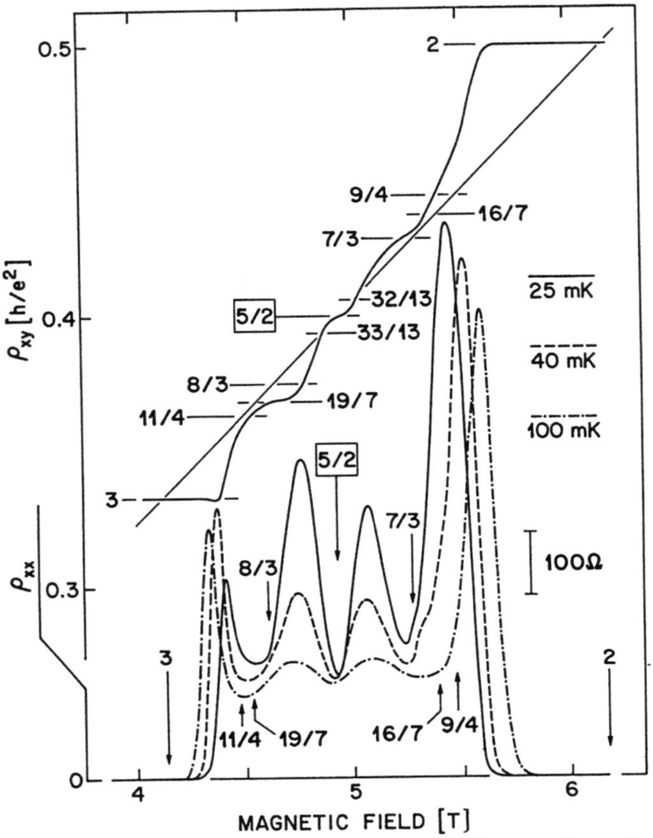

Fig. 2. Expanded view of ρ_{xx} and ρ_{xy} [enlarged section (a) from Fig. 1] at temperatures below 100mK. Filling factors are indicated.

FQHE plateau and therefore support our identification of the first even-denominator quantum number.

The development of the strong minimum in ρ_{xx} at $\nu=5/2$ only below 100mK implies a small associated energy scale. Although not shown in Fig. 2 the plateau observed in ρ_{xy} exhibits a similar temperature dependence as ρ_{xx}, disappearing altogether above 100mK. The development of the minimum in ρ_{xx} is noteworthy. The maxima immediately adjacent to the minimum rise up sharply as the temperature falls while the value at the minimum itself remains roughly constant. We suspect that this behavior represents a competition between the tendency for ρ_{xx} to rise because of decreasing scattering and the development of the zero-resistance state due to the formation of an energy gap. This general behavior has been observed in the development of weak odd-denominator fractional features[5].

Beyond the existence of the $\nu=5/2$ FQHE, the data shown in Fig. 2 are remarkable for the lack of any strong odd-denominator FQHE states. It is natural to anticipate the FQHE at 7/3 and 8/3 in analogy to the 1/3, 2/3, 4/3, and 5/3 states in the N=0 Landau level. No strong features are apparent in Fig. 2 at these filling fractions. Only broad minima, which shift considerably with temperature, are found in the vicinity. Aside from an apparent plateau off the classical Hall line at 19/7 and an inflection in ρ_{xy} near 7/3 the new state at $\nu=5/2$ represents the only conclusive evidence the FQHE in the second Landau level.

We have searched extensively for further evidence of even-denominator FQHE states. In the upper spin subband of the N=1 Landau level, $4>\nu>3$, a minimum in ρ_{xx} exists at $\nu=7/2$ flanked by minima above and below it. Only very weak structure has been observed in the Hall resistance at $\nu=7/2$. These features are broadly similar to those around $\nu=5/2$ filling at temperatures above 100mK. While these data are suggestive of an FQHE at $\nu=7/2$, higher quality samples are needed to conclusively identify fractional features in the range $4>\nu>3$.

By contrast, in the lowest Landau level $\nu<2$ extensive evidence for the traditional odd-denominator FQHE is present. Fig. 1 shows a broad, weak minimum in ρ_{xx} occurring around $\nu=1/2$. The shape of this minimum contrasts sharply with the families of odd-denominator fractions flanking it. Most importantly, down to 85mK, no associated feature has been found in the Hall resistance. In fact, ρ_{xy} follows the straight classical Hall line in this range. Similar weak minima are observed near $\nu=3/2$ and 3/4, but again without any associated features in ρ_{xy}. The range around $\nu=3/2$ has been studied down to 25mK, a temperature by which the $\nu=5/2$ FQHE is fairly well-developed, without evidence for a FQHE.

An interesting aspect of the $\nu=5/2$ FQHE data is the relatively small magnetic fields (~5T) at which it has been observed. Lower fields reduce the spin and Landau gaps as well as the electron correlation energies. New GaAs/AlGaAs heterostructures of low density and high mobility (typically $4x10^{10}cm^{-2}$ and $1.5x10^{6}cm^{2}/V/s$ respectively) have allowed us to study the FQHE in the lowest Landau level at magnetic fields comparable to those employed for observation of the $\nu=5/2$ FQHE in the higher density samples. In spite of these improvements there is still no evidence for even-denominator quantization at $\nu=1/2$ or 3/2 in the lowest Landau level as Fig. 3 reveals. It appears that the different structure of the electronic wavefunctions in the N=0 and N=1 Landau levels is a factor in the formation of half-filled fractional states such as $\nu=5/2$.

Even-denominator fractional quantization does not violate any known physical symmetry principle. LAUGHLIN'S[6] original explanation for the $\nu=1/m$ FQHE rests upon a specific choice of variational wavefunction to describe the relative motion of spin-polarized electrons in the lowest Landau level. Within that framework the antisymmetry of the wavefunction forces the restriction that m be odd. If the particles obeyed Bose rather than Fermi statistics m would have to be even[7,8]. A natural suggestion[1] then is that spin-reversed pairs of electrons are responsible for the FQHE at $\nu=5/2$. The low magnetic fields at which the 5/2-FQHE occurs might foster such spin reversal owing to the reduced Zeeman energy.

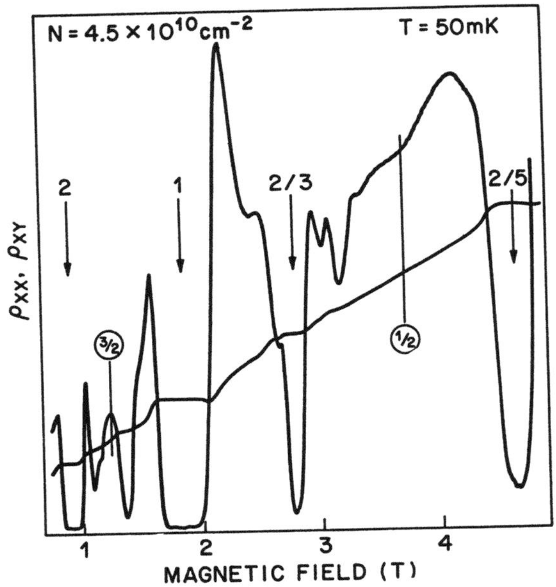

Fig. 3. Magneto-transport parameters at 50mK for low density sample described in text. Note absence of Hall plateaus at $\nu = 1/2$ and $3/2$.

Such spin reversal has already been predicted[9,10] to influence the quasiparticle energy gaps at low fields. Recently HALDANE and REZAYI[11] have shown that a half-integral FQHE *can* exist given an appropriate model interaction between electrons. They have proposed an unpolarized spin-singlet wavefunction whose stability is significantly enhanced in the second Landau level and by reduced Zeeman spin-flip energies at low magnetic fields. These are the conditions found in experiment. No quantitative estimates of the stability of the proposed state have been offered and the degree to which the model interaction represents reality has been criticized[12]. Nevertheless, HALDANE and REZAYI[11] have provided an existence proof for even-denominator fractional quantization and spin reversal is central to it.

A simple method for examining the spin states of 2D electron systems is provided by tilted-field experiments[13]. All orbital parameters of an ideal 2D system (Landau level spacings and degeneracies, coulomb energies, etc.) depend only upon the component of magnetic field perpendicular to the 2D plane while, by contrast, the spin Zeeman energy depends on the *total* magnetic field. A given Landau level filling factor ν occurs at higher total field when the sample is tilted. The scaling factor is $1/\cos\theta$ with θ measured from the normal to the 2D plane. Consequently, for fixed ν the spin flip energy increases with tilt. If the 5/2-FQHE state involves a significant number of reversed spins, flipped at the expense of liquid condensation energy, then tilting the sample should inhibit, or even quench, the state.

To examine the tilt-dependence of the 5/2-FQHE the sample is mounted on a spring-loaded paddle which can be rotated in a yoke via a string pulled from atop the cryostat[3]. Thus *in situ* rotation at low temperature is obtained. The tilt angle can be determined in several equivalent ways: measurement of the low-field Hall slope, the Shubnikov-de Haas periodicity or the field position of the 5/3 and 5/2 ρ_{xx} minima. Figure 4 shows diagonal resistivity data taken at 25mK for several tilt angles θ. For each angle the field range was adjusted to correspond to the filling factor range $3>\nu>2$. The $\theta=0$ data reveal the same structure shown in Fig. 2, with the arrow indicating the location of the $\nu=5/2$ feature. It is clear from Fig. 4 that as θ increases the $\nu=5/2$ minimum rapidly shrinks. This collapse of the ρ_{xx} feature is accompanied by a similar loss of the plateau in ρ_{xy} at $(h/e^2)/(5/2)$. The ρ_{xx} minima adjacent to the $\nu=5/2$ structure are either unaffected or actually strengthen with tilt.

Although it would be desirable to measure an activation energy associated with the 5/2-FQHE, this is not possible with the present samples owing to the roughly constant ρ_{xx} value at the minimum discussed above. Instead, we define the 'strength' of the 5/2-FQHE as the ratio $\Delta\rho_{xx}/<\rho_{xx}>$ of the amplitude of the minimum in ρ_{xx} to the average value of ρ_{xx} between the maxima and the minimum. Figure 5 shows how the temperature dependence of the strength of the 5/2-FQHE collapses as the tilt angle increases.

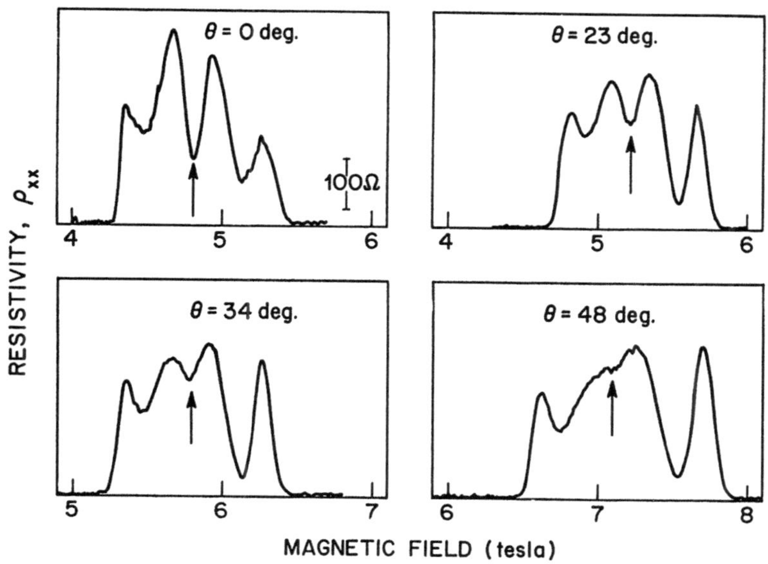

Fig. 4. Diagonal resistivity, ρ_{xx}, for filling factors $3>\nu>2$ at 25mK for several tilt angles θ. Arrows mark position of $\nu=5/2$ filling factor. Resistivity minima above and below the $\nu=5/2$ feature occur near $\nu=7/3$ and 19/7 respectively.

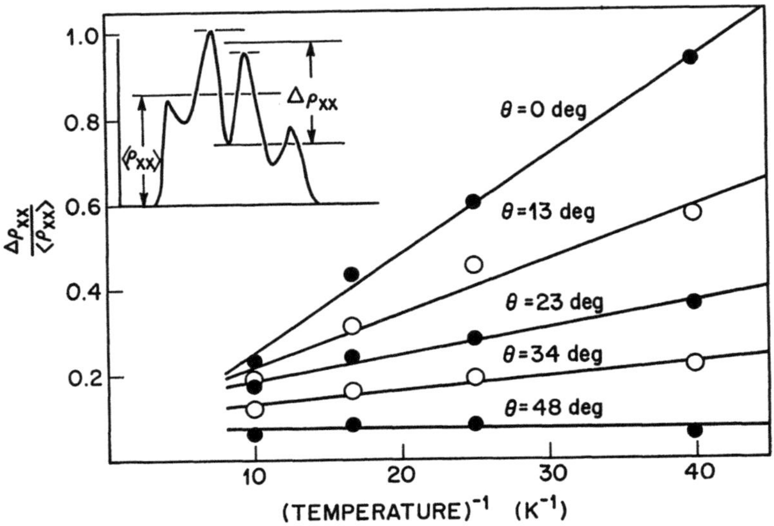

Fig. 5. Temperature dependence of the strength of the 5/2-FQHE for various tilt angles. Inset defines the strength ratio.

Tilting the sample effectively increases the spin-flip energy by the factor $1/\cos\theta$. This is equivalent to replacing the GaAs g-factor with the value $g_{Eff}=g/\cos\theta$ at constant perpendicular field. The collapse of the 5/2-FQHE may be viewed as a function of spin-flip energy by plotting the strength $\Delta\rho_{xx}/<\rho_{xx}>$ *vs.* the ratio g_{Eff}/g as shown in Fig. 6. From this perspective, the 5/2-state collapses rapidly with increased Zeeman energy, and appears only marginally stable even in a perpendicular field. The contents of Figs. 4-6 represent strong evidence for a 5/2-FQHE state with highly incomplete or zero spin polarization.

Tilting of a 2D sample causes effects beyond enhanced spin splittings, such as diamagnetic energy shifts and Landau level and electric subband mixings[14]. The mixings lead to a squashing of the electronic wavefunction perpendicular to the plane thereby creating a more ideal 2D system. This is expected to enhance the quasiparticle energy gaps[15,16]. Studies of the $\nu=2/3$ FQHE show strengthening of the ρ_{xx} structure upon tilting while the $\nu=1/3$ state remains largely unaffected[17,18]. In our sample (Fig. 1) we have found the activation energy of the $\nu=5/3$ state to be unchanged upon tilting to 34 deg. Suppression of these states by increased Zeeman splitting is not expected since the $\nu=1/3, 2/3$ and 5/3 FQHE states are all believed to be spin polarized quantum fluids. The new even-denominator $\nu=5/2$ FQHE behaves in a *qualitatively* different way although a full understanding of its tilted field dependence is lacking.

Given the currently available experimental evidence it appears likely that the existence of even-denominator fractional quantization poses no fundamental difficulty for the incompressible quantum liquid theory of the FQHE. The most recent data suggest two generalizations to the original approach of LAUGHLIN[6]

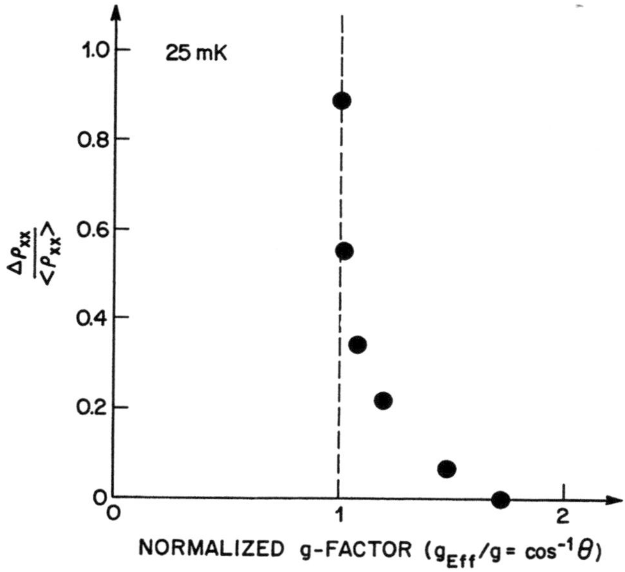

Fig. 6. Strength of 5/2-FQHE at 25mK *vs.* normalized g-factor defined as $g_{Eff}/g = 1/cos\theta$. Vertical dashed line indicates perpendicular field case.

are essential for understanding the $\nu=5/2$ FQHE. Primarily, the usual assumption of complete spin polarization of the electrons must be reconsidered. This is not surprising since typical electrostatic energies between 2D electrons in GaAs are significantly larger than the spin splittings at most magnetic fields of interest. Secondly, it appears that electron correlations in the second Landau level are substantially different from the lowest Landau level. It is the juxtaposition of both considerations which HALDANE and REZAYI[11] have shown may lead to stabilization of an unpolarized spin singlet state at $\nu=5/2$. While our data are consistent with such a state they cannot discriminate between it and other imaginable unpolarized wavefunctions.

It is a pleasure to thank my collaborators: R.L. Willett, H.L. Stormer, D.C. Tsui, A.C. Gossard, J.H. English, K.W. Baldwin, L.N. Pfeiffer, and K.W. West. Without their efforts this work would not have been possible.

1. R. Willett, J.P. Eisenstein, H.L. Stormer, D.C. Tsui, A.C. Gossard, and J.H. English, Phys. Rev. Lett. 59, 1776 (1987).
2. *The Quantum Hall Effect,* eds. R.E. Prange and S.M. Girvin (Springer-Verlag, New York, 1987).
3. J.P. Eisenstein, R.L. Willett, H.L. Stormer, D.C. Tsui, A.C. Gossard and J.H. English, to be published.

4. R.G. Clark, R.J. Nicholas, and A. Usher, Surf. Sci. 170, 141 (1986) and R.G. Clark, R.J. Nicholas, J.R. Mallett, A.M. Suckling, A. Usher, J.J. Harris, and C.T. Foxon, in *Proceedings of the Eighteenth International Conference on the Physics of Semiconductors, Stockholm, Sweden, 1986,* ed. O. Engstrom (World Scientific, Singapore, 1987), p.393.
5. R.L. Willett, H.L. Stormer, D.C. Tsui, L.N. Pfeiffer, K.W. West and K.W. Baldwin, Phys. Rev. B, to be published.
6. R.B. Laughlin, Phys. Rev. Lett. 50, 1395 (1983).
7. B.I. Halperin, Helv. Phys. Acta 56, 75 (1983).
8. Daijiro Yoshioka, Phys. Rev. B 29, 6833 (1984).
9. Tapash Chakraborty, Pekka Pietilainen and F.C. Zhang, Phys. Rev. Lett. 57, 130 (1986).
10. E.H. Rezayi, Phys. Rev. B *36,* 5454 (1987).
11. F.D.M. Haldane and E.H. Rezayi, Phys. Rev. Lett. 60, 956 (1988).
12. S.M. Girvin, A.H. MacDonald, D. Yoshioka, unpublished.
13. F.F. Fang and P.J. Stiles, Phys. Rev. 174, 823 (1968).
14. F. Stern and W.E. Howard, Phys. Rev. 163, 816 (1967).
15. A.H. MacDonald and G.C. Aers, Phys. Rev. B 29, 5976 (1984).
16. D. Yoshioka, J. Phys. Soc. Jpn. 53, 3740 (1984).
17. R.J. Haug, K. von Klitzing, R.J. Nicholas, J.C. Maan and G. Weimann, Phys. Rev. B 36, 4528 (1987).
18. D.A. Syphers and J. Furneaux, Surf. Sci. 196, 252 (1988).

The Influence of Coulomb Interactions on a 2DEG in High Magnetic Fields

R.J. Nicholas[1], *D.J. Barnes*[1], *R.G. Clark*[1], *S.R. Haynes*[1], *J.R. Mallett*[1], *A.M. Suckling*[1], *A. Usher*[1], *J.J. Harris*[2], *C.T. Foxon*[2], *and R. Willett*[3]

[1]Clarendon Laboratory, Parks Rd., Oxford, OX1 3PU, UK
[2]Philips Research Laboratory, Redhill, Surrey RH1 5HA, UK
[3]Massachusetts Institute of Technology, Cambridge, MA 02139, USA

Optical and thermal activation studies have been made of the magneto-excitations in a 2DEG in high magnetic fields. At integer Landau level occupancies it is found that both the 'bare' spin and cyclotron energies have considerable exchange contributions due to Coulomb interactions when measured by activation. For fractional occupancies the many body ground state at $\nu = 1/5$ has been shown to give a new hierarchy of states, and its excitation energies compared with the family arising from $\nu = 1/3$. Cyclotron resonance studies at 200mK where the carriers are condensed into the fractional ground state show no evidence of Coulomb interactions.

1. Introduction

Electron-electron interactions have been shown to play an important role in determining the properties of two-dimensional (2-D) systems. The earliest example of this was in the measurements of FANG and STILES [1], where it was found for silicon MOSFETs that the spin splitting in high magnetic fields could be enhanced by exchange interactions. A very pronounced example of this will be shown below for high mobility GaAs-GaAlAs heterojunctions, where the splitting is increased by a factor of over 15 [2,3] from that expected from the "bare" band structure value when the system is spin polarised. In addition, we have recently found that exchange can contribute significantly even to the simple Landau energy [2]. More dramatically the strong Coulomb force between electrons was found to give rise to a new correlated ground state which generates the Fractional Quantum Hall Effect (FQHE), observed at very low temperatures [4-6]. This arises from the formation of a quantum fluid [7-9], which has fractionally charged quasi-particles as the excitations of the system. Due to the very small intrinsic Zeeman splitting in GaAs, it is possible that the Coulomb interaction may lead to mixed spin states occurring for either the ground [10] or excited states [11] of the quantum fluid, as suggested recently [12] to explain much experimental data in both normal and tilted fields.

The scale of the Coulomb interaction between the electrons in high magnetic fields is the Coulomb Energy, $E_c = e^2/4\pi\epsilon\epsilon_0 l_c$ (l_c = Larmor radius), which is of order 51 $\sqrt{B}$ Kelvin. As the interaction is essentially two-dimensional it would be expected to depend only on the perpendicular component of field, which appears through l_c. In many cases however, such as the spin splitting, the exchange depends upon a population difference and can therefore act in a multiplicative fashion, through a form of positive feedback. In this case, the exchange energies can often become linearly proportional to B, and for spin splitting the exchange term can be determined by the total magnetic field and not just the perpendicular component. This is the basis of the co-incidence technique used to measure the g-factor in tilted fields [1,3]. Once the energy levels are well separated and the population difference has saturated, then the exchange energy will display its expected behaviour with a dependence only on the square root of the perpendicular field component [3].

Springer Series in Solid-State Sciences, Vol. 87
High Magnetic Fields in Semiconductor Physics II Editor: G. Landwehr

2. Excitations

Studies of the properties of a 2DEG in magnetic fields usually involve the different possible excitations of the system, which are created either optically or thermally. A typical example is the optical inter-Landau level transition known usually as cyclotron resonance. This can also be termed a magnetoplasmon excitation or magnetic exciton, with a wavevector $k \simeq 0$. For a filled ground state e.g. at an occupancy $\nu = 1$ ($\nu = nh/eB$), this produces a neutral particle consisting of a hole in the lower Landau level and an electron in the upper level, which can only be separated at finite k-vector, with a separation $r = kl_c^2$. Additional Coulomb energy is required to split up the two particles in order to generate free carriers, capable of giving rise to electrical conduction. Thus we may expect a set of dispersion curves for a generic excitation of the form [13]

$$E_m(k) = mh\omega_c + g\mu_B B\delta S_z + \Delta E_m(k) , \qquad (1)$$

where m is an integer, and $\Delta E_m(k)$ is of the order of the Coulomb energy E_c. The function ΔE_m will depend both upon the initial ground state of the system [13] (i.e. on ν) and the excitation being studied e.g. m = 1 for cyclotron resonance; m = 0, $\delta S_z = 1$ for spin-wave excitation.

A typical example of this sort of magnetoplasmon dispersion is shown in fig. 1, as derived by MACDONALD et. al. [14] for m = 1 at integer and fractional filling factors. Clearly $\Delta E_m(k) \to 0$ as $k \to 0$, in agreement with Kohn's theorem [15], which says that at long wavelengths an external perturbation only couples to the centre of mass motion, which is unaffected by electron-electron interactions.

A practical study of these excitation spectra is difficult, due to the relatively large scale of k, of order 1/100Å, at which changes occur. It is however possible to study the limits $k \to 0$, through optical absorption and $k \to \infty$, through the thermal excitation of dissociated electron-hole pairs, and both of these will be studied below for m = 1 and m = 0. One further possibility is that when there is disorder present, mixing may occur of the finite k states with those at k = 0, thus leading to shifts in position and changes in the linewidth of the resonance [14,16].

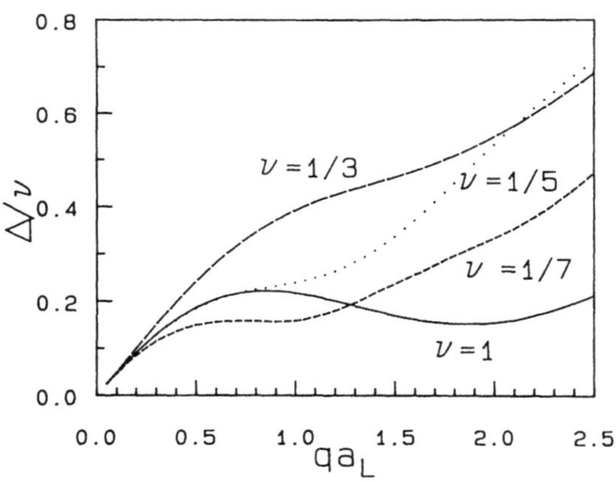

Fig. 1: Magnetoplasmon excitation energies measured from $\hbar\omega_c$ in units of νE_c at different occupancies. After MACDONALD et al [14].

3. Experimental

The samples studied were very high mobility GaAs/GaAlAs heterojunctions grown at the Philips Research Laboratories, Redhill [17]; conventional structures with spacer layers of between 400 and 1600 Å. The carrier concentrations (N_s) were varied over the range 0.22 to 2.5 x 10^{11} cm^{-2} using a red LED, and the mobilities were typically of order 1 x 10^6 cm^2/Vs at N_s = 1 x 10^{11} cm^{-2} with the mobility being approximately proportional to N_s. The transport measurements were made on lithographed Hall bars, with the exception of those where cyclotron resonance was studied simultaneously, where a rectangular geometry was used. The thermal activation measurements were made using calibrated germanium, carbon-glass and Speer resistors to cover the range 10K - 100mK.

Cyclotron resonance was performed at 200mK, using an optically pumped F.I.R. laser, with the sample mounted on a hollow extension to the mixing chamber of a dilution refrigerator. Optical access was via a lightpipe, passing through the mixing chamber, and extensive filtering was used to allow only the long wavelengths (typically 100μm) to reach the sample. The radiation was detected with a ground down Speer carbon resistor, which could also be used to measure the sample temperature. This was also checked by comparison with the transport results obtained from similar samples immersed directly in the mixture. Further details of this arrangement will be given in a future publication [18].

4. Activated Transport for Integer Filling

A typical method for studying energy gaps in semiconductors is to use temperature dependent conductivity, where

$$\sigma = \sigma_0 \exp(-\Delta E/2kT) . \qquad (2)$$

This expression holds exactly for $kT \ll \Delta E$, activation between sharp levels separated by ΔE, and a temperature independent mobility (μ). Since conduction involves the motion of independent electrons or holes, the corresponding energy gap, in the excitation picture, is that of separated carriers created at $k \rightarrow \infty$. In the case of Landau levels there can be a significant broadening, and activation is to a band of states of order 1meV (11K) wide [19-21] even in very high mobility samples, with some significant residual constant density between levels. Importantly however, only a narrow band of extended states will exist at the centre of each level. This is the origin of the quantum Hall effect [22]. A 'mobility edge' separates the extended and localised states, and one again expects a conductivity of the form of Eqation (2), with σ_0 known as the minimum metallic conductivity [23]. The extended state region may well contain only a small fraction of the total number of states [22,24].

An example of this is shown in fig. 2, for the activated minimum conductivity arising between two spin-split levels at occupancies of ν = 1 and 3. The curves show a typical form, bending over at high temperatures where $kT \approx \Delta E$, and at low temperatures when other conduction processes such as variable range hopping become significant. Similar plots are shown in fig. 3 for the Landau splitting at ν = 2.

The energy gaps deduced for the spin-split levels at occupancies ν = 1 and 3 are plotted in fig. 4 as a function of field, and hence of electron concentration, together with the data for the Landau splitting at ν = 2. The spin-splitting energies were both found to rise linearly with field at the same rate, 5K/T, tailing off at very low fields to a small negative offset of order 2K. This slope corresponds to a g-factor of 7.5, more than 15 times the bare g-factors in bulk GaAs (0.44)

[25] and low density heterojunctions (0.4) [26], which correspond to the k = 0 excitation as measured by E.S.R. This demonstrates the large spin exchange energy, which is comparable to, but a little larger than earlier measurements [3,27].

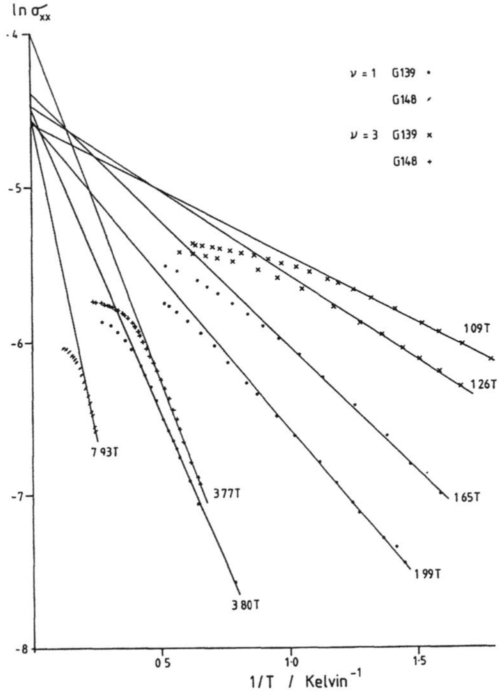

Fig. 2: Typical data for the activated conductivity minima at occupancies $\nu = 1$ and $\nu = 3$. The straight lines have slopes equal to $-\Delta E/2$, and give the activation energy.

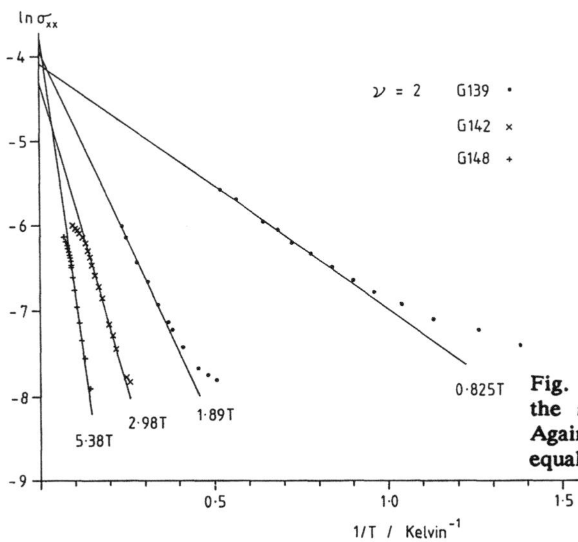

Fig. 3: Typical Arrhenius plots for the ν=2 activated conductivity minima Again, the straight lines have slopes equal to $-\Delta E/2$.

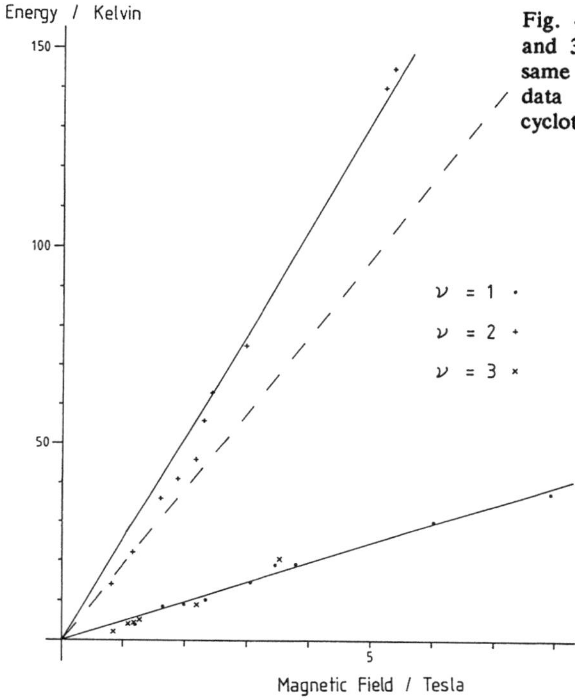

Fig. 4: Plots of ΔE vs. B for ν=1,2 and 3. The ν=1 and 3 data fit the same line, of slope 5K/T. The ν=2 data lie a similar amount above the cyclotron energy (dashed line).

The calculations of KALLIN and HALPERIN [13] suggest an exchange energy of $(\pi/2)^{\frac{1}{2}}E_c$ for a completely ionised spin flip excitation in a perfect 2-D system, for both ν = 1 and 3. In practice the finite z-extent of the wavefunction is expected to reduce this by approximately 40% [28], giving values approximately three times the largest values found here at high fields. This and the linear dependence on B suggest that there is still a significant overlap of the spin states leading to a field dependent spin population difference. This corresponds to the case considered originally by ANDO and UEMURA [29], who found an exchange energy (E_{ex}) of

$$E_{ex} = \sum_{N'}\sum_{q} \frac{V(q)}{\epsilon(q)} J_{NN'}(q)\ (n_{N'\uparrow} - n_{N'\downarrow})\ , \qquad (3)$$

where the overlap terms corresponding to exchange between electrons in different Landau levels can usually be ignored. A further factor which may reduce the magnitude of the exchange energy is disorder. This will introduce a broadening, or 'fuzziness' of the dispersion relation which will reduce the minimum energy required to excite the carriers. This may also be thought of as due to a reduction in the range of the electron-hole interaction which will be 'screened' at large separations.

The Landau splitting at ν = 2 was also found to increase approximately linearly with field, starting from a larger negative offset of order 7K, and rapidly exceeding the optically measured cyclotron energy (shown as a dashed line in fig. 4), by over 25% at high fields. The measured splitting is consistent with a comparable but slightly larger exchange energy, 7K/T, being added to the bare cyclotron energy. By 5.5T the exchange energy is almost 40K. For ν = 2 KALLIN AND HALPERIN [13] find that the exchange energy should be $(\pi/8)^{\frac{1}{2}}E_c$, which will be approximately 44K at 5.5T when a factor of 0.6 is introduced to account for finite z-extent. We thus conclude that for the Landau splitting there is very good agreement with theory. This is consistent with the much larger total splitting for this case, which will result in complete separation of the Landau levels with no overlap and hence the full expected exchange term. Thus at high fields we may conclude that we have measured the full ionisation energy of the m = 1 magneto-exciton.

At very low fields the activation energies fall below the energy gaps, because of the finite level broadening of the *extended* states. This has been modelled using both Gaussian and rectangular densities of mobile conducting states, to obtain the level widths separating the mobility edges [30]. The width found for ν = 2 (7K) was approximately three times that found at ν = 3 (2K). This may be related to the overlap of the two spin states at ν = 2, which may affect both the localisation and the broadening of the levels due to oscillatory screening [31], which is ineffective for a filled level. Cyclotron resonance on the same samples [32] shows that the ratio of the linewidths at ν = 2 and ν = 1 can be of order 5 or more.

5. Activation in the Fractional Regime

To date the main method for studying the excitations of the quantum fluid formed in the FQHE regime has been thermal activation, as discussed above for the integer regime. In this case however the excitations correspond to fractionally charged quasi-particles [7-12,33]. The main characteristics of the FQHE are the appearance of resistivity, and conductivity, minima at fractional occupancies, and associated Hall plateaux. A hierarchy exists of fractional ground states [7-9] based on the 'parent' state at ν = 1/3, which leads to the formation of 'daughter' states by correlation of the quasi-particles of the preceeding ground state. High order 'daughter' states up to ν = 6/13 have been reported, together with their 'hole-like' analogues [6]. A further hierarchy based on a 1/5 state should also exist, although there have been only limited reports of the FQHE at ν = 1/5 [6,34] and 4/5 [6,35] through the observation of weak resistivity minima and Hall plateaux.

Fig. 5(a) shows the development of FQHE structure in the resistivity of a very high quality heterojunction, G139, with a 1600 Å spacer layer, at temperatures down to 175 mK, measured using the M.I.T. Hybrid magnet [36]. The hierarchy based on 1/3 can be easily seen (2/5,3/7,4/9; 2/7,3/11), and there is also a well resolved ν = 1/5 minimum. Daughter states of the ν = 1/5 ground state are visible now as weak minima in ρ_{xx} at ν = 2/9 and 2/11. The 2/11 structure is best resolved at around 200mK (Fig. 5(b)), due to experimental difficulties in the extreme quantum limit, which are thought to be due to disorder and contacting problems. This prevents the observation of good Hall plateaux at very high fields (fig. 5(c)), and may also be related to the formation of a Wigner solid as reported recently [37].

The strengthening of the 1/5 structure at low temperature does not result in a decrease of the ρ_{xx} minimum except at the lowest temperatures. This is due to an overall increase in the background resistivity, thought to reflect a change in the quasi-particle mobility. To measure the activation energy in this case, the minimum was ratioed against a background resistivity (ρ^b), determined by interpolation between the resistivity peaks either side of the minimum [36] as shown in fig.6. The activation energy ($\Delta = \Delta E/2$) comes from the 'normalised' temperature dependence

$$\rho_{xx}/\rho^b \quad \propto \quad \exp(-\Delta/kT) \qquad (4)$$

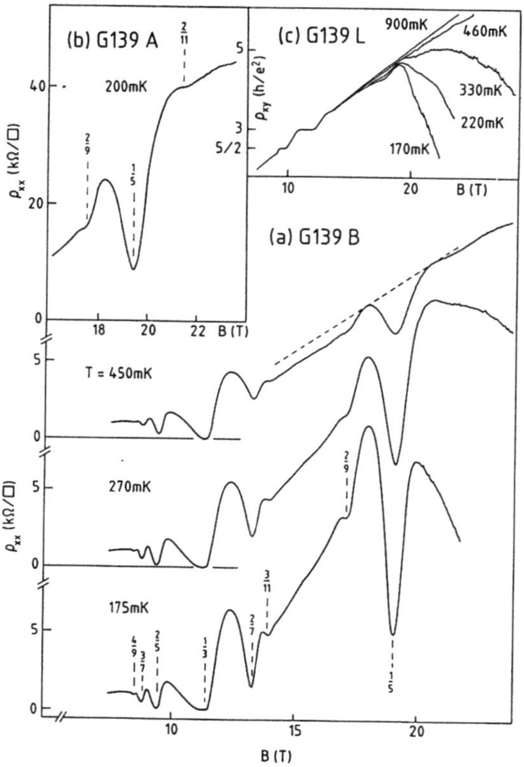

Fig. 5: Temperature dependence of ρ_{xx} vs. B for sample G139B.

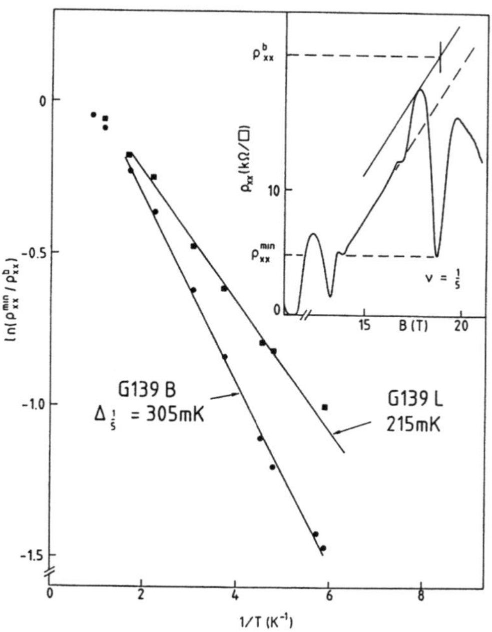

Fig. 6: Activation plot for the ν=1/5 minimum at 19T, using the ratio technique. The inset shows how the background, ρ^b is obtained at low temperatures.

as shown in fig. 6. The two samples studied give $\Delta \approx 0.22K$ and $0.3K$, broadly consistent with the temperature at which the minima first become visible. A weaker $\nu = 1/5$ minimum was also observed at lower field (9.8T) and its Δ estimated to be 50mK from the functional dependence of ρ_{xx} shown in fig. 6.

The activation energies of the 1/3 family of states (1/3,2/5,3/7) were also studied and are compared with the 1/5 results in fig. 7. The 1/5 gap is substantially smaller, even at 19T, than for 2/5 (3/5) at less than 10T. The energy gap should be the ionisation energy of an isolated quasi-particle pair, usually measured in units of the Coulomb energy $C(\nu)E_c$, corresponding to the $k \rightarrow \infty$ limit of the quasi-exciton dispersion. The consensus of theory [38] gives $C(1/3) \simeq 0.1$, $C(2/5) \simeq 0.05$, $C(1/5) \simeq 0.03$, assuming a completely spin polarised system with spin polarised excitation. Some calculations [10–12,39,40] suggest lower values for spin reversed excitations, which will be important at low fields [11] where the Zeeman energy is small. A further important reduction comes from the finite z-extent of the wavefunction, as well as the influence of disorder and overlap of the levels.

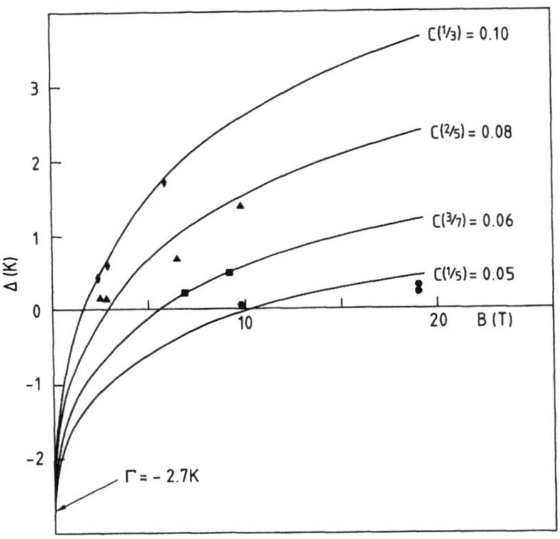

Fig. 7: Δ vs. B for G139 at $N_s = 9.5 \times 10^{10}$ cm^{-2}. One point is shown for $\nu=1/5$ at lower density. The solid curves show fits using the finite z-extent correction of ref. [28] as a correction to the $\sqrt{B}$ dependence for all fractional states, and a constant offset to Δ of $\Gamma = -2.7K$ to account for disorder.

Fig. 7 shows fits to the activation energies Δ for the groups of states p/q (1−p/q, 1+p/q, 2−p/q) at fixed electron concentration using finite z-extent calculations [28] (with $b^{-1} = 48$ Å), assumed to apply equally for all states. A fixed shift down of 2.7K was assumed to account for disorder, as suggested by MACDONALD et al. [41], and similar to the value deduced at $\nu = 1$ for this sample. The resulting values shown for $C(\nu)$ indicate clearly that $C(1/3) > C(2/5) > C(3/7) > C(1/5)$ in agreement with theory. The exact values depend upon the assumptions made about disorder and the wave-function correction, and the values for the higher order fractions are progressively less accurate. Complications may occur at low fields if the gaps become dominated by spin-reversed excitations [11,12,40] as will be discussed by MAKSYM et al and CLARK et al [12], or if disorder causes a significant overlap of the spin-states, as found at $\nu=1$. Both these effects would be expected to give a linear B dependence of the activation energy at low fields. Nevertheless the values of $C(\nu)$ deduced mainly from the high field data are in good agreement with theory.

The important conclusion however is the relative strength of the fractional states, and the existence of a new hierarchy based on 1/5.

6. Cyclotron Resonance in the Fractional Regime.

There have been reports [42,43] of cyclotron resonance studies as a function of occupancy which suggest that both minima [42] and maxima [43] in the linewidth may occur at fractional values corresponding to p/3. Neither of the reports to date have made measurements at temperatures below 1.4K and no corresponding fractional effects have been observed simultaneously in the resistivity components. In contrast we show in fig. 8 recordings of cyclotron resonance at the lowest temperatures yet reported, 200mK, at occupancies around 1/3. The sample resistivity was always recorded simultaneously and a series of typical recordings are shown in fig. 9, where the carrier concentration was controlled with a red LED. The 2/3 occupancy was moved progressively through a cyclotron resonance absorption at 5.55T, using a wavelength of $\lambda = 133.1\mu m$. The sample used comes from the same wafer, G139, as that studied in the previous section, and shows a very well resolved set of fractional resistivity minima corresponding to $\nu = 1/3, 2/3; 2/5, 3/5; 3/7, 4/7$, thus confirming the low sample temperature.

A systematic study was made of the linewidth and effective mass as a function of carrier concentration, at a fixed wavelength, around the fractional ground states at $\nu = 1/3$ and 2/3 as shown in figs. 10 and 11. The very narrow experimental half

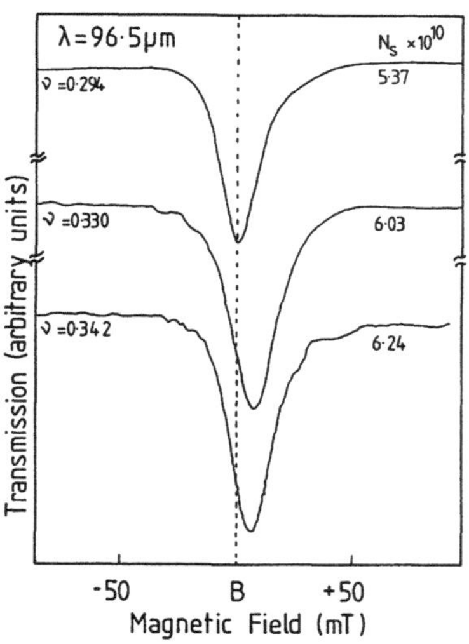

Fig. 8: Typical cyclotron resonance traces at 200mK close to $\nu = 1/3$.

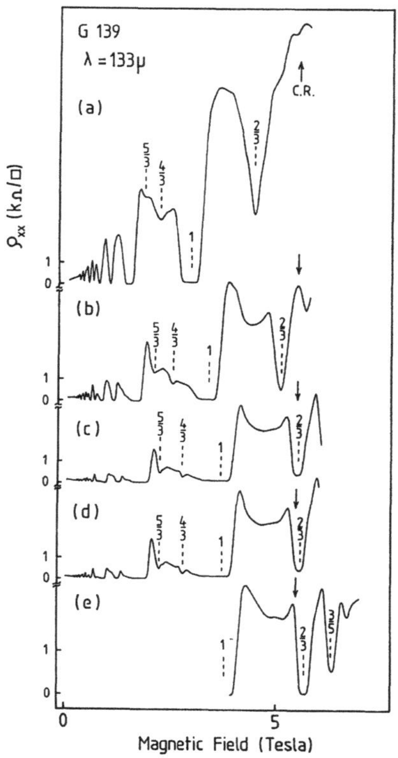

Fig. 9: Typical ρ_{xx} traces with a simultaneous resonance at $133\mu m$.

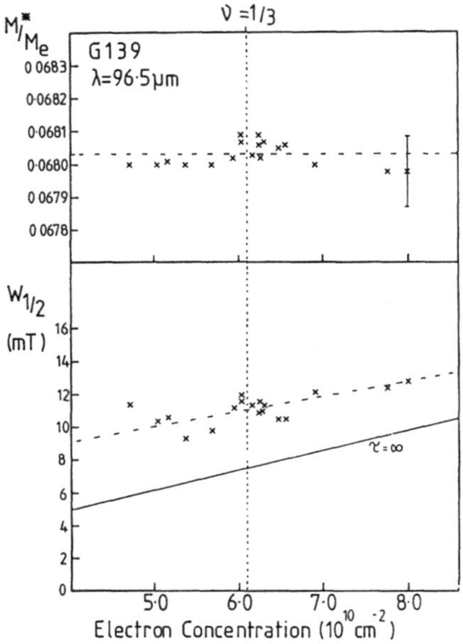

Fig. 10: The N_s dependence of the mass and linewidth at ν = 1/3.

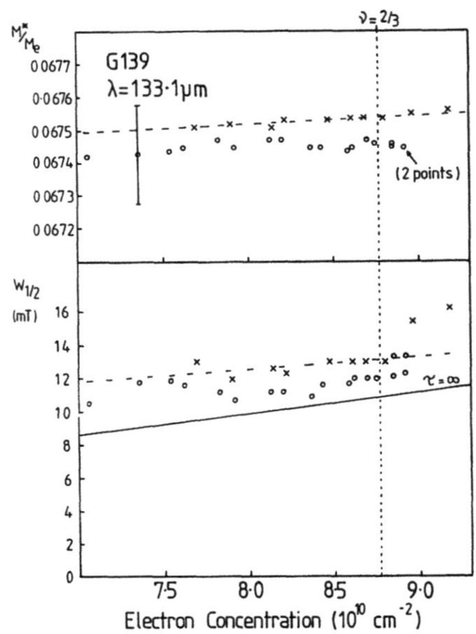

Fig. 11: The N_s dependence of the mass and linewidth at ν = 2/3.

widths at half-maximum, of order 10 mT, illustrate the exceptional quality of the heterojunctions. For such narrow resonances it is necessary to correct for the finite carrier concentration, using a Drude formalism [44]. This gives an experimental half width $\Delta B_{\frac{1}{2}}$ which is related to the true linewidth $\delta B_{\frac{1}{2}}$ by the relation

$$\Delta B_{\frac{1}{2}} = \delta B_{\frac{1}{2}} + m^*\omega_p/e \, , \qquad (5)$$

where $\omega_p = Z_o N_s e^2/m^*(1 + /\epsilon)$. The result of this is that there is an additive finite linewidth even for a perfect system where $\tau,\mu \rightarrow \infty$. This is shown by the solid lines in figs. 10 and 11.

The true linewidths $\delta B_{\frac{1}{2}}$ are of order 2 - 5 mT, with however a considerable amount of error due to the large constant term subtracted from the experimentally measurable quantity $\Delta B_{\frac{1}{2}}$. The conclusion from figures 10 and 11 however is that any influence of the fractional state upon the linewidth is at most of the order of 1mT, which is comparable to the errors. The only case where the data is suggestive of an effect is at ν = 1/3 for 96.5μm, where there could also be a very small influence on the effective mass. The original data for three occupancies around 1/3 is shown in fig. 8. In the two cases studied carefully (ν = 1/3,2/3) the effective mass has remained constant to a precision of ± 0.05%. On this scale a shift by 2Δ (the k = ∞ intra level energy gap, which should also be the Coulomb correction added to the inter Landau level excitation as k → ∞ [45]) would correspond to a change of 1.5%. These results are in distinct contrast to the earlier reports [42,43], where changes in linewidth of order 10 mT and more were reported even at 1.5K. It should be noted however that in order to obtain sufficient data it was necessary for these authors to mix data where both λ and N_s were varied.

Theoretically it has been suggested that many-body effects might influence the cyclotron resonance for both integer [16] and fractional [14,45] occupancies. In order for this to occur it is necessary to violate the assumptions of Kohn's theorem [15] by the introduction of spatial disorder which will mix the magnetoplasmon modes of finite wave-vector with those at k = 0. KALLIN and HALPERIN [16] have shown that for integer filling small shifts in resonance position of order 1% are possible. Recently SCHLESINGER et al[46] have reported shifts below $\nu = 1$, but no effects at fractional occupancy even down to 0.4K. SEIDENBUSCH et al [43] suggest that the occurrence of peaks in the linewidth at fractional occupancy could be due to oscillatory screening produced by the density of states gap formed on condensation of the ground state. This would be most pronounced for samples which show strong oscillatory effects at integer occupancy. In fact the same samples studied here do show a very pronounced peak at $\nu = 2$ ($\Delta B_{\frac{1}{2}}$ = 60mT: HOPKINS et al [32]), but no effect at $\nu = 1$, leading them to suggest that the oscillatory peak in linewidth is in fact related to localisation phenomena. Another possiblity suggested by KALLIN and HALPERIN [16] was that the broadening could come about by the rapid inelastic break up of the fractional states in the higher Landau level due to scattering, and this may also cause shifts in the resonance. Our data do not suggest any significant effect, with the possible explanation that these structures are 'too perfect', with not enough disorder to mix the various magnetoplasmon modes.

In conclusion we can say that we have no evidence for any influence on the FQHE ground state on the m = 1 magnetoplasmon excitation (cyclotron resonance) contrary to suggestions by some other workers.

7. References

1. F.F. Fang and P.J. Stiles, *Phys. Rev.* 174 823 (1968)
2. A. Usher, R.J. Nicholas, J.J. Harris, and C.T. Foxon, *Proc. Int. Conf. on Physics of Semiconductors, Warsaw, 1988, to be published* (1989)
3. R.J. Nicholas, R.Haug, K.vonKlitzing and G.Weimann, *Phys. Rev.*B37 1294(1988)
4. D.C. Tsui, H.L. Stormer and A.C. Gossard, *Phys. Rev. Lett.* 48 1559 (1982)
5. R.G. Clark, R.J. Nicholas, A. Usher, J.J. Harris, and C.T. Foxon, *Surf. Sci.* 170 141 (1986)
6. R. Willett, H.L. Stormer, D.C. Tsui, A.C. Gossard, J.H. English and K.W. Baldwin, *Surf. Sci.* 196 257 (1988)
7. R.B. Laughlin, *Phys. Rev. Lett.* 50 1395 (1983); *Surf. Sci.* 141 11 (1984)
8. F.D.M. Haldane, *Phys. Rev. Lett.* 51 605 (1983)
9. B.I. Halperin, *Phys. Rev. Lett.* 52 1583 (1984)
10. F.C. Zhang, and T. Chakraborty, *Phys. Rev.* B30 7320 (1984); *Phys. Rev.* B34 7076 (1986)
11. T. Chakraborty, P. Pietilainen and F.C. Zhang, *Phys. Rev. Lett.* 57 130 (1986)
12. P.A. Maksym, R.G. Clark, S.R. Haynes, J.R. Mallett, J.J. Harris and C.T. Foxon; and R.G. Clark, J.R. Mallett, S.R. Haynes, J.J. Harris, C.T. Foxon and P.A. Maksym, In Proc. Int. Conf. on Applications of High Magnetic Fields, Wurzburg, 1988 *to be published*
13. C. Kallin and B.I. Halperin, *Phys. Rev.* B30 5655 (1984)
14. A.H. McDonald, H.C.A. Oji, and S.M. Girvin, *Phys. Rev. Lett.* 55 2208 (1985)
15. W. Kohn, *Phys. Rev.* 123 1242 (1961)
16. C. Kallin and B.I. Halperin, *Phys. Rev.* B31 3635 (1985)
17. C.T. Foxon, J.J. Harris, R.G. Wheeler and D.E. Lacklison, *J. Vac. Sci. and Technol.* B4 511 (1986)

18. R.G. Clark, S.R. Haynes, R.G. Bendall, J.R. Mallett, A. Usher, A.M. Suckling, Y. Journeaux and H. Jones, *Cryogenics, to be published*
19. D. Weiss, K. vonKlitzing and V. Mosser, In Springer Series in Solid State Sciences 67 204 (1986)
20. T.P. Smith, B.B. Goldberg, P.J. Stiles and M. Heiblum, *Phys. Rev.* B32 2696
21. E.Gornik, In The Physics of the Two-dimensional Electron Gas ed. by J.T. Devreese and F.M. Peeters, NATO ASI Series B157 365 (1987)
22. H. Aoki, *Rep. Prog. Phys.* 50 655 (1987)
23. R.J. Nicholas, R.A. Stradling, and R.Tidey, *Solid State Commun.* 23 341 (1977)
24. G. Ebert, K. vonKlitzing, C. Probst, E. Schuberth, K. Ploog, and G. Weimann, *Solid State Commun.* 45 625 (1983)
25. C. Hermann and C. Weisbuch, *Phys. Rev.* B15 823 (1977)
26. M. Dobers, unpublished results, see G. Lommer, F. Malcher and U. Rossler, *Superlattices and Microstructures* 2 273 (1986)
27. T. Englert, D.C. Tsui, A.C. Gossard and C. Uihlein, *Surf. Sci.* 113 295 (1982)
28. F.C. Zhang and S. Das Sarma, *Phys. Rev.* B33 2903 (1986)
29. T. Ando and Y. Uemura, *J. Phys. Soc. Japan* 37 1044 (1974)
30. A. Usher, R.J. Nicholas, J.J. Harris and C.T. Foxon, *to be published*
31. T. Ando, *J. Phys. Soc. Japan* 43 1616 (1977)
32. M.A.Hopkins, D.Barnes, R.Nicholas, M.Brummell, J.Harris and C.T.Foxon, *Proc. Int. Conf. on Physics of Semiconductors, Warsaw, 1988, to be published* (1989)
33. R.G. Clark, J.R. Mallett, S.R. Haynes, J.J. Harris and C.T. Foxon, *Phys. Rev. Lett.* 60 1747 (1988)
34. E.E. Mendez, M. Heiblum, L. Chang and L.Esaki, *Phys. Rev.* B28 4886 (1983)
35. R.J. Nicholas, R.G. Clark, A. Usher, J. Mallett, A.M. Suckling, J.J. Harris and C.T. Foxon, In Springer Series in Solid State Science ed. G. Landwehr 71 146(1986)
36. J.R. Mallett, R.G. Clark, R.J. Nicholas, J.J. Harris and C.T. Foxon, *Phys. Rev.* B38 *in press*
37. E.Y. Andrei, G. Deville, D.C. Glattli, F.I.B. Williams, E. Paris and B. Etienne, *Phys. Rev. Lett.* 60 2765 (1988)
38. T. Chakraborty, *Phys. Rev.* B31 4026 (1985); R. Morf and B.I. Halperin, *Phys. Rev.* B33 2221 (1986); A.H. McDonald and S.M. Girvin, *Phys. Rev.* B33 4414 (1986); G. Fano, F. Ortolani and E. Colombo, *Phys. Rev.* B34 2670; F.D.M. Haldane and E.H. Rezayi, *Phys. Rev. Lett.* 54 237 (1985); A.H. McDonald and D.B. Murray, *Phys. Rev.* B32 2707 (1985); W.P. Su, *Phys. Rev.* B32 2617 (1985)
39. T. Chakraborty, *Phys. Rev.* B34 2926 (1986)
40. D. Yoshioka, *J. Phys. Soc. Japan* 55 3960 (1986)
41. A.H. McDonald, K. Liu, S.M. Girvin and P.M. Platzman, *Phys. Rev.* B33 4014
42. G.L. Rikken, H.W. Myron, G.J.Langerak and H.Sigg, *Surf. Sci.* 170 160 (1986)
43. W. Seidenbusch, E. Gornik and G. Weimann, *Phys. Rev.* B36 9155 (1987)
44. K.W. Chiu, T.K. Lee and J.J. Quinn, *Surf. Sci.* 58 182 (1976)
45. P. Pietilainen and T. Chakraborty, *to be published*
46. Z. Schlesinger, W.I. Wang and A.H. McDonald, *Phys. Rev. Lett.* 58 73 (1987)

Experimental Determination of Fractional Charge *e/q* in the FQHE and Its Application to the Destruction of States

R.G. Clark[1], *J.R. Mallett*[1], *S.R. Haynes*[1], *P.A. Maksym*[2], *J.J. Harris*[3], *and C.T. Foxon*[3]

[1]Clarendon Laboratory, University of Oxford, Oxford OX1 3PU, UK
[2]Department of Physics, University of Leicester, Leicester LE1 7RH, UK
[3]Philips Research Laboratories, Redhill, Surrey RH1 5HA, UK

1. Introduction

We have recently demonstrated [1] that the LAUGHLIN/ HALDANE prediction $e^*=\pm e/q$ for the charge e^* of quasiparticles excited across the energy gap Δ of FQHE ground states at $\nu=p/q$ [2,3], a new fundamental quantum of nature, is consistent with experiment. The experimental probe of e^* is $\sigma^c_{xx}=\sigma_{xx}(1/T=0)$ obtained from extrapolated values of ρ^c_{xx} determined from activation data defined by $\rho_{xx}=\rho^c_{xx}e^{-\Delta/kT}$. In a high quality, low density GaAs–GaAlAs heterojunction G139 ($n= 1 \times 10^{11}$ cm^{-2}, $\mu= 1 \times 10^6$ cm^2/Vs, 1600Å spacer) we found that $\sigma^c_{xx}=c(e/q)^2/h$ (independent of p) for some 13 fractional states at $\nu=p/q$ with q=3, 5, 7 and 9, where the numerical constant c= 0.91 ±0.11. This result is not only consistent with $e^*=\pm e/q$ but strongly indicative of the existence of a minimum quasiparticle conductivity [1]. In this paper, we report a study of σ^c_{xx} and Δ in a higher density heterojunction G156 ($n= 1.6 \times 10^{11}$ cm^{-2}, $\mu= 1.6 \times 10^6$ cm^2/Vs, 400Å spacer) at two angles of the B–field to the sample plane normal, $\theta=0^\circ$ and 47°. Our objectives are; (i) to examine the probe of e^* for fractional states with q=3, 5 and 7 in a different structure and geometry, (ii) to compare normal and tilted field activation energies to examine the nature of the quasiparticle energy gaps and (iii) to apply the e^* measurements for $\theta= 0^\circ$ and 47° to investigate what occurs when fractional states which are unpolarised at low field are 'destroyed' by the increased total field induced by tilt. Details of the FQHE ground state spin configurations for the N=0 Landau level are presented separately in these proceedings [4]. Here we consider aspects relevant to the quasiparticle charge.

2. Results

G156 resistivity data for $\theta=0^\circ$ and 47° are shown in Figs. 1 and 2. The experiment was performed using a dilution refrigerator in the Clarendon 20T Hybrid magnet. The sample current was 40.0 nA in the AC lock-in measurements and thermometry was provided by a Speer resistor calibrated and corrected for magnetoresistance by detailed in-situ studies. In Fig. 1(a), for the region $1<\nu<2$, 4/3, 5/3, 7/5 and 8/5 fractional states are resolved at 30mK. The structure is significantly altered on tilting the B–field to $\theta = 47^\circ$ as shown in Fig. 2(a), where it is qualitatively apparent that the 5/3 state is slightly strengthened, the 8/5 structure is completely destroyed, an 11/7 state is brought out, the 7/5 state is strengthened and 4/3 structure is virtually destroyed. For $\nu<1$ it would appear from Figs. 1,2(a) that in contrast, the 2/3, 3/5 and 4/7 states at high field remain essentially unaltered on increasing θ to 47°. The situation is quantified by an activation study to determine the parameters Δ and σ^c_{xx}. The development of fractional structure at selected temperatures is shown in Figs. 1,2(b) for $1<\nu<2$ and Figs. 1,2(c) for $\nu<1$. For the region $1<\nu<2$, activation plots ρ_{xx} v. 1/T are presented in Figs. 3 and 4 for $\theta = 0^\circ$ and 47°

Springer Series in Solid-State Sciences, Vol. 87
High Magnetic Fields in Semiconductor Physics II Editor: G. Landwehr

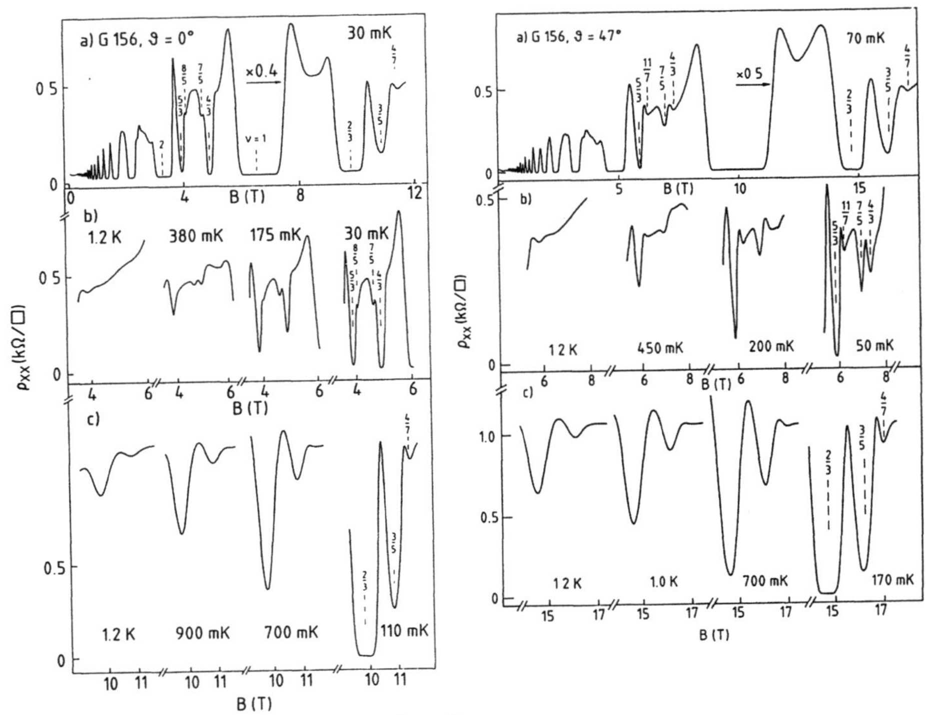

<u>Figs. 1,2</u> G156 transport data; $\theta=0^{\circ}$, 47°

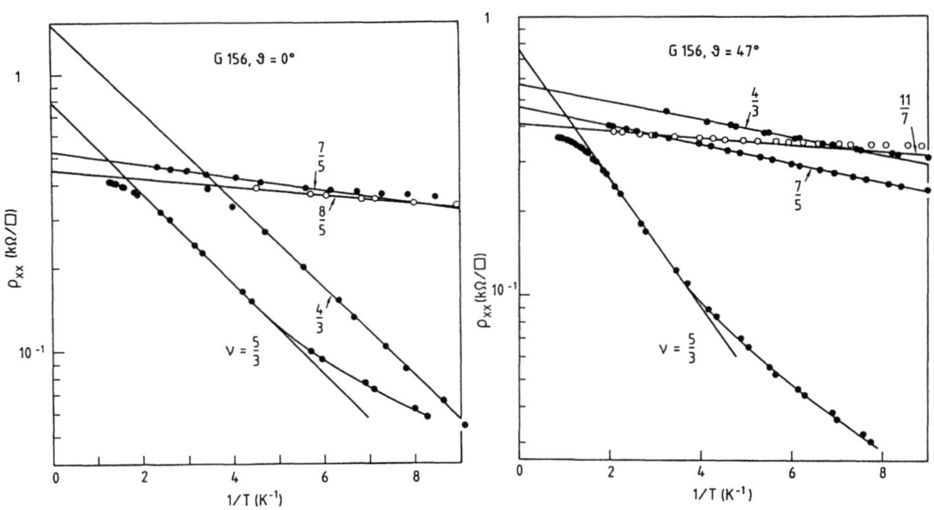

<u>Figs. 3 and 4</u> G156 activation plots for fractional states $1<\nu<2$ at $\theta=0^{\circ}$ and 47° (see also Fig. 4 of ref. 4)

respectively and in Figs. 5 and 6 for fractional states with $\nu<1$. Relevant parameters obtained from these figures are summarised in table 1, where we also show the low field spin configurations determined in ref. 4 where the notation is defined. Values of Δ are obtained from the raw data without subtraction of any modelled nonactivated components. Straight-line fits to the activated region are extrapolated to $1/T=0$.

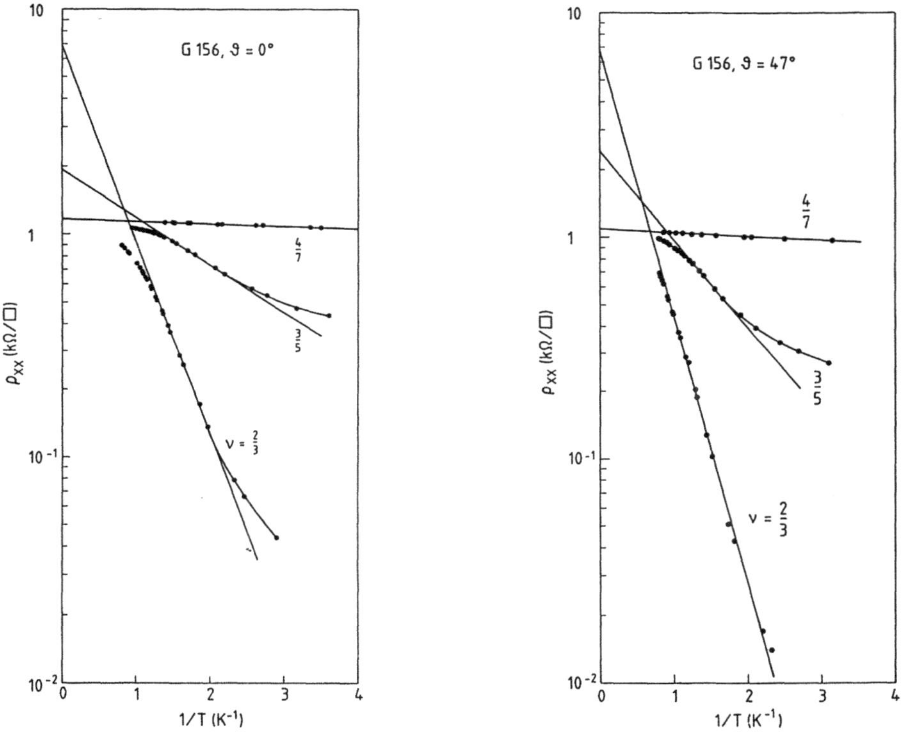

Figs. 5 and 6 G156 activation plots for fractional states $\nu<1$ at $\theta=0^\circ$ and 47°

Table 1. Low-B spin assignments [4] and G156 activation data with $\theta=0^\circ$ and 47°

ν	spin	B(T) $\theta=0^\circ$	B(T) $\theta\simeq47^\circ$	Δ(K) $\theta=0^\circ$	Δ(K) $\theta=47^\circ$	$\frac{\Delta(0^\circ)}{\cos47^\circ}$	ρ^c_{xx}(kΩ/□) $\theta=0^\circ$	ρ^c_{xx}(kΩ/□) $\theta=47^\circ$	$\sigma^c_{xx}(e^2/h)$ $\theta=0^\circ$	$\sigma^c_{xx}(e^2/h)$ $\theta=47^\circ$
4/3	↑↓	4.9	7.2	0.36	0.074		1.51	0.57	0.97/9	0.98/25
5/3	↑↑	3.9	5.8	0.375	0.53	0.55	0.79	0.76	0.79/9	0.76/9
7/5	↕	4.7	6.9	0.054	0.081	0.079	0.53	0.47	1.0/25	0.91/25
8/5	↑↓	4.1		0.034			0.45		1.1/25	
11/7	↑↑/↕		6.1		0.031			0.41		0.96/25
2/3	↑↓	9.8	14.5	1.99	2.75	2.92	6.8	6.8	1.02/9	1.02/9
3/5	↕	10.9	16.1	0.49	0.91	0.72	1.95	2.4	0.7/25	0.84/25
4/7	↑↓	11.4	16.9	0.023	0.035	0.034	1.18	1.13	0.73/49	0.7/49

3. Discussion

For $\theta=0^\circ$, G156 σ^c_{xx} values for p/q states 2/3, 4/3, 5/3, 7/5 and 8/5 are well-represented by the average $\sigma^c_{xx}= 0.98/q^2$ in units e^2/h. The 3/5 and 4/7 values 0.7/25 and 0.73/49 respectively are low, although the 3/5 extrapolation is subject to errors due to the form of the data (and the value is 0.84/25 at $\theta = 47^\circ$). When the field is tilted to $\theta=47^\circ$ the 2/3, 5/3, 3/5, 7/5 and 4/7 σ^c_{xx} values remain essentially locked at their $\theta=0^\circ$ values. For the 4/3 state, however, which is almost completely destroyed by the increased total field due to tilt angle, a dramatic and seemingly fundamental change occurs. σ^c_{xx} decreases from 0.97/<u>9</u> to 0.98/<u>25</u> e^2/h and the value of Δ is reduced from 0.36K at $\theta=0^\circ$ to 0.074K at $\theta=47^\circ$. In essence, the strength and intercept of the 4/3 state has been transformed from being very similar to the 5/3 state to being almost identical to the 7/5 data - see table 1 (solid/broken) underlined values. This is emphasised by plotting σ_{xx} v. 1/T corresponding to the Fig. 3 and 4 ρ_{xx} data and is presented as Fig. 4 of ref. 4. It would appear that in the destruction of the unpolarised state, the quasiparticle charge transforms from e/3 to e/5 at $\nu=4/3$ filling - an astonishing result. A similar effect is observed in the tilted-field annihilation of the 8/5 (spin-unpolarised) state and the subsequent emergence of an 11/7 state. The 11/7 σ^c_{xx} value is 0.96/<u>25</u> e^2/h.

The effect of the tilted field on Δ is also revealing. It can be seen from table 1 that, with the exception of the 3/5 state, $\Delta(\theta=47^\circ) \simeq \Delta(\theta=0^\circ)/\cos 47^\circ$. This indicates that the measured activation energies have a large Zeeman component in place of the expected Coulomb gap, which we discuss in ref. 4. This is an important result as it demonstrates that attempts to model theoretical Coulomb constants $C(\nu)$ to activation data must be reassessed, particularly at low B. SYPHERS et al [5] have also published activation data as a function of tilt angle, which they interpret in terms of the effect of the parallel field component in compressing the z-component of the wavefunction. In ref. 4 we show that this is not a dominant effect. In fact, we find that the data of ref. 5 are very well-represented by $\Delta(\theta) = \Delta(\theta=0^\circ)/\cos\theta$, which supports our finding of a <u>Zeeman</u> dominated gap, as does the linear B-dependence of Δ up to 12T reported by BOEBINGER et al [6a].

There is another point. In Fig. 1(b) it would appear from the 380mK ρ_{xx} data that the 5/3 state is significantly stronger than the 4/3 state, whereas at 30mK the 4/3 state appears to have emerged the stronger. This has been discussed as competition between states [6b]. In fact, the two states have roughly the same Δ's (table 1) and since we are now aware that they have the same σ^c_{xx} values due to the quasiparticle charge, their ρ^c_{xx} values <u>must</u> be different and the 4/3 state will develop at lower temperatures than the 5/3 state as seen in Fig. 3. We assert that 'competition' between states is simply a manifestation of their ρ^c_{xx} and Δ values and is not due to an 'interaction'.

4. Conclusion

On taking an average of G156 σ^c_{xx} values for all states in table 1 for $\theta=0^\circ$ and 47° <u>except</u> the $\theta=47^\circ$ 4/3 and 11/7 values which dramatically change character, we find that $\sigma^c_{xx}= 0.88\ (e/q)^2/h$ with standard deviation 0.14. In this average of 12 intercepts, the $\theta=0^\circ$ 3/5 and 4/7 values are low, but the remaining 9 are very consistent with our G139 study and provide compelling evidence for the use of σ^c_{xx} as a probe of e^*. This probe has been applied to investigate the destruction of spin-unpolarised ground states and we have observed a transformation from e/3 to e/5 for the quasiparticle charge at $\nu=4/3$ filling when the state is destroyed. Additionally, our comparison of normal and tilted-field activation energies demonstrates that measured gaps Δ have a large Zeeman component instead of the expected Coulomb dominance. This can be explained by $\Delta s \neq 0$ transitions from ground to excited states [4,7] ie. spin-reversed quasiparticles.

5. References

1. R.G. Clark, J.R. Mallett, S.R. Haynes, J.J. Harris and C.T. Foxon: *Phys. Rev. Lett.* 60 1747 (1988)
2. R.B. Laughlin: *Phys. Rev. B* 23 5632 (1981); *Phys. Rev. Lett.* 50 1395 (1983)
3. F.D.M. Haldane: *Phys. Rev. Lett.* 51 605 (1983)
4. P.A. Maksym, R.G. Clark, J.R. Mallett, S.R. Haynes, J.J. Harris and C.T. Foxon: this volume
5. D.A. Syphers and J.E. Furneaux: *Sol. Stat. Comm.* 65 1513 (1988)
6. G.S. Boebinger, H.L. Störmer, D.C. Tsui, A.M. Chang et al.:(a) *Phys. Rev. B* 36 7919 (1987):(b) *Phys. Rev. B* 32 4268 (1985)
7. T. Chakraborty, P. Pietiläinen, F.C. Zhang: *Phys. Rev. Lett.* 57 130 (1986)

Experimental σ_{xx} vs. σ_{xy} Scaling Diagram of the Fractional Quantum Hall Effect

J.R. Mallett[1], *R.G. Clark*[1], *J.J. Harris*[2], *and C.T. Foxon*[2]

[1]Clarendon Laboratory, University of Oxford, Oxford OX1 3PU, UK
[2]Philips Research Laboratories, Redhill, Surrey RH1 5HA, UK

1. Introduction

Recently, LAUGHLIN et al. [1] have proposed a σ_{xx} vs σ_{xy} scaling diagram for the hierarchical sequence of fractional quantum Hall effect (FQHE) ground states at filling factors $\nu=p/q$. The diagram contains 'delocalization' fixed points (denoted DF below) which determine the branching of flow trajectories to successively higher-order daughter states of the hierarchy and 'mobility' fixed (MF) points which account for the destruction of fractional states by disorder. Although flow lines in this theory correspond to T=0 length scale transformations, the effective sample size is related to an inelastic scattering length and the diagram can be probed experimentally by varying T. It is predicted [1] that DF points should scale as $(e^*)^2/\Gamma$ where e^* is the quasiparticle charge and Γ is a disorder parameter. Our objective is to establish the fixed points and flow topology experimentally to test the validity of scaling theory in the FQHE regime and the relation $e^*=\pm e/q$.

2. Procedure

As described in our preliminary work [2], we establish fixed points for the q=5, 7 and 9 hierarchical levels by studying $\sigma_{xx}(T)$ vs $\sigma_{xy}(T)$ for samples which, at the lowest temperatures reached (30mK) have (i) weak q=5 ρ_{xx}, ρ_{xy} structure with well-developed q=3 states, (ii) weak q=7 structure with well-developed q=3, 5 states and (iii) q=9 fractions resolved. For the region $1<\nu<2$ cases (i) and (ii) correspond to Fig. 1(a) and (b,c) respectively whilst for $\nu<1$ we study cases (ii), (iii) corresponding to Fig. 1(a) and (c). All samples are MBE-grown heterojunctions with Hall bar geometry. Temperature-driven conductivity flow is constructed from measured $\rho_{xx}(T)$, $\rho_{xy}(T)$ data over a grid of narrowly spaced magnetic field values, as reported for the less complicated integer regime [3].

3. σ_{xx} vs σ_{xy} Data

Results for sample G63 corresponding to the Fig. 1(a) situation ($q_{max}=5$ for $1<\nu<2$ and $q_{max}=7$ for $\nu<1$) are shown in Fig. 2 for selected temperatures from 4.2K down to 75mK. This is compared with results shown in Fig. 3 over the same temperature range for a different sample (G148) which has similar low-T ρ_{xx}, ρ_{xy} structure in the region $1<\nu<2$. Fully photoexcited G63 data corresponding to Fig. 1(b) ($q_{max}=7$ for $1<\nu<2$) are shown in Fig. 4 for temperatures between 800 and 35mK. This is compared with G139 results shown in Fig. 5 over the same temperature range, corresponding to Fig. 1(c) for the region $1<\nu<2$ for which $q_{max}=7$ also. G139 scaling data for a selected section of the region $\nu<1$ corresponding to Fig. 1(c), for which $q_{max}=9$, are shown on an expanded scale in Fig. 6. Arrows marked on the fixed field trajectories of Figs. 2–6 indicate flow as the temperature is lowered. Each fixed-field map is complimented by an isotherm plot and an example is shown in

Springer Series in Solid-State Sciences, Vol. 87
High Magnetic Fields in Semiconductor Physics II Editor: G. Landwehr

Fig. 7, which is the isotherm equivalent of the Fig. 2 mapping. DF/MF points fitted to the overall body of data are denoted ●/⊠ respectively and the experimental scaling diagram constructed from this study is shown in Fig. 8. Fixed points are numbered 1–13 in Figs. 2–6 and Fig. 8 for cross-reference.

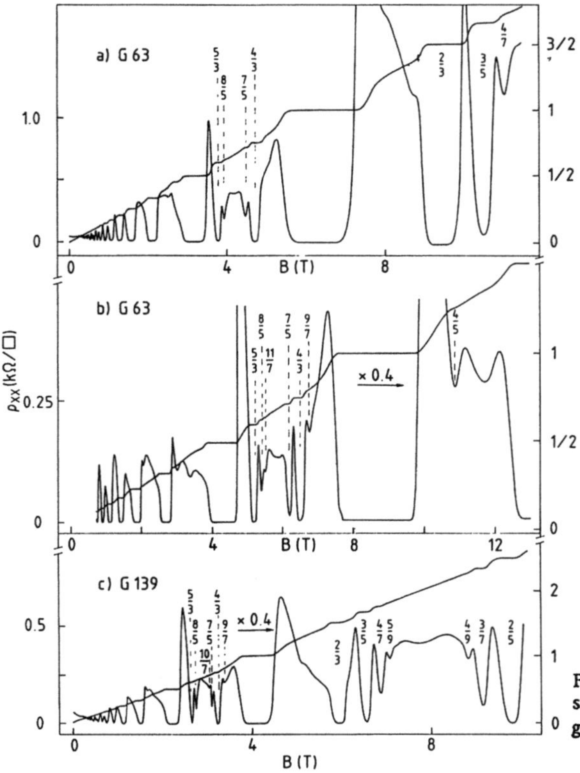

Fig. 1 Low-T transport data for successively higher-order FQHE ground states

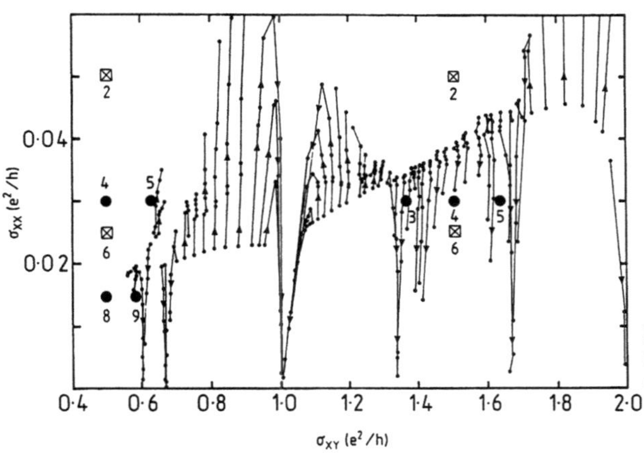

Fig.2 G63 scaling data ($n= 1.5 \times 10^{11}$ cm^{-2}) – see Fig. 1(a)

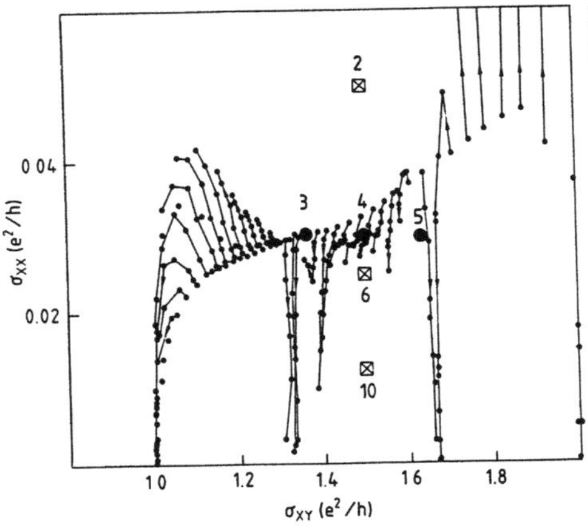

Fig.3 G148 scaling data ($n= 1.8 \times 10^{11}$ cm^{-2})

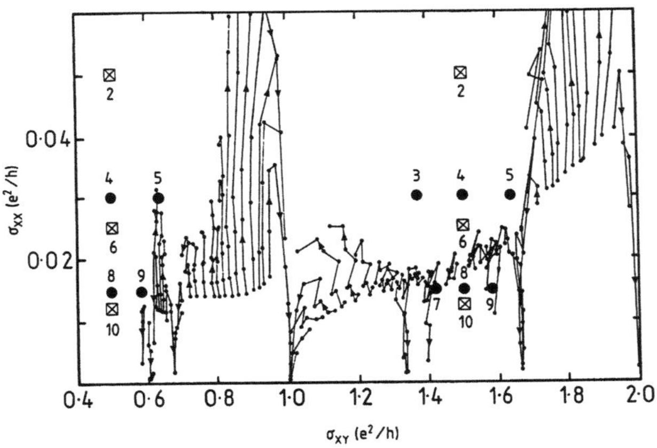

Fig.4 G63 scaling data ($n= 2.1 \times 10^{11}$ cm^{-2}) – see Fig. 1(b)

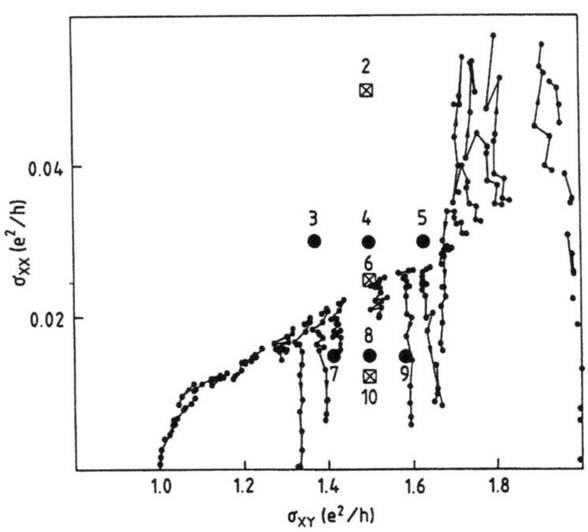

Fig.5 G139 scaling data ($n= 1 \times 10^{11}$ cm^{-2}) – see Fig. 1(c) $1<\nu<2$

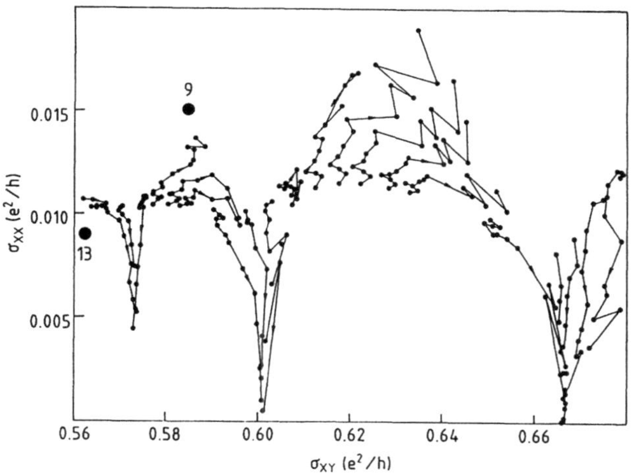

Fig. 6 G139 scaling data - see Fig. 1(c) $\nu<1$

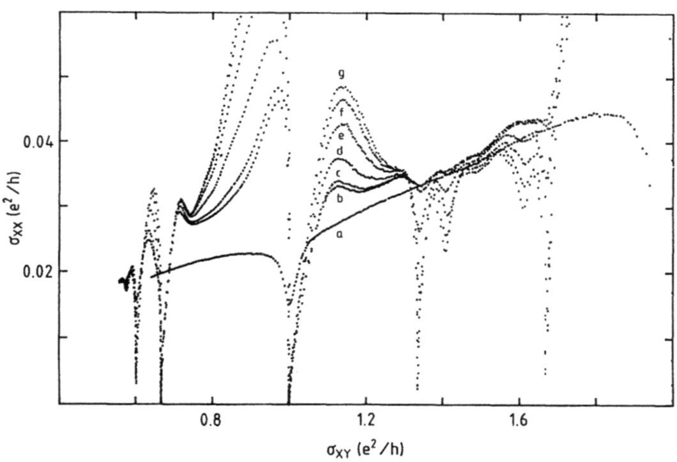

Fig. 7 G63 isotherm plot corresponding to Fig. 2 over the range 4.2K to 75mK (a–g)

4. Placement of Fixed Points

The detailed location of MF points 2, 6 and 10 in Fig. 8 at σ_{xx}= 0.05, 0.025 and 0.0125 e^2/h has been established by considering σ_{xx} values at σ_{xy}=1/2 (3/2) for some 70 sets of ρ_{xx}, ρ_{xy} vs B data for 14 samples. The critical parameter is the value of σ_{xx} for successive orders of p/q states (ie. q=5, 7 and 9) to become just observable at the lowest temperatures. The Fig. 2–6 data is consistent with these placements. For example, to observe q=3, 5 but not q=7 states, the high-T data must fall between MF points 2 and 6 of Fig. 8. Similarly, to observe q=3, 5, 7 but not q=9 fractions, the high-T data must fall between MF points 6 and 10. For the region $1<\sigma_{xy}<2$ this is strikingly verified by a comparison of Figs. 2, 4 (corresponding to Figs. 1(a) and (b) respectively). There are two factors constraining the placement of

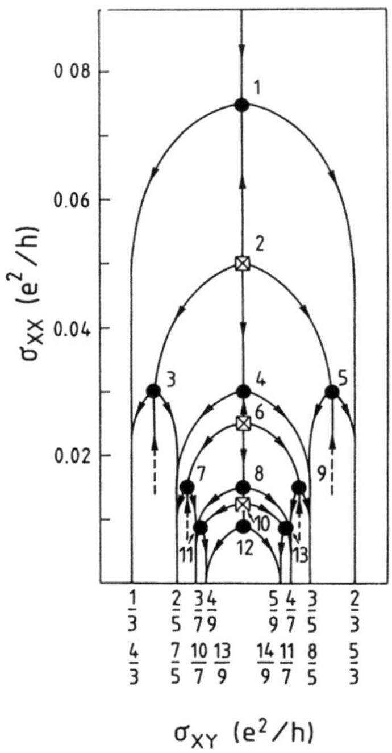

Fig. 8 **Scaling diagram**

DF points. With reference to Fig. 8 they occur at 'branching' of downward flow and represent the limit of upward flow. As a specific example, in the region $1<\sigma_{xy}<2$ of Fig. 2, branching of downward flow to σ_{xy}=4/3, 7/5 occurs at point 3 and to σ_{xy}=5/3, 8/5 at point 5. This is further verified by Fig. 3 and the isotherm plot, Fig. 7, although in Fig. 3 there is an 'anomalous' branching to the 11/7 as opposed to 8/5 state due to spin considerations discussed in these proceedings. For $\sigma_{xy}<1$, the high-T data falls below point 5 in the region $3/5<\sigma_{xy}<2/3$ in both Figs. 2 and 4 and hence point 5 must self- consistently determine the limit of upward flow in this area. Similar considerations apply to points 7, 8 and 9. In Fig. 2, for $\sigma_{xy}<1$ point 9 defines branching to σ_{xy}=3/5, 4/7 (see Fig. 1(a)). This is not shown on Fig. 4 as the 4/7 data falls outside our magnetic field range (see Fig. 1(b)). In Figs. 4 and 5, points 7 and 9 define branching to σ_{xy}=7/5, 10/7 and σ_{xy}=8/5, 11/7 although the 11/7 state in Fig. 1(b) (and hence Fig. 4) and the 10/7 state in Fig 1(c) (and Fig. 5) are only just observable, consistent with the high-T data at σ_{xy}=3/2 falling marginally below MF point 6 in Figs. 4 and 5. Points 7, 9 also represent the limit of upward flow when the high-T data falls below these points, which is verified for point 9 by the Fig. 6 data in the region $4/7<\sigma_{xy}<3/5$. Point 13 in Fig. 6 defines branching to σ_{xy}=4/7, 5/9.

5. Discussion

From the assignment of fixed points it would appear that a symmetry for the scaling diagram between σ_{xy}= 0−1 and 1−2 is compatible with the data. We have not attempted to place fixed points outside the region 1/3 (4/3)$<\sigma_{xy}<$2/3 (5/3) as the situation in the Landau level tails is complex, as can be seen from the steeply rising conductivity centred around σ_{xy}= 0.9, 1.1, and 1.9 e^2/h. There is also the complication of separate hierarchies emanating from ν=1/5, 4/5, 6/5 and 9/5. The flow pattern and fixed point scheme of LAUGHLIN et al. [1] is remarkably consistent with experiment and the parabolic trajectories characteristic of genuine scaling behaviour are clearly evident on the expanded σ_{xy} scale of Fig. 6. The (circular) DF points (3,4,5), (7,8,9) and (11,12,13) controlling branching to p/q states with q=5, 7 and 9 have been fitted to the overall body of data and are located at σ_{xx}= 0.03, 0.015 and 0.009 e^2/h respectively. DF point 1 in Fig. 8 is taken from our preliminary work [2] and must be regarded as tentative until further studies on samples which only show q=3 fractions have been carried out. The fitted DF points all scale as $1/q^2$ with respect to an (extrapolated) common integer fixed point at $\sigma_{xx} \simeq$ 0.75 e^2/h. From the prediction [1] that the DF points should scale as $(e^*)^2$, we conclude that the relation $e^*=\pm e/q$ for the quasiparticle charge is entirely consistent with the experimental scaling diagram of Fig. 8.

6. References

1. R.B. Laughlin, M.L. Cohen, J.M. Kosterlitz et al.: *Phys. Rev. B* 32 1311 (1985)
2. R.G. Clark, J.R. Mallett, A. Usher, A.M. Suckling, R.J. Nicholas, S.R. Haynes, Y. Journaux, J.J. Harris and C.T. Foxon: *Surf. Sci.* 196 219 (1986)
3. H.P. Wei, D.C. Tsui and A.M.M. Pruisken: *Phys. Rev. B* 33 1488 (1986)

The Spin Configuration of Fractional QHE Ground States in the N=0 Landau Level

P.A. Maksym[1], *R.G. Clark*[2], *S.R. Haynes*[2], *J.R. Mallett*[2], *J.J. Harris*[3], *and C.T. Foxon*[3]

[1]Department of Physics, University of Leicester, Leicester LE1 7RH, UK
[2]Clarendon Laboratory, University of Oxford, Oxford OX1 3PU, UK
[3]Philips Research Laboratories, Redhill, Surrey RH1 5HA, UK

1. Introduction

Current opinion is that FQHE ground states seen in the N=0 Landau level are fully spin polarized. No one has seriously questioned this, despite theoretical evidence to the contrary /1/ and observation of anomalies in the 1+(2>ν>1) fractions /2/. The purpose of the present work is to demonstrate that the spin behaviour is not so simple: fully polarized ground states are not favourable throughout the full range of filling factor and magnetic field. Furthermore, the measured activation energies have a large Zeeman component in place of the expected Coulomb gap.

The total energy of n electrons consists of a Coulomb term, E_c, and a Zeeman term: $E = E_c(s, B\cos\Theta) + g^*\mu_B Bs_z$. E_c depends on the spin quantum number s (spin squared = $\hbar^2 s(s+1)$)and the component of B normal to the 2DEG. For any particular s, the total energy is lowest when the z component of the spin, $s_z = -s$. Spin states other than fully polarized (s=n/2) are possible when the minimum value of E_c occurs at some s < n/2. However, they can only occur in practice if the Zeeman energy does not force the s=n/2 state to have the lowest total energy. This effect is studied experimentally and theoretically.

2. Theory

Finite size numerical studies of E_c have been done in the well known toroidal geometry /1,3/ which has the electrons in a periodic rectangular cell. As the aim is to study the physics, corrections for finite thickness and parallel component of B are not included. By making full use of symmetry,calculations are done for up to 6 electrons at 0+ (ν<1) filling. Electron-hole symmetry then gives results for up to 25-30 electrons at 1+ filling. Results for E_c are shown in fig 1. Clearly, minima in $E_c(s)$ occur at the spin assignments 1/3 ↑↑, 2/3 ↑↓, 2/5 ↑↓, 3/5 ⇵ where ↑↑, ⇵ and ↑↓ denote full, part and unpolarized. Corresponding results for the 1+ fractions are 5/3 ↑↑, 4/3 ↑↓, 8/5 ↑↓, 7/5 ⇵.

The ground states that prefer not to be fully polarized are characterized by a strong tendency for double occupation of Landau orbitals. This can be quantified by calculating a pair sum, Np, defined as the expectation value of $\Sigma_i n_{i\uparrow} n_{i\downarrow}$, where i labels the centre co-ordinate of individual orbitals ($X_i = 2\pi l^2 i/b$; b = cell height). Fig 1 shows that even for s=0 Np is small at ν = 1/q with q odd but much larger at intermediate fractions.

Springer Series in Solid-State Sciences, Vol. 87
High Magnetic Fields in Semiconductor Physics II Editor: G. Landwehr

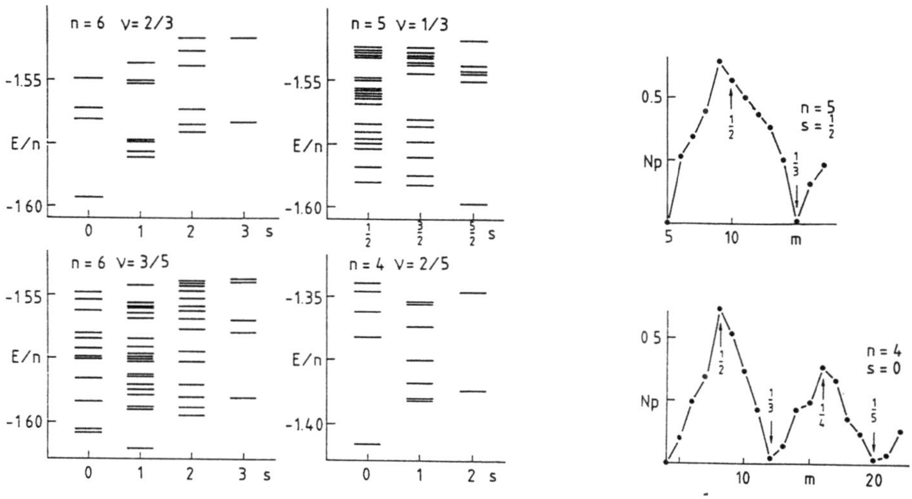

Fig 1: Computed energy levels (unit: $e^2/4\pi\varepsilon_0\ (2\pi/A)^{\frac{1}{2}}$) and pair sum Np.

The reason why double occupation is important can be seen from the pair correlation function, that is the expectation value of $\Sigma_{i\neq j}\ \delta(\underline{r}+\underline{r}_i-\underline{r}_j)/n(n-1)$. In 2nd quantization this is

$$n(n-1)\ g(\underline{r}) = \sum_{ss',j} \rho_{ss'}(j)\ G_{ss'}(j,\underline{r}) + \sum_{\substack{ss'\\ i,j\neq l,k}} g_{ijkl}\ \left\langle c^+_{si}\ c^+_{s'j}\ c_{s'k}\ c_{sl}\right\rangle ,$$

where $\rho_{ss'}(j) = \Sigma_k \left\langle n_{sk}\ n_{s'k+j}\right\rangle$.

Here ρ is the orbit centre correlation, G is the pair function for non-interacting electrons and the last term describes 'hopping'. The process $i\uparrow,\ i\downarrow \leftrightarrow i+p\uparrow,\ i-p\downarrow$ contributes to g(o,y) in proportion to - cos ($2\pi py/b$). Consequently the electrons tend to stay apart along y. In effect the system can reduce its energy by a kind of resonant mixing between paired and unpaired electrons. However this is not dominant so optimal configurations with pairs try to mix with optimal configurations of singly occupied orbitals to make the ground state. Symmetry forbids this at $\nu = 1/q$ with q odd /4,5/. Then Np is small and the ground state is insensitive to spin. For other ν, mixing is strong and E_c at $s<n/2$ is lower than at $s=n/2$. Calculations of ρ clearly illustrate these effects (Fig 2). At $\nu = 1/3$ the ground state $\rho \equiv \Sigma_{ss'}\ \rho_{ss'}$ is virtually independent of spin. Further, at small j there is almost no correlation for unlike spins. In contrast, the unlike spin correlation for the 1st excited state at $\nu = 1/3$, $s=\frac{1}{2}$ has peaks at j=0,2 and is small at j=1. This is characteristic of resonating pairs. Finally at $\nu = 2/3$ the ground state ρ are clearly different and for s=0 ρ is small at j=1 because pairing is important.

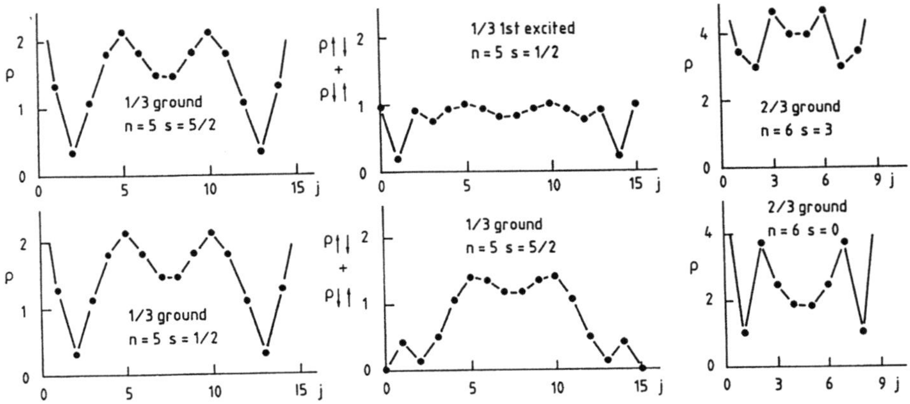

Fig 2: Orbit centre correlations ρ and $\rho_{\uparrow\downarrow} + \rho_{\downarrow\uparrow}$.

3. Experiment

The spin configuration of any fraction can be probed by changing the field at which it occurs. This is done either by tilting B and increasing its absolute value or by increasing n so the fraction occurs at higher B. Dramatic changes in conductivity are taken to indicate unpolarized states because changing the Zeeman energy then forces the system into a different spin state. Small changes imply a fully polarized state while changes at higher fields suggest a partly polarized state.

Fig 3 shows results for 2 angles and 2 concentrations. The experiments were done on high mobility GaAlAs - GaAs heterojunctions using a dilution

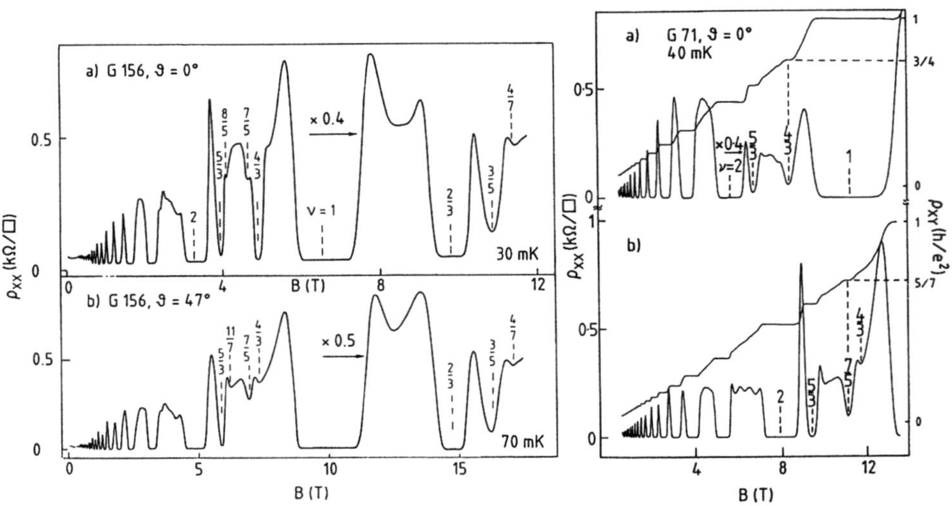

Fig 3: ρ_{xx} for sample G156 at $\Theta=0^{o}$ and 47^{o}($n=1.6\times10^{11}$ cm^{-2}) and G71 at n = (a) 2.7 and (b) 3.8 x 10^{11} cm^{-2}.

refrigerator in a 20T hybrid magnet and a.c. measurement techniques /6/. Except for an increase in the activation energy, tilting the field has little effect on the 0+ fractions. In contrast a 47° tilt almost destroys the 4/3 minima, completely destroys the 8/5, has little effect on 5/3 and 7/5 and brings out the 11/7. The effect of increasing n is similar to the effect of the tilt. This general picture has been confirmed by measurements on several other samples and is similar to anomalous behaviour reported in the literature /2/. The equivalence of changing tilt and increasing n is very important as it indicates the observed effects are due to the absolute value of B rather than its parallel component.

Activation plots for the 1+ fractions are shown in Fig 4. They confirm the trends seen in Fig 3, that is, tilt has little effect on 5/3 and 7/5 but a dramatic effect on 4/3. This is consistent with the theoretical spin assignments. Activation energies derived from Fig 4 obey the empirical relation $\Delta(47^\circ) \simeq \Delta(0^\circ)/\cos 47^\circ$. For example, at $\nu=5/3$ $\Delta(0^\circ)$ = 0.375K, $\Delta(47^\circ)$ = 0.533, $\Delta(0^\circ)/\cos 47^\circ$ = 0.550K. The implication is that Δ is linear in B and the Zeeman energy dominates. Again this is consistent with theory: Fig 1 shows that excitations with $\Delta s = \pm 1$ involve much smaller changes in E_C than excitations with $\Delta s = 0$. The behaviour of the intercepts of the activation plots is quite spectacular. As demonstrated elsewhere /7/, when σ_{xx} is measured in units of e^2/h the intercepts are close to the square of the fractional charge. Indeed at $\theta=0^\circ$ the 5/3, 4/3 intercepts are ~1/9 but at $\theta=47^\circ$ the 4/3 intercept is ~1/25! See details in /7/.

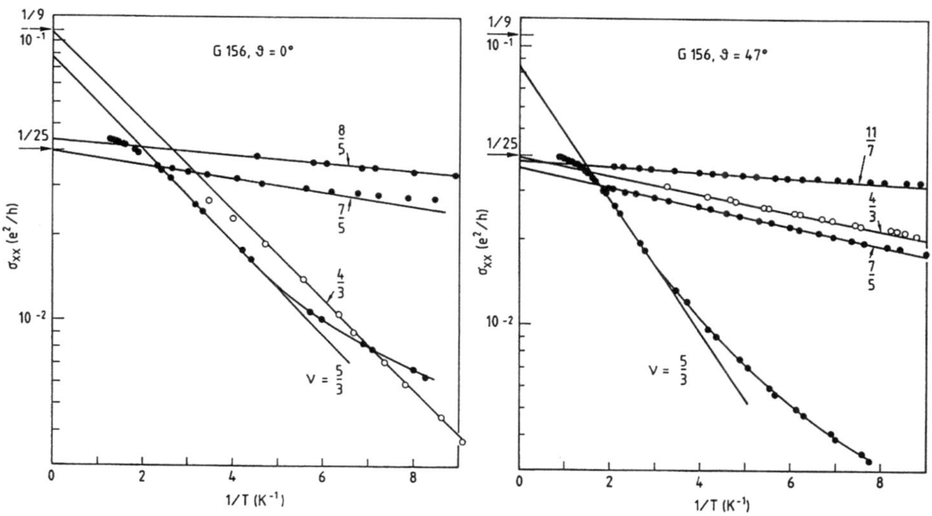

Fig 4: Activation plots for 1+ fractions in G156 at $\Theta=0^\circ$ and $\Theta=47^\circ$.

4. Conclusion

The experimental and theoretical results lead to the following spin assignments for the low field 1+ fractions: 5/3↑↑, 4/3↑↓, 8/5↑↓, 7/5↕. The cos θ scaling of the activation energies suggests that Zeeman effects dominate.

This implies that low B activation data cannot be fitted to a pure Coulomb gap and may account for observations of multiple activation energies /8/.

References

1. F.C. Zhang and T. Chakroborty: Phys. Rev. B30 7320 (1984)
2. R.G. Clark, J.R. Mallett et al: Surface Science 196 219 (1988)
   R.J. Haug, K.v. Klitzing et al: Phys. Rev. B36 4528 (1987)
3. D. Yoshioka: J. Phys. Soc. Japan 55 3960 (1986)
4. P.A. Maksym: submitted to Physical Review Letters.
5. P.A. Maksym: J. Phys. C.: Solid State Phys. 19 L247 (1986)
6. R.G. Clark, S.R. Haynes et al: submitted to Physical Review Letters.
7. R.G. Clark, J.R. Mallett et al: this volume; Phys. Rev. Lett. 60 1747(1988)
8. J. Wakabayashi, S. Kawaji et al: J. Phys. Soc. Japan 55 1319 (1986)

The Fractional Quantum Hall Effect with an Added Parallel Magnetic Field

J.E. Furneaux[1,*], *D.A. Syphers*[2,*], *and A.G. Swanson*[3,*]

[1]Naval Research Laboratory, Washington, DC20375, USA and Department of Physics, Boston University, Boston, MA02215, USA
[2]Department of Physics, Bowdoin College, Brunswick, ME04011, USA
[3]Department of Physics, Boston University, Boston, MA02215, USA
*Visiting Scientist Francis Bitter National Magnet Laboratory, MIT, Cambridge, MA02139, USA

The fractional quantum Hall states with fractions between 2 and 1 have been studied in the presence of an added parallel magnetic field. The fractional states in this range are observed to behave differently than those associated with fractions less than 1. Whereas there is a definite symmetry about 1/2 for the fractions less than 1, there is no correspondingly simple symmetry about 3/2 for the fractions between 2 and 1. The implications of these results for the hierarchical model and for calculations involving an exact numerical diagonalization for a small number of electrons are discussed.

We have been studying the effects of a parallel magnetic field on the fractional quantum Hall effect (FQHE) [1–3] for a number of years. Here we report some recent and somewhat preliminary results on the fractions between 2 and 1 taken at dilution refrigerator temperatures. We see behavior with added parallel field consistent with that seen previously [4,5], and we have some new results based to some degree on our extensive data obtained for fractions less than 1 [6]. For example, when comparing the fractions greater than 1 with those less than 1, we notice differences in the symmetries of both the resistivity maxima and minima about 3/2 and 1/2, respectively.

Figure 1 shows the effects of an added parallel magnetic field on the FQHE states associated with filling factors ($\nu = n_s h/eB$ where n_s is the two–dimensional electron gas (2DEG) density) between 2 and 1. It can be seen that all of the resistance maxima increase with added parallel field similar to the maxima observed for $1 > \nu > 1/2$ [2–6]. The maxima are not approximately constant between 3/2 and 1 as is observed for $\nu < 1/2$ [4,6]. It can also be seen that the hierarchical precedence of states between 4/3 and 7/5 seems to be broken as has been observed previously [5,7,8]. Whereas the 4/3 state is destroyed by an added parallel field, the 7/5 state is relatively unaffected similar to the behavior of the 5/3 state [9]. Therefore there is no symmetry about 3/2 which corresponds to that about 1/2. If we concede that the behavior of the $p/5$ states for $2 > \nu > 1$ is complicated and concentrate on the symmetry of the behavior of the 1/3 states there appears to be an approximate inversion symmetry about $\nu = 1$. The FQHE excitation gaps

Springer Series in Solid-State Sciences, Vol. 87
High Magnetic Fields in Semiconductor Physics II Editor: G. Landwehr

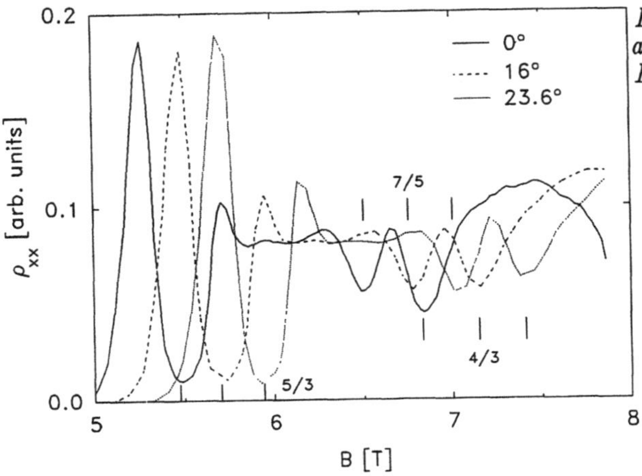

Figure 1 Magneto-resistance at 170 mK for three different $B_{\parallel}$

associated with 1/3 and 5/3 remain unchanged while the gap at 2/3 increases and the gap at 4/3 decreases as a parallel field is added.

We have examined this symmetry more quantitatively by analyzing the effects of the parallel magnetic field on the activation energies at 4/3. We calculate the first moment of the confinement wavefunction, $\langle z \rangle$, from n_s, $B_{\parallel}$, and the measured intersubband energy E_{10} using a fully self-consistent procedure [6], therefore calculating the added confinement due to $B_{\parallel}$ [2,3,10]. These calculations must be done self-consistently because perturbation theory gives grossly incorrect results for $B_{\parallel} > 1$ T. We then scale the results plotting Δ/Δ_0 as a function of $\langle z \rangle/\ell$, where Δ is the activation energy found from $\rho_{xx} = \rho_0 \exp(-\Delta/2T)$, $\Delta_0 = 0.1e^2/\epsilon\ell$, and $\ell = \sqrt{c\hbar/eB_{\perp}}$ is the cyclotron radius. We have found previously that this scaling gives a universal behavior for the effect of $B_{\parallel}$ on Δ for $\nu = 2/3$ [6]. Preliminary results confirm the inversion symmetry about $\nu = 1$ in that the scaled magnitudes of the change in Δ observed at 4/3 are the same as those at 2/3 for the same $\langle z \rangle/\ell$, but they are opposite in sign. This is a surprising result. Up to now, our parallel field investigations have relied on the hypothesis that the main part of the interaction is due to $B_{\parallel}$ causing a decrease in $\langle z \rangle/\ell$, due to the diamagnetic confinement of the wavefunction. Following the conventional FQHE theoretical calculations [11–13] this added confinement leads to a decrease in the finite size corrections to the FQHE excitation gaps of an amount which can explain the observed increases in Δ at $\nu = 2/3$ [2,3,6]. Even if there is an additional effect on the unconfined degrees of freedom due to $B_{\parallel}$ which leads to an increase in the gap energy at $\nu = 2/3$ and a decrease at 1/3, there should be a net increase in Δ/Δ_0 for the $p/3$ fractions for $2 > \nu > 1$ which is not observed. In fact if all of the $p/3$ FQHE states for $\nu < 2$ are considered, *i.e.*, the complete set for the lowest Landau level, these data indicate that there is no net effect due to the added parallel field and therefore no simple effect on Δ due to the change in $\langle z \rangle/\ell$ contrary to theoretical predictions [1-3,11–13].

In conclusion we find that the effects of $B_{\parallel}$ on the FQHE states for $2 < \nu < 1$ bring into question two important aspects of the conventional FQHE theory. First, as seen previously, the hierarchical model for the larger denominator fractions is brought into question by the behavior of the 4/3 and 7/5 states. Second, the calculations based on an exact numerical diagonalization for a small number of electrons, which lead to the finite size corrections, are brought into question in that the data are inconsistent with such a correction even when $\langle z \rangle / \ell$ is changed by up to 50% from 1.5 to 0.9 [6].

We would like to acknowledge the support of the Office of Naval Research (JEF), support of the Research Corporation through a William and Flora Hewlett Foundation Grant (DAS), support of the Exxon Educational Foundation (AGS), support from the staff of the Francis Bitter National Magnet Laboratory, and support from J. S. Brooks and his research group. The samples were provided by W. Wang of IBM and Columbia University with processing done by W. Tseng of NRL.

1. For a comprehensive review see The Quantum Hall Effect, ed. by R. E. Prange and S. M. Girvin (Springer, New York, 1987).
2. D. A. Syphers and J. E. Furneaux, Surf. Sci. 196, 252 (1988).
3. D. A. Syphers and J. E. Furneaux, Sol. St. Commun. 65, 1513 (1988).
4. R. J. Haug, K. von Klitzing, R. J. Nicholas, J. C. Maan, and G. Weimann, Phys. Rev. B36, 4528 (1987); Surf. Sci. 196, 242 (1988).
5. R. G. Clark, J. R. Mallett, A. Usher, A. M. Suckling, R. J. Nicholas, S. R. Haynes, Y. Journaux, J. J. Harris, and C. T. Foxon, Surf. Sci. 196, 219 (1988).
6. J. E. Furneaux, D. A. Syphers, and A. G. Swanson, To be Published.
7. R. G. Clark, R. J. Nicholas, M. A. Brummell, A. Usher, S. Collocott, J. C. Portal, F. Alexandre, and J. M. Masson, Sol. St. Commun. 56, 173 (1985).
8. G. S. Boebinger, A. M. Chang, H. L. Störmer, and D. C. Tsui, Phys. Rev. B32, 4268 (1985).
9. J. P. Eisenstein, private communication has seen the same lack of change in Δ at 5/3.
10. T. Ando, A. B. Fowler, and F. Stern, Rev. Mod. Phys. 54, 437 (1982).
11. A. H. MacDonald and G. C. Aers, Phys. Rev. B29, 5976 (1984).
12. F. C. Zhang and S. Das Sarma, Phys. Rev. B33, 2903 (1986).
13. D. Yoshioka, J. Phys. Soc. Jpn. 55, 885 (1986).

Plateau Formation by Force from Pinning Centres in the Fractional Quantum Hall Effect

H. Bruus [1], *O.P. Hansen* [2], *and E.B. Hansen* [2]

[1] Danish Institute of Fundamental Metrology, Building 322, Lundtoftevej 100, DK-2800 Lyngby, Denmark

[2] Physics Laboratory, University of Copenhagen, H.C. Ørsted Institute, Universitetsparken 5, DK-2100 Copenhagen Ø, Denmark

1. Introduction

Consider a QHE sample containing N electrons. As the flux is increased (decreased) relative to the mid-plateau value, $mN\Phi_0$, by a given number of flux quanta, the electron liquid responds by forming the same number V_v (V_a) of fractionally charged vortices (antivortices) [1,2,3]:

$$V_v = \frac{\Phi}{\Phi_0} - mN \qquad \text{for } \Phi > mN\Phi_0$$
$$V_a = mN - \frac{\Phi}{\Phi_0} \qquad \text{for } \Phi < mN\Phi_0 \quad . \tag{1}$$

The vortex (antivortex) states arise because hereby the system is able, in a maximal way, to preserve the favourable $\nu = \frac{1}{m}$-correlations. Let us first consider a vortex state. In each vortex region there is a deficit of $\frac{1}{m}$ of an electron charge [4]. The deficit charge is pushed out into the area between the vortices and it is precisely because of this feature that the local filling factor, ν, in this area can be maintained at the value $\frac{1}{m}$ [5]. In an antivortex state charge is removed from the area between the antivortices causing that $\nu = \frac{1}{m}$ in this area and giving rise to an excess of $\frac{1}{m}$ of an electron charge within each antivortex region.

The vortices (antivortices) are assumed to be pinned. Considering a pinned vortex with a transport current flowing by, we show that the pinning centre exerts the force $\vec{f}_P = \vec{\Phi}_0 \times \vec{j}$ on the electron liquid, where $\vec{j}$ is the two-dimensional current density, $|\vec{\Phi}_0| = \frac{h}{e}$, and where $\vec{\Phi}_0$ is parallel to the magnetic field. For an antivortex the sign is opposite, $\vec{f}_P = -\vec{\Phi}_0 \times \vec{j}$. By inclusion of the force from the pinning centres in a force balance equation for the electron liquid, we show that the Hall response, V_H, is independent of the magnetic field in an interval around $B_{\nu=\frac{1}{m}}$. The physical nature of $\vec{f}_P$ is discussed.

2. The Force from a Pinning Centre

We consider the situation where the magnetic field is larger that the mid-plateau field, but where the density of vortices is not so large that they perturb each other. Now consider one particular of these vortices. This vortex, we imagine, is pinned, but to a "pin" that we can move around. Initially the "pin" is located at some position a. Let $\vec{f}_P$ denote the force which the pinning centre exerts on the electron liquid flowing by. It is this force we must supply when, quasi-statically (carefully avoiding other vortices) we move the "pin", and thus the surrounding vortex, from a to another position b. In doing this we perform the work

Springer Series in Solid-State Sciences, Vol. 87
High Magnetic Fields in Semiconductor Physics II Editor: G. Landwehr

$$W = \int_a^b \vec{f}_P \cdot d\vec{s} \quad . \tag{2}$$

This work must be equal to the increase of energy. However, since the overall result of the quasi-static transport is the displacement of the charge $-\frac{e}{m}$ from b to a (equivalent to the displacement of the positively charged vortex from a to b), we have

$$W = -\frac{e}{m} \int_a^b \vec{E} \cdot d\vec{s} \quad . \tag{3}$$

Since (2) and (3) are valid for arbitrary a and b we conclude that

$$\vec{f}_P = -\frac{e}{m} \vec{E} \quad . \tag{4}$$

However, the electrical field, $\vec{E}$, can be related to the current density that would exist at the position of the vortex if the vortex was not there. Noting that in the region between vortices the filling factor is maintained at its canonical value $\nu = \frac{1}{m}$ [5], we conclude that the current density is given by the quantum Hall relation

$$j = \frac{1}{m} \frac{e^2}{h} |\vec{E}| \quad . \tag{5}$$

Using the geometry of ref. [2], equation (5) implies $-\frac{e}{m}\vec{E} = \vec{\Phi}_0 \times \vec{j}$, and consequently the validity of

$$\vec{f}_P = \vec{\Phi}_0 \times \vec{j} \quad . \tag{6}$$

The case where the density of vortices is so large that the individual vortex to a considerable extent experiences the velocity field of other vortices is quite complicated, and we shall not discuss it here.

Let w denote the width of the sample, it then follows from (1) and (6) that the pinning centres exert the force

$$F_P = \left(\frac{\Phi}{\Phi_0} - mN\right) \Phi_0 \frac{I}{w} \tag{7}$$

on the electron liquid.

3. Plateau Formation

Since, in the QHE-experimental set-up, the transverse current is zero, the total transverse force on the electron system must be zero:

$$0 = F_M + F_E + F_P \ , \tag{8}$$

where $F_M = -\Phi \frac{I}{w}$ is the Lorentz force, and where $F_E = +eN\frac{V_H}{w}$ is the force from the Hall field.

From (7) and (8) we obtain $\Phi \frac{I}{w} = Ne\frac{V_H}{w} + (\frac{\Phi}{\Phi_0} - mN)\Phi_0 \frac{I}{w}$ which implies

$$V_H = m\frac{\Phi_0}{e} I = m\frac{h}{e^2} I \quad . \tag{9}$$

Thus V_H comes out independent of the magnetic flux, i.e. as a Hall plateau.

4. The Physical Nature of the Pinning Force

Let us compare the two states of the electron liquid shown in fig. 1: the mid-plateau state which is without vorticity, and a vortex state containing one pinned vortex.

The velocity field and the density deviation of a vortex decay essentially exponentially [2,3]. Therefore, we assume, we can delimit a vortex region (the area within the dashed square in fig. 1(b)), outside which the flow of the liquid is unperturbed by the vortex. The linear extension of this vortex region is large compared to l , but small compared to the dimensions of the sample. The corresponding region, i.e. a region with the same location and shape, is displayed in fig. 1(a). In the mid-plateau state, fig. 1(a), the force balance for the electron liquid within the dashed region implies

$$\vec{f}_E^a + \vec{f}_M^a = 0 \quad , \tag{10}$$

where $\vec{f}_E^a$ is the electric force and $\vec{f}_M^a$ is the magnetic force. The force balance equation for the vortex region i.e. the dashed region i fig. 1(b) takes the form

$$\vec{f}_E^b + \vec{f}_M^b + \vec{f}_P = 0 \quad . \tag{11}$$

Since the average number of electrons in the vortex region is $\frac{1}{m}$ less than the average number of electrons within the dashed region in fig. 1(a) we have $\vec{f}_E^b = \vec{f}_E^a + \frac{1}{m}e\vec{E}$ which together with (4), (10) and (11) implies that

$$\vec{f}_M^a = \vec{f}_M^b \quad . \tag{12}$$

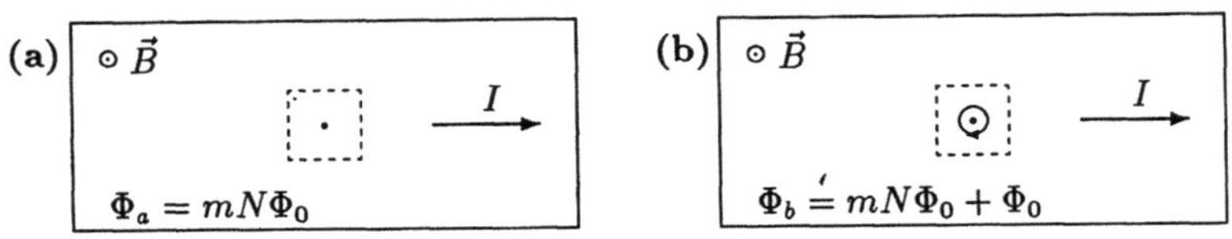

Figure 1: *The N-electron liquid at two different magnetic fluxes, Φ_a and Φ_b, in a sample with a pinning centre and a transport current I: the mid-plateau state (a) without vorticity ($\Phi_a = mN\Phi_0$) and a vortex state (b) containing one pinned vortex ($\Phi_b = mN\Phi_0 + \Phi_0$).*

Let A denote the total area of the sample. Noting that $B^b = B^a + \frac{\Phi_0}{A} \simeq B^a$ we conclude from (12) that the currents through the dashed regions in (a) and (b) are the same (a similar argument can be applied to the transition from the vortex state containing V_v vortices to a state containing $V_v + 1$ vortices). Although surprising, since the electron density is reduced in the vortex region, this conclusion can qualitatively be understood by inspection of fig. 2b.

As displayed in fig. 2c the magnetic force on the liquid in the vortex region is larger than the electric force. The physical significance of the pinning force, $\vec{f}_P$, is thus to balance the excess magnetic force.

Finally consider the case where the pinned vortex in fig. 1(b) is replaced by a pinned antivortex, the pinning centre now being a positively charged impurity. Again, by the

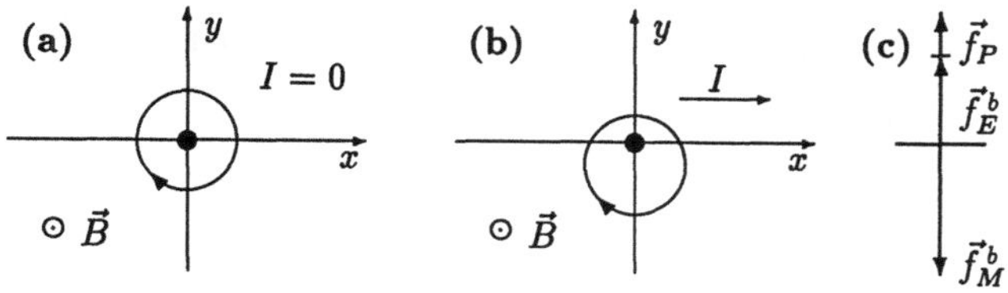

Figure 2: *In the currentless state we have rotational symmetry (a) of the density and the velocity field around the pinning centre. The passage of a transport current breaks this symmetry, and causes a displacement of the centre of the vortex relative to the pinning centre (b). Therefore the pinning centre (say a negatively charged impurity) repels the electron liquid above more strongly than below, thereby giving rise to a large current density in the upper vortex region. The forces on the electron liquid within the vortex region are shown in (c).*

same line of reasoning, we conclude that the currents through the two dashed regions are the same. However, in the antivortex case the pinning force balances an excess electric force.

5. References

1. S.M. Girvin: chap. 10 in *The Quantum Hall Effect* ed. by R.E.Prange and S.M. Girvin (Springer, Berlin, Heidelberg 1987)
2. H. Bruus, O.P. Hansen, and E.B. Hansen: Jour. Phys. C **21** L375 (1988)
3. H. Bruus, O.P. Hansen, and E.B. Hansen: to be published in Zeitschrift f. Physik
4. R.B. Laughlin: Phys. Rev. Lett. **50** 1395 (1983)
5. H. Bruus, O.P. Hansen, and E.B. Hansen: Proceedings of the 19. Int. Conf. on the Physics of Semiconductors, Warsaw August 1988

Different Behaviour of Integral and Fractional Quantum Hall Plateaus in GaAs-$Al_xGa_{1-x}As$ Heterostructures Under Back-Gating and Illumination

P.M. Koenraad[1], *F.A.P. Bloom*[1], *J.P. Cuypers*[1], *C.T. Foxon*[2], *J.A.A.J. Perenboom*[3], *S.J.R.M. Spermon*[3], *and J.H. Wolter*[1]

[1]Eindhoven University of Technology, P.O. Box 513, NL-5600 MB Eindhoven, The Netherlands
[2]Philips Research Laboratories, Redhill, Surrey RH1 5HA, UK
[3]High Field Magnet Laboratory, Toernooiveld, NL-6525 ED Nijmegen, The Netherlands

ABSTRACT

We measured the Shubnikov-de Haas resistivity of a 2-dimensional electron-gas (2DEG) in GaAs/AlGaAs heterojunctions and compared measurements in which for the same sample the electron density was increased to the same value either by back-gating or illumination. The electron-impurity interaction is different in these two cases. We observed that the integral and fractional Quantum-Hall (QH) plateaus behave quite differently in these two cases. We conclude that the integral Quantum-Hall Effect (IQHE) becomes stronger for stronger electron-impurity interaction while the fractional Quantum-Hall Effect (FQHE) is destroyed. This is consistent with the idea that the IQHE is due to electron-impurity interaction while the FQHE is based on electron-electron interaction.

1. INTRODUCTION

Recently HAUG et al. [1] showed that the width of the Quantum-Hall (QH) plateau and the asymmetry of the line shape of the Shubnikov-de Haas (SdH) resistivity ρ_{xx} in a GaAs-$Al_xGa_{1-x}As$ heterostructure can be influenced by applying a back-gate voltage. Although these experimental results could be consistently explained by assuming that the interaction between the 2DEG and attractive and repulsive scatterers is modified by a back-gate voltage, the question arises if these observations could not be also attributed to the change of the electron density.

With a positive back-gate voltage the electrons are pulled towards the back-gate (GaAs-region); the electron density increases whereas the number of scatterers remains constant. With illumination the electron density increases through both ionization of DX-centers and the excitation of carriers across the bandgap in GaAs. In this case, however, the number of positively charged donors in the AlGaAs increases and the number of negatively charged acceptors in the GaAs decreases [2]: the electrons are pulled towards the AlGaAs region. For the same electron density the interaction with positive scatterers is stronger in the case of illumination than in the case of a back-gate voltage. In this way for the same electron density we are able to study the influence of electron-impurity interaction on the IQHE and FQHE.

Springer Series in Solid-State Sciences, Vol. 87
High Magnetic Fields in Semiconductor Physics II Editor: G. Landwehr

2. EXPERIMENTS

The experiments were performed in a He^4 cryostat in a temperature range between 1.2 and 4.2 K and in magnetic fields up to 6 T and in a He^3/He^4 dilution refrigerator between 200 and 800 mK in magnetic fields up to 20 T. The sample was illuminated with a red-LED which was mounted inside the cryostat.

Fig. 1 presents measurements of ρ_{xx} of sample A under various back-gate and illumination conditions at 1.2 K. The initial electron density and mobilty are $1.22 \cdot 10^{11}$ cm^{-2} and $0.36 \cdot 10^6$ cm^2/Vs. For a back-gate voltage of +20 V these become $1.38 \cdot 10^{11}$ cm^{-2} and $0.41 \cdot 10^6$ cm^2/Vs respectively; after illumination during 9.2 μs with a red-LED these become $1.41 \cdot 10^{11}$ cm^{-2} and $0.40 \cdot 10^6$ cm^2/Vs. The sample was illuminated pulse by pulse until the same electron density was reached as for a back-gate voltage of +20 V. The most striking difference between cases b) and c) of Fig. 1 is the height of the N = 1 spin-up peak.

Measurements on sample B at 230 mk are shown in Fig. 2. The initial electron density and mobility are $1.27 \cdot 10^{11}$ cm^{-2} and $0.55 \cdot 10^{11}$ cm^2/Vs. For

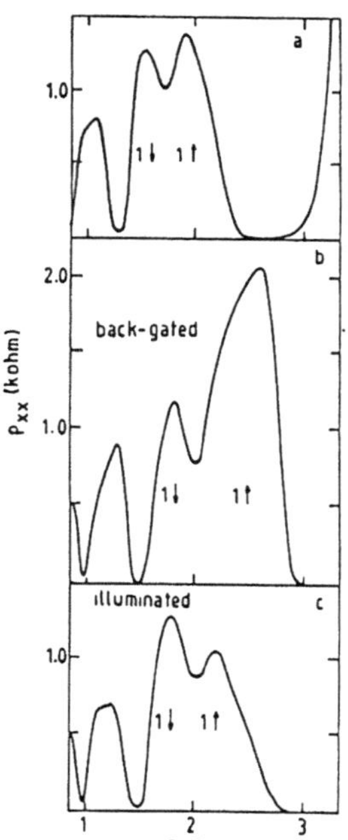

Fig. 1:

SdH-resistance of sample A at 1.2 K
a) without back-gating or illumination.
b) for a back-gate voltage of +20 V.
c) after 9.2 μs illumination.

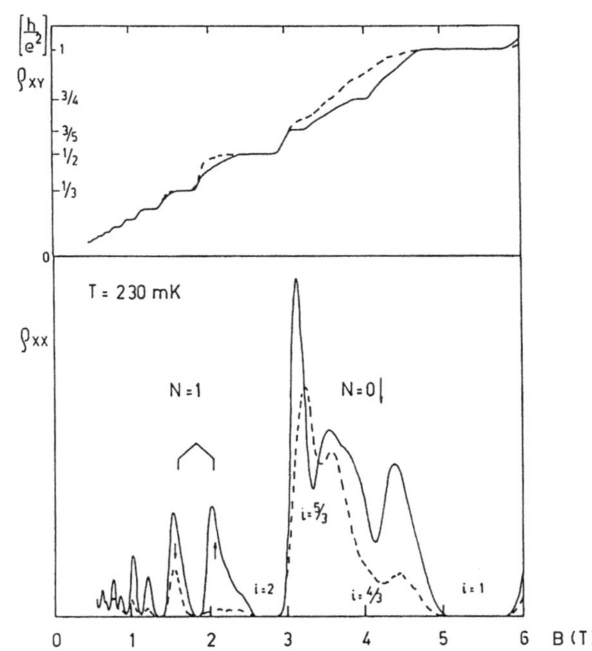

Fig. 2:

SdH- and QH resistance of sample B at 230 mK for a back-gate voltage of +25 V (solid line) and after 7.0 μs illumination (dashed line).

back-gating at +25 V these become $1.33 \cdot 10^{11}$ cm^{-2} and $0.60 \cdot 10^6$ cm^2/Vs and after 7.0 μs illumination $1.33 \cdot 10^{11}$ cm^{-2} and $0.58 \cdot 10^6$ cm^2/Vs. The solid and dashed lines represent the back-gated and illuminated sample respectively. The figure clearly shows that after illumination the spin splitted N = 1 peak and the N = 0 spin-down peak has become more asymmetric and that the IQH plateaus i = 1 and 2 have broadened. On the other hand, the fractional QH-plateaus i = 4/3 and 5/3 and the corresponding minima in ρ_{xx} have almost disappeared.

3. DISCUSSION

After illumination we observe a decreased spin-up peak and an almost constant spin-down peak, see Fig. 1. This proves that the asymmetry of ρ_{xx} is indeed caused by the interaction with the scatterers. The electron density is only effective through a better screening of the impurities and a higher Fermi energy of the electrons, [3].

Figure 2 shows that just as in Fig. 1 the asymmetry in the SdH-peaks and the width of the QH-plateaus increases when the interaction with the impurities becomes stronger. On the other hand, the FQH-plateaus and the corresponding minima in ρ_{xx} almost disappear. This is consistent with the idea that contrary to the IQHE the FQHE is an electron-electron interaction effect which is destroyed when the electron-impurity interaction becomes too strong. To our knowledge this is the first direct observation that in the same sample for the same electron density the IQHE and the FQHE respond completely differently to a change in the electron interaction with impurities; thus confirming that the IQHE and FQHE have a different physical origin.

4. CONCLUSION

In conclusion, we proved that the asymmetry in the SdH peaks is indeed dependent on the electron-impurity interaction. Furthermore, we showed for the first time in experiments carried out with the same sample that the IQHE and FQHE have a different physical origin.

REFERENCES

[1] R.J. Haug, K. v. Klitzing and K. Ploog, Phys. Rev. B35, 5933(1987)

[2] J.J. Harris, D.E. Lacklison, C.T. Foxon, F.M. Selten, A.M. Suckling, R.J. Nicholas and K.W.J. Barnham, Semi. Sci. Tech. 2, 783(1987)

[3] P.M. Koenraad, F.A.P. Blom, C.T. Foxon, G. Weimann and J.H. Wolter, to be published.

Temperature Dependence of Transport Coefficients of 2D Electron Systems at Very Small Filling Factors

R.L. Willett[1], *H.L. Stormer*[2], *D.C. Tsui*[3], *L.N. Pfeiffer*[2], *and K.W. West*[2]

[1]Massachusetts Institute of Technology, Cambridge, MA 02139, USA
[2]AT&T Bell Laboratories, Murray Hill, NJ 07974, USA
[3]Princeton University, Princeton, NJ 08544, USA

The temperature dependence of the transport coefficients was measured in three GaAs/AℓGaAs heterostructures of different quality at small Landau level filling factors ν. The highest quality specimen, as determined by the existence of strong fractional quantum Hall features, exhibits activated conduction as the filling factor approaches $\nu=0$. These characteristics are discussed in terms of magnetic freeze-out and Wigner solid formation.

The nature of two-dimensional electron systems in high magnetic fields at very small filling factors stands as an intriguing question because several electronic configurations may be realized at these physical extremes. The hierarchy of states of the fractional quantum Hall effect [1,2] may extend to very small ν but is predicted to be limited in real 2D systems. At small ν disorder may prohibit formation of these condensed electron fluids and magnetic freezeout or localization may predominate. It is also postulated that for sufficiently low disorder, crystallization to a 2D Wigner solid may occur [3].

We have extensively studied the temperature dependence of transport coefficients at small ν in several GaAs/AℓGaAs heterostructures. Two specimens were of similar densities but had low and high relative mobilities, and one high mobility low-density sample allowed experiments down to $\nu \sim 1/18$. For the two high-density samples we observe rather different temperature dependences at small ν. Both of these temperature dependences are in turn unlike the striking activated behavior observed in the high quality lowest-density specimen that displays very well developed FQHE features. These results are discussed with respect to magnetic localization or Wigner crystallization as possible explanations.

Our samples were single interface GaAs/AℓGaAs heterostructures. Their densities n and mobilities μ were: sample #1, $n \approx 1.5\times10^{11}$ cm^{-2} and mobility $\mu \approx 5\times10^{5}$ cm^{2}/Vsec, sample #2, $n \approx 1.6\times10^{11}$ cm^{-2} and $\mu \approx 5\times10^{6}$ cm^{2}/Vsec, and sample #3, $n \approx 4.5\times10^{10}$ cm^{-2} and $\mu \approx 1.7\times10^{6}$ cm^{2}/Vsec. A dilution refrigerator provided temperatures down to 80 mK in magnetic fields up to 31 T. Samples were submerged in He^{3}/He^{4} mixture adjacent to a calibrated carbon resistor with known magnetoresistance.

The longitudinal resistivity ρ_{xx} at four different temperatures for each of the three samples is displayed in Fig. 1. All three samples show marked temperature sensitivity at small ν. The sample of lowest mobility (#1) shows no FQHE minima at filling factors less than $\nu = 1/3$, instead ρ_{xx} increases dramatically as the temperature is lowered. In sample #2 the fractional minimum at $\nu = 2/7$ is well developed and at smaller filling factors ρ_{xx} increases with decreasing temperature. The lowest density

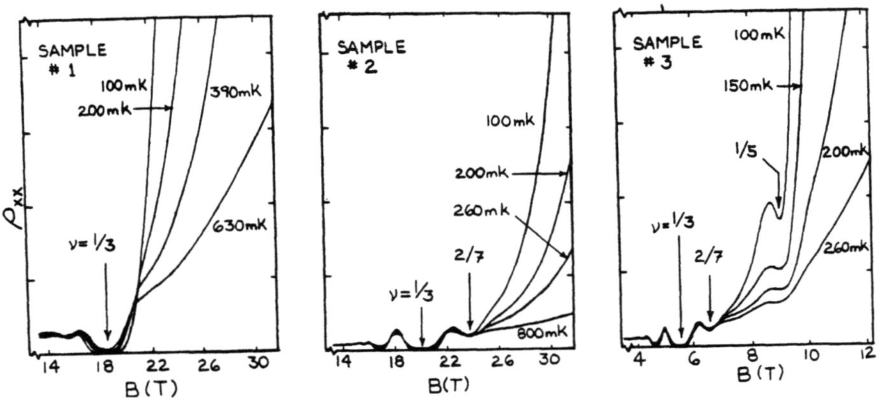

FIGURE 1. Longitudinal resistivities ρ_{xx} of three different two-dimensional electron systems at four temperatures.

sample #3, which allows examination of ρ_{xx} down to $\nu \approx 1/18$, develops well defined minima at $\nu = 2/7$ and $\nu = 1/5$; see Fig. 1. As in sample #2, ρ_{xx} increases strongly with decreasing temperature at filling factors less than $\nu = 2/7$.

In order to compare the temperature dependence at small ν for these three different 2D systems, we determined ρ_{xx} for five different filling factors in the range of observed temperature sensitivity. The results are shown in Fig. 2 on a semilog scale of ρ_{xx} versus T^{-1}. In the lowest mobility sample, the temperature dependence is not simply activated. Sample #2 likewise does not show an activated temperature dependence. It can be described phenomenologically by $\rho_{xx} = \rho_0 \exp(T_0/T)^{1/2}$ for the lowest three filling factors displayed.

In stark contrast, activated conduction is seen in the lowest-density sample #3. This is the heterostructure that displays a well developed series of FQHE minima at small ν. Activated conduction, described by $\rho_{xx} = \rho_0 \exp(E_a/T)$, is observed throughout the small filling factor range, from near $\nu = 2/7$ to the smallest observable ν. Activation over almost two decades of resistance is present. We further examined this sample by varying the 2D carrier density with a backside gate and determined activation energies, E_a, at different ν. These results are displayed in Fig. 3. As ν decreases, E_a increases for all three densities but with a peculiar crossover occuring at $\nu \sim 0.19$.

Although all three samples display marked increases in longitudinal resistivity at small ν, the different functional temperature dependences of ρ_{xx} on T suggest that different conduction mechanisms are present. Transport in sample #1 and sample #2 appears to be dominated by hopping conduction at small ν. The exponential $(T_0/T)^{1/2}$ dependence of sample #2 is suggestive of a variable range hopping mechanism proposed by ONO [4]. The T-dependence of ρ_{xx} of sample #3, which also shows a robust series of FQHE features down to small ν, differs substantially from samples #1 and #2. Its simple activated longitudinal resistance and its ability to support tenuous Coulomb interactions of high-order FQHE states suggest that sample #3's low ν behavior may reflect a different process. We therefore discuss the results from sample #3 with respect to magnetic freezeout or formation of a pinned 2D Wigner lattice.

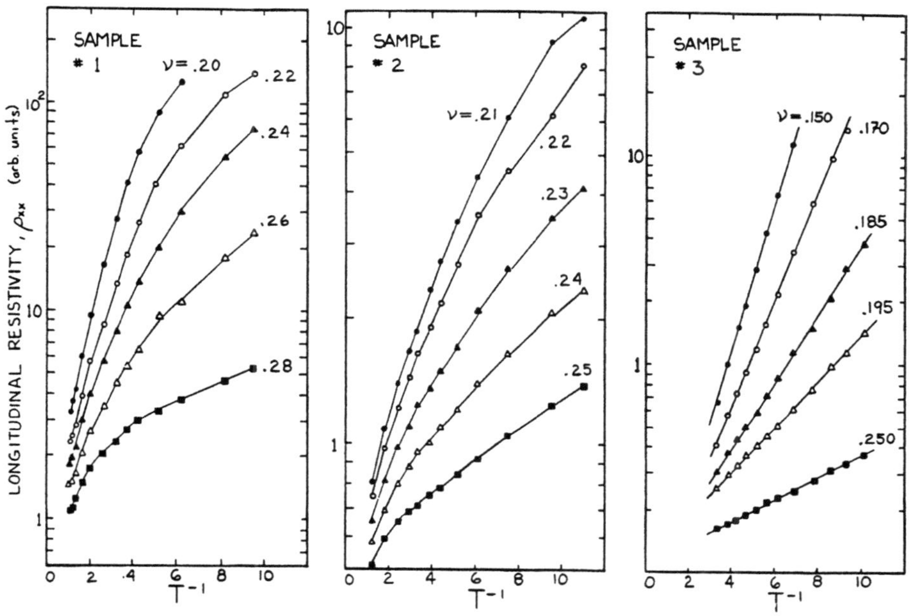

FIGURE 2. Temperature dependence of longitudinal resistivity ρ_{xx} at five different filling factors for the three samples of Figure 1.

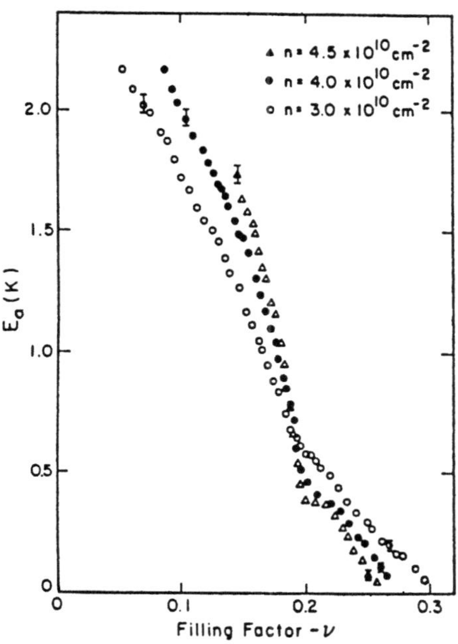

FIGURE 3. Filling factor dependence of activation energy E_a at three different carrier densities of sample #3.

In general activated behavior is not inconsistent with magnetic freezeout, and the energy scale of Fig. 3 is in agreement with a simple assessment of Landau level broadening from mobility, $\Gamma = h/\tau \sim 1$ K. However, the roughly linear behavior of E_a vs. ν (Fig.3) implies a peculiar density of states in the single particle localization model. The crossover behavior of E_a vs. ν does not have an obvious explanation within a localization model. Yet, improved screening at higher densities may shift the onset of temperature dependence to smaller ν, and this may be responsible for the density dependence of $E_a(\nu)=0$ in Fig. 3.

We also consider formation of a pinned Wigner lattice as an interpretation for our results. Theoretical estimates for the characteristic energies of a Wigner solid [5] far exceed the energy scale observed in our experiment. However, carrier correlation effects have not been included and are expected to lead to a considerable reduction of these energies. The trend of E_a as ν approaches the classical limit, $\nu \to 0$, is consistent with the monotonic energy increase expected for Coulomb energies in a Wigner solid, $E \sim \sqrt{n}$. Similarly, it is suggested that the onset of lattice formation should move to smaller ν as n increases [6], as is observed in $E_a(\nu) = 0$ for three different n. These results are remarkably consistent with features a Wigner lattice may possess.

In summary, our results show a marked difference in temperature dependence of ρ_{xx} at small Landau level filling factors for three samples. For the highest quality specimen, Wigner lattice formation is not inconsistent with many of these findings. However, magnetic freeze out can not be ruled out as the origin of these transport characteristics.

1. D.C. Tsui, H.L. Stormer, and A.C. Gossard. Phy. Rev. Lett. 48, 1559 (1982).
2. F.D.M. Haldane, Phys. Rev. Lett. 50, 1395 (1983).
3. Y.E. Lozovik and V.I. Yudson, JETP Lett. 22, 11 (1975).
4. Y. Ono, Jour. Phys. Soc. Jap. 51, 237 (1982).
5. Y. Yoshioka and P.A. Lee, Phys. Rev. B27, 4985 (1983).
6. P.M. Platzman, in The Physics of the Two-Dimensional Electron Gas, NATO ASI Series edited by T.F.J. Devreese and F.M. Peeters (Plenum Press, NY, London, 1987), p.87.

Electron Solid Formation at a Modulation Doped Heterojunction in a High Magnetic Field

F.I.B. Williams [1], D.C. Glattli [1], G. Deville [1], B. Etienne [2], E. Paris [2], and E.Y. Andrei [3]

[1] Service de Physique du Solide et de Résonance Magnétique, CEN-Saclay, F-91191 Gif-sur-Yvette Cedex, France
[2] Laboratoire de Microstructure et de Microélectronique, CNRS, 196, Av. H. Ravera, F-92220 Bagneux, France
[3] Department of Physics, Rutgers University, Piscataway, NJ 08854, USA

By means of a radiofrequency spectroscopy experiment which tests for shear rigidity by detecting a gapless magneto-phonon excitation branch, we are able to affirm that electrons at a high quality modulation doped GaAs/GaAℓAs heterojunction form a solid at low temperatures and high fields. We determine the phase diagram for Wigner-Seitz to Bohr radius ratio $1.6 < r_s < 2.5$; extrapolating it gives a zero temperature critical Landau level filling factor $\nu = 0.24 \pm 0.05$ and an infinite field melting temperature slightly lower, but of the same density dependence, as that for the classical Coulomb solid.

In zero magnetic field, the ground state of a two-dimensional electron gas at an interface of present GaAs/GaAℓAs heterojunctions is Fermi liquid like : the ratio $r_s = a/a_B$ of the characteristic Coulomb correlation energy $V_c = e^2/a$ to zero point kinetic energy $\hbar^2/ma^2$ is too low ($r_s \sim 2$) to expect a crystalline state (or Wigner Solid); $a = (\pi n_s)^{-1/2}$ is the Wigner-Seitz radius, n_s the density. But a magnetic field H channels the zero point energy into a cyclotron motion of extension $\ell_H = (\hbar c/eH)^{1/2}$, the electron delocalization can then be expressed simply by the spatial degeneracy of the cyclotron orbits centres, each of which shares an average $\nu = 2(\ell_H/a)^2$ electrons. One observes first a series of correlated quantum liquids (fractional quantum Hall effect) [1] and at higher field (smaller overlap of wavefunctions $\ell_H \ll a$ and larger spatial degeneracy $\nu \ll 1$) the orbit centres are expected to form a crystal [2] for $\nu < \nu_m$ if the temperature is low enough ($T \ll T_{mc} = V_c/\Gamma_{mc}$ the melting temperature of the classical Coulomb solid, $\Gamma_{mc} = 127 \pm 3$). The observation of such a Wigner solid is reported here [3], in high magnetic field, low temperature, in a novel type of high quality heterojunction.

The only intrinsic macroscopic property which distinguishes a solid from a liquid is its familiar elastic response to shear characterised by the shear modulus μ which vanishes at melting. Our experiment is then designed to test for the existence of elastic transverse (or shear) waves characterized by the gapless transverse phonon branch:

$$w_t(q) = (\mu/mn_s)^{1/2}q \qquad (1)$$

distinct from the longitudinal branch $w_\ell(q)$ which is common to both solid and liquid phases ($w_\ell \sim w_p = (\pi n_s e^2/\epsilon m)^{1/2} q^{1/2}$ the 2D plasmon frequency). In magnetic field shear and compressional modes are mixed by the Lorentz force : the upper hybrid branch presents a gap at $q = 0$: $w_+ = \left(w_c^2 + w_p^2\right)^{1/2}$ while the lower one $w_- = w_t w_p / w_c \sim u^{1/2} q^{3/2}$ remains gapless and vanishes at melting (here $w_c = eH/mc \gg w_t$).

Experimentally a low power (< 50pW) TEM wave is transmitted from a swept frequency (.01-2.GHz) source through a 50 Ohm 8μm wide stripline which meanders parallel to, and at a distance h=380nm from, the electron plane. The latter interacts with the spatial harmonics of the longitudinal field imposed by the meander line geometry. When the swept frequency crosses $w_-(pq_o)$ the small fraction of power absorbed is detected $q_o = 2/16\mu m$ and $1 < p < 15-2/hq_o$). In addition to r.f. measurements simultaneous low current density $j \sim .1 nAcm^{-1}$) transport measurements are made both to characterize the sample and to investigate the ρ_{xx} behaviour near melting (up to 20 MOhm $\square^{-1}$).

The samples are MBE grown bi-layer planar doped heterojunction with high mobility and low density (typically 1.2 $10^6 cm^2 V^{-1} s^{-1}$ at $n_s = 7\ 10^{10} cm^{-2}$) [4]. The first Si doping plane at 100nm from the inversion layer allows for low density, the second one at larger distance ~360nm cancels surface Fermi level effect without reducing the mobility. The sample is mounted on the r.f. meander line in the dilute phase in the plastic mixing chamber of a dilution refrigerator which gives 40mK in magnet environment.

A series of well resolved resonances are observed between .5 to 1.5Ghz at low temperature (T < .15Tmc) and high field ($\nu < 0.1$). The frequencies fit well a $(pq_o)^{3/2}$ law for $5 < p < 12$ and their absolute values compare well to $w_-(pq_o)$ when the parameters of the classical Coulomb solid of same density are used [5], as expected when $\ell_H/a \ll 1$. When the temperature increases or the magnetic field is lowered, the resonances become broader and their frequency decreases, Fig.1. Then they vanish. This is interpreted as the melting of the magnetically induced Wigner solid. The melting values (T_m, H_m) define the boundary of a phase diagram. Fig.2 shows that the results for different densities can be put in a single diagram with the universal units $t = T/T_{mc}$ and $\nu = en_s/H$ as expected for small r_s. A first order analysis of the data shows that the experimental values can be put on a single line (dashed line). The zero temperature extrapolated critical filling factor is $\nu = .24 \pm 05$. Three points around $\nu = 1/3$ are to be confirmed.

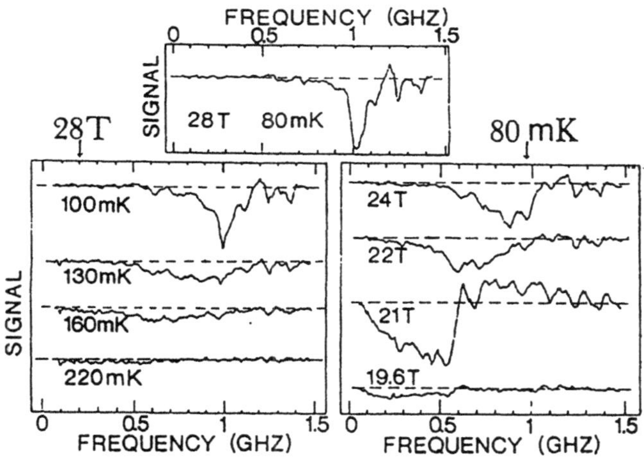

Figure 1 : *derivative of the absorption spectrum. A low frequency back-backgate voltage modulation (100Hz) is used. The density is n_s=.94.10^{11} cm^{-2} and ν = 3.92/H(Tesla) . The reduced temperature is t=T(mK)/560.*

Transport measurements are not a specific way to detect the solidification, as expected. ρ_{xx} shows no discontinuity at melting but strongly increases in the solid phase, when T is lowered or H is increased, see Fig. 3. Melting values of ρ_{xx} are in the range of 40 to 90 kOhm at high densities $n_s \sim 10^{11}$ and ~300kOhm for 4.5 10^{10}.

Until now the search for a nonlinear behaviour, related to the pinning of the Wigner solid, in the characteristic $V_{xx}(I)$ at electric

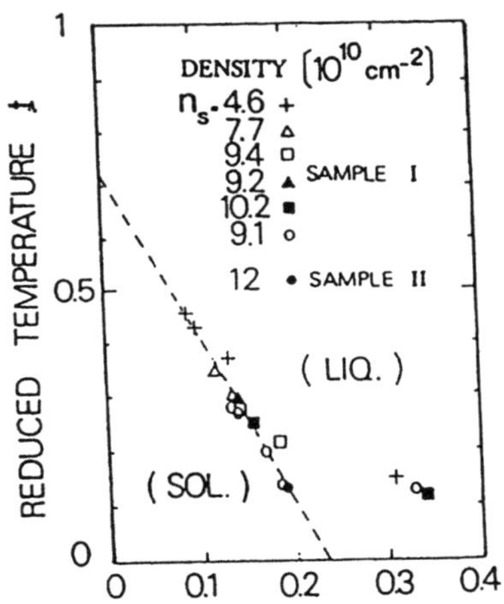

Figure 2 : *Phase diagram*

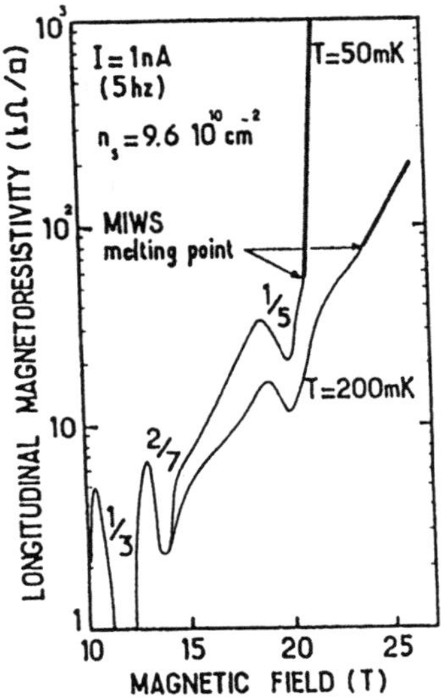

Figure 3 : ρ_{xx} *measurements at high field and low temperature. The arrows point the Magnetically Induced Wigner Solid (MIWS) transition determined by the onset of the radiofrequency resonance* $w_-(q)$ *directly related to shear rigidity. The heavier lines correspond to the solid phase.*

field down to $10\mu Vcm^{-1}$ well in the solid phase (for instance : v = .14 and t = .15) was negative. This indicates that the current still flows through remaining liquid droplets or along the sample boundaries, or that the depinning electric field threshold is still lower.

We are grateful to members of the CNRS-MPI High Field Magnet Laboratory at Grenoble, particularly J.C Vallier, H.J. Schneider-Munteau and J.C. Picoche. Inestimable technical help of J. Verrier and C. Chaleil and cryogenic counsel of A. Benoît and P. Pari were greatly useful. One of us (E.Y.A) benefits from NSF contract N° DMP87-05-977. A NATO grant facilitated international collaboration.

References

[1] D.C. Tsui, H.L. Störmer and A.C. Gossard, Phys. Lett. 48, 1559 (1982)

[2] J. Durkan, R.J. Elliot and N.H. March, Rev. Mod. Phys. 40, 812 (1968)

P.K. Lam and S.M. Girvin, Phys. Rev. B30, 473 (1984)

D. Levesque, J.J. Weiss and A.H. MacDonald, Phys. Rev. B30, 1056 (1984)

Yu. E. Lozovik, V.M. Fartzdinov and B. Abdullaev, J. Phys. C, L807 (1985)

[3] E.Y. Andrei, G. Deville, D.C. Glattli, F.I.B. Williams, E. Paris and B. Etienne, Phys. Rev. Lett. 60, 2765 (1988)

[4] B. Etienne and E. Paris, J. Physique (Paris) 48, 2049 (1987)

[5] L. Bonsall and A.A. Maradudin, Phys. Rev. B15, 1959 (1977)

K. Maki and X. Zotos, Phys. Rev. B28, 4349 (1984)

Part III

Heterostructures and Superlattices: Transport and Electronic Structure

Magnetic Oscillation of Many-Body Effects in Two-Dimensional Systems

T. Ando

Institute for Solid State Physics, University of Tokyo, 7-22-1 Roppongi, Minato-ku, Tokyo 106, Japan

A review is given of effects of strong magnetic fields on electron-electron interactions in two-dimensional systems with a special emphasis on their oscillations caused by the singular density of states. It includes the oscillatory enhancement of the g factor due to exchange effects, self-consistent screening which gives rise to a strong oscillation of broadening of Landau levels and cyclotron resonance, and oscillations of photoluminescence energy found recently.

1. Introduction

In a two-dimensional (2D) system in strong magnetic fields the orbital motion of electrons is completely quantized and the energy spectrum comprises discrete Landau levels. This singular density of states gives rise to varieties of interesting phenomena including the quantum Hall effect. In this paper a review is given on roles played by high magnetic fields in modifications of many-body electron-electron interactions, especially their magnetic oscillations.

The exchange interaction is a typical many-body effect and causes the enhancement of spin splitting or g factor and sometimes leads to a paramagnetic-ferromagnetic transition. In 2D systems in high magnetic fields, the exchange effect gives rise to a huge oscillatory enhancement of the g factor predicted theoretically and also observed experimentally in Si inversion layers and GaAs/AlGaAs heterostructures. This phenomenon is briefly discussed in Sec. 2.

The screening effect varies dramatically as a function of the Fermi-level position. When the Fermi level lies at the center of a Landau level, the screening is dominated by virtual low-energy intra-Landau-level excitations. Such intra-Landau-level screening depends on the broadening of the Landau level. Since scattering potentials are strongly screened by electrons in the system, the broadening and the screening should be determined self-consistently. When it lies in the tail region of Landau levels, on the other hand, only inter-Landau-level excitations are possible and the screening effect is reduced considerably. This gives rise to an oscillation of the broadening of Landau levels, which is discussed in Sec. 3.

Springer Series in Solid-State Sciences, Vol. 87
High Magnetic Fields in Semiconductor Physics II Editor: G. Landwehr

A well-known effect of many-body interactions is the reduction of the photoluminescence energy or the band gap in modulation-doped quantum wells. The exchange and correlation among electrons and interaction of a photo-excited hole with electrons reduce the photoluminescence energy considerably. In the presence of strong magnetic fields, such electron-electron and electron-hole interactions are modified, causing an oscillation of the luminescence energy. It will be shown in Sec. 4 that the oscillation is dominated by that of the self-energy shift of a hole.

2. Exchange Enhancement of g Factor

Suppose that the number of electrons with ↑ spin is larger than that of electrons with ↓ spin because of the spin-Zeeman splitting. Electrons with ↓ spin feel more repulsive force than those with ↑ spin, because electrons with the same spin cannot be close to each other due to the Pauli exclusion principle and electrons feel stronger repulsive force from those with different spins. The original splitting is enhanced by electron-electron interactions. In strong magnetic fields the difference in the number of electrons with ↑ and ↓ spins depends strongly on the position of the Fermi level. When the Fermi level lies at the middle of the gap between ↑ and ↓ levels as is schematically shown in Fig. 1, the difference becomes a maximum and the g factor has a maximum value larger than 2. When it lies midway between Landau levels with different level index N, the difference vanishes and the g factor has a minimum value 2.

This oscillatory g factor was first calculated in 2D systems in Si inversion layers [1]. It was shown that the g factor when enhanced can be as large as $3 \sim 4$ depending on magnetic fields and broadening of Landau levels. There have been various experimental attempts to determine the enhanced g factor. Most of such works have employed the fact that the Landau level separation and the spin splitting can be varied independently in the system. The spin splitting is determined by the total magnetic field, whereas the

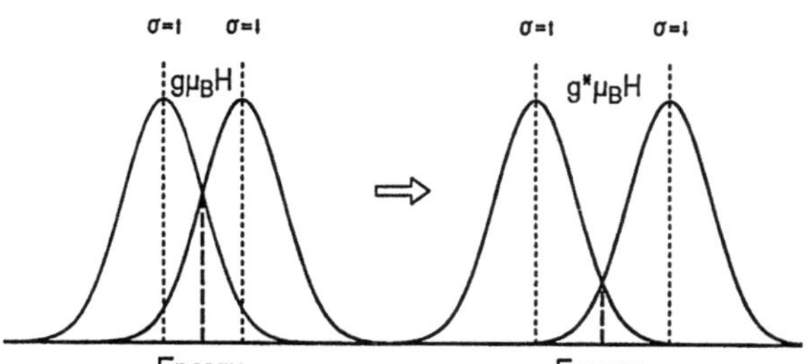

Fig. 1 A schematic illustration of density of states of spin-split Landau levels and enhancement of the spin splitting due to exchange effect

Landau level separation is determined by the magnetic field normal to the system. First experimental results which suggested the enhancement were provided by Fang and Stiles in 1968 [2]. Later, Englert and von Klitzing have provided convincing experimental proof that the spin splitting does show oscillatory behavior [3].

3. Broadening of Landau Levels – Self-Consistent Screening

In the self-consistent Born approximation (SCBA), which is known to work quite well in explaining various characteristic features of the transport in 2D systems [4], the broadening is given by

$$\Gamma_N^2 = 4\sum_{\mu} \int dz N_{\mathrm{i}}^{\mu}(z) \int \frac{dq}{(2\pi)^2} \left[\frac{v^{\mu}(q,z)}{\epsilon(q)}\right]^2 J_{NN}(q)^2, \tag{1}$$

where $N_{\mathrm{i}}^{\mu}(z)$ describes the distribution of the μth kind of impurities in the z direction perpendicular to the system, $v^{\mu}(q,z)$ is its potential, and $J_{NN}(q)$ is the usual form factor representing the Landau-level wave function. The dielectric function $\epsilon(q)$, calculated consistently within SCBA, shows a strong dependence on the broadening and also on the Fermi-level position. Therefore, the broadening and the dielectric function should be determined in a self-consistent way.

Such a self-consistent calculation was first performed in space-charge layers on Si surfaces and reported at a previous Conference in Würzburg in 1976 [5]. The calculation has shown that the level broadening exhibits a strong oscillatory behavior: When the Fermi level lies at the center of Landau levels, the broadening is small. When the Fermi level lies in the tail region, however, the broadening becomes even larger than the separation of adjacent Landau levels, *i.e.*, the density of states of Landau levels overlap with each other appreciably. Figure 2 schematically shows this oscillation. The broadening of the cyclotron resonance (CR), on the other hand, has turned out to show only a weak oscillation.

Similar calculations were recently performed for GaAs/AlGaAs single heterostructures [6,7]. Figure 3 gives an example of the results. The oscillation of the broadening Γ_N is clearly seen and appreciable also for the CR broadening. In the case of the heterostructures, the broadening has turned out to be strongly dependent on the thickness of the spacer which separates ionized donors from 2D electrons [7].

Figure 4 shows the broadening of the lowest Landau level as a function of the spacer thickness. When the thickness exceeds ~100 Å, the broadening is dramatically reduced. This means that scatterings from ionized donors play negligible roles and electron-electron interactions dominate the properties, when the spacer thickness is sufficiently large. Note that the relaxation time in the absence of a magnetic field does not vary so much as a function of the spacer thickness.

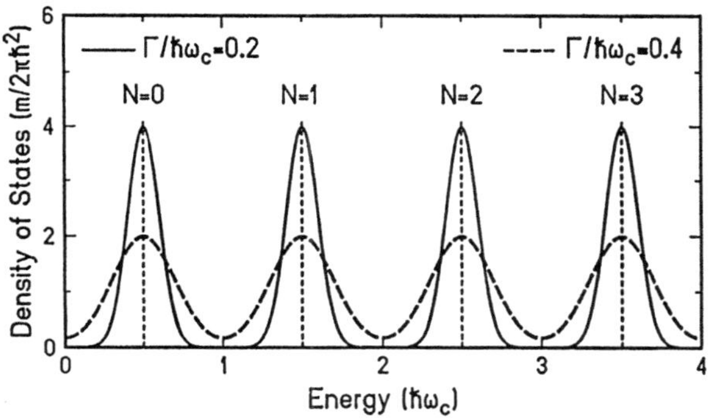

Fig. 2 A schematic illustration of the density of states in a 2D system in strong magnetic fields. When the Fermi level lies in the central part of the broadened Landau level, the broadening is small. When it lies in the tail region of the density of states, the broadening becomes large, causing overlapping of the density of states of adjacent Landau levels

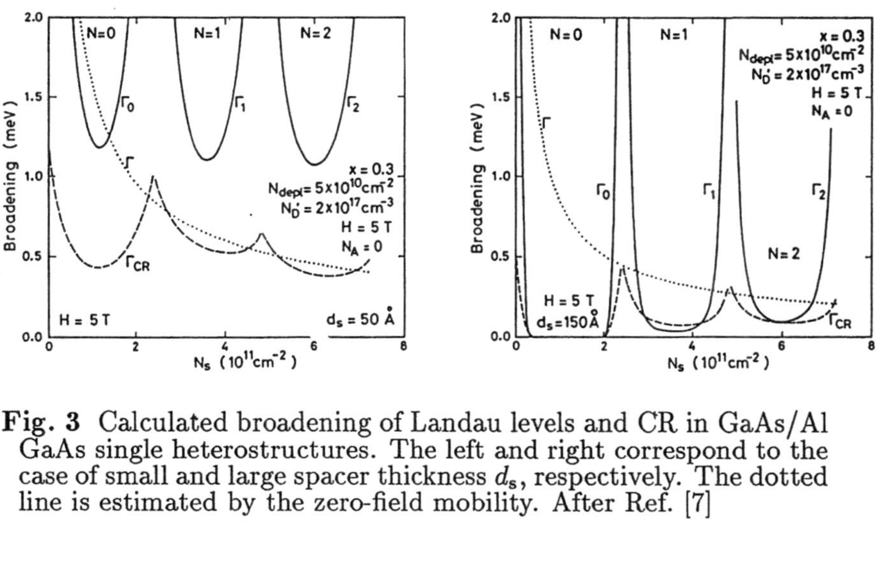

Fig. 3 Calculated broadening of Landau levels and CR in GaAs/Al GaAs single heterostructures. The left and right correspond to the case of small and large spacer thickness d_s, respectively. The dotted line is estimated by the zero-field mobility. After Ref. [7]

Experimental evidence of the oscillation of the broadening was provided by Kukushkin and Timofeev quite recently in Si inversion layers [8]. They have observed photoluminescence spectra corresponding to recombination of electrons in the inversion layer and holes trapped in acceptors in the vicinity of the Si/SiO_2 interface. The experiments have shown a beautiful Landau level density of states, whose width exhibits an oscillation consistent with the theoretical prediction.

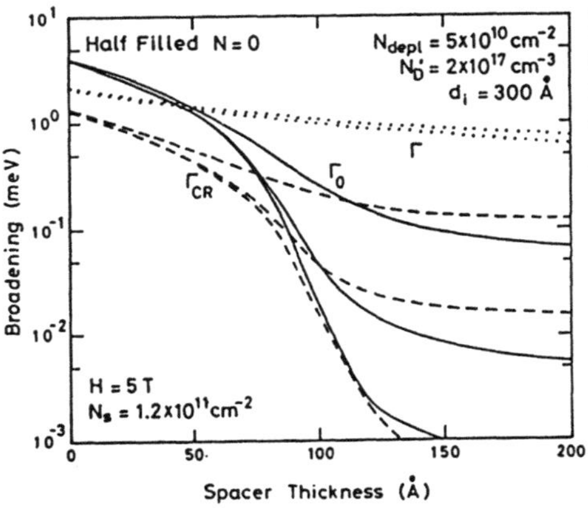

Fig. 4 Calculated broadening as a function of the spacer thickness in magnetic field $H = 5$ T for different values of background doping N_A ($N_A = 1\times10^{15}$ cm^{-3}, 1×10^{14}, and 0, from the top). The Fermi level lies at the center of the lowest Landau level. After Ref. [7]

Experimental situations are controversial in GaAs/AlGaAs heterostructures. Englert *et. al.* observed the CR broadening depending strongly on the Fermi-level position [9]. This result has not been confirmed by later experiments, however. Most convincing experiments have been provided by CR experiments of Heitmann *et. al.* in InAs quantum wells [10]. According to their experiments the amplitude of the oscillation can be as large as 2–6 times the typical broadening.

4. Photoluminescence in Modulation-Doped Quantum Well
4.1 Band-Gap Renormalization

Recent optical experiments such as photoluminescence [11–14] and absorption [15] have revealed a significant reduction of the energy gap in modulation-doped GaAs/AlGaAs quantum wells. This band-gap renormalization effect has been explained well by many-body effects in a 2D electron gas [16,17]

A theoretical analysis of photoluminescence spectra has been made using perturbation theory [16]. The many-body effects on the spectral function may be separated into the renormalization of electron and hole energies and the so-called vertex part. The vertex part is know to give rise to the Fermi-surface anomaly of absorption spectra. For sufficiently large electron densities, such vertex correction is not important and the shift of the luminescence energy is given in a good approximation by the sum of the renormalized quasi-particle energies at the top of the valence band and the bottom of the conduction band. [Adverse criticisms hurled at Ref. 16 by Ref. 17 without any calculations are unwarranted.]

Calculation of self-energy shifts has been made in the two different approximation schemes, diagramatic perturbations and a local density-functional method [16]. The dynamical dielectric function is calculated by adopting the plasmon-pole approximation. An example of the results are

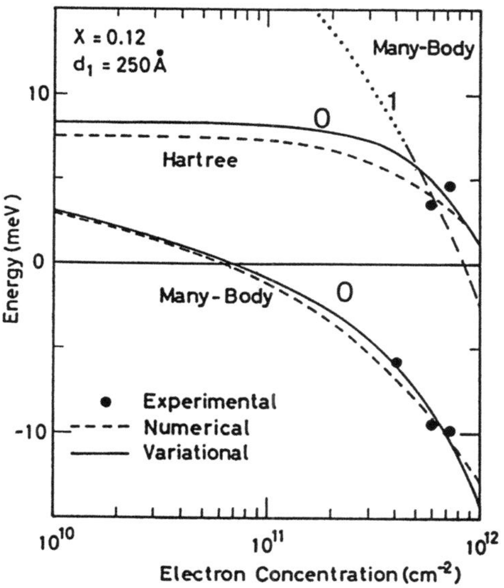

Fig. 5 Calculated photoluminescence energy in modulation-doped GaAs/AlGaAs quantum wells relative to the band gap of bulk GaAs. The curves labelled "Hartree" are modified by exchange-correlation as shown by the curves marked "Many-Body". The solid and dashed lines represent results calculated in different approximation schemes. After Ref. [16]

given in Fig. 5 together with experimental results [11]. We find a generally good agreement between the results of the different methods and between calculations and experiments.

The wave function of electrons and holes in a quantum well is primarily determined by the barrier potential and consequently the effect of the band-bending on the photoluminescence energy is cancelled to the lowest order between electrons and holes. As a matter of fact, the result of the Hartree approximation is nearly independent of the electron concentration except in the case of high concentrations. It can be concluded that the renormalization of photoluminescence energy reflects purely the strength of the electron-electron and electron-hole interactions.

4.2 Magnetic Oscillation of Photoluminescence Energy

When a strong magnetic field is applied, many-body renormalization of the photoluminescence energy can be modified through the change in the dielectric response of the system. Such effects were studied quite recently [18]. In the random-phase approximation (RPA), the quasi-particle self-energy for the electron Green's function Σ^e is given by a sum of the screened exchange (Σ^e_{xs}) and the Coulomb hole term (Σ^e_{CH}) as

$$\Sigma^{e}(N,\omega) = \Sigma^{e}_{xs}(N,\omega) + \Sigma^{e}_{CH}(N,\omega), \tag{2}$$

where the first and second term in the right hand side describe, respectively, the exchange interaction screened dynamically and the correlation effects, *i.e.* the creation of the Coulomb hole or electron-density deficiency around an electron through virtual density excitations. The two terms play complemental roles in reducing the electron energy. Explicitly, we have

$$\Sigma^{e}_{xs} = \sum_{N'}\sum_{\mathbf{q}} J^{2}_{NN'} F^{e}(q) \int_{-\infty}^{+\infty} \frac{d\omega'}{\pi} f_{F}(\omega') \frac{V(q)}{\epsilon(q,\omega-\omega')} \mathrm{Im} G^{e}_{N'}(\omega'), \tag{3a}$$

$$\Sigma^{e}_{CH} = -\sum_{N'}\sum_{\mathbf{q}} J^{2}_{NN'} F^{e}(q) \int_{-\infty}^{+\infty} \frac{d\omega'}{\pi} n_{B}(\omega') \mathrm{Im}\left[\frac{V(q)}{\epsilon(q,\omega')}\right] G^{e}_{N'}(\omega-\omega'), \tag{3b}$$

where $V(q) = 2\pi e^2/q\kappa$ with κ the background dielectric constant, f_F and n_B stand for the Fermi and Bose distribution, respectively, and F^e denotes the form factor for electron-electron interaction, which is evaluated by using the wave function corresponding to the ground subband in quantum wells. The electron Green's function G^e_N is evaluated by using SCBA to take account of impurity scatterings. A similar formula can be obtained for a photo-excited hole except that Σ^h_{xs} does not exist because of the negligibly small number of holes and that an appropriate form factor should be used for electron-hole interaction. The Coulomb-hole term Σ^h_{CH} describes the reduction of the hole energy by attracting a 2D electron cloud around the hole.

The dynamical dielectric function $\epsilon(q,\omega)$ of the 2D electron gas describes the screening effect on the exchange in Eq. (3a) and virtual excitations of density fluctuations in Eq. (3b). We employ the plasmon-pole approximation, which is known to work quite well in the absence of a magnetic field in spite of its simplicity. We have

$$\frac{1}{\epsilon(q,\omega)} = 1 + \alpha(q)\left[\frac{1}{\omega-\omega(q)+i\delta} + \frac{1}{\omega+\omega(q)+i\delta}\right], \tag{4}$$

where $\alpha(q)$ and $\omega(q)$ are determined so as to satisfy the sum rule and the Kramers-Kronig relation at $\omega = 0$. Explicitly, we have

$$\alpha(q) = \frac{\omega_p^2}{2\omega(q)} \quad \text{and} \quad \omega(q)^2 = \frac{\omega_p^2}{1-1/\epsilon(q,0)}, \tag{5}$$

where $\epsilon(q,0)$ is the static dielectric function and $\omega_p^2 = (2\pi e^2 N_s/m_e\kappa)q$ with m_e the electron effective mass and N_s the electron density per unit area. Figure 6 shows the calculated quasi-particle self-energies for a GaAs/Al$_x$ Ga$_{1-x}$As quantum well with $x = 0.23$ and $N_s = 4 \times 10^{11}$ cm^{-2} at zero temperature as a function of magnetic field. We employ standard parameters of bulk GaAs such as $m_e = 0.068m$, hole effective mass $m_h^* = 0.4m$, $\kappa = 12.6$, the well-width $L_W = 251$ Å, and the broadening $\Gamma = 1$ meV. We neglect the spin-slitting completely because the g factor of electrons and holes in GaAs is small and also for simplicity. The dashed, dash-dotted, and solid lines denote $-\Sigma^e_{xs}$, $-\Sigma^e_{CH}$, and $-\Sigma^e$, respectively. The hole self-energy is plotted by the solid line denoted by Σ^h.

Each of the self-energies exhibits clear oscillations as a function of the magnetic field. One of the most prominent features is that the phase of the oscillation of Σ^e_{xs} and Σ^e_{CH} is exactly opposite. When the Fermi level E_F sits on the Landau-level center, intra-Landau-level excitations give rise to a large polarization in the dielectric function ϵ. Large ϵ means importance of the Coulomb hole effect, *i.e.*, enhancement of $-\Sigma^e_{CH}$, and at the same time a reduction of the exchange interaction, $-\Sigma^e_{xs}$, due to large screening. As E_F approaches the edge of the density of states, on the other hand, such intra-Landau-level polarization is reduced and only small inter-Landau-level polarization remains, giving rise to enhancement of the exchange term and reduction of the Coulomb hole term. Therefore, there is a strong tendency for the cancellation of the oscillation between Σ^e_{xs} and Σ^e_{CH}. There exists another cancellation between Σ^e_{xs} and Σ^e_{CH}: Apart from the oscillation, $-\Sigma^e_{xs}$ decreases as H increases, while $-\Sigma^e_{CH}$ increases. Consequently the total self-energy shift $-\Sigma^e$ is almost independent of the magnetic field.

Such cancellation does not occur for holes, because only the Coulomb hole term is present. The self-energy shift of holes has a magnetic-field dependence similar to the Coulomb-hole term for electrons, *i.e.*, it exhibits a strong oscillation and increases almost linearly on the average as a function of magnetic field. Note that it is comparable to the shift of electrons and can be even larger in sufficiently strong magnetic fields. Therefore, the net magnetic-field dependence of the band-gap renormalization $-\Sigma^e - \Sigma^h$, especially its oscillation, is dominated by that of the self-energy of holes. In the limit of vanishing field, the gap renormalization approaches the value (about 14 meV) estimated in the absence of a field.

Perry *et. al.* have performed photoluminescence experiments in doped GaAs/AlGaAs quantum wells in large perpendicular magnetic fields and found unexpected oscillatory behavior of the luminescence energy [19]. Similar results have been obtained also in InGaAs/InP quantum wells by Skolnick and co-workers [20]. Figure 7 gives the photoluminescence energy measured from the band gap of bulk GaAs at the electron density corresponding to the experiments of Perry *et. al.* The open circles are experimental results at 2 K. The present calculation explains the overall feature of experiments very well. In particular, the calculation gives both period and phase of the oscillation in excellent accord with experiments. There appears a splitting of photoluminescence peaks observed experimentally above $H = 11$T, which is due to many-body enhancement of the g factor discussed in Sec. 2.

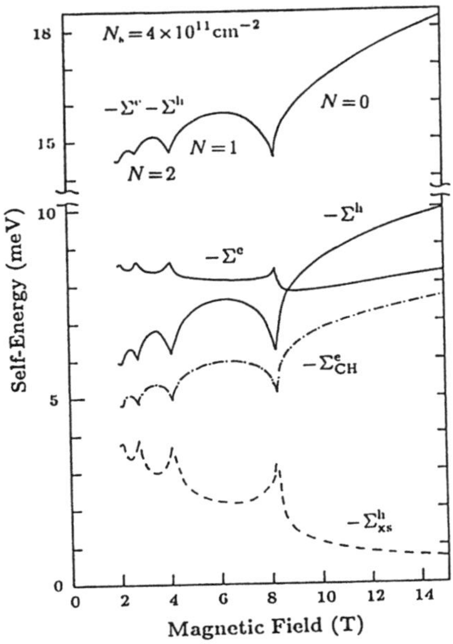

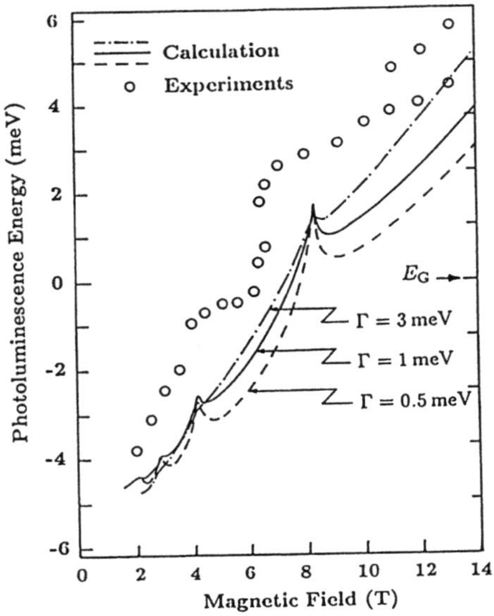

Fig. 6 Calculated quasi-particle self-energies as a function of the magnetic field in a modulation-doped GaAs/AlGaAs quantum well. After Ref. [18]

Fig. 7 Photoluminescence energy measured from the band gap of bulk GaAs as a function of the magnetic field in a modulation-doped GaAs/AlGaAs quantum well. The dashed, solid, and dash-dotted lines represent calculations for three different values of the broadening Γ, and open circles experimental results of Perry *et. al.*[19]. After Ref. [18]

There remains a slight discrepancy of the order of a few meV concerning the absolute value of calculated and observed photoluminescence energy. This is well in the range attributable to ambiguity of the parameters such as the band offset, the Al content x, the well width L_W, etc. A difference in shape of magnetic oscillations between calculations and experiments is more serious. A possible explanation is the oscillation of the broadening Γ discussed in Sec. 3. If we consider this effect, there is a crossover of the photoluminescence energy between small Γ and large Γ as a function of magnetic field. The sharp peaks near $H = 3.6$, 4.1, and 8.2 T tend to be broadened and the calculated energy becomes much closer to the observed step-like dependence.

Acknowledgments

The author wishes to thank Professor S. Katayama for collaboration in the work on magnetic oscillations of photoluminescence energy in modulation-doped quantum well. This work is supported in part by Special Distinguished Research from Ministry of Education, Science and Culture of Japan.

References

[1] T. Ando and Y. Uemura: J. Phys. Soc. Jpn. **37**, 1044 (1974).
[2] F.F. Fang and P.J. Stiles: Phys. Rev. **174**, 823 (1968).
[3] Th. Englert and K. von Klitzing: Surf. Sci. **73**, 70 (1978).
[4] T. Ando and Y. Uemura: J. Phys. Soc. Jpn. **36**, 959 (1974).
[5] T. Ando: J. Phys. Soc. Jpn. **43**, 1616 (1977).
[6] R. Lassnig and E. Gornik: Solid State Commun. **47**, 959 (1983).
[7] T. Ando and Y. Murayama: J. Phys. Soc. Jpn. **54**, 1519 (1985).
[8] I.V. Kukushkin and V.B. Timofeev: Surf. Sci. **196**, 196 (1988).
[9] Th. Englert, J.C. Maan, Ch. Uihlein, D.C. Tsui and A.C. Gossard: Solid State Commun. **46**, 545 (1983).
[10] D. Heitmann, M. Ziesmann and L.L. Chang: Phys. Rev. B **34**, 7463 (1986).
[11] A. Pinczuk, J. Shah, R.C. Miller, A.C. Gossard and W. Wiegmann: Solid State Commun. **50**, 735 (1984).
[12] M.H. Meynbadier, J. Orgonasi, C. Delalande, J.A. Brum, G. Bastard, M. Voos, G. Weimann and W. Schlapp: Phys. Rev. B **34**, 2482 (1985).
[13] D.A. Kleinman and R.C. Miller: Phys. Rev. B **32**, 2266 (1985).
[14] C. Delalande, G. Bastard, J. Orgonasi, J.A. Brum, H.W. Liu, M. Voos, G. Weimann and W. Schlapp: Phys. Rev. Lett. **59**, 2690 (1987).
[15] J.S. Lee, N. Miura and Y. Iwasa: Surf. Sci. **196**, 534 (1988).
[16] G.E.W. Bauer and T. Ando: Phys. Rev. B **31**, 8321 (1985); J. Phys. C **19**, 1537 (1986).
[17] S. Schmitt-Rink, C. Ell and H. Haug: Phys. Rev. B **33**, 1183 (1986).
[18] S. Katayama and T. Ando: to be published.
[19] C.H. Perry, J.M. Worlock, M.C. Smith and A. Petrou: *Application of High Magnetic Fields in Semiconductor Physics*, ed. by G. Landwehr (Springer, Heidelberg, 1986), p. 202.
[20] M.D. Skolnick: private communication.

n-i-p-i Doping Superlattices Under High Magnetic Fields

G.H. Döhler

Institut für Technische Physik, Erwin-Rommel-Str. 1
D-8520 Erlangen, Fed. Rep. of Germany

1. Introduction

It is well known, that the spatial separation between electrons and holes in doping superlattices results in quite remarkable pecularities of the electronic structure and the electrical and optical properties of semiconductors belonging to this class of materials /1,2/. Due to a strongly reduced overlap between electron and hole wave functions the recombination lifetimes may be enhanced by orders of magnitude compared with the uniform bulk material. Large deviations of the electron and hole concentrations in the n- and p-layers are metastable. The effective bandgap and also the energetic separation between the electric subbands become dynamically tunable. This tuning can be achieved by changing the carrier densities either by optical excitation or by carrier injection or extraction by means of selective n- and p-contacts /3,4/. Another important aspect when dealing with n-i-p-i doping superlattices is the role of impurities causing bound electronic states and, because of their random distribution, spatial potential fluctuations /1/. Although, in many cases, the impurities lead to undesired broadening of the energy levels, they can also lead to interesting properties, including metal-insulator transitions, in systems of appropriate design as pointed out in a recent paper /4/.

In spite of the fact that the properties of n-i-p-i doping superlattices can be modulated optically or electrically within a wide range it turns out, that high magnetic fields add an exciting new dimension for tuning the electronic properties of these materials. The effect of a magnetic field on the electronic structure does not only show up as a change of subband energies. Because of the conduction-valence band transitions, which depend sensitively on the overlap between electron and hole wave functions within the energy gap, we have a very delicate tool to monitor magnetic field induced changes of the wave functions even in the region far away from their origin. Such studies imply the possibility to obtain a lot of information which is not obtained in the simple variational calculations which are frequently used to solve similar problems. A variational approach may provide reasonably good results for the energy levels but will yield completely unsatisfactory approximations for the wave functions. These aspects and their implications will be discussed in section 2 of this paper.

In section 3 we will investigate some of the impurity related aspects of high magnetic fields applied to n-i-p-i doping superlattices, or to "hetero n-i-p-i's", an extension of the original concept in which compositional modulation is combined with modulation by periodic n- and p-doping. We will see, that variation of the

Springer Series in Solid-State Sciences, Vol. 87
High Magnetic Fields in Semiconductor Physics II Editor: G. Landwehr

magnetic field allows for a geometrical scaling of the structure, while electrical and/or optical tuning allows for depleting or filling up the densitiy of state distribution. Within this 2-dimensional parameter space we expect unique possibilities to study metal insulator transition of the Mott-Anderson and of the Mott-Hubbard type by optical and by transport measurements.

2. Effect of High Magnetic Fields on Energies and Transition Probabilities

In this section we will investigate the effect of a magnetic field which is oriented parallel to the layers or normal to them on the dynamically 2-dimensional states of an idealized n-i-p-i crystal (potential fluctuations due to random impurity distribution neglected).

2. 1. Magnetic Field Parallel to the Layers

We assume a n-i-p-i crystal consisting of a uniform semiconductor host material with band gap E_g^o, dielectric constant $\varkappa_o$ and effective masses m_e, m_{hh} and m_{lh} for electron, heavy and light holes near the band edges. Due to the periodic n- and p-doping in the superlattice direction, z, there is a periodic space charge

$$\rho(z) = e(n_D(z) - n_A(z) - n(z) + p(z)) \qquad (1)$$

due to ionized donors and acceptors as well as due to free electrons and holes, respectively. The space charge potential

$$v(z) = (4\pi e/\varkappa\ ) \int^{z} dz' \int^{z'} \rho(z'')dz'' \qquad (2)$$

with the superlattice period d superposed to the lattice-periodic potential $H_o = p^2/2m + V_o(\mathbf{r})$ of the host material. In the following we are using the effective mass approximation

$$H_o \approx p^2/2m_i \ ; \text{ with } i = e, hh, lh \qquad (3)$$

for electrons, heavy and light holes, respectively. Thus, the energies and Bloch functions for the bulk material become

$$\varepsilon_i(k) \approx \begin{cases} E_c + (\hbar^2/2m_e)k^2; & i = e \\ (-\hbar^2/2m_i)k^2; & i = lh,\ hh \end{cases} \qquad (4)$$

and

$$\varphi_{i,\mathbf{k}}(\mathbf{r}) \approx e^{i\mathbf{kr}} u_{i,\mathbf{k}_o}(\mathbf{r}); \qquad i = e,\ lh,\ hh\ . \qquad (5)$$

We introduce the magnetic field B in the xy-plane by the vector potential

$$\mathbf{A} = (0,-Bz,0) \qquad (6)$$

whence

$$\mathbf{B} = \mathrm{rot}\ \mathbf{A} = (B, 0, 0) \quad . \tag{7}$$

Thus, the effective mass Schrödinger equation to be solved reads

$$\{(1/2m_i)[p_x^2 + (p_y+eBz/c)^2 + p_z^2] + v(z) - \mathcal{E}^B_{i,\mu,\mathbf{k}_{\shortparallel}}\}\ \psi^B_{i,\mu,\mathbf{k}_{\shortparallel}}(\mathbf{r}) = 0 \quad . \tag{8}$$

In equ. (8) we have already taken into account that the carriers remain free with respect to their motion in the (x,y)-plane and therefore will be described by their momentum $\mathbf{k}_{\shortparallel}$ and the subband quantum number μ (we will neglect miniband dispersion $\varepsilon(k_z)$ due to interaction between neighbouring potential wells, assuming a sufficiently high and long-period space charge potential). Using the ansatz

$$\psi^B_{i,\mu,\mathbf{k}_{\shortparallel}}(\mathbf{r}) = e^{i(k_x x+k_y y)}\ u_{i,\mathbf{k}_o}(\mathbf{r})\ \zeta^B_{i,\mu}(z) \tag{9}$$

in equ. (8) allows one to separate the z-dependent part

$$\{-(\hbar^2/2m_i)\partial^2/\partial z^2 + v(z) + (e^2B^2/2m_ic^2)(z-z_{k_y})^2 - \mathcal{E}_{i,\mu}(\mathbf{k}_{\shortparallel}) + \hbar^2k_x^2/2m_i\}\ \zeta^B_{i,\mu}(z) = 0 \ ; \qquad i = e,\ hh,\ lh \tag{10}$$

where

$$z_{k_y} = \hbar k_y c/eB \quad . \tag{11}$$

A good approximate solution of equ. (10) can be obtained easily if the n- and p-layer are zones of uniform n- and p-doping, respectively, with thicknesses fulfilling the condition $n_D d_n = n_A d_p$. In this case equ. (10) becomes the Schrödinger equation for harmonic oscillators with energy levels given by

$$\mathcal{E}_{i,\mu,\mathbf{k}_{\shortparallel}} = \begin{cases} E_g^0 - 2V_0 + (\hbar^2/2m_e)\{k_x^2 + (\omega_{p,e}/\omega_e^B)^2k_y^2\} + \hbar\omega_e^B(\mu+1/2); & i = e \\ -(\hbar^2/2m_i)\{k_x^2 + (\omega_{p,i}/\omega_i^B)^2k_y^2\} + \hbar\omega_i^B(\mu+1/2); & i = hh,\ lh, \end{cases} \tag{12}$$

with

$$2V_0 = (\pi e^2/2\varkappa_0)\ (n_D d_n^2 + n_A d_p^2) \tag{13}$$

and

$$\omega_i^B = (\omega_{p,i}^2 + \omega_{c,i}^2)^{1/2} \quad . \tag{14}$$

Compared with the B = 0 case where the subband spacing is given by the (formal) plasmon energies

$$\hbar\omega_{p,i} = (4\pi e^2\hbar^2/\varkappa_0)^{1/2}\begin{cases}(n_D/m_e)^{1/2}\ ; & i = e \\ (n_A/m_i)^{1/2}\ ; & i = hh,\ lh\end{cases} \tag{15}$$

the subband spacing is now renormalized by the Landau energies

$$\hbar\omega_{c,i} = \hbar eB/m_i c \quad . \tag{16}$$

Also, the energies are shifted by the (diamagnetic) term $\hbar(\omega_i^B - \omega_{i,p})/2$. In GaAs with $n_D = 4 \times 10^{17} cm^{-3}$ and B = 15T we find, for instance,

$\hbar(4\pi e^2 n_D/\varkappa_0 m_e)^{1/2} \approx 25$ meV $\approx \hbar eB/m_e c$, or $\hbar\omega_e^B \approx 35$ meV.

The wave functions are also renormalized and shifted into the z-direction.

$$\zeta_{i\mu}^B(z) = (m_i\omega_i^B/\pi\hbar)^{1/4}(2^n n!)^{-1/2} H_\mu\big((m_i\omega_i^B/\hbar)^{1/2}(z-\Delta_{i,k_y}-z_{in})\big)\cdot$$
$$\exp[-(m_i\omega_i^B/2\hbar)(z-\Delta_{i,k}-z_{in})^2] \quad , \tag{17}$$

where

$$z_{in} = nd \; ; \quad i = e \quad \text{and} \quad (n-\tfrac{1}{2})d < z < (n+_2)d \tag{18}$$

and

$$z_{in} = (n+\tfrac{1}{2})d \; ; \quad i = hh,\, lh \quad \text{and} \quad nd < z < (n+1)d \quad . \tag{19}$$

The shift of the wave functions in z-direction

$$\Delta_{i,k_y} = z_{k_y}(\omega_{c,i}/\omega_i^B)^2 = (\hbar k_y/m_i)\omega_{c,i}/(\omega_i^B)^2 \tag{20}$$

is a consequence of the Lorentz force experienced by particles with a momentum $\hbar k_y$ perpendicular to the B-field. The Lorentz force is also the reason for the increase of the effective mass with respect to the in-plane motion perpendicular to the B-field (see kin energy term in Eq. (12)).

The renormalization of the subband energies has been investigated by Maan et al. /5/ in magnetoresistance experiments. These authors were able to demonstrate that the conductance showed quantum oscillation as a function of magnetic field in quantitative agreement with the expected behavior.

From equs. (17) to (19) we can calculate the changes of overlap and the resulting changes of the conduction to valence subband transition probabilities. A comparison between experiment and theory is particularly interesting as we are investigating with high sensitivity the changes of the *wave function* induced by the B-field. Usually, only the changes of the *energies* induced by a perturbation are observed. For the $\mu = 0$ electron light-hole transition we find for the change of transition probability reduction factors with and without B-field (assuming $n_D = n_A$, $d_n = d_p$, $m_{lh} \approx m_e$)

$$w_{oo}^B = w_{cv}^{bulk} \exp[-m_e\omega_e^B d^2/8\hbar] \tag{21}$$

and

$$w_{oo} = w_{cv}^{bulk} \exp[-m_e\omega_{p,e}d^2/8\hbar] \quad (22)$$

respectively. We have used Equ. (35) from Ref. 6 in order ot obtain Eqs. (21) and (22).

For our previous example ($\hbar\omega_{p,e}$ = 25 meV, B = 15 T) we find for d ≈ 50 nm a red shift of the luminescence of about 85 meV. The transition probability reduction factor which is ≈ 10^{-16} at B = 0 becomes ≈ $4 \cdot 10^{-9}$ at B = 15 T, i.e., the transition probabilities for luminescent or absorptive transitions change by a factor ≈ 250 by the applications of the B-field.

So far we have neglected the shift Δ_{i,k_y} of the wave functions which occurs for carriers with a finite momentum $\hbar k_y$ perpendicular to the B-field. From (11) and (20) we see, that the sign of this shift depends on the direction of the y-component of the motion of the carriers. Also it has opposite sign for electrons and holes. We can easily calculate the effect on the transition probabilities for the symmetric case discussed before. Equation (21) has to be modified by replacing d/2 by $d/2 - \Delta_{e,k_y} + \Delta_{lh,k_y} \approx d/2 - 2\Delta_{e,k_y}$. We find

$$w_{oo}^{B}(k_y) = w_{cv}^{bulk} \exp[-m_e\omega_e^B(d/2-2\Delta_{k_y})^2/2\hbar] \,, \quad (23)$$

whereas (21) remains unchanged.

For our previous values and assuming a reasonable electron concentration of $n^{(2)} = 5 \cdot 10^{11}\text{cm}^{-2}$ we find for the recombination probability of electrons and light holes with the Fermi momentum $k_{F,y} = 2^{1/4}(2\pi n)^{1/2}$ (the factor $2^{1/4}$ is a consequence of the change of effective mass, m_y)

$$w_{oo}^{15T}(k_{F,y} = 2 \cdot 10^6 \text{ cm}^{-1}) \approx 120\, w_{oo}^{B=0} \approx 3 \cdot 10^4\, w_{oo}^{15T}(k_y = 0) \ .$$

Thus, we find, that the *strong reduction* of transition probability for carriers of zero momentum perpendicular to the B-field, which is due to the potential becoming steeper, can be *largely overcome* by the Lorentz force driving electrons and holes towards each other for carriers with momentum close to the Fermi momentum. For our specific example the overall transition probability remains unchanged if $n^{(2)} \approx 10^{11}\text{cm}^{-2}$, but changes by about +2 (−2) orders of magnitude for $n^{(2)} \approx 5 \cdot 10^{11}\text{cm}^{-2}$ ($2 \cdot 10^{10}\text{cm}^{-2}$), if a B-field of 15 T is switched on.

We have discussed here only the simplest case in order to illustrate the expected phenomena. An ideal system for the observation of the carrier density dependent effect of a parallel magnetic field on the interband transition probabilities is the PbTe n-i-p-i /7/. The high field case, $\hbar\omega_{c,i} \geq \hbar\omega_{p,i}$, is not only realized at relatively low B-fields, but also the assumption of free electrons and holes is justified for electrons and holes as well. The present results have only to be extended to the (nearly symmetric with respect to electrons and holes) multi-valley band structure of PbTe /8/. It is clear, that the results will change qualitatively for other doping profiles, such as δ-doping /1,8/ or in hetero - n-i-p-i's /2/. Of particular interest is the situation, if the carriers are no longer moving perpendicular to the B-field. This can be due to the formation of Landau orbitals under a B-field having also a z-component or due to localization by impurity potentials or spatially random potential fluctuations. These cases, for which we expect qualtitative changes will be discussed in sections 2.2 and 3.

2.2. Magnetic Field Normal to the Layers

We chose the gauge

$$\mathbf{A} = (By/2, -Bx/2, 0) \quad . \tag{24}$$

This is also the appropriate gauge for a system with additional potentials v(x,y) which we will treat in section 3. Replacing **p** by (**p**−e**A**/c) in the effective mass Schrödinger equation now yields, instead of our previous eq. (8)

$$\{(1/2m_i)(p_x^2+p_y^2+p_z^2) + \frac{m_i\omega_c^2}{8}(x^2+y^2) + \hbar\omega_c L_z + v_o(z) - \varepsilon_{i,\mu,n,L_z}\}\,\psi^B_{i,\mu,n,L_z}(\mathbf{r}) = 0 \quad . \tag{25}$$

Eq. (25) reflects the full quantization of the system by the formation of Landau orbits in the (x,y)-plane with quantum numbers n and L_z and the confining superlattice potential $v_o(z)$ in the z-direction. Eq. (25) can be separated into a z-dependent and a (x,y)-dependent part. For the ground state we find (assuming again uniform doping in the n- and p-layers)

$$\psi^B_{i,o,o,o} = N^{-1/2}\exp\{-\frac{m_i\omega_{c,i}}{2\hbar}(x^2+y^2)\}\exp\{-\frac{m_i\omega_{p,i}}{2\hbar}(z-z_{in})^2\};\quad i = e,\ lh,\ hh \tag{26}$$

and

$$\varepsilon_{i,o,o,o} = \begin{cases} E_c - 2V_o + \hbar\omega_{c,e}/2 + \hbar\omega_{p,e}/2; & i = e \\ -\hbar\omega_{c,i}/2 - \hbar\omega_{p,e}/2 \quad ; & i = hh,\ lh \quad . \end{cases} \tag{27}$$

where the degeneracy of the Landau levels per layer is $(eB/\pi\hbar c)$-fold, as known from any 2-D system. From (2b) we see, that the interband transition probabilities should not be affected by the normal B-field, as the z-dependence of the wave function isn't.

2.3. Arbitrary Orientation of the Magnetic Field

Without loss of generality (except for the case of an anisotropic band structure of the host material) we can choose a magnetic field

$$\mathbf{B} = (B_x,\ 0,\ B_z) \quad . \tag{28}$$

Having the results of 2.1 and 2.2 in mind, one might first expect that the system is now fully quantized due to the B_z-induced Landau level formation, but with the space charge induced subband energies renormalized by the field component B_x

parallel to the layers. Actually, however, the system is now 1-dimensional, as the carriers are still free to move in the x-direction, although experiencing a Lorentz force due to the z-component of the B-field.

The B-field (28) can be derived from the vector potential

$$\mathbf{A} = (-B_z y,\ -B_{x1} z,\ B_{x2} y) \tag{29}$$

with

$$B_{x1} + B_{x2} = B_x \quad . \tag{30}$$

Inserting the ansatz

$$\psi_{i,\mu,n}(\mathbf{r}) = N^{-1/2} e^{ik_x x} \Phi_n(y) \zeta_\mu(z) \tag{31}$$

into the effective mass Schrödinger equation yields

$$\{\hbar^2 k_x^2/2m_i - (\hbar^2/2m_i)(\partial^2/\partial y^2 + \partial^2/\partial z^2) + (m\omega_y^2/2)(y + \Delta_{k_y}^2) + (m\omega_z/2)z - i\hbar\omega_{x1} z(\partial/\partial y) + i\hbar\omega_{x2} y(\partial/\partial z) - \mathcal{E}_{i,\mu,n}(k_x)\}\varphi_n(y)\zeta_\mu(z) = 0 \tag{32}$$

with

$$\Delta_{i,k_x} = (\hbar k_x/m_i)(\omega_{c,i}/\omega_{y,i}^2) \quad , \tag{33}$$

$$\omega_{c,i} = eB_z/m_i c \ ; \quad \omega_{x1,i} = eB_{x1}/m_i c \ ; \quad \omega_{x2,i} = eB_{x2}/m_i c \ , \tag{34}$$

$$\omega_{y,i}^2 = \omega_{c,i}^2 + \omega_{x2,i}^2 \ , \tag{35}$$

$$\omega_{z,i}^2 = \omega_{p,i}^2 + \omega_{x1,i}^2 \quad . \tag{36}$$

The ansatz

$$\Phi_o(z)\ \zeta_o(z) = \exp\{-(m_i\omega_{yi}/2\hbar)(y+\Delta_{i,k_x})^2\}\ \exp\{-(m\omega_{z,i}/2\hbar)(z-z_{in})^2\}; \quad i = e,\ hh,\ lh \tag{37}$$

turns out to be the solution for the lowest index subbands, if the gauge of B_x is chosen such that

$$\omega_{x1}\omega_z = \omega_{x2}\omega_y \quad . \tag{38}$$

From the definitions for the frequencies (38) it follows, that this is equivalent to

$$B_{x2}/B_{x1} = \omega_c/\omega_p \quad . \tag{39}$$

The energies become

$$\mathcal{E}_{i,o,o}(k_x) = \begin{cases} E_g^o - 2V_o + (\hbar\omega_{y,i} + \hbar\omega_{z,i})/2 + (\hbar^2/2m_i)(\omega_{x1,i}/\omega_{y,i})^2 k_x^2 ; & i = e \\ -(\hbar\omega_{y,i} + \hbar\omega_{z,i})/2 - (\hbar^2/2m_i)(\omega_{x1,i}/\omega_{y,i})^2 k_x^2 ; & i = hh,lh \end{cases} \quad (40)$$

The expression for the 1-dimensional subband reduces to the 0-dimensional Landau system (27) in the limiting case of $B_x \to 0$. Note, that in this case the renormalized effective mass for the motion in x-direction in (40) diverges, i.e., $m_i^{eff} = m_i(\omega_{y,i}/\omega_{x1,i})^2 \to \infty$. This implies the transition from the 1-dimensional density of states, which exhibits logarithmic singularities at the discrete energy levels $(n+1/2)\hbar\omega_c+(\mu+1/2)\hbar\omega_z$ to the δ-function-density of states of the Landau level system discussed in 2.2.

From (37) and (33) we see that a finite momentum in x-direction shifts the wave function in y-direction. The z-dependent part of the wave function, however, remains centered at the middle of the n- and the p-layers. Thus, the k_y-dependent *increase* of the transition probabilities, discussed in section 2.1 will not occur, if, in addition to the magnetic field parallel to the layer, B_x, there is also a magnetic field, B_z, normal to them (at least, if it is strong enough to achieve the magnetic quantum limit).

For our previous example this means, that the transition probabilities, which are enhanced for carriers the Fermi level by the field B_x = 15 T by a factor ≈ 120, compared with the field-free case will drop to a value which is even by a factor of ≈ 20 lower than the field-free case value if a B_z-field of 9 T is added.

3. Effects of High Magnetic Fields in the Case of Localized States

So far we have assumed that the carriers are moving free in the (xy)-plane as long as no magnetic fields are applied. In particular, we found for a Fermi system that the transition probabilities may increase upon the application of a field parallel to the layers, instead of the expected reduction if the density is high enough. An additional field normal to the layers, however, may turn the increase back into a strong reduction, due to the formation of Landau levels. In this section we will investigate the consequences of disorder which causes localization of the states in the (xy)-plane.

3.1 Magnetic Fields Parallel to the Layers

We expect that the effect of localization on the magnetic field dependence will be weak, if the random potential, which results from the disorder, is small. If, however, the spatial potential fluctuations are so strong that the classical motion in the ground state of a localized state corresponds to kinetic energies at which the k_y-dependence from eq. (23) leads to strong *increase* of the interband transition probabilities, it is of interest to know whether this *increase* also applies to bound states. In the following we will treat a simple model which can be solved analytically. We assume, that, in addition to the parabolic space charge potential $v_o(z) = (m_i\omega_{i,p}^2/2)\, z^2$ there is also a parabolic potential

$$v_o(x,y) = (m_i\omega_{xo}^2/2)x^2+(m_i\omega_{yo}^2/2)y^2 \quad . \tag{41}$$

The ground state of this system will be renormalized by a magnetic field parallel to the plane, B_x. Using the gauge

$$\mathbf{A} = (0,\ -B_{x1}z,\ B_{x2}y)\ ;\quad B_{x1}+B_{x2} = B \tag{42}$$

and the ansatz

$$\psi_o(\mathbf{r}) = N^{-1/2}\, e^{-(m_i\omega_{xi}^2/2\hbar)x^2}\, e^{-(m_i\omega_{yi}^2/2\hbar)y^2}\, e^{-(m_i\omega_{zi}^2/2\hbar)z^2} \tag{43}$$

we find that (43) is the ground state solution of the effective mass Schrödinger equation with

$$\varepsilon_o = \begin{cases} E_g^o-2V_o+(\hbar\omega_{xi}+\hbar\omega_{yi}+\hbar\omega_{zi})/2\ ; & i = e \\ -(\hbar\omega_{x,i}+\hbar\omega_{y,i}+\hbar\omega_{z,i})/2\ ; & i = hh,\ lh \end{cases} \tag{44}$$

given by

$$\begin{aligned} \omega_{x,i}^2 &= \omega_{xo,i}^2 \\ \omega_{y,i}^2 &= \omega_{yo,i}^2\,\{1+[(eB/m_ic)/(\omega_{p,i}+\omega_{yo,i})]^2\} \\ \omega_{z,i}^2 &= \omega_{p,i}^2\,\{1+[(eB/m_ic)/(\omega_{p,i}+\omega_{yo,i})]^2\} \quad . \end{aligned} \tag{45}$$

This result shows that the localization increases in the two directions perpendicular to the B-field, whereas the wave function is unaffected in the direction of the B-field. By comparison of (45) and (43) with (17) (for μ=0) and (21) we see that the transition probabilities *decrease* with increasing B_x-field, though less strongly as in the case of $k_y = 0$. Specifically we find for $\hbar\omega_{yo}/2 = \varepsilon_F(5\cdot 10^{11}\ \mathrm{cm}^{-2}) = 15$ meV in our previous example

$$\omega_{z,i} = 1.1\ \omega_{p,i} \tag{46}$$

and

$$w_{oo,loc}^{1ST} \approx 0.25\, w_{oo}^{B=o} \approx 2\cdot 10^{-3}\ w_{oo}^{1ST}(k_y = k_{F,y}(5\cdot 10^{11}\ \mathrm{cm}^{-2}) \tag{47}$$

if conduction and valence band states are localized. The last number in (47) demonstrates the dramatic difference betweeen extended and localized states with respect to interband transition probabilities, quite analoguous to the results obtained by the reduction of dimensionality introduced by an additional B-field component normal to the layer. This difference in transition probabilities may be a useful tool to distinguish experimentally between extended and localized states.

In this section we have only treated the ground state for a simple model. We expect, however, that the results will not change qualitatively if more realistic situations are assumed.

3.2 Magnetic Field Perpendicular to the Layer

If we assume the same potential $v_o(z)$ due to the space charge potential and $v_o(x,y)$ due to disorder as in section 3.1, with the magnetic field, however, oriented in z-direction, we obtain, with the gauge for the vector potential given in (24), the same results (43) and (44) for the ground state solution of the effective mass Schrödinger equation. The values of ω_x, ω_y, and ω_z, however, are now given by

$$\omega_{x,i} = \omega_{y,i} = [\omega_{o,i}^2 + (eB/m_i c)^2]^{1/2}$$
$$\omega_{z,i} = \omega_{p,i} \quad . \tag{48}$$

These results confirm the expectation that a magnetic field normal to the layers will not influence the z-dependent part of the ground state wave function, also in the case that one is dealing with states localized in the (xy)-plane. The results will not change drastically if $v_o(x,y)$ is no longer a harmonic potential, as long as it is weak compared with $v_o(z)$. A case of particular interest is, of course, the Coulomb potential of an impurity atom /10/. One has to be aware that a solution obtained, say for an impurity at the center of a uniformly doped n- or p-layer, is of limited relevance for a realistic system. For the realistic system we have to take into account that the potential $v(\mathbf{r})$ is not simply the superposititon of the average potential of the uniform space charge back-ground, $v_o(z)$, and the potential of a specific impurity at $\mathbf{R}_i$.

$$v_i(\mathbf{r}) = e^2/(\varkappa_o\,|\mathbf{r}-\mathbf{R}_i|) \quad . \tag{49}$$

In addition, there are strong spatial fluctuations, due to the random distribution of the impurities in the doping layers. There are, however, a number of possibilities to create systems in which the potential can be treated as a superposition of a one-dimensional and a point charge potential, if the requirement of a parabolic shape of $v_o(z)$ is given up. An example of such a structure is proposed in Fig. 6 of Ref. 4. The quantizing field due to the impurities in the n_1- and p_1 layers is nearly uniform within the (very wide) quantum well in between. If the average distance within the (xy)-plane between the impurity atoms is sufficiently large, the potential for the effective mass Schrödinger equation can be described appropriately by the superposition of a triangular potential $v_o(z)$, the Coulomb potential (49) and a vector potential

$$\mathbf{A}_{\mathbf{R}_i} = (B/2)\,(y - y_i\,,\, -x + x_i\,,\, 0) \quad . \tag{50}$$

Such a system is of particular interest, as it is uniquely suited for the study of metal insulator transitions. This is due to the fact that it can be dynamically tuned within a two-dimensional parameter space. The (two-dimensional) carrier density can be tuned from zero to "high" values such that not only the impurity band is occupied but also the Hubbard band, corresponding to doubly occupied impurities, by variation of optical excitation and/or external potential applied between selective contacts to the n_1- and p_1 - layers /4/. By variation of the magnetic field B_z on the other hand the overlap between neighbouring impurity wave functions can be tuned over a wide range. In particular, the (2-dimensional) doping density can be chosen such that the system is on the metallic side of the Mott-Hubbard metal/insulator transition /10/ at $B_z = 0$, but goes through the transition with increasing B_z. The photoluminescence spectra due to recombination of electrons from the impurity- and Hubbard band with free holes in the uppermost hole subband monitor the density of states as a function of Fermi level position and bandwidth. Similarly, a somewhat differently designed system allows transport measurements above and below the Mott-Hubbard (or also the Mott-Anderson) transition /4/.

4. Conclusions

We have shown that the effects of a magnetic field on doping superlattices are quite dramatic. The band to band transition probabilities may *decrease* or *increase* in the presence of a magnetic field parallel to the layers by orders of magnitude, depending on the momentum of the carriers perpendicular to the magnetic field. By changing the direction of the field from parallel to normal to the layers the system changes its character from 2-dimensional and finally to 0-dimensional. In the case of states which are localized at B = 0 no increase of the transition probabilities is expected under finite B-fields, but only a more or less strong decrease. Magnetic field studies of n-i-p-i doping superlattices turn out to be a sensitive tool for the study of the wave functions and metal-insulator transitions.

References

1. G.H. Döhler: Phys. Stat. Sol. 52, 79 and 533 (1972); G.H. Döhler: J. Vac. Sci. Technol. 16, 851 (1979)
2. For a recent review see: G.H. Döhler: IEEE QE-22, 1682 (1986); or G.H. Döhler: CRC Review in Solid State and Material Sciences 13 (1986)
3. G.H. Döhler, G. Hasnain, and J.N. Miller: Appl. Phys. Lett. 49, 704 (1986)
4. G.H. Döhler: C.Y. Fong, ed., Plenum Publishing Company, N.Y., in press
5. J.C. Maan, Th. Englert, Ch. Uihlein, H. Künzel, A. Fischer, and K. Ploog: Solid State Commun. 47, 383 (1983)
6. G.H. Döhler and P.P. Ruden: Phys. Rev. B30, 59332 (1984)
7. G.H. Döhler and P.P. Ruden: Surface Science 142, 474 (1984)
8. D.L. Mitchell and R.F. Wallis: Phys. Rev. 151, 581 (1966); G. Martinez, M. Schlüter, and M.L. Cohen, Phys. Rev. B11, 651 (1975)
9. H.P. Hjalmarson: J. Vac. Sci. Technol. 21, 524 (1982); E.F. Schubert, Y. Horikoshi, and K. Ploog: Phys. Rev. B32, 1085 (1985)
10. Y. Yafet, R.W. Keyes, and E.N. Adams: J. Phys. Chem. Solids 1, 137 (1956)
11. N.F. Mott and E.A. Davis: Electronic Processes in Non-Crystalline Materials, Oxford (1979)

Hot Electron Magnetotransport in $Al_xGa_{1-x}As$-GaAs Samples of Different Geometry

R.J. Haug, K. von Klitzing, and K. Ploog

Max-Planck-Institut für Festkörperforschung, Heisenbergstr. 1, D-7000 Stuttgart 80, Fed. Rep. of Germany

The size dependence of the magnetotransport properties of $Al_xGa_{1-x}As$-$GaAs$ heterojunctions has been studied in high electric and magnetic fields. For the breakdown of the quantum Hall effect an increase of the critical drift velocity with decreasing sample width was found in Hall bar geometries, whereas in Corbino geometries a slight decrease was observed. In the regime of the breakdown step-like structures were observed for all sample sizes. A new instability was seen for filling factors of $\nu < 2$ at electric fields of $\approx 200V/cm$.

1. INTRODUCTION

The behaviour of a two-dimensional electron gas (2DEG) at low temperatures is highly non-ohmic due to the non-equilibrium electron energy distribution in electric fields. In quantizing magnetic fields interesting new effects have been found [1,2], which are not fully understood to date. In recent experiments [2 - 5] on very narrow structures astonishing results were obtained on the 'breakdown' of the quantum Hall effect first observed in relatively broad samples by Ebert et al. [1].

We have studied the dependence of these hot electron effects upon the geometry of the samples. The structures used were $Al_xGa_{1-x}As$-$GaAs$ heterojunctions with 4K mobilities between $30,000 cm^2/Vs$ and $300,000 cm^2/Vs$ and carrier concentrations of about $n_s = 3 \cdot 10^{11} cm^{-2}$. We compared Corbino and Hall bar samples of different widths prepared from one wafer by standard photolithographic techniques. The sample widths varied between $500\mu m$ and $10\mu m$. In this regime the effects of the edge depletion should not be significant and the electrical and lithographic widths should be equal.

2. THE BREAKDOWN IN DIFFERENT GEOMETRIES

Experiments on the breakdown of the quantum Hall effect have been performed on both relatively broad ($\approx 400\mu m$) [1,6] and very narrow structures ($< 5\mu m$) [2 - 5]. Here we report studies of the intermediate range. In the Corbino devices the current is measured as a function of the voltage applied to the sample. At a certain critical voltage the current rises abruptly from zero to a finite value. This onset of a finite conductivity is more abrupt in the wider devices, in agreement with a recent calculation [7]. In similar experiments on Hall bar samples produced from the same wafer an abrupt rise in the voltage was observed as the current was varied.

In order to analyze these results independently of the geometry and magnetic field we have calculated the drift velocities from the measured critical voltages (in Corbino devices) or critical currents (in Hall bar samples). The critical drift velocity v_d in a Hall bar sample is given by

$$v_d = \frac{E_H}{B} = \frac{R_H I_c}{B\,w} \quad , \tag{1}$$

Springer Series in Solid-State Sciences, Vol. 87
High Magnetic Fields in Semiconductor Physics II Editor: G. Landwehr

where B is the magnetic field, E_H the critical Hall field, $R_H = h/\nu e^2$ the Hall resistance, I_c the critical current through the sample and w the width of the Hall bar. For the critical drift velocity v_d in a Corbino device we assume

$$v_d = \frac{U_{DS}}{B\ b} \tag{2}$$

where U_{DS} is the critical source-drain voltage and b the width of the Corbino ring.

Figure 1 shows the dependence of critical drift velocities upon the sample width for 3 Corbino samples and 3 Hall bar devices for a filling factor of $\nu = 2$, i. e. for the splitting between the two lowest Landau levels which is the best resolved splitting. For the broader samples the critical drift velocities are comparable, consistent with the results in [8]. The critical current densities in the Hall bar samples increase from $j_c = 0.76A/m$ to $j_c = 1.1A/m$ with decreasing sample width. This increase could be due to the growing importance of the edge states, or the decreasing influence of inhomogeneities.

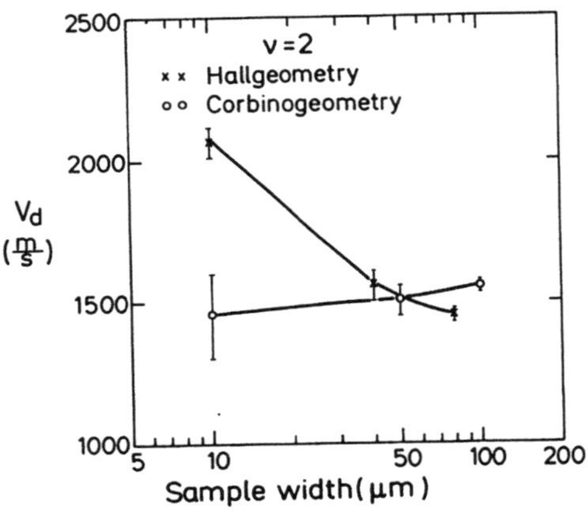

Fig. 1
The critical drift velocity v_d as a function of the sample width for different Corbino and Hall bar samples at a filling factor of $\nu = 2$ (B = 6.7T)

One can think of two possible explanations for the remarkably different behaviour of the Hall bar and Corbino samples. Firstly, the Corbino samples are produced by evaporation and annealing of the ring contacts and lateral diffusion of the contact material could lead to a narrowing of the effective sample width. For the critical drift velocities in the two geometries to be comparable one would have to assume a lateral diffusion of about $2\mu m$. This value would seem unrealistically high, but no studies of the lateral diffusion of AuGeNi contacts have been reported to date. Secondly, the potential and current distributions in the quantum Hall regime are totally different for the two geometries. In a Hall bar the two singularities in the corners always produce some dissipation, whereas in the Corbino geometry there are no dissipative points. The additional heating of the electron gas in the dissipative regions of the Hall bar samples would be expected to lead to smaller critical drift velocities. This is contrary to our observations, possibly due to the different role of edge states in the Hall bar and Corbino geometries.

3. STEPLIKE STRUCTURES

In the breakdown of a very narrow constriction Bliek et al. [2] observed step-like structures which they attributed to a new quantum effect. We observe similar effects in most of our samples, even the very broad structures. Figure 2 shows the dependence of the current in a 500μm wide Corbino sample upon the magnetic field for different applied voltages (near a filling factor of $\nu = 1$). The voltage is varied from 2.5V to 3.8V in steps of 0.1V. One can see the expected narrowing of the minimum and increase of the minimum current with increasing voltage. Some steps in the curves are also distinguishable.

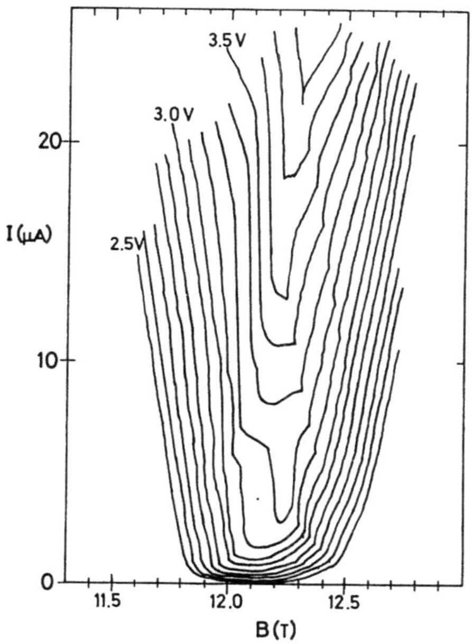

Fig. 2
The current I as a function of magnetic field B near a filling factor of $\nu = 1$ for source-drain voltages from 2.5V to 3.8V in steps of 0.1V. The Corbino ring has a width b of 500μm.

The current measured at these step-like structures varies linearly with the applied voltage. It is difficult for us to imagine that these structures in our broad samples are connected to the number of flux quanta in the sample as proposed in [2] or that collective behaviour of all the electrons is responsible for this effect [9]. A mechanism based on localized states at impurities, as proposed in [10], seems to us to be more realistic. A connection with the quantized conductivity of point contacts [11] could also be possible via electric field dependent depletion of small parts of an inhomogeneous sample.

4. A NEW TYPE OF INSTABILITY

In studying the breakdown at a filling factor of $\nu = 1$ we found a new type of instability at higher electric fields. Figure 3 shows the current as a function of applied voltage for 3 Corbino samples of different width b. At a critical voltage (for example 0.83V in the $b = 100\mu m$ sample) the breakdown of the quantum Hall effect can be observed. At about twice this value a second instability occurs. The critical voltages for this instability correspond to electric fields between

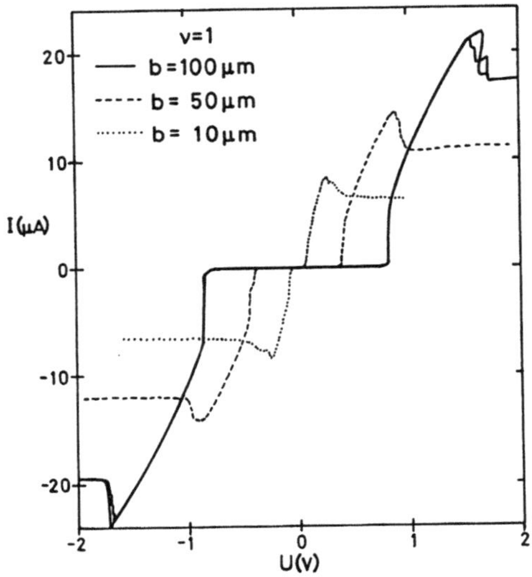

Fig. 3
The dependence of the current I upon the applied voltage U for Corbino devices of different width b at a magnetic field of B = 12.5T, corresponding to a filling factor of $\nu = 1$.

150V/cm for $b = 100\mu m$ and 220V/cm for $b = 10\mu m$. In all three samples we find a single value of $j_s = 0.034 A/m$ for the current density at electric fields above this new critical value. This instability can be observed over filling factor range $\nu < 2$, but is more pronounced at higher magnetic fields. Some indications of this instability were also found in the Hall bar samples. We attribute this effect to the formation of current or potential domains. Kurosawa et al. [12] showed that such instabilities can easily occur in a high magnetic field.

5. CONCLUSION

We have studied the breakdown of the quantum Hall effect for sample sizes from macroscopic to mesoscopic dimensions and found an increasing difference between Corbino and Hall bar geometries. The different critical drift velocities are thought to be due to the totally different potential and current distributions in the two geometries. A new instability observed at higher electric fields is attributed to the formation of current and potential domains.

ACKNOWLEDGEMENT

We gratefully acknowledge the financial support of the Bundesministerium für Forschung und Technologie, Federal Republic of Germany.

1. G. Ebert, K. von Klitzing, K. Ploog, and G. Weimann, J. Phys. C 16, 5441 (1983)
2. L. Bliek, G. Hein, V. Kose, J. Niemeyer, G. Weimann, and W. Schlapp, in High Magnetic Fields in Semiconductor Physics, G. Landwehr (Ed.), Springer Berlin (1987), p. 113
3. L. Bliek, E. Braun, G. Hein, V. Kose, J. Niemeyer, G. Weimann, and W. Schlapp, Semicond. Sci. Technol. 1, 110 (1986)
4. J. R. Kirtley, Z. Schlesinger, T. N. Theis, F. P. Miliken, S. L. Wright, and L. F. Palmateer, Phys. Rev. B34, 5414 (1986)
5. P. G. N. de Vegvar, A. M. Chang, G. Timp, P. M. Mankiewich, J. E. Cunningham, R. Behringer, and R. E. Howard, Phys. Rev. B36, 9366 (1987)

6. M. E. Cage, R. F. Dziuba, B. F. Field, E. R. Williams, S. M. Girvin, A. C. Gossard, D. C. Tsui, and R. J. Wagner, Phys. Rev. Lett. 51, 1374 (1983)
7. J. K. Jain, and S. A. Kivelson, Phys. Rev. B37, 4276 (1988)
8. F. Kuchar, G. Bauer, G. Weimann, and H. Burkhard, in Proc. 17th Int. Conf. Phys. Semicond., World Scientific Singapore (1984), p. 275
9. L. Eaves, and F. W. Sheard, in High Magnetic Fields in Semiconductor Physics, G. Landwehr (Ed.), Springer Berlin (1987), p. 118
10. J. K. Jain, and S. A. Kivelson, Phys. Rev. Lett. 60, 1542 (1988)
11. B. J. van Wees, H. van Houten, C. W. J. Beenakker, J. G. Williamson, L. P. Kouwenhoven, P. van der Marel, and C. T. Foxon, Phys. Rev. Lett. 60, 848 (1988)
12. T. Kurosowa, H. Maeda, and H. Sugimoto, J. Phys. Soc. Jpn. 36, 491 (1974)

p-Type GaAs-(GaAl)As Heterostructures in Tilted Magnetic Fields: Theory and Experiments

W. Heuring[1], *E. Bangert*[1], *G. Landwehr*[1], *G. Weimann*[2], *and W. Schlapp*[3]

[1]Physikalisches Institut der Universität Würzburg, Am Hubland, D-8700 Würzburg, Fed. Rep. of Germany
[2]Walter-Schottky-Institut, D-8046 Garching, Fed. Rep. of Germany
[3]Forschungsinstitut der Deutschen Bundespost, D-6100 Darmstadt, Fed. Rep. of Germany

I. Introduction

In a p-type GaAs-(GaAl)As-heterostructure the inversion symmetry is broken giving rise to a complex subband structure. If a strong magnetic field is applied perpendicular to the interface, a complicated set of Landau levels (LLs) arises. The single levels have no longer a well defined spin character and level crossing can occur. This has been confirmed by magnetoresistance experiments [1]. In a magnetic field perpendicular to the interface well resolved Shubnikov-de Haas oscillations can be observed at temperatures below 1K . The complex oscillatory patterns are well explained by the LL calculations performed by Bangert [2]. In order to put the model to a more stringent test, magnetoresistance measurements were performed in which a substantial longitudinal magnetic field was present. For a comparison with theory the previous calculations were extended by incorporating both a parallel and a transverse component of the magnetic field.

II. Results

The 4x4-Luttinger-Kohn-Hamiltonian was modified taking account of an additional parallel field and diagonalized in a system of wavefunctions ϕ_N, the eigenfunctions in the simple perpendicular case (N=Landau quantum number). The Hamiltonian mixes the wave functions ϕ_N with $\phi_{N\pm1}, \phi_{N\pm2}, \phi_{N\pm3}$ and $\phi_{N\pm4}$. Thus, N is no longer a good quantum number.

The calculations were performed for a carrier concentration $p = 2.55x10^{11}/cm^2$. As the parallel field component we chose $B_p = B_y$ ($B_x = 0$). In order to compare our theoretical results with experimental data, the energy levels were calculated over a wide range of B_p at such perpendicular fields $B_s = B_z$, at which the Landau level filling factor $\nu = 2\pi l^2 p$ (l is the magnetic length) is integer for a given carrier concentration p. The results are shown in Fig.1, where the energies of the 10 highest levels are plotted versus the parallel magnetic field B_p. It can be seen that the levels are significantly shifted due to B_p and that levels 8 and 9 are crossing.

It should be mentioned that this behavior cannot easily be explained as a consequence of enhanced spin splitting due to the parallel field component, because the LLs have no defined spin character and the wave functions are mixed with each other. The theoretical predictions were compared with results from Shubnikov de Haas (SdH) measurements. If the Fermi energy is located midway between two adjacent LLs, activated behavior of the longitudinal resistance R_{xx} can be measured:

Springer Series in Solid-State Sciences, Vol. 87
High Magnetic Fields in Semiconductor Physics II Editor: G. Landwehr

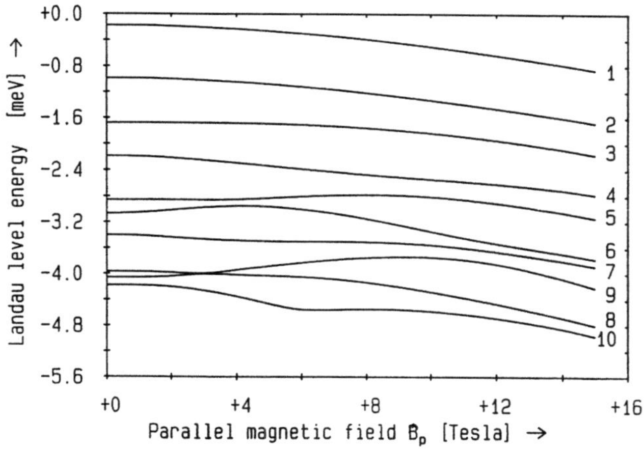

Figure 1: Energy of the 10 uppermost Landau levels as a function of parallel magnetic field B_p. The perpendicular component of the magnetic field is the same for all levels: $B_s = 5.2$ Tesla. A crossing is observed for levels 8 and 9.

$$R_{xx} = const * exp(-E_a/2kT) \quad .$$

E_a is the distance between the two neighbouring mobility edges of the LLs involved. At high perpendicular fields, i. e. at low filling factors, the mobility edges are positioned very close to the center of the LLs so that the measured activation energies are nearly identical with the actual energy difference of the levels. At a filling factor $\nu = 2$ ($B_s =$ 5.2 Tesla) the distance between levels 2 and 3 decreases with increasing parallel magnetic field as can be seen from Fig.1 and 4. At a filling factor 3 ($B_s = 3.6$ Tesla) the difference of energy levels 3 and 4 is varying only within a few percent over a wide range of B_p but increases significantly at $B_p > 10$ Tesla (Fig. not shown).

In our experiments we have investigated two different samples with carrier concentrations of $2.6x10^{11}/cm^2$ and very high mobilities of 120000 cm^2/Vs. at 1.3 K in tilted magnetic fields up to 14 Tesla. With the first sample, #1394 we studied the qualitative behavior at filling factors 2 and 3. The results are plotted in Fig.2 and 3. Fig.2 shows the resistance of the SdH minimum at filling factor 2 as a function of perpendicular magnetic field B_s for various parallel components B_p. B_p was calculated from the angle Φ between applied total field B and the sample plane. The amplitude of the oscillation decreases with increasing B_p, indicating that the distance between the two neighbouring LLs becomes smaller, as theoretically predicted. In Fig. 3 the SdH minima corresponding to filling factor 3 are plotted versus total magnetic field. The amplitude of this oscillation is nearly constant at low parallel field components and becomes more and more pronounced at $B_p > 10$ Tesla indicating that the LL difference is increasing. Another sample, #1458, was also studied quantitatively. The resistance of the SdH minimum at filling factor 2 was measured over a wide temperature range at four certain parallel field components. Analysing the data with respect to the above equation

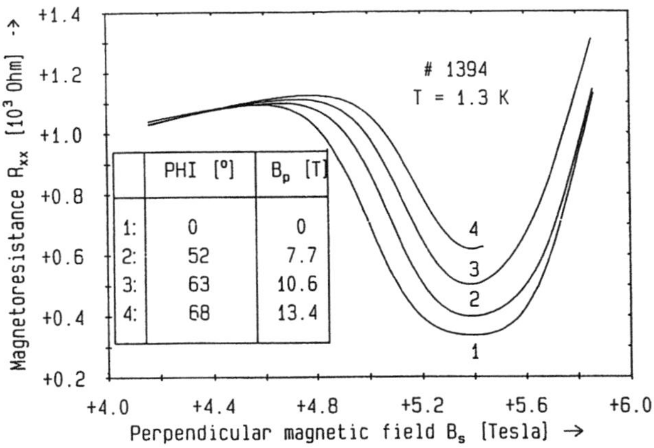

Figure 2: Shubnikov-de Haas minimum at a filling factor 2 for different angles Φ between applied total field B and the parallel component $B_p = B * sin(\Phi)$. The amplitude of the oscillation decreases with increasing B_p as predicted by the calculations.

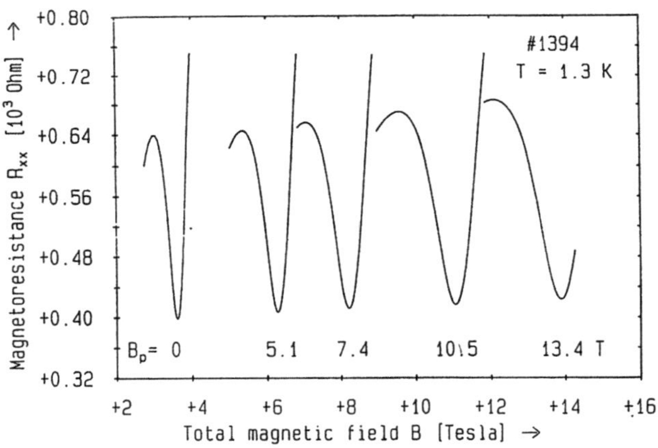

Figure 3: Shubnikov-de Haas minimum at filling factor 3 (B_s = 3.6 Tesla) and different parallel components. The B_p in Tesla are : 0 (curvel), 5.1 (2), 7.4 (3), 10.5 (4) and 13.4 (5). The amplitude of the oscillation is slightly increasing at high parallel fields.

the corresponding activation energies E_a were obtained and plotted versus B_p in Fig.4 (x-symbols). The solid line in Fig.4 represents the difference between levels 2 and 3 from Fig.1. The experimentally deduced activation energies are found to be in good agreement with the theoretical data.

In conclusion, we have shown by a rigorous test the validity of our subband calculations.

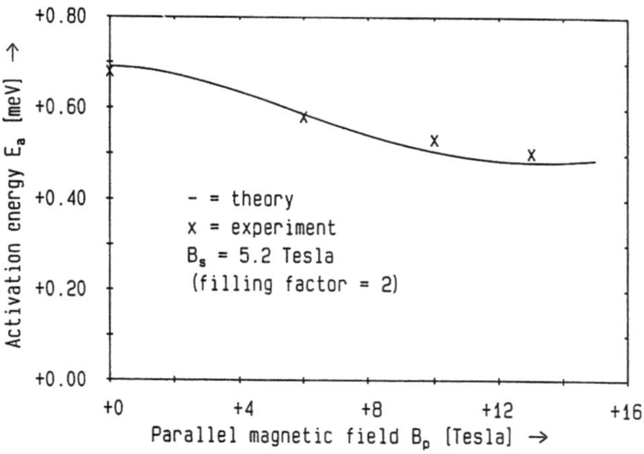

Figure 4: Activation energies at a filling factor 2 with different parallel magnetic components (x-symbols). The hard line represents the calculated energy difference between LLs 2 and 3 from Fig.1.

Acknowledgements

This work was supported by the Deutsche Forschungsgemeinschaft. The calculations were made at the Rechenzentrum, Universität Würzburg.

References:

1. G. Remenyi, G. Landwehr, W. Heuring In: **High Mágnetic Fields in Semiconductor Physics**, ed. by G. Landwehr, p. 166 (Springer 1987)
2. E. Bangert, G. Landwehr: Surf. Sci. **170**, 593 (1986)

Tilted Field Magnetotransport Experiments on Germanium Bicrystals

M. Kraus[1], *H. Mrotzek*[1], *N. Steinmetz*[1], *E. Bangert*[1], *G. Landwehr*[1], *and G. Remenyi*[2]

[1]Physikalisches Institut der Universität Würzburg, Am Hubland, D-8700 Würzburg, Fed. Rep. of Germany

[2]CNRS, CRTBT, and Max-Planck-Institut für Festkörperforschung, Hochfeldmagnetlaboratorium, F-38042 Grenoble, France

1. Introduction

Previous magneto-transport experiments on p-type space charge layers adjacent to the grain boundary of Ge-bicrystals revealed the existence of a quantized, 2-dimensional system. Shubnikov-de Haas (SdH) experiments allowed the identification of two electric subbands, which can be attributed to light and heavy holes [1]. The background magnetoresistance suggested that the g-factor of the holes should be much larger than 2. The determination of the g-factor of a 2D-system is sometimes possible by studying quantum oscillations in tilted magnetic fields [2]. Because the Landau splitting depends on the normal component of the magnetic field and the spin splitting on the total field, the relative magnitude of the splittings varies with the tilt angle and gives rise to interference effects. These were actually observed in previous experiments [3], although no spin splitting of individual SdH peaks was found due to level broadening. Results of recent SdH- experiments in fields up to 23 T on high quality bicrystals produced in Würzburg are in good agreement with previous subband calculations [1]. Again interference effects were found in tilted magnetic fields. They can be explained on the basis of numerical model calculations. The reason for the interference effects is that spin splitting, even if unresolved, gives rise to amplitude variations as a function of the tilt angle. The assigned g^* factors of the light and heavy holes are considerably larger than 2, they are close to 6.

2. Experimental

Germanium bicrystals were grown using the Czochralski pulling technique. The starting point for the crystal growth was two [100] seeds cut from a high quality germanium single crystal and tilted along the [001] axis. The tilt angle Θ between the two seeds was 15 degrees. From simple geometrical considerations, dislocation pipes perpendicular to the pulling direction are supposed to be formed with a distance $D = a/2sin(\Theta/2)$ where a is the lattice constant. Two kinds of samples with parallel or with perpendicular dislocation pipes were prepared (see fig.1). The magnetotransport experiments in the temperature range from 1.5K to 4.2K were performed in magnetic fields up to 23 Tesla.

Springer Series in Solid-State Sciences, Vol. 87
High Magnetic Fields in Semiconductor Physics II Editor: G. Landwehr

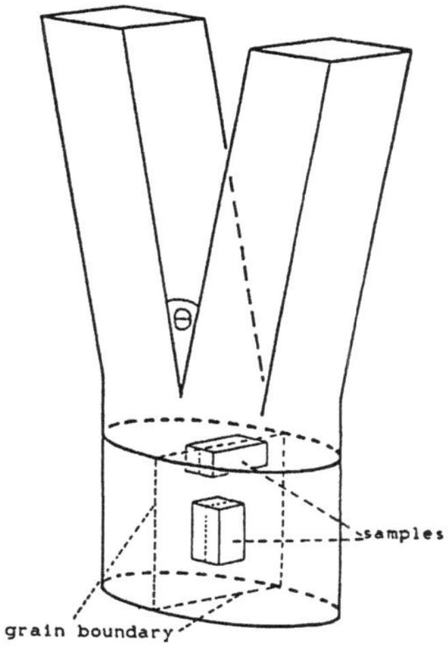

Fig.1. Bicrystal.

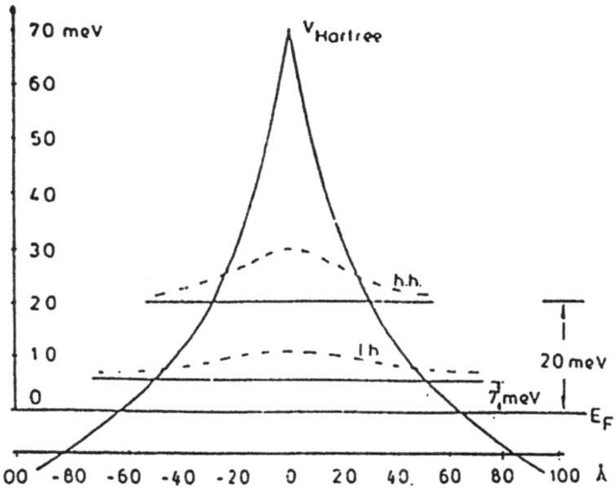

Fig.2. Hartree potential.

3. Subband Calculations

To obtain detailed insight into the band structure of the p-type inversion layer in a bicrystal, self consistent calculations were performed [4]. Contrary to the well-known situation in MOSFETs, the Hartree potential shows inversion symmetry and the spin degeneracy of the bulk states is not lifted. Fig.2 shows the calculated Hartree-potential for a total carrier concentration of $4 * 10^{12} cm^{-2}$. For this concentration two occupied subbands are expected.

4. Experimental Results

The magnetoresistance for different tilt angles θ between the normal vector of the 2D-plane and the magnetic field is shown in fig.3. On top of a strong background magnetoresistance two SdH-oscillations with low and high frequency are superimposed. As expected for 2D-systems, the SdH-extrema are shifted to higher magnetic fields following a $B/cos(\theta)$ law for tilt angles $0° < \theta < 20°$. However, the amplitudes of the SdH-oscillations vanish at distinct angles ($\theta = 25°$ and 57.5° for the high and $\theta = 58°$ and 72° for the low frequency oscillation).

5. Discussion

From theoretical considerations and from the observed, relatively low mobilities, it is clear that the Landau levels formed in a magnetic field are broadened quite substantially in the bicrystal system. To understand the angular dependence of the SdH-amplitudes, it is essential to consider the shape of the density of states (DOS). In a tilted magnetic field, where the Landau splitting $\Omega_c = \hbar eB/m_c$ depends on the normal component of the magnetic field and the spin splitting on the total field, the relative magnitude of

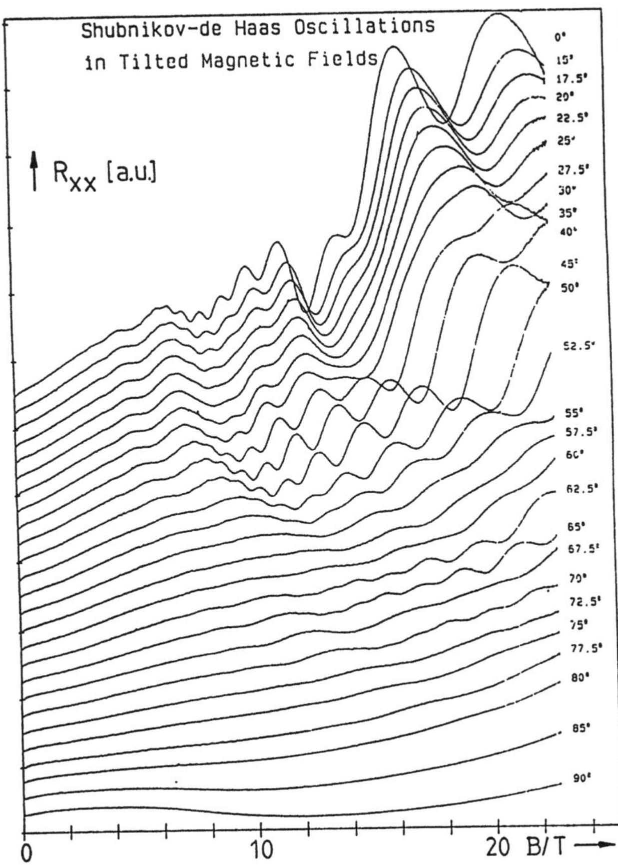

Fig.3. Magnetoresistance for different tilt angles.

the splittings varies with the tilt angle (fig.4). For relatively strong broadening of the Landau levels and particular tilt angles the situation may occur that all energy levels are equally spaced (fig.4b), resulting in a constant DOS. Employing the energy spectrum for 2d-systems in the free charge carrier approximation [2]:

$$E_{i,l,M,\theta} = E_i + \hbar\omega_c cos\theta(l + \tfrac{1}{2} \pm \tfrac{M}{2cos\theta}); \qquad M = \tfrac{1}{2}g^*\tfrac{m_c}{m_o}$$

with E_i being the particular subband energy and θ the tilt angle between the normal of the 2d-system and the magnetic field, the energy levels belonging to one particular subband are equally spaced if $M/cos\theta$ is an integer multiple of 1/2. If at least two such angles θ can be found, the spin factor can be determined. Together with the effective masses of the charge carriers, obtained from the temperature dependence of the oscillation amplitudes, it is possible to assign g*-factors to the charge carriers:

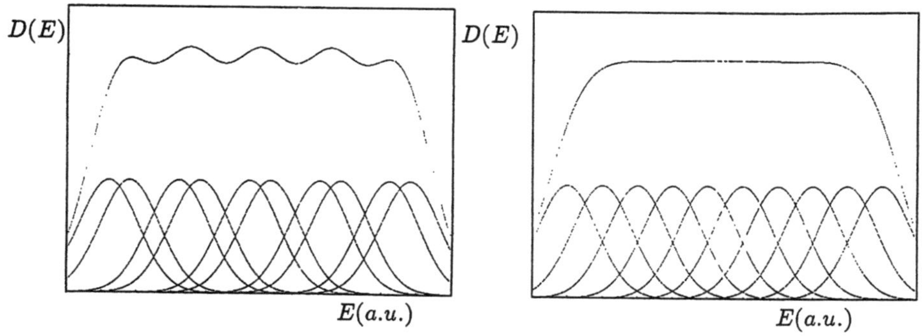

Fig.4. Part a shows the strongly broadened energy levels grouped in pairs. This leads to an oscillating density of states. Part b covers the case where all energy levels are equally spaced, resulting in a constant density of states.

6. Numerical Simulation

With the parameters obtained from the experiment (table 1), a numerical simulation based on a formula given by Roth and Argyres [5] was performed. The series expansion of the magnetoresistance can be restricted to the fundamental oscillation, because a spin splitting is not explicitly resolved. The result of the simulation is shown in fig.5. The numerical data match very well with the experimental results shown in fig.3. The extinctions of the SdH-oscillations occur at the same angular positions as in the experiment.

table 1.	carrier conc. $[10^{12} cm^{-2}]$	effective mass $[\frac{m_c}{m_o}]$	g*-factor
heavy holes	3.75	0.44	6.1
light holes	0.75	0.27	5.9

7. Conclusion

A two-dimensional charge carrier system has been investigated by magnetotransport experiments. The tilting of the magnetic field can be considered as an essential tool to gain insight into the energy level structure of the 2d-system. Despite the strong broadening of the energy levels, values for the g*- factors of the two subbands could be determined. Their values are much larger than the value for free charge carriers, indicating that the interaction between the charge carriers plays an important role. This conclusion is also supported by the large background magnetoresistance. The cyclotron masses obtained from the experiment agree well with the theoretical values [4] and the numerical calculation for the magnetoresis-tance shows good agreement between theory and experiment.

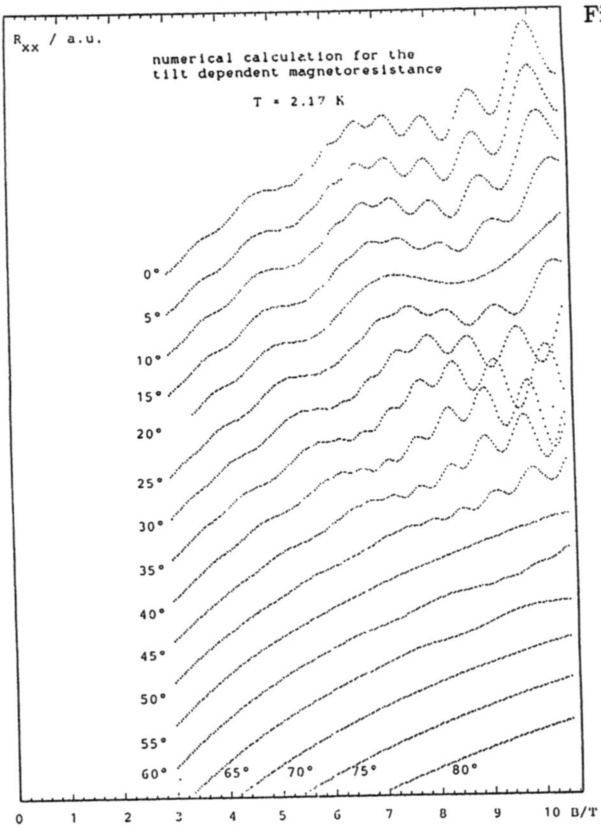

Fig.5. Numerical calculation.

8. Acknowledgements

The authors would like to thank Mr. H.Koubek for growing the bicrystals. The support by Stiftung Volkswagenwerk is gratefully acknowledged.

9. References

[1] G.Landwehr,E.Bangert,S.Uchida Solid-State El. **28**, Nos. 1/2,pp. 171-180, (1985)

[2] See: T.Ando, A.B.Fowler, and F.Stern, Rev. of Modern Physics 54, 551 (1982)

[3] S.Uchida, G.Remenyi, G.Landwehr : In High Magnetic Fields in Semiconductor Physics, ed. by G.Landwehr, Springer Series in Solid State Sciences 71, 353 Springer (1987)

[4] G.Landwehr,E.Bangert: In Two Dimensional Systems, Heterostructures and Superlattices, edited by G.Bauer, F.Kuchar, and H.Heinrich Springer Verlag,(1984)

[5] L.M.Roth,P.N.Argyres: In Semiconductor and Semimetals, Vol.1, ed. by R.K.Willardson and A.C.Beer, Academic Press (1966)

Parallel and Perpendicular Field Magnetotransport Studies of MBE Grown GaAs Doping Superlattices and Slab Doped InSb Formed by Selective Doping with Silicon

R. Droopad[1], *S.D. Parker*[1], *E. Skuras*[1], *R.A. Stradling*[1], *R.L. Williams*[1], *R.B. Beall*[2], *and J.J. Harris*[2]

[1]Blackett Laboratory, Imperial College, Prince Consort Road, London SW7 2BZ, UK

[2]Philips Research Laboratories, Cross Oak Lane, Redhill, Surrey RH1 5HA, UK

Selective doping in the form of slabs or spikes of dopants offers new possibilities for low-dimensional structures. The results for spike-doped superlattices in GaAs can be interpreted in terms of a nearly free-electron model with the electrons occupying a Fermi surface which has 'neck' and 'belly' orbits for the magnetic field applied perpendicular to the doping planes and with magnetic break-down occurring at high fields with B parallel to the doping planes. The carrier concentration is varied by applying hydrostatic pressure and the change in orbit area with carrier concentration is consistent with this model.

Carrier confinement is reported for the first time in selectively doped InSb. A large number of subbands are occupied in slab-doped samples. As the carrier mobilities for the same doping level are an order of magnitude higher than with GaAs, very striking dimensional effects are observed.

1. Introduction to Spike or δ-Doped GaAs

Extremely high local concentrations of silicon donors can be obtained with MBE by the spike or δ-doping technique in which growth is interrupted by shutting off the Ga beam while at the same time opening a silicon shutter. The areal densities which can be achieved by this technique ($\leq 2 \times 10^{13}\,cm^{-2}$), exceed by between one and two orders of magnitude those found in GaAs:GaAlAs heterostructures. The Munich group [1,2] have made extensive magneto-transport measurements on δ-doped layers and from Fourier analysis of the Shubnikov-de Haas effect have shown very pronounced two-dimensional behaviour with up to six subbands occupied.

The sum of the occupancies of the subbands was found to be accurately equal to the doping level until a concentration of $2 \times 10^{13}\,cm^{-2}$ was reached. At this point the carrier concentration saturated with no further increase being found with increasing doping. A concentration of $2 \times 10^{13}\,cm^{-2}$ corresponds to a bulk doping level of $\sim 10^{20}\,cm^{-3}$ if the dopant is localised on a single atomic plane. At this concentration the Fermi energy is extremely high even when many subbands are filled and it seemed reasonable that the saturation in the free carrier concentration as a function of doping might arise from occupancy of localised states associated with higher order conduction bands. However, it was also noted that the relative occupancies of the lowest (i = 0) and first excited subband (i = 1) did not fit self-consistent calculations unless it was assumed that the silicon diffused significantly away from the initial dopant plane. Approximately 50Å diffusion was required at a doping level of $5 \times 10^{12} cm^{-2}$ and 200Å at $2 \times 10^{13} cm^{-2}$ to fit the experimental data. These areal densities would then both correspond to a bulk density of $10^{19} cm^{-3}$ which is close to the value at which silicon ceases to substitute solely on the gallium site and act as a donor in bulk growth.

In order to determine the extent to which resonant impurities were playing a role in the δ-doped layers, high pressure measurements were undertaken as a collaborative project between Imperial College and the Technical University at Munich [3]. The effect of the application of high pressure was unexpected in that the first order effect was an increase in amplitude rather than a very pronounced shift in position of the Shubnikov-de Haas oscillations. The increase in strength is associated with an improvement in the mobility of the carriers. This effect is particularly pronounced in the case of lowest (i = 0) subband. In the case of the results shown in Fig. 1 for sample PD14 (grown in the Imperial College MBE Facility) the increase in mobility approaches an order of magnitude for the i = 0 subband at 18 kbar. In contrast, the decrease in carrier concentration is about 25% for i = 0 subband with proportionally greater falls for the higher order subband. An identical effect was seen with samples grown at Stuttgart. The key to the understanding of this effect lies in the diffusion of the dopants coupled with a consideration of the extent and symmetry of the wavefunctions in the z-direction for the different subbands. The i = 0 subband has an antinode at the original doping plane and for $n = 5 \times 10^{12} cm^{-2}$ and no dopant diffusion, self-consistent calculations show that $\langle z^2 \rangle \sim 25$Å. In contrast $\langle z^2 \rangle \sim 75$Å for the i = 1 subband and, because there is a node in the wavefunction at z =0, the probability of finding the electron within 20Å of the doping plane is extremely low. Suppose that the dopants diffuse a distance of 50Å (a typical distance for $n = 5 \times 10^{12} cm^{-2}$) and that, for simplicity, the diffusion profile is a slab of total extent 100Å. The extent of localised electronic states associated with higher order conduction band minima is likely to be < 10Å as the binding energy to the minima which

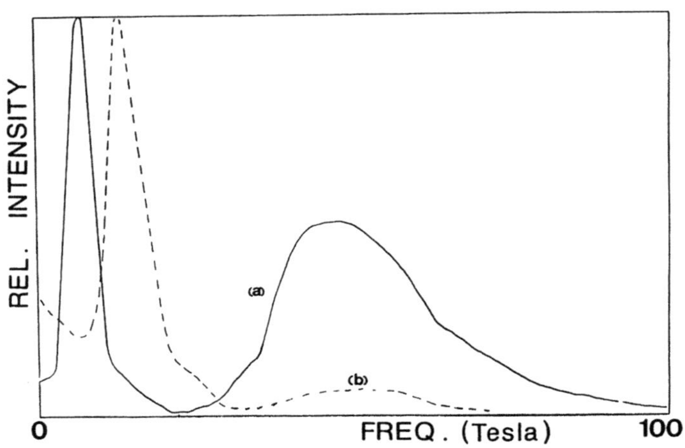

Fig. 1 Fourier analysis of the Shubnikov-de Haas measurements for a single plane spike doped GaAs sample grown at ICST showing the increase in amplitude of the i = 0 contribution on applying hydrostatic pressure. (a) 18kbar (b) 10kbar

give them their character is < 100 meV. Consequently, these deep levels will have eigenvalues which are a function of distance z and will have a minimum value when z = 0 as this is where the potential energy is a minimum. Thus the impurities will become neutralised first at z = 0 as pressure is increased and then a neutralised zone will spread outwards from the z = 0 plane with increasing pressure. For the simple model chosen, when 50% of the donors are neutralised, this zone will extend ~ 25A from z = 0. The carriers in the i = 0 subband will then move in a zone where there are no ionised impurity scatterers and a spatial separation of scatterers and carriers will have been achieved, giving the mobility enhancement observed. In contrast, although the carriers in the i = 1 subband will sense the fall in the number of scatterers, the scattering sites which have disappeared are from a region which is remote from where the carriers are located. Although these ideas remain to be put on a quantitative footing, they can readily explain the features observed and provide corroborative evidence for diffusion of the dopant impurities which was first deduced from relatively small changes in relative occupancy of the lower subbands compared with predictions of self-consistent theory calculated with the assumption that the dopants were located on a single atomic plane.

2. Spike doped superlattices

Evidence that the diffusion of Si deduced in [3] is of the correct order is provided by studies of a δ-doped GaAs:Si superlattice with 80Å period grown by MBE at the Philips Research Labs at Redhill.

In the absence of an applied hydrostatic pressure, no Shubnikov-de Haas oscillations were observed up to magnetic fields of 13T. With an applied hydrostatic pressure of 16 kbar, however, oscillations became obvious, Fig. 2(a). The Fourier transform power spectrum for this pressure is shown in Fig. 2(b).

Figure 3 (a) shows Shubnikov-de Haas data at 12.5 kbar obtained with the magnetic field in the plane of the sample surface. Oscillations are again observed, with two frequencies immediately apparent.

The appearance of multiple frequencies in this geometry points to some coupling between the spike-doped wells, since a completely two-dimensional system would be expected to show only a diamagnetic shift and depopulation of the subbands. The associated Fourier spectrum is shown in Figure 3(b).

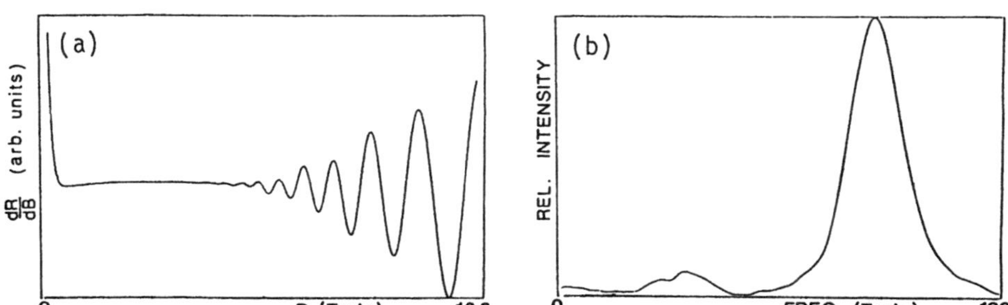

Fig. 2 (a) Shubnikov-de Haas oscillations, at 16kbar, with magnetic field along the surface normal (b) Associated Fourier transform power spectrum

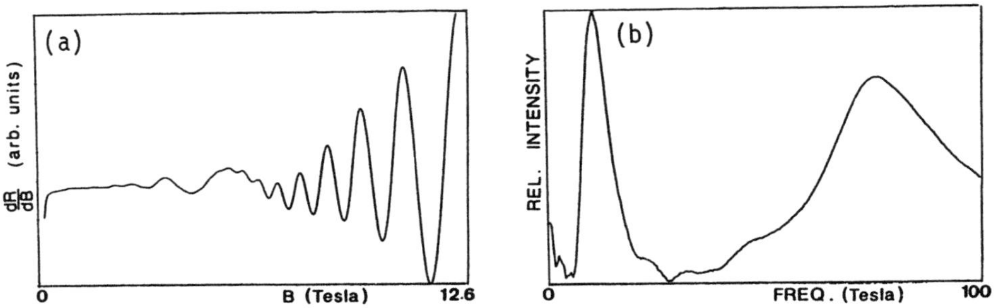

Fig. 3 (a) Shubnikov-de Haas oscillations, at 12.5 kbar, with the magnetic field parallel to the sample surface (b) Associated Fourier transform power spectrum

The data may be analysed in terms of the dimensions of the superlattice Fermi surface depicted in Fig. 4, which may be constructed as a slight distortion of the free electron Fermi sphere across the superlattice zone boundaries.

For a magnetic field of arbitrary direction, the Shubnikov-de Haas oscillations measure the extremal cross-sections of the Fermi surface in a plane perpendicular to the field direction. For a magnetic field normal to the sample surface, the frequencies observed would be expected to correspond to the 'Belly' orbit and the 'Neck' orbit in Fig. 4. A simple geometrical construction gives

$$A_{Belly}=A_{Neck}+\pi K_B^2, \qquad A=\pi(3\pi^2N)^{2/3},$$

where N is the effective carrier density. For the 'Neck' concentration of $6.7\times10^{17}\,cm^{-3}$, observed at 16 kbar, we would predict a 'Belly' concentration of $3.6\times10^{18}\,cm^{-3}$, which is to be compared with an observed concentration of $3.4\times10^{18}\,cm^{-3}$. The agreement is clearly excellent.

To understand the appearance of two frequencies in the data with B parallel to the surface, we assume that at high magnetic fields a break-

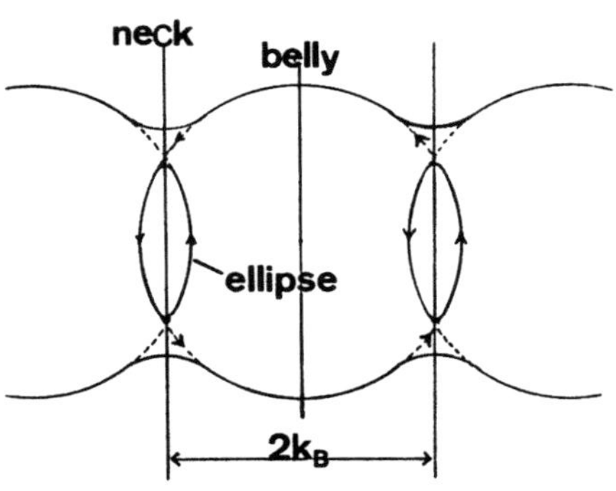

Fig. 4 Superlattice Fermi surface for a Fermi energy in the second mini band

down of Bragg reflection from the superlattice periodicity occurs, so that the dotted 'belly' orbit as well as the low field 'elliptic' orbit of Fig. 4 becomes possible. This phenomenon is exactly analogous to that of Magnetic Breakdown in the de Haas-Van Alphen oscillations of HCP metals [4]. The condition for magnetic breakdown to occur is $\hbar\omega_c$ = 11 meV at 6T, and E_F = 150 meV. Consequently $E_G \sim 40$ meV. A simple geometrical argument based on Fig. 4 gives the relation between the extremal areas $A_{ellipse}$ and A_{belly} as,

$$A_{ellipse} = 2\{\Theta A_{belly}/\pi - K_B(A_{belly}/\pi - K_B{}^2)^{1/2}\}, \cos\Theta = \{\pi K_B{}^2/A_{belly}\}^{1/2}.$$

Figure 5 shows the variation of the 'ellipse' concentration with 'belly' concentration, as the latter is varied with applied hydrostatic pressure. No 'ellipse' concentration is observed below $N_{belly} = 2\times10^{18}\,cm^{-3}$ in good agreement with theory, indicating that a minimum doping level of $1.5\times10^{12}\,cm^{-2}$ is required to populate the elliptic orbit at this superlattice periodicity. Extensions of the nearly free electron model are required to include zone boundary effects.

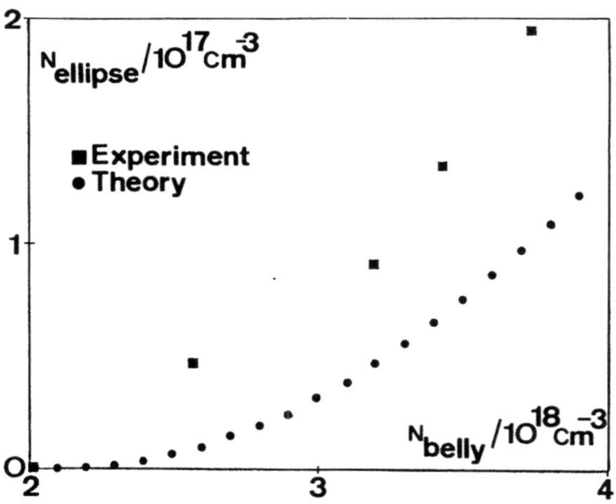

Fig 5 Variation of $N_{ellipse}$ with N_{belly} as the carrier concentration is varied using hydrostatic pressure

3. Shubnikov-de Haas in Slab-doped InSb

(a) Electrical properties of MBE InSb

InSb epilayers of nominal thickness 1μm were grown on GaAs substrates. GaAs provided better electrical isolation than InSb with the additional advantages of cheapness, better structural quality and easier cleaning procedures. The penalty lies in the large lattice mismatch (14%) between the two materials which gives an extremely high density of misfit dislocations at the interface. However, TEM studies show that the density of threading dislocations falls to relatively low levels ($< 10^9\,cm^{-2}$) after distances of the order of 1/4 μm. The prime purpose of the series of samples grown was to perform studies of the incorporation and possible amphoteric behaviour of silicon dopants while at the same time providing information on how the electrical quality of the material was affected by the defects present. Earlier MBE studies of InSb on GaAs [5,6] have suggested a severe degradation of the mobility at distances of less than 1 μm from the interface and also a slow conversion from p-type to n-type character with increasing distance. With this in mind the silicon was introduced in a slab 850Å in thickness whose depth was varied in different samples. At this thickness good depth resolution could be achieved while at the same time allowing the electron gas to remain approximately three-dimensional and ensuring that the slab conductance was considerably greater than that of the background undoped material at the silicon doping levels achieved ($\leq 3 \times 10^{18} cm^{-3}$). Nevertheless, some systematic corrections to the measured electrical parameters were undoubtedly present because (1) of the parallel conduction from the undoped material (2) the two-dimensional nature of the electron gas because the confinement energies are large for InSb even for thicknesses of the order of 1000A because of the low effective mass of this material (3) boundary scattering because the mean free path in bulk material at the carrier concentrations and mobilities concerned is more than an order of magnitude greater than the slab thickness (1.2 μm at $3 \times 10^{18} cm^{-3}$). Nevertheless the electrical measurements over a wide range of doping levels (1×10^{17} to $3 \times 10^{18} cm^{-3}$) gave values of mobility identical to the bulk values. Shubnikov-de Haas measurements gave the same carrier concentrations as low-field Hall measurements within experimental error and the Shubnikov-de Haas oscillations tended to persist to rather lower magnetic fields than with bulk samples substantiating the measurement of high mobilities in the slab-doped material. The Shubnikov-de Haas measurements did, however, show very striking two-dimensional behaviour, with strong differences in the results for the magnetic field applied perpendicular to the slab and in the plane of the slab, as will be discussed in the next section. The low field electrical measurements also showed that (1) while 100% activation of silicon as a donor was achievable at low growth temperatures, at higher growth temperatures $N_A/N_D \sim 0.5$ (2) the 77K mobility at the InSb/GaAs interface was more than a factor of two lower than when the slab was positioned 0.5μm from the interface (3) there was a detectable fall in the mobility as the slab approached the surface, although this may have been due to the loss of $\sim 5 \times 10^{11} cm^{-2}$ carriers to surface states rather than an increase in the scattering centres.

3(b). The Shubnikov-de Haas Effect with Slab-Doped InSb

Figure 6 shows the very pronounced Shubnikov-de Haas oscillations which are observed when the magnetic field B is applied perpendicular to a slab doped with a high carrier concentration ($\sim 3 \times 10^{18} cm^{-3}$ i.e. $2.5 \times 10^{13} cm^{-2}$). It should be realised that at this carrier concentration many subbands are occupied because of the low effective mass of the electrons and because the slabs are relatively wide (it is estimated on a parabolic model that approximately ten bands are occupied - a more appropriate non-parabolic calculation has not been attempted yet). Above 6T the diameter of the cyclotron orbits becomes less than the slab thickness and hence the system becomes effectively three-dimensional. The carrier concentration deduced from the period of the high frequency peaks agrees well with the carrier concentration deduced from low field Hall measurements. Although in Fig. 6 the low field peaks appear to form a single series, expansion of the field scale in this region reveals the presence of several distinct series as would be expected because of the occupancy of a large number of subbands.

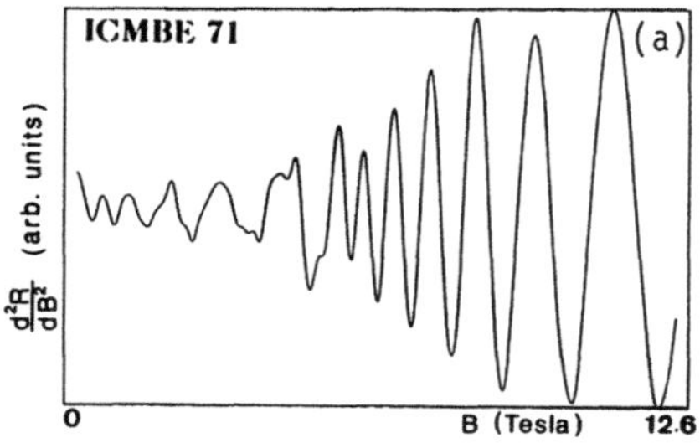

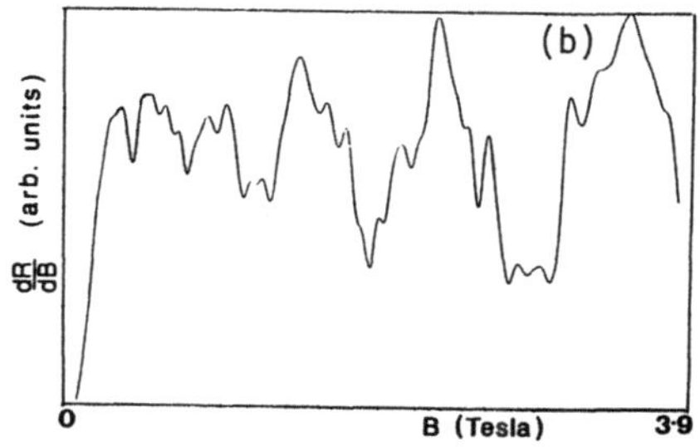

Fig 6(a) Second derivative of the magnetoresistance against magnetic field for a heavily doped slab 850Å wide (n = 3 x 10^{18} cm^{-3}) within InSb with B perpendicular to the slab showing quasi-3D Shubnikov-de Haas oscillations above 6T and 2-D behaviour below this field

Fig 6(b) shows an expanded low field plot of the first derivative of the magnetoresistance showing complicated structure due to the occupancy of many subbands

Figure 7 shows the results of applying the field parallel to the slab. The high field peaks are in almost identical positions to those found in the perpendicular orientation. However, the low field peaks are radically different in character and extend to remarkably low fields as is illustrated dramatically in the differential recordings. Oscillatory terms start at approximately 0.2T. There are approximately seven peaks which are not part of the high field series. Instead of having the familiar 1/B dependence on field the separation of these peaks increases only slowly with field. Qualitatively this striking difference between peaks observed with B in the two orientations reproduces at lower carrier concentrations as can be seen from Fig. 8.

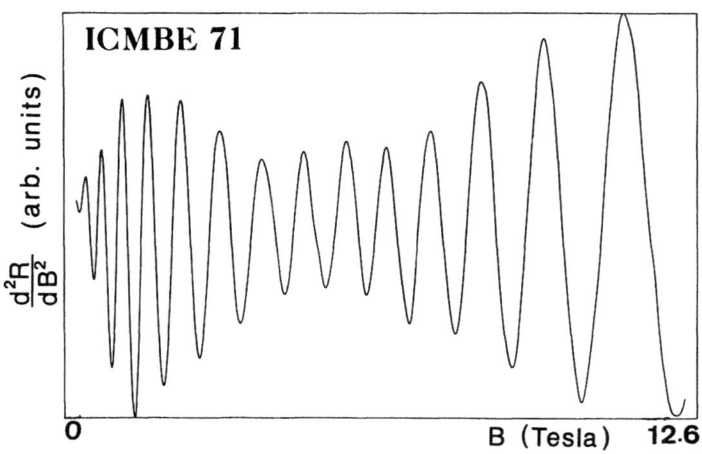

Fig 7 shows the second derivative of the magnetoresistance against magnetic field for the same InSb sample used for Fig. 6 with B applied parallel to the plane of the slab. Again quasi-3D behaviour is observed above 6T but radically different behaviour is observed at low fields

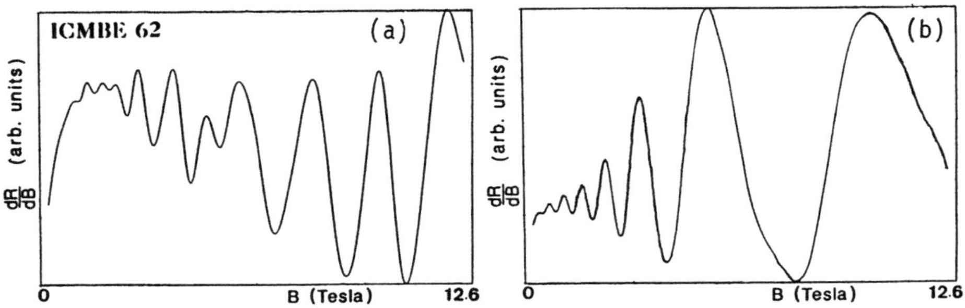

Fig. 8 shows the first derivative of magnetoresistance for a slab doped at 2×10^{17} cm^{-3} (a) B perpendicular to the slab (b) B in the plane of the slab

The likely explanation for this behaviour lies in the diamagnetic emptying of the higher order subbands with increasing field in the parallel orientation, i.e. the seven peaks correspond to the emptying of seven subbands. It should be noted however that the difference between $<z^2>$ for the different subbands is not great so these subbands may be close to the top of the well produced by the slab. Complicating a more quantitative fit may be (i) the modification of the density of states by skipping orbits formed at the edge of the well (the mean free path is more than an order of magnitude greater than the well width) (ii) possible size effects and a jump in the scattering time when the diameter of the cyclotron orbits become less than the thickness of the slab. Whether (i) or (ii) is more significant depends on how smooth the effective potential barrier is at the edge of the slab-doped region. In detailed calculations non-parabolicity, spin-splitting and the real form of the confining potential, including depletion effects at the edges which will depend on the background contamination in the 1 μm layers, will have to be included. Spin splitting is very evident in the recordings for the 2×10^{17} cm^{-3} sample but is not apparent with the higher carrier concentration.

These results are only in the preliminary stage but are presented to demonstrate the exceptionally good electrical quality of the InSb samples. Spike and selective doping has already produced some fascinating results with GaAs, as shown in the earlier sections. The mobilities achieved in InSb for the same carrier concentrations are already an order of magnitude higher. The potential for striking results for quantum wires and dots is obvious.

References

1. F. Koch, A. Zrenner, M. Zachau: Two Dimensional Systems, Springer Ser. Solid-State Sci., 67, 175 (1986)
2. A. Zrenner, H. Reisinger, F. Koch, K. Ploog, J.C.Maan: Phys. Rev. B33, 5607 (1986)
3. A. Zrenner, F. Koch, R.L. Williams, R.A. Stradling: To be published
4. M.H. Cohen, L.M. Falicov: Phys. Rev. Lett. 7, 231 (1961)
5. M. Yano, T. Takara, M. Kimatu: Phys. Stat. Solidi (a) 54, 707 (1979) Jap. J. App. Phys. 18, 387 (1979)
6. A.S. Noreika, M.H, Francombe, C.E.C. Wood: J. App. Phys. 52, 7416 (1981)

Magnetotransport on HgTe/CdTe Superlattices Grown by LAMBE

L. Ghenim[1], *R. Mani*[1], *J.R. Anderson*[1], *and J.T. Cheung*[2]

[1]Joint Program for Advanced Electronic Materials, Department of Physics, University of Maryland, and Laboratory for Physical Sciences, College Park, MD 20742, USA

[2]Rockwell Science Center, Thousand Oaks, CA 91360, USA

Low temperature Shubnikov-de Haas oscillations and the Hall effect have been studied on a HgTe/CdTe superlattice grown by laser MBE. The results indicate that three subbands in the quantum well contribute to the conductivity. A lineshape fit of the SdH data also indicates that the intermediate SdH component has an electron effective mass ratio which is five times larger than the mass ratio of the lower frequency SdH component.

1 Introduction

The inverted band structure of semimetallic HgTe makes the HgTe/CdTe superlattice (SL) a type III system and a promising candidate for infrared detectors /1/. This SL system is superior to the bulk (Hg,Cd)Te alloy for infrared applications partly because box quantization of the quantum wells makes the band gap of the SL increase with decreasing HgTe layer thickness without depending upon alloying effects as in the bulk alloy. High quality HgTe/CdTe SL samples /2/ which show the integer quantum Hall effect (IQHE) and large Shubnikov-de Haas (SdH) oscillations have already revealed new semiconductor physics such as antilocalisation even though there have been only a few comprehensive transport studies of this unique type III system. Previous investigators /2/ have suggested multi-subband conduction to explain the complicated SdH pattern observed in this system but they were unable to characterize these carriers due to weak SdH oscillations. Our aim is to provide further insight into the band structure of the HgTe/CdTe system and also determine the relations between subband occupancy and SdH frequency in this multi-subband system. Thus, we compare Hall effect measurements with SdH oscillations observed using field modulation techniques. The SdH oscillations have been fitted with the Best Recursive Fit (BRF) technique /3/ in order to separate the oscillatory contributions of different subbands and determine the effective masses and scattering temperatures associated with these subbands.

2 Experiment

The HgTe/CdTe SL used in this study was grown by laser MBE /4/. The low growth temperature used in this technique is believed to produce less interdiffusion between the HgTe and the CdTe layers. Thus, these samples are superior to conventional MBE material. The sample consists of 12 periods of 90A HgTe and 40A CdTe on a (100) CdTe substrate. The magnetotransport measurements were made in the Van der Pauw configuration in magnetic fields, B, up to 70kOe at temperatures, T, below 20K. The SdH oscillations were enhanced using standard field modulation techniques. In order to simplify the fit to the SdH data, we have assumed that the oscillations obey the following semi-empirical formula which is valid even for this narrow-gap system in the high temperature, low field limit:

Springer Series in Solid-State Sciences, Vol. 87
High Magnetic Fields in Semiconductor Physics II Editor: G. Landwehr

$$\rho = \rho_o + \Sigma_i A_i \exp(-\lambda(T+T_D^{i})m^*_i/mB) \sin(2\pi F_i/B + \delta_i). \quad (1)$$

Here, m^*_i/m is the electron effective mass ratio, T_D^i is the Dingle temperature, F_i is the SdH frequency, δ_i is the infinite field SdH phase of the i'th SdH component and λ=146.9 kOe/K. We point out that the Dingle temperature /5/ measures the broadening, Γ, of the individual Landau levels, $\Gamma=\pi k_B T_D$, and is related to the single particle lifetime, τ_s, through $\tau_s=\hbar/(2\pi k_B T_D)$.

3 Data and Discussion

In Fig. 1, we have shown a trace of the SdH oscillations observed in this HgTe/CdTe SL at T=1.55K. The data implies more than two SdH components since the figure does not show a simple beating pattern in 1/B. In order to extract the SdH frequencies, we have plotted in Fig.2 the $d\rho_{xx}/dB$ extrema positions in 1/B vs integers; the half-cycle plot shows three linear regions with increasing 1/B which correspond to SdH frequencies F_1=238kOe, F_2=65kOe, and F_3=12kOe respectively. These frequencies are temperature independent to within experimental accuracy for T<20K.

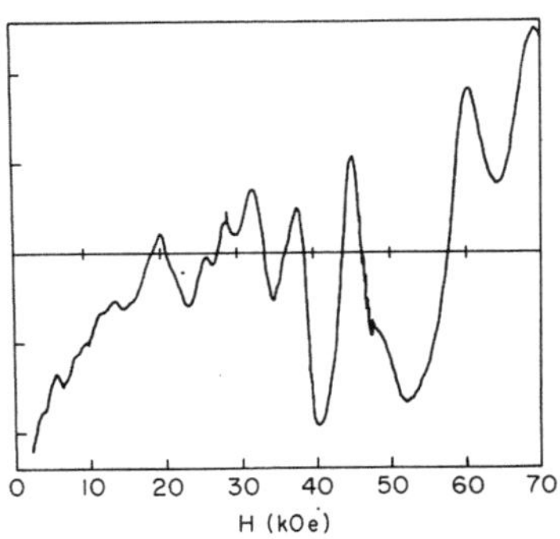

Fig. 1

SdH oscillations $d\rho_{xx}/dB$ vs the magnetic field H at 1.55K for a HgTe/CdTe SL.

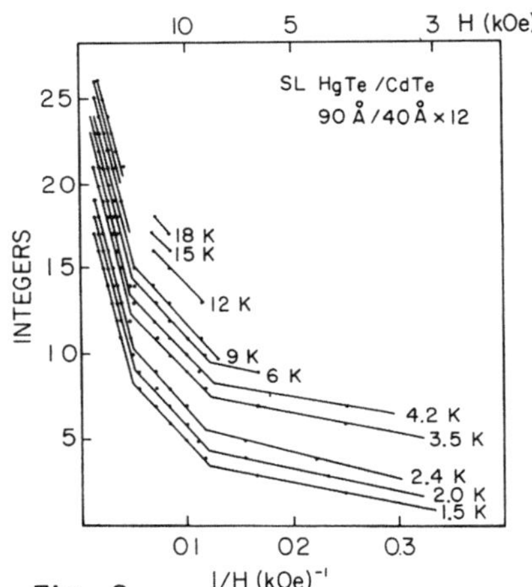

Fig. 2

Plot of the $d\rho_{xx}/dB$ extrema positions in 1/B vs integers for temperature up to 18K.

In Fig. 3, we have shown the lineshape fit to the SdH data over the field range 3-20kOe (Fig. 3a), 8-20kOe (Fig. 3b), and 25-60kOe (Fig. 3c) for T=1.5 K. The temperature dependence of the exponential terms in eqn.1 were used to measure the effective masses and the Dingle temperatures (see Table 1). We were unable to determine the mass for the i=1 SdH component partly because oscillation associated with this component occurred in the high field, quantum Hall regime. We point out that the i=2 SdH component (F=66kOe) has an effective mass ratio which is a factor of five times larger than m^*/m for the i=3 SdH component (F=12kOe). These results are consistent with our observation that SdH peaks associated with the F=66kOe

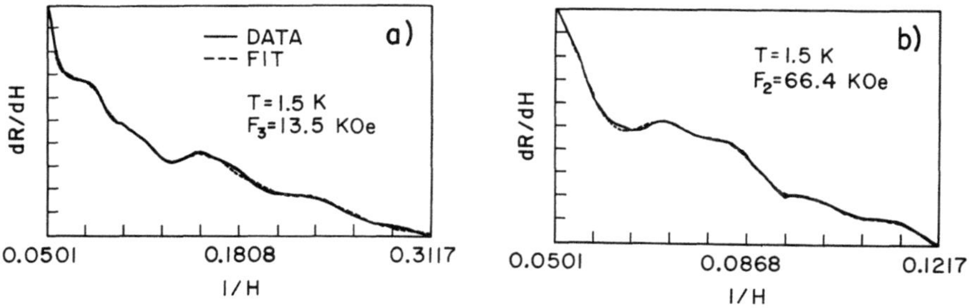

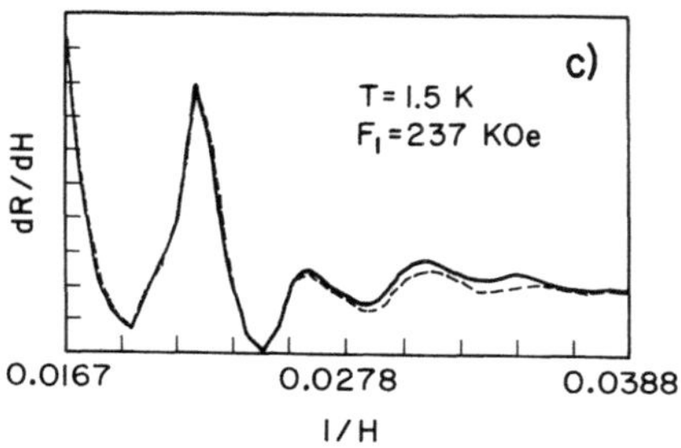

Fig. 3

Lineshape fit to the SdH data over the field range a) 3-20kOe, b) 8-20kOe, c)25-60kOe.

component are damped more strongly with temperature than the peaks due to the other two components. We estimate possible 20% error in our mass estimates due to approximating the SdH oscillations by simple damped sinusoids and also due to random errors. The Dingle temperatures indicate that sharp Landau levels are associated with these SdH components, i.e. $\hbar\omega \gg k_B T_D$.

Low field Hall effect measurements indicated a linear-in-magnetic-field Hall voltage which is independent of temperature for T<20K. The Hall mobility, μ, increased from 48,000cm^2/V-s at T=77K to 50,000cm^2/V-s at T=4.2K.

The SdH frequencies do not measure the individual subband populations in systems with many occupied subbands when the Fermi energy oscillates with the magnetic field and the degeneracy is independent of subband index /6,7/. As originally pointed out by Vinter and Overhauser /6/, the highest SdH frequency measures the total density rather than the population of the lowest subband at high magnetic fields when $\hbar\omega^1 \gg k_B T_D{}^1$. Thus, we have determined the subband populations given in Table 1 as follows: First, the highest SdH frequency measures the total density, i.e. $N_S = 2eF_1/\hbar = 1.1\times10^{12}$cm^{-2}; Second, we assume the lowest SdH frequency measures the occupation of the i=3 subband, $N_3(B) = N_3(0) = 2eF_3/\hbar$, since the large effective mass ratio of the i=2 subband produces a quasi-static Fermi level over the range of fields (B < 10kOe) where oscillations associated with the i=3 subband are observed. Third, the average population of the i=1 and i=2 subband differ from their zero-field value over the range of fields where oscillations associated with these subbands are observed when i=3 subband becomes completely depleted of its carriers. Thus, $N_2(B) = N_2(0) + \delta N_3$ and $N_1(B) = N_1(0) + (1-\delta)N_3 = N_s - N_3(0) - N_2(B)$. Here, $\delta \sim N_2(B)/N_1 \sim$

F_2/F_1 measures the fraction of carriers transferred from the i=3 subband to the i=2 subband in the extreme quantum limit for the i=3 subband.

Using these assumptions, we have found that the population of the i=1 subband is roughly 30% smaller than the value which would be obtained by assuming the highest frequency measures the i=1 subband density rather than the total density. The subband positions with respect to the Fermi energy, given in Table 1, were calculated from the measured density and m^*/m neglecting non-parabolicity. We point out that the second and third subbands are equidistant from E_F although their mass ratios differ by a factor of five; this feature is difficult to understand in the context of conventional band structure models /8/.

Table 1:

i	F(kOe)	m^*/m	T_D(K)	$N_S(cm^{-2})$	E_F-E_i (meV)
1	237	----	---	$8x10^{11}$	--
2	66	0.05	0.7	$3x10^{11}$	15
3	13	0.01	5.5	$6x10^{10}$	18

In summary, our study of the SdH effect in the HgTe/CdTe SL has revealed, for the first time, that three subbands are occupied in this type III system. We have found that m^*/m for the i=2 subband is roughly five times larger than m^*/m for i=3 subband although the i=2 and i=3 subbands are close together in energy. We are currently investigating diamagnetic SdH oscillations in the $B_{//}$ configuration to check the self-consistency of our results and also determine the spatial extent of the wavefunctions.

4 Acknowledgements

We would like to acknowledge many useful discussions with Dr. D. R. Stone and Dr. W. B. Johnson. This research was supported in part by the Defense Advanced Research Projects Agency (DARPA) and (A.R.O.) Army Research Office under grant DAAG29-85-K-0052.

5 Literature

1. J. Schulman, T. McGill: Appl. Phys. Lett. 34, 663 (1979).
2. N. P. Ong, J. Moyle, J. Bajaj, J. T. Cheung: J. Vac. Sci. Tech. A5, 3079 (1987); J. Moyle: Ph.D Thesis (unpublished).
3. J. R. Anderson, P. Heimann, W. Bauer, R. Schipper, D. R. Stone: Conf. Ser. Inst. Phys. No. 39, 81 (1978).
4. J. T. Cheung, G. Nizawa, J. Moyle, N. P. Ong, B. M. Paine, T. Vreeland: J. Vac. Sci. Tech. A4, 4 (1986).
5. R. B. Dingle: Proc. Roy. Soc. A211, 517 (1952); T. Ando, A. B. Fowler, F. Stern: Rev. Mod. Phys. 54, 437 (1982).
6. B. Vinter, A. Overhauser: Phys. Rev. Lett. 44, 47 (1980).
7. H. L. Stormer, Z. Schlesinger, A. Chang, D. C. Tsui, A. C. Gossard, W. Wiegmann: Phys. Rev. Lett. 51, 126 (1983).
8. Y. R. Lin-Liu, L. J. Sham: Phys. Rev. B32, 5561 (1985); G. Bastard: Surf. Sci. 170, 426 (1986).

Quantized Particle Motion in High Magnetic Fields

J. Riess

Centre de Recherches sur les Tres Basses Temperatures,
CNRS, BP 166X, F-38042 Grenoble Cedex, France

1. Introduction

Recently the dynamics of charged particles in an electric field was discussed from a novel point of view by investigating the evolution of the characteristic structure of the phase $\eta(x,y,...)$ of the wave functions $\psi(x,y,..) = |\psi(x,y,..)|\exp(i\eta(x,y,..)$ [1]. It was shown for a wide class of Hamiltonians, that this structure behaves according to general laws. In the presence of disorder these laws can change dramatically leading to entirely non-classical particle dynamics. This has been illustrated in the case of one-dimensional conductance without magnetic field [1]. In the present paper we show that analogous non-classical behaviour is possible in high magnetic fields. We give an example of a two-dimensional system, where disorder leads to quantized, discontinuous motion of particles between distinct sites. This may give new insight into the mechanisms of Hall conductance in disordered systems.

We consider charged particles (charge q, mass m) on an infinitely long strip in x-direction with finite width L_y, subject to a magnetic field $\mathbf{B} = (0,0,B) = \mathrm{curl}(0,Bx,0)$ in z-direction and a homogeneous electric field E_y in y-direction. In addition a static potential $V(x,y) = V(x) + V^1(x,y)$ is present, where V(x) has a first derivative, which (piecewise) can be considered as approximately constant over the range of the magnetic length $\lambda = |\hbar c/qB|^{1/2}$. $V^1(x,y)$ is a disorder potential which for the moment is not further specified, except that it has to break the homogeneous symmetry in y-direction.

The Hamiltonian has the form

$$H = \left(\frac{1}{2m}\right)\left\{\left[\frac{\hbar}{i}\frac{\partial}{\partial x}\right]^2 + \left[\frac{\hbar}{i}\frac{\partial}{\partial y} - \frac{q}{c}[Bx + \phi(t)/L_y]\right]^2\right\} + V(x,y). \tag{1}$$

Here the electric field E_y is expressed by the time-dependent vector potential $A_y(t) = -cE_y t = \phi(t)/L_y$. Since we neglect edge effects, we choose periodic boundary conditions in y-direction: $\psi(x,y,t) = \psi(x,y + L_y,t)$.

2. Absence of Disorder

We first treat the case $V^1(x,y) = 0$. Here the adiabatic solutions ψ of the time-dependent Schrodinger equation, i.e., the eigenfunctions of H, where $\phi(t)$ is considered as a parameter, can be obtained approximately by expanding V(x) around the centre of each Landau orbital. For simplicity we neglect derivatives of V(x) of order higher than one (but this assumption is not essential for our main conclusions). A straightforward calculation gives

Springer Series in Solid-State Sciences, Vol. 87
High Magnetic Fields in Semiconductor Physics II Editor: G. Landwehr

$$\psi^n{}_p(x,y) = \exp(i2\pi py/L_y)\chi^n{}_p(x), \text{ p integer,} \tag{2}$$

where $\chi^n{}_p(x)$ is the product of a Hermite polynomial and a Gaussian $g_p(x) = \exp[-(x-x_p)^2/(2\lambda^2)]$, which is centred at

$$x_p = [chp/qBL_y)] - \phi/(BL_y) - [mc^2/(q^2B^2)]\left.\frac{dV(x)}{dx}\right|_{x=x_p}. \tag{3}$$

The corresponding energy is

$$E^n{}_p = \hbar\omega(n + 1/2) + V(x_p) + (m/2)\left[(c/qB)\frac{dV(x_p)}{dx_p}\right]^2. \tag{4}$$

Generally the x_p-dependence of the last term can be neglected with respect to $V(x_p)$. The ϕ-dependence of the individual energies $E_p(\phi)$ (we omit the band index n) is then given by the term $V[x_p(\phi)]$. This means that the levels $E_p(\phi)$ intersect whenever V(x) is not monotonically increasing or decreasing. This is illustrated in Fig. 1, where V(x) in the form of a smooth barrier is assumed. The individual energies $E_p(\phi)$ are usually not periodic with ϕ, in contrast to the total spectrum, which is periodic with period hc/q (each energy curve is periodically reproduced after a ϕ-shift of hc/q). Note that the wave vector $k_y{}^p = 2\pi p/L_y$ is constant along each individual energy curve $E_p(\phi)$, i.e., it is constant in time.

3. Influence of Disorder

As a next step we admit a disorder potential $V^1(x,y)$. We consider a case, where the unperturbed levels $E_p(\phi)$ intersect and are discrete, as in Fig. 1. In the presence of the perturbation V^1 the energy levels anti-cross in accordance with the Wigner-von Neumann theorem [2]. Now each individual perturbed level is periodic with period hc/q (Fig. 2). From this it follows on general grounds [1], that the phase winding numbers

$$W_x(\phi) = (1/2\pi)\int_0^{L_y} \eta(x,y;\phi)dy, \tag{5}$$

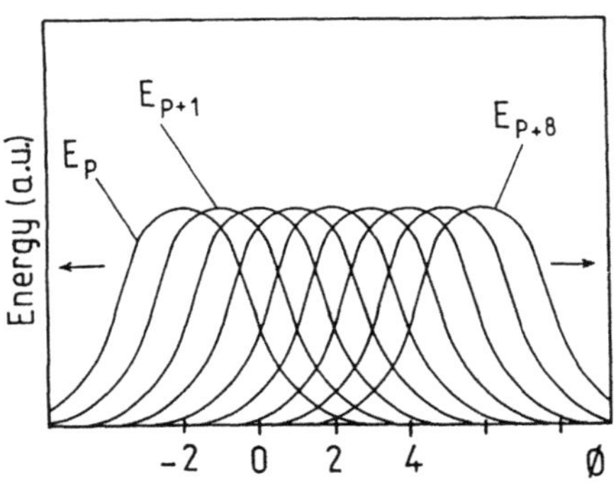

Fig. 1 Energy levels $E_p(\phi)$ in the presence of a smooth potential barrier V(x) in the absence of disorder V^1. (ϕ in units of hc/q)

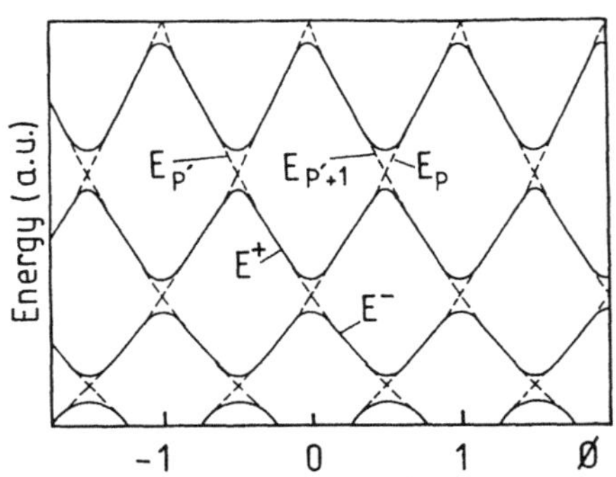

Fig. 2 Energy levels $E^+(\phi)$, $E^-(\phi)$,... in the presence of weak disorder V^1 (full lines) corresponding to the unperturbed levels $E_p(\phi)$, $E_{p'}(\phi)$ (dashed lines).

of the perturbed orbitals are no longer independent of ϕ, in opposition to the unperturbed case, where $W^p{}_x(\phi) = W^p = p = L_y k_y/2\pi$. In the presence of $V^1(x,y)$ these numbers (which are integers) obey the law

$$W_x(\phi + 1) = W_x(\phi) + 1 \quad (\phi \text{ in units of } hc/q). \tag{6}$$

This implies a quantized change of $W_x(\phi)$ in integer steps in each ϕ-interval of length hc/q. This change is associated with the passage of phase singularities through the physical domain (the two-dimensional strip) in a way analogous to the previously discussed systems [1]. Physically this corresponds to non-classical behaviour.

In the present example the perturbed orbitals can be obtained explicitely as follows: Let us consider the intersection of two unperturbed levels $E_p(\phi)$ and $E_{p'}(\phi)$. We choose the origin of ϕ such, that this happens at $\phi = 0$ (Fig. 2). We further assume

$$|V^1{}_{p,p'}| = |\langle\psi_p|V^1|\psi_{p'}\rangle| = f|E_p(1/2) - E_{p'}(1/2)|, \quad f << 1 \tag{7}$$

(called case of weak disorder). In this case the perturbed eigenfunctions $\psi^+(\phi)$, $\psi^-(\phi)$ and eigenvalues $E^+(\phi)$, $E^-(\phi)$ can be calculated approximately in a neighbourhood of $\phi = 0$ in the usual way [3] as a linear combination of the two unperturbed functions $\psi_p(\phi)$ and $\psi_{p'}(\phi)$. This calcuation has to be repeated for each level anti-crossing with a different pair of unperturbed functions. In the case of homogeneous disorder (where the diagonal elements $V_{p,p}$ can be neglected), the calculations give the following qualitative behaviour of the normalized perturbed function $\psi^+(\phi) = \Sigma c_p{}^+(\phi)\psi_p(\phi)$ (ϕ expressed in units of hc/q): for $-1/2 < \phi < 0$ we have $|c^+{}_{p'}| \simeq 1$, for $\phi \simeq 0$ both $c^+{}_{p'}$ and $c^+{}_p$ are appreciably different from zero (but the former decreases and the latter increases). For $0 < \phi < 1/2$ one has $|c^+{}_p| \simeq 1$, for $\phi \simeq 1/2$ both $c^+{}_p$ and $c^+{}_{p'+1}$ are appreciably different from zero, for $1/2 < \phi < 1$ we have $|c^+{}_{p'+1}| \simeq 1$, etc.. For ψ^- an analogous result holds with p' and p interchanged.

Physically this means that the perturbed orbital $\psi^+(x,y,t)$ describes a particle which is located at $x_{p'}(t)$ during the time interval $-\tau/2 < t < 0$, where $\tau = |h/(qE_yL_y)|$. At $t \simeq 0$ the particle disappears at $x_{p'}$ and reappears at x_p, where it remains during the time $0 < t < \tau/2$, then it disappears and reappears again at $x_{p'+1}(t)$ ($= x_{p'}(t-\tau)$), etc.. The adiabatic solution $\psi^+(x,y,t)$ thus describes a wave packet moving with the small, classical drift velocity $v = cE_y/B$ (because of (3)), but which is alternately localized in a small neighbourhood of one of the two fixed sites $x_p(0)$ and $x_{p'}(0)$. After each time interval $\tau/2$ the particle changes site discontinuously by disappearing at the first site and simultaneously reappearing at the second site. The distance passed with the classical velocity v $= cE_y/B$ in one half period $\tau/2$ is very small (of the order of 10^{-11}cm for B = 6 Tesla and L_y = 1 cm). But *the distance* $|x_{p'} - x_p|$ *passed non-classically* (i.e., in the discontinuous way just described) is $|p - p'|$ times larger. This means, it *is larger by many orders of magnitude*. The time integral over the period τ of the total velocity is zero for each of the perturbed orbitals, i.e., the single-particle current has no dc-component, in contrast to the unperturbed orbitals, which describe particle motion in accordance with classical mechanics.

4. Discussion

The non-classical dynamics induced by disorder in high magnetic fields has its analogy in zero magnetic field, as we have already mentioned. For example consider a one-dimensional loop threaded by a time-dependent magnetic flux $\phi(t)$.

Here, in the presence of disorder, the adiabatic wave functions (perturbed Bloch-functions) also describe a current oscillation with the same frequency $1/\tau$ [1,4]. This common behaviour is a result of the general structure of the corresponding Hamiltonians (which in both cases belong to the same subclass characterized by n = 1 in equation (13) of ref. [1]).

If the electric field E_y is increased above a threshold value $E_y{}^{th}$, Zener tunnelling across the minigaps occurs. The wave functions then behave in a non-adiabatic way, tending more and more to the unperturbed functions $\psi_p(x,y,t)$, which correspond to classical particle motion with the constant drift velocity $v = cE_y/B$. In the case of "weak disorder" (condition (7)) the threshold field can roughly be estimated from the Zener tunnelling probability [5] by setting the exponent in the expression for the tunnelling probability equal to one. With (7) one obtains

$$qE^{th}{}_y L_y = 2\pi^2 f|V^1{}_{p,p'}| \simeq \sqrt{2}\pi^2 fd\ \exp[-(x_p - x_{p'})^2/(2\lambda)^2], \tag{8}$$

where

$$d = (1/L_y)\int V^1 \exp[i2\pi y(p - p')/L_y]dy. \tag{9}$$

(In homogeneous disorder d is considered independent of x.) Assuming the upper limits $f \simeq 0.2$, $d = 0.1$ eV and $dV(x)/dx = 10^6$ eVcm^{-1} one obtains from (4), (7) and (8) $E_y{}^{th}L_y \simeq 10^{-5}$ V (corresponding to $x_p - x_{p'} \simeq 6\lambda$). This threshold value might be considered as an upper limit in the so-called weak disorder case (defined by (7)) in semiconductor heterostructures.

In the strong disorder case, i.e., where $f > 0.2$, the threshold field between adiabatic and non-adiabatic behaviour is expected to be much higher than $E_y{}^{th}$ calculated from (8), where f is set 0.2. In this case the adiabatic functions generally are linear combinations of more than two unperturbed orbitals. In any case, in the presence of disorder, the adiabatic functions always obey the general relations developed in [1], in particular (6). This implies the passage of phase singularities through the physical domain, which is associated with non-classical change of the momentum k_y, hence with non-classical motion of the localization centres. The adiabatic particle motion is therefore always non-classical in the presence of disorder. Our estimation of the threshold field $E_y{}^{th}$ shows that such non-classical processes may not be negligeable in high magnetic fields.

REFERENCES

1. J. Riess: Phys. Rev. B38, 3133 (1988)
2. J. von Neumann and E. Wigner: Phys. Z. 30, 467 (1929)
3. See e.g. L. D. Landau and E. M. Lifshitz: In Quantum Mechanics, Course of Theoretical Physics, Vol.3 (Pergamon, London, Paris 1959), sec.76
4. See e.g. D. Lenstra and W. van Haeringen, J. Phys. C14, 5293 (1981)
5. C.Zener, Proc. R. Soc. (London) A137, 696 (1932)

Connection Between Low and High Temperature Magneto-transport Measurements in GaAs/GaAlAs Heterojunctions

T. Rötger [1], G.J.C.L. Bruls [1], J.C. Maan [1], P. Wyder [1], K. Ploog [2], and G. Weimann [3]

[1] Max-Planck-Institut für F-38042 Grenoble Cedex, France
[2] Max-Planck-Institut für Festkörperforschung, Heisenbergstr. 1 D-7000 Stuttgart 80, Fed. Rep. of Germany
[3] Forschungsinstitut der Deutschen Bundespost beim Fernmeldetechnischen Zentralamt, D-6100 Darmstadt, Fed. Rep. of Germany

Magnetotransport measurements in several GaAs/GaAlAs heterojunctions show a linear magnetoresistance at 30K <T<120K. At lower temperatures where the quantum Hall effect is observed the magnetoresistance is found to be proportional to the magnetic field times the derivative of the Hall voltage with respect to the field, which can be seen as the extension of the linear magnetoresistance to the quantum regime. A relation between Hall voltage and magnetoresistance valid for all samples and temperatures is given.

The quantization of the Hall resistance to the values h/ie^2 (h Planck's constant, e the electron charge and i an integer) at low temperatures (Quantum Hall Effect, QHE) is a spectacular phenomenon in magnetotransport in the two-dimenensional electron gas. The existence of plateaux of finite width is believed to be due to localization as a consequence of inhomogeneities in the sample [2]. It is well known that in metals inhomogeneities in a sample give rise to a term in the magnetoresistance which is linear in the magnetic field (linear magnetoresistance, LMR) [3]. As the two-dimensional electron gas (2DEG) in GaAs/GaAlAs heterojunctions exhibits a metallic behaviour (non-activated transport), it is interesting to investigate whether these materials also show a LMR in a similar way as normal metals. Such a linear magnetoresistance is clearly observable only at higher temperatures when quantum oscillations have disappeared. Since both the LMR at high temperature and the QHE at low temperature are explained by sample inhomogeneities, the question arises whether there is a connection between these two phenomena. LMR is believed to be due to local changes in Hall resistance ρ_{xy}, which then contribute to the longitudinal magnetoresistance ρ_{xx}. Therefore we investigate whether in the case of the 2DEG in GaAs/GaAlAs heterojunctions a similar link between ρ_{xx} and ρ_{xy} can be established, which is valid both at low temperature (QHE regime) and at high temperature (LMR regime).

ρ_{xx} and ρ_{xy} of several Hall-bar shaped GaAs/GaAlAs heterostructure samples of different carrier densities n and mobilities μ (see Table 1) at temperatures from T = 1.4 to $\simeq$ 120 K are measured. At low temperatures all samples show the usual Shubnikov-de

Springer Series in Solid-State Sciences, Vol. 87
High Magnetic Fields in Semiconductor Physics II Editor: G. Landwehr

Table 1

Carrier densities n and low-field mobilities μ_0 at 4.2 K of the samples used. Conductivities extrapolated to $B = \infty$ (σ_∞), (see figure 1) and proportionality constant α_0.

Sample	n [10^{11} cm^{-2}]	μ [10^4 cm^2/Vs]	σ_∞ [e^2/h]	α_0
1	1.78	22.7	0.008	0.069
2	6.46	13.8	0.05	0.059
3	5.07	39.7	0.04	0.065
4	2.95	7.0	0.02	0.070

Haas oscillations and quantum Hall plateaux, which disappear gradually as the temperature is increased. At temperatures above $\simeq$ 40 K, only a smooth monotonic increase in the magnetoresistance is observed (Fig. 1, inset).

It is customary [4] to describe the magnetoresistance in the form of a power series

$$\rho_{xx}/\rho_0 = 1 + \alpha(\mu B) + \beta(\mu B)^2, \tag{1}$$

where $\rho_0 = 1/ne\mu$ is the zero field resistance and α and β dimensionless coefficients. In the high-field approximation ($\mu B \gg 1$), eq. (1) can be rewritten as

$$\sigma_{xx} = \alpha\sigma_{xy} + \beta\sigma_0, \tag{2}$$

where $\sigma_0 = 1/\rho_0$ is the zero field conductivity. We therefore plotted the data in a σ_{xx} vs σ_{xy} plot (Fig. 1). The figure shows how the oscillatory behaviour at low temperatures evolves for T $\gtrsim$ 50 K into a strictly linear relation between σ_{xx} and σ_{xy}, given by eq. (2). This linear dependence is observed in all samples studied for filling factors $\nu \le 2$. With further increase of the temperature the behaviour of the conductivity remains linear, only the slope α of the curves changes as a function of T. In all cases β is so small that the quadratic term in (1) is only a small contribution to the magnetoresistance.

At low temperatures, where quantum oscillations become observable, the simple linear relation (1) between σ_{xx} and σ_{xy} is no longer obeyed. However, Chang and Tsui [5] have found that experimentally

$$-\frac{d\rho_{xy}}{dn}\,n = \frac{d\rho_{xy}}{dB}\,B = \frac{\rho_{xx}}{\alpha}. \tag{3}$$

Neglecting the quadratic term in B in eq. (1) this empirical relation is a generalization of the simpler relation (2), also applicable in the regime of quantum transport. The occurrence of relation (3) was examined by Vagner and Pepper [6], and attributed to the presence of long-range potential fluctuations due to defects [7].

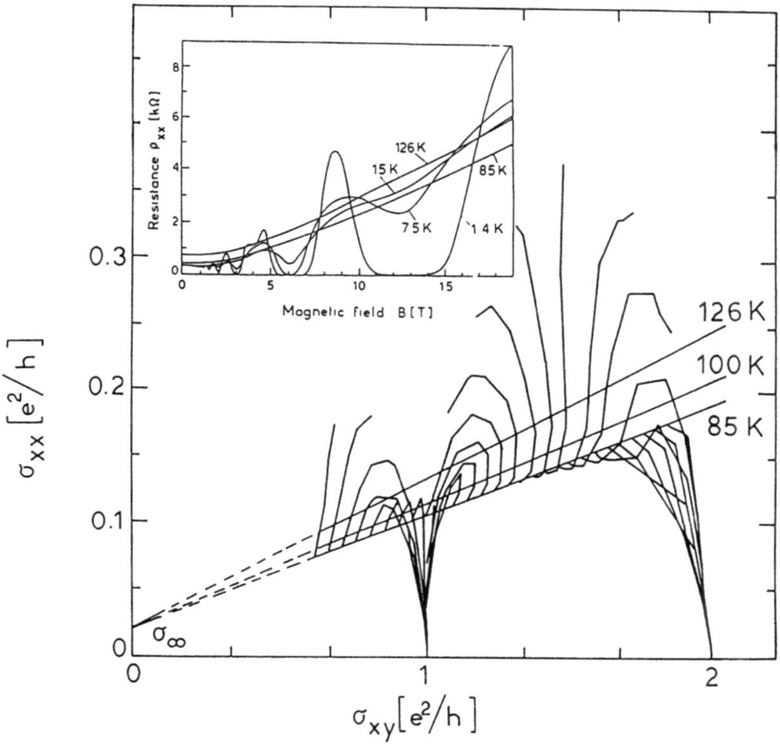

Fig. 1

a) Magnetoresistance ρ_{xx} of sample 4 at temperatures from 1.4 to 126 K. (inset)
b) Conductivities σ_{xx} vs σ_{xy} (calculated from ρ_{xx} and ρ_{xy}). For temperatures between 1.4 and 85 K, magnetoconductivities evolve from quantum oscillations to a linear behaviour. At these lower temperatures, the σ_{xx}/σ_{xy} points for the same magnetic field, but at different T are connected. At higher T, the magnetoconductivities follow the linear relation $\sigma_{xx} = \alpha(T)\ \sigma_{xy} + \sigma_\infty$. To demonstrate this behaviour, we connected σ_{xx}/σ_{xy} points at the same temperature (85 K, 100 K, 126 K) as a function of the magnetic field.

In Fig. 2, α, as determined from either eq.(2) or (3) is shown as a function of temperature. One sees that at low temperature, α is approximately constant and has, within the precision with which eq. (3) is obeyed, the same value $\alpha \simeq 0.07$ for all our samples. At high temperatures, α rises proportionally to $T^{3/4}$ for all samples, although the proportionality factor differs from sample to sample. By comparing the behaviour of α with that of μ we guess that α is proportional to $\sqrt{\mu_0/\mu(T)}$, where μ_0 is the low T and $\mu(T)$ the mobility at that temperature.

The inset of Fig. 2 shows this dependence of α for all samples and temperatures, and despite the very different mobilities and densities all results fall on the same curve, given by

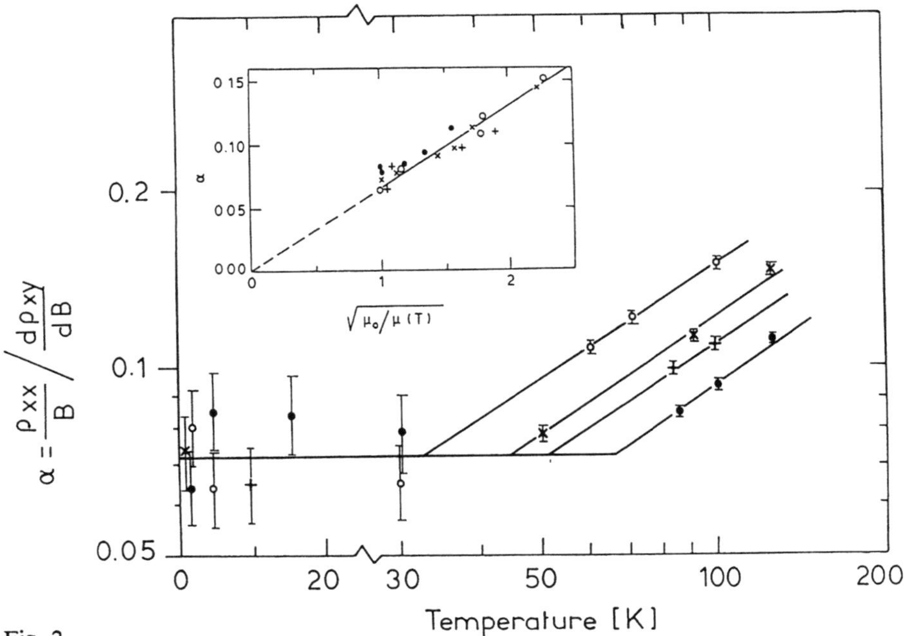

Fig. 2

The proportionality constant α as a function of temperature for sample 1: ×, 2: +, 3: ○, 4: • In the inset: α vs $\sqrt{\mu_0/\mu(T)}$ with symbols for different samples as in Fig. 1.

$$\alpha = \alpha_0 \sqrt{\frac{\mu_0}{\mu(T)}} \tag{4}$$

with $\alpha \simeq 0.07$ for all investigated densities, mobilities, and temperatures. Note that this constant proportionality factor is valid both at low and high temperatures. The universality of this relation is striking in two respects, namely the fact that all samples have the same proportionality constant and the same dependence on the mobility. We cannot offer a satisfactory explanation for these rather puzzling facts but only give some commentary remarks on these two aspects.

The width of the potential fluctuations is characterized by the Landau level width, which for the case of short-range scatterers at low T is given by [9]

$$\Gamma_0 = \frac{\hbar e}{m^*} \sqrt{\frac{2}{\pi} \frac{B}{\mu_0}}, \tag{5}$$

where Γ_0 is the intrinsic low T Landau level width, and m^* the effective mass (0.068 m_0 [free electron mass] for GaAs). At higher temperatures phonons become thermally

excited giving rise to additional broadening. Supposing that also at higher T a similar relation between the field and the mobility (now determined by phonon scattering) and the Landau level width exists as given by eq. (5) [10], eq. (4) reads $\alpha \propto \Gamma/\Gamma_0$, i.e. α is proportional to the relative broadening due to phonons with respect to the intrinsic broadening Γ_0 at low T, given by eq. (5).
The second puzzling observation is the apparent universality of the constant $\alpha_0 \simeq 0.07$ in eq. (4). In Si MOSFETs Tausendfreund and v. Klitzing [11] have measured the proportionality between the minimal values of $d\rho_{xy}/dn$ and ρ_{xx} in the QHE plateaux, and the corresponding proportionality factor in their case can be estimated from their results as being $\alpha_0 \simeq 0.17$, as compared to 0.07 in GaAs/GaAlAs heterostructures. These numbers seem to suggest a relation to the effective masses in these materials, which are 0.19 and 0.068, respectively, which would be a very striking observation.
In conclusion, we have shown that the 2DEG in GaAs/GaAlAs heterostructures shows a linear magnetoresistance which is dependent on temperature and sample mobility as $\sqrt{\mu_0/\mu(T)}$. Eq. (3) is thereby valid for all investigated temperatures and mobilities using eq.(4) for α with the same proportionality factor $\alpha_0 \simeq 0.07$. This very general connection between ρ_{xx} and $d\rho_{xy}/dB \cdot B$ both in the QHE and in the LMR regime shows that the same inhomogeneities are responsible for both behaviours, which are thus clearly connected.
Although many questions remain unanswered, we believe that our data of LMR at high temperatures may contribute to a better understanding of the transport properties of the 2DEG in high magnetic fields.

References

1. K. v. Klitzing, G. Dorda and M. Pepper, *Phys. Rev. Lett.* **45**, 494 (1980).
2. H. Aoki, *Rep. Prog. Phys* **50**, 655 (1987).
3. G. J. C. L. Bruls, J. Bass, A. P. van Gelder, H. van Kempen and P. Wyder, *Phys. Rev.* **B32**, 1927 (1985).
4. H. Weiss, in *Semiconductors and Semimetals* Vol. 1, eds.: R. K. Willardson and A. C. Beer (Academic Press, New York, San Francisco, London, 1966), p. 315.
5. A. M. Chang and D. C. Tsui, *Solid State Commun.* **56**, 153 (1985).
6. I. D. Vagner and M. Pepper, *Phys. Rev.* **B37**, 7147 (1988).
7. D. Bergers and J. Hajdu, *Solid State Commun.* **20**, 683 (1976).
8. F. Stern, in *Proceedings of the International School on Physics and Applications of Quantum Wells and Superlattices*, in *ASI Series* (Plenum Press, New York, 1988).
9. T. Ando and Y. Uemura, *J. Phys. Soc. Jpn.* **36**, 959 (1974).
10. M. A. Hopkins, R. J. Nicholas, P. J. Barnes, M. A. Brummell, J. J. Harris and C. T. Foxon, to be published.
11. B.Tausendfreund and K.von Klitzing,*Surface Sci.* **142**,220, (1984).

Hybrid Magneto-electric Quantisation in Quasi-2D Systems

W. Zawadzki

Institute of Physics, Polish Academy of Sciences,
PL-02-668 Warsaw, Poland

Quantisation of electron energies in quasi-2D systems is considered, in which an external magnetic field is perpendicular to the gradient of electric potential (crossed-field configuration). The hybrid magneto-electric quantisation in the case of the rectangular well, the triangular well and the parabolic well is calculated analytically and it is shown that the hybrid quantisation offers a continuous transition between 2D and 1D systems as the magnetic field increases.

1. INTRODUCTION

Most of the experimental and theoretical work on two-dimensional systems in the presence of a magnetic field is concerned with the configuration in which the magnetic field is perpendicular to interfaces confining electrons or holes. In this geometry the magnetic field is parallel to the gradient of electric potential (i.e. to the electric field). The electron energies are easily quantised since parallel fields act independently: the electric confinement quantises the motion in the z direction while the magnetic field quantises the motion in the x-y plane. As a result, one deals with ladders of Landau levels belonging to each electric subband. Such a system is basically zero-dimensional since no free motion exists in the presence of parallel fields.

In the following we shall be concerned with a less frequently investigated configuration in which an external magnetic field is perpendicular to the gradient of confining potential, i.e. to the electric field. In most cases this corresponds to a situation in which the magnetic field is parallel to interfaces of a semiconductor system. In the crossed-field configuration the effects of both fields may not be separated either physically or mathematically. Instead, one deals with hybrid magneto-electric states. As we shall show, the crossed field geometry offers a possibility to realize a continuous transition between a quasi-2D system (electric limit) to a quasi-1D system (magnetic limit).

Various crossed-field configuration possess some general features, which become apparent in all cases. However, it is of interest to investigate specific cases, both because of their different technological feasibility as well as the variety of mathematical methods used for their description. We begin with a rectangular infinite well in a parallel field, which has simplifying symmetry properties. Then we continue with an asymmetric triangular well realized in MOS structures. The third, parabolic well case is particularly simple and it allows one to calculate not only analytical form of the energy dispersion, but also the density of states. This is used to illustrate the above-mentioned continuous transition between quasi-2D and quasi-1D systems.

Springer Series in Solid-State Sciences, Vol. 87
High Magnetic Fields in Semiconductor Physics II Editor: G. Landwehr

2. RECTANGULAR WELL IN PARALLEL MAGNETIC FIELD

In this section we investigate the electron behaviour in an infinitely high rectangular potential well in the presence of a magnetic field perpendicular to the gradient of the confining potential, i.e. parallel to the "interfaces". The considered case is well suited to illustrate a continuous transition between electric and magnetic quantisation since the two limiting cases are commonly known. The description is based on the work of ZAWADZKI [1]. The initial eigen-energy equation in the one-band effective mass approximation reads

$$[\frac{1}{2m^*}P^2 \pm \frac{1}{2}g^*\mu_B B + U(y)]\psi = E\psi \quad , \tag{1}$$

where $\vec{P} = \vec{p} + e\vec{A}$ is the kinetic momentum, $\vec{A}$ is the vector potential of the magnetic field, m* is the effective mass, g* is the spin Landé factor, μ_B is the Bohr magneton. U(y) describes the confining potential. We take the magnetic field in the z direction and use the Landau gauge: $\vec{A} = [-By, 0, 0]$. The x and z variables may then be separated by looking for solutions of the form: $\psi(r) = \exp(ik_x x + ik_z z)f(y)$. Considering an infinitely high rectangular well of width a and taking the zero of energy at the bottom of the well we obtain (1) in the form

$$[-\frac{\hbar^2}{2m^*}\frac{d^2}{dy^2} + \frac{m^*\omega_c^2}{2}(y - y_o)^2]f(y) = E'f(y) \quad , \tag{2}$$

where $\omega_c = eB/m^*$ is the cyclotron frequency, $y_0 = k_x L^2$ represents the centre of oscillations, $L = (\hbar/eB)^{1/2}$ is the magnetic radius, and $E' = E \pm \frac{1}{2}g^*\mu_B B - \hbar^2 k_z^2/2m^*$. Due to the presence of the well the solutions of the differential equation (2) are subject to the boundary conditions $f(\pm a/2) = 0$. In Fig. 1 we show schematically the potential of the rectangular well and the potential due to the magnetic field (for the electron at the centre, i.e. for $y_o = 0$). It is clear that the electron behaviour depends on the relation between the well width and the magnetic parabola width. At weak fields (L>a) it is the rectangular well that confines the electron motion (the electric case), while at strong fields (L<a) it is the magnetic field that confines the motion (the magnetic case).

The general behaviour of energies as functions of average electron position y_o for the high-field case L << a is known. For y_o near the centre of the well one deals with flat Landau levels, whose energies begin to increase

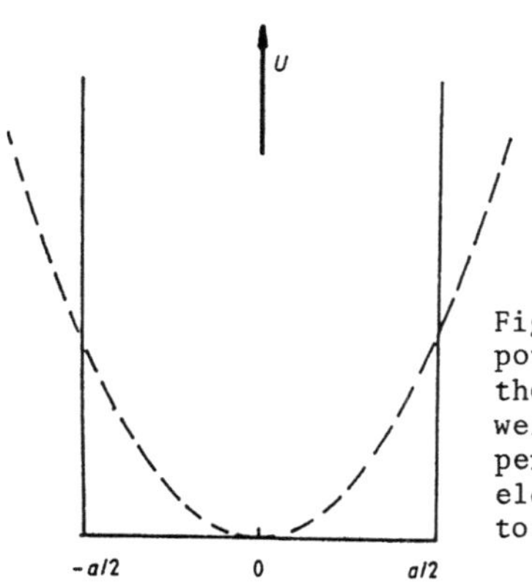

Fig.1 Electric and magnetic potentials for the electron in the centre of a rectangular well and a magnetic field perpendicular to the gradient of electric potential (parallel to interfaces).

as y_o approaches the limits $\pm a/2$. It is clear from the symmetry considerations that $E(y_o)$ is a symmetric function of y_o. Since both intra-band and inter-band magneto-optical transitions occur at sub-band extrema, and the Shubnikov-de Haas magnetoresistance maxima are also related to these points, we limit our considerations to the physically interesting case of $y_o = 0$. This simplifies the problem because both the potential in (2) and the boundary conditions are then symmetric with respect to the centre of the well.

We introduce a dimensionless variable $q = y\sqrt{2}/L$ and a dimensionless eigenvalue $\beta = E'/\hbar\omega_c = E/\hbar\omega_c \mp \nu - k_z^2L^2/2$ where $\nu = g^*m^*/4m_o$. Equation (2) reads then (for $y_o = 0$)

$$\frac{d^2f}{dq^2} - (\frac{1}{4}q^2 - \beta)f = 0 \tag{3}$$

with the boundary conditions

$$f(\pm q_o) = 0 , \tag{4}$$

where $q_o = a/L\sqrt{2}$. It is seen that q_o is the only important parameter for the problem. Equation (3) represents the standard form of the parabolic cylinder differential equation (with the changed sign of β, see [2]). Together with the boundary conditions (4) it describes a symmetrically constrained quantum harmonic oscillator. Our aim is to find the $\beta(q_o)$ dependence.

The semi-classical WKB approach is not expected to give good results in our case. The reason is that the semi-classical phase is not well defined. For the infinite rectangular well each turning point contributes the phase $\pi/2$, so that the total phase is π. On the other hand, for the parabolic well the total phase is $\pi/2$. Thus, one expects the WKB procedure to be useful only in the two limiting cases: $q_o \ll 1$ (phase π) or $q_o \gg 1$ (phase $\pi/2$), whereas it is the intermediate case $q_o \simeq 1$, which is of interest. In order to determine the eigenvalues we use the method of DEAN [3]. First, it can be observed that certain portions of the well known harmonic oscillator functions between their symmetric zero values satisfy both equation (3) and the boundary conditions (4). This allows one to assign specific values of q_o to the eigenvalues $\beta_i = i + \frac{1}{2}$, using the tabulated values of symmetric zeros of the Hermit polynomials. The intermediate $\beta_i(q_o)$ values may then be found by interpolation. In addition DEAN has found the development of $\beta_i(q_o)$ in power series of q_o by employing the developments of even and odd parabolic cylinder functions in powers of q (which contain eigenvalues β) and equating each term to zero at $\pm q_o$ points. His results may be summarised as follows

$$\beta_i(q_o) = \frac{c_i}{q_o^2} + b_2^{(i)}q_o^2 + b_4^{(i)}q_o^6 + \dots , \tag{5}$$

where

$$i = 0, 1, 2, \dots ,$$
$$c_i = [(i + 1)/2]^2\pi^2 ,$$
$$b_2^{(i)} = 1/12 - 1/8c_i ,$$
$$b_4^{(i)} = 1/720c_i - 5/192c_i^2 + 7/128c_i^3 ,$$

for both even and odd i. (We have changed the signs in agreement with the previous convention that β is positive for positive energies.) Numerical values of c_i, $b_2^{(i)}$, $b_6^{(i)}$ for $0 \leqslant i \leqslant 5$ may be found in [3]. The development ceases to be valid for higher values of q_o. In this range, however, $\beta_i(q_o)$ should extrapolate to the eigenvalues of the unconstrained harmonic oscillator: $\beta_i = (i + \frac{1}{2})$, respectively.

In Fig. 2 we plot $\beta_i(q_o)$, as calculated from (5) (full curves) and extrapolated (chain curves). At high q_o one obtains the usual Landau levels. The magnetic radii of the higher states are larger: $L_i = (2i+1)^{1/2}L$, so that higher Landau states are more strongly affected by the well confinement. As a consequence, at a constant magnetic field (fixed L), the energy separations between levels increase for the decreasing well width. In the dimensionless scale of figure 2 the spin levels are given simply $\beta_i^{\pm} = \beta_i \pm \nu$, with ν defined above. Thus the ratio of the orbital splitting to the spin splitting can be tuned as a function of q_o, since the orbital splittings depend on the well width and the spin splittings do not.

The development (5) and the plots in Fig. 2 are not convenient if one is interested in weaker magnetic fields (small q_o values), since the β_i become very large. It is then useful to multiply (5) by $\hbar\omega_c/(\hbar^2\pi^2/2m^*a^2) = q_o^2(4/\pi^2)$. This gives

$$\eta_i^{\pm} = \frac{4}{\pi^2}(c_i + b_2^{(i)}q_o^4 + b_6^{(i)}q_o^8 + \dots \pm \nu q_o^2) + \frac{k_z^2a^2}{\pi^2} , \tag{6}$$

where $\eta_i^{\pm} = E^{\pm}/(\hbar^2\pi^2/2m^*a^2)$ is again the dimensionless energy (in electric units). The spin splitting has been explicitly included now as the last term in the parentheses. It can be seen that at $B = 0$ (i.e. $q_o = 0$) and $k_z = 0$ equation (6) gives $\eta_i = (i + 1)^2$, which is the exact result for an infinite square well. Thus the expansion (6) is exact in the low-q_o limit. In Fig. 3 we plot the $\eta_i^{\pm}$ as functions of q_o^2 (i.e. linear in B) for $k_z = 0$ and $\nu = 0.177$, which corresponds to InSb band-edge parameters: $m^*=0.0138\ m_o$ and $|g^*| = 51.3$. At low q_o the results may be regarded as the Zeeman effect on electrically quantised levels. As a whole, Fig. 3 shows a continuous

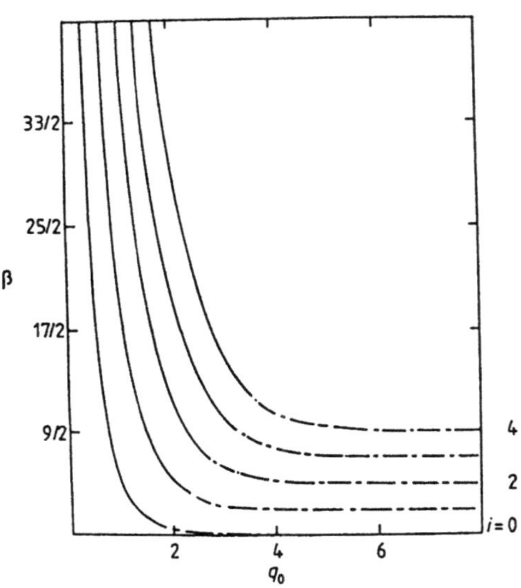

Fig.2 Energies of hybrid magneto-electric levels (in magnetic units) versus $q_o = a/L\sqrt{2}$ for the electron in the centre of a rectangular well. At high q_o the Landau levels are obtained (after [1]).

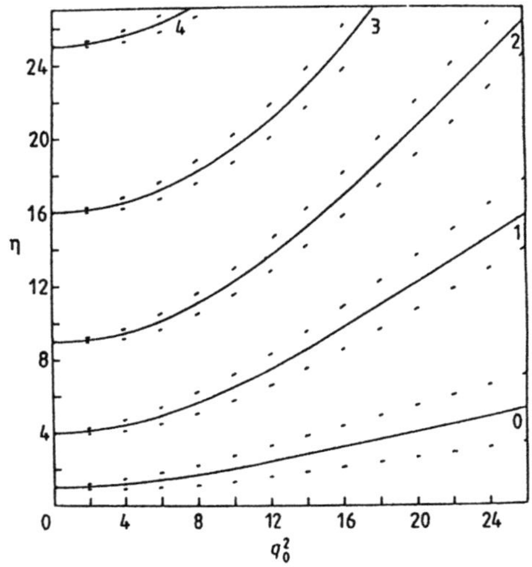

Fig.3 Energies of hybrid magneto-electric levels (in electric units) versus q_0^2 showing the continuous transition from electric to magnetic quantisation. The broken curves indicate spin levels for $\nu = g^*m^*/4m_o = 0.177$ (after [1]).

transition between electric and magnetic quantisation in the crossed-field configuration. At high q_o values the η_i become linear functions of q_o^2 (the Landau level range). The continuous transition from the unequally spaced electric levels to the equally spaced Landau levels allows one to realize almost arbitrary combinations of inter-level energies. The selection rules for optical transitions between the hybrid magneto-electric states and a generalisation of the above procedure for the case of nonparabolic band in narrow-gap semiconductors are described in ref. [1].

3. TRIANGULAR WELL IN PARALLEL MAGNETIC FIELD

This situation is realized experimentally in MOS structures (whose potential can be approximated by a triangular well) with magnetic field parallel to the metal-oxide interface. The case in question is more complicated mathematically than that of the rectangular well since it does not have the mirror symmetry. On the other hand the electron behavior in crossed uniform electric and magnetic fields (i.e. our problem without the barrier) is quite well understood, also in narrow-gap semiconductors (cf. [4,5]). The following description is based on the work of CRASEMANN et al [6] and ZAWADZKI et al [7].

The initial eigenvalue equation reads

$$\left[\frac{1}{2m^*}(\vec{p} + e\vec{A})^2 \pm \frac{1}{2}g^*\mu_B B + eFy\right]\psi = E\psi \tag{7}$$

for $y \geqslant 0$. For $y < 0$ the potential is assumed to be infinitely large. Taking $\vec{A} = [-By,0,0]$ and separating the variable in the usual way one obtains the one-dimensional equation

$$\left[-\frac{\hbar^2}{2m^*}\frac{d^2}{dy^2} + \frac{m\omega_c^2}{2}(y-y_o)^2\right]f(y) = \beta_\pm f(y) \tag{8}$$

where $\omega_c = eB/m^*$; $y_o = \alpha/m^*\omega_c^2$; $\alpha = \hbar k_x\omega_c - eF$; $\beta_\pm = \lambda_\pm + \alpha^2/2m^*\omega_c^2$; and $\lambda_\pm = E - \hbar^2k_x^2/2m^* - \hbar^2k_z^2/2m^* \pm (1/2)g^*\mu_B B$. The wave function is subject to the boundary conditions $f(y) = 0$ for $y = 0$ and $y = +\infty$. Dividing (8) by $\hbar\omega_c$ and introducing a dimensionless variable $q = y\sqrt{2}/L$, where $L = \sqrt{\hbar/eB}$, one obtains

$$\left[\frac{d^2}{dq^2} - \frac{1}{4}(q-q_o)^2 - a\right]f(q) = 0 \tag{9}$$

with

$$f(q) = 0 \text{ for } q = 0 \text{ and } q = +\infty. \tag{10}$$

Here $q_o = y_o\sqrt{2}/L$ and $a = -\beta_\pm/\hbar\omega_c$. Quantity q_o is the only parameter in the problem. One can also shift the variable $q' = q-q_o$ and the boundary conditions become $f(q') = 0$ for $q' = -q_o$ and $q = +\infty$. The task is to find eigenvalues a as functions of q_o, the energy E can then be found using the above relations. Equation (9) represents the parabolic cylinder equation with asymmetric boundary conditions. There exist the following analytic approximations for the eigenvalues:

$$a_n \simeq -(n+\tfrac{1}{2}) - \frac{1}{\sqrt{2\pi}}\frac{1}{n!}q_o^{2n+1}\exp(-q_o^2/2) \qquad (q_o \gg 1) \quad ,$$

$$a_n \simeq -(2n+\tfrac{3}{2}) + \frac{\sqrt{2}}{\pi}\frac{\Gamma(n+\tfrac{1}{2})}{n!}(2n+1)q_o \qquad (|q_o|<1) \quad ,$$

$$a_n \simeq -\tfrac{1}{4}q_o^2 - [\tfrac{3}{4}\pi(n+\tfrac{3}{4})q_o]^{2/3} \qquad (q_o \ll -1) \quad ,$$

The first relation is asymptotic [8], the third results from the WKB semiclassical quantization, the second has been derived by DEAN [3], who considered (9) in some detail.

In Fig.4 we show $E_n(k_x)$ dependences for $k_z = 0$. The increase of energies for negative k_x (or y_o) is explained in Fig. 5, the increase of energies for positive k_x (or y_o) is due to the increasing potential energy eFy_o in constant electric field (cf.(7)). The last increase is absent for F = 0, and one deals in this case with magnetic surface levels (flat energies for large positive y_o, cf. Fig.6). Finally, Fig. 7 shows energy shifts of electric levels in increasing magnetic field (called the diamagnetic shift). The influence of the band's nonparabolicity on the above behavior has been discussed in ref. [7].

4. PARABOLIC WELL IN PARALLEL MAGNETIC FIELD

In practice parabolic potential wells are realized in n-i-p-i structures. The case of a parabolic well in a parallel field is particularly simple and it can be quantized in analytical form (cf. e.g. [9]). We will use it here to calculate the density of states and to show how the transition from a 2D system (electric limit) to a 1D system(magnetic limit) occurs.

The initial eigenenergy equation reads (spin is omitted),

$$\left[\frac{1}{2m^*}(\vec{p}+e\vec{A})^2 + \frac{m^*\omega^2}{2}y^2\right]\psi = E\psi \quad , \tag{11}$$

where $m^*\omega^2/2$ characterises the parabolic well. Taking $\vec{A} = (-By,0,0)$ and separating the variables in the usual way one obtains

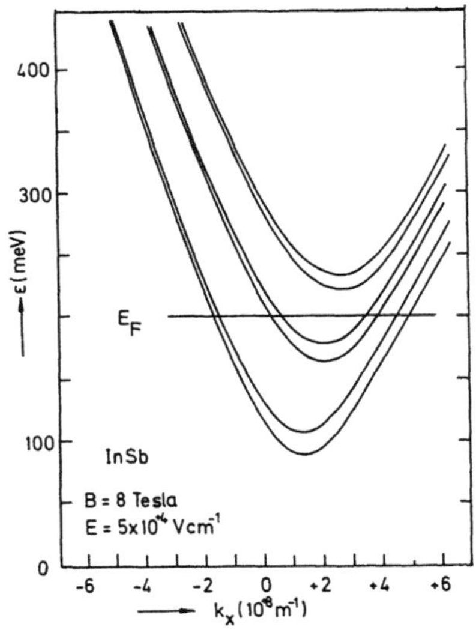

Fig.4 Eigen-energies of electron in triangular well and parallel magnetic field versus k_x calculated for InSb band parameters (including nonparabolicity of the band, after [7])

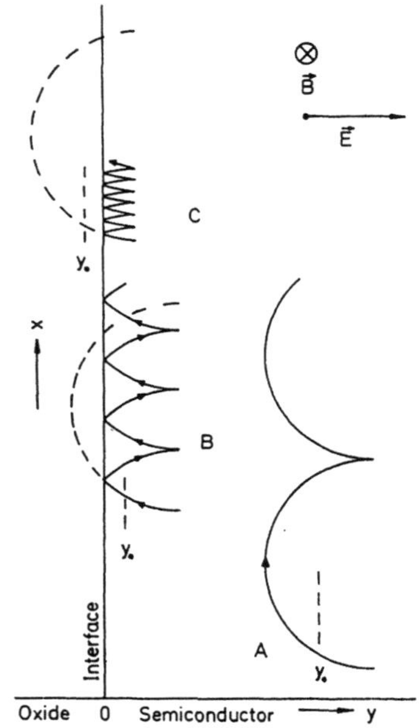

Fig.5 Classical orbits of electrons in a triangular well and parallel magnetic field for different electron positions y_0 with respect to the barrier (schematically, after [7]).

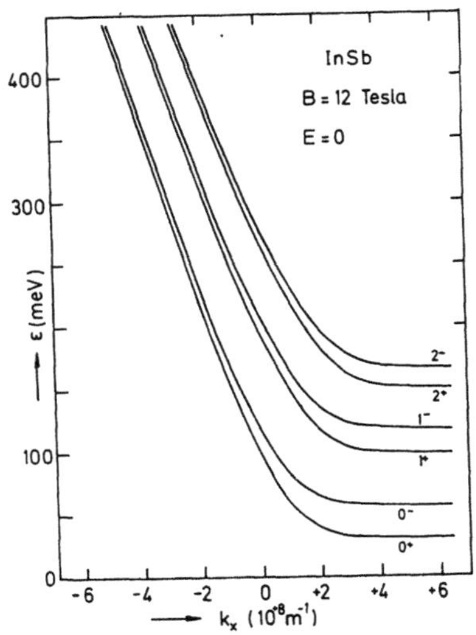

Fig.6 Energies of magnetic surface levels versus k_x calculated for InSb band parameters (including nonparabolicity, after [7]).

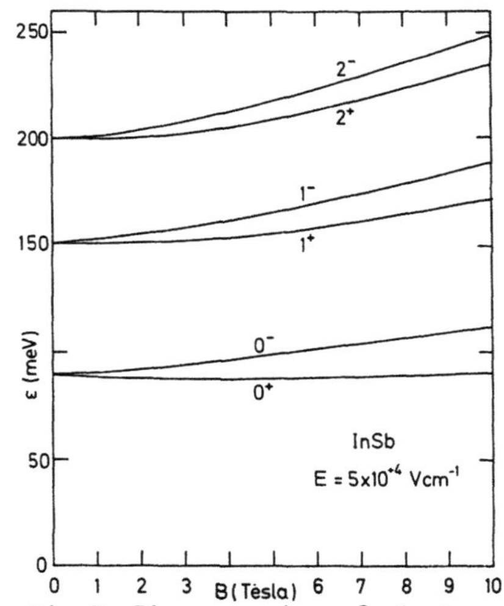

Fig.7 Eigenenergies of electron in a triangular well and parallel magnetic field versus magnetic field (the diamagnetic shift), calculated for InSb band parameters (including nonparabolicity, after [7]).

$$(-\frac{\hbar^2}{2m^*}\frac{d^2}{dy^2} + \frac{m\Omega^2}{2}y'^2)f(y') = E'f(y') \quad , \tag{12}$$

where $\Omega^2 = \omega^2 + \omega_c^2$ is the effective frequency, $y' = y - y_o$ in which $y_o = \hbar k_x \omega_c / m\Omega^2$. The quantization gives

$$E = \hbar\Omega(n + \frac{1}{2}) + \frac{\hbar^2 k_x^2}{2m^*}\frac{\omega^2}{\Omega^2} + \frac{\hbar^2 k_z^2}{2m^*} \quad , \tag{13}$$

where $n = 0,1,2,\ldots$
For $\omega_c \ll \omega$ equation (13) describes a 2D electric subband and for $\omega_c \gg \omega$ it describes a 1D magnetic one.

The density of electron states for the subband n is given by

$$\rho_n(E) = \sum_{k_x k_z} \delta(E - \lambda_n - \alpha k_x^2 - \beta k_z^2) \quad , \tag{14}$$

where we have introduced the notation: $\lambda_n = \hbar\Omega(n + 1/2)$; $\alpha = (\hbar^2/2m^*)\cdot(\omega^2/\Omega^2)$; $\beta = \hbar^2/2m^*$. The summation in (14) can be replaced by an integration in the standard way. If the electric field is weak (small ω), the center of magnetic oscillations y_o should be within the crystal, which gives $|k_x| < L_y m\Omega^2/2\hbar\omega_c$. For ω not too small (which is our case) the condition that the center of oscillations be within the crystal should be replaced by the condition that the electron be within the parabolic potential well. For the ground state this corresponds to replacing $L_y/2$ by $(m\omega/\hbar)^{1/2}$. It is seen from (14) that the integration over k_z gives a nonvanishing contribution only if $E - \lambda_n - \alpha k_x^2 \geq 0$, so that the integration over k_x is limited to the region $\alpha k_x^2 \leq E - \lambda_n$. The integration over k_z is carried out using the well known properties of the δ-function. In the integration over k_x one should consider two cases:

1) $\alpha k_x^2(\max) < E - \lambda_n$, where $k_x(\max) = L_y m^*\Omega^2/2\hbar\omega_c$ (for small ω), or $k_x(\max) = (m^*/\hbar\omega)^{1/2}\Omega^2/\omega_c$ (for not too small ω). In this case the integration over k_x gives finally

$$\rho_n(E) = \frac{L_x L_z}{(2\pi)^2}\frac{2}{\sqrt{\alpha\beta}}\arcsin\left(\frac{\alpha}{E-\lambda_n}\right)^{1/2} k_x(\max) \quad . \tag{15}$$

In particular, for $k_x^2(\max) \ll \alpha/(E-\lambda)$ one has $\arcsin y \simeq y$ and the usual 1D DOS for electrons in a magnetic field $\rho_n(E) \sim (E-\lambda_n)^{-1/2}$ is obtained.

2) $\alpha k_x^2(\max) \geq E - \lambda_n$. In this case the limiting points of integration are $k_x = \pm[(E-\lambda_n)/\alpha]^{1/2}$ and (15) gives

$$\rho_n(E) = \frac{L_x L_z}{(2\pi)^2}\frac{1}{(\alpha\cdot\beta)^{1/2}}\pi \quad , \tag{16}$$

i.e. the usual 2D density of states independent of energy.

For given ω and ω_c the two above cases are determined by the magnitude of $(E-\lambda_n)/\alpha$, i.e. by the energy E of interest. For not too small ω the density of states of the lowest subband is

$$\rho_o(E') = A(1+x^2)^{1/2}\arcsin\left[\frac{1/2}{E' - (1+x^2)/2}\right]^{1/2}\left(1 + \frac{1}{x^2}\right)^{1/2} \quad , \tag{17}$$

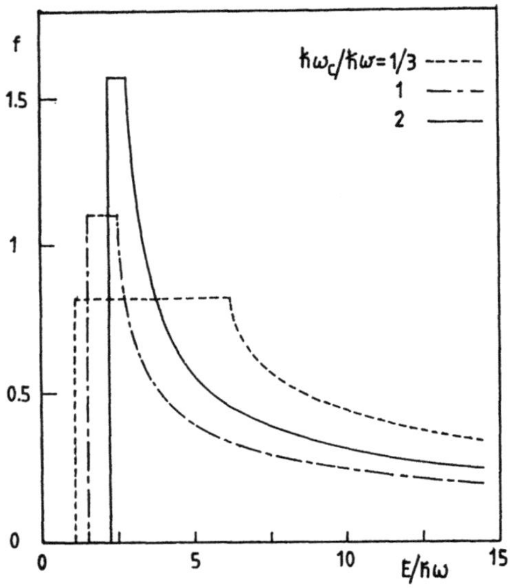

Fig.8 The density of states (not normalized) for electron in a parabolic well and parallel magnetic field versus $E/\hbar\omega$ for different values of $x = \omega_c/\omega$. The transition from 2D (electric limit) to 1D (magnetic limit) is seen.

where $E' = E/\hbar\omega$ and $x = \omega_c/\omega$. For fixed x and $(1+x^2)/2<E'<E'_{cr}$, such that the argument t of arcsin t is larger than 1, the DOS remains constant. The function $f(E') = \rho_o(E')/A$ for different x is plotted in Fig. 8. One can see how the 2D DOS (electric limit, $x < 1$) goes over into the 1D DOS (magnetic limit, $x > 1$) as $x = \omega_c/\omega$ increases. The transition between 2D and 1D cases occurs when the width of plateau becomes smaller than the distance to the next subband. This occurs for $x = 0.8$, i.e. for $\omega_c \approx \omega$, which could be expected. One can also see from (17) that the width of plateau may not become narrower than 1/2 (in $E/\hbar\omega$ units). This means that the strict magnetic limit may be obtained only for $\omega \rightarrow 0$.

5. EXPERIMENTAL EXAMPLES

Below we quote two experimental results showing hybrid magneto-electric quantisation in crossed-field configuration. There are certainly more to come.

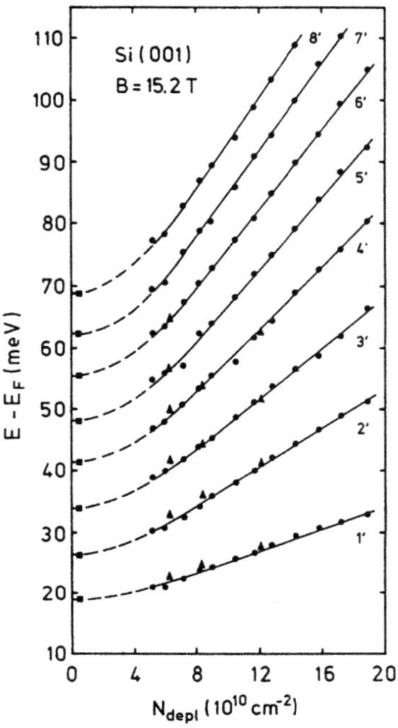

Fig.9 Energies of hybrid magneto-electric states in the accumulation and inversion layers of Si at constant magnetic field as functions of depletion charge density (after [10]).

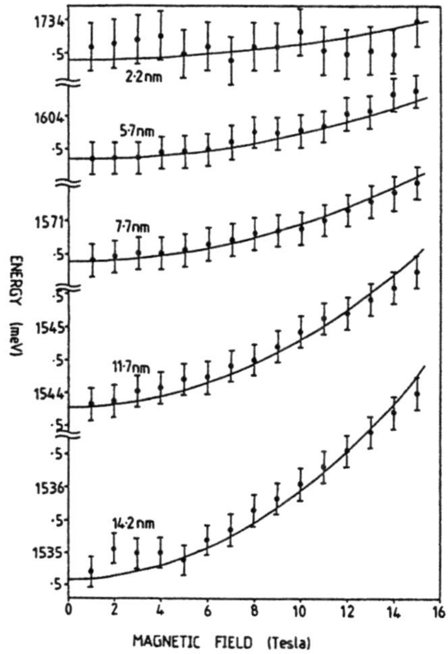

Fig.10 The diamagnetic shifts of optical transitions between electric subbands in GaAs - GaAlAs quantum wells versus intensity of parallel magnetic field (after [11]).

Literature

1. W.Zawadzki; Semicond. Sci. Technol. 2, 550 (1987)
2. J.C.P. Miller: Handbook of Mathematical Functions ed. by M. Abramowitz and I.A. Stegun, (Dover Publ. New York 1965) p. 685
3. P. Dean: Proc.Camb.Phil.Soc. 62, 277 (1966)
4. W. Zawadzki, Surface Science 37, 218 (1973)
5. W. Zawadzki, S. Klahn and U. Merkt: Phys. Rev. Lett.55, 983 (1985)
6. J.H. Crasemann, U. Merkt and J.P. Kotthaus: Phys.Rev. B28, 2271 (1983)
7. W. Zawadzki, S. Klahn and U. Merkt: Phys. Rev. B33, 6916 (1986)
8. E.A. Kaner, N.M. Makarov and I.M. Fuchs: Zh. Eksp. Teor. Fiz. 55, 931 (1968) [Sov. Phys. - JETP 28, 483 (1969)]
9. J.C. Maan: in Lecture Notes in Physics vol 177, ed. G. Landwehr (Springer, Berlin 1983) p. 163
10. U. Kunze: Surface Science 196, 374 (1988)
11. N.J. Pulsford, J. Singleton, R.J. Nicholas and C.T.B. Foxon, Journ. de Physique 48 (Supplement to N 11), 231 (1987)

Part IV

Heterostructures and Superlattices: Optics

Classification of Magneto-excitons in Quantum Wells

L.J. Sham

Department of Physics, B-019, University of California, San Diego, La Jolla, CA92093, USA

In an undoped quantum well, the exciton lines dominate in the optical spectra whether the magnetic field is present or not. The theory of magneto-exciton by Yang and Sham which includes the effect of the degenerate valence band edge is reviewed here with the emphasis on explaining the classification of the lines in order to facilitate interpretation of experiments.

1. INTRODUCTION

In an undoped quantum well, the presence of the exciton lines in the optical spectra is more clearly seen than in a bulk semiconductor [1].Thus, exciton plays a prominent role in the exploitation of the optical properties for the device applications [2].A high magnetic field normal to the planes of the quantum well splits the electron and hole subband structure into a number of discrete Landau levels. The resultant optical spectra involving valence to conduction band transitions are dominated by the magneto-exciton lines. The study of the energy and magnetic field strength dependence then furnishes details of the exciton properties. A number of experiments have been carried out on the magneto-optical properties of undoped GaAs-AlGaAs quantum wells [3-7]. The complex valence band structure greatly complicates the interpretation of the line spectra. Here, I would like to do more than recount the theory of YANG and SHAM [8] which includes the valence band complex in the theory of magneto-excitons. In terms of the angular momentum concept, the subband-to-subband, and Landau-level-to-Landau-level exciton lines can be classified by a physically apparent scheme. It is hoped that this will facilitate the interpretation of the experimental spectra.

2. EXCITONS IN A MAGNETIC FIELD

The theory of the magneto-exciton in the two-dimensional hydrogenic model has been given by AKIMOTO and HASEGAWA [9]. One important feature which is independent of the quadratic dispersion of the electron and hole is the non-existence in the optical spectrum of unbound electron-hole pairs in the presence of the normal magnetic field. The lines in the finite field are all excitons, which reduce to the optically active exciton states in zero field. This is easy to see in the equation for the exciton developed below in the symmetric gauge, where the normal magnetic field provides a quadratic potential well in the x-y plane and the Coulomb attraction between the electron and hole lifts the degeneracy of the Landau levels. The complex valence band states, however, require changes in the angular momentum classification of the hydrogenic model and give rise to new lines in the spectrum.

Springer Series in Solid-State Sciences, Vol. 87
High Magnetic Fields in Semiconductor Physics II Editor: G. Landwehr

The z-axis is chosen to be normal to the interface planes of the quantum well. The magnetic field $\mathbf{B}$ points along the z-axis. The coordinates of a particle are represented by $(\mathbf{r}, z)$, the vector in the x-y plane and the z component. Suffices e and h denote the electron and hole respectively. The multiband effective mass approximation is used, with the electron wave function expressed in terms of the basis $u_s, s = \pm 1/2$, the Bloch wave function at the bottom of the conduction band of the bulk semiconductor material of the well (e.g. GaAs in the GaAs/AlGaAs well) and with the hole wave function in terms of the basis, v_m, m =3/2, 1/2, -1/2, -3/2, the fourfold degenerate Bloch waves at the top of the bulk valence bands. The isomorphism of the hole basis to the spin 3/2 states will be exploited.

We first motivate a rather unconventional choice of the exciton wave function because of the convenience in evaluating the optical transition matrix element.

3. THE EXCITON WAVE FUNCTION

The exciton wave function is defined in terms of the electron and hole bases by

$$\Psi^*(\mathbf{r}_h, \mathbf{r}_e, z_h, z_e) = \sum_m \Psi_{sm}(\mathbf{r}_h, \mathbf{r}_e, z_h, z_e) u_s^* v_m \quad . \tag{1}$$

This wave function is the complex conjugate of the usually defined exciton wave function. Thus, in the limit of no electron-hole attraction, the wave function (1) reduces to the product of the hole wave function times the complex conjugate of the electron wave function. Under optical excitation,

$$H_{ext} = \mathbf{k} \cdot \mathbf{A}_{ext}, \tag{2}$$

with $\mathbf{k} = \frac{1}{i}\frac{\partial}{\partial \mathbf{r}} + \frac{e}{\hbar c}\mathbf{A}$, the transition matrix element is given by

$$M = \sum_m \langle u_s | \mathbf{k} \cdot \mathbf{A}_{ext} | v_m \rangle \int dr^2 \int dz \Psi_{sm}(\mathbf{r}, \mathbf{r}, z, z) \quad . \tag{3}$$

From the dipole matrix element of the Bloch waves, the selection rule in terms of spin momentum conservation can be deduced,

$$s - m = m_p \ , \tag{4}$$

if one defines the momentum of the photon m_p to be ± 1 for polarization $\sigma\pm$ and to be zero for the polarization along the z axis (π polarization). This is in keeping with the rules in Ref. 1. The ease of visualizing the valence to conduction band transition is one of the reasons for using the "spin" index m of the valence band rather than the conventional change of sign for the spin momentum of the holes.

The other reason is the preservation of the LUTTINGER [10] Hamiltonian for the valence bands. Thus, we choose the signs for the wave vector operators:

$$\mathbf{k}_h = \frac{1}{i}\frac{\partial}{\partial \mathbf{r}_h} + \frac{e}{\hbar c}\mathbf{A}(\mathbf{r}_h),$$

$$\mathbf{k}_e = \frac{1}{i}\frac{\partial}{\partial \mathbf{r}_e} - \frac{e}{\hbar c}\mathbf{A}(\mathbf{r}_e), \tag{5}$$

with the proton charge e.

From (3), additional selection rules can be deduced. For a symmetric quantum well, the parity with respect to z_e and z_h must be the same. Selection rules governing the functional dependence on the x and y coordinates will be given later.

4. THE EXCITON HAMILTONIAN

The exciton Hamiltonian consists of the electron part, the hole part and the Coulomb attraction. In terms of the electron and valence subband bases described above, the Hamiltonian has the form:

$$H_{ex} = \begin{bmatrix} W_+ + \frac{3}{2}\kappa B & -S & R & 0 \\ -S^\dagger & W_- + \frac{1}{2}\kappa B & 0 & R \\ R^\dagger & 0 & W_- - \frac{1}{2}\kappa B & S \\ 0 & R^\dagger & S^\dagger & W_+ - \frac{3}{2}\kappa B \end{bmatrix} , \tag{6}$$

where:

$$\begin{aligned} W_\pm &= \frac{\hbar^2}{2m}k_{ez}^2 + V_e(z_e) - g_e\mu_B B_S + \frac{\hbar^2}{2m}(\gamma_1 \mp 2\gamma_2)k_{hz}^2 \\ &+ V_h(z_h) + \frac{\hbar^2}{2m_c}(k_{ex}^2 + k_{ey}^2) + \frac{\hbar^2}{2m}(\gamma_1 \pm \gamma_2)(k_{hx}^2 + k_{hy}^2) \\ &- u(\mathbf{r}_e - \mathbf{r}_h, z_e - z_h) , \\ u &= e^2/\epsilon\sqrt{|\mathbf{r}_e - \mathbf{r}_h|^2 + (z_e - z_h)^2} , \end{aligned} \tag{7}$$

$$\begin{aligned} &R = -\tfrac{\hbar^2}{2m}\sqrt{3}\bar{\gamma}k_{h-}^2, \quad \bar{\gamma} = \tfrac{1}{2}(\gamma_2 + \gamma_3) , \\ &R_A = \tfrac{\hbar^2}{2m}\sqrt{3}\mu k_{h+}^2, \quad \mu = \tfrac{1}{2}(\gamma_3 - \gamma_2) , \\ &S = \sqrt{3}\tfrac{\hbar^2}{m}\gamma_3 k_{hz}k_{h-}, \quad k_{h-} = k_{hx} - ik_{hy} . \end{aligned}$$

The γ's and κ are the Luttinger parameters for the valence band, g_e is the electron g factor, m_c is the conduction band effective mass and m is the free electron mass. The κ terms on the diagonal represent the valence band Zeeman splitting. The diagonal terms $W_\pm$ contain, in order, the z-components of the electron and hole motion with effective square well potentials, the kinetic energy terms in the x-y plane and the electron-hole attraction. The off-diagonal terms are all hole terms from the Luttinger Hamiltonian.

The expression for R is called the cylindrical approximation since it preserves the rotation symmetry about the z-axis for the Hamiltonian. For the full Luttinger expression, the anisotropic correction R_A must be added to the R terms. We shall neglect the anisotropic terms and discuss the justification below.

5. EQUATION OF MOTION IN RELATIVE COORDINATES

To use the concept of the angular momentum about the z-axis, it is convenient to adopt the symmetric gauge,

$$\mathbf{A} = \mathbf{B} \times \frac{\mathbf{r}}{2}. \tag{8}$$

In analogy to the LAMB [11] transformation for the hydrogen atom in a magnetic field, YANG and SHAM [8] introduced the transformation for the exciton wave function:

$$\Psi(\mathbf{r}_h, \mathbf{r}_e, z_h, z_e) = e^{ie\chi/\hbar c} e^{i\mathbf{K}\cdot\mathbf{r}} \psi(\mathbf{r}_h - \mathbf{r}_e, z_h, z_e) \quad . \tag{9}$$

The gauge transformation with

$$\chi = \mathbf{B} \times \mathbf{r}_h \cdot \mathbf{r}_e/2 \tag{10}$$

changes the vector potential terms in both the electron part and the hole part of the Hamiltonian to $\mathbf{A}(\mathbf{r})$, as functions of the relative coordinates in the x-y plane:

$$\mathbf{r} = \mathbf{r}_h - \mathbf{r}_e. \tag{11}$$

The resultant wave function as shown on the right of (9) can be written as a product of a plane wave with wave vector $\mathbf{K}$ in the interface plane and a function of the relative coordinates in the x-y plane. From (3) for the optical transition matrix element, it is clear that $\mathbf{K}$ has to vanish. The effective Hamiltonian for the wave function $\psi(\mathbf{r})$ has the same form as (6) but with the wave vector operators now given by

$$\begin{aligned} \mathbf{k}_h &= \frac{1}{i}\frac{\partial}{\partial \mathbf{r}} + \frac{e}{\hbar c}\mathbf{A}(\mathbf{r}), \\ \mathbf{k}_e &= \frac{1}{i}\frac{\partial}{\partial \mathbf{r}} - \frac{e}{\hbar c}\mathbf{A}(\mathbf{r}). \end{aligned} \tag{12}$$

6. ANGULAR MOMENTUM ABOUT THE Z-AXIS

The wave vector operators satisfy the commutation relations:

$$\begin{aligned} [k_{hx}, k_{hy}] &= -i/R_c^2 \ , \\ [k_{ex}, k_{ey}] &= i/R_c^2 \ , \end{aligned} \tag{13}$$

where $R_c = \sqrt{\hbar c/eB}$ is the cyclotron radius. Thus, the Hamiltonian can be expressed in terms of the operators [10,12]

$$\begin{aligned} a_h &= k_{h-} R_c/\sqrt{2} \ , \\ a_e &= k_{e+} R_c/\sqrt{2} \end{aligned} \tag{14}$$

and their Hermitian conjugates. They obey the commutation relations

$$\begin{aligned} \left[a_i, a_j^\dagger\right] &= \delta_{ij} \ , \\ [a_i, a_j] &= 0. \end{aligned} \tag{15}$$

In terms of polar coordinates $\mathbf{r} = (r, \phi)$, the z-component of the angular momentum operator is given by

$$L = \frac{1}{i}\frac{\partial}{\partial\phi} = a_h^\dagger a_h - a_e^\dagger a_e \; . \tag{16}$$

Thus, $a_h^\dagger$ and a_e raise the angular momentum quantum number and a_h and $a_e^\dagger$ lower it. An alternate way to see the role of the angular momentum raising and lower operators is from the following expression in terms of the polar coordinates:

$$k_{h\pm} = e^{\pm i\phi}\left(\frac{1}{i}\frac{\partial}{\partial r} \pm \frac{1}{r}\frac{\partial}{\partial\phi} \pm i\frac{r}{2R_c^2}\right) ,$$

$$k_{e\pm} = e^{\pm i\phi}\left(\frac{1}{i}\frac{\partial}{\partial r} \pm \frac{1}{r}\frac{\partial}{\partial\phi} \mp i\frac{r}{2R_c^2}\right) \; . \tag{17}$$

By examining the exciton Hamiltonian in terms of these operators, we see that the diagonal terms conserve the angular momentum L. The term S, by containing a_h, lowers L by one, and by being in the Hamiltonian matrix elements (3/2, 1/2) and (-1/2, -3/2), raises the spin m by one. In the cylindrical approximation, R, by containing two factors of a_h, lowers L by two, and by being in the matrix elements (3/2, -1/2) and (1/2, -3/2), raises m by two. The Hamiltonian then conserves the sum of the angular and "spin" momentum in the z direction. The exciton wave function may be written in the form

$$(\psi_{\ell-2,s,3/2}, \psi_{\ell-1,s,1/2}, \psi_{\ell,s,-1/2}, \psi_{\ell+1,s,-3/2}). \tag{18}$$

The suffix ℓ denotes the angular dependence $exp(i\ell\phi)$ for the s, m component of the exciton wave function. This form of the angular dependence is equally applicable in the zero field case [13].

By examining the equation for small r, it is straightforward to show that

$$\psi_{\ell s m}(r, \phi) \sim r^{|\ell|} e^{i\ell\phi}. \tag{19}$$

From the expression for the transition matrix element, (3), the selection rule $\ell = 0$ follows.

If the electron-hole attraction is neglected, each wave function component is a product of the harmonic wave functions for the electron and hole with Landau indices n_e and n_h, which are eigenvalues of $a_e^\dagger a_e$ and $a_h^\dagger a_h$ respectively. The relative angular momentum of the electron-hole pair is given by $\ell = n_h - n_e$.

The anisotropic term in R_A which contains two factors of $a_h^\dagger$ and occurs in matrix elements (3/2, -1/2) and (1/2, -3/2) raises the total angular momentum by four and therefore connects the states with ℓ values differing by 4. In general, this term is unimportant, as in the hole Landau level case [14].

7. CLASSIFICATION OF THE EXCITON STATES

An exciton state may be specified by six quantum numbers, μ, s, ν, j, n, M. In the limit of vanishing electron-hole attraction, the state reduces to the electron subband μ with spin s and hole subband ν with spin j. The hole spin j is well defined at zero field for the zero in-plane wave vector state where the Luttinger Hamiltonian is diagonal. At a finite field, the j state will have a mixture of m components. The value of j is chosen to be the value of m which by continuity is the dominant component at low fields. The $j = \pm 3/2$ states are sometimes referred to as the heavy hole excitons and $j = \pm 1/2$ the light hole excitons. YANG [15] has found that for fields up to 16 Tesla in a symmetric quantum well, the dominant component of the j state is $m = j$, a medium-strength component is $m = j+2$ or $m = j-2$, which is sometimes comparable with the "dominant" component, and the remaining two components are weak, and often negligible. The index n is the principal quantum number of the exciton states and M is the azimuth quantum number in the cylindrical approximation, related to ℓ in (18) by

$$M = \ell - j - 1/2. \tag{20}$$

The advantage of M over ℓ is that M is the angular momentum quantum number of the dominant component of the wave function, just as in the hole Landau level case [16].

8. EXCITONS IN THE MAGNETO-OPTIC SPECTRA

We are now in a position to enumerate the possible exciton lines in the optical spectrum in a symmetric quantum well. Since the selection rules derived above are for a component of the exciton wave function, for each allowable component there are four optically active exciton states, which we shall classify into a strong line, a medium line and two weak lines, bearing in mind that the medium line can be comparable in intensity to the strong line.

Consider the transition from the top valence subbands with the four different j's to the two lowest nearly degenerate conduction subbands with $s = \pm 1/2$. Since both set of bands have even parity in the z direction, parity conservation is obeyed.

8.1 Polarization $\sigma+$

For the $\sigma+$ polarization, $m_p = 1$, two sets of transitions are allowed: from $m = -3/2$ to electron spin down, $s = -1/2$ and from $m = -1/2$ to electron spin up, $s = 1/2$. For the first set, the orbital angular momentum rule says that the $m = -3/2$ component of the exciton wave function must have zero orbital momentum. From (18) with $\ell = -1$, the orbital momentum values for the four components are in order $(-3, -2, -1, 0)$. Thus, $j = -3/2$ states with $M = 0$, which have the strong component at $-3/2$, give a series of strong down-spin heavy-hole excitons, HHE↓ns, n being the principal quantum number and s denoting the zero orbital angular momentum. The states $j = 1/2, M = -2$ have medium strength at the component $-3/2$ and give rise to the series of up-spin light-hole excitons, LHE↑3d-,..., where d- denotes $M = -2$. The states $j = 3/2, M = -3$ and $j = -1/2, M = -1$ yield two series of weak lines, HHE↑nf- and LHE↓np-.

For the transitions, $m = -1/2 \rightarrow s = 1/2$, the $m = -1/2$ component of the exciton wave function must have zero orbital angular momentum. From (18) with $\ell = 0$, the angular momentum values of the four components are $(-2, -1, 0, 1)$. The strong series is $j = -1/2, M = 0$, or LHE↓ns. The medium series is $j = 3/2, M = -2$, or HHE↑nd-. The two weak series are LHE↑np- and HHE↓np+.

8.2 Polarization $\sigma-$

With the photon angular momentum $m_p = -1$, possible transitions are $m = 3/2 \rightarrow s = 1/2$ and $m = 1/2 \rightarrow s = -1/2$. For $m = 3/2$, from (18) with $\ell = 2$, the angular momentum values for the four components of the wave function are (0,1,2,3). This set of transitions has a strong series for $j = 3/2, M = 0$, denoted by HHE↑ns, a medium series for $j = -1/2, M = 2$, denoted by LHE↓nd+, and two weak series for $j = 1/2, M = 1$ and $j = -3/2, M = 3$, denoted by LHE↑np+ and HHE↓nf+.

For $m = 1/2$, from (18) with $\ell = 1$, the angular momentum values are $(-1, 0, 1, 2)$. Thus, the strong series is given by $j = 1/2, M = 0$, or LHE↑ns, the medium series given by $j = -3/2, M = 2$, HHE↓nd+, and the weak series given by $j = 3/2, M = -1$ and $j = -1/2, M = 1$, i.e. HHE↑np- and LHE↓np+.

In the $\sigma-$ spectra, the two lowest lines have much weaker magnetic field dependence than the higher lines and have commonly been interpreted as the ground state excitons HHE↑1s and LHE↑1s. In the calculations of YANG and SHAM [8], the field dependence of LHE↑1s is so strong that we found it more appropriate to assign this state to the third line up in the energy spectrum. The state HHE↓3d+ has a component at $m = 1/2$ comparable in strength to the $-3/2$ component and has its energy between LHE↑1s and HHE↑1s. Therefore, we assigned the old LHE↑1s line to HHE↓3d+. The energy dependence of these two states on the magnetic field is also similar to the corresponding valence subband Landau levels.

8.3 Polarization π

The two set of transitions have the same values for s and m. The possible series are the same as for the $m = \pm 1/2$ cases described above.

9. DISCUSSION

We review here a theory of the excitons confined to a well in a normal magnetic field taking into account the valence band structure. The features of the magneto-optic spectra are then richer than the two-dimensional hydrogen model. Within the cylindrical approximation, the angular momentum concept can still be used in conjunction with the spin angular momentum to classify the exciton states. The exciton lines reduce, as the magnetic field strength goes to zero, to the exciton states in the zero field limit. There are no unbound electron-hole pairs in finite magnetic field. Even though the electron cyclotron energy is the largest energy, the field dependence of the exciton energy does not follow the hydrogenic model. The valence band mixing effect gives rise to another set of optically active excitons.

ACKNOWLEDGEMENTS

This work is supported by NSF Grant No. DMR 85-14195.

LITERATURE

1. For a review, see R.C. Miller and D.A. Kleinman: J. Lumin. 30, 520 (1985)
2. D.A.B. Miller: Surf. Sci. 174, 221 (1986)
3. J.C. Maan, G. Belle, A. Fasolino, M. Altarelli and K. Ploog: Phys. Rev. B 30, 2253 (1984)
4. S. Tarucha, H. Okamoto, Y. Iwasa and N. Miura: Solid State Comm. 52, 815 (1984)
5. W. Ossau, B. Jakel, E. Bangert, G. Landwehr and G. Weimann: Surf. Sci. 174, 188 (1986)
6. D.C. Rogers, J. Singleton, R.J. Nicholas, C.T. Foxon and K. Woodbridge: Phys. Rev. B 34, 4002 (1986)
7. F. Ancilotto, A. Fasolino, and J.C. Maan: Superlattices and Microstructures 3, 187 (1987)
8. S.-R. Eric Yang and L.J. Sham: Phys. Rev. Lett. 58, 2598 (1987)
9. O. Akimoto and H. Hasegawa: J. Phys. Soc. Jpn. 22, 181 (1967)
10. J.M. Luttinger: Phys. Rev. 102, 1030 (1956)
11. W. Lamb: Phys. Rev. 85, 259 (1952)
12. This is similar to the impurity case, given by A.H. MacDonald and D.S. Ritchie: Phys. Rev. B 33, 8336 (1986)
13. B. Zhu and K. Huang: Phys. Rev. B 36, 8102 (1987)
14. S.-R. Eric Yang, D.A. Broido, and L.J. Sham: Phys. Rev. B 32, 6630 (1985)
15. S.-R. Eric Yang: Ph.D. Thesis, University of California, San Diego (1986)
16. L.J. Sham: In High Magnetic Fields in semiconductor Physics, ed. by G. Landwehr (Springer, Berlin, Heidelberg 1987) p.288

Mixing of Magnetoexcitons in Quantum Wells

G.E.W. Bauer

Philips Research Laboratories, P.O. Box 80000,
NL-5600 JA Eindhoven, The Netherlands

The mixing of magnetoexcitons due to the complicated valence band structure in III-V semiconductors is considerably modified by a quantum well potential. Several examples are discussed from a theoretical point of view. Most experimental findings are well explained by a recently developed effective mass theory and the Luttinger parameters of the bulk band structure.

1. Introduction

In spite of the importance of magneto-optical studies in unravelling the electronic structure of bulk semiconductors and in determining effective mass parameters [1], the usefulness of a magnetic field as an instrument in optical studies of quantum wells has been demonstrated only rather recently [2-5]. The number of theoretical approaches which take into account the degeneracies of the valence band is also quite limited [6-10], which stems mainly from the computational complexity of the problem. Recently we submitted a review of the effects of exciton mixing on optical spectra [11]. The present paper addresses some properties of magnetoexcitons which have not been dealt with in depth in [11], *i.e.* the exciton spin-splittings as a function of well widths, the combined effect of electric and magnetic fields, and the properties of ground state excitons in magnetic fields above 10 tesla. Our attention is limited here to quantum wells grown in the [001] crystal direction, though other growth orientations are of interest too [11,12]. Only external fields normal to the interfaces are considered.

2. Method

We treat a Wannier exciton in a quantum well in the effective mass approximation, representing the valence band structure by the Luttinger Hamiltonian. The exciton wave function then becomes

$$\Psi^{\mu}_{\mathrm{ex}}(\vec{r}_e,\vec{r}_h)=\frac{\Omega}{\sqrt{A}}\sum_{MM'}F^{\mu}_{MM'}(\vec{\rho},z_e,z_h)\overset{c}{\psi}_M(\vec{r}_e)\hat{K}\overset{v}{\psi}_{M'}(\vec{r}_h), \tag{1}$$

where Ω is the total volume of the system and A the area of the quantum well. The arguments in Eq. (1) are the coordinates of electron and hole $\vec{r}_e=(x_e,y_e,z_e)$ and $\vec{r}_h=(x_h,y_h,z_h)$ while the difference coordinates in the plane of the well $\vec{\rho}=(x_e,y_e,0)-(x_h,y_h,0)$. $\overset{c}{\psi}_M$, $\overset{v}{\psi}_{M'}$ denote the Bloch functions of conduction and valence band electrons at the direct band gap, where M and M' are magnetic quantum numbers, *i.e.* $M=(-1/2,1/2)$ for the conduction band and $M'=(-3/2,-1/2,1/2,3/2)$ for

Springer Series in Solid-State Sciences, Vol. 87
High Magnetic Fields in Semiconductor Physics II Editor: G. Landwehr

the valence band. $\hat{K}$ is the time inversion operator which transforms the Bloch function of the valence band electron into that of a hole. We disregard the exchange interaction and comment briefly on this approximation in the discussion (Ch. 6). The electron label may then be omitted and the functions $F^{\mu}_{M'}$ are the elements of a 4-component vector which diagonalizes the following Hamiltonian (in atomic units):

$$\mathscr{H}_{ex} = \begin{bmatrix} A_{-3/2} & B & C & 0 \\ B^* & A_{-1/2} & 0 & C \\ C^* & 0 & A_{1/2} & -B \\ 0 & C^* & -B^* & A_{3/2} \end{bmatrix} + [-\frac{2}{\varepsilon|\vec{r}_e - \vec{r}_h|} \pm \frac{g_e}{2}\gamma]\,\mathbf{I}_4, \tag{2}$$

where

$$A_{\pm 3/2} = -\frac{m_0}{m_e}\frac{\partial^2}{\partial z_e^2} + V_c(z_e) - (\gamma_1 - 2\gamma_2)\frac{\partial^2}{\partial z_h^2} + V_v(z_h)$$

$$+(\gamma_1 + \gamma_2 + \frac{m_0}{m_e})(-\nabla_\rho^2 + \frac{\gamma^2\rho^2}{4}) + (-\gamma_1 - \gamma_2 + \frac{m_0}{m_e})\gamma L_z \pm (3\kappa + \frac{27}{4}q)\gamma, \tag{3}$$

$$A_{\pm 1/2} = -\frac{m_0}{m_e}\frac{\partial^2}{\partial z_e^2} + V_c(z_e) - (\gamma_1 + 2\gamma_2)\frac{\partial^2}{\partial z_h^2} + V_v(z_h)$$

$$+(\gamma_1 - \gamma_2 + \frac{m_0}{m_e})(-\nabla_\rho^2 + \frac{\gamma^2\rho^2}{4}) + (-\gamma_1 + \gamma_2 + \frac{m_0}{m_e})\gamma L_z \pm (\kappa + \frac{1}{4}q)\gamma, \tag{4}$$

$$B = -2i\sqrt{6}\,\gamma_3 k^- \frac{\partial}{\partial z_h}, \tag{5}$$

$$C = \sqrt{3}(\gamma_2 + \gamma_3)(k^-)^2 + \sqrt{3}(\gamma_2 - \gamma_3)(k^+)^2, \tag{6}$$

$$k^{\pm} = \frac{i}{\sqrt{2}} e^{\pm i\phi}\{\frac{\partial}{\partial\rho} \pm \frac{i\partial}{\rho\partial\phi} \pm \frac{\gamma\rho}{2}\}. \tag{7}$$

Here $\gamma_1, \gamma_2, \gamma_3, \kappa$ and q are the Luttinger valence band parameters, m_e and g_e are the effective mass and g-factor of the conduction band electron and $\mathbf{I}_4$ denotes the 4×4 unit matrix. L_z is the operator for the exciton angular momentum normal to the interfaces, k^+ and k^- are raising and lowering operators, γ denotes the magnetic field and ∇_ρ^2 is the 2-dimensional Laplacian.

The components of the exciton envelope function can be expanded as follows:

$$F^{\mu}_{M'}(\vec{\rho}, z_e, z_h) = \sum_{nmij} c^{\mu M'}_{nmij}\, \zeta^e_i(z_e)\zeta^{M'}_j(z_h) R_{nm}(\vec{\rho}), \tag{8}$$

where $\zeta^e_i(z_e)$ and $\zeta^{M'}_j(z_h)$ are the subband envelope functions of electrons and holes and $R_{nm}(\vec{\rho})$ is a member of a set of in-plane basis functions with angular momentum m.

It is a good approximation [6] to assume axial symmetry such that the total angular momentum $\mathrm{M} = M - M' + m$ is a good quantum number. Excitons with

$M = -1, 0, 1$ couple to a light field with σ^-, π, σ^+ polarization. The heavy-hole s-type excitons ($M' = \pm 3/2, m = 0$) belong to the irreducible representation Γ_7 while the light-hole s-type excitons ($M' = \pm 1/2, m = 0$) transform as Γ_6 [11]. The eigenvalue problem is solved numerically as detailed in Ref. 11.

Finally the values of the empirical parameters of the theory have to be filled in. Two parameter sets will be considered in the following. Set I [13] uses results of a fit of magnetoexciton spectra of bulk GaAs [1] while set II [14] has been found to represent the effective mass anisotropy as observed in quantum wells grown in different crystal directions. The g-factors κ and q are adopted from [1].

3. Spin-Splitting of Ground State Excitons

In order to get a feeling for the physics involved, let us first disregard all complications which are not essential for an understanding of the effects (but may be important to obtain quantitative results). To this end, the cubic anisotropy in the plane of the quantum well and the electron spin is neglected. The most important basis functions in the expansion (8) turn out to be those with low radial quantum numbers in the envelope function, ($n = 0,1,2$ and $n = |m|$), while higher ones, *i.e.* $2s,3p,4d$ *etc.*, may be disregarded for the moment. It is helpful to consider first a zero-order problem in which the the non-diagonal operators B,C and the kinetic energy of the motion in the plane of the well are neglected. Some energy levels of this model system are shown in Figs. 1a and 1b as a function of magnetic field for a 160Å quantum well. $h^{\uparrow}/h^{\downarrow}$ and $\ell^{\uparrow}/\ell^{\downarrow}$ denote the character of the valence band as heavy $M' = \pm 3/2$ and light $M' = \pm 1/2$. The hole subband index is given as a subscript (only the first electron subband is of interest here) and the character of the radial envelope function is indicated in brackets. Only excitons with the same symmetry as the ground states are plotted in Figs. 2a and 2b, *i.e.* states which are allowed to mix in the presence of the

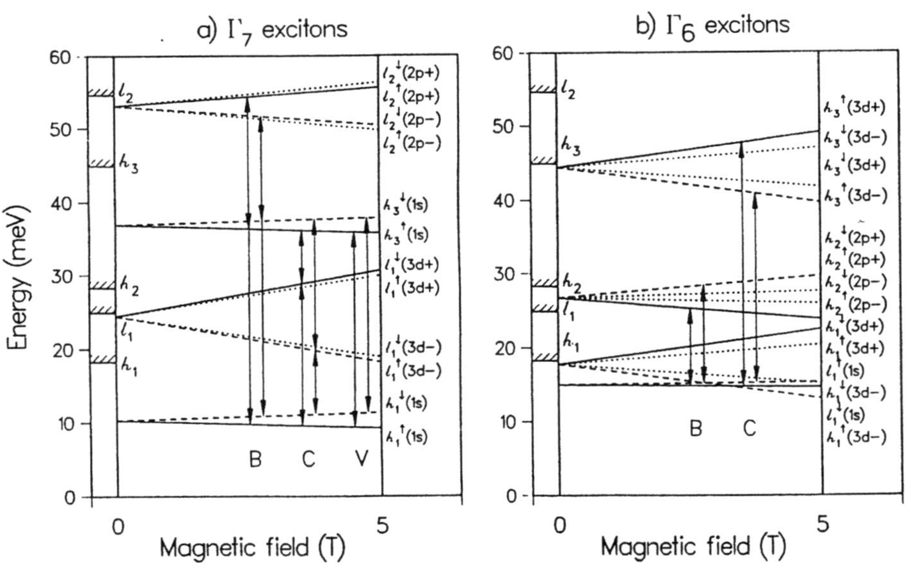

Figs. 1a and 1b: Exciton spin-splittings in quantum wells for the model system described in the text.

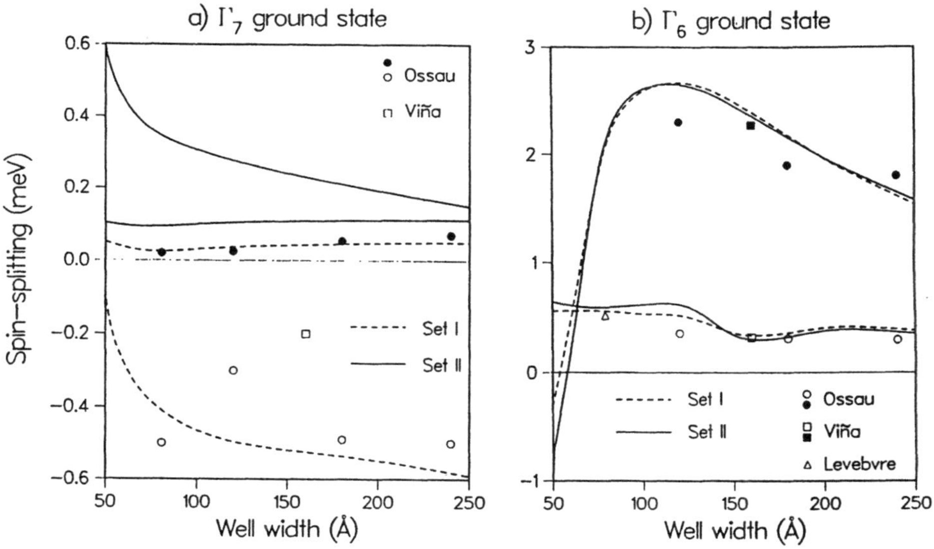

Figs. 2a and 2b: Exciton spin-splittings in $GaAs|Al_{0.3}Ga_{0.7}As$ quantum wells as a function of well width. The solid markers represent the experimental results for 10 tesla while the open symbols refer to a magnetic field of 1 tesla. The experimental values are from Ossau et al. [15], *Lefebvre* et al. [16], *and Viña* [17]. *The experimental error is estimated to be about* ± 0.1 *meV* [15,17]. *Dashed and solid curves give the theoretical results for the parameter sets I and II. The curves which indicate small and almost constant splittings represent the low-field behaviour (1 T) while the others correspond to 10 T.*

non-diagonal operators. Dashed/solid lines indicate excitons which may couple to σ^-/σ^+ polarized light. Excitons belonging to the dotted lines do not couple to the others (or a light field) and are given only for the sake of completeness. The splitting of the "unmixed" *s*-type excitons in Figs. 1a and 1b reflects the angular spin momentum of the Bloch bands, *i.e.* arises solely from the valence band *g*-factors κ and q. On the other hand, excitons with envelopes of higher angular momentum show a considerable splitting due to the L_z operator and the large difference in effective mass of electrons and holes.

Let us now switch on the mixing interactions between the different exciton states as indicated by the vertical arrows in Figs. 1a and 1b. The labels B and C refer to the non-diagonal operators in Eqs. (2) while V is the electron-hole Coulomb interaction. For the present parameters we have $B > C > V$. The theoretical results for the exciton spin-splittings as obtained for the full model are plotted in Figs. 2a and 2b as a function of well widths. The results can be understood as follows. Concentrating on Fig. 1a it can be seen that the excitons $\ell_2^{\downarrow}(2p\text{-})$ and $\ell_1^{\uparrow}(3d\text{-})$, to which the $h_1^{\uparrow}(1s)$ state couples, both decrease in energy as the field increases. The coupling operators B and C both tend to push $h_1^{\downarrow}(1s)$ to lower energies, thus cancelling the effect of the hole g-factor to a large extent. At lower and intermediate fields the experimental and theoretical spin splittings are therefore very small. At high magnetic fields the effect of the coupling wins over the Zeeman term and the spin splitting changes sign. The field at which the crossing occurs turns out to be quite sensitive to the choice of parameters. The experimental results are best represented by parameter set I. The crossing occurs at much higher fields (> 20T) for parameter set II. This could be an indication that the κ parameter consistent with the zero-field parameters of [14] should be smaller than the value of 1.2 from Ref. 1. The effect of mixing dominates the Zeeman splitting even at zero field for $h_3(1s)$ [10,11] which is a direct consequence of the relative proximity of $\ell_2(2p)$.

The situation for the light hole ground state is more complicated. By disregarding the kinetic energy in the plane, $h_1(3d)$ and $\ell_1(1s)$ cross, which does not happen in wide wells because the diamagnetic shift of $h_1(3d)$ is much larger. The arrows which indicate the couplings with $h_1(3d)$ are therefore omitted in Fig. 1b. The B and C mixings now have an opposite effect, but B is larger and wins. The non-diagonal coupling consequently enhances the pure Zeeman splitting considerably. The small-field splitting in Fig. 2b is slightly larger for narrow wells. This effect is due to the change of energy ordering of $\ell_1(1s)$ and $h_1(3d)$ at $\simeq 125$Å. A fine structure in the anticrossing behaviour beween these states [11] is omitted here. For narrower wells, $\ell_1(1s)$ is strictly speaking no longer the ground state but a weak Fano resonance with the $h_1(nd)$ continuum. At high magnetic fields all exciton levels are well separated and the mixing between $h_1(3d)$ and $\ell_1(1s)$ increases. At 10 T and for small well widths the Γ_6 spin-down (σ^-) exciton ground state consists mainly of $h_1^{\uparrow}(3d\text{-})$, which explains the sudden drop of the spin-splitting in this regime.

Lefebvre *et al.* [16] proposed a modification of the Luttinger parameters in quantum wells in order to explain the experimental spin-splittings. The good agreement between experiment and theory in Figs. 2a and 2b allows us to conclude that such an assumption is not necessary.

4. Electric and Magnetic Fields

It has been shown very recently by Viña *et al.* [18] that a weak magnetic field can be employed advantageously as a spectroscopic tool for the interpretation of optical spectra. A previous study on electric field induced couplings [19] remained inconclusive with respect to the role of d-type excitons. However, by the application of a magnetic field of 0.5 T the oscillator strengths of excited exciton states are sufficiently enhanced for higher angular momentum states to be clearly resolved. In Figs. 3a and 3b the measured exciton energies and transition oscillator strengths as a function of an applied electric field are compared with theoretical results. From the good agreement it can be concluded that *e.g.* the filled squares correspond to the third excited state $\Gamma_6^{(3)}$ which at low electric fields consists mainly of $h_1^{\uparrow}(3d\text{-})$.

5. High-Field Behaviour

The numerical accuracy of the ground state exciton energies using the present expansion basis set has been found to be better than 0.1 meV for fields up to 10 tesla [11]. Although this value is expected to deteriorate it is interesting to look at results for higher fields.

In Figs. 4a and 4b the ground state exciton energies and oscillator strengths of a 160 Å quantum well is plotted up to 30 tesla. The effect of an electric field of 20 kV/cm is also shown. Mendez *et al.* [20] reported a crossing of heavy and light exciton ground states at 20 T which shifts to higher magnetic fields when applying an electric

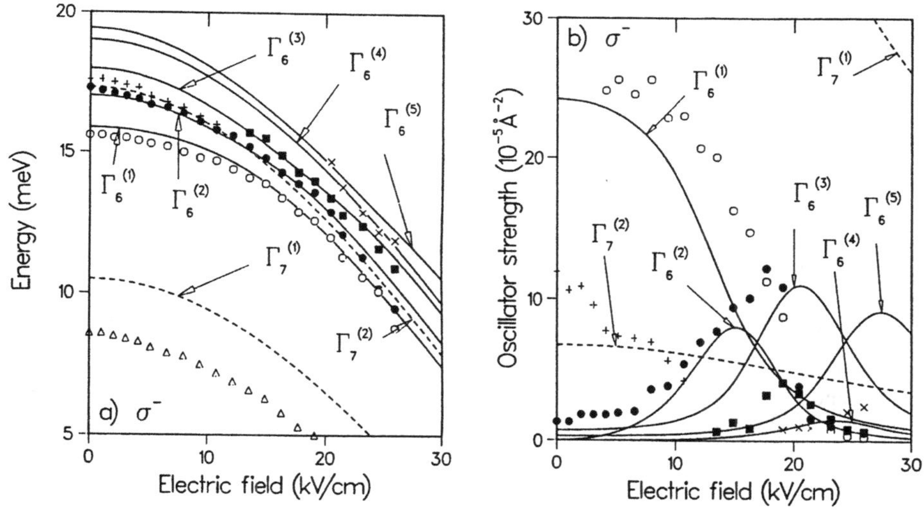

Figs. 3a and 3b: *Exciton energies and oscillator strengths of a 160Å $GaAs/Al_{0.32}Ga_{0.68}As$ quantum well as a function of an electric field (from [18]). A constant magnetic field of 0.5 T is applied. Theoretical results have been obtained with parameter set II (see Ch. 2).*

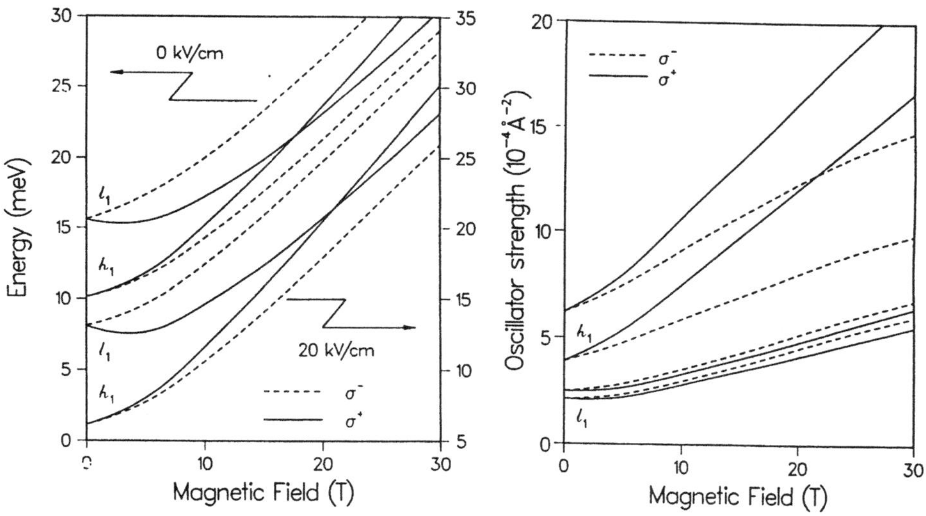

Figs. 4a and 4b: *Exciton ground state energies and oscillator strengths of a 160Å $GaAs/Al_{0.32}Ga_{0.68}As$ quantum well as a function of electric and magnetic fields as calculated with parameter set I. The fine structure in the results for the light hole exciton and finite fields (see Figs. 3a and 3b) has been omitted here.*

field, which agrees well with the present theory using parameter set I (see Ch. 2). This result is easily understood by the mass dependence of the confined Stark effect: as the electric field increases, the heavy-hole exciton shifts faster to lower energies than the light-hole exciton.

6. Discussion

The theory of the magneto-optical properties of intrinsic direct-gap quantum wells appears to be sound in so far as various experimental results are well explained. An effect which we have not considered hitherto is the exchange interaction which couples the electron and hole spins. Bauer *et al.* [21] claimed to have observed an exchange splitting of the heavy-hole exciton in quantum wells of up to a few meV, which is difficult to reconcile [22] with most bulk measurements. Van Kesteren *et al.* [22,23] have determined exciton exchange splittings of the order of 1 μeV in type II quantum wells. The trends in the experiments can be explained well by an effective mass theory [25] similar to [22], but with a much smaller bulk exchange integral. A strong magnetic field can be used to investigate the exchange coupling by letting heavy- and light-hole excitons cross (Ch. 5), although experimental results for a 160 Å quantum well could not detect a significant interaction [20].

Acknowledgements

I am indebted to Professor T. Ando for his guidance, to Drs. W. Ossau and L. Viña for unpublished experimental data and to Professor M.F.H. Schuurmans for his support and critical observations.

References

1. K. Hess, D. Bimberg, N.O. Lipari, J.U. Fishbach, M. Altarelli: Proceedings of the 13th International Conference on the Physics of Semiconductors, ed. by F.G. Fumi (North-Holland, Amsterdam, 1977).
2. J.C. Maan, G. Belle, A. Fasolino, M. Altarelli, K. Ploog: Phys. Rev. B30, 2253 (1984).
3. N. Miura, Y. Iwasa, S. Tarucha, H. Okamoto: Proceedings of the 17th International Conference on the Physics of Semiconductors, ed. by J.D. Chadi and W.A. Harrison (Springer, New York, 1985).
4. W. Ossau, B. Jäkel, E. Bangert, G. Landwehr, G. Weimann: Surf. Science 174, 188 (1986).
5. D.C. Rogers, J. Singleton, R.J. Nicholas, C.T.B. Foxon, K. Woodbridge: Phys. Rev. B34, 4002 (1986).
6. G.E.W. Bauer and T. Ando: Proceedings of the 18th International Conference on the Physics of Semiconductors, Stockholm, 1986, ed. by O. Engström, (World Scientific, Singapore, 1987).
7. F. Ancilotto, A. Fasolino, J.C. Maan: J. Superlattices Microstruct. 3, 187 (1987).
8. S.R. Eric Yang and L.J. Sham: Phys. Rev. Lett. 58, 2598 (1987).
9. G.E.W. Bauer and T. Ando: Phys. Rev. Lett. 50 , 601 (1987); J. Phys. (Paris) C5, 215 (1987).
10. G.E.W. Bauer and T. Ando: Phys. Rev. B37, 3130 (1988).
11. G.E.W. Bauer and T. Ando: Phys. Rev. B, in press.
12. L.W. Molenkamp, G.E.W. Bauer, R. Eppenga, C.T.B. Foxon: Phys. Rev. B, in press.
13. U. Ekenberg and M. Altarelli: *Phys.Rev.***B35**,7585(1987).

14. L.W. Molenkamp, R. Eppenga, G.W. t'Hooft, P. Dawson, C.T.B. Foxon: Phys. Rev. B, in press.
15. W. Ossau, B. Jäkel, E. Bangert, G. Weimann: The Basic Properties of Impurity States in Superlattice Semiconductors, ed. by C.Y. Fong, (Plenum, New York, 1988); W. Ossau, private communication.
16. P. Lefebvre, B. Gil, P. Lascaray, H. Mathieu, D. Bimberg, T. Fukunaga, and H. Nakashima, Phys. Rev. B37, 4171 (1988).
17. L. Viña, private communication.
18. L. Viña, G.E.W. Bauer, M. Potemski, J.C. Maan, E.E. Mendez, W.I. Wang: submitted to Phys. Rev. B.
19. L. Viña, R.T. Collins, E.E. Mendez, W.I. Wang: Phys. Rev. Lett. 58, 832 (1987).
20. E.E. Mendez, L. Viña, J.A. Brum, W.I. Wang, E. Isaacs, X. Wang: Bull. Am. Phys. Soc. 33, 364 (1988).
21. R. Bauer, D. Bimberg, J. Christen, D. Oertel, D. Mars, J.N. Miller, T. Fukunaga, H. Nakashima: Proceedings of the 18th International Conference on the Physics of Semiconductors, Stockholm, 1986, ed. by O. Engström (World Scientific, Singapore, 1987).
22. Y. Chen, B. Gil, P. Lefebvre, H. Mathieu: Phys. Rev. B37, 6429 (1988).
23. H.W. van Kesteren, E.C. Cosman, F.J.A.M. Greidanus, P. Dawson, K.J. Moore, C.T.B. Foxon: Phys. Rev. Lett. 61, 129 (1988).
24. H.W. van Kesteren, E.C. Cosman, P. Dawson, K.J. Moore, C.T.B. Foxon: Proceedings of the 19th International Conference on the Physics of Semiconductors, Warsaw, 1988, ed. by J. Kossut, to be published.
25. B. Rejaei Salmassi and G.E.W. Bauer, to be published.

High Magnetic Fields as a Tool to Study the Optical Properties of Quantum Wells and Superlattices

J.C. Maan, M. Potemski, and Y.Y. Wang

Max-Planck-Institut für Festkörperforschung, HML,
PB 166X, F-38042 Grenoble Cedex, France

I.INTRODUCTION

The application of high magnetic fields has proven to be a very fruitful technique for studying the fundamental properties of semiconductors. The effect of a field is to quantize the band structure, to split spin degenerate levels, and to quantize the orbital motion. In this paper we will briefly describe three different phenomena, which are observable only because of this threefold quantization. In the next chapter we will study the luminescence in GaAs/GaAlAs quantum well under high excitation in magnetic fields. The quantization of energy by the field serves to create richer spectra and to obtain more detailed information about the many-particle states. In chapter III we will study relaxation between spin split Landau levels in similar samples, and we will show that the discrete nature of spin split levels in two-dimensional systems leads to a bottleneck in the energy relaxation. Finally in chapter IV we show results of theoretical calculations of the energy levels in quasi-periodic Fibonacci superlattices, and show that the orbital quantization which can be varied by the field, can lead to self-similarity in the energy dispersion.

II *ENERGY LEVEL STRUCTURE IN QUANTUM WELLS IN A MAGNETIC FIELD UNDER HIGH EXCITATION*

In this chapter we will discuss experimental results of interband luminescence under high excitation in GaAs/GaAlAs quantum wells in high magnetic fields. This luminescence comes from the filled discrete magnetic levels (Landau levels), at high carrier density. Comparing these results with low radiation intensity excitation experiments, which measure interband absorption between empty electron and hole states, dressed with the Coulombic interaction (excitons), the change of the many-particle to the two-particle level structure can directly be observed.

Experimentally the luminescence experiments were performed, using high intensity excitation (excitation intensities up to 500 kW/cm-2 from a pulsed nitrogen laser pumped Rhodamine 6G dye laser, with pulse-length ~10ns) on high quality GaAs quantum wells (thickness 14nm<d<3nm) sandwiched between $Ga_{1-x}Al_xAs$ layers (0.26<x<0.45). In the high excitation luminescence spectra we observe emission from filled states. The level structure when the states are empty (excitonic or two particle spectrum) is measured with conventional excitation spectroscopy (~ absorption). A few representative high excitation (HE) spectra are shown in fig.1, together with an excitonic spectrum at the same field. The position of the peaks in the (HE) luminescence as well as of the excitonic spectra are shown in fig.2. The excitation spectrum shows several excitonic transitions involving the hole and conduction band Landau levels. These magneto-excitonic states have been extensively studied in the past [1-4], and are known to result from the interaction between hole and electron landau levels under the influence of Coulomb interaction. The HE luminescence at the lowest powers shows a single peak coinciding with the exciton ground state. As the exciting power is increased, this single peak is broadened, initially shifts slightly to lower energies and develops a shoulder at the high energy side. At full power, the high energy side shows a regular pattern of equally

Springer Series in Solid-State Sciences, Vol. 87
High Magnetic Fields in Semiconductor Physics II Editor: G. Landwehr

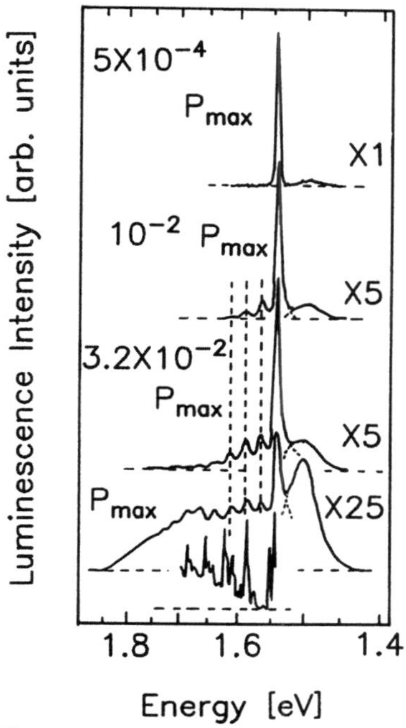

Fig.1
The luminescence intensity at 18T in a 9 nm GaAs quantum well for different powers of the exciting light. To compare the actual intensities the peak heights should be multiplied by the factor indicated. The lower trace is a normal excitation spectrum, shown for comparison.

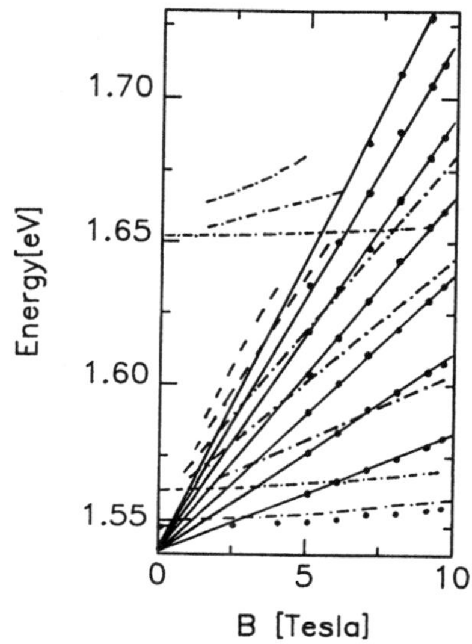

Fig.2
Magnetic field dependence of the luminescence at P_{max} (circles) in a 9nm GaAs quantum well; the drawn line for the ground state is a guide to the eye and for the excited states a Landau ladder. The dashed lines are the measured main transitions in the excitation spectra.

spaced peaks. With dashed lines we have shown the position of the maxima which as a function of excitation intensity hardly change position. On the other hand, the HE luminescence spectra are notably different from the low power excitation spectrum (see fig.2), which demonstrates the change of a two-particle to an N-particle excitation. In particular, the second peak in luminescence occurs at an energy position where there is no transition visible in the excitation spectrum. In fig. 2 the positions of the HE luminescence peaks are plotted as a function of the magnetic field and compared with the excitation spectra. The excited states of the HE spectra exhibit indeed a very regular Landau level structure, characterized by a single reduced mass and a linear field dependence for all transitions. All these excited states extrapolate, at zero field, to the same energy. On the other hand, the ground state shows a parabolic field dependence and is at higher energy at zero field than this extrapolation point.

On the basis of the theoretical and experimental understanding of highly excited semiconductors, the following phenomena are expected: a reduction of the fundamental band gap, an increase in the effective mass (band gap and effective mass renormalization), a disappearance of the exciton because of screening and/or overlap between excitonic wavefunctions and a relaxation of the selection rules [5-8]. In two-dimensional systems, this bandgap renormalization in term of energy (meV) is higher than in 3D [7,9]. All these effect are observed in our data as we will explain below.

From our data, we can deduce the renormalized band gap as the extrapolation to zero field of the excited states. This method is applicable when at least the second level is observed in luminescence which is around 10T and for powers ~5 of P_{max}. These conditions correspond to densities of $\sim 10^{12}\ cm^{-2}$ and higher. As peak positions do not change with increasing power, we must conclude that the renormalization does not change anymore with density, and that the main change has already occurred at lower densities. In this density region a density dependent band gap was found by Tränkle et al. [9] and Delalande et al. [10]. The renormalized band gap as a function of the well thickness is shown in Fig. 3. It is found to depend on well width in a similar way to the exciton binding energy and is found to be roughly two times the exciton binding energy determined from the excitation spectra. A dependence on well thickness as observed here was not observed in ref. [9], but is theoretically predicted by Bauer et al. [11].

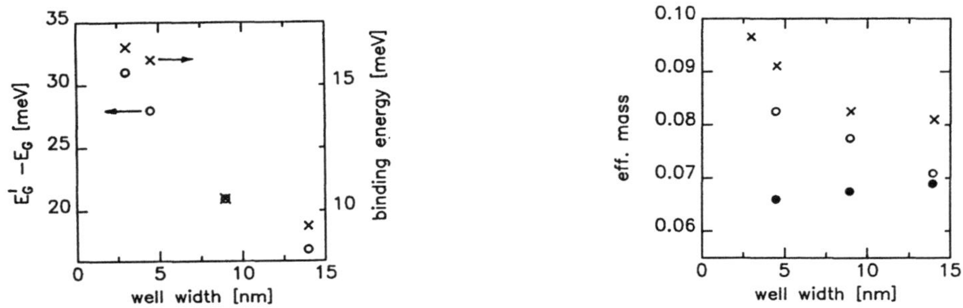

Fig.3
Measured renormalized bandgap and exciton binding energy as a function of the well width (left). Right, reduced effective mass as a function of the well width. (x) the HE renormalized mass, (o) reduced mass from the 2s excitonic and (o) from the 3s excitonic transitions.

From the field dependence of the excited states transitions, we can obtain the effective mass for interband transitions and in fig.4 the results are shown together with the reduced mass obtained from the excitation spectra. The slope of these latter transitions are partially influenced by their excitonic nature, but for the higher excited states may be assumed to represent an upper limit for the free carrier reduced mass. The HE mass is definitely increased compared to that at low excitation. Such a clear mass renormalization has not been observed before, but has theoretically been predicted in ref. [7].

The disappearance of the exciton can be very clearly seen in the behaviour of the 1s-2s transition. Due to the electron-hole pair correlation the 1s-2s splitting in a two-dimensional exciton is 0.9 times the exciton binding energy. We observe this splitting to be 12meV in the 4.5nm sample. This splitting disappears if there is no binding energy, and in fig.2 this effect can very clearly be seen as the drastic reduction of the splitting between the ground and the first excited state under high excitation. Therefore all electron-hole correlation seems to have disappeared for the excited states, and they all are rigidly shifted downward by the renormalization of the bandgap. On the other hand the ground state has hardly moved in energy with the high excitation, shows a parabolic magnetic field dependence and its intensity is enhanced with respect to the other peaks. This behaviour is typical for an exciton and it seems that for the ground state electron-hole correlation remains important, implying the coexistance of excitons with the electron hole plasma. This is a rather surprising observation, and if confirmed by other experiments, could necessitate a rethinking of the properties of a 2D high density plasma. We have verified to what extend spatial and temporal inhomogeneities could be invoked. We observed that the luminescence at all energies comes at the same time (within 1ns) and from the same position in the sample (within 60 μm diameter). These data although still needing further investigations show however that much new information about highly excited 2D systems can be obtained using a magnetic field that leads to a much richer spectrum.

III *OPTICAL PUMPING IN QUANTUM WELLS AT HIGH FIELDS*

In a high magnetic field in a two-dimensional electronic system, the spin split Landau levels are fully quantized, i.e. the density of states between them is in principle zero. In three dimensions the momentum in the direction of the magnetic field is conserved and the density of states is never zero. In 3D there are therefore many energy relaxation paths, involving various combinations of momentum, energy and magnetic moment. On the other hand in two dimensions, the relaxation of carriers between spin split levels can only take place by the simultaneous change in energy and magnetic moment. It is interesting to note that the two-dimensional spin split band states are in some sense analogous to Zeeman split levels in atoms or localized impurities in semiconductors which are also discrete and which are known to have long spin relaxation times. On the other hand in quantum wells the states are very different from those in atoms because they are band states and the particles are free. The discrete nature of the states permits a selective optical excitation of particles with one spin orientation only. This process,which in atomic physics is known as optical pumping [12], can be studied with conventional excitation spectroscopy.

In a high magnetic field, the heavy hole exciton ground state in quantum wells are split into two components and these two components, designated as peak A and B, are both observed in luminescence, although with opposite circular polarization of the emitted light (see fig.4, inset). It is important to note that the higher energy peak B remains always well observable, even at the lowest excitation power (< 1 mW/cm^2), thus showing that these states are not in thermal equilibrium. This implies that the relaxation time between the states is comparable to the recombination time. Since the relaxation between those states involves an spin-flip this fact is an indication that indeed this spin relaxation time is long. We can now measure an excitation spectra on each of this two

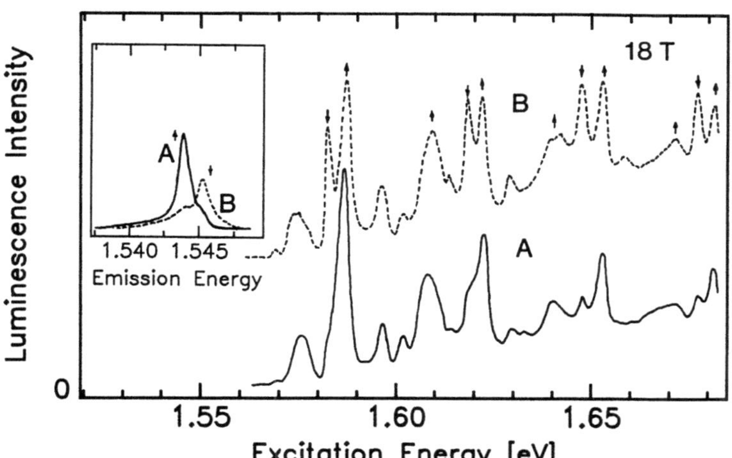

Fig.4
In the inset the A and B luminescence peaks, which are σ^+ and σ^- polarized respectively. The intensity dependence of A resp. B as a function of the σ^+ polarized exciting light energy. The arrows indicate the electron spin orientation of the transitions involved.

peaks, but always with the same circular polarization of the exciting light, so that the same transitions are excited. The results of this experiment are shown in fig. 4 and it can be seen that although the exciting light is absorbed in the same states the intensity of the peaks are different in the two spectra. This fact demonstrates a selective relaxation from the excited states to either A or B, and which relaxation path is preferred is a property of the excited state.

It is important to realize that the observed transitions involve electron Landau levels with pure spin-up or spin-down character and hole Landau levels with a mixed spin-orbit character. States which can be excited in one polarization, may involve electron levels with either spin-up or spin-down character, i.e. the polarization dependence is not simply linked to a certain spin orientation. On the other hand, the two components of the ground state, the luminescence peaks A and B, are purely spin-up in one polarization and spin-down in the other polarization [1-4]. The hypothesis to explain the data is that the excited states involving a certain conduction band spin orientation preferentially relax to that ground state which has the same orientation of the electronic spin, and do not change their magnetic moment during the relaxation. The thermalization via the hole levels is thought to occur without difficulties because through the spin orbit coupling the holes can relax easily by interaction with the lattice or impurities. The electron, however, can only change its spin through a magnetic interaction which is more improbable.

In order to prove this hypothesis we have compared the observed high magnetic field transitions with the theoretical calculations of the interband transitions, which are allowed in this polarization (see fig.5). From the identification of the transitions we can attribute the conduction band spin component for each of them. In turn, it can be seen in fig.4 that, in a given excitation spectrum, those peaks are enhanced which involve the same spin components of conduction band states both in absorption and in emission.

To be more quantitative we can evaluate the intensity ratio between the luminescence peaks A and B using rate equations. A simplified scheme of the processes which determine this intensity ratio is shown in fig.6. The relevant processes can be described in the following way. Carriers are

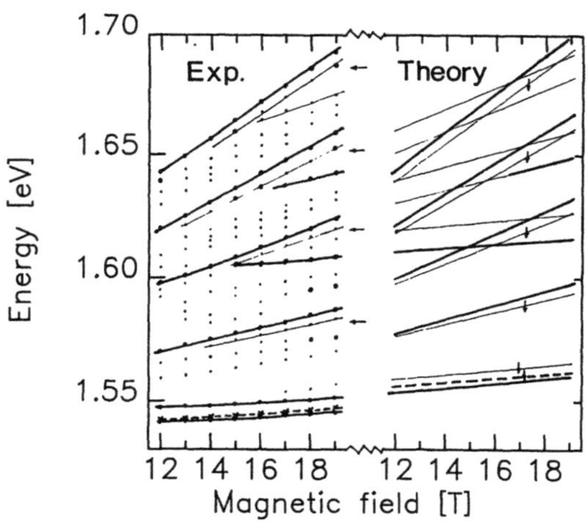

Fig.5
Energy of the maxima in the σ^+ excitation spectra as a function of the magnetic field (left). The size of the data points reflects the intensity of the peaks. The arrows indicate peaks that are enhanced when the excitation spectra are measured on peak B (indicated by the dashed line). Right, the calculated field dependence of the most intense transitions in σ_+ polarization. The arrows indicate state with the same CB spin as the σ^- ground state, indicated by the dashed line.

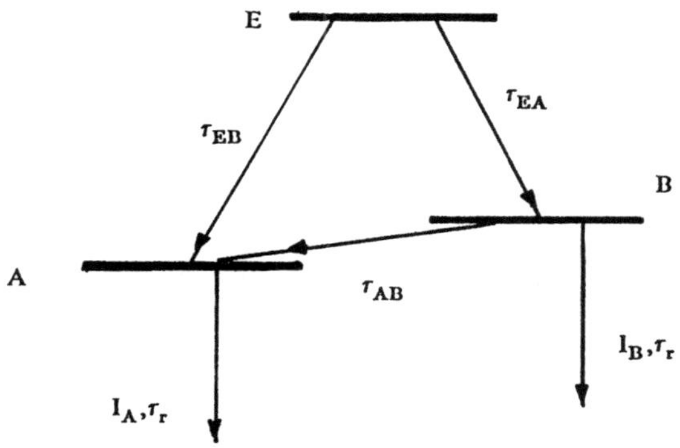

Fig.6
Schematic level structure showing the relaxation and recombination processes involved.

created optically in some arbirary state E. From this state they relax toward the states A and B, with characteristic relaxation times τ_{EA} and τ_{EB}. The ratio between these times describes to which state the relaxation takes place preferentially (the spin conserving process). When the carriers have relaxed down to the state A, which is the lowest in energy, they can only recombine radiatively thus giving rise to the luminescence I_A. Carriers which relax down to state B, can either relax further to A, which implies a electron spin-flip, or can recombine directly, thus giving rise to the luminescence I_B. The intensity ratio $R=I_A/I_B$ will therefore depend also on the ratio of the recombination lifetime τ_r toward relaxation time between A and B, τ_{BA}. With this model we can write:

$$R = \frac{I_B}{I_A} = \frac{\dfrac{\tau_{BA}}{\tau_r}}{\left(\dfrac{\tau_{EB}}{\tau_{EA}}\right)\left(1+\dfrac{\tau_{BA}}{\tau_r}\right) + 1} \quad . \tag{1}$$

Since there are two ratios of characteristic times which deterimine I_A/I_B, it is impossible to evaluate these charateristic times directly from a measurement of the intensity ratio which involves only one particular excited state E. However, the same equation is valid for different states E, with of course different values for τ_{EB}/τ_{EA} depending on the nature of this state, but with the same value for τ_{BA}/τ_r since this latter ratio has no relation with the excited state. To estimate the relaxation times, we choose for E a state that gives rise to a maximal value of the intensity ratio. It can be seen from eq.(1) that $R \le \tau_{BA}/\tau_r$, the equal sign occurring when τ_{EB}/τ_{EA} is zero, i.e. when all carriers excited in E relax to B only. We have found that experimentally R=4 as a maximal value, therefore $\tau_{BA}/\tau_r \ge 4$, i.e the spin-relaxation time between the two components of the spin split heavy hole ground states, is at least four times longer than the recombination lifetime. Using τ_{BA}/τ_r equal 4, when R=0.1, which is the lowest value measured for excitation of carriers in another state E, we obtain $\tau_{EB}/\tau_{EA} = 10$, i.e. spin conserving relaxation for this state is at least 10 times as efficient as the spin-flip process.

We have shown that at high magnetic field in two dimensions the conduction band spin-flip is effectively suppressed. A consequence of this effect is that the intensity of excitation spectra at high fields may not necessarily reflect the absorption. To obtain the real spectra in fact the sum of the spectra A and B, on both luminescence peaks should be taken. A second implication is that it is very easy to create a population inversion between the lowest two spin split heavy hole exciton levels, by exciting in a state with the same electron spin as the upper heavy hole spin split level. We have indeed observed such a population inversion experimentally. Finally, this effect leads to new spectroscopic technique to select states, which have the same electron spin orientation. In combination with the well-known polarization dependence this leads to a very powerful spectroscopical tool.

IV *MAGNETIC LEVELS IN FIBONACCI SUPERLATTICES*

In this chapter we will discuss the energy level structure of a particular type of non-periodic superlattice, namely a Fibonacci superlattice [13,14] in a magnetic field, applied parallel to the layers. In this configuration the electrons orbit in a plane perpendicular to the layers of the superlattice, with an orbit radius $l=\sqrt{\hbar/eB}$. Therefore the length scale on which the non-periodicity is "seen" by the electrons can be varied. Due to this non-periodicity the energy depends on the position of cyclotron orbit and we will show that this orbit center dispersion, can be self-similar at certain values of the magnetic field.

The Schrödinger equation that describes the motion of the electrons in a potential and a magnetic field B in the x-direction, can be written as:

$$\left[-\left(\frac{d^2}{du^2}\right)+(u-u_0)^2+v(u)\right]\Psi(u)=E\Psi(u) \quad . \tag{2}$$

In this equation the gauge is chosen a A=(0,-zB,0), the wavefunctions in the x and y direction are plane waves with wavevector k_x and k_y respectively. The energy dispersion in the direction of the field (k_x) is unaffected by the potential and we will discuss the properties only at zero k_x wavevector. Furthermore the potential v and the eigenvalues E are normalized to the cyclotron energy $\hbar eB/m$, and the length $u=z/l$ to the magnetic length $u_0=\hbar k_y/eB$, the center coordinate of the cyclotron orbit. This Schrödinger equation has become dimensionless with these units of energy and length and therefore it possesses a "trivial" scaling property, namely, that different samples in different fields will have the same properties provided that the normalized units are the same. In the following we wish to discuss a "non-trivial" scaling in which self-similarity is found in the same sample but at different fields.

In the absence of the potential eq.(1) is the well-known harmonic oscillator equation with eigenvalues $E=(N+1/2)$, with N the Landau level quantum number, for all orbit center coordinates u_0. This degeneracy, which is due to translation symmetry will be broken when $v\neq0$ and in this case the Landau levels will broaden into Landau bands. If v is a periodic potential, as is the case for a "real superlattice" it has been shown that for energies within the superlattice miniband width also flat Landau levels exist [15-17]. For a non-periodic structure this is no longer the case and we expect dispersive (=orbit center dependent) Landau bands. In the following it is exactly this dispersion that we will be interested in.

We consider a quasi-periodic superlattice which is built in the following way. We generate a Fibonacci sequence w_n with as elementary building blocks "a" and "b" following the prescription:

w_1="a", w_2="b".

$w_n=w_{n-2}+w_{n-1}$ if n is even

and

$w_n=w_{n-1}+w_{n-2}$ if n is odd

leading to: ="...ababbabb..." . (3)

It is easy to see that the length of the sequence w_n (the number of elements "a" and "b") corresponds to the nth Fibonacci number [18]. Therefore the ratio w_n/w_{n-1} tends to $\tau=\left(\sqrt{5}+1\right)/2$, the golden mean, as this is a property of the Fibonacci sequence. From this procedure of generation one can also see the self-similarity of the sequence, namely by replacing,

"ab"→"a"

"abb"→"b" (4)

one obtains the sequence w_{n-2} from w_n. Similarly the transformation,

"b"→"a"

"ab"→"b" (5)

transfers w_n into the inverse of w_{n-1} (the sequence w_{n-1} read from right to left). To translate this sequence w_n into a one-dimensional potential we take "a" as a barrier with width d_a and height v_0 and "b" a well with width d_b. Two successive "b" in the sequence are a well with $2d_b$ wide. This comb-like potential, with irregularly spaced teeth, is shown in fig.1 in the bottom half. In order to obtain self-similarity in the length scale it is easy to show that d_a/d_b has to be equal to τ. Since $B\equiv l^{-2}$, scaling of τ in the length implies a scaling of τ^2 in the field, and thereby a of $1/\tau^2$ in the energy.

With the potential constructed in this way we have calculated the energy levels of eq.(2) numerically. The results are shown in fig.7a and 7b., and were calculated with parameters corresponding to GaAs/GaAlAs superlattices, with layer thicknesses of 1.12 and 1.68 nm to have $d_a/d_b\simeq\tau$ and at the same time an integer number of lattice planes per layer. The barrier height in the calculation was 0.2eV and the corresponding range of the fields were up to 20T. It can be seen from fig.7 that the energy levels all show a dispersion in the cyclotron orbit position, which is a direct consequence of the non-periodicity of the system. However, the striking result is, that this dispersion for a given field B and at τ^4 times this field is the same when the reduced units, introduced in eq.(2), are used for the energies and lengths. For $B\tau^2$ the levels are seen to be antiself-similar, with respect to those of B, which means self-similar if the coordinate system is inverted . These observations are illustrated more clearly in fig.8 which shows the calculated density of states for several values of the magnetic field. It can be noted that only scaling with τ^{2n} with n an integer leads to self-similarity. It can now also be seen clearly that the τ^2 scaling leads to the same density of states as that at B, because for the density of states the inversion of the coordinate system has no consequences. For values of the field with different powers of τ no similarity is observed.

There are of course limitations on the field and sample parameters, for which the self-similar behaviour descibed above is observed. It is for instance obvious that for infinitely high potential barriers, this behaviour would not be observable. In this case there would be no coupling between the successive wells and therefore the level stucture would only depend on the individual wells, and not on their quasi-periodic arrangement. It can be shown that only for parameters, which allow a treatment of the potential in eq.(2) as perturbation, the previously described properties can be observed. The main reason for the scaling property being that the eigenvalues depend on the average value of the potential within the magnetic length. It turns out that the Fibonacci potential has the remarkable property that a scaling in the length by τ^n leads to a scaling of $1/\tau^{2n}$ in the average potential. This is exactly the same scaling property that is required by the magnetic Hamiltonian (2), and it is this identical scaling behaviour that is responsible for the self-similarity in the level structure as shown in fig.7 and 8.

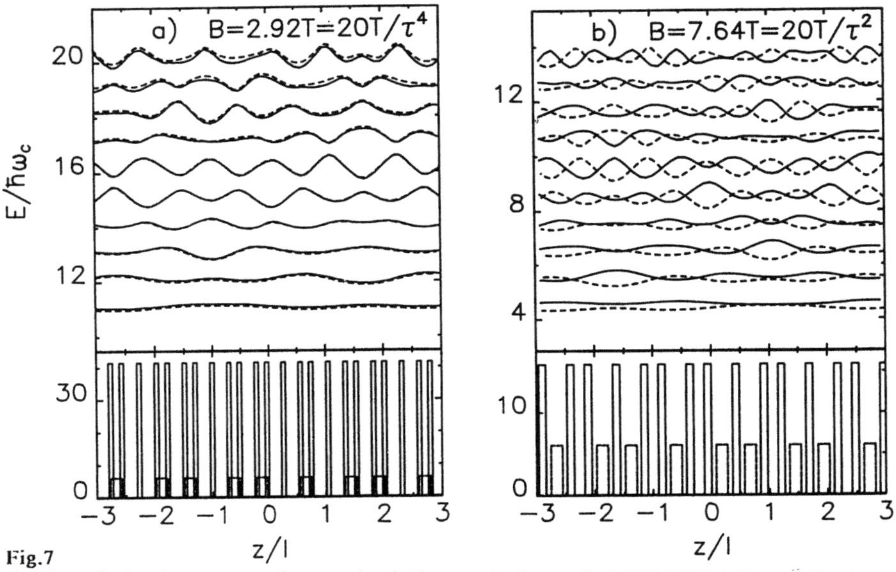

Fig.7

a) Magnetic level structure of a quasiperiodic superlattice at $B=2.92T=20T/\tau^4$ (drawn lines) and at 20T (dashed lines) as a function of the cyclotron orbit center coordinate (top) at zero wavevector for the motion in the direction of the field. In the lower part, a corresponding section of the Fibonacci potential at the same fields. Energies are all in units of the cyclotron energy and lengths in units of the magnetic length.

b) As in a) but for $B=7.64T=20T/\tau^2$.

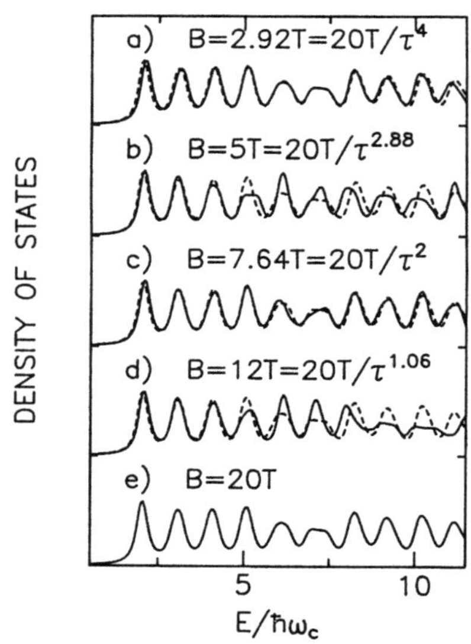

Fig.8

Calculated density of states at zero wavevector for the motion in the direction of the field. The energy scale is in units of the cyclotron energy for all curves. The zero of the energy scale for all but the lower curve is linearly shifted and aligned with the first peak. For comparison the dashed curves, all representing the result at 20T, are drawn on the other results.

Note that only for even powers of τ self-similarity is observed.

V *SUMMARY*

In the previous chapters we have shown that the quantization of the band structure, the lifting of spin degeneracy and the quantization of the orbital motion, which are three important consequences of the application of a magnetic field, can be used to study very different physical problems. We hope that this work has contributed to the feeling that there are still many interesting problems that can be investigated in the domain of the application of high magnetic fields in semiconductor physics.

Acknowledgement

We would like to thank G.Weimann and K.Ploog for the gift of the high quality samples without which this work would not have been possible. We would further like to thank A.Fasolino for performing the calculations of the interband transitions, and for many clarifying discussions. Finally we would like to acknowledge P.Wyder for his interest in this work.

REFERENCES

1. G. E. W. Bauer and T. Ando, Phys. Rev. Lett. **50**, 601 (1987).
2. S. R. Eric Yang and L. Sham, Phys. Rev. Lett. **58**, 2598 (1987).
3. F. Ancilotto, A. Fasolino, and J. C. Maan, J. Superlatt. Microstr. **3**, 187 (1987).
4. J. C. Maan,in "Physics and Applications of Quantum Wells and Superlattices",ed. E.E.Mendez and K. von Klitzing, ASI series B:Physics Vol 170, Plenum, New York,(1987)
5. P. Nozieres, Physica **117/118B**, 16 (1983).
6. T. M. Rice, J. C. Hensel, T. G. Philips, G. A. Thomas, Solid State Physics 32 (1977).
7. S. Schmitt - Rink, C. Ell, and H. Haug, Phys.Rev. **B33**, 1183 (1986).
8. R. C. Miller, D. A. Kleinman, O. Munteanu, and W. T. Tsang, Appl. Phys. Lett. **39**, 1 (1981).
9. G. Tränkle, H. Leier, A. Forchel, H. Haug, C. Ell, and G. Weimann, Phys. Rev. Lett. **58**, 419, (1987).
10. C. Delalande, G. Bastard, J. Orgonasi, J. A. Brum, H. W. Liu, M. Voos, G. Weimann, W. Schlapp, Phys. Rev. Lett. **59**, 2690, (1987).
11. G.E.W. Bauer, Proc. 19th Int. Conf. on the Physics of Semiconductors, Warsaw, (1988)
12. C. Hermann and G. Lampel, Ann. Phys. Fr. **10**, 1117, (1985)
13. R. Merlin, K. Bajema and R. Clarke, Phys. Rev. Lett. **55**, 1768, (1985)
14. M. Kohmoto, B. Sutherland and C.Tang, Phys. Rev. **35**, 1020 (1987)
15. G. Belle, J. C. Maan and G. Weimann, Solid State Commun, **56**, 65 (1985)
16. T.Duffield, R. Bhat, M. Koza, D. M. Hwang, P. Grabbe and S. J. Allen, Phys. Rev. Lett. **56**, 2724 (1986)
17. J. C. Maan, in: Festkörperprobleme (Advances in Solid State Physics), Vol.27, 137 (1987)
18. M. R. Schroeder, "Number Theory in Science and Communication", Springer series in Information Sciences, Vol. 7, Springer, Berlin (1986)

Far Infrared Magneto-optical Studies of Shallow Impurities in GaAs/AlGaAs Multiple-Quantum-Well Structures

B.D. McCombe, A.A. Reeder, J.-M. Mercy, and G. Brozak

Department of Physics and Astronomy, SUNY at Buffalo,
Buffalo, NY 14260, USA

1 Introduction

Infrared and far infrared (FIR) magnetospectroscopy has been an important tool for the study of semiconductors for more than 30 years. With advances in materials growth techniques high quality, thin semiconductor films and repeated structures with compositional control in the growth direction on the scale of one or two monolayers, and well-controlled selective doping can be produced. These techniques have been used to grow structures both for basic physical studies and for novel device applications. Detailed understanding of the nature of the impurity states in these structures is important in both areas.

Over the past few years there has been considerable theoretical and experimental interest in the effects of confinement on shallow donor impurities in AlGaAs/GaAs quantum well (QW) structures [1-6]. Far infrared magnetospectroscopic studies have provided the most detailed information about donors in GaAs QWs. Results [4,7] for Si donors doped in the centers of the GaAs wells in MQW structures with the magnetic field along the growth direction are in good agreement with variational calculations [8].

In these lecture notes we review several recent FIR spectroscopic investigations of shallow impurities in GaAs/AlGaAs MQW structures: 1) Shallow donors in the centers of wide quantum wells in the presence of a magnetic field parallel to the layers; 2) MQW structures doped in the centers of the barriers; 3) MQW structures doped with donors at the "top" and "bottom" edges of the GaAs QWs; and 4) MQW structures doped in the well centers with Be acceptors.

2 Theoretical Background

In the effective mass approximation the Hamiltonian for a hydrogenic donor in a quantum well structure with a uniform magnetic field applied along the growth direction (z-direction) is (in dimensionless units) [8]

$$H = -\nabla^2 - \frac{2}{[\rho^2+(z-z_i)^2]^{\frac{1}{2}}} + \gamma L_z + \gamma^2\rho^2/4 + V_W(z), \qquad (1)$$

where ρ is the position of the electron in the x-y plane, z_i is the position of the impurity ion, L_z is the projection of the orbital angular momentum operator along the z direction, and $\gamma = (\hbar\omega_c/2)Ry^*$ is a dimensionless measure of the magnetic field. Energy is measured in units of Ry* (the effective Rydberg) = 5.83 meV for GaAs, and length in units of

Springer Series in Solid-State Sciences, Vol. 87
High Magnetic Fields in Semiconductor Physics II Editor: G. Landwehr

a_o* (the effective Bohr radius) = 98.7Å for GaAs; $\omega_c = eB/m^*c$ is the cyclotron frequency with m* the electron effective mass. For an isolated quantum well of width L with the origin taken at the center of the well

$$V_W(z) = 0 \qquad |z| < L/2 ,$$
$$V_W(z) = V_o \qquad z < -L/2 \text{ or } z > + L/2 . \tag{2}$$

Variational solutions to this Hamiltonian [8] give results consistent with physical intuition. For an impurity at the well center the binding energy (and transition energies) increase with decreasing well width, since the electron on average is forced to be closer to the attractive positive ion. For constant well width the binding energy decreases as the impurity is moved from the well center to the edge, and decreases further as it is moved into the barrier (with the electron remaining in the well). With the positive ion near the interface or in the barrier material the average position of the electron is farther away from the positive ion due to the barrier repulsion, thus reducing the magnitude of the potential energy.

With the magnetic field along the quantum well axis the z-component of orbital angular momentum is conserved, (1); thus, irrespective of the position of the impurity, the electric dipole selection rules for light propagating along the magnetic field (Faraday geometry) are $\Delta m = \pm 1$, where m is the magnetic quantum number. The ground state always corresponds to m = 0; hence the final state of an allowed transition must have $m = \pm 1$.

For an impurity in the well center the lowest lying excited states are $2p_x$, $2p_y$ and $2p_z$, in the atomic orbital notation, all of which are degenerate for a 3D system ($V_W(z) \equiv 0$). The confining potential of the quantum well removes this degeneracy, separating the $2p_z$ state to higher energies and leaving $2p_x$ and $2p_y$ states degenerate. For sufficiently narrow wells the $2p_z$ state becomes the lowest impurity state associated with the first excited confinement subband [9].

With B along the z-axis eigenfunctions of (1) are linear combinations of $2p_x$ and $2p_y$ with eigenvalues of $L_z = \pm h$, and the strongest allowed transitions from the 1s state involve these two states. Transitions to the $2p_z$ state are not allowed in this geometry. However, with $\vec{B} \perp \hat{z}$ the situation is considerably different as discussed below.

3 Experimental Details

The AlGaAs/GaAs MQW samples used in these investigations where grown by MBE at Cornell University, the Naval Research Laboratory, and AMOCO Research Laboratories. Quantum well widths were between 100Å and 450Å and barrier widths between 125Å and 150Å. Barrier composition was nominally between x = 0.22 and x = 0.3. One set of samples was selectively doped with Si donors. The barrier- and center-doped sample was nominally uniformly doped over the central 1/3 of the AlGaAs barriers at $4 \times 10^{16} cm^{-3}$ and the central 1/3 of the 30 GaAs wells at $1 \times 10^{16} cm^{-3}$. The 450Å well-width sample was doped over the central 200Å of the GaAs wells at $5 \times 10^{15} cm^{-3}$. The other samples (top edge-doped, bottom edge-doped and center-doped 210Å wells and top edged-doped 140Å wells and 375Å wells) were all doped over the appropriate 1/3 of the GaAs wells at $1 \times 10^{16} cm^{-3}$. "Top-edge and "bottom-edge" are referred to the growth direction with the substrate down. The other set of samples was doped with Be acceptors at $3 \times 10^{17} cm^{-3}$ over the central one third of the GaAs wells. A "bulk" 3μm thick epi-layer uniformly doped at $3 \times 10^{16} cm^{-3}$ was also investigated for comparison. All sample substrates were wedged at 6° in order to minimize multiple internal reflection effects.

Magnetotransmission data were obtained with two separate Fourier transform interferometric spectrometers, a SPECAC model 40.000 polarizing interferometer and a BOMEM Model DA3.02, in conjunction with a 9T superconducting magnet system and light-pipe optics. The BOMEM was used primarily for collecting spectra at frequencies above 80 cm^{-1} with a 4.2K Ge:Ga photoconductive detector, while the SPECAC was used primarily in the spectral region between 10 and 80 cm^{-1} with a Si composite bolometer operated at 1.6K. Measurements were made at temperatures between 4.2K and 50K with samples in He exchange gas. For the donor doped samples spectra taken at a given magnetic field after illumination for a short time with visible light were ratioed either to a zero field background or, in some cases, to a background obtained at the same magnetic field immediately after cool down in the "dark". (All donor-doped samples exhibited a persistent photo-effect.) Persistent photo effects were not observed in the acceptor-doped samples. In addition, the acceptor absorption features are only weakly dependent on magnetic field. Thus it was necessary to use a rotatable sample holder, which permitted either the active sample or a reference sample (a section of GaAs substrate) to be placed in the beam in-situ at low temperature. Transmittance data in this case were obtained by dividing the sample spectrum by the reference spectrum. All samples were wedged at ~6° to avoid multiple internal reflection effects.

4 Results and Discussion

4.1 Donors

We consider first the interesting case of a magnetic field applied parallel to the layers, ($\vec{B} \parallel \hat{x}$). In this situation the cylindrical symmetry is broken, and m is no longer a good quantum number. For example, the appropriate degenerate 2p states which would be taken in linear combination to make up the eigenfunctions of L_x in a three-dimensional situation ($|2p_y\rangle \pm i|2p_z\rangle$) are no longer degenerate; the confining potential in the z-direction removes this degeneracy. However, for appropriate samples and range of magnetic fields, measurements of FIR absorption with light propagating perpendicular to the magnetic field (Voigt geometry) and polarized perpendicular to the magnetic field (along the y-direction) can probe the $2p_z$ state. There will be substantial admixture of $|2p_y\rangle$ into $|2p\rangle$ when the magnetic field energy ($\frac{1}{2}\hbar\omega_c$) is comparable to the splitting between $2p_z$ and $2p_{x,y}$; in the high field limit, $\frac{1}{2}\hbar\omega_c \gg E(2p_z)-E(2p_{x,y})$, the confinement becomes unimportant and the 3D limit is approached, both for the selection rules and the energies of the states.

Experimental results for the "$1s-2p^+$" transition frequency vs. magnetic field for a center-doped 450Å well-width MQW sample are shown in Fig. 1.

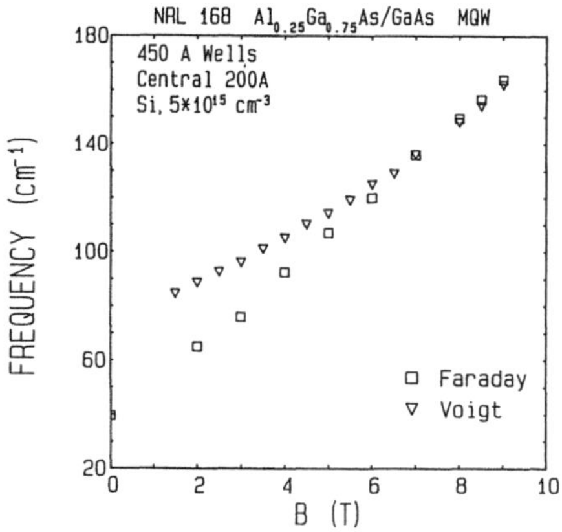

Fig. 1 Plot of the positions of the transmission minima at 4.2K for the "$1s\text{-}2p^+$" transitions of a MQW sample with 450Å well widths vs. Magnetic field for two geometries: 1) magnetic field along the growth direction (normal to the layers), Faraday geometry; and 2) perpendicular to the growth direction (in the plane of the layers), Voigt geometry.

Both Faraday and Voigt geometry measurements are shown. In the Faraday geometry the positions of the $1s\text{-}2p^+$ transition are in good agreement with previous variational calculations [9]. In the Voigt geometry the lowest field at which an absorption line is detectable (~1% absorption) is 1.5T. At this field the final state is mostly $2p_z$-like. As the field is increased the absorption strength increases to ~4% as more of the $2p_y$ state is admixed. The behavior is qualitatively as expected. A comparison of these experimental results with a detailed variational calculation shows reasonable agreement, including the crossover of transition energy vs. magnetic field curves at high fields [10].

Transmission data obtained from the barrier and center-doped MQW sample at a magnetic field of 9T are shown in Fig. 2(a) [11]. Three distinct features are observed. The weakest line at the highest energy, line 3, is a 1s-2p(m=+1) transition associated with donors located at the center of the wells. The two remaining strong lines are not normally observed in samples doped in the GaAs wells only. Line 1 is approximately 10cm^{-1} above cyclotron resonance at high fields, and line 2 appears at intermediate energies. Line 2 has a slightly larger slope vs. Magnetic field than line 1. Studies as a function of the angle of the magnetic field with respect to the growth direction show that all three lines scale approximately with the component of the field along the growth direction, and are thus confinement-related. As shown in Fig. 2(b), both lines 1 and 2 rapidly lose intensity as the temperature is increased from 4.2K to 15K while line 3 (donors at the well centers) remains essentially unchanged.

Line 1 is due to intraimpurity transitions from the m=0 ground state to the m=+1 first excited state of electrons in the well bound to their parent impurity positive ions near the centers of the barriers. Extrapolation to zero field gives a value of 20±4 cm^{-1} in reasonable agreement with recent calculations for barrier impurities in MQW samples [12]. These calculations also provide qualitative agreement between the relative absorption strengths of the well-center and barrier-center absorption lines (calculated at B=0) and the observed relative absorption strengths of lines

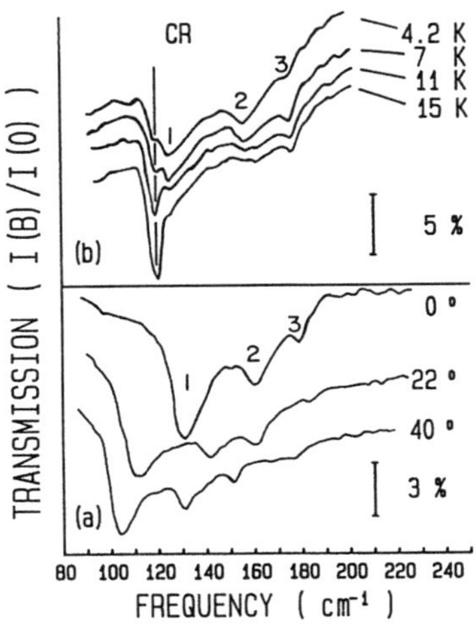

Fig. 2. a) Angular dependence of the absorption features in a barrier and center-doped sample (MBE414) at 9T. b) Temperature dependence for field normal to the sample at 9T. Transmission data are the ratio of the detected signal at the indicated magnetic field to the detected signal at zero magnetic field.

3 and 1. For MQW samples there is a singularity in the density of states vs. position along the growth direction for impurities in the centers of the barriers, as well as for impurities in the centers of the wells; and there is an increased oscillator strength for barrier impurity transitions [12]. Recent studies of several samples with a large range of barrier doping densities has also shown that the intensity of line 1 scales with doping in the barrier [13]. The temperature dependence of line 2 is very similar to that of line 1 (both indicative of a very weakly bound initial state); and, in addition, the observed resonance energy vs. B is in reasonable agreement with calculations [12] for transitions from the m=0 ground state to the second excited state with m=+1 for electrons in the wells bound to impurities at the center of the barriers. Thus a plausible assignment of this line is to this higher excited state transition for barrier impurities.

As part of a systematic extension of earlier work to determine the possible redistribution of Si impurities during growth, several edge-doped samples have been investigated. Magnetotransmission results for two 210Å well-width samples, one nominally doped uniformly over the "top" 1/3 of each well and one over the "bottom" 1/3, along with similar measurements on a nominally center-doped 210Å well-width sample are shown in Fig. 3. All other parameters of these samples are nominally identical; the top- and bottom-edge doped samples were grown sequentially during the same growth run. Low temperature photoluminescence measurements of the heavy hole exciton show that all samples are of high quality.

There are clear systematic differences among the samples as well as important similarities. The sharp feature in the center-doped sample at $162cm^{-1}$ is due to 1s-2p (m=+1). This sample shows a low energy tail extending down to $110cm^{-1}$. The edge-doped samples show dramatically different behavior. The bottom edge-doped sample exhibits an onset of absorption near $160cm^{-1}$, a weak broad minimum near $143cm^{-1}$, and a long asymmetric tail extending below $100cm^{-1}$. The top edge-doped sample, on the other hand, shows an inverted line profile with a transmittance edge at low

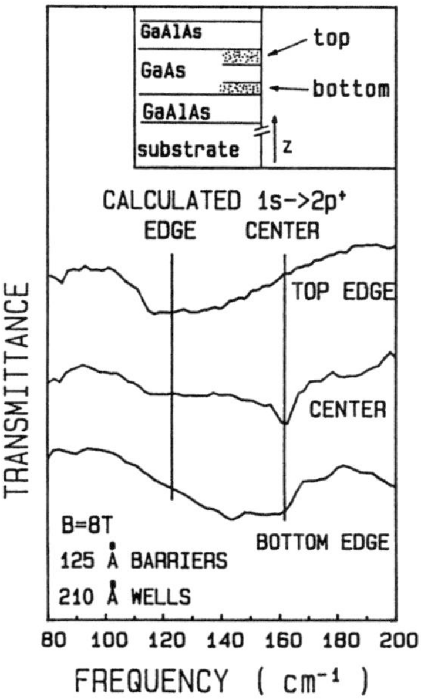

Fig. 3 Comparison of experimental transmission line profiles at 8T and 4.2K for "top-" edge "bottom" edge, and center-doped MQW samples having 210Å well widths. Two vertical lines denote the calculated positions for impurities at the edges and at the centers of the GaAs QWs for parameters appropriate to these structures. The inset illustrates the definition of "top" and "bottom"

frequencies near $110cm^{-1}$, a broad absorption band from $120\text{-}150cm^{-1}$, and a tail extending to frequencies above $160cm^{-1}$.

The weak absorption in the center- and bottom edge-doped samples at low frequencies (around $110cm^{-1}$) is apparently due to residual positive ions in the barrier weakly binding electrons in the well. The remaining line shape features are attributed to redistribution of Si dopant along the direction of growth. In the bottom-doped sample there is a delay in incorporation of Si resulting in a low density of impurities at the interface and a peak in the distribution displaced toward the well center. The peak in the distribution accounts for the weak transmission minimum near $143cm^{-1}$, and the tail of the distribution reaching the well center accounts for the high frequency edge. For the top-edge-doped sample delayed incorporation and displacement of the profile along the growth front leads to a tail of impurities extending into the AlGaAs barriers; this leads to the low energy onset and weak minimum near $115cm^{-1}$. The remaining broad absorption is attributed to a peak in the impurity distribution between the intended abrupt onset in doping, at 1/6 of the distance between the well center and the top edge, and the top edge of the well. Thus these measurements provide information about redistribution of donors during growth on a scale of roughly 10-20Å.

4.2 Acceptors

The valence bands of GaAs-like materials are much more complex than the simple s-like conduction band; thus a description of the acceptor states is considerably more complicated than that presented above for shallow donors. It is helpful to consider the bulk states prior to introducing the confining quantum well potential.

An acceptor in bulk GaAs in lowest order can be thought of as a spin 3/2 particle in a Coulomb potential, essentially a one particle spin 3/2 atom. The total angular momentum, F, of the "atom" is obtained by adding the total spin of the hole (3/2) and the orbital angular momentum of the hydrogenic envelope (L) according to the usual quantum mechanical rules for the addition of angular momenta; thus F takes interger-spaced values between L+3/2 and |L-3/2|. The valence band, however, has cubic symmetry which leads to splittings of the acceptor states of a spherically symmetric potential. The acceptor states are given an irreducible representation of the valence band point group to identify them uniquely. The lowest acceptor states are: $1s_{3/2}\Gamma_8$, $2p_{3/2}\Gamma_8$, $2s_{3/2}\Gamma_8$, $2p_{5/2}\Gamma_8$, $2p_{5/2}\Gamma_7$, and $2p_{1/2}\Gamma_6$.

The energy levels of 3D hyrogenic acceptors have been treated theoretically by several workers. In the simplest treatment [14] only spin-orbit coupling was considered; a later extension of this work [15] included the cubic term as a perturbation. KIRKMAN et al. [16] used a "zeroth order" wave function approach to determine the allowed optical transitions for acceptors in GaAs. For the transitions relevant to this discussion ($1s_{3/2}\Gamma_8$ to $2p_{5/2}\Gamma_8$ and $2p_{5/2}\Gamma_7$) the strongest absorption lines in the Faraday geometry arise from $\Delta m_J=\pm1$ and weaker transitions from $\Delta m_J=\pm3$.

The only reported calculation of acceptor states in GaAs/AlGaAs quantum wells are those of MASSELINK et al. [17]. These workers used the Luttinger Hamiltonian with the addition of a confining potential V(z), (2) where V_o represents the valence band discontinuity, and W is the well width.

The presence of the quantum well potential reduces the symmetry of the valence band to D_{2d}. In this reduced symmetry the J=3/2 valence band states are split into "heavy" ($m_J=\pm3/2$) and "light" ($m_J=\pm1/2$) hole bands. "Heavy" and "light" refer to the curvature effective mass in the confinement direction. Additionally, the $1s_{3/2}\Gamma_8$ acceptor ground state and the $2p_{5/2}\Gamma_8$ excited state are split into $1s_{3/2}(\Gamma_6+\Gamma_7)$ and $2p_{5/2}(\Gamma_6+\Gamma_7)$, respectively. The calculations were done only for the s-like ground state and s-like excited states. Results show that as the quantum well width decreases from infinity (bulk) to 100Å, the acceptor ground state splits by about 2.5 meV, and the binding energy increases by about 4 meV. Since the acceptor ground state wave function in the bulk is much less extended than the donor ground state wave function, the percentage effect over this range of well-width is much smaller for acceptors. Confinement effects on the excited states are much greater due to their more extended wave functions. Raman [18] studies of transitions between s-like states are in reasonable agreement with calculations. There are no reported calculations of the p-like acceptor excited states; the dominant features in infrared absorption measurement are due to transitions between the s-like ground states and high p-like states.

We have carried out far infrared transmission and magnetotransmission studies of Be acceptors in GaAs/$Al_{1-x}Ga_xAs$ (x=0.3) quantum wells with widths between 300Å and 100Å, as well as thick (~3µm) Be-doped GaAs epitaxial layer to provide a basis for understanding the quantum well results.

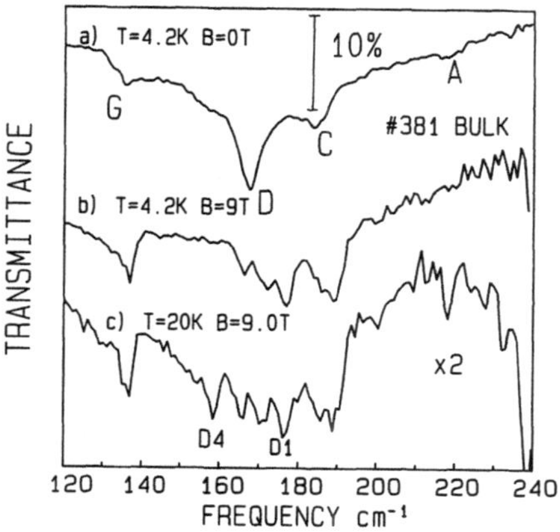

Fig. 4 Absorption spectra of 3μm of Be ($3\times10^{16}cm^{-3}$) doped GaAs at a) 0T and 4.2K b) 9T 4.2K and c) 9T 20K. Note: at high magnetic field heating the sample causes a feature D4 to appear.

The transmission spectrum of 3μm of Be-doped ($3\times10^{16}cm^{-3}$) GaAs at zero field and 4.2K is shown in Fig. 4 (Trace a). Several transmission minima are evident at 135, 167, 184 and 220 cm^{-1}. These features are labeled G, D, C and A, respectively, consistent with the literature. The relative separation of the features G, D, C, and A is consistent with the photoconductive peaks observed by KIRKMAN et al [16]. These lines arise from transitions between the $1s_{3/2}\Gamma_8$ ground state and $2p_{3/2}\Gamma_8$, $2p_{5/2}\Gamma_7$ and $2p_{3/2}\Gamma_6$ states, respectively. In the presence of a magnetic field the D line moves to higher energies and splits into several components, the C feature also moves up slightly in energy. At 20K, new lines appear at frequencies below the D feature (see Figs. 4b, 4c). At lower magnetic fields a similar temperature dependence was observed but with proportionally smaller splittings. The appearance of lower energy features associated with the D line at 20K at 9.0T is due to splitting of the ground state $1s_{3/2}\Gamma_8$ in the magnetic field. When k_BT is comparable to the field splitting, the higher energy states of the ground states multiplet are populated, and the absorption intensity for transitions originating in these states becomes appreciable. With the g-values given Kirkman, we cannot explain the temperature dependence of the absorption found in the present work. Although the magnitudes of the splittings are consistent with the magnitude of the calculated g-values, the sign of the g-values appear to be in error; i.e., the ordering of the levels of either the ground state ($1s_{3/2}\Gamma_8$) or excited state ($2p_{5/2}\Gamma_8$) multiplet is incorrect.

In Fig. 5, traces a), b), c), d) and e), the zero field transmission spectra of the bulk, 300Å, 200Å, 150Å and 100Å samples are shown. A series of absorption lines which move to higher energies as the well width decreases is evident. Although the C and D features are much sharper in the quantum well than would be expected for a bulk sample with the same dopant density, they are still quite broad; and it is difficult to identify fine structure, as was the case with bulk sample. As the quantum well width decreases from 300Å to 100Å the C and D features move to higher energies (Fig. 5). This is due to the quantum well confinement. Thus, the energy of all the states will increase at a rate that depends on the Bohr radius of the state and the orientation of the lobes of its wave function

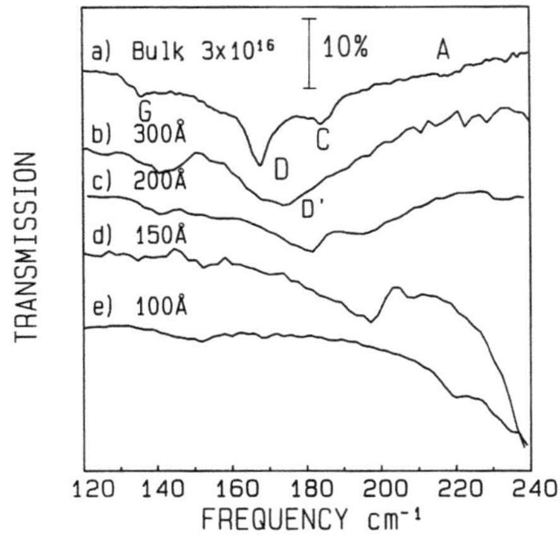

Fig. 5 Zero field transmission spectra taken at 4.2K of a) Be doped GaAs 3μm thick; and GaAs/AlGaAs Quantum Wells Be doped over the center 1/3 at 3×10^{17} cm^{-3} with widths b) 300Å c) 200Å d) 150Å and e) 100Å.

relative to the confinement axis. The movement to higher energies of the dominant absorption feature in the confined structures with decreasing well width is quite similar to that calculated by MASSELINK et al [16] for the $1s_{3/2}\Gamma_8 \rightarrow 2s_{3/2}\Gamma_8$ transitions (parity forbidden for infrared dipole transitions). However, only qualitative comparisons can be made in the absence of detailed calculations for the excited p-like states.

The dominant absorption feature labeled D′ in Fig. 5(b) displays a magnetic field dependence, approximately 1 cm^{-1}/T, similar to the highest component of the D line of the bulk acceptor. This line splits into 2 resolvable components at high magnetic field. Thus it is likely that the D′ feature in the 200Å well-width sample is due to $1s_{3/2}\Gamma_6$ to $2p_{5/2}\Gamma_6$ or $1s_{3/2}\Gamma_7$ to $2p_{5/2}\Gamma_7$ transitions. However, it is not presently clear why only 2 components are observed at high fields, since the splitting of the ground state by confinement should be very small for a 200Å well. The magnitudes of the magnetic field-dependent shifts are much smaller than would be inferred from the calculations of Ref. 17. Liu et al. have reached similar conclusions from studies of Landau level to acceptor recombination in these same samples [19].

Acknowledgements
The authors are grateful for expert MBE growth of the samples used in these studies by J. Ralston and G. Wicks (Cornell), W. Beard and J. Comas (NRL) and F. Chambers and G.P. Devane (AMOCO). Thanks are also due B.V. Shanabrook, E. Glaser and D.M. Larsen for stimulating discussions. This work was supported in part by the Office of Naval Research, the National Science Foundation through a grant to the National Nanostructure Facility at Cornell University, the Army Research Office, and the University at Buffalo through the center for Electronic and Electro-optic Materials.

References

1. G. Bastard, Phys. Rev. B 24, 4714 (1981).
2. C. Mailhiot, Y.C. Chang and T.C. McGill, Phys. Rev. B 26, 4449 (1982): R.L. Greene and K.K. Bajaj, Solid State Commun. 45, 825 (1982).
3. B.V. Shanabrook and J. Comas, Surf. Sci. 142, 504 (1984).

4. N.C. Jarosik, B.C. McCombe, B.V. Shanabrook, J. Comas, J. Ralston and G. Wicks, Phys. Rev. Lett. 54, 1283 (1985).
5. R.J. Wagner, B.V. Shanabrook, J.E. Furneau, J. Comas, N.C. Jarosik and B.D. McCombe in Proceedings of the Tenth Int'l. Symposium on Gallium Arsenide and Related Compounds, Inst. Phys. Conf. Ser. 74 (Inst. Phys., Bristol-London, 1985).
6. B.V. Shanabrook, J. Comas, T.A. Perry and R. Merlin, Phys. Rev. B 28, 7096 (1984).
7. J.-M. Mercy, N.C. Jarosik, B.D. McCombe, J. Ralston and G. Wicks, J. Vac. Sci. Technol. B4, 1011 (1986).
8. R.L. Greene and K.K. Bajaj, Phys. Rev. B 31, 913 (1985).
9. R.L. Greene and K.K. Bajaj, Phys. Rev. B 31, 4006 (1985).
10. G. Brozak, B.D. McCombe and D.M. Larsen, to be published.
11. J.-M. Mercy, B.D. McCombe, W. Beard, J. Ralston and G. Wicks, in Proc. of the Seventh Int'l. Conference on the Electronic Properties of Two-Dimensional Systems (EP2DS-VII), Surf. Sci. 196, 334 (1988).
12. R.L. Greene and P. Lane, Phys. Rev. 34, 8639 (1986); and R.L. Greene, private communication.
13. E. Glaser, B.V. Shanabrook, R.L. Hawkins, W. Beard, J-M. Mercy, B.D. McCombe and D. Musser, Phys. Rev. B36, 8185 (1987).
14. A. Baldereschi and N.O. Lipari, Phys. Rev. B8, 2697 (1973).
15. A. Baldereschi and N.O. Lipari, Phys. Rev. B9, 1525 (1974).
16. R.F. Kirkman, R.A. Stradling and P.J. Lin-Chung, J. Phys. C11, 419 (1978).
17. W.T. Masselink, Yia-Chung Chang and H. Morkoc, Phys. Rev. B32, 5190 (1985).
18. D. Gammon, R. Merlin, W.T. Masselink and H. Morkoc, Phys. Rev. B33, 2919 (1986).
19. X-C. Liu, A. Petrou, A.L. Moretti, F.A. Chambers and G.P. Devane, J Superlattice and Microstructures, 4(2), P141 (1988).

The *H*-Band Luminescence of p-Type GaAs-(GaAl)As Heterostructures in High Magnetic Fields

W. Ossau[1], *T.L. Kuhn*[1], *E. Bangert*[1], *and G. Weimann*[2]

[1]Physikalisches Institut der Universität Würzburg, Am Hubland, D-8700 Würzburg, Fed. Rep. of Germany
[2]Walter Schottky Institut der Technischen Universität München, D-8046 Garching, Fed. Rep. of Germany

We have investigated p-type GaAs-(GaAl)As heterojunctions by low temperature photoluminescence experiments in magnetic fields up to 9.5 T. The H-band luminescence splits in two well resolved lines denoted as e- and d-line respectively. From the energy shift and the splitting-behaviour, we conclude that the e-line is emitted by the recombination of a (3D)-flat-band electron with a (2D)-hole confined in an excited subband near the interface. In the case of the d-line the electron stems from a donor.

1. INTRODUCTION

In p-type GaAs/(AlGa)As heterojunctions an unusual new luminescence band has been observed in the spectral range between the emission of bound excitons and donor-acceptor pairs, referred to as either the H-band /1/ or the M-band /2/. No convincing explanation for the origin of this band has been offered in these publications, however, both agree that the band is correlated with the interface. Yuan et al. /1/ suggest that the H-band stems from the recombination of conduction band electrons with photo-created holes trapped near the interface, whereas Alferov et al. /2/ postulate a transition from donors to acceptor like surface states. Both papers report a dramatic blue shift of the band with excitation intensity (3-4 meV/decade), they disagree in the observed radiative lifetime, differing more than three orders of magnitude.

On the other hand 't Hooft et al. /3/ observed a luminescence band centered at 1.509 eV that showed a very small dependence on the excitation density (0.5 meV/decade). The lifetimes measured by these authors are as short as those published by Yuan et al. (1-2 ns). However, 't Hooft et al. supposed, that due to the observed difference in the blue-shift their luminescence line and the H-band did not correspond to the same recombination mechanism. They suggested, similar to Alferov and co-workers, a donor-acceptor pair recombination, with the impurities positioned on either side of the heterojunction interface.

Springer Series in Solid-State Sciences, Vol. 87
High Magnetic Fields in Semiconductor Physics II Editor: G. Landwehr

2. EXPERIMENTAL RESULTS

The p-type GaAs/(AlGa)As-heterojunctions used in this study were grown with MBE on semi-insulating (001)-GaAs substrates. An undoped (AlGa)As spacer was inserted between the GaAs and the doped (AlGa)As layers to separate the two-dimensional carrier gas from the ionized impurities. Beryllium was used as a dopant to achieve a 2D-hole gas. We have investigated samples with different doping levels, aluminium contents of the (GaAl)As layers ($0.5 < x < 0.65$), and spacer-thicknesses (9.5 nm $< d_s <$ 53 nm) in order to observe the dependence of the photoluminescence spectrum on these parameters. An argon ion laser as well as a pumped CR-599 dye laser with pyridine II as a dye were used for the excitation. Magnetic fields up to 9.5 T were generated with a superconducting split-coil magnet. A variable temperature cryostat insert allowed sample temperatures between 1.4 and 300 K. The luminescence was dispersed through a 1 m Jobin-Yvon spectrometer and detected with a cooled photomultiplier.

In Fig. 1 we plotted photoluminescence spectra of the H-band for different laser excitation powers. The spectra we obtained from our high quality samples show that the H-band splits into two well resolved lines. For simplicity we denote the low energetic component as d-line and the high energy component of the H-band as e-line. The intensity of the d-line exceeds that of the e-line for low excitation and therefore

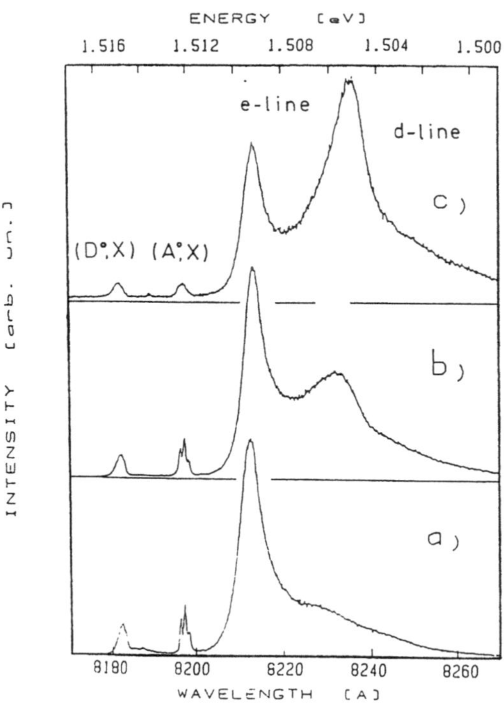

Fig. 1 *Luminescence spectra of the H-band for three different excitation densities I. The H-band splits into two well resolved lines labelled as e-line and d-line. ($I_a : I_b : I_c = 250 : 5 : 1$)*

may have simulated the drastic linear shift of an unresolved H-band with laser excitation density as observed by Yuan et al. and Alferov et al., respectively. The e- and d-line indeed show a noticeable blue-shift with increasing laser intensity but not so pronounced as published by these authors. The amount of this shift shows a dependence on the growth parameters, as can be seen from Fig. 2, where we plotted the energy shift of the e-line as a function of excitation power for three heterojunctions. The main difference of the samples is in the thickness of the spacer. As we excite with photon energies below the bandgap of the (GaAl)As layers, the different thicknesses of the (AlGa)As layers should not influence the excitation conditions near the interface. The observed dependence of the shift on the spacer-thickness is consistent with our model for the H-band recombination and will be discussed later. The mean shift of the photon energy with excitation density is similar to that reported by 't Hooft and co-workers, therefore we believe that all of the literature cited above deals with the same recombination process. The enhanced blue-shift observed previously /1,2/ seems to be caused by unresolved e- and d-lines in samples of not so high quality.

Another important result of Fig. 2 is that for low excitation the energy shift of the e- and d-line vanishes, which allows one to characterize different heterojunctions by this line. The energy difference ΔE of the e-line to the fundamental gap of GaAs varies between 7 and 15 meV. From our study of 8 p-

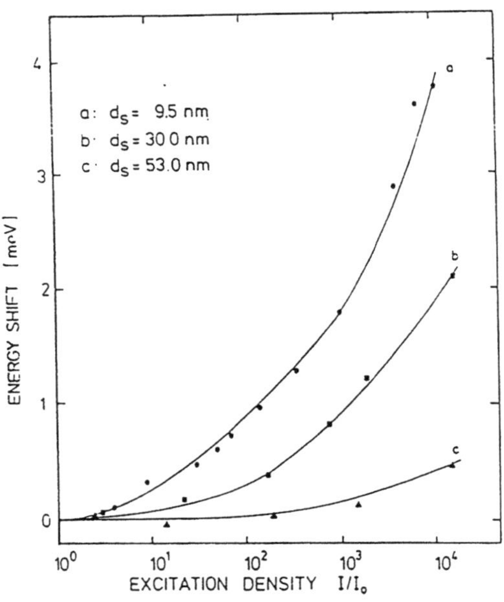

Fig. 2 *Energy shift of the e-line as a function of excitation density with the spacer-thickness d_s as a parameter. The photon energy of the exciting laser is well below the bandgap of the (GaAl)As-layers.*

type samples we conclude that ΔE increases with increasing doping level, increasing aluminium concentration and decreasing spacer-thickness.

In a magnetic field perpendicular to the interface the e-line splits into four components with two $\sigma+$ and two $\sigma-$ polarized lines which show a pronounced shift to higher energies with increasing field strength. This energy shift is greater than that of bound exciton lines (e.g. A^0,X represented in Fig. 3 by squares). Therefore the components of the e-line merge with that of the acceptor bound excitons at high magnetic fields as can be seen in Fig. 3. After the crossing of the lines the acceptor bound excitons are lower in energy and therefore the intensity of this luminescence is greater than that of the H-band. This makes the determination of the energetic position difficult and may have caused the incorrect interpretation of Alferov et al. that the energy shift of the H-

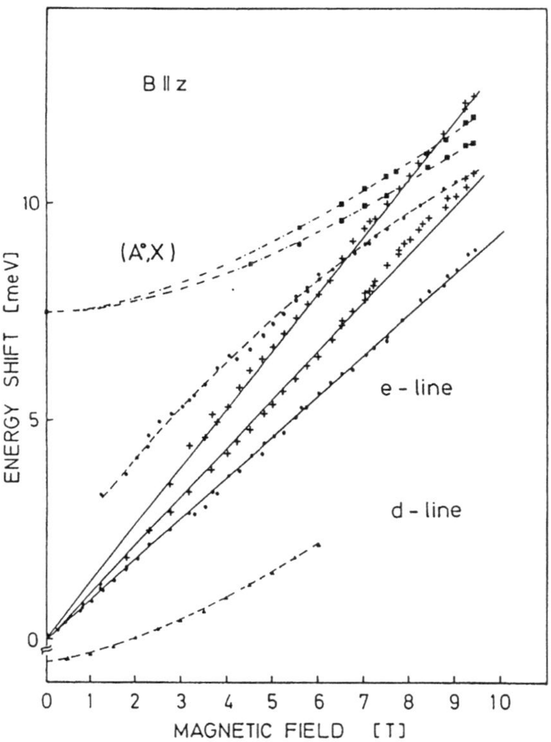

Fig. 3 *Photon energy of the H-band and the acceptor bound excitons as a function of magnetic field strength. The crosses represent σ−, dots σ+ polarization of the e-line. The squares indicate the shift and splitting of the acceptor bound exciton. The triangles represent the shift of the d-line. For the sake of clarity the d-line is shifted upwards on the energy scale.*

band in magnetic fields shows a sudden change of the slope for certain field strength. As all of the investigated samples show nearly the same behaviour we will discuss the results for the sample with the greatest distance of the H-band from the bandgap. For this sample the splitting is well resolved, even for magnetic fields up to 9.5 T. As the mean energy shift of the e-line in a magnetic field is roughly the same as observed by Alferov et al. for their M-band, we think that both lines should be attributed to the same recombination mechanism.

For the e-line we observe a linear shift and splitting with magnetic field for the three low energy lines. The fourth component behaves in a qualitatively different way and has another origin, as can be seen by extrapolating the nonlinear curvature in Fig. 3 to B = 0 T. For a magnetic field orientation parallel to the interface the e-line shows no splitting. For this field direction the shift of the remaining line still increases linearly with magnetic field strength with a slope of 0.84 meV/T, nearly identical to that of the first Landau-level of a GaAs conduction band electron (0.86 meV/T).

The pronounced shift in a magnetic field, as well as the line shape and the blue-shift distinguish the e-line from the luminescence lines observed first by Künzel and Ploog /4/. In the spectral region in discussion a series of sharp lines, up to 50, have been detected in MBE grown GaAs layers, and they are attributed to defect-bound excitons and defect-complex acceptors /5/.

To prove the conjecture that the H-band is correlated to the interface we have varied the angle Θ between the direction of the magnetic field and the interface normal. In Fig. 4 the splitting of the three low energetic components is plotted as a function of cos(Θ) for a total magnetic field strength of 9 T. The splitting of the e-line is a linear function of that cosine. This means the splitting depends on the effective magnetic field strength perpendicular to the interface.

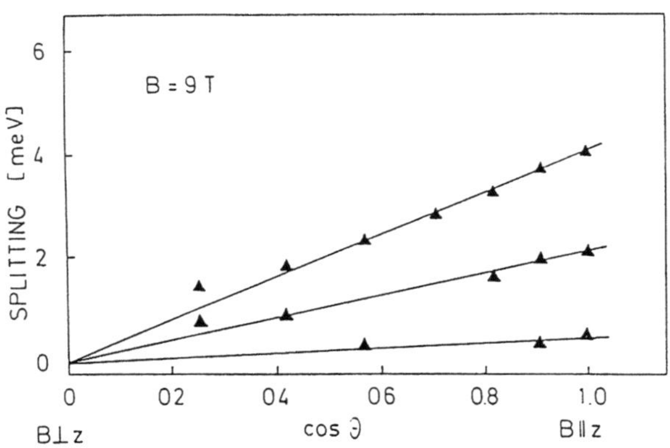

Fig. 4 *Splitting of the e-line as a function of the cosine of the angle between the directions of the magnetic field and growth. Magnetic field strength B = 9 T.*

Considering all the experimental results, we conclude that the e-line luminescence is caused by the recombination of a free conduction band electron in the flat-band region with a hole confined in a subband near the interface /6/.

1. From the energy shift in a magnetic field it is obvious that both particles of the recombination are not bound together via Coulomb interaction. Excitonic binding would result in a nonlinear - diamagnetic - shift, much smaller than that observed.

2. The splitting of the e-line is determined by the effective magnetic field strength normal to the interface. This clearly indicates that at least one of the states involved in the transition has two-dimensional character.

3. The shift in the parallel field case is equal to that known for the lowest electron Landau-level. This means that the electron involved in the transition is free and has three -dimensional character. Therefore the two-dimensional behaviour of the luminescence line must be caused by the hole, bound to the interface.

Electric subbands of two-dimensional hole gas in a magnetic field were calculated in a self-consistent Hartree-approximation /7/. As the valence band structure of GaAs is degenerate the calculations are rather complex. The subband energies in these calculations are scaled relative to the Fermi-energy and therefore not easily applied to our experiments, because the position of the Fermi-energy is not known exactly for the photo-excited samples studied here. Generally, undoped GaAs MBE samples are of p-type. This means the Fermi-energy is located near the ground state of the acceptor at T = 0 K. Considering the photo-excitation and the sample temperature the Fermi-level is shifted a few meV closer to the valence band. Accepting this, the final state of the H-band recombination fits well with the second heavy hole subband. The three linearly shifting components of the e-line then are the emitted by the recombination of electrons from the lowest Landau-level of the conduction band with holes of the first three Landau-levels of the second subband. The fourth component, seen in high magnetic fields (Fig.3) may be related to the first light hole subband only 1 meV apart, taking into account the parameters of the heterojunction discussed here. If the high energetic component arises from a light-hole subband, this will explain the increase of the intensity with magnetic field. In a magnetic field we have a strong mixing of heavy and light hole states especially for excited subbands. Therefore we have relaxed selection rules compared to nondegenerate bandstructures. The strong mixing of the subband state in addition allows the observation of interband transitions from the first electron Landau-level of the conduction band to the second and third Landau-level of the hole subband.

That we mainly observe transitions into the second heavy hole subband and not into the first or third and so on can be explained as follows. The electron and hole involved in the transition are separated in real space. The corresponding wavefunctions, however, have a finite overlap. Because the recombination between conduction band electron and subband hole depends on this overlap, a transition to the highest subband is not probable. In Fig. 5 we have compared the squared wavefunctions of the first three heavy hole subbands. The wavefunctions of the excited subbands are much more extended, this means the overlap of these states with the wavefunction of the electrons in the flatband region of the conduction band is much greater than the overlap produced by the very confined first hole subband. Unfortunately subbands with indices greater than two have energetic positions closer to the gap and therefore merge with the bound exciton lines and are difficult to be resolved or totally reabsorbed.

As the excited subbands are energetically not far from the valence band in the flat-band region, the photo-created holes in these subbands are occupied with valence band electrons by a slight increase of the sample temperature. This will lead to a drastic decrease in the luminescence intensity, which is indeed observed. At temperatures near 15 K the H-band luminescence intensity is too weak to be resolved from the background.

If we subtract the energy shift of the conduction band electron - as for instance obtained from the measurements with the magnetic field applied parallel to the interface - from the energy shift measured with fields normal to the interface we obtain the shift and splitting of the hole subbands involved in the transition. In Fig. 6 we have compared this result with the Landau-levels calculated in the self-consistent Hartree approximation. We obtain good agreement in the slopes between calculated and measured data. However, up to now the

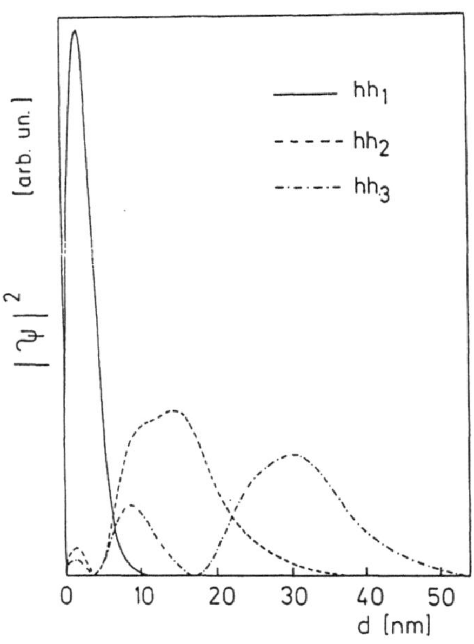

Fig. 5 *Squared wavefunctions of the first three heavy-hole subbands as a function of the distance from the interface.*

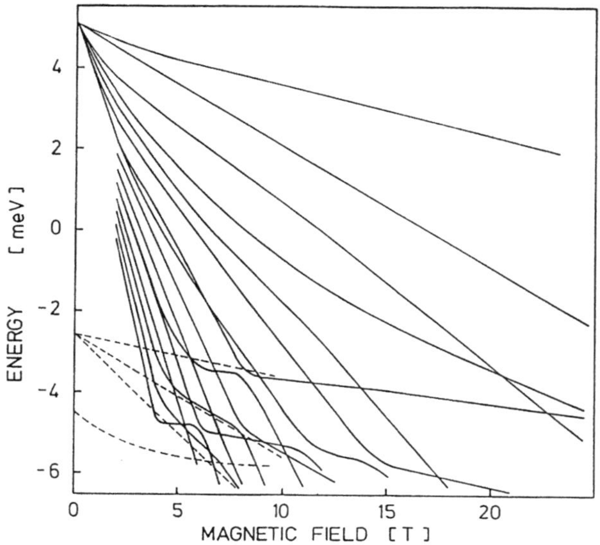

Fig. 6 *Landau levels of the hole subbands as a function of magnetic field (unbroken lines). (This Landau-level scheme was initially planed to interpret SdH data, therefore the higher subband Landau-levels are not drawn in detail). The dashed lines represent the experimental data, shifted vertically to compare the slopes.*

calculated Landau-level scheme of excited subbands has not been very accurate, caused by the low convergence of the numerical calculations. The similarities of experimental and theoretical data, nevertheless, provide insight into the Landau-level structure of excited subbands, usually not accessible with cyclotron resonance measurements.

The d-line at the low energy side of the H-band cannot be a transition into the first hole subband. Such a transition would be located closer to the impurity related luminescence (e,A^0 and D^0,A^0). As the observed line shows a distance of about 4 meV to the H-band in nearly every sample studied, we believe that this recombination arises from a donor to subband transition. This is confirmed not only by the observed intensity variation with excitation density, but also by the shift of this line in a magnetic field. It can be seen in Fig. 3 that the shift of the peak position with magnetic field strength is nonlinear and slightly greater than that of the acceptor bound excitons, as is expected for a donor to subband transition.

The fact that both components of the H-band have a different origin can also be seen from the measured lifetime and the intensity variation with magnetic field strength. In a magnetic field perpendicular to the interface the intensity of the e-line increases with increasing field strength, whereas that of the d-line shows a dramatic decrease. Therefore we are not able to observe the shift and splitting pattern of this line at high magnetic fields. The decrease of intensity with increasing field strength of the d-line is consistent with the proposed recombination model. In a magnetic field the wavefunction of a donor shrinks perpendicular to the direction of field. Therefore the overlap of the wavefunctions is reduced, resulting in a decreased luminescence intensity.

The energetic positions calculated for subbands of a two-dimensional hole gas are strongly dependent on the doping level and the spacer-thickness. In the performed luminescence experiments the photo-created holes are collected in the electrical subbands near the interface. This influences the hole density and the energetic position of the subbands. Therefore the H-band shifts with excitation density. The amount of this shift is dependent on the bending of the band-edge near the interface and therefore influenced by the growth parameters. To explain the observed result quantitatively carrier concentration measurements under illumination and subband calculations are being persued.

In addition we have performed lifetime measurements /8/ of the H-band luminescence. We confirm the short lifetimes reported by Yuan et al. and 't Hooft et al. For the e-line we obtain 0.8 ns, indicating that the transition occurs between free particles. The measured lifetime for the donor-subband transition (d-line) is about 8 ns and increases with increasing wavelength. The longer lifetime compared to that obtained for the e-line and the dependence on the wavelength confirms that a bound state could be involved in the d-line luminescence. In no case could we observe lifetimes in the range of μs as reported by Alferov et al. previously.

The experiments of Alferov et al. /2/ are performed on unintentionally doped samples with a high concentration of residual impurities. Caused by the smaller distance between impurities in samples of low purity, recombinations between donors and acceptors located on either side of the interface seems to be a good explanation for the observed lifetimes in the μs-range. However, the fact that the H-band as well as the M-band shows the same drastic shift of the photon-energy in a magnetic field suggests the same recombination mechanism for both bands. The difference in the observed lifetimes for the H-band /1,3,8/ and the M-band /2/ by more than three orders of magnitude is not yet understood.

Acknowledgement - The authors wish to thank G. Landwehr for useful discussions. This work was supported by the Deutsche Forschungsgemeinschaft.

References:

/1/ Y.R. Yuan, M.A.A. Pudensi, G.A. Vawter and J.L. Merz, J. Appl. Phys. 58, 397, 1985

/2/ Z.I. Alferov, A.M. Vasil'ev, P.S. Kop'ev, V.P. Kochereskho, I.N. Ural'tsev, A.L.Efros, D.R. Yakovlev, JETP Lett., 43, 569, 1986

/3/ G.W. 't Hooft, W.A.J.A. van der Poel, L.W. Molenkamp, C.T. Foxon, Appl. Phys. Lett. 50, 1388, 1987
/4/ H. Künzel, K. Ploog, Appl. Phys. Lett. 37,416, 1980
/5/ L. Eaves, M.S. Skolnick, D.P. Halliday, J.Phys. C 19, L445, 1986
/6/ W. Ossau, E. Bangert, G. Weimann, Solid State Commun., 64, 711, 1987
/7/ E. Bangert, G. Landwehr, Superlattices and Microstructures, 1, 363, 1985
/8/ W. Ossau, G. Mayer, A. Forchel, G. Weimann, to be published

Spectral Blue-Shifts in Optical Absorption and Emission of the 2D Electron System in the Magnetic Quantum Limit

D. Heiman[1], *B.B. Goldberg*[1], *A. Pinczuk*[2], *C.W. Tu*[2,*], *J.H. English*[2,**], *A.C. Gossard*[2,**], *D.A. Broido*[3], *M. Santos*[4], *and M. Shayegan*[4]

[1]Massachusetts Institute of Technology, National Magnet Lab., Cambridge, MA 02139, USA

[2]AT&T Bell Labs., Murray Hill, NJ 07974, USA

[3]Department of Physics, Boston College, Chestnut Hill, MA 02167, USA

[4]Department of Electrical Engineering, Princeton University, Princeton, NJ 08544, USA

The 2D electron system in GaAs quantum wells was investigated using optical absorption and photoluminescence measurements in the bandgap region, for magnetic fields $B \leq 30$ T and temperatures $T \geq 0.4$ K. Several optical anomalies appear as spectral blue-shifts for both increasing field and decreasing temperature, and are related to electron-electron interactions. The anomalies observed at the 2D Landau filling factors $\nu \simeq 1$ and $\nu < 2/3$ have very different field and temperature dependencies. The one near $\nu = 1$ is associated with changes in population of the lowest spin-split Landau level. The ones at higher fields are observed only for $T < 3$ K, concurrent with the fractional quantum Hall effect.

1. INTRODUCTION

Two-dimensional electron systems under high magnetic fields have been well-characterized through transport experiments [1,2]. The integer quantum Hall effect [3] has stimulated a surge of interest in 2D systems in general, and the discovery of the fractional quantum Hall effect (FQHE) [4] exposed novel many-particle electron-electron interactions [1]. The FQHE is observed at high fields and low temperatures as a manifestation of condensation of the 2D electron gas into a quantum fluid. Departures from the ground state configuration of this incompressible fluid appear as quasiparticle excitations having a fractional charge [5]. Quasiparticles are induced through changes in the filling factor, from field or density changes, and appear at local defects and impurities.

Unlike magnetoresistance experiments that probe the 2D electron system mostly on a macroscopic scale, optical measurements examine local as well as extended regions by virtue of single quantum events. Magneto-optics is likely to reveal new physics of the many-body condensate not accessible with transport experiments. Near-bandgap optical studies have been reported for silicon MIS transistors and GaAs quantum wells. Magneto-optics on Si showed a shift in the recombination photoluminescence (PL) at $\nu = 7/3$ for $T < 2$ K [6]. A spectral doublet was first reported [7] at filling factor $\nu \simeq 1$ and $\nu < 2/3$ in GaAs [8,9], and then in Si at $\nu = 7/3$ and 8/3 in Si [10].

High-field experiments using optical absorption and emission have been applied to both undoped [11] and doped [12-15] quantum wells, but in contrast to those studies, our work has focused on higher-mobility GaAs layers

Springer Series in Solid-State Sciences, Vol. 87
High Magnetic Fields in Semiconductor Physics II Editor: G. Landwehr

(μ=1 to 5x10^5 cm^2/V-sec), lower temperatures (T≲0.05 K) and larger fields (B≲30 T). Experiments at these extrema require small-bore magnets and He-3 cryostats or dilution refrigerators, which have only recently become accessible with fiber-optic systems, developed at the Francis Bitter National Magnet Lab [16]. This combination of unique experimental technique and high-quality sample material has uncovered several unusual effects in the magnetic quantum limit ν<1.

We present results of the intrinsic optical transmission and emission due to transitions between conduction and valence Landau levels in GaAs-(Al,Ga)As multiple quantum wells (MQW). Several remarkable anomalies have been observed in the PL from the 2D electron system in GaAs, where spectral blue-shifts occur for increasing field or decreasing temperature [9]. In the anomaly at ν≃1 we see a discontinuous shift of the intrinsic PL peak to higher energy for increasing B. The same behavior has recently been observed in absorption [17]. The spectral shift is about 1 meV at T=1 K and decreases to near zero by T~10 K. Subsequent calculations that include many-electron effects thru the self-energy and exchange interaction have reproduced some qualitative features of this blue-shift [18]. A second anomaly occurs in the regime of the FQHE for ν<2/3 — a spectral doublet develops in the intrinsic emission. The magnitude of the splitting, ~0.6 meV, is independent of T and B, and persists over a wide (10 T) field range. Of particular interest is the relative intensities of the two spectral components which is a strong function of T, similar to that of the measured ρ_{xx} magnetoresistance minima at the ν=2/3 FQHE. Also, in a low-density sample, the intrinsic PL feature shifts to higher energy with decreasing T in a manner similar to the thermal activation of the ν=1/3 magnetoresistance minima.

Simultaneous transport and PL measurements were made on n-type, modulation-doped GaAs-(Al,Ga)As multiple quantum well structures (30-50 layers). Well widths were either 175 or 200 A, with 25 to 85 periods. The samples had carrier densities between n=1.4 and 6x10^{11} cm^{-2} and mobilities of 1.3 - 4x10^5 cm^2/V-sec, and all showed well-developed FQHE ρ_{xx} minima at either ν=2/3 or 1/3. Samples for optical transmission were cemented to (Al,Ga)As substrates with cyanoacrylate before jet-etching to reduce strain upon cooling [19]. A fiber-optic apparatus [16] was used to access the He^3 cryostat and high-field magnet, and polarization was set by placing a circular polarizer next to the sample. The power was limited to 10^{-6} W over an area 10^{-2} cm^2 to avoid heating.

2. RESULTS

2.1 Basic Transitions

Before examining the optical anomalies we first describe the allowed selection rules for the electron-hole (e-h) Landau level transitions. Figure 1 shows the transitions involving the lowest electron Landau level in the vicinity of ν=1. (Details of the transitions in this region and at higher filling factors are given in Refs. 17 and 19.) The points represent energy positions of the main absorption peaks for a sample with carrier density n = 3.9 x 10^{11} cm^{-2} having ν=1 at B = 16.3 T. The inset shows the initial valence level (z-component of angular momentum m_j, and harmonic oscillation index n in the absence of mixing, Sham Ref. 20) and final conduction level for the three transitions:

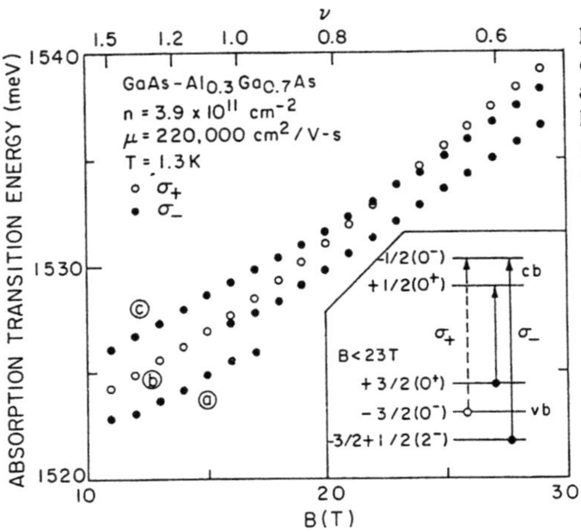

Fig. 1 Lowest Landau transition energies versus B, obtained from absorption spectra of GaAs-(Al,Ga)As MQW; closed (open) circles are for σ- (σ+) polarizations; the inset shows initial and final states

{a} +3/2 (n=2) ⟶ +1/2;
{b} -3/2 (n=-1) ⟶ -1/2;
and {c} -3/2 (n=1), +1/2 ⟶ -1/2.

Since the electron g-value is negative (g = -0.44) the lowest electron level is taken to be +1/2 (spin-up). Transition {b} involves the lowest hole Landau level of the -3/2 hh which does not couple to any other states, while {c} is from a hole state which was originally -3/2 hh at B=0, but the strong coupling at high fields makes the +1/2 lh component dominant. Note the crossing of {b} and {c} at B=22 T due to valence level crossing. (See Ref. 20 for complete descriptions of the theory.)

Slopes of the transitions in Fig. 1 are not straightforward to interpret because they are not simply related to the electron and hole masses. Complications arise from confinement effects that lead to mixing of hole states, and the mixing itself varies with magnetic field. In addition to a calculation of the hole levels, Landau level broadening and Coulomb corrections must also be included [19]. We also note that neither the B=0 transition nor the B=0 intercept is equal to the undoped bandgap due to many-body renormalization [21].

The observed polarizations of the absorption features are consistent with the selection rules of the previous assignments. In transitions {a} and {c} the electron loses one quantum of angular momentum in going from the valence to conduction levels which is taken up by the absorption of a σ- photon. In PL the polarization is almost totally σ- above B=8 T (ν=2). This is due to the photoexcited hole having relaxed to the highest hole level (+3/2), and is <u>not related to the population of the electron spin state</u>.

Near ν=1 we find that the lowest Landau level does <u>not</u> have a spin splitting much larger than the level broadening. In fact, recent optical studies of the Fermi level jumps indicate a large broadening (~2 meV at B=16 T) [19]. The spin splitting appears to be small, as determined from analysis of both the magnitude of the absorption and the difference between transition energies. A significant overlap of the spin states gives rise to the sizeable absorption of transition {a} for $\nu>1$ at low temperature, which would be absent if all the electrons were in the lowest spin-split level. Comparing the calculated hole levels (including uncertainties) with measurements we can place an upper bound of 3 on the exchange enhancement of the electron g-factor at $\nu=1$.

2.2 Anomaly at $\nu\simeq1$

Figure 2 shows the PL spectrum for $\nu\simeq1$ of the sample with $n=3.9\times10^{11}$ cm^{-2}. The dominant spectral features on the right are ascribed to recombination of an electron in the 2D electron gas with a photoexcited hole, since the PL peak energy coincides with the onset of absorption. (A smaller defect-related peak lies about 4 meV below the intrinsic peak.) The arrows indicate a new spectral peak that emerges for fields higher than B=16 T. Notice the line is shifted to higher energy than the original line by about 1 meV. For increasing fields just above B=16 T there is an intensity transfer from the lower- to the higher-energy line. We point out that these "multiplet features" are unlike the "nonlinear" tuning [12] observed at larger filling factors. Similar behavior is seen in samples with n=1.8, 3.5 and 6.0×10^{11} cm^{-2}, where instead of two well-separated PL peaks, only an abrupt shift to higher energies is observed, but it occurs at the same filling factor $\nu\simeq1$.

This blue-shift is also clearly observed in the absorption spectra [17], Fig. 3. The $+3/2 \rightarrow +1/2$ transition peak at lower energies denoted by vertical bars shows a discontinuous shift to higher energy. As in the PL results, the absorption anomaly is seen at $\nu\simeq1$ in samples with different carrier densities, n=3.9 and 6.0×10^{11} cm^{-2}. The transition energies plotted versus filling factor in the left hand side of the figure exemplify the connection to the quantum limit. The discontinuity seems to occur for ν slightly less than 1, and the two peaks coexist only over a small range of filling factor $\Delta\nu\sim0.1$.

The temperature dependence of the $\nu\simeq1$ absorption anomaly [17] is exhibited in Fig. 4. For $\nu>1$ (B=14 T is displayed) the energy position of the line is reasonably independent of temperature. On the other hand, for

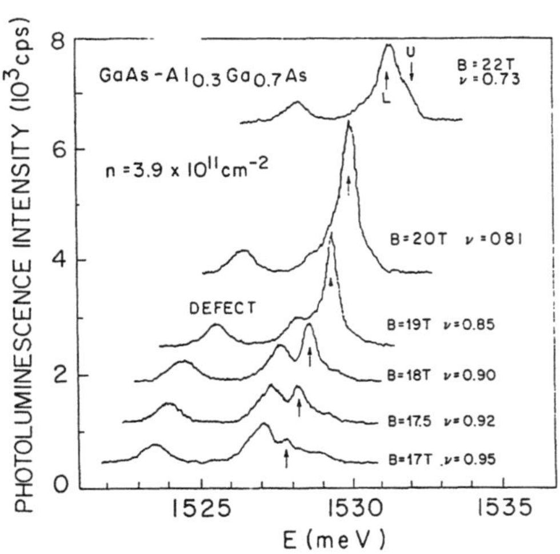

Fig. 2 Emission spectra in the region $\nu\sim1$ at various magnetic fields for GaAs-(Al,Ga)As MQW; sample has not been thinned for transmission; arrows show the new transition; from Ref. 9

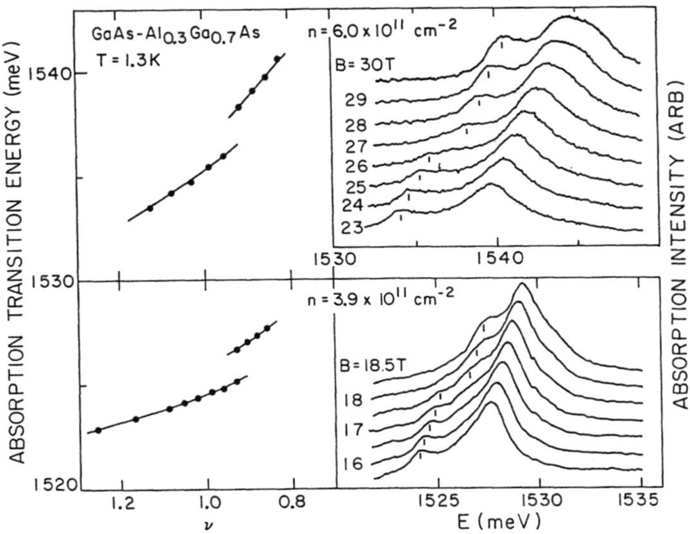

Fig. 3 Transition energies and absorption spectra near ν=1, for samples with two electron densities of GaAs-(Al,Ga)As MQW

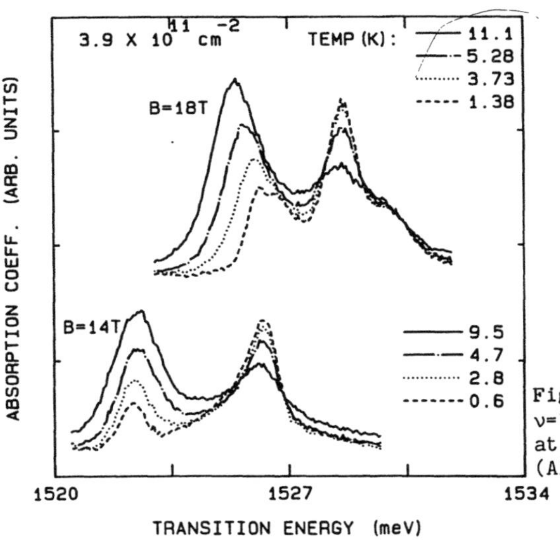

Fig. 4 Absorption spectra near the ν=1 anomaly for various temperatures at two magnetic fields in GaAs-(Al,Ga)As MQW, from Ref. 17

ν<1 (at B=18 T) the peak shifts dramatically to higher energy for decreasing temperature. Thus the blue-shift disappears for T>10 K, an energy (kT) which is of the same magnitude as the spin splitting.

2.3 Emission Anomaly at $\nu<2/3$

The anomaly for $\nu<2/3$ is quite different from the one at $\nu\simeq1$. Their only common characteristic is that for increasing field a new PL peak appears at higher energy. The energies of the PL doublet components are plotted as a function of B in Fig. 5. The stronger component is represented by the solid circles. In the vicinity of $\nu=2/3$ the splitting is ~0.6 meV and independent of field, and the two components persist over wide field range ~10T, in contrast to the $\nu\simeq1$ anomaly. These two traits are not consistent with level crossing, such as resonant polaron coupling (cyclotron energy equal to LO phonon energy) which occurs at B=22 T, or to crossing of cyclotron level with inter-subband levels. Polaron coupling results in a sharp magnetic-field-dependent level splitting which is not observed.

We next look at the B- and T-dependence of the relative spectral weights of the doublet components. Figure 6 displays the PL spectra, where we clearly observe a spectral doublet. Vertical bars indicate the intensities and energies of the two components determined by curve-fitting the spectra to the sum of two Gaussians with equal linewidths. At the lowest temperature T=0.4 K, Fig. 6(a) shows that for increasing B the higher-energy component increases in intensity relative to the lower-energy component. Note that equal intensities occur between B=24 and 25 T. On the right hand side, at a higher temperature T=0.8 K, the intensity crossover occurs at a higher field, closer to B=27 T. The ratio of intensities I_L/I_U (lower divided by upper) is plotted as a function of B for various temperatures in the top of Fig. 7. The field dependencies are similar for T=0.4, 0.8 and 1.5 K, but are shifted to higher fields for higher temperatures. The crossover condition $I_L/I_U=1$ occurs at the field corresponding to $\nu=2/3$ for the lowest temperature T=0.4 K.

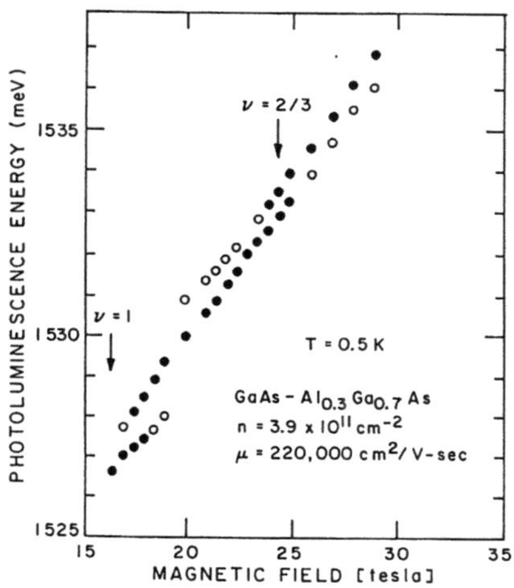

Fig. 5 Transition energies near $\nu=2/3$ versus B, obtained from emission spectra of GaAs-(Al,Ga)As MQW; closed (open) circles represent strong (weak) intensities; from Ref. 9

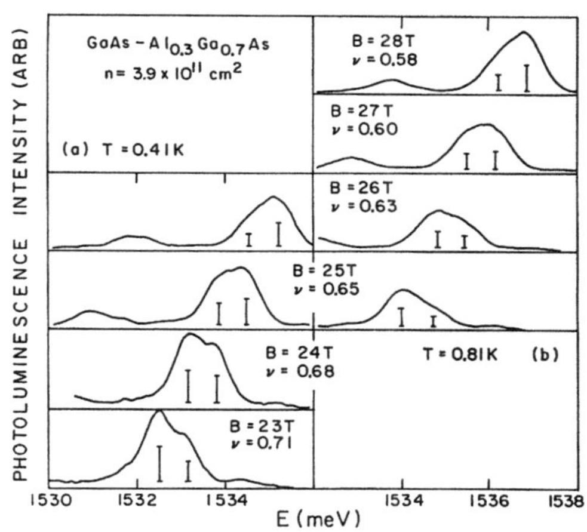

Fig. 6 Emission spectra near $\nu=2/3$ at various magnetic fields, for two temperatures in GaAs-(Al,Ga)As MQW; the vertical bars give energies and intensities (x1/2) obtained by curve-fits to the sum of Gaussians; from Ref. 9

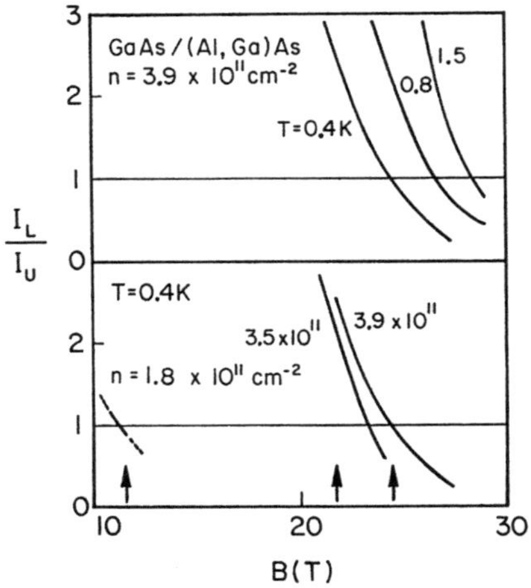

Fig. 7 Intensity ratio I_L/I_U of emission doublet near ν=2/3 for GaAs-(Al,Ga)As MQW; the upper curves show T-dependence of one sample, and the lower-half shows samples having three different densities; arrows denote position of ν=2/3 magnetoresistance minima; from Ref. 9

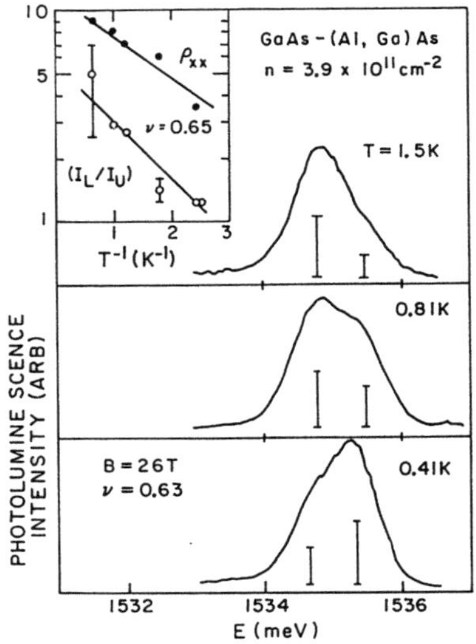

Fig. 8 Emission spectra near ν=2/3 for three temperatures of GaAs-(Al,Ga)As MQW; the inset displays magnetoresistance at ν=2/3 and intensity ratio I_L/I_U on a log scale versus 1/T; from Ref. 9

Two other samples having different electron densities also show the same anomaly near ν=2/3, exhibited in the lower half of Fig. 7. The sample with $n=3.5\times10^{11}$ cm^{-2} has a similar field dependence of the intensity ratio, although the condition $I_L/I_U=1$ is shifted from ν=2/3 as shown by the arrow. In the lowest mobility sample with $n=1.8\times10^{11}$ cm^{-2} we could not accurately curve-fit the data; however, a temperature-dependent shoulder was observed and its intensity was roughly estimated. (Unfortunately, the intrinsic peak at ν=1/3 in this sample was obscured by the defect band which is close in energy at these fields.) Thus, in widely differing samples where we see a spectral doublet there seems to be a coincidence between the optical anomaly and the filling factor ν=2/3.

One of the most significant features of these results is the <u>scale</u> of the temperature dependence of the ν<2/3 optical anomaly, which is similar to that of the FQHE. Figure 8 shows the intensity crossover at fixed field for three temperatures. The inset compares I_L/I_U with the activated behavior of the minima in ρ_{xx} at ν=2/3. Magnetotransport has an activation energy Δ/2=0.44 K, which is about a factor of 4 lower than in a comparable mobility single heterostructure [22]. The optical results show a similar slope. This is not to demonstrate that the optical intensity ratio is activated, but merely to show it has the same energy scale. In fact, the slope of I_L/I_U is only weakly dependent on field between B=21 and 29 T, whereas the activation energy in ρ_{xx} is maximum at ν=2/3 and decreases to zero for a change of field of ΔB=±3 T.

2.4 Absorption Spectra at ν<2/3

Optical transmission measurements were made in the region of the ν<2/3 anomaly. Strain-free mounting of the micron thick sample was difficult to obtain, and the broadened PL doublet appeared as a single peak. In Fig. 9 we show the PL and absorption spectra at three temperatures for B=27 T. Between T=0.55 and 1.3 K the PL peak shifts to higher energy by 0.3 meV, about one-half the doublet splitting in an unprocessed sample. In contrast, the peak in the absorption does not shift. The inset shows the Stokes shift, defined as the difference in energies between the absorption and emission peaks. At high temperatures the shift is relatively field-independent at about 1.2 meV, and arises from many-electron renormalization of the final state of the PL transition [27]. As the temperature is lowered, the Stokes shift at higher fields is reduced by the blue-shift of the PL. Notice that the field at which deviation starts is large at high T, but moves to lower B at lower T, identical in behavior to the unprocessed sample (Figs. 5 and 7). At the lowest temperature, the onset again occurs near ν=2/3. Aside from the quantitative features, the key attribute here is that the blue-shift only appears in the final state interaction, unlike the ν≃1 anomaly.

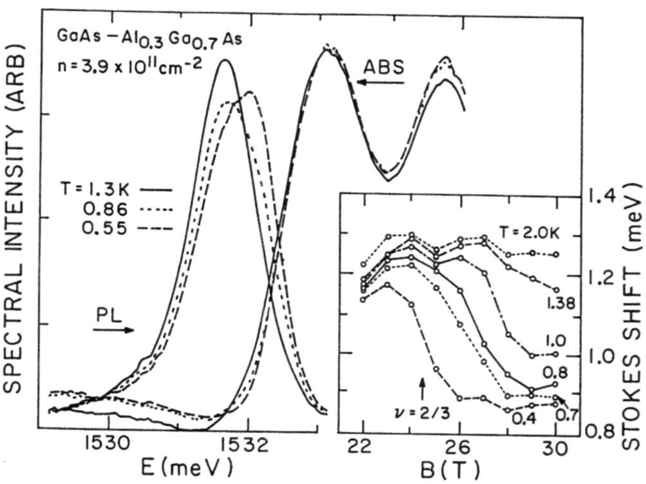

Fig. 9 Spectral intensities near ν=2/3 of emission (PL) and absorption (ABS), for three temperatures of GaAs-(Al,Ga)As MQW; the inset displays the Stokes shift (ABS minus PL) versus B and T

2.5 Anomaly at ν~1/3

A low-density sample with $n=1.4\times10^{11}$ cm^{-2} and $\mu=400{,}000$ cm^2/V-sec was investigated in the region of ν=2/3 and 1/3. Unlike the other samples, the ρ_{xx} minima at ν=2/3 was quite weak, but had a strong minima at ν=1/3.

In addition to the direct effect of the lower carrier density, differences might be attributed to the smaller screening of Coulomb (exciton) interactions. Figure 10 displays the shift in the transition energy of the main PL line as a function of ν at different temperatures. This shift was

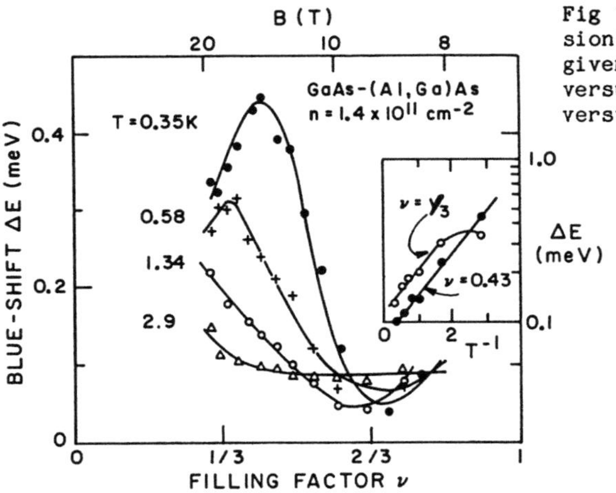

Fig 10 Spectral shift ΔE of emission peak (transition energy at given T minus energy at T=4.2K) versus ν; the inset shows log ΔE versus 1/T

defined relative to the spectral position at T=4.2 K. Below ν~2/3 the peak shifts to higher energy for both increasing field and decreasing temperature. In the inset we plot the log of the shift versus 1/T for ν=0.4 and 1/3. Activated-like behavior appears above T~0.5 K, with an activation energy 0.6 K. This is one-half the ρ_{xx} activation of 1.0 K at ν=1/3. The structure in the field-dependence of the blue-shift accompanies freeze-out in ρ_{xx} and will be presented at a later time.

3. DISCUSSION

It is interesting to compare the optical anomaly at ν<2/3 in the present work with the results reported for optical emission from Si inversion layers at ν=7/3 and 8/3 [6,10]. In the Si experiments the photoexcited holes reside on acceptors that may not lie in the 2D layer, and consequently may have smaller perturbation on the electron-electron correlations. The dependence of emitted photon energy on filling factor was then related to the quasi-particle gap of the FQHE. In our case, however, the doublet splitting is not used to estimate the energy gap of the FQHE because of the large perturbation of the hole.

Optical transitions depend significantly on the local environment of the minority carrier, the photoexcited valence band hole in the present case. Screening of the positively charged hole alters the local charge density of the 2D electron system by redistributing the correlated electrons (or equivalently quasiparticles of the quantum fluid). For ν<1 there is enhanced screening [23,24] because the electrons can change their state within the lowest Landau level. Further changes are caused by condensation of the electron into an incompressible many-particle state. The resulting modification in the local electron distribution contributes to: (1) the many-particle correlation energy of the 2D electron system [25-28]; (2) Coulomb corrections to the transition energy (e.g. exciton binding energy) [18,29,30]; and (3) changes in oscillator strength (matrix element) of the transition. Spectral shifts are effected by (1) and (2), while spectral intensity is effected by (3). In the FQHE regime the response of the hole should be treated like the cases of charged impurities and disorder [31-33]. A theoretical model should take these into account.

A spectral blue-shift that is qualitatively similar to the observed B- and T-dependence at $\nu \simeq 1$ has been observed in cyclotron resonance (CR) [34]. The collective magnetoplasmon modes calculated by Kallin and Halperin [30] give rise to a red-shift in the CR at $\nu=1$. An increase in the screening which occurs for $\nu<1$ reduces the coupling of the magnetoplasmon modes and leads to a blue-shift. The effect is strengthened for both increasing B and decreasing T as the screening increases.

We have shown that the spectal blue-shifts (splittings) observed in the bandgap optical transitions at $\nu \simeq 1$ and $\nu<2/3$ in 2D electron systems have some similarities, but differ remarkably in their magnetic field and temperature dependence. The spectral anomaly at $\nu \simeq 1$ appears in both the initial and final states, and seems to be related to the spin gap at $\nu \simeq 1$ [17]. Its temperature dependence is associated with an energy gap much larger than 1 K. On the other hand, the spectral anomaly for $\nu<2/3$ only appears in the final state. The optical changes in the regime of the FQHE occur on a smaller temperature scale, similar to the activation energy of the magnetoresistance of the FQHE. In all cases, the optical blue-shifts appear at $\nu<1$ for increasing field or decreasing temperature.

Acknowledgements

We thank H.L.Stormer, L.J. Sham, D.C. Tsui, P. Hawrylak, B.I. Halperin, J.E. Furneaux, and P.M. Platzman for many helpful discussions, and J.P. Valladares, B. Brandt and L. Rubin for assistance. B.B. Goldberg acknowledges support of an MIT Bantrell postdoctoral fellowship. The Francis Bitter National Magnet Lab is supported by the National Science Foundation DMR-8511789. The Princeton authors acknowledge support by NSF grant No. ECS-8553110 and a grant by the New Jersey Commission on Science and Technology.

References

*Present address: University of California, San Diego, CA 92093.
**Present address: University of California, Santa Barbara, CA 93106.

1. The Quantum Hall Effect, ed. R.E. Prange and S.M. Girvin (Springer-Verlag, New York, 1987); and references cited therein.
2. High Magnetic Fields in Semiconductor Physics, ed. G. Landwehr, Sol. State Sci. vol. 71 (Springer-Verlag, Berlin, 1988).
3. K. von Klitzing, K. Dorda and M. Pepper, Phys. Rev. Lett. 45, 494 (1980).
4. D.C. Tsui, H.L. Stormer and A.C. Gossard, Phys. Rev. Lett. 48, 1559 (1982); H.L. Stormer, A.M. Chang, D.C. Tsui, J.C.M. Hwang, A.C. Gossard and W. Weigmann, Phys. Rev. Lett. 53, 997 (1984).
5. R.B. Laughlin, Phys. Rev. Lett. 50, 1395 (1983).
6. I.V. Kukushkin and V.B. Timofeev, JETP 43, 387 (1986).
7. D. Heiman, Bull. Am. Phys. Soc. 32, 577 (1987).
8. B.B. Goldberg, D. Heiman, A. Pinczuk, C.W. Tu, A.C. Gossard, and J.H. English, Surf. Sci. 196, 209 (1988).
9. D. Heiman, B.B. Goldberg, A. Pinczuk, C.W. Tu, A.C. Gossard and J.H. English, Phys. Rev. Lett. 61, 605 (1988).
10. I.V. Kukushkin and V.B. Timofeev, Surf. Sci. 196, 196 (1988).
11. J.C. Maan, G. Belle, A. Fasolino, M. Altarelli, and K. Ploog, Phys. Rev. B30, 2253 (1984).

12. C.H. Perry, J.M. Worlock, M.C. Smith and A. Petrou, Ref. 2, p. 202.
13. J.S. Lee, N. Miura, and Y. Iwasa, Surf. Sci. 196, 534 (1988).
14. W. Ossau, B. Jakel, and E. Bangert, Ref. 2, p. 213.
15. D.C. Rogers, J. Singleton, R.J. Nicholas, and C.T. Foxon, Ref. 2, p. 223.
16. D. Heiman, Rev. Sci. Instrum. 56, 684 (1985); E.D. Isaacs and D. Heiman, Rev. Sci. Instrum. 58, 1672 (1987).
17. B.B. Goldberg, D. Heiman, A. Pinczuk, C.W. Tu, A.C. Gossard, and J.H. English, (to be published in Phys. Rev.).
18. T. Uenoyama and L.J. Sham (to be published).
19. B.B. Goldberg, D. Heiman, M.J. Graf, D.A. Broido, A. Pinczuk, C.W. Tu, A.C. Gossard and J.H. English, (submitted to Phys. Rev. Rap. Commun.).
20. L.J. Sham, Ref. 2, p. 288; D.A. Broido and L.J. Sham, Phys. Rev. B31, 888 (1985); E. Bangert and G. Landwehr, Superlattt. Micostruct. 1, 363 (1985); M. Altarelli, Festkorperprobleme XXV, 381 (1985); G. Bastard, ibid., p. 389.
21. A. Pinczuk, J. Shah, R.C. Miller, A.C. Gossard and W. Wiegman, Sol. State Commun. 50, 735 (1984); G.E.W. Bauer and T. Ando, Phys. Rev. B31, 8321 (1985).
22. G.S. Boebinger, H.L. Stormer, D.C. Tsui, A.M. Chang, J.C.M. Hwang, A.Y. Cho, C.W. Tu, and G. Weimann, Phys. Rev. B36, 7919 (1987).
23. T. Ando and Y. Murayama, J. Phys. Soc. Japan 54, 1519 (1985).
24. S. Das Sarma, Sol. State Commun. 36, 357 (1980); and Phys. Rev. B23, 4592 (1981).
25. R. Sooryakumar, A. Pinczuk, A.C. Gossard, D.S. Chemla and L.J. Sham, Phys. Rev. Lett. 58, 1150 (1987).
26. M.S. Skolnick, J.M. Rorison, J.K. Nash, D.J. Mowbraz, P.R. Tapster, S.J. Bass and A.D. Pitt, Phys. Rev. Lett. 58, 2130 (1987).
27. A.E. Ruckenstein and S. Schmitt-Rink, Phys. Rev. B35, 7551 (1987); A.E. Ruckenstein, S. Schmitt-Rink and R.C. Miller, Phys. Rev. Lett. 56, 504 (1986).
28. Y.C. Chang and G.D. Sanders, Phys. Rev. B32, 5521 (1985).
29. S.-R. Yang and L.J. Sham, Phys. Rev. Lett. 58, 2598 (1987).
30. C. Kallin and B.I. Halperin, Phys. Rev. B31, 3635 (1985); ibid., B30, 5655 (1984).
31. F.C. Zhang, V.Z. Volovic, Y. Guo and S. Das Sarma, Phys. Rev. B32, 6920 (1985).
32. E.H. Rezayi and F.D.M. Haldane, Phys. Rev. B32, 6924 (1985).
33. A.H. MacDonald, K.L. Liu, S.M. Girvin and P.M. Platzman, Phys. Rev. B33, 4014 (1986).
34. Z. Schlesinger, W.I. Wang, and A.H. MacDonald, Phys. Rev. Lett. 58, 73 (1987).

Resonant-Subband Landau-Level Coupling in a Two-Dimensional Electronic System: Depolarization Effect and Dependence on Carrier Density

K. Ensslin, D. Heitmann, and K. Ploog

Max-Planck-Institut für Festkörperforschung, Heisenbergstr. 1
D-7000 Stuttgart 80, Fed. Rep. of Germany

For a quantitative understanding of the properties of a two-dimensional electron gas the exact knowledge of subband energies and wavefunctions is very important [1]. Experimentally one usually applies optical techniques such as Raman scattering [2,3] and resonant-subband-Landau-level-coupling (RSLC) spectroscopy [4-6]. Using Raman scattering the presence of strong bandgap radiation causes quasi-accumulation conditions and fixes the two-dimensional (2D) carrier density N_s . So far RSLC was investigated as a function of bandgap illumination [5,6] or back-gate voltage [5], which both strongly influence the depletion charge N_{depl} of an $Al_xGa_{1-x}As - GaAs$ heterostructure. In particular it was not clear whether the energies, determined from RSLC experiments, are depolarization shifted as in charge density excitation experiments [3]. Therefore it was difficult to compare experiments and results of self-consistent calculations [7,8] in a satisfactory manner. Here we prepared modulation doped $Al_xGa_{1-x}As - GaAs$ heterostructures with front-gates which allowed us to change N_s in a controlled and reproducible way [9,10]. This gave us the possibility to study RSLC as a function of N_s, while all other parameters remained essentially unchanged. The agreement between our experimental results and a self-consistent calculation is excellent and clearly reveals that the energies, determined from RSLC experiments, are depolarization shifted.

The samples were grown by molecular beam epitaxy and have the following structure: 2.1-μm buffer layer and active GaAs, 18-nm spacer of $Al_xGa_{1-x}As$, 52-nm Si-doped $Al_xGa_{1-x}As$ (x=0.36, $N_d \approx 5 \cdot 10^{17} cm^{-3}$) and a 8-nm GaAs cap layer. The samples were mesa etched, defining gate areas of $\approx$ 3 mm diameter. 1000Å Al_2O_3 [11] followed by 50Å NiCr were evaporated onto the gate area. Outside the gate area contacts were made to the 2D channel. Cyclotron resonance (CR) was measured with a Fourier transform spectrometer which was connected via a waveguide system to a 14.5 T superconducting magnet. Spectra were measured at fixed N_s and magnetic field B. The normal of the sample was slightly tilted ($\alpha \approx 5^o$) with respect to the magnetic field to couple the motion of the electrons perpendicular and parallel to the 2D system. This is a precondition to observe RSLC. The temperature was normally 2.2 K and the resolution of the spectrometer was set to 0.1 cm^{-1}. A gate voltage V_g was applied between the gate and the channel contacts. N_s was measured in situ under the conditions of the experiment from magnetocapacitance measurements [9].

Figure 1 shows experimental spectra measured at fixed B=13.78 T for different V_g. With decreasing V_g and therefore decreasing N_s the subband separation $E_{10} = E_1 - E_0$ (E_i is the energy of subband i) decreases and crosses the condition $E_{10} = \hbar\omega_c$ ($\omega_c = e \cdot B/m^*, m^*$ is the bandstructure mass), where the well known RSLC occurs [4].

High Magnetic Fields in Semiconductor Physics II Editor: G. Landwehr

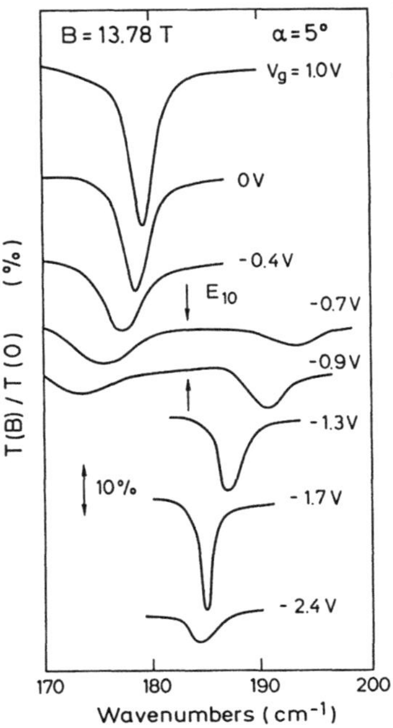

Figure 1 : *Normalized transmission $T(B)/T(B=0)$ of unpolarized FIR radiation measured at fixed B=13.78 T for various gate voltages V_g. For $V_g \approx -0.8V$ the two resonances have approximately the same absorption strength. The arrows indicate the position of E_{10}.*

Although so far it was more usual to change $\hbar\omega_c$ (which means B) instead of E_{10} (which means N_s) we choose this way of presenting the data, because in these plots some additional effects on the CR beyond simple one particle effective mass theory become clearly appearent. Several investigations [10,12,13] have shown that m^* at least depends on nonparabolicity , filling factor dependent effects, localization of the carriers and interaction between transitions from different Landau and spin levels. Our presentation offers the possibility to separate all these features from the experimental results to get the unperturbed RSLC.

Figure 2 shows a measurement for B=13.78 T. Each data point corresponds to the resonance position of one CR. The full lines are the results of a calculation following Ando [14]. First one solves Poisson and Schrödinger equation self-consistently in the direction perpendicular to the 2D layer. Then one diagonalizes the Hamiltonian containing all quantum numbers involved in this problem (subband quantization, Landau quantization and center coordinate) for a convenient choice of the basis. Convenient means, that the Hamiltonian is diagonal with respect to the motion in the x-y plane if we neglect couplings between different Landau levels of the same subband. The calculation was carried out in the effective mass approximation. For the last part of the calculation increasing (decreasing) the value of m^* means decreasing (increasing) the energy postion of the theoretical curves in fig. 2, respectively. The splitting of the energy itself scales essentially with the absolute energies $E_{10} \approx \hbar\omega_c$. We used a value of $m^*/m_0 = 0.071$ (corresponding to the nonparabolicity at B=13.78 T) for this last part of the calculation which gives satisfactory agreement between theory and experiment.

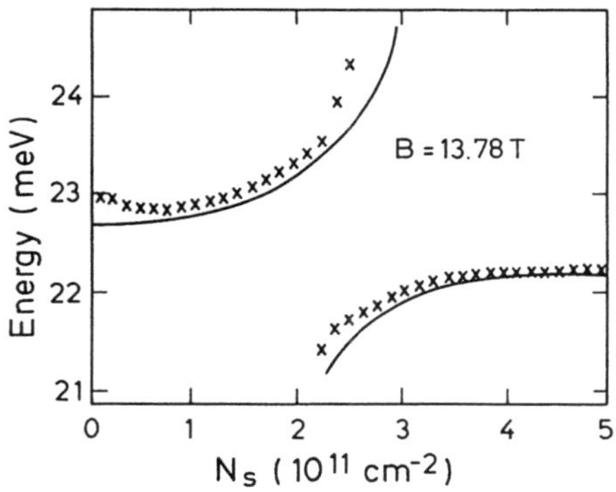

Figure 2 : *Experimental resonance positions (×) versus N_s for B=13.78 T. Results of a self-consistent calculation are indicated by full lines. Comparison with measurements of the untilted sample reveals that the deviations of the experimental points from theory at small N_s are caused by nonparabolicity effects.*

The data (fig. 1) shows that in the regime, where we can observe two resonances, the development of the amplitudes and the linewidths is essentially symmetric to the N_s value of minimum splitting. So we determined the N_s value where both resonances have the same amplitude. For this N_s we took the mean value of the energetic positions of the two resonances, which gave us the experimental value of E_{10} (fig. 3).

The only free parameter of our self-consistent calculation is the concentration of ionized acceptors N_A in the GaAs buffer layer. Varying N_A shifts in particular all values of E_{10} by a fixed amount leaving the slope of the E_{10} versus N_s curve mainly unchanged [7,8].To explain the observed slope of the E_{10} (N_s) curve, it is necessary to include many body effects like exchange interaction, exciton and, most important, depolarization effect. Figure 3 shows the theoretical curves together with the experimental points. For the parametrization of the exchange energy we used the same result as [7]. Following Ando [15] we considered exciton and depolarization effect which shift the subband separation as $\tilde{E}_{10} = E_{10} \cdot (1 + \alpha_{11} - \beta_{11})^{1/2}$ where α_{11} (depolarization effect, increases the observed energy) and β_{11} (exciton effect, decreases the observed energy but is much smaller than the depolarization effect) are complicated functions

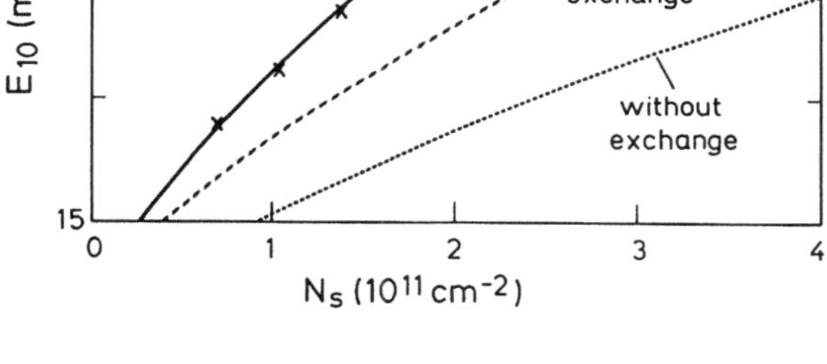

Figure 3 : *The experimental results for the subband separation E_{10} are marked by crosses (×). Calculated subband separations are shown as follows : dotted line = without exchange interaction, dashed line = including exchange interaction, full line = including exchange interaction, exciton and depolarization effect.*

of the self-consistently calculated wavefunctions in z-direction. One clearly sees that the experimental points are best described when all effects are included in the theory. Usually the z-component of the incoming light excites the electrons and a depolarization shifted intersubband resonance is observed [3]. Although our sample is slightly tilted with respect to the incoming light the z-component is negligible because of the high refractive index of GaAs. The reason for this depolarization shifted resonance is the fact that even a small tilt angle leads to a coupling of the motion of the carriers perpendicular and parallel to the 2D layer. The out-of-plane component of the Lorentz force excites the electrons in the z-direction, leading to a charge density excitation.

In summary we have shown that there is nearly perfect agreement between subband separations determined from RSLC experiments and self-consistent calculations, if one includes the depolarization effect in the calculation.

We acknowledge financial support by the Bundesministerium für Forschung und Technologie, Federal Republic of Germany.

References

1. T. Ando, A. B. Fowler, and F.Stern: Rev. Mod. Phys. **54**, 437 (1982)
2. G. Abstreiter and K. Ploog: Phys. Rev. Lett. **42**, 1308 (1979)
3. A. Pinczuk, J. M. Worlock, H. L. Störmer, R. Dingle, W. Wiegmann, and A. C. Gossard: Solid State Commun. **36**, 43 (1980)
4. Z. Schlesinger, J. C. M. Hwang, and S. J. Allen: Phys. Rev. Lett. **50**, 2098 (1983)
5. G. L. J. A. Rikken, H. Sigg, C. J. G. M. Langerak, H. W. Myron, J. A. A. J. Perenboom, and G. Weimann: Phys. Rev. B **34**, 5590 (1986)
6. A. D. Wieck, J. C. Maan, U. Merkt, J. P. Kotthaus, K. Ploog, and G. Weimann: Phys. Rev. B **35**, 4145 (1987)
7. F. Stern and S. Das Sarma: Phys. Rev. B **30**, 840 (1984)
8. T. Ando: J. Phys. Soc. Jpn. **51**, 3893 (1982)
9. K. Ensslin, D. Heitmann, H. Sigg, and K. Ploog: Surf. Sci. **196**, 263 (1988)
10. K. Ensslin, D. Heitmann, H. Sigg, and K. Ploog: Phys. Rev. B **36**, 8177 (1987)
11. Al_2O_3 suppresses leakage currents, idea of J. Richter
12. F. Thiele, U. Merkt, J. P. Kotthaus, G. Lommer, F. Malcher, U. Rössler, and G. Weimann: Solid State Commun. **62**, 841 (1987)
13. E. Batke, H. L. Störmer, A. C. Gossard, and J. H. English: Phys. Rev. B **37**, 3093 (1988)
14. T. Ando: Phys. Rev. B **19**, 2106 (1979)
15. T. Ando: Z. Physik B **26**, 263 (1977)

Quasi-Two-Dimensional Shallow Donors in a High Magnetic Field

S. Huant[1], *W. Knap*[2,*], *R. Stepniewski*[1,*], *G. Martinez*[1], *V. Thierry-Mieg*[3], *and B. Etienne*[3]

[1]SNCI/CNRS, BP 166X, F-38042 Grenoble Cedex, France
[2]GES/USTL, Place Eugène Bataillon, F-34060 Montpellier Cedex, France
[3]L2M/CNRS, 196, Avenue Henri Ravera, F-92220 Bagneux, France

1. Introduction

Hydrogenic donors in confined semiconducting materials are worth studying because they exhibit interesting similarities with the hypothetical two-dimensional (2D) hydrogen atom /1/. A detailed understanding of their properties is also required for the material doping in view of device applications. Far-infrared magneto-spectroscopy - especially in high magnetic fields - has long proved to be a valuable tool for the study of impurity states in bulk semiconductors and also more recently in quasi-2D materials /2,3/. In the present communication, we summarize our magneto-optical studies carried out on the Silicon impurity in GaAs-GaAlAs multi-quantum-wells (MQW). Special emphasis is given to the optical phonon energy region where 2D bound resonant polarons are observed.

2. Samples and experimental details

The samples are MBE grown and either intentionally δ-doped (with growth interrupt during doping) with Si ($N_d = 10^{10}\,cm^{-2}$) at various selected places in the MQW sequence (150 periods) or uniformly doped ($N_d = 8 \times 10^{15}\,cm^{-3}$). The well and barrier thicknesses are 100 Å. The experiments have been performed with magneto-transmission (MT) and photoconductivity (MPC) techniques (with ohmic contacts on the samples) at low temperatures (2 K < T < 20 K) using a far-infrared laser (λ= 33 μm - 70μm) and magnetic fields up to 20 T. Photoluminescence experiments have also been made.

3. The real dopant profile and the role of acceptor states

MT spectra such as shown in Fig. 1 can be used to extract the real doping profile /2/ within an appropriate fitting procedure /3/. This was done for the samples used in this study and is discussed elsewhere /4/. For the δ-doped samples, it was confirmed that the dopant profiles are much sharper than those obtained with the standard doping procedure without growth

* Permanent address: Institute of Experimental Physics, Warsaw University, Hoza 69, 00-681 Warsaw, Poland

Springer Series in Solid-State Sciences, Vol. 87
High Magnetic Fields in Semiconductor Physics II Editor: G. Landwehr

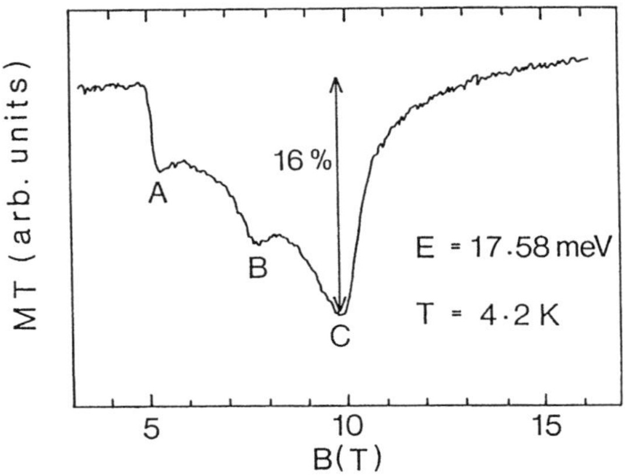

Fig. 1 MT spectrum of the uniformly doped MQW. Peaks A and C are the 1s → $2p^+$ transitions for on-well-centre and on-barrier-centre impurities respectively /5/. Peak B is discussed in the text.

interrupt. Typically 30 % of the dopant atoms introduce themselves at the intended location. The remaining part is incorporated after the growth restart and gives rise to a dopant tail in the growth direction in such a way that residual doping ($N_d \approx 10^9 cm^{-2}$) is present at the barrier centres in samples which are nominally δ-doped at the well centres.

However, this analysis - although very precise in principle - provides us only with semi-quantitative information because of the presence of Carbon and Silicon related acceptor states which are clearly identified in our luminescence experiments. The role they possibly play on the donor properties is discussed in more detail in Ref.4. For the purpose of this work, let us simply note that we think they are responsible for the additional peak B in Fig.1. No explanation for this peak is found within the one-impurity theory /5/. Indeed, some donors in the barriers are ionised in order to compensate for the acceptor states. Then, there is a finite probability for the electrons weakly bound to the remaining neutral donors in the barriers to feel the attractive fields of these ionised donors. This increases their binding energy, as was shown in bulk GaAs /6/. Thus, peak B is due to the 1s → $2p^+$ transitions for these "deeper" shallow donors in the barriers. It is worth noting that peak B has almost the field position expected for interface impurities /5/.

4. Bound resonant polarons in two dimensions

Resonant polaron coupling occurs in polar semiconductors when the magnetic field dependent energy, like the cyclotron energy ("free" polaron) or the 1s → $2p^+$ intradonor energy ("bound" polaron) /7/, is close to the optical phonon energy $\hbar\omega_{LO}$. Recently, 2D resonant free polarons have been extensively studied /8/. The experimental results have long been controversial because of the key role played by free carrier screening and/or Landau level occupation effects which mask most of - if not all - the dimensionality effects. The complexity of analysing the data has also continued the controversy, as was shown in Ref.9.

In our lightly doped MQW, no carrier concentration related problem occurs since all the electrons are frozen on impurity states at low temperatures. Resonant polarons are observed in MPC measurements well below the phonon energies where the phonon response function still behaves non-resonantly /9/. The high signal-to-noise ratio provided by the MPC technique in an energy range where the samples are opaque in MT measurements is exhibited by the spectra shown in Fig.2. It should, however, be kept in mind that it is not possible to compare directly the relative intensities of the peaks in MPC measurements /10/. For instance, peak C was not seen in MPC measurements: this point is not yet understood.

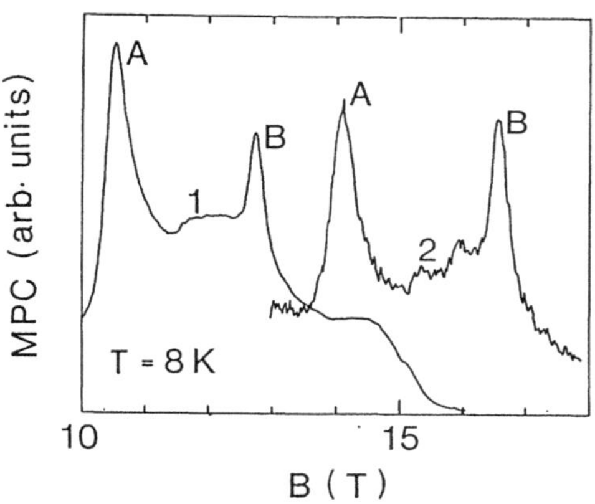

Fig 2 Typical MPC spectra for the nominally on-well-centre δ-doped MQW. 1: E= 26.05 meV; 2: E= 31.06 meV. Tilting the field by an angle θ with respect to the normal of the layers confirms that the whole structure is confinement related.

The field positions of peaks A and B are collected in Fig.3 for a on-well-centre δ-doped sample, together with the bulk data of Ref.7. The deviations from linearity at E > 25 meV are essentially due to resonant polarons which also cause the hybridisation of the 1s → $2p^+$ transitions into two polaron branches /10/. It is immediatly seen in Fig.3 that resonant polarons for on-well-centre impurities and bulk impurities have comparable magnitude, but that they are almost twice as strong for the "deeper" barrier impurities. This point is very interesting since it is directly connected with the reduced dimension. Indeed, we have shown in a simple second-order perturbation calculation that bound resonant polarons have a magnitude roughly proportional to the square root of the oscillator strength for the 1s → $2p^+$ transitions /10/. This strength is strongly position dependent: it is increased by a factor of approximately 3 from the well centre to the interface /5/, thus resulting in a stronger polaron coupling for interface impurities (and for "deeper" barrier impurities which have almost the same binding energy).

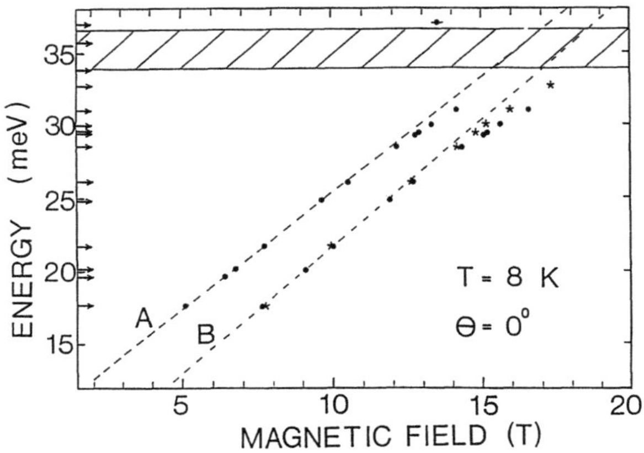

Fig. 3 Field positions of the MPC peaks of Fig. 2 at various laser lines shown by arrows. For some of them close to the phonon energies, no resonance could be detected. The stars are the data for the bulk material /7/. The Reststrahlen band is hatched.

This entails the remarkable feature that quasi-2D bound resonant polarons can be monitored by the impurity position, and also by all the MQW parameters (well width...) since the bare impurity properties depend upon all these parameters /5/. Thus, the effect of the reduced dimension on the Fröhlich polaron is clearly stressed in our experiments.

References

1. G. Bastard, Phys. Rev. B24, 4714 (1981).
2. B.D. McCombe, N.C. Jarosik, and J.M. Mercy, Springer Series in Solid State Sciences 71, 238 (1987) [Preceeding Würzburg Conference], and references therein.
3. S. Huant et al., Solid State Commun. 65, 1467 (1988).
4. S. Huant et al., in Proc. of 4th Int. Conf. Superlattices, Microstructures and Microdevices, Trieste, 1988, to be published in Superlattices and Microstructures.
5. P. Lane and R.L. Greene, Phys. Rev. B33, 5871 (1986); R.L. Greene and P. Lane, Phys. Rev. B34, 8639 (1986).
6. D. Paget and P.B. Klein, Phys. Rev. B34, 971 (1986).
7. H. Sigg, H. Bluyssen, and P. Wyder, Solid State Commun. 48, 897 (1983).
8. See e.g. C.J.G.M. Langerak et al., submitted to Phys. Rev. B, and references therein.
9. K. Karraï, S. Huant, G. Martinez, and L.C. Brunel, Solid State Commun. 66, 355 (1988).
10. S. Huant, W. Knap, G. Martinez, and B. Etienne, accepted for publication in Europhys. Lett.

Two-Dimensional Magnetoplasmons in Gated $Al_xGa_{1-x}As$-GaAs Heterojunctions

M. Tewordt[1], *E. Batke*[1], *J.P. Kotthaus*[1], *G. Weimann*[2], *and W. Schlapp*[2]

[1]Institut für Angewandte Physik, Universität Hamburg, D-2000 Hamburg 36, Fed. Rep. of Germany

[2]Forschungsinstitut der Deutschen Bundespost, D-6100 Darmstadt, Fed. Rep. of Germany

The magnetoplasmon dispersion of inversion electrons in gated $Al_xGa_{1-x}As$-GaAs heterojunctions is investigated at high magnetic fields ($B \leq 12T$) and low electron densities ($0 < N_s < 4 \cdot 10^{11} cm^{-2}$) in transmission with Fourier spectroscopy. The magnetoplasmon resonance position at $B \gtrsim 2.5T$ is well described by the classical local dispersion. Magnetoplasmon resonance line shape anomalies are discussed in terms of a magnetic field dependent line width partially caused by an inhomogeneous surface charge density.

Many experimental studies on cyclotron excitation in two-dimensional (2D) space charge layers are devoted to the interrelation between the high-frequency conductivity and the integer or fractional Quantum Hall Effect [1-4]. In this context we present transmission studies of the wave vector dependent high-frequency conductivity of inversion electrons on GaAs in high magnetic fields $B \leq 12T$ at low temperatures $T \approx 2K$. The samples are modulation doped $Al_xGa_{1-x}As$-GaAs heterojunctions with spacer (28nm), spike-doped $Al_{0.3}Ga_{0.7}As$ layer (28nm) and GaAs cap layer (10nm). A continuous semitransparent NiCr front gate ($R_g \approx 1k\Omega/\square$) serves to vary the surface charge density N_s over a wide range ($0 \leq N_s \leq 4 \cdot 10^{11} cm^{-2}$). An additional metal grating of period a=885nm is prepared on top of the gate to couple the perpendicularly incident unpolarized far-infrared (FIR) radiation to the 2D plasmon resonances at finite wave vector $q=2\pi/a$. We determine N_s from SdH oscillations of the magnetocapacitance and the magnetoresistance. Figure 1 shows the four point magnetoresistance obtained in gate voltage variation for different magnetic field strengths. We find well defined zero resistance states at integer filling factors $f=2\pi l^2 N_s=1,2,...$ (l: cyclotron radius). Please, note the onset of fractional quantization at f=2/3 and 5/3, respectively.

In Fig. 2(a) we show the relative change in transmission $-\Delta T/T = (T(0)-T(N_s))/T(0)$ measured at B=11.5 T for different gate voltages. The wave vector independent cyclotron resonance (CR) dominates compared to the magnetoplasmon resonance (MPR) at $q=2\pi/a$ observed at higher energy. Both the CR and MPR decrease in strength with decreasing gate voltage, and the magnetoplasmon shifts towards the CR. In the inset of Fig. 2(a) we compare CR (ω_c) and MPR (ω_{mp}) positions. The closed line is calculated from the known N_s with the local MP dispersion

$$\omega_{mp}^2 = \omega_c^2 + \omega_p^2, \tag{1}$$

where

$$\omega_p^2 = \frac{N_s e^2}{m^* \varepsilon_0} \frac{q}{\varepsilon(1+\coth(qd))} \tag{2}$$

is the screened local plasmon frequency [5] with $m^*=0.07m_0$ and $\varepsilon=12.8$. The agreement between experiment and theoretical dispersion is satisfactory.

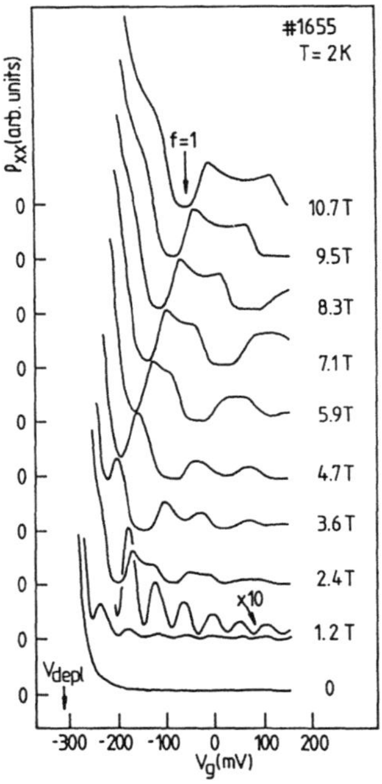

Fig. 1: Four point magnetoresistance of a square shaped $Al_xGa_{1-x}As$-GaAs heterojunction at 2K versus gate voltage for different magnetic fields. Measurements are performed at a constant current I=1μA. V_{depl} indicates the voltage necessary to completely deplete the 2D channel. The arrow indicates the filling factor f=1

Note that we observe no structure, e. g., at f=1 which is reached at V_g=-70 mV. Figure 2(b) shows $-\Delta T/T$ measured at V_g=-120 mV for different magnetic field strengths. As predicted by (1) the MPR approaches ω_c with increasing B. With decreasing B the MPR halfwidth increases, and at B=0 no resonance could be identified. In Fig. 3 we show the amplitude and full width at half maxima (FWHM) of the CR and the MPR. We will restrict the discussion to fields B=0 and B>2.5T where we can neglect nonlocal corrections in the MP dispersion [5]. In Fig. 3(a) ω_p is determined from (1) with the experimental values ω_c and ω_{mp} and is in reasonable agreement with the calculated screened local plasmon frequency ω_p=23.5 cm^{-1}. Cyclotron masses $m^*=eB/\omega_c$ shown in Fig. 3(b) increase with increasing B due to the nonparabolicity of the GaAs conduction band [3]. The B-dependence of the MPR amplitude and FWHM is in contrast to the expectation for a homogeneous classical 2D electron gas with constant phenomenological scattering time τ_p. Within this model assuming linearly polarized incident FIR radiation the MPR amplitude is largest at B=0. This is due to the fact that the plasmon is sensitive to both the left and the right circular polarization. With increasing B the plasmon becomes CR-like and will be less sensitive to the CR-inactive polarization. Thus, with increasing B, we expect a drop in MPR amplitude reaching at high fields about 50% of the amplitude at B=0. The observed increase in MPR amplitude with increasing B we attribute to inhomogeneity and a field dependent scattering time τ_p. Although τ_p and the scattering time τ_c that describes the CR lineshape are not equivalent the difference is expected to be small for the q of interest here [5]. Thus, qualitatively the FWHM of the MPR and the CR should exhibit a similar B-dependence, which indeed is confirmed experimentally. However, $\tau_p \approx \tau_c$ implies $\Delta\omega_p \approx \Delta\omega_c/2$ if we neglect saturation effects [5]. The experiment shows $\Delta\omega_p \approx 2\Delta\omega_c$ indicating the importance of inhomogeneity in N_s. In Fig. 4(a) we calculated MPR line shapes assuming linearly polarized incident FIR radiation and a Gaussian-type density profile as indicated in the inset and τ_p independent

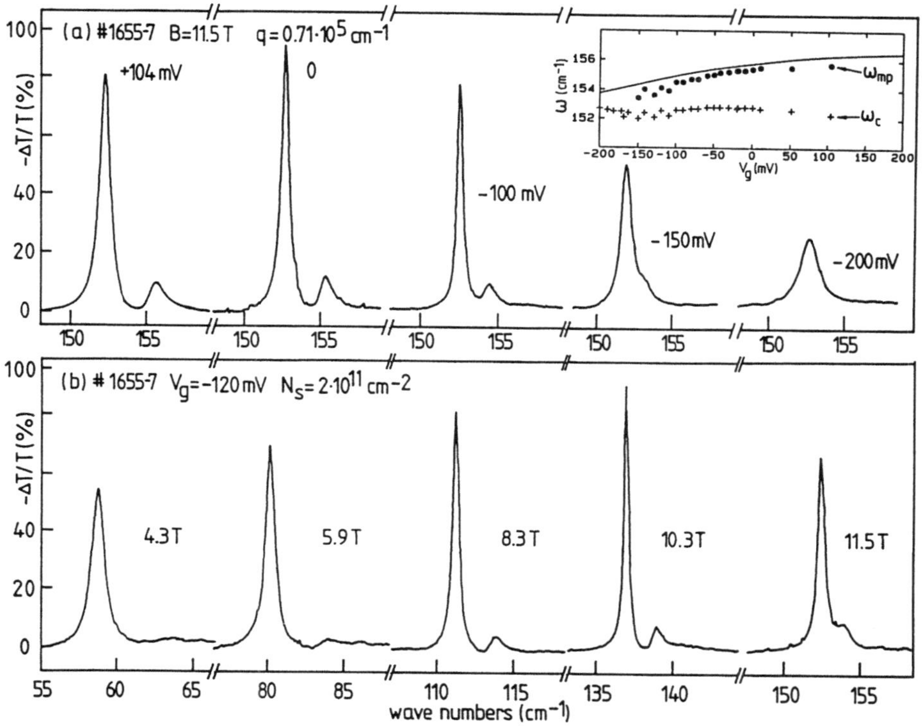

Fig. 2: Relative change in transmission $-\Delta T/T$ at (a) fixed B for various values of V_g and (b) fixed V_g and various values of B. The inset in (a) summarizes the resonance positions ω_c and ω_{mp}. The closed line is calculated from (1)

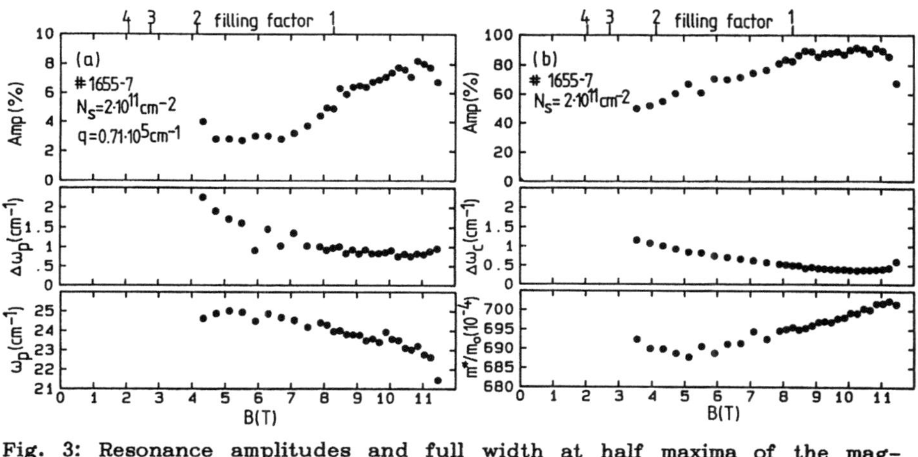

Fig. 3: Resonance amplitudes and full width at half maxima of the magnetoplasmon resonance (a) and CR (b), respectively. m^* is the cyclotron mass, ω_p is determined from (1)

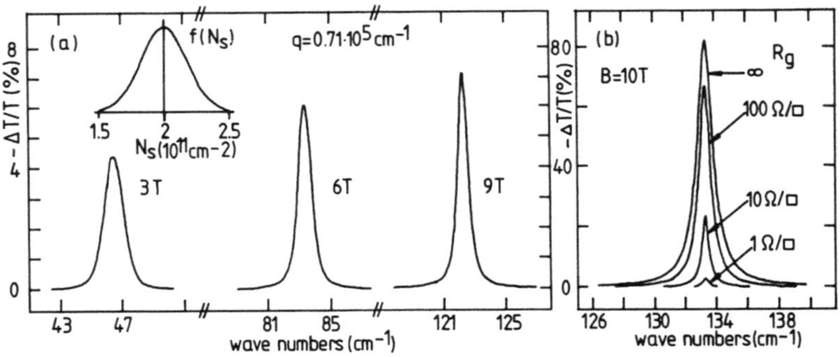

Fig. 4: (a) Calculated magnetoplasmon resonances for different B assuming linearly polarized incident FIR radiation and a Gaussian-type density profile as indicated. (b) Calculated CR line shapes for different effective gate conductivities. Conversion of the incident FIR radiation to purely CR active polarization via the grating coupler is assumed. The scattering times are $\tau_p=\tau_c=20\cdot10^{-12}$s

of B. Qualitatively the amplitude and the FWHM variation with B can be reproduced demonstrating the importance of inhomogeneity in N_s.

As observed previously [5] CR in samples with grating coupler exhibits smaller FWHM and enhanced amplitudes, exceeding 50%, compared to CR in samples without grating coupler (Figs. 2, 3). We attribute this enhancement to near field effects of the grating coupler. In Fig. 4(b) we show calculated CR line shapes for different effective gate conductivities assuming ad hoc that the grating near field is circularly polarized. The introduction of an effective gate conductivity is justified since the FIR field will feel an average gate conductivity from the low conductivity continuous gate and the high conductivity grating stripes. With increasing effective gate conductivity we find a sharpening of the CR line shape reflecting the suppression of saturation by a highly conducting gate. Since we measure a static average gate conductivity of order 100Ω/□ for the direction perpendicular to the grating we can only explain the observed cyclotron amplitudes if we assume that the grating coupler effectively converts the incident radiation to CR active circular polarization. In the quantitative evaluation of CR in the presence of a grating coupler we therefore assume such polarization conversion. A quantitative theory that explains this electrodynamic effect is highly desirable. In summary we find that the B-dependence of the MPR line width can be used to investigate N_S-inhomogeneities in gated heterojunctions. Perhaps such inhomogeneities are responsible for not seeing quantum effects that correlate with discrete filling factors.

Financial support of the Deutsche Forschungsgemeinschaft is gratefully acknowledged.

REFERENCES

1. T. Englert, J. C. Maan, C. Uihlein, D. C. Tsui, and A. C. Gossard, Solid State Commun. 46, 545 (1983)
2. D. Heitmann, M. Ziesmann, and L. L. Chang, Phys. Rev. B 34, 7463 (1986)
3. E. Batke, H. L. Störmer, A. C. Gossard, and J. H. English, Phys. Rev. B 37, 3093 (1988)
4. W. Seidenbusch, E. Gornik, and G. Weimann, Phys. Rev. B 36, 9155 (1987)
5. E. Batke, D. Heitmann, and C. W. Tu, Phys. Rev. B 34, 6951 (1986)

Magneto-optical Study of Excitons Localized Around 2D Defects of BiI_3 in Pulsed High Magnetic Fields up to 47T

K. Watanabe[1], *S. Takeyama*[2], *T. Komatsu*[3], *N. Miura*[2], *and Y. Kaifu*[3]

[1]Department of Physics, Yamanashi University, Kofu 400, Japan
[2]Institute for Solid State Physics, University of Tokyo, Tokyo 106, Japan
[3]Department of Physics, Osaka City University, Osaka 558, Japan

1. Introduction

BiI_3 is a typical layered crystal, and its intrinsic optical absorption edge is an allowed indirect exciton transition/1/. Very sharp absorption lines have been observed below the edge. The origin of these sharp lines is understood to be associated with exciton states localized around various types of 2D defects formed in this crystal. We labelled these lines P, Q, R, S, T and W. The lines originate from the following three different kinds of 2D defects. (1) The origin of the P line was assigned to an exciton state associated with a polytypic structure contained in the crystal with small concentration/2/. (2) The origin of the Q, R, S and T lines has been attributed to exciton states localized around the 2D plane of a stacking fault naturally formed during crystal growth/3/. (3) The W line is induced by a crystal deformation by shear stress. The W line is considered to be associated with a deformation type 2D defect, however, the electronic origin of the W line has not been so clear up to now. In this paper, we present detailed magneto-absorption (MA) spectra of those lines, and discuss the characteristic magnetic field effects of the excitons localized around the 2D defects. This is the first systematic magneto-optical study of all these excitons using very high magnetic fields for the purpose of a general understanding.

2. Experimental procedure

Samples with typical absorption lines were carefully selected among many as-grown single crystals/4/. The MA spectra are measured by using a spectrometer, an optical multichannel analyzer, and an optical fiber system/5/. A pulsed magnetic field up to 47T is generated with 8msec duration. The spectra were obtained at the flat part near the top of the magnetic field in both $[B \parallel z, k \parallel z]$ and $[B \perp z, k \parallel z]$ configurations, where B is the magnetic field, k is the wave vector of the incident light, z is the axis chosen for the direction of the layer stacking. The spectral resolution of the system is estimated to be less than 0.2meV.

3. Experimental results and discussion

Figure 1(a) and (b) shows MA spectra of the P, Q, R, S and T lines in both $[B \perp z, k \parallel z]$ and $[B \parallel z, k \parallel z]$ configurations. The P line splits into two components, which are denoted P_1 and P_2, in magnetic fields of $[B \perp z]$. The energy shifts of P_1 and P_2 are proportional to B^2. In $[B \parallel z]$, the P line also splits into two lines, but the shifts in $[B \parallel z]$ are much smaller than those in $[B \perp z]$. The shifts of the Q, R, S and T lines are very small in

Springer Series in Solid-State Sciences, Vol. 87
High Magnetic Fields in Semiconductor Physics II Editor: G. Landwehr

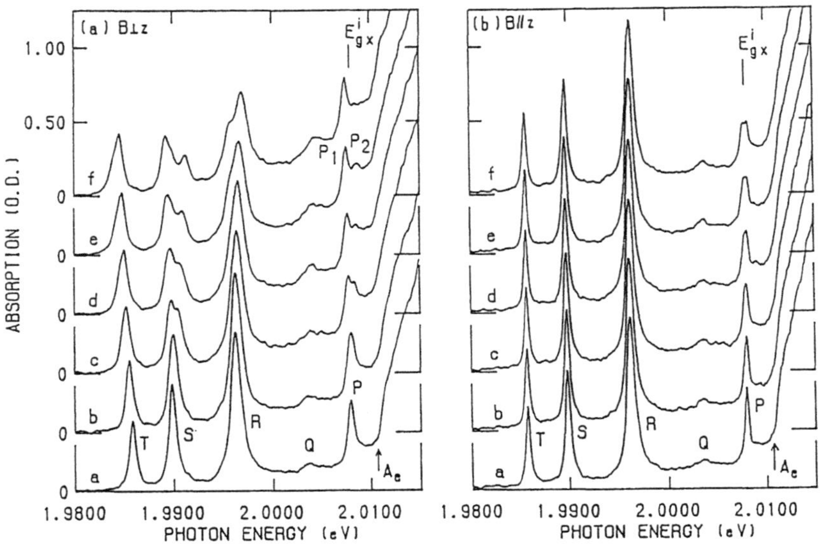

Figure 1(a) MA spectra at 4.2K in [B ⊥ z, k ∥ z] configuration: spectrum a, B=0T; b, 17.3T; c, 26.0T; d, 30.9T; e, 34.9T, f, 40.4T. A_e denotes the threshold of the indirect exciton absorption edge, and E^i_{gx} denotes the indirect exciton energy. (b) MA spectra at 4.2K in [B ∥ z, k ∥ z] configuration: spectrum a, B=0T; b, 15.6T; c, 22.2T; d, 29.6T; e, 35.9T; f, 40.9T. The scale of the axis is the same as in (a).

[B ∥ z]. In [B ⊥ z], the shifts of the lines are much larger than those in [B ∥ z], and proportional to B^2. Figure 2(a) and (b) shows MA spectra around the W line in both [B ∥ z, k ∥ z] and [B ⊥ z, k ∥ z] configurations. The W line splits into two components in [B ⊥ z], which are indicated by W_1 and W_2. Two peaks newly appeared on both sides of the W line in the magnetic field orientation [B ⊥ z], and are indicated by W_3 and W_4. The W_2 and W_3 lines shift to the higher energy side, and the W_1 and W_4 lines shift to the lower energy side with increasing magnetic field. The shifts of the four lines are proportional to B^2 as shown in Fig.3. The absorption intensities of the W_3 and W_4 lines are also proportional to B^2. In [B ∥ z], the shift of the W line is smaller than that in [B ⊥ z]. In [B ∥ z], the energy shifts of the P, Q, R, S, T and W lines are less than 0.2meV at 40T. On the other hand, for [B ⊥ z], larger shifts occur than for [B ∥ z], which are proportional to B^2, and the appearance of new lines is observed. These lines show very anisotropic spectra concerning the magnetic field direction with respect to the z axis.

The characteristic magnetic field effects on the P, Q, R, S, T and W lines can be interpreted by the cationic exciton model. According to this model, the bulk excitons correspond to the excitation from $6s^2$ to 6s6p Bi cation states, and the excitons at the band edge consist of four states of triplet origin named ϕ_3, ϕ_4, ϕ_6 and ϕ_{10} /6/. The transitions from the ground state to ϕ_3, ϕ_6 and ϕ_{10} are optically allowed by singlet state mixing

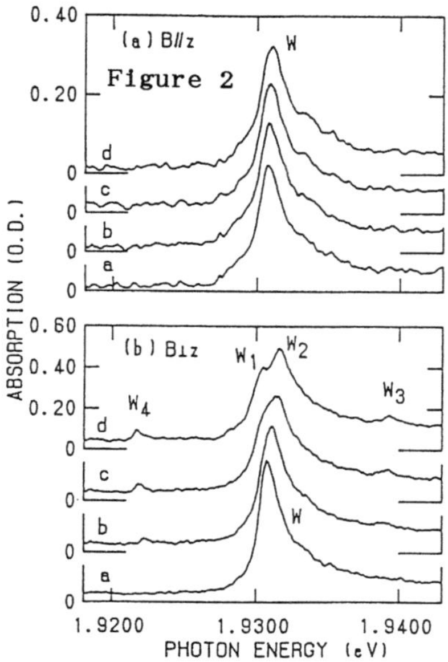

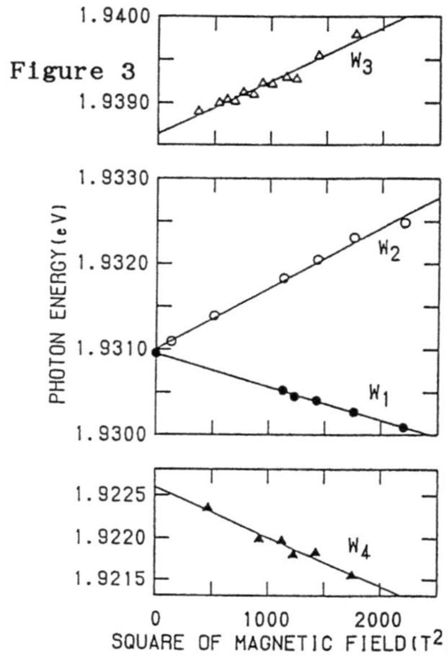

Figure 2(a) MA spectra at 4.2K in [B ∥ z, k ∥ z] configuration: spectrum a, B=0T; b, 22.0T; c, 35.6T; d, 41.8T. (b) MA spectra at 4.2K in [B ⊥ z, k ∥ z] configuration: spectrum a, B=0T; b, 21.7T; c, 30.5T; d, 35.1T.

Figure 3 Magnetic field dependence of the photon energies of the W_1(open triangles), W_2(open circles), W_3(solid circles) and W_4(solid triangles) lines at 4.2K.

because of spin-orbit interaction. The ϕ_6 state is a (x+iy)-like, and ϕ_{10} state is a (x-iy)-like. These two states are degenerate, and the transitions to the states are allowed in [E ⊥ z], where E is electric field of the incident light. The ϕ_3 state is a z-like, and lies on the higher energy side of ϕ_6 and ϕ_{10}. The transition to the ϕ_3 is allowed in [E ∥ z], therefore, it is forbidden in the present configuration. The ϕ_4 state is a pure triplet, therefore, the transition is optically forbidden. This state lies at the lower energy side of ϕ_6 and ϕ_{10}. From the calculation which takes into account the linear Zeeman Hamiltonian as a perturbation, a large anisotropy of the magnetic field effect of the cationic excitons is estimated. In the magnetic field of the [B ⊥ z] configuration, there is no diagonal Zeeman matrix element in the cationic states. However, spin Zeeman off-diagonal matrix elements result in a mixing between the ϕ_4 and ϕ_{10} and a mixing between the ϕ_3 and ϕ_6 states. By these mixings, the transitions from the ground state to ϕ_3 and ϕ_4 become partially allowed in the magnetic field, and the absorption intensities of the transitions are proportional to B^2. The energies of ϕ_3 and ϕ_{10} states shift to the higher energy side, and those of ϕ_4 and ϕ_6 shift to the lower energy side with increasing magnetic field. These shifts are also proportional to B^2. On the other hand, in [B ∥ z], there is no off-diagonal Zeeman matrix element among the four states, therefore, no mixing takes place among these states. However, since diagonal orbital Zeeman matrix elements exist in the [B ∥ z] configuration, the energies of ϕ_6 and ϕ_{10} shift to the higher and lower energy side depending on the signs of the diagonal elements, respectively.

This model can also be applied to the indirect exciton, and the magnetic field effect of the indirect exciton interpreted by this model/7/. The photon energy of the P line is very close to the indirect exciton energy. The magnetic field behavior of the P_1 and P_2 lines is quite similar to that of the indirect exciton states ϕ_6^1 and ϕ_{10}^1, respectively/7/. From these facts, we confirmed that the state of the P line is a localized state around a polytypic disorder having the same nature as the indirect exciton state. The magnetic field effects on the Q, R, S and T lines have been interpreted by the model of cationic excitons localized around the deformed unit cell of a stacking fault plane/8/. It is a characteristic of exciton states that a spin triplet state lies on the lower energy side of its singlet state. This phenomenon is clearly observed in the W line as well. It is probable that the transfer energy of the bulk exciton is modified at the deformation type defect. This results in a formation of a new exciton state localized around the defect, which is observed as the W line. Comparing the magnetic field behavior of the W line to the above calculation, it is concluded that the origin of the W_1, W_2, W_3 and W_4 corresponds to the perturbed states of the ϕ_6, ϕ_{10}, ϕ_3 and ϕ_4 states, respectively.

It is concluded that all P, Q, R, S, T and W lines are exciton transitions localized around 2D defects. The very small shifts of these excitons for [B ∥ z] suggests that the radii of the excitons are very small. The small radii of the excitons bring about the clear observation of a localization effect of the excitons. The magnetic field effect of these excitons is a typical example of that of small radius excitons. In the present system, the mixing among cationic exciton states by spin Zeeman effect and the exchange interaction between electrons and holes play a more important role than the internal motions of the excitons in the magnetic field. The energy shifts of these 2D localized excitons in magnetic fields are very small compared with those of Wannier excitons in usual semiconductors. Nevertheless, clear magnetic field effects on the excitons were observed. This is due to the application of very strong pulsed magnetic fields and also due to the sharpness of the exciton absorption lines.

References

1. Y.Kaifu, T.Komatsu: J.Phys.Soc.Jpn. 40, 1377 (1976)
2. T.Karasawa, T.Komatsu, Y.Kaifu: Solid State Commun. 44, 323 (1982)
3. Y.Kaifu, T.Komatsu, T.Aikami: Nouvo Cimento 38B, 449 (1977)
4. T.Komatsu, Y.Kaifu: J.Phys.Soc.Jpn. 40, 1062 (1976)
5. N.Miura, T.Goto, S.Takeyama, K.Nakao, T.Sakakibara, F.Herlach: J.Magn.Magn.Mater. 54-57, 1409 (1986)
6. Y.Kaifu: J.Luminescence 42, 61 (1988)
7. K.Watanabe, T.Komatsu, S.Takeyama, Y.Iwasa, N.Miura, Y.Kaifu: J.Phys.C 20, 6315 (1987)
8. T.Komatsu, Y.Kaifu, S.Takeyama, N.Miura: Phys.Rev.Lett. 58, 2259 (1987)

Magneto-optical and Magneto-transport Investigations of a Wide Modulation Doped (InGa)As/InP Quantum Well

D.G. Hayes [1;2], *M.S. Skolnick* [2], *L. Eaves* [1], *L.L. Taylor* [2], *and S.J. Bass* [2]

[1]Department of Physics, University of Nottingham, Nottingham NG7 2RD, UK
[2]Royal Signals and Radar Establishment, Great Malvern NG7 2RD, UK

Photoluminescence measurements in magnetic fields up to 10 T on a 80 nm wide modulation doped (InGa)As/InP quantum well are reported. Magneto-luminescence from three electron subbands is observed. Magneto-transport measurements (with $\underline{B}$ both perpendicular and parallel to the plane of the well) also demonstrate occupancy of three subbands. Discontinuities in the Landau level transition energies observed in photoluminescence at certain field values are interpreted in terms of charge redistribution effects between the subbands.

Magneto-luminescence measurements on 2D electron gases in several material systems (Si [1], GaAs/(AlGa)As [2], (InGa)As/InP [3,4]) have recently attracted considerable interest. In this paper we concentrate on the magneto-optical and transport properties of a wide (80 nm) n-type quantum well (QW) which is modulation doped on one side. The structure comprises the following layers in order of growth from the Fe-doped semi-insulating (or n^+-doped) InP substrate: (i) 200 nm, nominally undoped InP buffer; (ii) 80 nm, lattice-matched (InGa)As quantum well, [In] = 0.53; (iii) 10 nm, InP spacer; (iv) 40 nm, 3.0×10^{17} cm^{-3} S-doped top layer.

Hall measurements at 2 K gave a sheet density $n = 3.7 \times 10^{15}$ m^{-2} and mobility $\mu = 6.8$ $m^2/V.s$ in the dark. Under illumination n increased to 7.2×10^{15} m^{-2} ($\mu = 12$ $m^2/V.s$) due to persistent photoconductivity effects [5].

Figure 1 shows the magnetic field-dependence of the photoluminescence (PL) at 4.2 K. At B = 0, three peaks are observed which are attributed to three occupied subbands in the electronic potential well. The two lowest subbands, E(0) and E(1), are localised at the well interfaces, whereas the higher subband E(2) is continuous across the well. The occupancies and energies of the E(0) and E(1) bound states suggest that they are predominantly localised at opposite sides of the well. The photo-created holes thermalise to their lowest energy state near the centre of the well and hence overlap strongly with E(2) which thus dominates the PL spectrum at B = 0 [6,7].

The magnetic field-dependence of the PL confirms that all three lines are associated with bound two-dimensional electronic states of the well as they evolve into Landau level structure at fields as low as 1 T for $\underline{B} \| \underline{z}$, where $\underline{z}$ is perpendicular to the plane of the well. For $\underline{B} \perp \underline{z}$ no Zeeman splittings are observed. The narrowing of the lines with increasing field is due to the removal of free carrier broadening. Figure 2 shows the Landau level splittings. The full lines represent slopes of $\frac{1}{2}\hbar\omega$ and $3/2\hbar\omega$ where $\omega = eB/m^*$ and $m^* = 0.044$ m is the effective mass of electrons in

Springer Series in Solid-State Sciences, Vol. 87
High Magnetic Fields in Semiconductor Physics II Editor: G. Landwehr

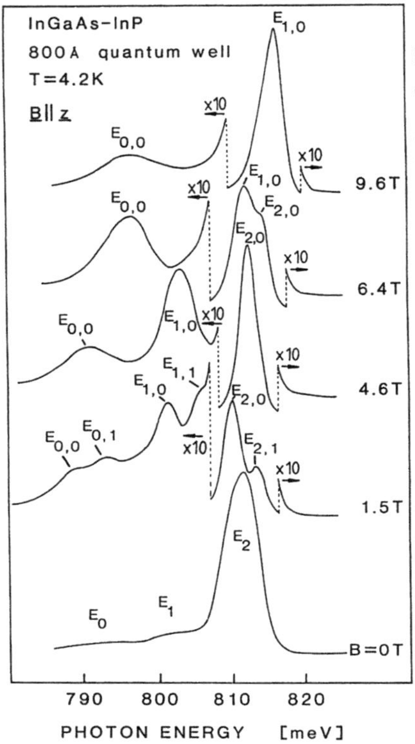

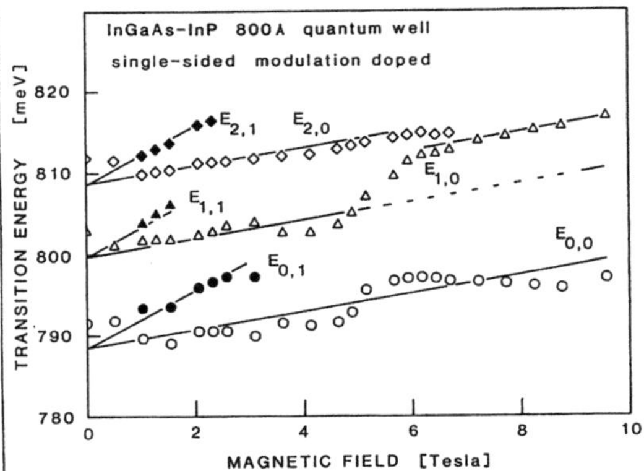

Fig. 2 (above) The magnetic field-dependence of PL peak energies. The solid lines represent slopes of $\frac{1}{2}\hbar\omega$ (open symbols) and $3/2\hbar\omega$ (full symbols) for the three subbands. Notation for Landau levels is as in Fig. 1

Fig. 1 (above) PL spectra (intensity in arbitrary units) at several magnetic fields. Note that $E_{m,n}$ in this figure corresponds to E(m,n) in the text

(InGa)As [8] and E(m,n) corresponds to the n-th Landau level of the m-th subband. The E(2,1) line is weak at 1.5 T and disappears above 2 T as it passes through the Fermi energy, E_F, which we estimate to correspond to a photon energy of 814 meV at B = 0 (see Fig. 2). By extrapolating the solid line in Fig. 2 to B = 0, we deduce that the three bound states are 5, 14 and 25 meV (± 1 meV) below E_F. Using $m^* = 0.044\,m$, the total sheet density deduced from these three subbands is $8.1 \times 10^{15}\ m^{-2}$. The deviation of the observed peak position from a linear dependence is particularly noteworthy. Between 4 and 6 T, the energy of the E(1,0) line moves up rapidly relative to that of E(2,0), whilst the strong E(2,0) line decreases in intensity and vanishes for B ≥ 7 T. E(0,0) also shows an energy upshift between 4 and 6 T. These upshifts are directly related to the depopulation of the spatially extended E(2,0) state as it passes through E_F. The redistribution of electrons into the lower energy bound states E(0) and E(1) renormalises the confining potentials at the well interfaces and leads to the observed upshifts of the E(0,0) and E(1,0) states [7,9]. Note that above 6 T, E(1,0) increases as $\frac{1}{2}\hbar\omega$ and follows the extrapolated variation of E(2,0). The strong intensity of E(1,0) above 6 T suggests that the E(1,0) state now extends right across the well. Figure 2 shows that E(0,1) will depopulate above 6 T when it

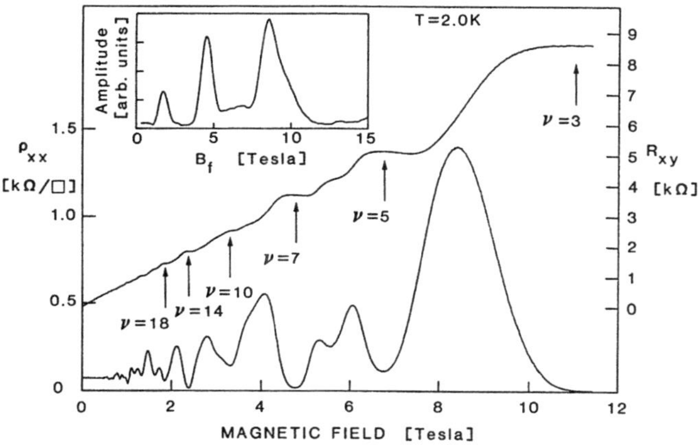

Fig. 3 Magneto-transport results showing the SdH oscillations (axis on left) and the QH plateaux (axis on right). The inset is a Fourier transform of the low field SdH oscillations discussed in the text.

crosses E(1,0). This additional transfer of charge (from E(0) to E(1)) leads to the increase in separation between E(0,0) and E(1,0) above 6 T.

Figure 3 shows the Shubnikov-de Haas (SdH) oscillations and the Quantum Hall (QH) effect under continuous illumination at 2 K. The inset shows the Fourier transform of the SdH oscillations up to 2.5 T (no spin splitting resolved) and reveals three distinct series with fundamental fields $B_f = [\Delta(1/B)]^{-1}$ of 1.7, 4.5 and 8.5 T for E(2), E(1) and E(0) subbands respectively. With m* = 0.044 m, these give Fermi energies of 4.5, 11.8 and 22.4 meV, in fairly good agreement with those deduced from the magneto-luminescence. Further confirmation of the population of three electronic subbands is obtained from $B\perp z$ magneto-resistance measurements [10], where depopulation of E(2) and $\overline{E}(\overline{1})$ at 3 T and 5 T is observed. The Landau level filling factors ν as a function of field can be deduced directly from the resistances (from $\nu = h/e^2R_{xy}$) of the QH plateaux, and are shown on Fig. 3. The total sheet density n is then given by $n = \nu eB(min)/h$, where B(min) is the field value of the corresponding minima of the SdH oscillations. A value for $n = 8.1 \times 10^{15}\ m^{-2}$ is deduced in this way in excellent agreement with that found from the PL. The sheet density can also be deduced from the low field (< 2.5 T) SdH data since

$$n = \sum_{1,2,3} \frac{2eB_f}{h},$$

giving a value of $7.1 \times 10^{15}\ m^{-2}$, in good agreement with the values obtained from the PL and QH results discussed above. Fig. 4, showing the Landau level energies and Fermi energy variation with B, is constructed from the subband energies deduced from PL and from the filling factors deduced from Fig. 3. The observation of clear QH plateaux in Fig. 3 is seen to be consistent with the fields corresponding to well-defined jumps of E_F between Landau levels. Note, however, that the construction neglects charge redistribution and renormalisation of the subband energies, which are particularly important above 6 T. Such effects increase E(0) and E(1) as shown in Fig. 2, and explain the absence of the ν = 4 plateau due to the close coincidence of the opposing spin states of E(0,1) and E(1,0) at about 8 T.

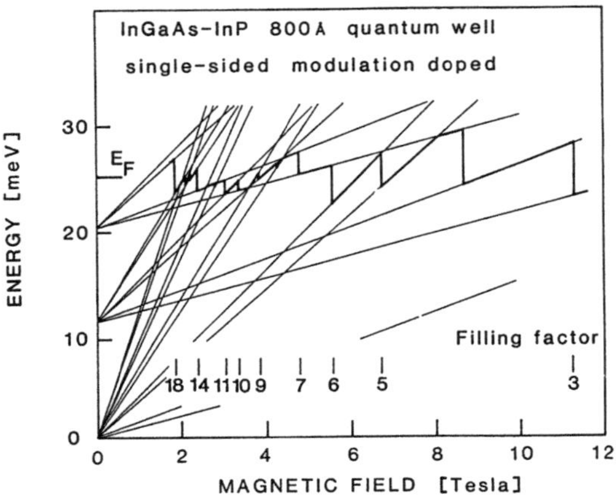

Fig. 4 Construction of variation of Fermi energy with magnetic field, neglecting charge redistribution effects. The subband energies are taken from Fig. 2 and the jumps in E_F from Fig. 3

We conclude that even in a relatively complicated system in which three electronic subbands are occupied [11], the analysis of complementary magneto-optical and magneto-transport measurements provides a detailed understanding of the nature of the electron states.

REFERENCES

1. I. V. Kukushkin and V. B. Timofeev: Sov. Phys. JETP Lett., 43, 499 (1986)
2. M. C. Smith et al: Proc. 17th ICPS, San Francisco 1984, (Springer) p.549
3. M. S. Skolnick et al: Solid State Commun., in press
4. M. S. Skolnick et al: Phys. Rev. Lett., 58, 2130 (1987)
5. M. J. Kane et al: J. Appl. Phys., 60, 657 (1986)
6. F. Meseguer, J. C. Maan and K. Ploog: Phys. Rev. B, 35, 2505 (1987)
7. T. Rotger et al: J. Physique, 48, C5-389 (1987)
8. C. K. Sarkar et al: J. Phys. C, 18, 2667 (1985)
9. P. E. Simmonds et al: Solid State Commun., to be published
10. J. C. Portal et al: Solid State Commun., 43, 907 (1982)
11. M. Razeghi et al: Appl. Phys. Lett., 48, 712 (1986)

Temperature and Magnetic Field Dependence of the Lifetime of Resonantly Excited 2D Carriers in Magnetic Fields up to 25 T Studied Using Picosecond Time-Resolved Photoluminescence

T.T.J.M. Berendschot[1], *H.A.J.M. Reinen*[1], *P.C.M. Christianen*[1], *H.J.A. Bluyssen*[1], *and H.P. Meier*[2]

[1]Research Institute for Materials and High Field Magnet Laboratory, University of Nijmegen, Toernooiveld, NL-6525 ED Nijmegen, The Netherlands

[2]I.B.M. Zürich Research Laboratories, Säumenstrasse 4, CH-8803 Rüschlikon, Switzerland

The effect of confinement on the dynamics of carriers in multi-quantum well structures (MQWS) has been studied extensively in the past [1]. The confinement changes the density of states, the electron-phonon interaction and the magnitude of excitonic effects. It was shown to have consequences for the intraband scattering rates [2] and the efficiency of radiative recombination through spontaneous [3] or stimulated emission [4] processes.

Recently DUBARD et al. [5] have shown that the recombination rate is enhanced considerably due to stimulated emission, at high excitation intensities and low temperatures. In their pump-probe experiment with 200 fs pulses they found that at 15 K the band-to-band absorption recovered within 10 ps. Time-resolved detection of the edge-emitted luminescence showed that the occurrence of a fast radiative recombination could be associated with substantial light amplification.

The stimulated emission rate depends on the density of states and the effective carrier temperature. Applying a magnetic field to the MQWS fully quantises the carrier motion, resulting in quasi-zero-dimensional carrier states. As a result of the complete quantisation of the carrier motion both the density of states and the effective carrier temperature are affected [6].

In this paper we present a study of the lifetime of two-dimensional carriers, optically excited to within three times the optic phonon energy of the band-edge, at high excitation intensities, in magnetic fields up to 25 T and at bath temperatures T_b of 4.2 and 77 K. We find that at 4.2 K and at zero magnetic field the carriers recombine within 15 ps, in accordance with ref. 5. However, with increasing magnetic field the carrier lifetime increases, until at 12 T it is enhanced by a factor of 4 over the B=0 value. For fields above 14 T the lifetime again decreases slightly. If T_b is increased to 77 K the carrier lifetime increases by 25% but the overall magnetic-field dependence remains unchanged.

The experiments were carried out on an MBE-grown modulation-doped quantum well structure consisting of 10 unintentionally-doped GaAs QW's of 5 nm width between 80 nm GaAlAs barriers. The barriers are composed of two 30 nm undoped spacer layers which clad a doped center 20 nm thick. The carrier density per quantum well was found to be 6.1×10^{11} cm^{-2} from Shubnikov-de Haas measurements. Optical excitation was achieved by lightpulses (duration 2 ps, repetition rate 82 MHz, emission wavelength from 690 to 830 nm) from a synchronously pumped c.w. dye-laser. The pulses were focused to a 20 μm diameter spot on the sample surface. For time-resolved detection of the luminescence radiation a light-gate, consisting of a nonlinear optical crystal, a delay line and a monochromator with a photomultiplier tube were used [7]. Magnetic fields up to 25 T were provided by one of the hybrid magnet systems of the University of Nijmegen.

To study the time evolution of the quantum well luminescence, carriers were directly excited in the GaAs layers with an excess energy of about 100 meV. Figure 1 shows the time evolution of the quantum well luminescence for three values of the magnetic field, 0, 4 and 14 T at T_b=4.2 K. The time dependence was measured at the low energy side of the luminescence spectra within one optic phonon energy (36 meV) of the 2D band-edge. The rise-time of the luminescence at B=0 is within the temporal resolution (∿ 5 ps) of the system, but increases to about 20 ps at B=14 T. Values of the decay-time for different magnetic field strengths and at T_b=4.2 K and 77 K, obtained from measurements similar to those in Fig. 1, are given in Fig. 2. At B=0 and 4.2 K we find a decay time less than 15 ps, which increases to 65 ps at B=12 T. For B>14 T a slight decrease to 50 ps at B=20 T is observed. For T_b=77 K the decay-time increases by about 25% with respect to 4.2 K but shows a similar magnetic-field dependence up to B=15 T. Measurements for several different excitation wavelengths give similar results, showing that the observed magnetic field dependence is not an absorption effect.

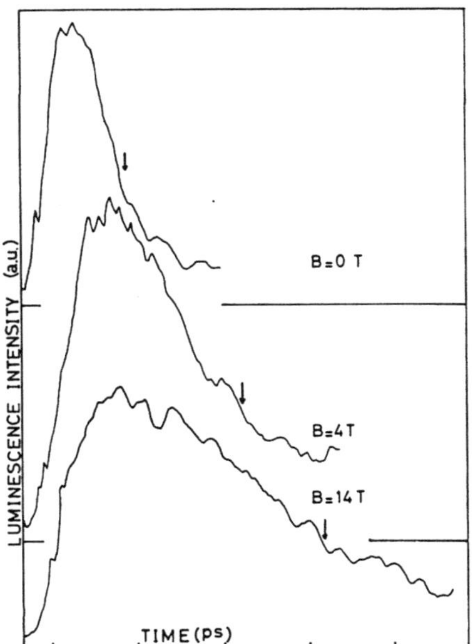

Figure 1 Time evolution of the quantum well luminescence for B=0,4 and 14 T and T_b=4.2 K, measured to within one optic phonon energy of the band-edge

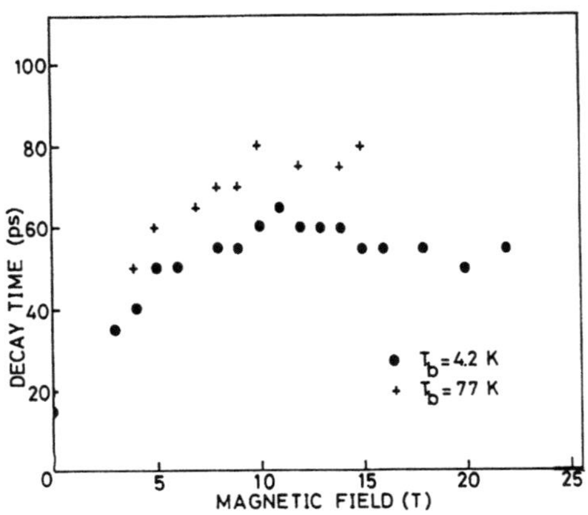

Figure 2 The luminescence decay-time versus magnetic field for T_b=4.2 K (dots) and T_b=77 K (crosses)

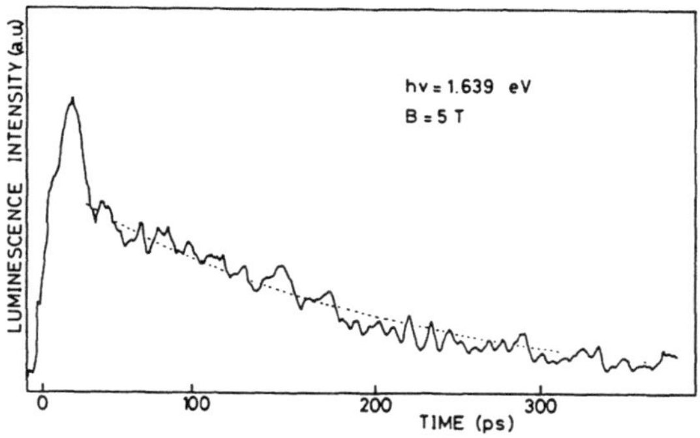

Figure 3 Time evolution of the quantum well luminescence for B=5 T and T_b=4.2 K. The dashed curve is given by exp(−t/τ) where τ=200 ps

In Fig. 3 the decay of the luminescence is followed for a much longer time (400 ps) at B=5 T and T_b=4.2 K. After reaching the maximum, a rapid decay of 35 ps is followed by a much slower one of 200 ps. The latter time appeared to be independent of the magnetic field. Moreover, a measurement at 300 K and B=0 showed no rapid processes, and only a slow decay-time of order 200 ps.

The fast rise-time at B=0, as observed in Fig. 1, is due to excitation of electrons with an excess energy of 100 meV, i.e. well below the energy of the L- and X-valleys. Therefore no inter-valley scattering can occur. A fast rise time of the Γ valley luminescence has also been observed recently in InP by SHAH et al. [8]. In InP, where no inter-valley scattering can occur when an excitation wavelength of 610 nm is used, a rise time of about 2 ps was found. This rise time was assigned to the cooling of the thermalised electrons and holes, since the integrated luminescence intensity depends on the carrier temperature. The increase of the rise time with magnetic field observed here may therefore be due to the previously observed effect of the magnetic field on the cooling of 2D carriers [6,9].

Previous experiments on GaAs-GaAlAs QW's have shown that recombination rates can be enhanced by stimulated emission at carrier temperatures under 100 K [5]. The T_b-dependence of the luminescence decay rates at B=0 in our work suggests that our ultra-short recombination times are also due to stimulated emission. Under the almost resonant excitation conditions, carriers can cool very rapidly to a carrier temperature around 100 K, where stimulated emission is significant [5]. The application of a magnetic field perpendicular to the plane of the QW's completely quantises the carrier motion (producing quasi-zero-dimensional states) leading to a considerable change in the cooling rate of the 2D carriers. Previously the cooling rate in QW's has been found to decrease with increasing B up to ∿ 10 T, and then to increase as the field is increased further, possibly due to the increasing importance of acoustic-phonon emission [6]. The stimulated emission rate is suppressed by high carrier temperatures [11], and so the observed variation of the decay-time of the fast component of the luminescence is probably due to the magnetic-field dependence of the carrier cooling rate. A decrease in the stimulated emission with increasing magnetic field has also been reported for bulk GaAs [12]. However, in that case only the shift of the stimulated emission peak in a magnetic field was explained, but not the decrease.

Finally we attribute the slow decay observed in Fig. 3 to recombination by spontaneous emission. After the fast decay of the luminescence directly after excitation, the carrier density is reduced to a value where stimulated emission is no longer dominant. The lifetime of a few hundred ps is in agreement with earlier reported lifetimes in quantum wells, where spontaneous emission is the main recombination process [4]. At T_b=300 K spontaneous emission is the main recombination process, since stimulated emission is not important at high carrier temperatures. The decay time of the luminescence at 300 K is comparable to the slow decay observed at 4.2 K after 35 ps.

In summary we have shown preliminary results on the lifetime of resonantly excited 2D carriers at high excitation intensities and in magnetic fields up to 25 T. We have found that: (1) the observed short lifetimes are due to a large stimulated emission rate and (2) the magnetic field dependence of both the rise time and decay time of the quantum well luminescence can be qualitatively understood as an effect of the magnetic field on the carrier cooling.

Acknowledgement - We thank A.F. van Etteger for assistance with the laser equipment. Thanks are also expressed to dr. J. Singleton for a critical reading of the manuscript. Part of this work was supported by the "Stichting voor Fundamenteel Onderzoek der Materie" (FOM) with financial support of the "Nederlandse Organisatie voor Wetenschappelijk Onderzoek" (NWO).

1. Jagdeep Shah: IEEE J. of Quantum Electron. QE-22, 1728 (1986) and references therein
2. F.A. Riddoch and B.K. Ridley: J. Phys. C 16, 6971 (1983)
3. N. Holonyak, R.M. Kolbas, R.D. Dupuis and P.D. Dapkus: IEEE J. Quantum Electron. QE-16, 701 (1980)
4. E.O. Göbel, H. Jung, J. Kuhl and K. Ploog: Phys. Rev. Lett. 51, 1588 (1983)
5. J. Dubard, J.L. Oudar, F. Alexandre, D. Hulin and A. Orszag: Appl. Phys. Lett. 50, 821 (1987)
6. R.W.J. Hollering, T.T.J.M. Berendschot, H.A.J.M. Reinen, H.J.A. Bluyssen and P. Wyder: Proc. of the Int. Conf. on the Physics of Semiconductors, ed. by O. Engström, Stockholm (1986), page 1323
7. R.W.J. Hollering, T.T.J.M. Berendschot, H.J.A. Bluyssen, P. Wyder, M. Leys and J. Wolter: Solid State Commun. 57, 527 (1986)
8. Jagdeep Shah, Benoit Deveaud, T.C. Damen, W.T. Tang, A.C. Gossard and P. Lugli: Phys. Rev. Lett. 59, 2222 (1987)
9. J.F. Ryan, R.A. Taylor, A.J. Turberfield and J.M. Worlock: Surface Science 170, 511 (1986)
10. A.J. Turberfield: Solid State Electron. 31, 387 (1988)
11. K. Hess, A. Vojak, N. Holonyak, R. Chin, P.D. Dapkus: Solid State Electron. 23, 585 (1980)
12. J.L. Shay and W.D. Johnston Jr.: Phys. Rev. B 6, 1605 (1972)

New Magnetically Tunable Far-Infrared Solid State Lasers

E. Gornik[1,*], *K. Unterrainer*[1], *M. Helm*[1], *C. Kremser*[1], *and E.E. Haller*[2]

[1]Institut für Experimentalphysik, Universität Innsbruck, Technikerstraße 25/4, A-6020 Innsbruck, Austria

[2]Lawrence Berkeley Laboratories, Berkeley, CA 94720, USA

*)present address: Walter–Schottky–Institut, TU–München, Am Coulombwall, D–8046 Garching, FRG

The basic principles for achieving population inversion and stimulated emission in the far infrared from p–Ge is discussed. In the heavy–light hole lasing mode a broad gain region is found resulting in a broad multimode spectrum due to intracavity modes. By attaching external plates of Ge a single mode operation is realized. The obtained powers are in the W range with linewidths of ≈ 0.2 cm^{-1}. A single mode magnetically tunable coherent source is achieved with the light hole cyclotron resonance laser. With external mirrors a tuning range from 20 cm^{-1} to 120 cm^{-1} with magnetic fields between 1 and 6 T is achieved. The intensity of the single mode is in the order of mW, the linewidth below 0.2 cm^{-1}.

1. Introduction

It is a rather old idea to use the population inversion of hot carriers in bulk semiconductors to generate coherent sub–mm and mm radiation. The first proposal came in 1958 from KRÖMER /1/ who suggested a negative mass amplifier (NEMAG). A concept for a magnetically tunable laser based on radiative transitions between Landau levels was proposed in the early 1960's by LAX /2/ and WOLFF /3/. LAX has suggested a cyclotron maser in n–InSb pointing out that the nonparabolicity causes unequally spaced Landau levels. Wolff proposed in addition that a sudden decrease in relaxation time above a certain energy level terminates the Landau ladder. Thereby effectively eliminating upward transitions by electrons excited into higher levels. Nearly a decade later spontaneous cyclotron emission in a semiconductor was observed for the first time by GORNIK /4/ in 1972 and by OTSUKA's group /5/ in 1973.

It is quite remarkable that it took another decade of intense investigations of hot carrier phenomena until a solid state laser in the far infrared spectral region could be realized: A cyclotron resonance laser based on light hole transitions in p–Ge was first demonstrated by IVANOV and VASILEV /6/ in 1983, the light–heavy hole laser in p–Ge by ANDRONOV et al. /7/ and CHEBOTAREV and MURZIN /8/ in 1984.

This success would not have been possible without the intense theoretical study of highly anisotropic carrier distribution functions resulting from optical phonon emission by several groups /9–18/. It was the pioneering work of VOSILYUS and LEVINSON /10,11/ and by MAEDA and KUROSAWA /12,13/ which found inverted carrier distributions in the case of streaming for crossed electric and magnetic fields. In the following, it was the Russian theoretical work around ANDRONOV and POZELA /19/ which tried to obtain a bulk high frequency negative differential resistivity which would provide an active medium for the realization of the cyclotron resonance maser. The basic idea was that a scattering free transit of carriers up to the energy of the strong scattering threshold (optical phonon emission) can provide not only an inverted hot carrier distribution but also a mechanism for a dynamic negative differential resistance (for reviews see Ref. 20 and 21).

Springer Series in Solid-State Sciences, Vol. 87
High Magnetic Fields in Semiconductor Physics II Editor: G. Landwehr

In this paper the present status of the solid state far infrared lasers is reviewed. After a brief description of the basic pump mechanism for the p–Ge lasers the properties of the light–heavy hole laser are discussed. Finally, recent results on the cyclotron resonance laser are presented.

2. Pumping Mechanisms

In high purity semiconductors at low temperatures carriers are strongly scattered when accelerated in external fields up to the optical phonon energy. They gain the energy practically collisionless from the field. If we consider a p–type semiconductor with a light and heavy hole mass m_l and m_h respectively, the conditions for streaming are met at different electric (E) and magnetic (B) fields. The situation is best described in a schematic representation of hole trajectories in the $v_z = 0$ plane in crossed E and B fields. The center P of the cyclotron orbits is (0, –E/B). In Fig. 1a the situation when the heavy holes are streaming and the light holes are accumulated is sketched. For a given E/B ratio the center P is the same for both types of carriers. However, the velocity corresponding to the optical phonon energy ($\hbar\omega_{OP}$) is quite different: $v_{OP}{}^{l,h} = \sqrt{2\hbar\omega_{OP}/m^*{}_{l,h}}$ since it depends on the effective mass. From Fig. 1a it is evident that several Landau levels lie within the optical phonon energy for light holes which leads to an accumulation of light holes in a limited area of velocity space. The heavy holes are streaming: They are accelerated collisionless to the optical phonon energy and scattered back to the origin. The resulting distribution function is schematically shown in Fig. 1b. As a result of the accumulation the light hole distribution is inverted while the heavy holes are distributed nearly homogeneously up to the phonon energy.

The situation is best described by a dimensionless parameter $\xi_{h,l} = v_{OP}{}^{h,l}/(E/B)$ which is the ratio between v_{OP} and the orbit center velocity E/B. The most efficient pumping process for light hole inversion is present when $1 < \xi_h < 2$ and $\xi_l > 2$.

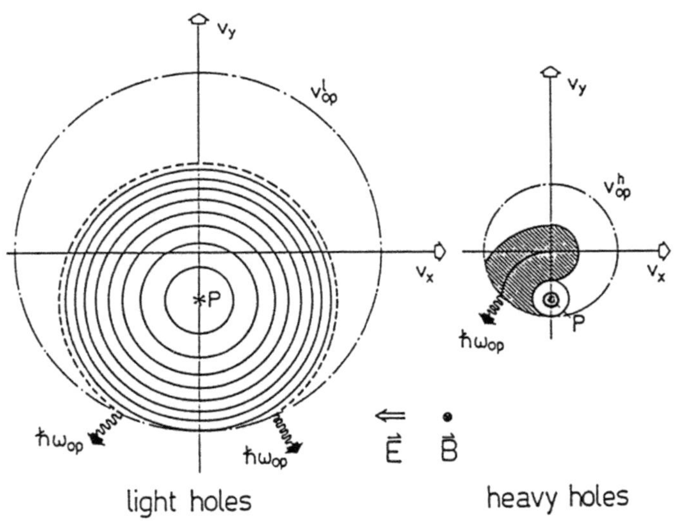

Fig. 1a: Schematic representation of hole trajectories in the $v_z = 0$ plane for a crossed field configuration (B = 1T, E = 1.4 kV/cm) simulating a situation $\xi_l > 2$ and $\xi_h \approx$ 1.4. The orbits for light holes correspond to Landau level, the shaded area for heavy holes to the streaming region

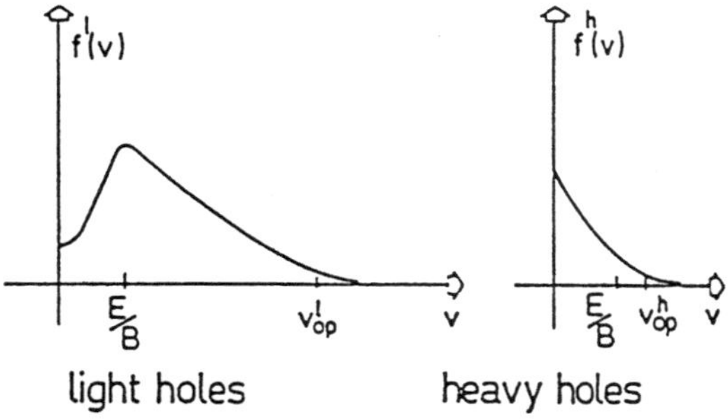

Fig. 1b: Schematic form of the light and heavy hole distribution function for the situation in Fig. 1a

According to a classical calculation for p–Ge by ANDRONOV et al. /21/ with $m^*_h = 0.35\ m_0$, $m^*_l = 0.043\ m_0$ and $\hbar\omega_{op} = 37$ meV the optimum condition of the ratio E/B for stimulated emission is $\xi_h \approx 1.4$ and $\xi_l = 2.85$. In this case stimulated emission covers the emission range $\lambda = 40$ to $400\ \mu$m, where the λ–position for the maximum gain shifts to lower values (higher energy) with increasing B (or E).

In terms of a laser pumping process we can sketch a four level pumping scheme which illustrates the process quite well (Fig. 2a). The streaming process is represented by the ground level at $v_h \approx 0$ and the upper level at $v_h = v_{op}{}^h$. Scattering transfers heavy holes either directly into the light hole accumulation region or by an emission of one optical phonon to $v_h = 0$: After consecutive tunneling transitions to the light hole band they are accelerated to the accumulation region by the electric field. After the radiative transition light hole – heavy hole the heavy holes are transferred by the electric field back to the streaming process. In Fig. 2b the same four step process is shown within the valence band bandstructure for clarity. Detailed calculations of the gain coefficient under the above situation have been performed by POZELA et al. /22/ already before the observation of stimulated emission.

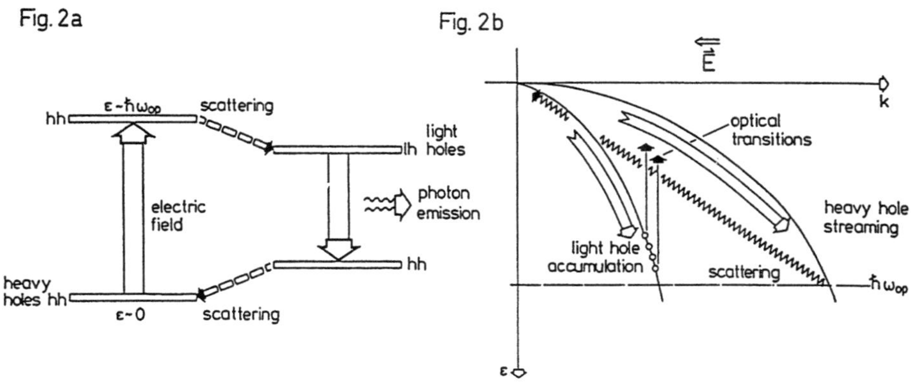

Fig. 2a: Schematic four level pumping scheme for the heavy–light hole laser process

Fig. 2b: Band structure of p–Ge. The different pumping, scattering and optical transitions are indicated for the lasing process

3. Light hole – heavy hole laser oscillation

Experimental studies started from the observation of spontaneous emission in p–Ge /23–26/ in 1981/82 and were followed only a few years later by the first success in observing stimulated emission by the Russian group around ANDRONOV /27/ and the subsequent observation by KOMIYAMA et al. /28/ and HELM et al. /29/.

In this chapter a study of the stimulated emission from p–Ge in the light–heavy hole mode which was performed in Innsbruck is presented. By investigating samples with different dimensions with a broadband (p–Ge) and narrowband (n–GaAs) detector the intensity, the spectrum and the mode behaviour is obtained.

Three different samples were used for the experiments, all cut from a Ga–doped Ge–crystal with $p = N_A - N_D = 1.2 \times 10^{14} cm^{-3}$. The sample dimensions are given in table 1:

TABLE 1 – Sample dimensions

Sample	d	a	l	(in mm)
1	3.7	2.4	46.9	
2	4.2	3.15	43.1	
3	5.2	3.75	48.4	

The distance between the electrical contacts is d, l is the sample length in the direction of the magnetic field (in our case the (111)–direction), and a is the width. Electrical contacts were made by evaporating Indium (with some gold) and alloying at 400°C. All faces of the samples were polished and are parallel within 30", which turned out to be crucial for the achievement of laser oscillation. Voltage pulses of up to 1000 V with 1 μs duration were applied to the samples with a low impedance high power pulse generator. The typical current through the samples was 200 A, in accordance with the relation $j = 1/2\ pev_{op}^h$, where j is the current density, and $1/2\ v_{op}^h$ is the approximate drift velocity of the heavy holes under "streaming motion" conditions /30/. The pulse repetition rate was chosen to be around 5 Hz. For rates higher than 10 Hz lasing breaks down due to the sample heating. The experiments were performed at 4.2 K, so the sample was immersed in liquid He at the center of a superconducting solenoid. To check the range of electric and magnetic fields where lasing can be achieved, the radiation was detected with a broadband Ge:Ga–detector with peak sensitivity at about 110 μm. A magnetic field tunable narrowband GaAs–detector was used for the detailed analysis of the emission spectrum. It was placed in the center of a second magnet in the cryostat below the sample. The Zeeman–split 1s – $2p^+$ shallow donor line is used for detection /31/. This line is tunable between 45 cm^{-1} and 130 cm^{-1} for magnetic fields of 1 T to 6.5 T. Due to the fact that the GaAs contains only one chemical species of shallow donor, the absorption line is extremely narrow (0.25 cm^{-1}), and the GaAs can be used as a high resolution spectrometer.

Figure 3 shows typical traces of stimulated emission from sample 2, recorded with a Ge:Ga–detector. The detector signal due to stimulated emission is a factor of $10^2 - 10^3$ above the level of spontaneous emission. The region of lasing extends from somewhat less than B = 1 T for the lowest electric field of $E \simeq 1000$ V/cm up to 2 T for $E \simeq 2000$ V/cm. This corresponds to $1.3 < \xi_h < 1.9$. These data agree quite well with the results of other authors /7,21,28,32–34/. Comparing the magnitude of the signal with the signal obtained from a FIR gas laser and a InSb–Landau source, the emitted power was estimated to reach 500 mW.

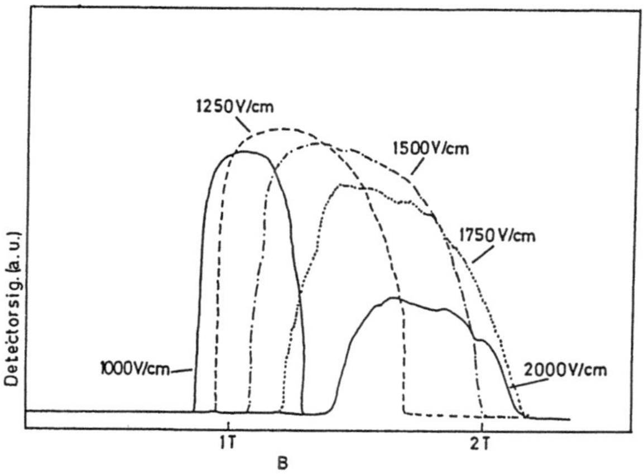

Fig. 3: Ge:Ga–detector signal from sample 2 as a function of magnetic field for different electric field values

To analyze the spectrum of the stimulated emission the electric and magnetic fields applied to the sample were set to a fixed value, while the magnetic field of the GaAs–detector was tuned. Using the 1s–2p⁺ shallow donor transition, the frequency of the detected radiation is linearly related to the magnetic field over a wide range. Figure 4 shows three spectra from sample 2, for three different values of the electric and magnetic field. The spectra consist of 10 – 20 lines with a regular mode spacing of about 1.4 cm^{-1}. Their positions remain fixed with changes in the applied fields. In the high–gain region some additional lines appear, disturbing the regular structure. The measured widths are about 0.4 cm^{-1}. We conclude that the real linewidth of the emission is of the same order of magnitude as the detector linewidth (0.25 cm^{-1}). By

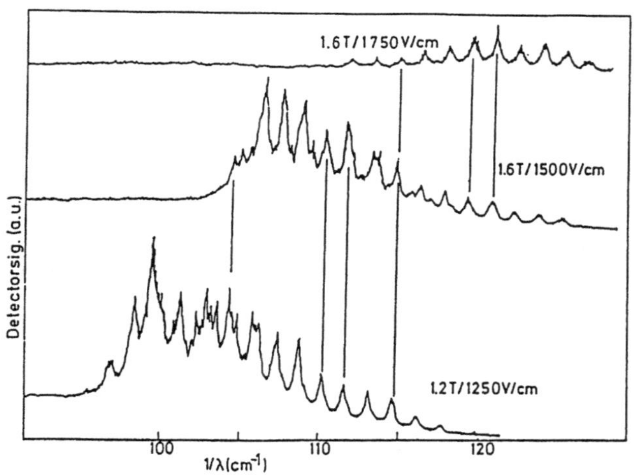

Fig. 4: Spectrum of stimulated emission from sample 2 for different electric and magnetic fields

varying the electric and magnetic fields it is possible to obtain lasing between 75 μm and 110 μm. There is a general trend towards emission at shorter wavelengths for both higher E– and B–fields (see Fig. 2). In contrast to ANDRONOV's /34/ and KOMIYAMA's /33/ work, no long–wavelength–lasing (150 – 250 μm) could be found, not even at lower magnetic fields.

ANDRONOV et al. /7/ and KOMIYAMA et al. /21/ suggested that the lasing modes are due to total reflections at the sample surfaces, since without mirrors the quality factor of the cavity is far too low to allow longitudinal or transverse modes to oscillate (the Ge surface has 36 % reflectivity at normal incidence). According to Fig. 5 the observed mode spacings are 1.74 and 1.20 cm^{-1} for samples 1 and 3, respectively, and 1.40 cm^{-1} for sample 2 (not shown). Comparing the spacings with the dimensions of the samples it appears to be very likely that the width, a, is the crucial parameter determining the mode spacing (see Table 1).

The mode spacings can be explained by considering the transverse resonance condition given by $k_x n = kn \sin\theta = M\pi/a$, where n is the refractive index, M is an integer, and θ is the angle between k and the longitudinal axis of the crystal (see insert in Fig. 5). The analogous longitudinal resonance condition is not considered, because it would yield a very narrow mode structure which is not observed in the experiment. In addition, we have the geometrical condition $\tan\theta = Na/l$ (N integer), which ensures that the path of the ray is closed within the sample. This leads to a mode spacing (in wave numbers) $\Delta v = (2an \sin\theta)^{-1}$. θ must satisfy the condition $14.5^0 < \theta < 75.5^0$, because 14.5^0 is the angle of total reflection for germanium. One can assume that modes with the smallest possible θ have the lowest losses, because they undergo the fewest reflection. Thus, for each sample the mode spacing is determined first by choosing the angle θ with the smallest possible N, and then this angle is used in the expression for Δv. This procedure gives a mode spacing Δv of 1.77 cm^{-1} (N = 6), 1.41 cm^{-1} (N = 5), and 1.14 cm^{-1} (N = 4), for samples 1, 2 and 3, respectively, in good agreement with the measured values. This interpretation is confirmed by the work of ANDRONOV et al. /36/, who reported emission of a single line by using a very thin (a = 0.7 mm) sample, where the mode spacing is as wide as the gain region.

After attaching external mirrors to sample 3 (see insert in Fig. 6), the region of electric and magnetic fields where lasing occurs stays roughly the same. The detector signal is also of the same order of magnitude with and without mirrors. The observed spectra, however, exhibit a dramatic change. Figure 6 shows spectra of sample 3 for different electric and magnetic fields, recorded with the GaAs detector. A considerably reduced number of lines appears with wider mode spacing and higher amplitude, accompanied by some other less dominant structure. Varying the electric and magnetic fields allows us to switch on and off lines on the long or short wavelength side of the spectrum due to a change in the gain spectrum. In this way it was possible to obtain operation of a single line (lower part of Fig. 6). The emission wavelength is 87.3 μm (114.5 cm^{-1}), the adjacent line at 83.3 μm (120.0 cm^{-1}) is just at threshold. As is seen from Fig. 6 the single line contains approximately the same total intensity as the three lines together. However, one cannot exclude the coexistence of several cavity modes within this line, because their spacing would be below the resolution of the detector. The spectra observed by ANDRONOV et al. /34/ from samples with external mirrors showed a very irregular and complicated structure. KOMIYAMA and KURODA /32/ reported some changes in the spectrum and were able to reduce the number of oscillating lines by sandwiching a thin layer of TPX between the sample and one mirror.

The mode spacing in Fig. 6 is equal to $\Delta v = 5.5\ cm^{-1}$, which is too large for fundamental longitudinal or transverse cavity modes. A possible explanation is the formation of an intracavity interference filter by the two Ge plates /35/. The polarization characteristics and the effects of uniaxial stress on the laser oscillations were studied by KOMIYAMA /33/. The gain of the laser is found to be rather small in the range of 5 x 10^{-3} to $10^{-2}cm^{-1}$ which explains the necessary high quality factor of the cavity.

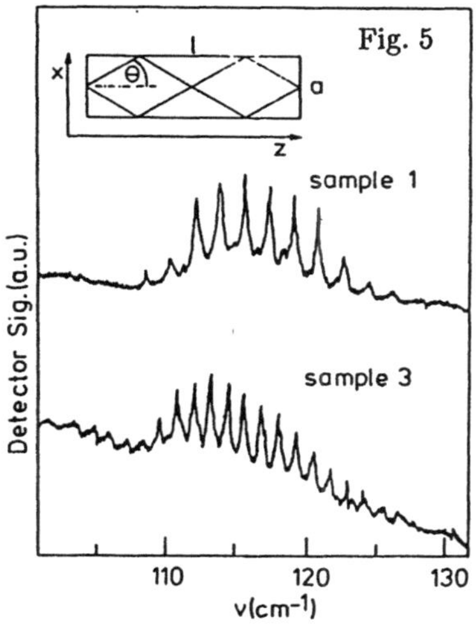

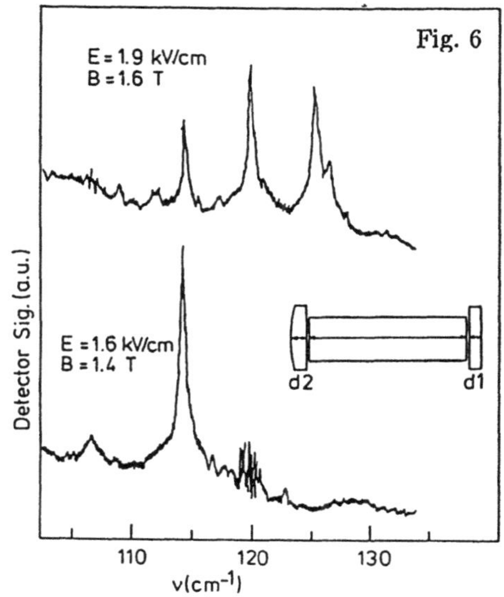

Fig. 5: Spectrum of stimulated emission for different electric and magnetic fields. Sample 1: E = 1.8 kV/cm, B = 1.4 T. Sample 3: E = 1.9 kV/cm, B = 1.7 T. The insert schematically shows the ray path in the Ge crystal

Fig. 6: Emission spectra from sample 3 after attaching external mirrors for different electric and magnetic fields. The insert shows the cavity configuration

4. Light hole cyclotron resonance laser

Stimulated cyclotron emission of light holes was discovered during the investigation of spontaneous emission in crossed fields in an extremely pure Ge sample ($p \approx 10^{13} cm^{-3}$) with the help of a narrowband GaAs detector /6/. A strong emission well above the spontaneous level was observed at certain E and B fields. The requirement of parallel sample faces and the threshold behaviour imply that a lasing effect is present. An analysis of the frequency spectrum reveals a linear relation between frequency and field according to $\omega = eB/m^*$ with an effective mass 0.0464 m_0.

In the intracavity mode a rather broad emission spectrum is observed with samples of active length up to 1 cm /6,36/. A remarkable breakthrough was achieved with samples of 4 to 6 cm length when external resonators were applied /37–39/. The resonators consist of two copper mirrors (isolated by mylar foils) which were attached to the end faces. The light is coupled out by a 1 mm hole in one of the copper mirrors.

Spectra obtained by the group of MURZIN /38,39/ are shown in Fig. 7. A single mode operation is found. The single laser line is tuned linearly by the magnetic field. The observed linewidth is spectrometer limited. The intensity of the cyclotron laser lines is in the order of several mW thus about 2 orders of magnitude smaller than for the heavy–light hole laser.

In Fig. 8 the tuning characteristics for two different materials are shown. The shaded area demomstrates the range of electric and magnetic fields in which lasing is obsrved.

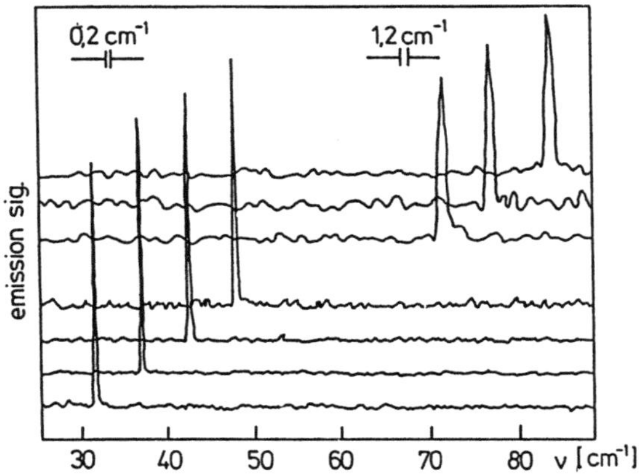

Fig. 7: Emission spectrum in the light hole lasing mode with external mirrors and hole outcoupling after Ref. 38

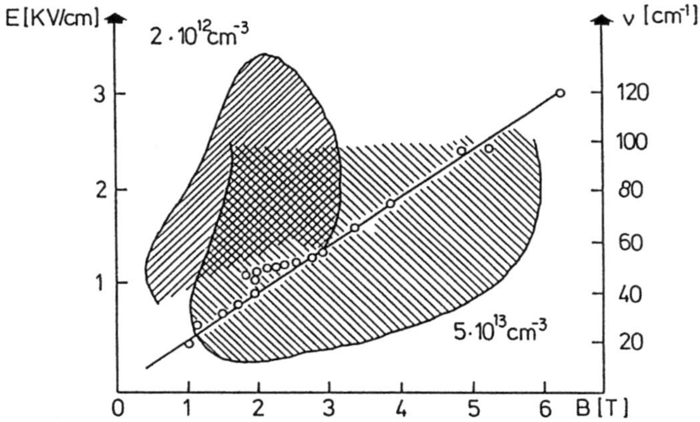

Fig. 8: Electric and magnetic field range of light hole cyclotron laser action (shaded areas) for two different sample materials. The tuning range of the single mode laser as a function of magnetic field is shown by the open circles. The solid line indicates an effective mass of $m^* = 0.046\ m_0$ after Ref. 38

A rather wide range from 20 cm^{-1} to 100 cm^{-1} is found for a material with a hole concentration of 5 x $10^{13}cm^{-3}$, which seems to be the upper limit of concentration where the cyclotron mode works. For a doping of 2 x $10^{12}cm^3$ lasing is only observed between 20 and 60 cm^{-1}.

The explanation of the pumping mechanisms is not settled yet. In the low magnetic field range between 1 and 2 T the same pumping process as discussed in Fig. 2 is most likely responsible for the inversion within the light hole Landau levels as is evident from Fig. 1b. A theoretical explanation has been given by KOZLOV et al. /40/ and D'YAKONOV and PEREL /41/, who suggest a tunneling transition between the heavy and light hole band near $v_{l,h} \approx 0$ to be responsible for the built–up of the light hole inversion.

The lasing process in the higher magnetic field region above 4 T has to be explained in a different way since the streaming condition for heavy holes is not fulfilled any more. Also the heavy holes are accumulated. In this situation a mechanism involving intersubband hole transitions and an interaction of light and heavy hole Landau levels in regions of crossing is suggested /42/. As a result a hybridization of light and heavy hole wave functions takes place leading to an inversion within the light hole Landau levels.

The total gain for transitions between Landau levels is in the order of 0.3 cm^{-1} /36/, due to the rather small inversion of the light hole population within the Landau levels. The matrix elements for inter–Landau level transitions are known to be quite large. Limiting factors of the lasing mode are on the one hand the impurity limited spontaneous linewidth of cyclotron resonance transitions and on the other hand the nonradiative transition by phonons and an inter–Landau–level Auger–effect /31/. Both limitations are avoided in high purity samples.

Acknowledgement: The work is partly supported by the Jubiläumsfonds der Österreichischen Nationalbank, Vienna, and the European Research Office, London.

Literature

1. H. Krömer: Phys. Rev. 109, 1856 (1958)
2. B. Lax: In Quantum Electronics, (Columbia University Press 1960) p. 428
3. P.A. Wolff: Physics 1, 147 (1964)
4. E. Gornik: Phys. Rev. Lett. 29, 595 (1972)
5. K.L. Kobayashi, K.F. Komatsubara, E. Otsuka: Phys. Rev. Lett. 30, 702 (1973)
6. Yu.L. Ivanov, Yu.V. Vasiljev: Sov. Tech. Phys. Lett. 9, 264 (1983)
7. A.A. Andronov, I.V. Zverev, V.A. Kozlov, Yu.N. Nozdrin, S.A. Pavlov, V.N. Shastin: Sov. Phys.–JETP Lett. 40, 804 (1984)
8. A.P. Chebotarev, V.N. Murzin: Sov. Phys.–JETP Lett. 40, 1005 (1984)
9. T. Kurosawa: J. Phys. Soc. Jpn 21, Suppl. 424 (1966)
10. I.I. Vosilyus, I.B. Levinson: Sov. Phys.–JETP 25, 672 (1967)
11. I.I. Vosilyus: Sov. Phys. Solid State 11, 755 (1969)
12. T. Kurosawa, H. Maeda: J. Phys. Soc. Jpn 31, 668 (1971)
13. H. Maeda, T. Kurosawa: Proc. 11th Int. Conf. Phys. Semicond., (Warsaw 1972) p. 602
14. Ya.I. Al'ber, A.A. Andronov, V.A. Valov, V.A. Kozlov, A.M. Lerner, I.P. Ryazantseva: Sov. Phys.–JETP 45, 539 (1977)
15. T. Kurosawa: Solid State Commun. 24, 357 (1977)
16. A.A. Andronov, V.A. Kozlov, L.S. Mazov, V.N. Shastin: Sov. Phys.–JETP Lett. 30, 551 (1979)
17. T. Kurosawa: J. Physique 42, Suppl. 10 C7–377 (1981)
18. A.A. Andronov, E.P. Dodin, Z.F. Krasil'nik: Sov. Phys. Semicond. 16, 133 (1982)
19. A.A. Andronov, Yu.K. Pozela: In Hot Electrons in Semiconductors: Streaming and Anisotropic Distributions in Crossed Fields and Inverted Distributions of Hot Electrons in Semiconductors, both in Russian ed. by Andronov and Pozela, (Gorky 1983)
20. S. Komiyama, T. Kurosawa, T. Masumi: In Topics in Applied Physics, Hot Electron Transport in Semiconductors, Vol. 58, ed. by Reggiani (Springer–Verlag 1985) p. 177
21. A.A. Andronov et al.: Physica 134B, 210 (1985)
22. Yu.K. Pozela, E.V. Starikov, P.N. Shiktorov: Sov. Phys. Semicond. 17, 566 (1983) and Phys. Lett. 96A, 361 (1983)
23. Yu.L. Ivanov: Sov. Phys.–JETP Lett. 34, 515 (1981)
24. S. Komiyama: Phys. Rev. Lett. 48, 271 (1982)
25. V.I. Gavrilenko, V.N. Murzin, S.A. Stoklitskii, V.P. Chebotarev: Sov. Phys.–JETP Lett. 35, 97 (1982)

26. A.A. Andronov, V.I. Gavrilenko, O.F. Grishin, V.N. Murzin, Yu.N. Nozdrin, S.A. Stoklitskii, A.P. Chebotarev, V.N. Shastin: Sov. Phys.–Dokl. 27, 932 (1982)
27. A.A. Andronov, I.V. Zverev, V.A. Kozlov, Yu.N. Nozdrin, S.A. Pavlov, V.N. Shastin: Sov. Phys.–JETP Lett. 40, 804 (1984)
28. S. Komiyama, N. Iizuka, Y. Akasaka: Appl. Phys. Lett. 47, 958 (1985)
29. M. Helm, K. Unterrainer, E. Gornik, E.E. Haller: Solid State Electron. 31, 759 (1988)
30. S. Komiyama: Adv. Phys. 31, 255 (1982)
31. E. Gornik: Physica 127B, 95 (1984)
32. S. Komiyama, S. Kuroda: Solid State Coomun. 59, 167 (1986)
33. S. Komiyama: In Proc. of the 18th Int. Conf. on the Physics of Semiconductors (World Scientific, Singapore 1987), p. 1641
34. A. Andronov, A. Muravjev, I. Nefedov, Y. Nozdrin, S. Pavlov, V. Shastin, Y. Mityagin, V. Murzin, S.A. Stoklitskii, I. Trofimov, A. Chebotarev: In Proc. of the 18th Int. Conf. on the Physics of Semiconductors (World Scientific, Singapore 1987), p. 1663.
35. K. Unterrainer, M. Helm, E. Gornik, E.E. Haller, J. Leotin: Appl. Phys. Lett. 52, 564 (1988)
36. Yu.B. Vasilyev, Yu.L. Ivanov: In Proc. of the 18th Int. Conf. on the Physics of Semiconductors (World Scientific, Singapore 1987), p. 1659
37. Yu.L. Ivanov (private communications)
38. I.M. Melnitczuk, Yu.A. Mityagin, V.N. Murzin, S.A. Stoklitskii, I.E. Trofimov: Kratkie Soobshcheniya po Fizike 12. 3 (1987), in Russian
39. Yu.A. Mityagin, V.N. Murzin, S.A. Stoklitskii, and I.E. Trofimov: JETP Lett. 46, 144 (1987).
40. A. Kozlov, L.S. Mazow, I.M. Nefedov, M.R. Zabolotskikh: Sov. Phys.–JETP Lett. 37, 170 (1983)
41. I. D'yakonov, V.I. Perel: Sov. Phys.–JETP 65, 200 (1987)
42. S.A. Stoklitskii: to be published in Russian

Part V

Tunneling in Heterostructures

Resonant Tunnelling Devices in a Quantising Magnetic Field

L. Eaves [1], E.S. Alves [1], M. Henini [1], O.H. Hughes [1], M.L. Leadbeater [1], C.A. Payling [1], F.W. Sheard [1], G.A. Toombs [1], A. Celeste [2], J.C. Portal [2], G. Hill [3], and M.A. Pate [3]

[1] Department of Physics, University of Nottingham, Nottingham NG7 2RD, UK
[2] SNCI-CNRS, F-38042 Grenoble and LPS, INSA, F-31077 Toulouse, France
[3] Department of Electronic Engineering, University of Sheffield, Sheffield S1 3JD, UK

1. INTRODUCTION AND ABSTRACT

High magnetic fields, applied either parallel or perpendicular to the barriers, are extremely useful for investigating the electrical properties of resonant tunnelling structures [1-11]. By considering the current-voltage characteristics of a variety of double barrier structures (DBS) based on n-(AlGa)As/GaAs with differing well and barrier thicknesses, we show how quantising magnetic fields can be used to study tunnelling processes in which transverse momentum is conserved; the charge distribution in the device, particularly the build-up of electronic charge in the quantum well at resonance; intrinsic bistability; non-resonant elastic and inelastic scattering processes which do not conserve transverse momentum; ballistic transport involving electrons tunnelling into hybrid magneto-electric states of wide quantum wells.

2. RESONANT TUNNELLING IN A QUANTISING FIELD $\underline{B}||\underline{J}$

An idealised symmetric DBS under bias V is shown schematically in Fig. 1. We consider first the case when the emitter and collector electrodes are heavily doped n^+ GaAs layers with Fermi energy E_F. The GaAs quantum well (width w) is an undoped layer of the same material and supports a quasibound state at energy E_r. For simplicity, we represent the screening charges in n^+ regions adjacent to the (AlGa)As barriers (each of width b) and the charge, Q, due to electrons occupying the well, as sheets of infinitesimal thickness. If the quasibound level is very narrow, conservation of energy and transverse momentum means that an emitter electron can only tunnel into the well if its longitudinal kinetic energy $\hbar^2 k_0^2/2m^*$ is equal to

$$\hbar^2 k_0^2/2m^* = E_r + e\phi - eV, \tag{1}$$

as shown in Fig. 1. We take $E_r > E_F$, so that at low temperatures when the Fermi surface is sharp, a finite bias is required before emitter states with $k_x^2 = k_0^2$ are occupied and resonant tunnelling into the well can occur. Tunnelling out of the well then takes place into empty states in the collector contact. The onset of resonant tunnelling corresponds to $k_0^2 = k_F^2$ and occurs at a voltage of $V_{th} = 2(E_r - E_F)/e$. For voltages up to threshold, the electronic charge Q in the well is zero. However, at higher bias voltages, since $\phi = V/2 + Q/2C$, it can be shown [10] that

Springer Series in Solid-State Sciences, Vol. 87
High Magnetic Fields in Semiconductor Physics II Editor: G. Landwehr

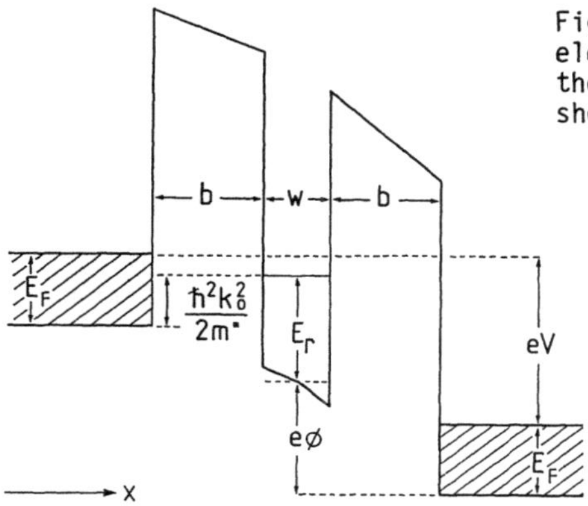

Fig. 1 (above) Spatial variation of electron potential energy through the double-barrier structure, showing bound state and Fermi levels

$$Q = \frac{Q_m e}{2E_F} \frac{V - V_{th}}{1 + \alpha} , \qquad (2)$$

where $Q_m = ek_F^2 T_e/2\pi(T_e + T_c)$, $\alpha = Q_m e/2CE_F$, $C = \varepsilon_r\varepsilon_0/(b + \frac{1}{2}w + \lambda)$, λ is the screening length in the emitter and collector (assumed zero in Fig. 1), and T_e and T_c are the transmission coefficients of the emitter and collector barriers.

For $\underline{B}$ applied perpendicular to the barriers ($\underline{B}||\underline{J}$), the transverse kinetic energy of the electrons is quantised into Landau levels. Since resonant tunnelling involves electrons with transverse kinetic energy up to $E_F - (\hbar^2 k_0^2/2m^*)$, the magneto-oscillations occur when [10]

$$E_F - \hbar^2 k_0^2/2m^* = (n + \tfrac{1}{2})\hbar\omega_c \quad n = (0,1,2,\ldots), \qquad (3)$$

where $\omega_c = eB/m^*$. Using Eqs. (1) and (2) this may be rewritten

$$V - V_{th} = [2\hbar(1 + \alpha)B/m^*](n + \tfrac{1}{2}). \qquad (4)$$

Expressing the periodicity of the magneto-oscillations in terms of a fundamental field $B_{fr} = [\Delta(1/B)]^{-1}$ we obtain

$$B_{fr} = \frac{m^*}{2\hbar} \frac{V - V_{th}}{1 + (L/a_0)T_e/(T_e + T_c)} , \qquad (5)$$

where $L = 2b + w + 2\lambda$ and $a_0 = 10$ nm is the Bohr radius for GaAs. Note that $B_{fr} = 0$ at the threshold voltage.

Let us consider how this relation compares with the observed magneto-oscillations of a DBS (Structure 1) comprising the following layers in order of growth from the n^+ substrate, doped at 2×10^{18} cm^{-3}: (i) 2×10^{17} cm^{-3} n^+GaAs buffer layer, 1 μm thick; (ii) undoped (AlGa)As barrier, [Al] = 0.33, 8.8 nm; (iii) undoped GaAs well, 5.6 nm; (iv) (AlGa)As barrier, [Al] = 0.33, 8.8 nm; (v) 2×10^{17} cm^{-3}, n^+GaAs, 1 μm top layer. Figure 2a shows the I(V) characteristics (B = 0) for a 200 μm diameter mesa. There is a threshold around 70 mV. In the region of negative differential conductivity beyond the peak at 170 mV the current breaks into high frequency oscillations [9]. Typical magneto-oscillations at a fixed bias of 170 mV are shown in the inset of Fig. 2a. The plot of B_f versus V is shown in Fig. 2b. At low voltage ($\lesssim$ 100 mV) the magneto-oscillations are effectively independent of V. Here the small current is non-resonant and the magneto-oscillations correspond to Landau levels passing through the Fermi energy of the n^+ emitter contact, when $(n + \frac{1}{2})\hbar\omega_c = E_F$. This gives $B_f = E_F m^*/e\hbar = 8$ T for a doping level of 2×10^{17} cm^{-3}, as is observed. Above threshold, the resonant current grows rapidly with the appearance of a new series of oscillations whose B_f increases linearly with V, in agreement with Eq. 5. For structure 1, $\lambda \simeq 8$ nm and $T_e \lesssim T_c$. This gives a rate of change of B_f with V of ~0.11 T mV^{-1}, in good agreement with experiment (0.1 T mV^{-1}).

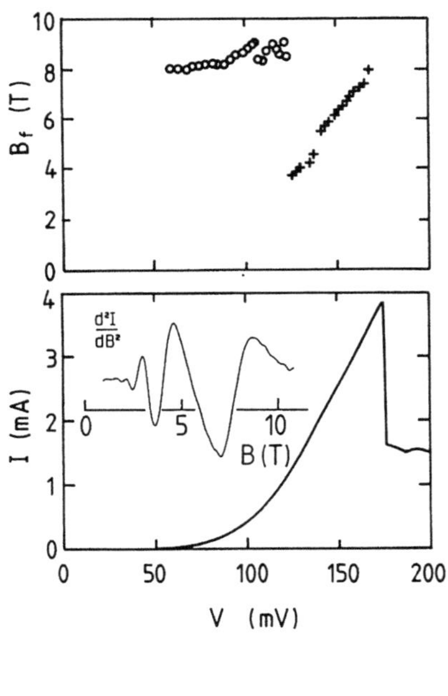

Fig. 2 (a) The I(V) curve of structure 1 in the region of the first resonance. The inset shows the magneto-oscillations in I at V = 170 mV (b) the fundamental field $B_f = [\Delta(1/B)]^{-1}$ versus V

3. INTRINSIC BISTABILITY

The charge build-up in the well at resonance can lead to intrinsic bistability in the I(V) characteristics of DBS structures due to electrostatic feedback [10-13]. In order to enhance the charge build-up and hence the voltage range of the bistability, we have prepared and investigated a structure (structure 2) in which the two barriers have different widths [14]. It comprises the following layers, in order of growth from the n^+ substrate: (i) 2 µm of 2 x 10^{18} cm^{-3} GaAs; (ii) 50 nm of 1 x 10^{17} cm^{-3} GaAs; (iii) 50 nm of 1 x 10^{16} cm^{-3} GaAs; (iv) 3.3 nm of undoped GaAs; (v) 8.3 nm of undoped (AlGa)As, [Al] = 0.4; (vi) 5.8 nm of undoped GaAs; (vii) 11.1 nm of undoped (AlGa)As, [Al] = 0.4;(viii) 3.3 nm of undoped GaAs; (ix) 50 nm of 1 x 10^{16} cm^{-3} GaAs; (x) 50 nm of 1 x 10^{17} cm^{-3} GaAs; and (xi) 0.5 µm of 2 x 10^{18} cm^{-3} GaAs top contact.

The I(V) characteristics of this device (200 µm diameter mesa) at 4 K are shown in Fig. 3. There is a pronounced difference between forward and reverse bias (forward bias = substrate positive, thick emitter barrier, thin collector barrier). In reverse bias, shown in Fig. 4, electrons are injected from the emitter into the well through the thinner barrier. The small tunnelling probability through the thicker collector barrier retains the charge in the well and enhances the electrostatic feedback. We believe that Fig. 3 provides an unambiguous observation of intrinsic bistability (see refs. 12, 13 concerning the controversy over previous

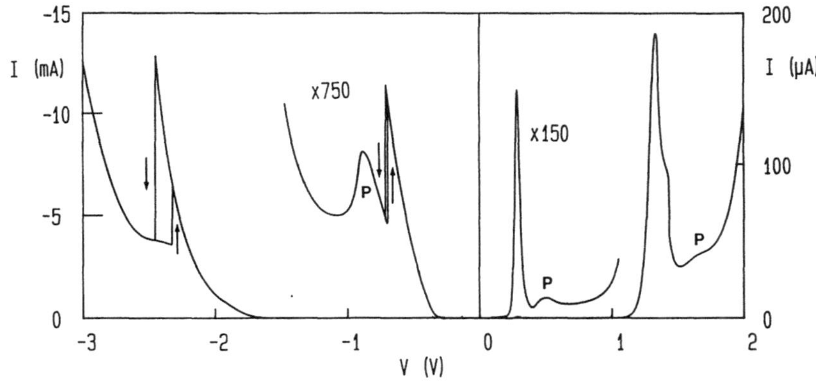

Fig. 3 The current-voltage characteristics at 4 K of a 200 µm diameter mesa fabricated from the asymmetric double barrier structure 2 described in the text. The LO phonon-assisted peaks are marked P. Note the very strong LO phonon-assisted peak at -0.85 V

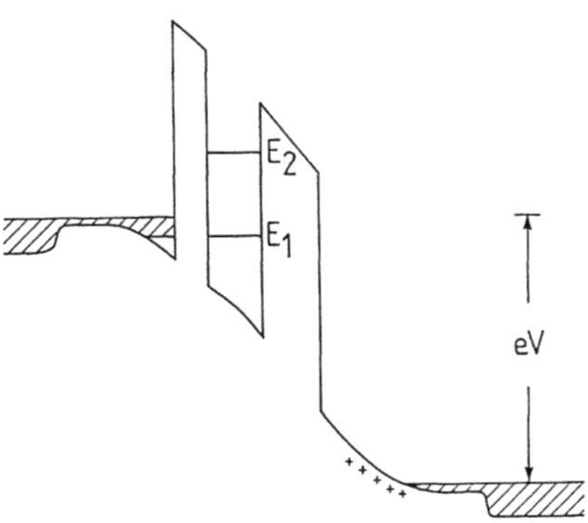

Fig. 4 (above) Spatial variation of electron potential through the asymmetric double barrier structure 2 in reverse bias

reports). Due to the light doping in the contact layers and the large threshold voltage, a two-dimensional electron gas (2DEG) is formed in the emitter accumulation layer. Resonant tunnelling occurs only when the applied voltage brings the bound state energy of this 2DEG into resonance with the bound state of the quantum well. The light doping also gives rise to a significant voltage drop across the collector depletion layer. This depletion layer voltage magnifies the electrostatic feedback effect. The sheet density of the 2DEGs in the emitter accumulation layer (n_e) and the quantum well (n_w) can be determined from the magneto-oscillations in the tunnel current and capacitance, shown in Fig. 5 for various fixed bias voltages. The origin of the high frequency series is discussed in section 4. In forward bias, there is a single low frequency series which gives $n_e = 2eB_{fe}/h$ where $B_f^{-1} = \Delta(1/B)$. n_e (forward) increases steadily with V since little charge builds up in the well, even at resonance. In reverse bias, Fourier transformation of the oscillations reveals two series corresponding to n_e and n_w. In reverse bias, over the voltage range of the first resonance (-0.325 to -0.7 V), n_e remains constant and n_w increases steadily so that $n_w \simeq n_e$ at the peak in current at -0.7 V.

This corresponds to $n_w = 2 \times 10^{11}$ cm^{-2}, significantly higher than in previous studies [2,8,12]. The large charge build-up in the well at resonance in reverse bias is fully consistent with our observation of the intrinsic bistability effect.

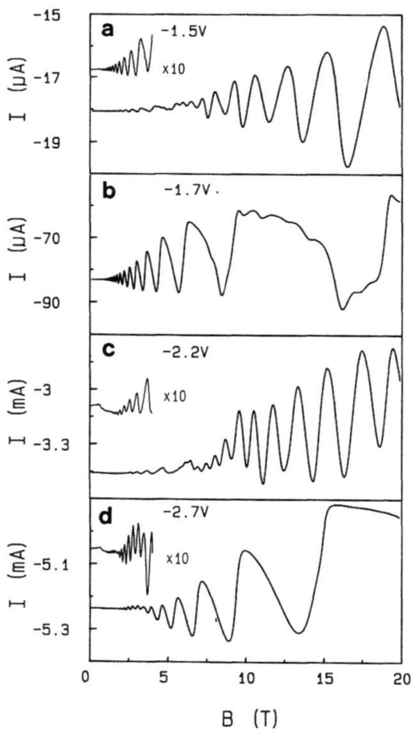

Fig. 5 (right) Magneto-oscillations in the tunnel current of structure 2 at various fixed reverse biases

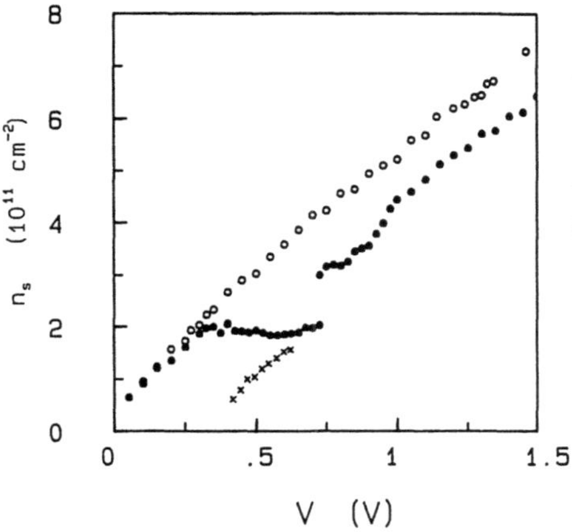

Fig. 6 Variation of the 2DEG sheet density n versus applied voltage V for structure 3.
n_e (Forward bias = o)
n_e (reverse bias = ●)
n_w (reverse bias = x)
The data were obtained from magneto-capacitance measurements for B||J

4. TUNNELLING ASSISTED BY ELASTIC AND INELASTIC SCATTERING

It can be seen from Fig. 3 that there are subsidiary maxima, marked P, beyond the main resonant peak in the I(V) curves of structure 2. This feature is due to tunnelling of electrons, assisted by LO phonon emission, from the emitter contact into the quantum well [3,6,11]. If an electron traverses the barriers without scattering, which is assumed to be the case for the majority of electrons contributing to the main resonant peak, the k-vector component, $k_\perp$, perpendicular to the tunnelling direction is conserved. The emission of an LO phonon or an elastic scattering process violates this conservation rule. The LO phonon-related peak and the contribution to the valley current due to elastic scattering can be revealed more clearly by applying a magnetic field B||J. This quantises the energy of motion in the plane of the barriers so that the energies of electrons in the accumulation layer of the emitter and in the quantum well are given by

Accumulation layer: $E = E_0 + (n + \frac{1}{2})\hbar\omega_c$

Quantum well: $E = E_1 + (n' + \frac{1}{2})\hbar\omega_c$

where n and n' are the Landau level quantum numbers, and E_0 and E_1 are the lowest bound state energies in the emitter and well respectively. The conservation of $k_\perp$ for B = 0 corresponds to the requirement that $p = n' - n = 0$ in the presence of the magnetic field, so that in both cases resonant tunnelling occurs at an applied voltage for which $E_0 = E_1$.

Figure 7 shows the effect of a quantising magnetic field B||J on the I(V) characteristics of structure 3, which comprises the following layers in order of growth from the n^+ substrate: (i) 2 μm of 2 x 10^{18} cm^{-3} GaAs buffer layer; (ii) 50 nm of 2 x 10^{16} cm^{-3} GaAs; (iii) 2.5 nm of undoped GaAs; (iv) 5.6 nm of undoped (AlGa)As, [Al] = 0.4; (v) 11.7 nm of undoped GaAs; (vi) 5.6 nm of undoped (AlGa)As, [Al] = 0.4; (vii) 2.5 nm of undoped GaAs; (viii) 50 nm of 2 x 10^{16} cm^{-3} GaAs; and (ix) 0.5 μm of 2 x 10^{18} cm^{-3} GaAs top contact layer.

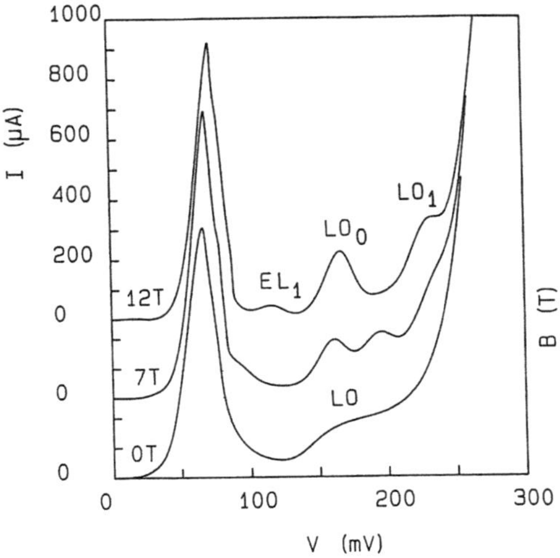

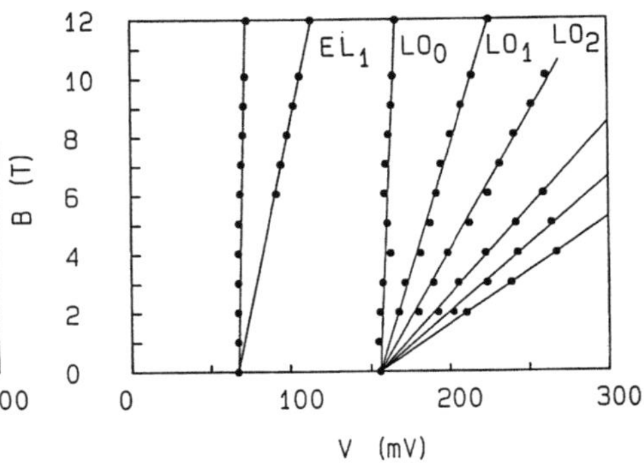

Fig. 7 (left) The I(V) characteristics around the first resonance at 4 K and at various magnetic fields $\underline{B}||\underline{J}$ for a 100 μm diameter mesa fabricated from structure 3

Fig. 8 (right) Fan chart showing the magnetic field dependence ($\underline{B}||\underline{J}$) of the peaks in I(V). The elastic and inelastic (LO phonon) scattering processes giving rise to the oscillations are discussed in the text

The following features in Fig. 7 are noteworthy. Firstly, the magnetic field increases the amplitude of the main LO phonon-related peak. Secondly, a weak secondary peak (EL_1) emerges from the main resonant peak in the I(V) curve as the magnetic field is increased. Similar subsidiary peaks also evolve from the LO phonon feature with increasing B. Thirdly, the magnetic field increases the peak/valley ratio. The fan chart in Fig. 8 shows the evolution of the magnetoquantum peaks in the I(V) characteristics. Transitions for which $k_\perp$ (or n) is not conserved are governed only by energy conservation. Therefore, for elastic or LO phonon emission we have

$$E_0 = E_1 + \frac{p\hbar eB}{m^*} \text{ or } E_0 = E_1 + \frac{p\hbar eB}{m^*} + \hbar\omega_L \;. \qquad (6)$$

The chart shows two clearly identifiable groups of lines. In the limit B = 0, peak EL_1 extrapolates back to the main resonant peak at 70 mV. The peaks LO_p extrapolate back to the LO phonon satellite at 160 meV. The main resonant peak, which shifts only very slightly with B, corresponds to tunnelling processes for which n is conserved. The peak marked EL_1 corresponds to non-resonant tunnelling involving an elastic (or quasi-elastic) scattering-induced transition from the n = 0 Landau level in the emitter contact to the n = 1 Landau level in the well. Such a transition could be caused by scattering due to ionised impurities or interface roughness (elastic) or by acoustic phonon emission (quasi-elastic). At sufficiently high magnetic fields (> 10 T) only the n=0 Landau level of the emitter accumulation layer is occupied. The peaks marked LO_p then correspond to transitions of electrons from the n = 0 Landau level in the emitter to the pth Landau level in the well, together with the emission of an LO phonon. Similar transitions, involving LO phonon emission and a change in Landau quantum number, were reported recently [6] in triple barrier devices.

It is interesting to note that the effect of the magnetic field is to suppress the (quasi-) elastic scattering induced-transitions at certain voltages below and above EL_1 and to enhance them at other voltages, corresponding to the peak EL_1. In zero magnetic field such processes are allowed energetically for all voltages beyond the main resonant peak in the tunnel current. At these voltages the lowest energy bound state in the well is below the energy of the electrons in the emitter. However, a large magnetic field quantises the electronic motion in the plane of the barriers, thus giving rise to sharp peaks in the densities of states. Therefore, energy conservation allows elastic scattering into the well only at certain voltages (EL_1) and inhibits it elsewhere. This explains the enhancement of the peak/valley ratio with increasing B (see Fig. 7).

Figure 9 shows, in more detail, the forward bias I(V) characteristics of the asymmetric double barrier device, structure 2, at B = 0 and in the presence of a longitudinal magnetic field. At 0 T the phonon satellite peak has a "flat top" structure. Application of a magnetic field clearly resolves two phonon peaks at voltages (470 and 530 mV) which do not shift with increasing B. They must arise from optical phonon emission with no change in Landau level number. We attribute the two components to the two LO phonon modes in (AlGa)As. Structure due to the elastic scattering process (EL_1) and LO phonon emission accompanied by a change of Landau quantum number is also observed, but is complicated by the overlapping of two different phonon series. The differences in voltage between the phonon satellites and the main resonant peak in Figs. 7 and 9 are considerably larger than the LO phonon energy, $\hbar\omega_L$. This is because only a fraction (~30%) of the total applied voltage V is dropped between the

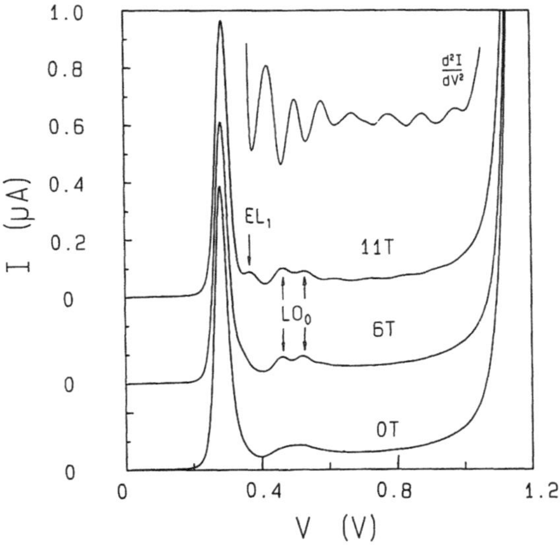

Fig. 9 The I(V) characteristic of structure 2 around the first resonance in fixed bias at 4 K and at various magnetic fields B||J for a 200 μm diameter mesa. The top curve is d^2I/dV^2 at 11 T, revealing additional Landau structure.

emitter contact and the quantum well, with a large portion of the voltage being dropped across the collector barrier and depletion layer in the collector contact. The potential distribution across the device varies with applied bias. Estimates of the potential difference V_1 between the emitter contact and the well can be obtained over a range of applied voltage V by noting that the separations between the main resonance and EL_1 and between the peaks LO_p and LO_{p+1} corresponds to $eV_1 = \hbar\omega_c$. Using this procedure, we can estimate the energy of the LO phonons involved in the phonon-assisted tunnelling process. For structure 3 we obtain $\hbar\omega_L$ = 35 meV, in good agreement with the LO phonon energy of GaAs. For structure 2, the two phonon peaks correspond to energies of $\hbar\omega_{L1}$ = 35.5 meV and $\hbar\omega_{L2}$ = 48 meV. These values are in good agreement with the energies of the GaAs-like and AlAs-like LO phonon modes of (AlGa)As with Al = 0.4 [15]. A possible explanation for the observation of two phonon modes in structure 2 but only one in structure 3, is that structure 2 has a narrower well so the wavefunction of the bound state in the well penetrates further into the (AlGa)As barrier. Tunnelling electrons could then couple more effectively to the AlAs-like mode.

An additional type of LO phonon-related magneto-oscillations occurs when electrons resonantly tunnel into an upper subband of the quantum well (e.g. E_2 in Fig. 4). In this case strong magneto-oscillations with a large B_f (~70-100 T) can occur by LO-phonon assisted transitions of the resonantly tunnelling electron out of E_2 into the higher Landau levels of E_1 according to the relation [9]

$$E_0 \simeq E_2 = E_1 + n\hbar\omega_c + \hbar\omega_L,\ n = \text{integer}. \tag{7}$$

For the asymmetric device (structure 2) these resonances correspond to the strong oscillations observed above ~7 T at V = -2.2 V (Fig. 5c), close to the peak of the second reverse bias resonance in Fig. 3.

5. TUNNELLING INTO HYBRID MAGNETO-ELECTRIC ORBITS IN WIDE QUANTUM WELLS WITH $\underline{B}\perp\underline{J}$

Figure 10 shows the conductance G at 4.2 K and B = 0 for a DBS (structure 4) with a 120 nm quantum well. It comprises the following layers, in order of growth from the n^+ substrate: (i) 2 µm, 2 x 10^{18} cm^{-3} n^+GaAs buffer layer; (ii) 50 nm, 2 x 10^{16} cm^{-3} GaAs; (iii) 2.5 nm undoped GaAs; (iv) 5.6 nm undoped (AlGa)As, [Al] = 0.4; (v) 120 nm undoped GaAs; (vi) 5.6 nm undoped (AlGa)As, [Al] = 0.4; (vii) 2.5 nm undoped GaAs; (viii) 50 nm, 2 x 10^{16} cm^{-3} GaAs; (ix) 0.5 µm, 2 x 10^{18} cm^{-3} n^+GaAs top contact.

In Fig. 10, the beating of the oscillations in G at high voltage arises from the interference of electron waves scattered from the two interfaces of the collector barrier. The effect of a magnetic field is to modify the motion of the electrons in the well, as shown schematically in Fig. 11. Note that the electrons are accelerated by a large, uniform electric field $\underline{E}$ in the well.

The electrons can tunnel through the emitter barrier into two distinct types of state: (a) those associated with "skipping" orbits which interact with the emitter barrier only - these correspond to cycloidal motion under the action of crossed $\underline{E}$ and $\underline{B}$; (b) "traversing" states in which the electron wave is repeatedly reflected off both barriers. (The two types

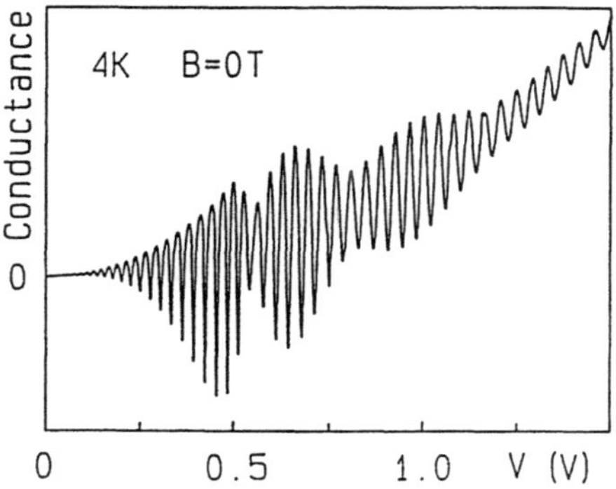

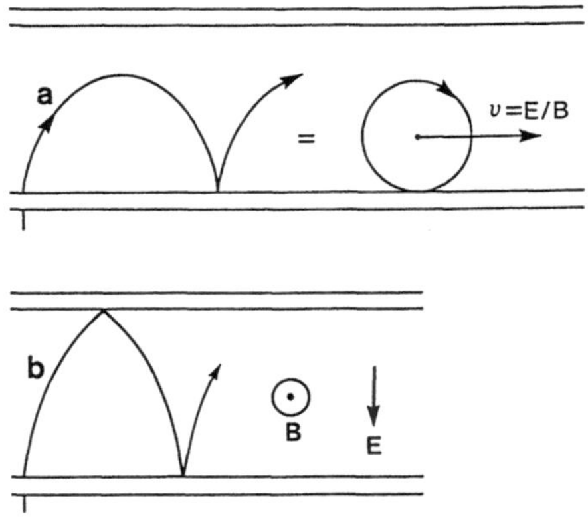

Fig. 10 (left) The differential conductance G = dI/dV for 100 μm mesa of structure 4, B = 0, T = 4 K.

Fig. 11 (right) Schematic diagram of electrons tunnelling into magneto-electric orbits in crossed fields.

are also called "magneto-electric" states). The transition between the two types of state has recently been related to the quenching of the Hall effect in small structures [16].

Electrons tunnel into these states from the Fermi sea of the emitter accumulation layer 2DEG with conservation of energy and $k_\perp$. At low temperatures, the sharp cut-off in the emitter distribution function at $k_\perp = \pm k_F$ means that oscillatory structure appears in the conductance as V or B is varied [17]. The resonance condition can be obtained by treating states (a) and (b) with the WKB approximation [11,17]. The values of $k_\perp = \pm k_F$ give rise to two distinct series of oscillations from each type of magneto-electric state. However, the different tunnelling probability for $\pm k_F$ in the presence of a magnetic field means that only one $k_\perp$ value is observed for certain values of V and B. Typical magneto-oscillations arising from tunnelling into magneto-electric orbits are shown in Fig. 12.

Above 6 T, tunnelling occurs into type (a) orbits. The two series of oscillations (0-2 T and 2-5 T) correspond to type (b) with $k_\perp = \pm k_F$. Note that at high B($\gtrsim$15 T) the tunnel current is quenched as the large Lorentz force deflects electrons back into the emitter. The fan chart in Figure 13 shows the evolution of the type a and b orbits for structure 5. As B → 0, the b-orbit resonances evolve into the electric subband resonance of the quantum well at finite V, whereas the a-orbit resonances tend to V → 0 as B → 0.

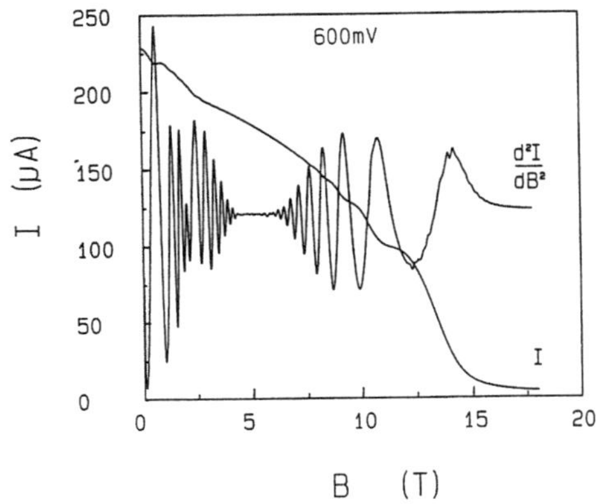

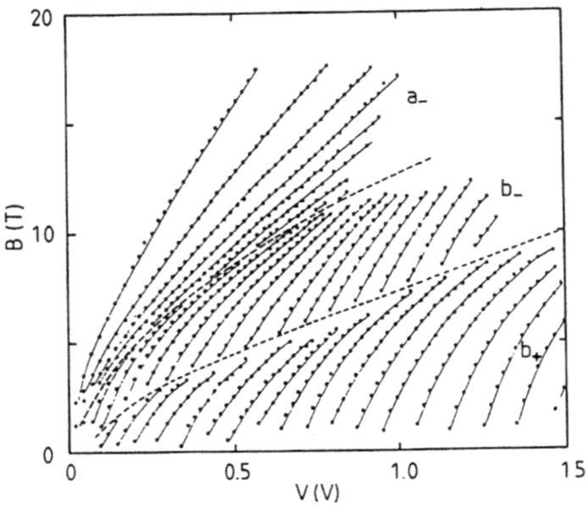

Fig. 12 (left) Plots of I(B) and d^2I/dB^2 at V = 600 mV, T = 4 K for a 100 μm diameter mesa of structure 4, well thickness 120 nm, showing oscillatory structure due to tunnelling into magneto-electric states

Fig. 13 (right) Fan chart showing the positions of the maxima in the current in B-V space for structure 5 (similar to structure 4 except that the well thickness is 60 nm). The dashed lines define the regions corresponding to different types of orbit. The solid lines are guides

REFERENCES

1. E. E. Mendez, L. Esaki and W. I. Wang: Phys. Rev. B33, 2893 (1986)
2. V. J. Goldman et al: Phys. Rev. B35, 9387 (1987)
3. V. J. Goldman et al: Phys. Rev. B36, 7635 (1987)
4. E. E. Mendez: Int. School of Solid State Devices Research, NATO ASI Series B, 170, Erice, Sicily (1987)
5. V. J. Goldman et al: J. Physique 48, C5-463 (1987)
6. H. Bando et al: Proc. 18th Int. Conf. on Low Temperature Physics, Kyoto 1987. Jap. J. Appl. Phys. 26 (Supplement 26-3) (1987)
7. C. A. Payling et al: J. Physique 48, C5-289 (1987)
8. C. A. Payling et al: Surf. Sci. 196, 404 (1988)
9. L. Eaves et al: Appl. Phys. Lett. 52, 212 (1988); cf. G. A. Toombs et al: GaAs and Related Compounds, Inst. Phys. Conf. Ser. 91, 581 (1987)
10. F. W. Sheard and G. A. Toombs: Appl. Phys. Lett. 52, 1228 (1988)
11. L. Eaves et al: Physics and Technology of Sub-Micron Structures, to be published in Springer-Verlag Solid State Series (1988)
12. V. J. Goldman et al: Phys. Rev. Lett. 58, 1256, ibid. 1623 (1987)
13. T. C. L. G. Sollner: Phys. Rev. Lett. 59, 1622 (1987)
14. E. S. Alves et al: Electronics Lett. (in press)
15. O. K. Kim and W. A. Spitzer: J. Appl. Phys. 50, 4362 (1979)
16. C. W. J. Beenakker and H. van Houten: Phys. Rev. Lett. 60, 2406 (1988)
17. B. R. Snell et al: Phys. Rev. Lett. 59, 2806 (1987)

A Wigner Function Study of Magnetotunneling

*N.C. Kluksdahl, A.M. Kriman, and D.K. Ferry**

Center for Solid State Electronics Research,
Arizona State University, Tempe, AZ 85287, USA

Abstract The Wigner function formalism for quantum transport is used to study the magnetoconductance of single-barrier tunneling structures. We use a Gaussian wave packet to numerically simulate a single electron in two spatial dimensions, and study the skipping orbits of the electron wave packet resulting from the combined interactions of the magnetic field and the tunneling barrier. We find that transmission through a barrier leads to a packet that is split into different skipping orbits.

With the increasing use of high magnetic fields in the experimental study of quantum semiconductor systems (GOLDMAN et. al. [1], VON KLITZING et. al. [2], AHARONOV et. al [3], LEADBEATER et. al. [4], CHAN et. al. [5]), it is important that a quantum mechanically correct modeling technique be used to support the experimental work, and to gain an understanding of the physics which underlies the observed phenomena.

Recently, the Wigner function formalism has been used to model quantum electronic transport in semiconductor tunneling structures (RAVAIOLI et. al [6], KLUKSDAHL et. al. [7], FRENSLEY [8]). The Wigner function formalism is a phase-space representation; although real, it differs from a true phase-space distribution in that it is not strictly positive (WIGNER [9]). Nevertheless, the Wigner function has many attractive features and has been widely studied (MOYAL [10], LEVINSON [11], IAFRATE et. al [12], BARKER [11]). The Wigner function and its equation of motion are fully quantum mechanical. While the phase-space representation is not strictly positive, the integral over either of the conjugate coordinates yields a true probability distribution.

The Wigner function is the Wigner-Weyl transform of the density matrix. For a pure state, the Wigner function is

$$f_w(x,p) = \frac{1}{2\pi\hbar} \int dy\, e^{ipy/\hbar}\, \psi^*(x + \frac{y}{2})\, \psi(x - \frac{y}{2}). \tag{1}$$

The equation of motion for the Wigner function is found by taking the Wigner-Weyl transform of the quantum Liouville equation. In the absence of a magnetic field, the equation of motion is

$$\frac{\partial}{\partial t} f_w(x,p,t) + \frac{p}{m}\frac{\partial}{\partial x} f_w(x,p,t) + \Theta\cdot f_w(x,p,t) = 0, \tag{2}$$

where

* This work supported by the U.S. Office of Naval Research.

$$\Theta \cdot f_w(x,p,t) = \frac{1}{2\pi\hbar} \int dP \int dy\ \sin(2Py)\ [V(x+\tfrac{y}{2})-V(x-\tfrac{y}{2})]\ f_w(x,p+P,t). \quad (3)$$

Nonclassical effects arise from the nonlocality of this potential operator.

For magnetic fields, the Hamiltonian is more complicated, and thus the transformation to an equation of motion is slightly more difficult. Fortunately, a formalism is available (MOYAL [10]) which permits easy transformation. For an arbitrary Hamiltonian H, the equation of motion is

$$\frac{\partial}{\partial t} f_w(x,p,t) = \int d\eta \int d\xi\ S(p,x|\eta,\xi)\ f_w(\eta,\xi,t), \quad (4)$$

where

$$S(p,x|\eta,\xi)=\frac{i}{\hbar} \int d\tau \int d\theta\ [H(\eta+\tfrac{\hbar\theta}{2},\xi-\tfrac{\hbar\tau}{2})-H(\eta-\tfrac{\hbar\theta}{2},\xi+\tfrac{\hbar\tau}{2})]e^{i[\tau(\eta-p)+\theta(\xi-x)]}. \quad (5)$$

We consider the Hamiltonian

$$H(p,x) = \frac{1}{2m} [p + eA(x)]^2 + V(x), \quad (6)$$

where A(x) is the vector potential. The p^2 and potential terms of the Hamiltonian yield the same contributions as (2). The terms linear in A yield the classical electromagnetic correction of (2). The quadratic terms in A are odd in the dummy variables ξ and τ, and vanish in the integral over $d\tau$. After simplifying, we obtain

$$\frac{\partial}{\partial t} f_w(x,p,t) + \frac{1}{m} (p + eA(x)) \cdot \frac{\partial}{\partial p} f_w(x,p,t) = -\Theta\cdot f_w(x,p,t). \quad (7)$$

The quantum effects for a static uniform magnetic field enter strictly through interaction with the quantum nonlocality of the potential.

We simulate (7) by numerically integrating this time-evolution equation, as in earlier work (KLUKSDAHL et. al. [7]). For tunneling studies, the potential V is uniform in the $\hat{y}$ and $\hat{z}$ directions and has an abrupt quantum barrier in the $\hat{x}$ direction. We use the gauge $A = \frac{1}{2}(Bx\hat{y} - By\hat{x})$ for a field along the $\hat{z}$ direction (parallel to the barrier). Thus, we discretize in the 4 dimensions $\hat{x}$, $\hat{y}$, $\hat{p}_x$, and $\hat{p}_y$. The initial wave function is a Gaussian wave packet in position, which has a Wigner representation which is Gaussian in both position and momentum (CARRUTHERS et. al. [14]).

The initial wave packet is given a velocity perpendicular to the quantum barrier. Figure 1a) shows the wave packet, prior to the tunneling event, approaching the barrier. As the system evolves in time, the wave packet accelerates, initially along $\hat{y}$, to follow a cyclotron orbit. This acceleration is reflected in the asymmetry of the wave packet. As the packet begins to tunnel through the quantum barrier, Fig. 1b), it has gained a significant $\hat{y}$ velocity. In Fig. 1c), the portion of the wave packet which tunneled through the barrier has split, in contrast to zero-field tunneling, in which a Gaussian packet leads to a unimodal transmitted packet. Analysis in momentum space shows that the near-barrier peak has velocity almost entirely in the $\hat{y}$ direction, while the other peak has a velocity oblique (about 60 degrees) to the barrier. The reflected pulse is small because the barrier is weak (0.3 eV, 1 nm).

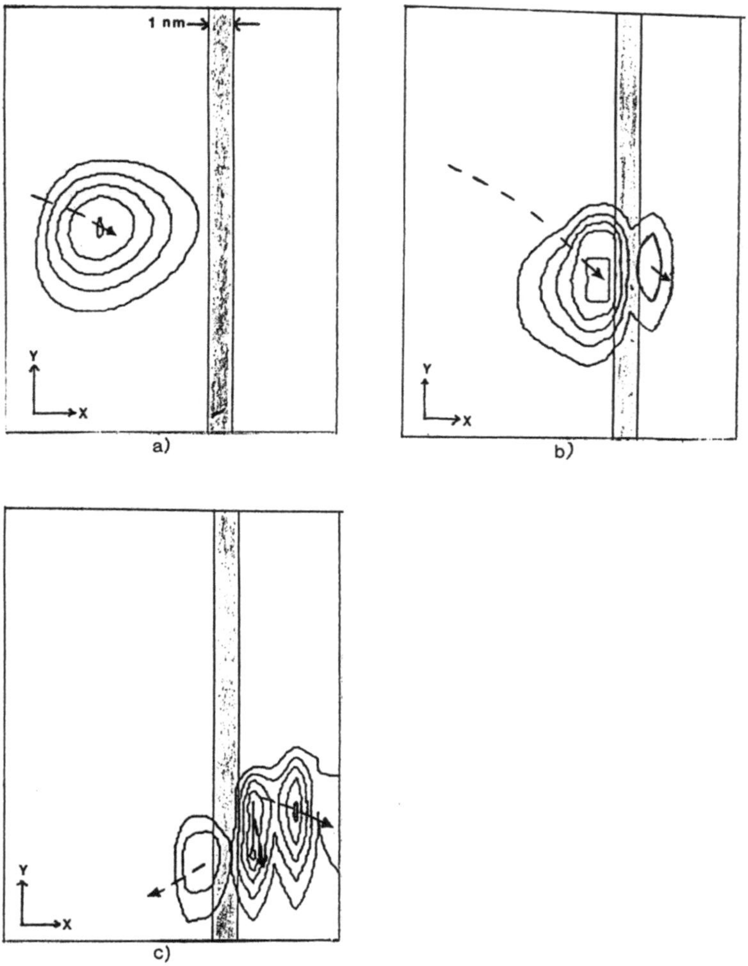

Figure 1. Evolution of a Gaussian wave packet in the presence of a magnetic field. The magnetic field is perpendicular to the plane of the plot, and the wave packet is illustrated with a contour plot. Barrier is indicated by the shaded region. a) the initial wave packet. b) the wave packet starting to tunnel. c) the wave packet has split into 3 components after tunneling.

Both of the transmitted peaks represent distributions of skipping orbits. Skipping orbits arise classically from repeated reflections off a barrier in a magnetic field. Quantum tunneling through the barrier causes the electron to scatter into both shallow and distant skipping orbits (orbits with centers to the left and right of the barrier's right edge). The momentum spread of the electron wave packet leads to a range of skipping orbits.

We have extended the Wigner formalism for modeling quantum electronic transport to include magnetic field effects. The magnetic field enters the equation of motion through classical Boltzmann Equation terms. Early studies indicate that the magnetic field causes electron wave packets to enter skipping orbits during interaction with the quantum barriers. Interaction with the barrier splits the transmitted wave packet into quantized skipping states. The system has no constraint on which skipping state (one with guiding center on each side of the barrier) is excited, therefore the tunneling electron excites both possible orbits. This agrees with experimental work of Eaves *et al.*[4]

1. V.J. Goldman, D.C. Tsui, J.E. Cunningham: Phys. Rev. B 35, 9387 (1987).
2. K. von Klitzing, G. Dorda, M. Pepper: Phys. Rev. Lett. 45, 494 (1980).
3. Y. Aharonov, D. Bohm: Phys. Rev. 115, 485 (1959).
4. M.L. Leadbeater, L. Eaves, P.E. Simmonds, G.A. Toombs, F.W. Sheard, P.A. Claxton, G. Hill, M.A. Pate: Sol. St. Elec. 31, 707 (1988).
5. K.S. Chan, L. Eaves, D.K. Maude, F.W. Sheard, B.R. Snell, G.A. Toombs, E.S. Alves, J.C. Portal, S. Bass: Sol. St. Elec. 31, 711 (1988).
6. U. Ravaioli, M.A. Osman, W. Pötz, N.C. Kluksdahl, D.K. Ferry: Physica 134B, 36 (1985).
7. N.C. Kluksdahl, W. Pötz, U. Ravaioli, D.K. Ferry: Superlattices and Microstructures 3, 41 (1987).
8. W.R. Frensley: Phys. Rev. Lett. 57, (1986) 2853.
9. E. Wigner: Phys. Rev. 40, 749 (1932).
10. J.E. Moyal: Proc. Cambridge Phil. Soc. 45, 99 (1949).
11. I.B. Levinson: Sov. Phys. JETP 30, 362 (1970).
12. G.J. Iafrate, H.L. Grubin, D.K. Ferry: Phys. Lett. 87A, 145 (1982).
13. J.R. Barker: J. Phys. C 6, 2663 (1973).
14. P. Carruthers, F. Zachariason: Rev. Mod. Phys. 55, 245 (1983).

Surface-Field Induced InAs Tunnel Junctions in High Magnetic Fields

*U. Kunze**

Institut für Elektrophysik, Technische Universität Braunschweig, und Hochmagnetfeldanlage der Physikalischen Institute der Technischen Universität Braunschweig, D-3300 Braunschweig, Fed. Rep. of Germany

Tunneling from an electron inversion layer into bulk valence states is studied in Yb/oxide/degenerate p-InAs tunnel junctions. Magneto-tunneling characteristics yield the effective electric field for Zener transitions and demonstrate a striking difference to Esaki diodes.

1. Introduction

In Yb/oxide/InAs tunnel junctions on degenerately doped p-type substrate, the high electric surface field associated with the low work function of the metal electrode influences an electron inversion layer at the InAs/oxide interface. Due to the high doping level the depletion layer is sufficiently thin that Zener tunneling occurs between the localized electron subband and the extended bulk valence states. If the thickness of the oxide barrier is small the top metal provides a low-resistive tunneling contact to the two-dimensional (2D) electrode. As illustrated by Fig. 1, elastic tunneling occurs sequentially /1/. At forward bias V_{pn} between the p-type bulk and the n-type layer equal to the sum of $V_p=(E_v-E_{Fp})/e$ and $V_n=(E_{Fn}-E_0)/e$, the elastic tunneling interval goes to zero and, hence, a negative differential conductance arises as in Esaki diodes. However, the reduced dimensionality of the cathode of the surface-field-induced tunnel (SFIT) junction causes some characteristic features in the current-voltage (I-V) characteristics which are better understood by means of magneto-tunneling experiments.

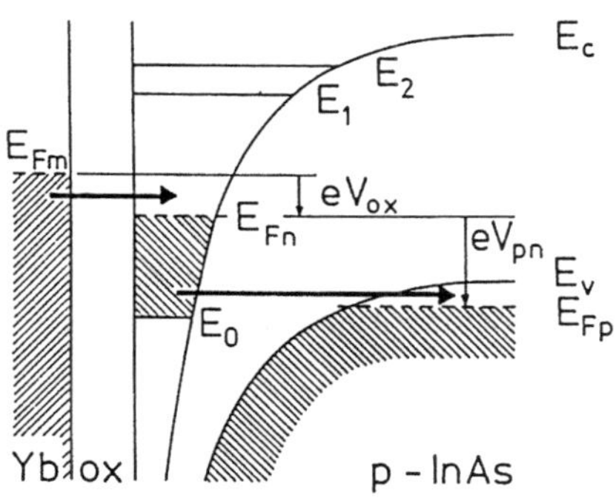

Fig. 1. Schematic energy-band diagram of a SFIT junction under forward bias conditions. Partial voltage drops across the oxide, V_{ox}, and across the depletion layer, V_{pn}, open energy intervals for elastic tunneling transitions as indicated by solid arrows.

*Present address: Institut für Technische Physik, Universität Erlangen, D-8520 Erlangen, Fed. Rep. of Germany

Springer Series in Solid-State Sciences, Vol. 87
High Magnetic Fields in Semiconductor Physics II Editor: G. Landwehr

2. Experimental Results and Discussion

The junctions used for the present experiments were prepared on p-type InAs(001) wafers with hole concentration of 1.9×10^{18} cm^{-3} at T=77 K. Thin oxide barriers were formed by thermal oxidation at 100 °C for 30 min in pure O_2. Subsequently Yb with an Al capping layer was evaporated. Circular dots of 44 μm diameter were defined by standard lithography. Finally, the area surrounding the contacts was insulated with an evaporated Al_2O_3 coating.

Low-temperature (T=4.2 K) measurements of the I-V curves and their second derivative d^2I/dV^2 have been performed with magnetic fields up to B=15 T transverse and longitudinal with respect to the tunneling direction. Typical results for the influence of magnetic field on the tunneling current are shown in the Figs. 2(a) and 2(b). The reduction of the current arises from decreasing the effective electric field in the transverse case of crossed electric and magnetic fields, and from increasing the effective energy gap in the longitudinal case where wave-vector conservation is replaced by quantum-number conservation in tunneling between magnetic quantum levels. The tunneling current in Esaki diodes in the presence of transverse and longitudinal magnetic fields has been calculated by WEILER et al. /2/ using WKB solutions to the two-band model. Assuming that the 2D character of the SFIT cathode does not seriously alter the WKB transmission coefficients we apply the results of /2/

$$I_{trans} \propto \exp[-(\pi E_g^2/2\hbar eF_{eff})(m^*/2E_g)^{1/2}] \qquad (1)$$

$$I_{long} \propto FB \sum_n (E_n/E_g)^4 \exp[-(\pi E_n^2/2\hbar eF)(m^*/2E_g)^{1/2}], \qquad (2)$$

where $F_{eff}=F(1-E_gB^2/2m^*F^2)^{1/2}$, $E_n=[E_g^2+4E_g(n+1/2)\hbar\omega_c]^{1/2}$, and $\hbar\omega_c=eB/m^*$. Using the InAs parameters E_g=418 meV and $m^*=0.025\ m_o$ for tunneling into the light-hole valence band, the only unknown parameter is the built-in electric field F. A fit of (1) and (2) to the experimental peak currents yields $F=(2.65\pm0.08)\times10^7$ V/m for the transverse and $F=(2.49\pm0.07)\times10^7$ V/m for the longitudinal case. This amounts to approximately 59% and 55%, respectively, of the depletion field at the InAs/oxide interface, which has been calculated from $F_{depl}\cong[2eN_A/\varepsilon(E_g/e+V_p+V_n-V_{pn})]^{1/2}$ at an internal bias voltage $V_{pn}=V_n$. In view of the fact that tunneling occurs inside the depletion layer where the depletion field is only some fraction of its maximum value at the interface and where the surface field is screened by the inversion charge, this result seems reasonable.

The B=0 characteristic in Fig. 2 strongly resembles those of double-barrier resonant tunnel devices /3/ in that the current starts to increase at a certain threshold voltage more or less close to the peak voltage. This is completely different from Esaki diodes whose differential conductance gradually decreases as the applied voltage runs from reverse to forward bias /4/. The threshold characteristic of SFIT junctions is the result of transverse momentum conservation which predicts the current peak position at B=0 at an internal bias V_{pn} where the bulk Fermi level is aligned with the bottom of the subband /1/. This prediction can be examined by tunneling spectroscopy of subband Landau levels /5/. Figure 3 shows magnetooscillations in d^2I/dV^2 that arise in longitudinal magnetic fields whenever a

Landau level is swept through the bulk Fermi level as bias is increased. A plot of the bias position of the oscillation minima versus magnetic field yields a fan diagram which converges onto the bias position of the peak current as expected for $V_{pn}=V_n$.

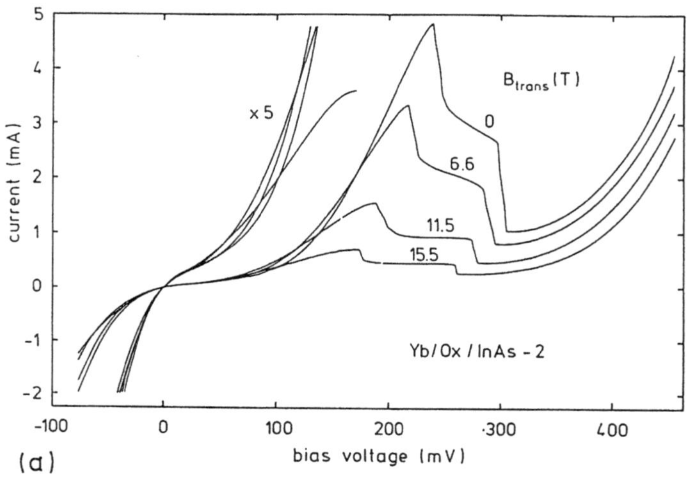

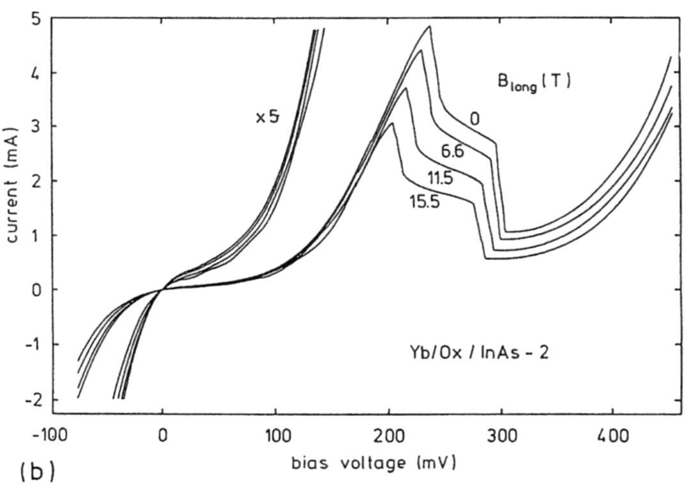

Fig. 2. Current-voltage characteristics of a surface-field induced tunnel junction at T=4.2 K with different magnetic fields applied (a) transverse and (b) longitudinal with respect to the current flow

Magnetooscillations in d^2I/dV^2 as a function of magnetic field (Fig. 4) are periodic in 1/B, and their period $\Delta(1/B)$ is inversely proportional to the areal density of states that are enclosed in the $\vec{k}$ plane by the con-

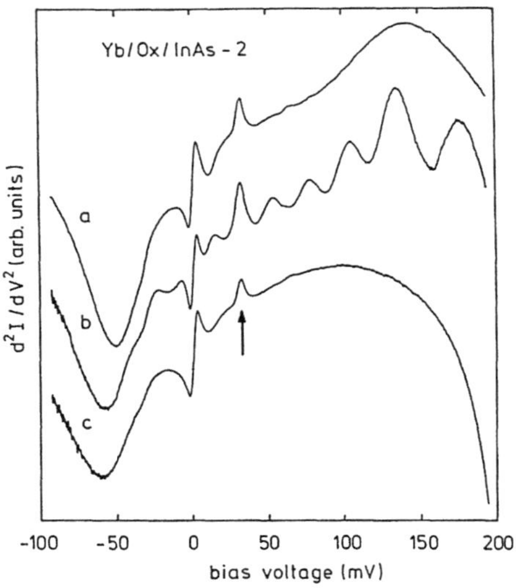

Fig. 3. Second derivative d^2I/dV^2 as a function of applied bias voltage with (a) zero magnetic field, (b) longitudinal field B=9.8 T, and (c) transverse field B=9.8 T. The arrow indicates the position of the structure due to phonon-assisted tunneling.

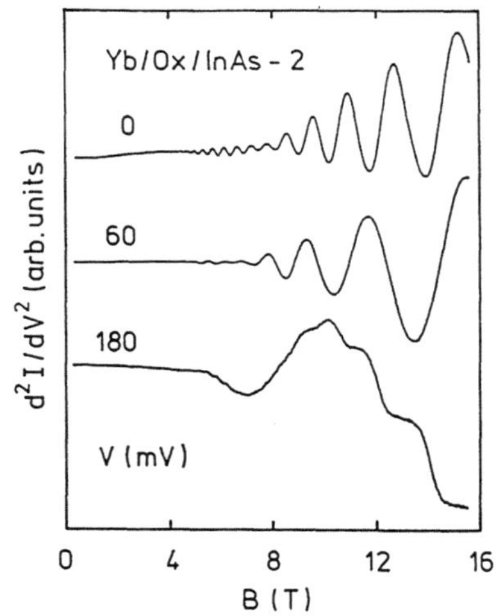

Fig. 4. Second derivative d^2I/dV^2 as a function of longitudinal magnetic field with applied bias V as parameter

stant-energy contour given by $E-E_o=e(V_n-V_{pn})$ /5/. At V=0 the period gives the density of occupied states $N_s^o=(3.7\pm0.1)\times10^{12}$ cm^{-2}. A remarkable detail of the oscillations in Fig. 4 is the small amplitude about B=8 T reminescent of beating, which is probably due to unresolved spin splitting. At large forward bias (and reverse, not shown in Fig. 4) a second oscillation of short period shows up. At these biases the differential resistance of the subband-to-bulk junction is small and an appreciable fraction of the modulation voltage drops across the oxide. Hence the oscillation can be attributed to the change of the metal-to-subband-junction resistance that arise from moving of subband Landau levels through the metal Fermi level which is shifted with respect to the subband Fermi level by $-eV_{ox}$. From the oscillation period at V=180 mV we obtain an areal state density which exceeds N_s at V=0 by only 10%. Since at forward biases also the decrease of N_{depl} leads to an increase of N_s we may conclude that V_{ox} is only a small fraction of the applied voltage. However, a more accurate analysis requires an exact knowledge of the areal density of charged interface states which also depends on bias voltage.

References

1. U. Kunze and W. Kowalsky: to appear in Appl. Phys. Lett.
2. M.H. Weiler, W. Zawadzki, and B. Lax: Phys. Rev. 163, 733 (1967).
3. V.J. Goldman, D.C. Tsui, J.E. Cunningham, and W.E. Tsang: J. Appl. Phys. 61, 2693 (1987).
4. H.P. Kleinknecht: Solid-State Electron. 2, 133 (1961); E.O. Kane: J. Appl. Phys. 32, 83 (1961).
5. U. Kunze and G. Lautz: Surf. Sci. 142, 314 (1984).

Part VI

Transport in Submicron Structures

Magnetoconductance in Lateral Surface Superlattices*

D.K. Ferry[1], *G. Bernstein*[1,+], *R. Puechner*[1], *J. Ma*[1], *A.M. Kriman*[1], *R. Mezenner*[1], *W.-P. Liu*[1], *G.N. Maracas*[1], *and R. Chamberlin*[2]

[1]Center for Solid State Electronics, Arizona State University, Tempe, AZ 85287, USA

[2]Department of Physics, Arizona State University, Tempe, AZ 85287, USA

Abstract. Microfabrication through the use of electron-beam lithography has allowed the creation of nanostructures with interesting and novel properties, and it is now possible to create potential-induced lateral surface superlattices. The magnetoconductance in such structures can be expected to be significantly different from that of normal quasi-two-dimensional systems. Here, we discuss the theoretical understanding and experimental studies that show the presence of the Aharonov-Bohm effect, universal conductance fluctuations, and structure in V_H and σ_{xx} with h/e and 2h/e periodicity in the flux coupled through each quantum well of the superlattice. At high V_{SD}, we have seen evidence for negative differential conductivity, which might be attributable to Bloch oscillations.

1. INTRODUCTION

Since the introduction of integrated circuits in the late 1950's, the number of individual transistors that can be placed upon a single circuit has approximately doubled every three years. Many commercial companies are experimenting with the production of chips with critical dimensions of 0.1 µm, and university laboratories have produced individual devices with gate lengths as short as 25 nm [1,2]. The creation of devices whose spatially important scales may be only a few tens of nanometers opens the door to the study of many new and important physical effects, some of which have been described earlier [3,4]. Indeed, it can rightfully be said that it will be impossible to understand fully the operation of these devices without a full understanding of these newly appearing physical effects.

In the reduction of individual feature sizes of a transistor, used as the basic building block for ULSI, the critical length (e.g. the gate length or a depletion length [5]) is becoming so small that it is approaching the *coherence length* of the electrons. Over the past several years, it has become evident that this latter length is not the wavelength of the electron itself, but is the *inelastic mean free path*. It has been known for some time that in modern device structures, this latter length can be more than 1 µm at low temperatures in both Si and GaAs devices. The

---

* This work was supported in part by the U.S. Office of Naval Research and the Army Research Office

+ Present address: Department of Electrical Engineering, Notre Dame University, Notre Dame, Indiana, USA

Springer Series in Solid-State Sciences, Vol. 87
High Magnetic Fields in Semiconductor Physics II Editor: G. Landwehr

consequence is that such small devices must now be treated as quantum mechanical objects, and many phenomena, ignored in the normal classical and semi-classical treatments of semiconductor devices, become important. This has led to studies of quantum behavior, many of which are poorly understood at best, in device structures.

In very dense arrays of next generation ULSI chips, the devices will appear as if they are a superlattice of structures imposed upon the basic semiconductor material [6]. We are already facing near-neighbor interactions which change the basic behavior of the architecture [7]. While such behavior is detrimental to the standard mode of operation of these chips, it opens the door to creation of parallel methods of operation through the *lateral surface superlattice* (LSSL) [8], a structure which was first studied by BATE [9]. Many groups are now studying coupling mechanisms in such arrays through the idea of quantum boxes in a variety of structures such as we have suggested [10,11]. The inclusion of a LSSL in a high-electron mobility transistor structure has been pursued by our group [12-14] and others [15]. The idea of these arrays of quantum boxes is that the electrons can be localized in the boxes by the potential barriers introduced by a grid gate, and the superlattice is developed when the carriers lying in the localized boxes can interact with each other and sense the periodicity.

In this paper, we will describe the study of LSSL structures and their magnetoconductance. First, in the next section, we review the work on the Bloch oscillation structure and discuss its relation to resonant tunneling and the formation of a Stark ladder of states. We then turn to the properties of larger LSSL structures and their magnetoconductance at low temperatures. Structures which have been prepared on modulation-doped heterojunction layers exhibit universal conductance fluctuations and the Aharonov-Bohm effect, while devices in which the LSSL is placed over a normal MESFET-type layer (highly doped epitaxial layer on lightly doped substrate) show more interesting structure in the Hall voltage and σ_{xx} that appears to have periodicities related to both h/e and 2h/e flux units coupled through each cell of the LSSL.

2. BLOCHFETS

In a few samples, prepared on one wafer of modulation-doped heterojunction GaAlAs/GaAs material, we have seen negative differential conductivity [12-14]. In the best case, the drop-off in current was complete for a gate voltage near to the pinchoff value, as expected from theoretical considerations [16]. The relative difficulty in observing the effect, and lack of its observation by other groups [15] clouds the interpretation of these experiments. We can understand the difficulties by understanding what the process of LSSL involves. The LSSL is induced by the periodic gate potential (we show such a gate structure in the next section) of the grid gate. This potential variation, induced by the differences in the gate bias and the pinned surface potential at the GaAlAs surface between the grids, must be transferred to the inversion layer of the modulation-doped structure. The use of delta-doping, as in the work of [15], means that the potential variation is induced in a material layer satisfying Laplace's equation, rather than Poisson's equation, and it is almost impossible to transfer the potential variation to the inversion layer. Thus, it is necessary to have uniform doping through the GaAlAs (except for the spacer layer adjacent to the heterojunction).

The Bloch oscillation effect is very closely related to the process known as sequential resonant tunneling [17]. The occurrence of Bloch oscillations refers to electrons which cycle through the entire minizone of the reciprocal lattice, and are thus localized in real space. In sequential resonant tunneling, the bias shifts the bands of adjacent quantum wells. In this latter process, the current rise occurs as empty states in the minibands are swept past filled states of the adjacent well. The negative resistance occurs when the current drops to zero as the bands no longer overlap, so that the electrons are localized in each quantum well. The shift in energy of the adjacent wells is the miniband energy width ε_o, which is related to the bias as ε_o=eV, where V is the potential drop *per well in the linear direction between source and drain.* If an electron subsequently tunnels by an inelastic process, the energy loss required to couple to the non-overlapping minibands is just eV, and this corresponds to a frequency $\omega= \varepsilon_o/\hbar=eV/\hbar=eFd/\hbar$, where F=V/d is the electric field. This frequency is identically the Bloch frequency, which illustrates the close connection between the Bloch oscillation and the sequential resonant tunneling process. This connection, and the possibility of direct radiation of this frequency, has been discussed previously [18]. When the minibands are moved further apart, the overlapping of wavefunctions from adjacent wells is strongly reduced, and the bands are narrowed to essentially just single energy levels. This chain of energies, one level per well, shifted linearly down the array by eV is the form that the Stark ladder takes in these structures. In either the localized Stark ladder, or the coupled cosinusoidal minibands, the onset of an a.c. signal results in steps in the current (or photo-absorption) due to the mixing of the a.c. signal with harmonics of the Bloch frequency [19], and this has recently been seen in photo-absorption [20]. To date, LSSLs have been fabricated with periodicities of 150-180 nm [12-15]. For this periodicity, the minibands are quite narrow, and this is thought to be one of the major reasons for the difficulty in observing the effect. Any modest inhomogeneity in the potential between source and drain will produce a voltage fluctuation much larger than the miniband width, and may occlude the results.

3. MAGNETOCONDUCTANCE IN LSSL

The inelastic mean free path, or phase-breaking length, arises in yet another context in regard to the field of localization. Here, the localization length describes the extent of the phase coherence of an electron in a random network, where the randomization can be provided either by the use of random site energies, such as in a random alloy on a crystalline lattice, or in randomization of lattice and bond lengths, such as in a truly disordered material, *or in our quantum boxes.* The localization phenomena appear differently in various dimensional structures, but are the behavior responsible for an important phenomenon -- universal conductance fluctuations. In systems which have constrained lateral dimensionality, and in which the length of the major conduction path is smaller than the important phase-breaking length, the conductance exhibits fluctuations, with regard to variations of the chemical potential or a magnetic field, of the order of

e^2/h. The fluctuations themselves are external characterizations of internal quantum interference of the phase variations of a few possible conductance paths, which in turn relate to the various "channels" for conduction. As the chemical potential is varied, or as an external magnetic field is varied (which also shifts individual site energies), it is thought that various channels are switched on or off, changing the phase interference effects within the sample. This phase interference is very much related to a comparable interference effect -- the Aharonov-Bohm effect, in which conducting rings, which enclose a magnetic field, show quantum interference effects in the end-to-end transmission, provided the ring dimensions are small compared to the phase-breaking length.

In general, the system is more complicated than this. In a two-dimensional periodic potential, with an applied magnetic field, the Hamiltonian is given by

$$H = \frac{1}{2m}(\mathbf{p} - e\mathbf{A})^2 + V[\cos(gx) + \cos(gy)] \; ,$$

where $g=2\pi/a$, and a is the periodicity of the potential. In fact, the Schrödinger equation with this Hamiltonian can not be solved in general [21]. There are two characteristic lengths in the problem: the magnetic length $l_c=(h/eB)^{1/2}$ and the lattice periodicity a. Only when the ratio of these lengths satisfies the condition of rationality: $eBa^2/h=p/q$, where p and q are integers, can the problem be solved. This leads to a "fractal" set of allowed energy levels [22,23], and the breakup of Landau levels into a set of subbands [24]. It is unlikely that just exciting (or de-exciting) one more mode of the quasi-one-dimensional conductor would create the fractal conductance found for the universal conductance fluctuations (indeed, some studies of carefully turning on modes show uniform steps [25]). Rather, it is more likely that these latter are related to the complicated energy states allowed in this system. We can probe this with our LSSL.

We can illustrate the above possibility by constructing a realistic Anderson model for our superlattice. Localization phenomena can arise either from energy fluctuations on a well defined lattice or from random lattices. The Anderson model treats the localized regime in a tight-binding Hamiltonian in which the individual site energies ε_i are assumed to be randomly distributed over a range [-W,W], and the sites are coupled to nearest neighbors by a (non-random) overlap energy V_{ij}. This leads to the Hamiltonian

$$H = \sum_i \varepsilon_i a_i^+ a_i + \sum_{i,j} V_{ij} a_i^+ a_j \; ,$$

where a^+ and a are the creation and annihilation operators for an electron at site i or j. The ratio of the range of site energies W to the overlap energy V (W/V) gives a measure of the degree of disorder in the system. A numerical evaluation can be achieved by evaluating the Green's functions in the site representation and using these to evaluate the Kubo formula [25,26]. We chose to illustrate the latter, as it gives a straight-forward method to evaluate the universal conductance fluctuations and the results are expressible in terms of the Landauer multi-channel conductance formula as well. Results of this study have yielded such a fractal structure in other than one-dimensional conductors [4]. The magnetic field variation is shown in Fig. 1. The apparent periodicity arises from the microscopic loops of area a^2 that exist in this lattice.

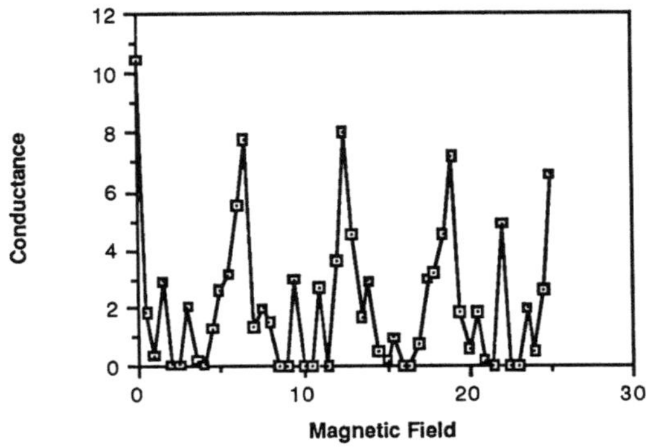

Fig. 1 The conductance, in units of e^2/h, is plotted as a function of the magnetic field, in units of $h/2\pi ea^2$. This plot is for values of W=2V, periodic boundary conditions, and a Fermi Energy E=V.

The amplitude of the oscillations and fluctuations shown in Fig. 1 are quite sensitive to the level of the Fermi energy, but seem to be independent of the exact details of the model. For example, the same results are obtained if the magnetic field is applied only to the localized region, and not to the contacts. Again, the dominant fluctuation effect arises from the stochastic interference of many electron wave packet trajectories traversing the sample; i.e. the switching of channels in the multi-channel formulation. It is important to note here that the Green's functions fully incorporate all the transverse modes relevant to the multi-channel Landauer formula [27]. While observing the periodic oscillations in magnetoconductance would require some 165 megaGauss, if *a* were 0.25 nm for a regular lattice, only 2000 Gauss is required in a typical LSSL with a periodicity of 150 nm. We also note in Fig. 1, that the conductance goes to zero at several values of the magnetic field, which is usually an indication of the quantum Hall effect. Here, we think that the structures are showing behavior similar to that calculated by Hofstadter [22], with a magneto-conductance showing fractal behavior.

3.1 High Electron Mobility Structures

In Fig. 2, we show the layout of the samples used to investigate this effect. The first structures were placed on modulation-doped GaAlAs/GaAs heterojunction devices. Here, the GaAlAs was doped uniformly to 10^{18} cm^{-3} except for a 20 nm spacer layer adjacent to the heterojunction. The overall thickness of this layer was 35 nm. The devices were operated quite near to pinchoff, so as to insure that the superlattice potential was induced in the inversion layer. In Fig. 3, we show the source current, which essentially is σ_{xx}, for one of these devices. It is clearly evident that there are universal conductance fluctuations present, as well as larger amplitude oscillations. The general structure of the curve is similar to that of Fig. 1, so that the effects expected seem to be here. The Aharonov-Bohm oscillations correspond to a size length (for a square ring) of approximately 150 nm, whereas we feel the periodicity of the grid is closer to 160 nm.

The density of carriers is thought to be only of the order of 2-4x10^9 cm^{-2}, which corresponds to filling the lowest miniband less than half full. On the other hand, we can estimate the inelastic mean free path from the amplitude of the fluctuations, and this gives approximately 1 μm, so that coherence is being maintained over

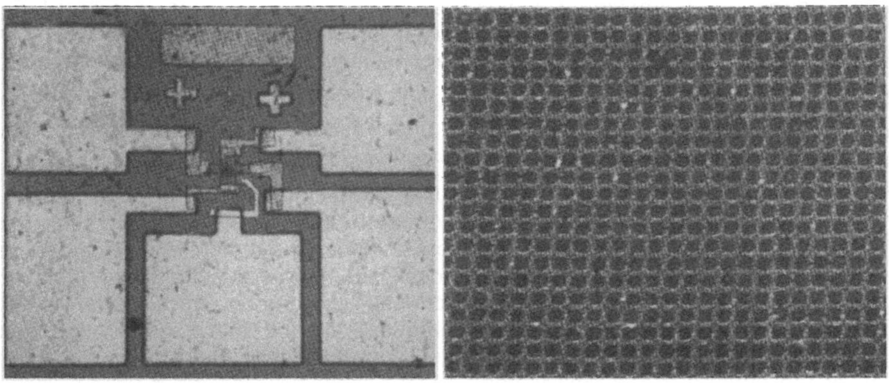

Fig. 2 The device structure (a) and the latera surface superlattice gate grid (b) used in the experiments discussed here. The periodicity ranged from 150 nm to 180 nm.

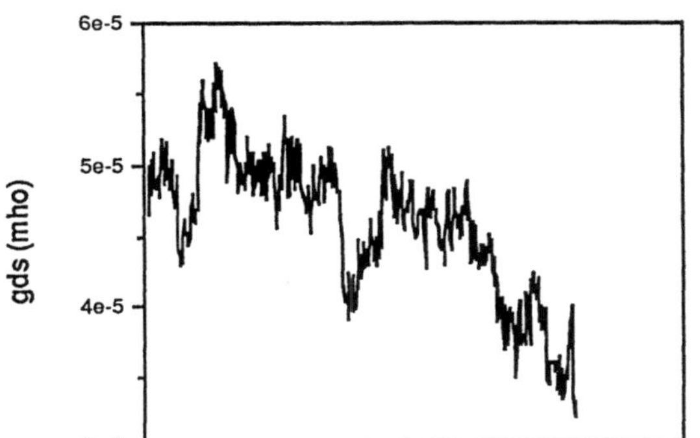

Fig. 3 The conductance, or source current, along the source to drain direction is shown for a MODFET. The presence of universal conductance oscillations and a larger period oscillation is apparent. The periodicity is apparently related to the Aharonov-Bohm effect in the lateral surface superlattice.

many cell lengths. The peaks in Fig. 1 correspond to electrons basically hopping around a ring formed by four adjacent quantum wells. On the other hand, the periodicity found in the experiment seems to be related more to the ring formed by the inside dimensions around a ring in the gate grid.

3.2 MESFET Structures

We have also incorporated the LSSL gates into structures fabricated on material normally used for MESFETs. Here, the active layer is a 60 nm epitaxial layer grown by vapor phase epitaxial on a lightly doped substrate. The epi-layer is doped to 1.5×10^{18} cm^{-3}. It has been demonstrated previously that such layers will show a quasi-two-dimensional behavior at low temperatures near pinchoff [28]. The devices measured here show a variety of phenomena, that are not fully understood by us at present. In particular, we note that at the highest magnetic fields used here, 0.9T, we still have $\omega_c\tau_{el}<1$, where τ_{el} is the elastic scattering relaxation time, so that we do not expect to see magnetic quantum effects. In Fig. 4a, we show the source current (σ_{xx}) and Hall voltage for one sample, which clearly indicates the presence of universal conductance fluctuations in the LSSL device. In Fig. 4b, we plot these parameters for another device on a compressed scale, so that the presence of the negative magnetoresistance at low magnetic fields can be seen. From this clear evidence of weak localization, we can estimate an inelastic mean free path to be 1-2 μm. Other effects can also be seen in this latter figure. The source current shows sharp drops at regular values of the magnetic field, and sharp changes in the Hall voltage are often correlated with these. However, the sign of the change in the Hall voltage seems random, although it usually is an increase in this quantity.

In contra-distinction to the high electron mobility samples, where we had less than one electron per quantum well, the electron density in these samples is

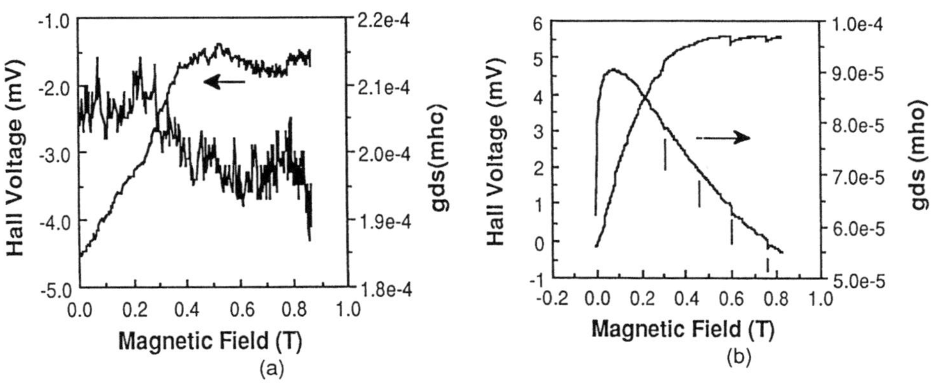

Fig. 4 Source current and Hall voltage for two MESFETs at low temperatures. In (a), we illustrate the presence of universal conductance fluctuations in these samples. In (b), we show on an expanded scale the jumps in conductance and Hall voltage that are seen in these samples.

thought to be about 3-5x10^{11} cm^{-2}, so that there may be as many as 100 electrons per quantum well. There are certain to be complicated many-body corrections within each well. Nevertheless, it appears from Fig. 4b, that the basic periodicity in the conductance is about 1500 Gauss (±100 Gauss), which corresponds to about 165 nm periodicity, which is quite close to the periodicity thought to exist in the grid. We remark that the h/e structure could relate to the half-band width, since the period of the Landau orbit diverges at an energy corresponding to a half-filled miniband. This follows from the divergence in the two-dimensional density of states at this energy [16]. Moreover, we can estimate a new inelastic mean free path near these jumps from the negative magnetoresistance in the source current following the jump. This value is only about 0.3 μm, which is much reduced from the value estimated at zero magnetic field, and is about twice the grid periodicity. While we do not understand the jumps apparent in the Hall voltage appearing at 2h/e periodicities, this could correspond to sweeping the Fermi level entirely through a miniband with the magnetic field. On the other hand, if the complicated band structure mentioned above is invoked, Thouless has speculated on the possibilities of jumps in the Hall conductance due to mixing of Landau levels and/or subbands [24], and this should occur at preferential values of the number of flux quantum per well. Moreover, the coupling of various orbits by Bragg scattering from the superlattice should occur at field values such that $\Phi=2n\Phi_o$ [29].

4. SUMMARY AND CONCLUSIONS

In each of the LSSL structures, we have chosen to plot the source conductance, which relates to the transverse conductance σ_{xx}, and the Hall voltage rather than the Hall resistance. Particularly in the case of the MESFET samples, structure in the source conductance often had no counterpart to that occurring in the Hall voltage, so that the former would have introduced a structure in the Hall resistance (V_H/I_S) that did not occur in the Hall voltage. In mesoscopic systems, it is well known now that great care must be taken in four-terminal measurements as the quantum interference phenomena are quite non-local. In this manner of presenting the data, we find that in low density (high electron mobility) systems, we observe universal conductance fluctuations as well as Aharonov-Bohm type oscillations with h/e periodicity in areas closely related to the basic periodicity of the LSSL. On the other hand, in dense systems (the MESFET samples), in addition to the universal conductance fluctuations, we observe structure characterized by sharp changes in the conductance and Hall voltage. This latter structure has components with both h/e and 2h/e periodicities, whose origins are not well understood.

Particularly in the case of the MESFET samples, where the scattering remains high, and the Landau levels are not expected to be fully formed, it is interesting that the quantum effects apparently still appear, not only in phase interference phenomena such as universal conductance fluctuations, but also in effects periodic in the basic flux quantum h/e. For both sets of samples, we are apparently at the "low field" limit where the two-dimensional periodic potential dominates the magnetic contribution to the energy because of the heavy scattering. Future attempts to elucidate the roles played by each of these interactions are expected to shed more light on these unexpected and unusual observations.

References

1. G. Bernstein and D. K. Ferry: Superlatt. and Microstruc. **2**, 147 (1986)
2. J. Ryan, Master's Thesis, Arizona State University (unpublished)
3. J. R. Barker and D. K. Ferry: Sol.-State Electronics **23**, 519 (1981)
4. D. K. Ferry: In **Physics and Technology of Submicron Structures**, Ed. by H. Heinrich, G. Bauer, and F. Kuchar (Springer-Verlag, Heidelberg, in press)
5. D. K. Ferry, L. A. Akers, and E. W. Greeneich: **ULSI Microelectronics** (Prentice-Hall, Englewood Cliffs, N.J., 1988)
6. J. R. Barker and D. K. Ferry: Sol.-State Electronics **23**, 531 (1981)
7. D. K. Ferry: In **Quantum Transport in Semiconductors**, Ed. by H. L. Grubin, C. Jacoboni, and D. K. Ferry (Plenum Press, New York, 1988)
8. J. R. Barker and D. K. Ferry: In **Proc. Int. Conf. on Cybernetics and Society** (IEEE Press, New York, 1979) p. 762
9. R. T. Bate: Bull. Am. Phys. Soc. **22**, 407 (1977)
10. G. J. Iafrate, R. K. Reich, and D. K. Ferry: Surf. Sci. **113**, 485 (1982)
11. Y. Tokura and K. Tsubaki: Appl. Phys. Letters **51**, 1807 (1987)
12. G. Bernstein and D. K. Ferry: Superlatt. and Microstruc. **2**, 373 (1986)
13. G. Bernstein and D. K. Ferry: Z. Physik B **67**, 449 (1987)
14. D. K. Ferry, G. Bernstein, and W.-P. Liu: In **Physics and Technology of Submicron Structures**, Ed. by H. Heinrich, G. Bauer, and F. Kuchar (Springer-Verlag, Heidelberg, in press)
15. Y. Jin, D. Mailly, F. Carcenac, B. Etienne, and H. Laundis: Microelectr. Engr. **6**, 195 (1987)
16. R. K. Reich, R. O. Grondin, and D. K. Ferry: Phys. Rev. B **27**, 3483 (1983)
17. F. Capasso, K. Mohammedi, and A. Y. Cho: Appl. Phys. Letters **48**, 478 (1986)
18. R. F. Kazarinov and R. A. Suris: Sov. Phys. Semiconductors **5**, 707 (1971)
19. R. O. Grondin, W. Porod, J. Ho, D. K. Ferry, and G. J. Iafrate: Superlatt. and Microstruc. **1**, 183 (1985)
20. E. E. Mendez, F. Agullo-Rueda, and J. M. Hong: Phys. Rev. Letters **60**, 2426 (1988)
21. See, e.g., G. H. Wannier: Phys. Stat. Sol. (b) **88**, 757 (1978)
22. D. R. Hofstadter, Phys. Rev. B **14**, 2239 (1976)
23. D. J. Thouless, M. Kohmoto, M. P. Nightingale, and M. den Nijs: Phys. Rev. Letters **49**, 405 (1982)
24. D. Thouless: Phys. Repts. **110**, 279 (1984)
25. D. J. Thouless and S. Kirkpatrick: J. Phys. C **14**, 235 (1981)
26. P. A. Lee and D. Fisher: Phys. Rev. Letters **47**, 882 (1981)
27. M. Büttiker: Phys. Rev. Letters **57**, 1761 (1977)
28. M. Pepper: J. Phys. C **10**, L173 (1977)
29. A. B. Pippard: Phil. Trans. Roy. Soc. (London) **256A**, 317 (1964)

Conductance Fluctuation Phenomena in Submicron Width High Mobility GaAs/AlGaAs Heterojunctions

J.P. Bird[1], *A.D.C. Grassie*[1], *M. Lakrimi*[1], *K.M. Hutchings*[1], *J.J. Harris*[2], *and C.T. Foxon*[2]

[1]Physics Division, University of Sussex, Brighton BN1 9QH, UK
[2]Philips Research Laboratories, Cross Oak Lane, Redhill, Surrey RH1 5HA, UK

1. Introduction

Recent advances in fabrication technology mean that it is now possible to manufacture devices in which transport is coherent or even ballistic. This has resulted in the observation of a rich variety of quantum mechanical phenomena which were previously inaccessible in larger systems due to the effects of self-averaging [1]. An important consequence of coherence in metallic systems is the prediction that the conductance should fluctuate as a function of magnetic field by an amount e^2/h independent of sample size or degree of disorder [2]. If however the phase coherence length L_Φ is shorter than the sample length L, then averaging effects will reduce the amplitude of the fluctuations. Such behaviour has now been observed in a wide variety of metallic systems in good agreement with theory (see ref. [1]).

The theory of Universal Conductance Fluctuations applies well to a two terminal measurement on a metal. Electronic motion in metallic systems is typically diffusive, i.e. the mean free path Le is less than any of the dimensions, in stark contrast to high mobility semiconductor structures in which motion can be quasi-ballistic (Le > Width) or even ballistic (Le > L). Recent experiments have therefore focussed on the applicability of the U.C.F. theory to high mobility devices with constrasting results [3-5].

In this paper we present the results of measurements on a sample of geometrical width 1 μm and length 30 μm, fabricated from a high mobility GaAs/AlGaAs heterojunction grown by M.B.E. at the Philips Research Laboratories at Redhill. As the temperature is reduced we observe possible transitions from diffusive to quasi-ballistic scattering and to phase-coherent transport. However, the amplitude of the phase-coherent fluctuations is found to differ from the value predicted for diffusive systems [2], but is in good agreement with estimates based on quasi-ballistic transport [5].

2. Experiment

The samples used are patterned using a wet etch technique described elsewhere [6]. The effect of sidewall depletion means that the actual conducting width W will be much less than the geometrical width.

After mounting to the base of a dilution unit the sample is cooled in the dark to 4.2 K. Four terminal magnetoresistance measurements are made using a low frequency A.C. technique in the temperature range 4.2 to 0.04K, with the voltage across the sample kept below 4 μV to avoid spurious heating effects.

Springer Series in Solid-State Sciences, Vol. 87
High Magnetic Fields in Semiconductor Physics II Editor: G. Landwehr

3. Results

Typical magnetoresistance traces at three temperatures are shown in Fig.1. At low temperatures fluctuations of up to 100% in the resistance are observed. As the temperature is increased the amplitude of the fluctuations decreases, while the magnetoresistance trace retains the same general shape. The measurements are reproducible (within the noise level of < 0.5%), as long as the sample is kept below 4.2 K. On the thermal cycling a new set of fluctuations are observed with similar characteristics.

An important feature of the traces is that below 4.2 K there is no background due to weak localisation or interaction effects, so that in the following analysis we do not subtract a background from the data. Furthermore, the maximum field available (1.2 T) and the sheer size of the fluctuations mean we are unable to directly calculate transport parameters such as the scattering lengths Le and $L\Phi$, the thermal diffusion length L_T, the mobility μ, or the diffusion coefficient D.

Plotting the r.m.s. fluctions of the dimensionless conductance, δg, as a function of temperature we observe it is approximately constant below 90 mK (see Fig.2), suggestive of a transition to phase-coherent transport. However, the observed amplitude of fluctuation, 0.15, is in disagreement with the predicted value of 0.729 for diffusive systems [2].

Recently Beenakker and van Houten [5] have calculated the correction to δg in high mobility systems due to the effect of boundary scattering. The correction arises because electrons which scatter off the walls enclose zero flux, so a greater field is required to change the flux through the sample by an amount h/e, and is reflected in an enchanced value for the field ΔBc. They find that in the presence of boundary scattering,

$$(\delta g)^2 = (4\pi/3)(L_T^2 L\Phi/L^3) \qquad L\Phi \gg L_T \quad , \tag{1a}$$

$$(\delta g)^2 = 6(L\Phi/L)^3 \qquad L\Phi \ll L_T \quad . \tag{1b}$$

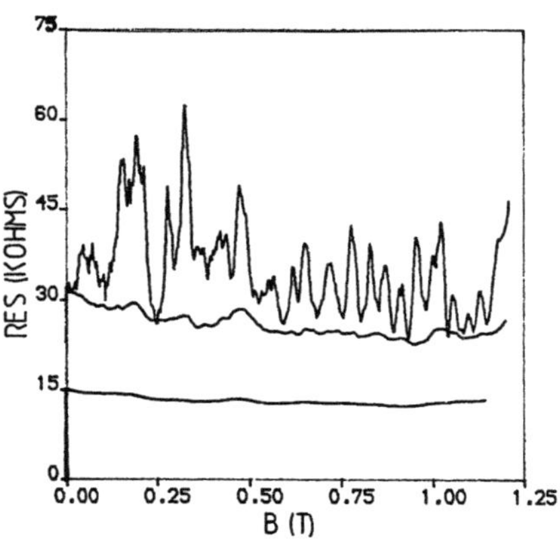

Fig.1. Resistance fluctuations at (from top to bottom) 0.058K, 1.1K and 4.2K.

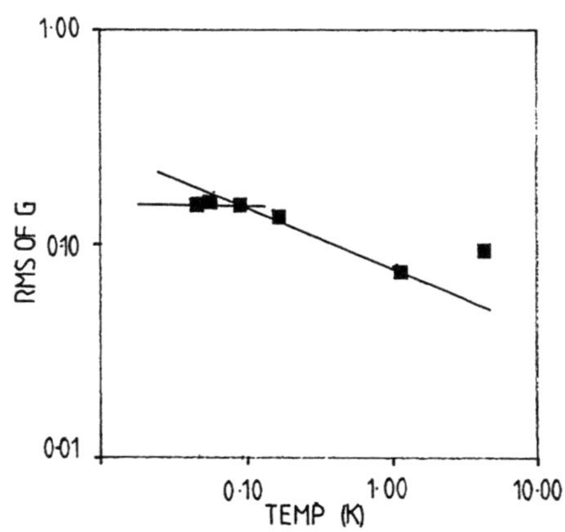

Fig.2. The r.m.s. conductance fluctuation, in units of e^2/h, as a function of temperature.

If we assume that at 90 mK, L_Φ is equal to the sample length, i.e. 30 μm and take our value for δg to be 0.15 then we can extract various transport parameters. From eqn.(1a) we find L_T = 2.2 μm, D = 0.055 m^2/s and W = 0.4 μm. We know from previous work that samples with widths under 0.7 μm show no Shubnikov de-Haas oscillations below 1 T, so the extrapolated value for W would therefore seem a reasonable estimate.

Further evidence for quasi-ballistic behaviour is provided by an analysis of the correlation function $F(\Delta B)$. The fluctuations are correlated over a field range ΔBc which is a measure of the field required to change the flux through the sample by an amount h/e and is given by,

$$\Delta Bc = \text{Constant } (h/e)/WL_\Phi \qquad L \gg L_\Phi \qquad (2)$$

The constant depends on the form of the scattering (see refs. [2] and [5]).

In Fig.3 we plot the correlation function at four different temperatures. Unexpectedly we find that the correlation field increases from 4.2 to 1.1 K and suggest that this behaviour is due to an enhancement associated with a transition from diffusive to quasi-ballistic transport. Another unusual feature is the oscillations seen at low temperatures. Although their amplitude changes with temperature the positions of the peaks and troughs does not. This oscillatory behaviour has been reported previously in similar structures by Timp et al. [4] and is presumably related to fundamental periodicities in the magnetoresistance resulting from flux quantisation.

Finally we calculate the Fourier spectra of the data, the results of which are shown in Fig.4. The spectra consist of a series of peaks which decrease in amplitude as the temperature is increased. Similar results have been found in n^+-GaAs wires by Taylor et al. [7] who associate the different peaks with the flux enclosed by different backscattered electrons, illustrating the connection between weak localisation and conductance fluctuations. While a weak localisation approach may well describe systems in which ensemble averaging dominates it will not apply to those in which transport is phase-coherent, which we believe to be the case here.

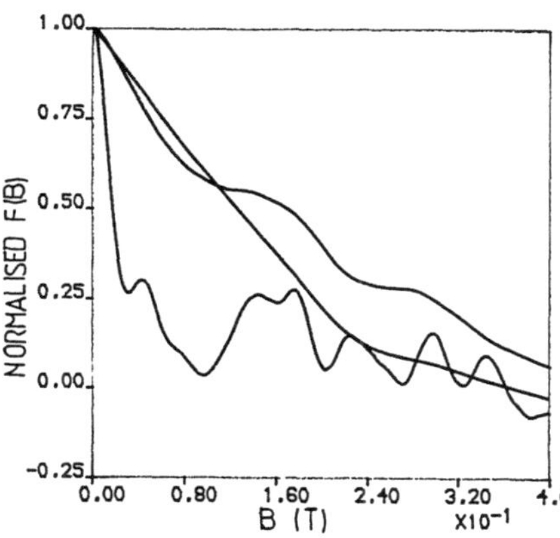

Fig.3. The Normalised correlation function at (from top to bottom at 0.08T) 4.2K, 1.1K and 0.068K.

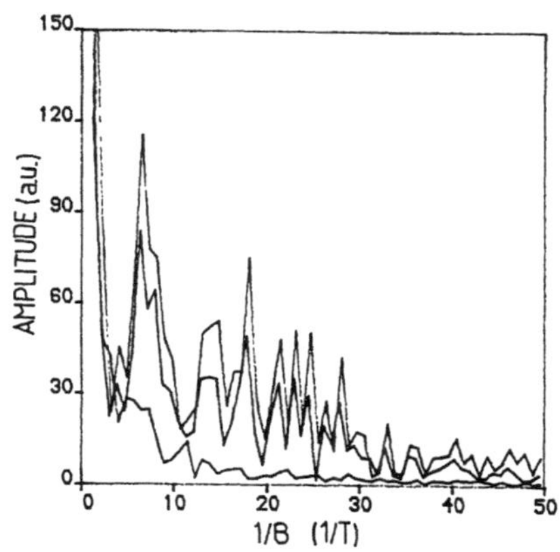

Fig.4. The Fourier spectra of the fluctuations at (from top to bottom) 0.058K, 0.170K and 1.1K.

We propose that the peaked structure in the spectra actually results from a flux quantisation effect in which electrons coherently scatter around regions of inhomogeneity and from Fig.4 we find areas between 2000 to 14000 nm^2.

4. Conclusions

We have measured the four terminal magnetoresistance of a high mobility heterostructure and find that the amplitude of fluctuation differs from the value predicted and obtained in a two terminal measurement on a metal. Because we are unable to directly measure transport parameters our conclusions remain tentative and other explanations exist (a temperature independent coherence length, for example, could explain the saturation of δg at low temperatures). A direct measurement of L_Φ is therefore required to substantiate our conclusions and this work is currently in progress.

References

1. S. Washburn and R.A. Webb: In Advances in Physics, 35, 375 (1986).
2. P.A. Lee, A.D. Stone and H. Fukuyama: In Phys. Rev. B. 35, 1039 (1987).
3. T.J. Thornton et al: In Phys. Rev. B. 36, 4514 (1987).
4. G. Timp et al: In Phys. Rev. Lett. 59, 732 (1987).
5. C.W.J. Beenakker and H. van Houten: In Phys. Rev. B. 37, 6544 (1988).
6. K.M. Hutchings et al: to be published (1988).
7. R.P. Taylor et al: In Surface Science, 196, 52 (1988).

New Magnetotransport Phenomenon in a Two-Dimensional Electron Gas in the Presence of a Weak Periodic Submicrometer Potential

D. Weiss [1], *K. v. Klitzing* [1], *K. Ploog* [1], *and G. Weimann* [2]

[1]Max-Planck-Institut für Festkörperforschung, Heisenbergstr. 1, D-7000 Stuttgart 80, Fed. Rep. of Germany

[2]Walter-Schottky-Institut, TUM, D-8046 Garching, Fed. Rep. of Germany

A new type of magnetoresistance oscillation periodic in 1/B is observed when the carrier density N_s of a two-dimensional electron gas is weakly modulated with a period smaller than the mean free path of the electrons. In high mobility AlGaAs-GaAs heterojunctions the carrier density is periodically modulated by illumination of the samples with two interfering laser beams at liquid helium temperatures. The modulation arises due to the persistent photoconductivity of the samples at low temperatures. The experiments show that the period of the additional quantum oscillation is determined by the separation a of the interference fringes. This period corresponds to Shubnikov-de Haas oscillations where only the electrons within the first reduced Brillouin zone with $|k| < \pi/a$ contribute.

1. Introduction

At low temperatures the magnetoresistance of a degenerate two-dimensional electron gas (2-DEG) exhibits the well-known Shubnikov-de Haas (SdH) oscillations periodic in 1/B. Lateral confinement of the 2-DEG on a submicrometer scale leads to deviations from the typical 1/B periodicity of these SdH-oscillations [1]. Within a naive picture these deviations occur when the cyclotron orbit of an electron is larger than the width of the confining potential [2]. An intermediate situation between a purely two-dimensional system and a quasi-one-dimensional quantum wire arises if the confining potential of the quantum wire is replaced by a weaker periodic potential with a length scale smaller than the mean free path of the electrons. The experiments show that under this condition a new type of magnetoresistance oscillation becomes visible with a periodicity depending on the period of the superimposed potential.

In selectively doped AlGaAs-GaAs heterostructures a persistent increase in the two-dimensional electron density is observed at temperatures below T=150K if the device is illuminated with infrared or visible light. This phenomenon is usually explained on the basis of the properties of DX-centers which seem to be related to a deep Si donor [3]. The increase in the electron density depends on the photon flux absorbed in the semiconductor so that a spatially modulated photon flux generates a modulation in the carrier density. In our measurements a holographic illumination of the heterostructure at liquid helium temperatures is used to produce a periodic potential with a period on the order of the wavelength of the interfering beams. In Fig.1 the interference of two plane light waves which create a periodic modulation of the ionized donors and therefore of the carrier density is shown schematically.

The existence of a periodically modulated carrier density in such structures has been demonstrated by Tsubaki, Sakaki, Yoshino, and Sekiguchi [4] who measured the anisotropy of the resistivity parallel and perpendicular to the interference fringes at 90K. Tsubaki and coworkers stated that the smallest possible period for finding an anisotropic resistivity is four or five times as large as the thickness of the AlGaAs layers on top of the 2-DEG. Since we have created smaller fringe periods we have observed nearly isotropic resistivity at zero magnetic field. In the presence of a magnetic field,

Springer Series in Solid-State Sciences, Vol. 87
High Magnetic Fields in Semiconductor Physics II Editor: G. Landwehr

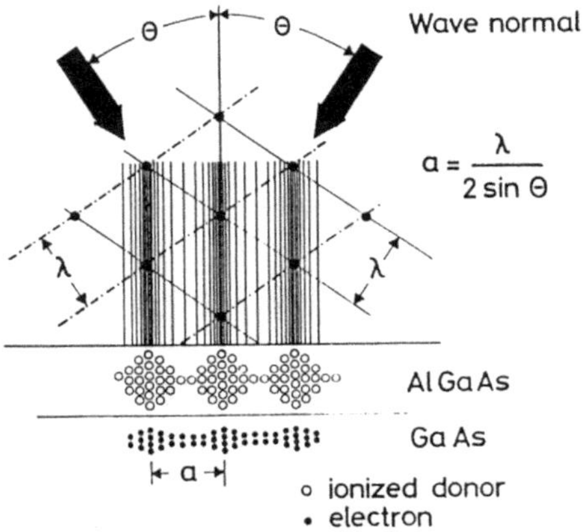

Figure 1: Sketch of the spatial modulation of the concentration of ionized donors in the AlGaAs layer and of electrons in the 2-DEG produced by holographic illumination using two interfering laser beams with wavelength λ. The interference pattern created is shown schematically

however, a new oscillatory phenomenon occurs demonstrating clearly that a periodic modulation of the 2-DEG is present.Details will be given in the following sections.

2. Experimental

The samples used in the experiments were conventional AlGaAs-GaAs heterostructures grown by molecular beam epitaxy [5] with carrier densities between $1.5 \cdot 10^{11}cm^{-2}$ and $4.3 \cdot 10^{11}$ cm^{-2} and low temperature mobilities ranging from $0.23 \cdot 10^{6}cm^{2}/Vs$ to $1 \cdot 10^{6}cm^{2}/Vs$. Illumination of the samples increases both the carrier density and the mobility at low temperatures. The heterojunctions discussed in the following sections consist of a semi-insulating GaAs substrate, followed by a $1\mu m - 4\mu m$ thick undoped GaAs buffer layer, an undoped AlGaAs spacer (6nm-33nm), Si-doped AlGaAs (33nm-84nm), and an undoped GaAs toplayer ($\approx$ 22nm). We have chosen an L-shaped geometry (sketched on the right hand side of Fig.2) to investigate the magnetotransport properties parallel and perpendicular to the interference fringes. Such a mesa structure was produced using standard photolithographic and etch techniques. Ohmic contacts to the 2-DEG were formed by alloying AuGe/Ni layers at 450°C. Some of the samples investigated have an evaporated semi-transparent NiCr front gate (thickness $\approx$ 8nm) or a back gate, respectively, in order to vary the carrier density after holographic illumination.

The experiment was carried out using either a 5mW HeNe laser (λ = 633nm) or a 3mW Argon-Ion laser (λ = 488nm) both linearly polarized. The experimental realization of the holographic illumination is shown schematically on the left hand side of Fig.2. The laser system was mounted on top of the sample holder which was immersed in liquid helium (4.2K) within a 10-Tesla magnet system. The laser beam which was expanded to a diameter of 40mm entered the sample holder through a quartz window and a shutter. The shutter ensures well-defined illumination times of the sample down to 25ms. Short exposure times are important to prevent jumping of the fringes; therefore exposure times between 25ms and 100ms were typically chosen. The mirrors which split the laser beam into two coherent waves, are located close to the device and are arranged in such a way that an interference pattern with a period a is generated at the surface of the device. An aperture mounted above covers the sample from direct illumination. The period a of the fringes created in this way depends on the wavelength λ of the laser and the incident angle Θ (see Fig.1): $a = \lambda/2sin\Theta$. With the wavelengths used in this experiment and $\Theta = 56^o$ we obtain periods of 382nm and 294nm for the interference pattern. Care was taken to minimize stray light originating from components inside

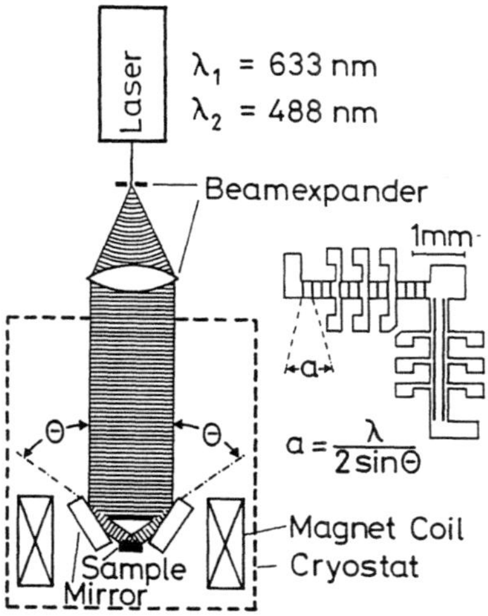

Figure 2: Schematic experimental set-up (left hand side) and top view of the L shaped sample geometry where the interference pattern is sketched

the sample holder by coating them with an absorbing layer. In order to reduce wavefront distortion due to rough surfaces of the optical components, all mirrors, lenses, and windows have a high quality polished surface. The wavefront distortion specification of these components is better than $\lambda/10$ where λ is the wavelength of a HeNe-laser ($\lambda = 633$nm).

3. Magnetotransport Measurements

After holographic illumination of the sample which was carried out at 4.2K we have measured the resistivities $\rho_\perp$ (pependicular to the interference fringes) and $\rho_{||}$ (parallel to the fringes) at cryogenic temperatures as a function of the magnetic field. The magnetic field was perpendicular to the plane of the 2-DEG. In addition, the Hall resistances $R_H^\perp$ and $R_H^{||}$ have been measured. The use of the indices $\perp$ and $||$ is illustrated in Fig.3. The resistivities and Hall resistances are measured applying a constant current (1μm-10μm) and measuring the voltage drop between potential probes along and perpendicular to the direction of current flow, respectively. A typical result obtained for a fringe period of a =382nm is shown in Fig.3. At magnetic fields above 0.5T both $\rho_\perp$ and $\rho_{||}$ show the well known SdH-oscillations with a periodicity $\Delta(1/B)$ inversely proportional to the 2-DEG carrier density N_s. Below 0.5T pronounced additional oscillations dominate $\rho_\perp$ while weaker oscillations with a phase shift of 180° relative to the $\rho_\perp$ data are visible in the $\rho_{||}$ measurements. The oscillations in $\rho_\perp$ are accompanied by a remarkable positive magnetoresistance at very low magnetic fields. The Hall resistances ,however, show no additional structure at magnetic fields below 0.5T even at the highest resolution used. Deviations from the clear linear behaviour of $R_H^\perp$ and $R_H^{||}$ occur only at magnetic fields above 0.5T, and are connected to the SdH oscillations starting to be resolved. The slightly different slopes of $R_H^\perp$ and $R_H^{||}$ are due to a small difference in the carrier density of about 4%. The additional oscillations in $\rho_\perp$ and $\rho_{||}$ are perfectly periodic in 1/B. This is demonstrated in the inset of Fig.4 where the oscillation index m is plotted versus the inverse magnetic field. The points in the inset correspond to minima in $\rho_\perp$. The oscillations in Fig.4 are obtained for a fringe period of 382nm. The behaviour of

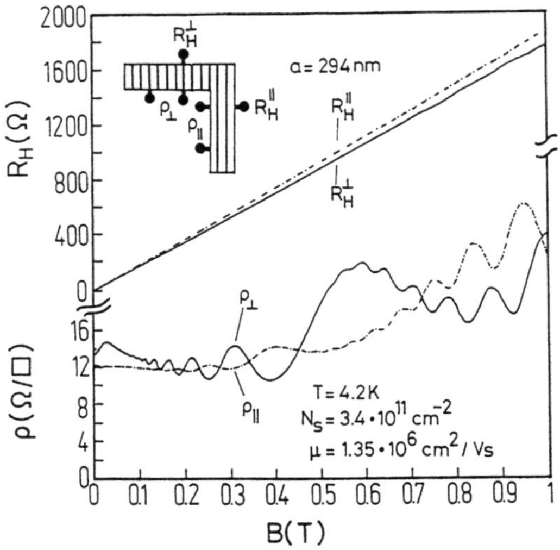

Figure 3: Magnetoresistivity ρ and Hall resistance R_H parallel and perpendicular to the interference fringes. The inset defines the use of the indices $\perp$ and $\|$ with respect to the interference fringes shown schematically

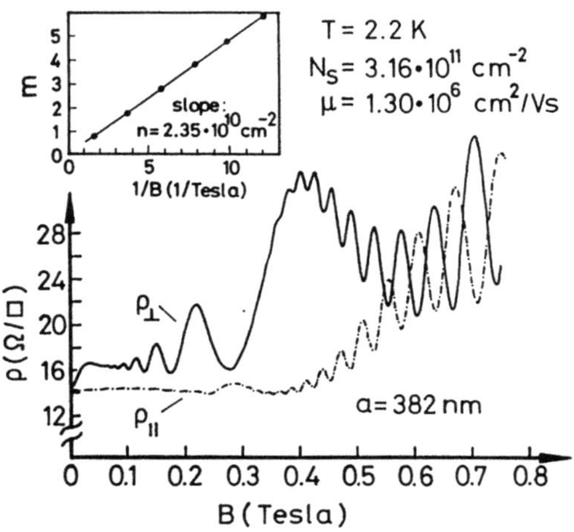

Figure 4: Magnetoresistivity ρ parallel and perpendicular to the interference fringes. The inset displays the 1/B dependence of the additional oscillations where the points correspond to minima in $\rho_\perp$

these additional oscillations of $\rho_\perp$ and $\rho_\|$ as displayed in Fig.4 is essentially the same as those shown in Fig.3. Since the additional oscillations in $\rho_\perp$ are perfectly periodic in 1/B the periodicity can be expressed as a carrier density n. Analogous to SdH-oscillations one obtains (spin splitting not resolved)

$$n = \frac{e}{\pi\hbar}(\Delta\frac{1}{B})^{-1} \quad . \tag{1}$$

Here e is the elementary charge, $\hbar$ is Planck's constant and $\Delta(1/B)$ is the difference between adjacent minima (or maxima) of $\rho_\perp$ or $\rho_\|$ on a 1/B scale. The additional oscillations in Fig.4 correspond to a carrier density $n = 2.35 \cdot 10^{10}\mathrm{cm}^{-2}$ while the periodicity of $\rho_\perp$ in Fig.3 gives $n = 3.2 \cdot 10^{10}\text{cm}^{-2}$. This result suggests that a shorter period a results in higher values for n as long as the carrier densities N_s are approximately equal.

Up to now we have not completely ruled out the possibility that the additional oscillatory structures are due to an occupation of the next higher subband. Temperature dependent measurements of $\rho_\perp$ shown in Fig.5 demonstrate that the additional oscillations are much less temperature dependent than SdH oscillations. The SdH oscillations have a strong temperature dependence since their amplitudes depend on the Landau level seperation $\hbar\omega_c$ as compared to the thermal energy k_BT. Therefore the additional oscillations do not originate from the Landau level seperation $\hbar\omega_c$ as one would expect also for electrons in higher subbands.

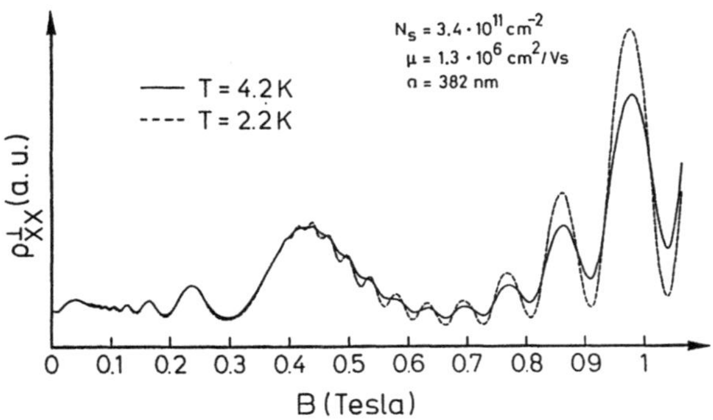

Figure 5: $\rho_\perp$ versus B for two temperatures. The SdH oscillations exhibit the expected strong temperature dependence, while the additional oscillations remain nearby unchanged for a change in temperature of 2K

By means of Fig.6 we wish to demonstrate that the additional oscillations are due to the existence of a periodically modulated charge pattern. After the first holographic illumination of the sample, $\rho_\perp$ displays pronounced additional oscillations. The interference pattern created by a second illumination should be shifted with respect to the first one due to vibrations of the experimental set up. The periodic modulation should therefore be smeared out and the additional oscillatory structure should disappear. This effect is visible in Fig.6 where the amplitude of the additional oscillations is drastically reduced after a second illumination with 500ms duration. We have found that continous

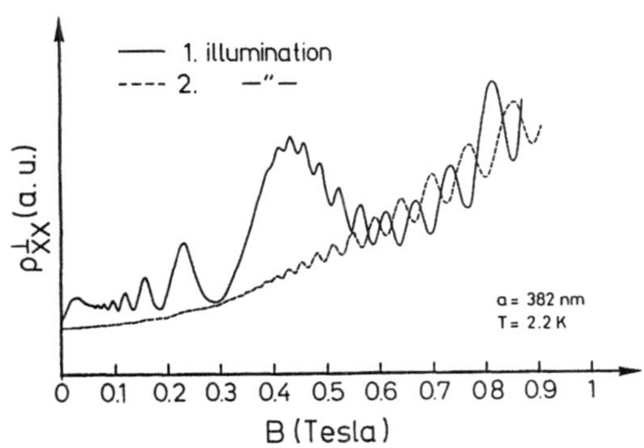

Figure 6: $\rho_\perp$ versus B. The amplitude of the additional oscillation is drastically reduced after a second holographic illumination

illumination always destroys the oscillations indicating that a proper fringe pattern is essential for observing the effect.

From the SdH-oscillations (using Eq.1) as well as from low field Hall measurements an average carrier density N_s of the periodically modulated 2-DEG can be obtained. The carrier density N_s and mobility μ we have measured parallel and perpendicular to the interference pattern are equal within 5%. Since, in addition, we do not observe additional structure in the higher magnetic field regime of $\rho_\perp$ and $\rho_{||}$ due to a second carrier density, we conclude that we have only a weak modulation of the carrier density. This means that the amplitude of the additional periodic potential introduced by holographic illumination is small compared to the Fermi energy of the system. A stronger potential modulation, achievable e.g. by holographic illumination of a photoresist coated sample and followed by suitable etching techniques leads to a clear anisotropy of the resistivity parallel and perpendicular to the grating [8].

The period of the additional oscillations depends on the carrier density N_s. This is shown in Fig.7 where the carrier density has been varied after holographic illumination. The data in Fig.7 were obtained applying a voltage to a semi-transparent gate on top of the sample. Changing the carrier density N_s results in a change of the periodicity of the additional oscillations in $\rho_\perp$ and $\rho_{||}$; as N_s increases, $\Delta(1/B)$ decreases and therefore n increases, too. In Fig.7 less additional oscillations are resolved compared to the magnetoresistivity data shown i.e. in Fig.4. This we contribute to the lower mobility of the sample and will be discussed within the next section.

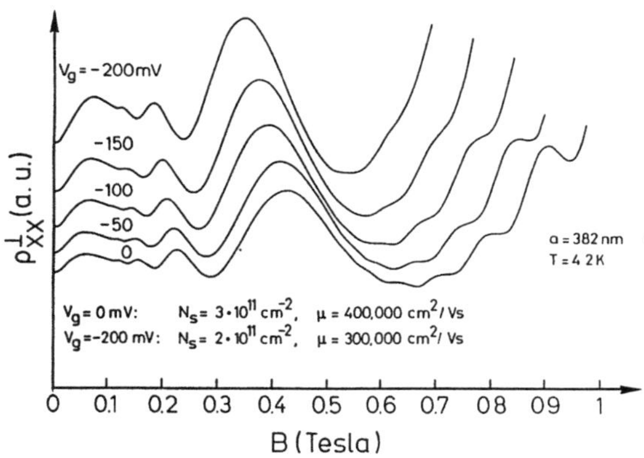

Figure 7: Gate voltage dependence of the magnetoresistivity $\rho_\perp$. Reducing the carrier density N_S results in an increasing $\Delta(1/B)$ value and therefore in a reduced carrier

4. Discussion

The experiments described in the previous sections have been carried out using samples with different mobilities and carrier densities. In addition we have changed the carrier density applying a voltage to a front- or backgate, we have varied the illumination times and we have used two periods a=294nm and a=382nm, respectively. Nevertheless, the results can be summarized in a simple plot. In Fig.8 the carrier density n determined by the periodicity (Eq.1) of the additional oscillations is plotted as a function of the carrier density N_s. The lines in Fig.8 are calculated curves without any free parameter based on the assumption that the period is simply given by the condition that the classical cyclotron orbit is equal to an integer number of fringe periods a (visualized in the inset of Fig.8):

$$R_c = \frac{m^*}{e}\frac{v_F}{B} = \frac{a}{2}(m+\varphi) \qquad m = 1,2,3,\ldots \quad . \tag{2}$$

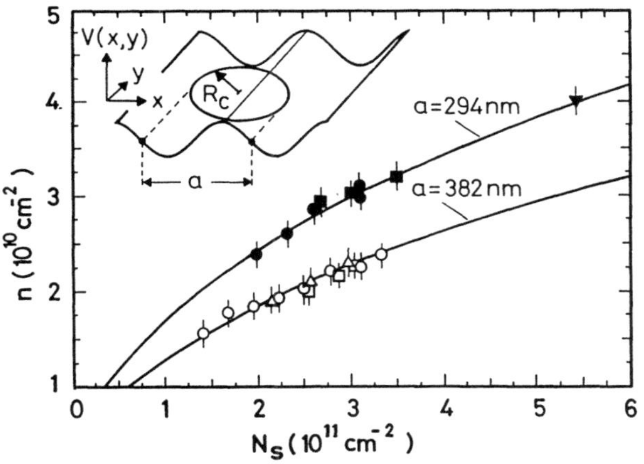

Figure 8: n versus N_s. Full symbols correspond to a laser wavelength λ=488nm, open symbols to λ=633nm and different symbols represent different samples. The solid lines are calculated using the condition that the cyclotron orbit diameter $2R_c$ is equal to an integer multiple of the interference period a, as sketched in the inset

R_c is the cyclotron radius, v_F is the electron velocity at the Fermi energy, m^* is the electron effective mass and φ is a phase factor. Substituting for v_F using $v_F = \hbar k_F/m^*$ we can rewrite Eq.2 to obtain

$$\Delta\frac{1}{B} = e\frac{a}{2\hbar k_F} \tag{3}$$

calculating n using Eq.1 and the relation $k_F = (2\pi N_s)^{1/2}$.

$$n = \frac{2k_F}{\pi a} \quad . \tag{4}$$

This simple picture is in excellent agreement with the experimental data. The model described above requires an elastic free path l_e at least as long as the perimeter of the cyclotron orbit. This agrees with our finding that the number of oscillation periods depends on the mobility of the sample; the higher the mobility the more oscillations are observable. From the mobility of the sample used in the measurements of Fig.4 a mean free path of 12μm can be calculated in agreement with the estimated cyclotron orbit perimeter of 11μm for the largest cyclotron orbit which produces a structure in the $\rho_\perp$ curve at about 0.054T (m=9 in Fig.4). If maxima in the $\rho_\perp$ curves are interpreted as maxima in the scattering rate then the experimentally determined phase shift φ=0.17$\pm$0.06 in Eq.4 implies that the strongest interaction of the cyclotron motion with the periodic potential occurs if the diameter of the cyclotron orbit is 0.17a larger than an integer multiple of the period a. On the other hand minima occur when φ=$-$0.25$\pm$0.06. It should be mentioned that this effect seems to be similar to the magnetoacoustic resonance [6,7]; the detailed mechanism however which results in minima in $\rho_\perp$ while at the same magnetic field $\rho_{||}$ displays maxima can not be explained within this picture at the moment.

The period of the additional oscillations can be described in a more quantum mechanical picture. The additional periodic potential introduces a new reduced Brillouin zone scheme with the period of the reciprocal lattice vector $2\pi/a$. This is illustrated in

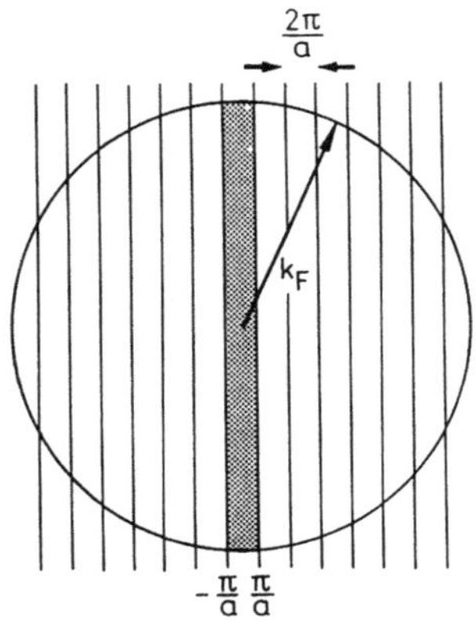

Figure 9: New Brillouin zone structure obtained by superimposing a one-dimensional periodic potential with period a on a free 2-DEG. The states in the first Brillouin zone are displayed shaded. The carrier density required to fill the first Brillouin zone is equal to the carrier density n deduced from the additional oscillations

Fig.9 where the first Brillouin zone is displayed shaded. Since π/a is small compared to k_F in our system, the area A (in k-space) of the occupied states in the first Brillouin zone is given to good approximation by

$$A = 4k_F\frac{\pi}{a} \quad . \tag{5}$$

The carrier density n required to fill the first Brillouin zone can now easily be calculated bearing in mind that one state in k-space occupies the area $(2\pi/L)^2$ (L = length of the sample) and that each state can be occupied by two electrons with opposite spin:

$$n = N/L^2 = \frac{2k_F}{\pi a} \quad . \tag{6}$$

This result is the same as that which we have already obtained in Eq.4. Our finding is that the carrier density which fills the first reduced Brillouin zone is equal to that connected with the additional oscillations. Equation 6 describes correctly the experimentally observed periodicity, however, without explaining the mechanism leading to the oscillatory behaviour of $\rho_\perp$ and $\rho_{||}$. The solution of this problem may be connected to the magnetic breakdown [9],[10].

5. Summary

In summary we have reported a new type of magnetoresistance oscillation, the periodicity of which can be equally well described either by the condition that the classical cyclotron orbit equals an integer multiple of the periodic potential period or in terms of the carrier density required to fill the first reduced Brillouin zone of the modulated system. The detailed mechanism, however, which leads to the additional oscillatory structure remains to be solved.

6. Acknowledgements

We would like to thank D. Heitmann for stimulating suggestions and R. R. Gerhardts, A. MacDonald, P. Středa, and S. Bending for very helpful discussions.

7. References

[1] K.-F. Berggren, T. J. Thornton, D. J. Newson, M. Pepper *Phys.Rev.Lett.* **57**, 1769 (1986)

[2] H. van Houten, B. J. van Wees, E. J. Mooij, G. Roos, K.-F. Berggren *Superlattices and Microstructures* **3**, 497 (1987)

[3] E. F. Schubert, J. Knecht, K. Ploog, *J.Phys.* **C18**, L215 (1985)

[4] K. Tsubaki, H. Sakaki, J. Yoshino, Y. Sekiguchi *Appl. Phys. Lett.* **45**, 663 (1984)

[5] E. F. Schubert, K. Ploog, *Appl. Phys.* **A33**, 63 (1984)

[6] A. B. Pippard *Phil. Mag.* **2**, 1147 (1957)

[7] M. H. Cohen, M. J. Harrison, W. A. Harrison *Phys. Rev.* **117**, 937 (1960)

[8] M. Hundhausen, T. Ichiguchi, Y. Shiraki *Appl. Phys. Lett.* **53**, 110 (1988)

[9] M. H. Cohen, L. Falicov *Phys. Rev. Lett.* **7**, 231 (1961)

[10] A. MacDonald & P.Středa private communication

Quantisation of Resistance in One-Dimensional Ballistic Transport

D.A. Wharam, T.J. Thornton, R. Newbury, M. Pepper, H. Ahmed, J.E.F. Frost, D.G. Hasko, D.C. Peacock, D.A. Ritchie, and G.A.C. Jones*

Cavendish Laboratory, University of Cambridge, Cambridge CB3 0HE, UK

*Also at GEC Hirst Research Centre, Wembley, Middlesex HA9 7PP, UK

1. Introduction

Recent experiments by VAN WEES et al [1] and WHARAM et al [2] on short, narrow constrictions defined in the two-dimensional electron gas (2DEG) of a GaAs-AlGaAs heterostructure have shown that such a constriction possesses a quantised resistance,

$$R = h/2ie^2 \qquad (1)$$

where i is the number of one-dimensional subbands occupied. The existence of the quantised resistance is clearly illustrated in Fig. 1 which shows the gate characteristic of a narrow split gate FET. As the gate voltage becomes more negative the electronic confinement increases and the one-dimensional energy levels pass successively through the Fermi level giving rise to the quantised steps in resistance between plateaux.

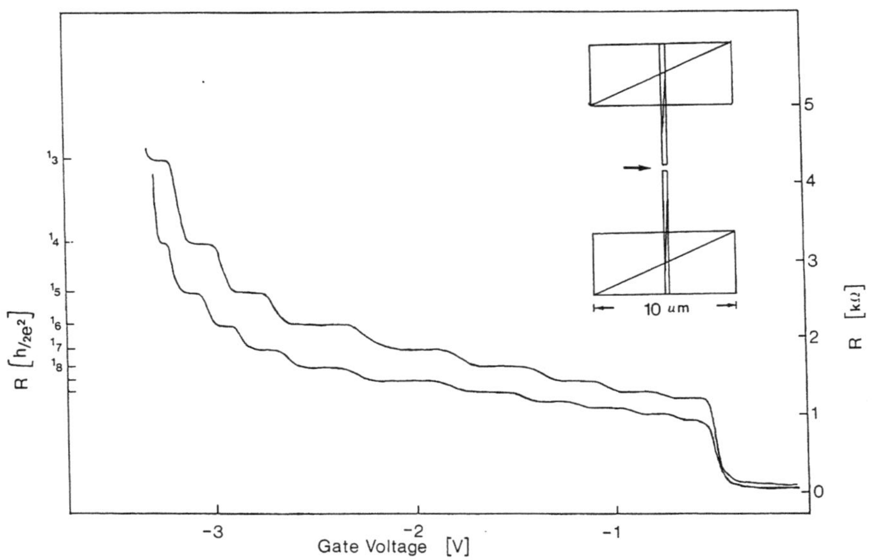

Figure 1. The channel resistance at T ~ 100 mK is plotted as a function of gate voltage for two different carrier concentrations induced by illumination. The existence of a quantised resistance is clearly illustrated. The inset shows a schematic diagram of the split gate structure and the arrow indicates the direction of current flow [2].

Springer Series in Solid-State Sciences, Vol. 87
High Magnetic Fields in Semiconductor Physics II Editor: G. Landwehr

2. Device Fabrication

All of the devices we have studied have been fabricated using high resolution 60 kV electron beam lithography to define a pattern in a 2000 Å PMMA resist layer on the surface of a GaAs-AlGaAs heterostructure. Subsequent evaporation and lift-off of 500 Å of AuPd defined the Schottky barrier which acted as the gate. A typical structure is shown in the inset of Fig. 1. The low temperature mobility of the electrons in the 2DEG was generally between than 2.5 10^5 and 10^6 $cm^2V^{-1}s^{-1}$ and the carrier concentration was between 2 and 5 10^{11} cm^{-2} yielding a mean free path in excess of 2 µm. Typically the fabricated split gate structures had lengths shorter than 0.3 µm and channel widths of the order of 0.5 µm; at the low temperatures at which our experiments were conducted electrons will pass through the channel ballistically and the one-dimensional subband structure of the channel will become apparent.

3. Theoretical Background

The value of quantised resistance given in (1) above has now been derived by various different theoretical approaches all of which have led to the same quantised value [3,4,5,6]. This result may also be readily derived from the Landauer multi-channel fomula [7], $g = \mathrm{Tr}\, tt^*$, where g, the dimensionless conductance is determined for the case of perfect transmission in the absence of scattering between channels, and essentially states that the current contributed by each subband is the same; a consequence of the exact compensation that exists in one dimension between the density of states at the Fermi energy and the Fermi velocity.

In the presence of a transverse magnetic field the value of measured quantised resistance is changed according to,

$$R(B) = h/2i(B)e^2 \tag{2}$$

since i, the number of occupied subbands, is no longer merely dependent upon the electric confinement produced by the split gate. BERGGREN et al [8,9] have shown using a parabolic confining potential well that the essential effects af a transverse field are twofold. Firstly the subband dispersions are flattened via the introduction of an effective magnetic mass, and secondly the subband energy separations are increased with increasing field. This latter effect leads to the phenomenon of magnetic depopulation; as the magnetic field is increased the subband energies pass successively through the Fermi energy and we again observe transitions from one quantised value of resistance to another. This depopulation is further accelerated by the associated increase in the density of states with increasing field.

Furthermore one would expect the application of a magnetic field should also lift the spin degeneracy, which has been assumed in the derivation of (1) above, and give rise to additional plateaux at values of resistance $R = h/2(i+1/2)e^2$ independent of the orientation of the field. However recent experimental evidence [10] suggests that the spin splitting becomes anisotropic in one dimension due to an anisotropy in the enhancement of the Landé g-factor.

4. Experimental Results

The phenomenon of magnetic depopulation discussed in the previous section is clearly shown by the data of Fig. 2. As the magnetic field increases the electric confinement required to depopulate a subband decreases resulting in the steps in resistance shifting to less negative values of applied gate voltage. A variational model for this depopulation of one-dimensional subbands has recently been developed [11] and has allowed us to extract important device parameters, such as the effective channel width and channel carrier concentration, which are in excellent agreement with both the lithographic details and the material parameters.

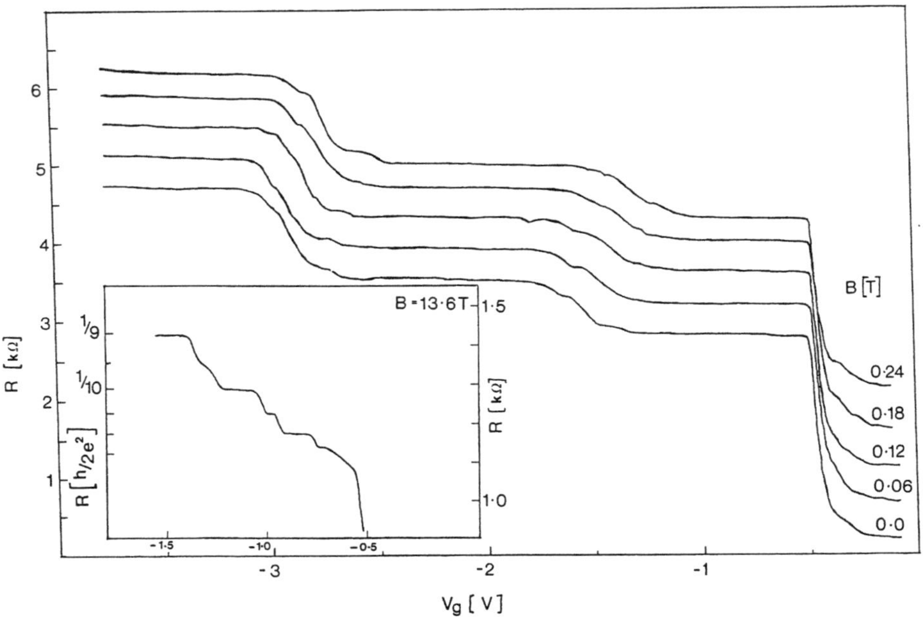

Figure 2. The channel resistance is plotted as a function of gate voltage for various values of transverse field. The mobility of the device at the temperature of measurement, 100 mK was 6.1 10^5 $cm^2V^{-1}s^{-1}$. The curves are offset vertically for clarity. The inset figure demonstrates the effect of a large parallel field, B = 13.6 T [2]. The lifting of the spin degeneracy gives rise to additional plateaux at $R = h/21e^2$, $h/23e^2$, and an incipient plateaux at $h/19e^2$.

The quantised conductance of such magneto-electric subbands has also been studied by VAN WEES et al [12] who in addition observe the onset of spin-splitting at high transverse magnetic fields (B ~ 2 Tesla). However we have only been able to observe spin-splitting in the parallel field orientation and then only at very high magnetic fields i.e. B > 10 Tesla. The additional plateaux introduced by the lifting of the spin degeneracy can be clearly seen in the inset of Fig. 2. Presumably the discrepancy between these observations may be explained by the anisotropic enhancement of the Landé g-factor as found by SMITH et al [10].

More recently we have fabricated double split gate structures in order to investigate the addition of two ballistic resistances in series [13]. Two split gates defined to be 0.3 μm long and 0.3 μm wide were fabricated a distance 1.0 μm apart on the surface of a GaAs-AlGaAs heterostructure which had a low temperature mean free path of 4.6 μm; thus electronic motion through the channel may be regarded as truly ballistic. Independent voltage control of each split gate pair enabled us to vary the width of each narrow constriction separately. The quantised structure for each gate separately was well preserved.

Intuitively we expected that, in the absence of scattering, when the subband structure of both constrictions became identical then an electron would pass ballistically through both constrictions and the value of quantised resistance observed would be that determined by the number of occupied subbands in either of the split gate constrictions. This is of course equivalent to the notion that the ballistic resistance is independent of length. In practice we found that the measured resistance included an anomalous series resistance which complicated the measurement of the quantised resistance. Unique determination of the number of occupied subbands however proved to be possible through the quantisation of the steps between plateaux. The quantised resistance of the combined channel, after subtraction of the anomalous series resistance, is shown in Fig. 3 in a three-dimensional representation as a function of the two applied gate voltages. The quantisation of resistance is maintained, the number of occupied subbands being determined by the narrower of the two constrictions (i.e. that with fewer occupied subbands). Furthermore the proximity of the two split gates leads to a perturbation of the confining potential in the constriction which may be seen by the behaviour of the point of inflection between plateaux four and five.

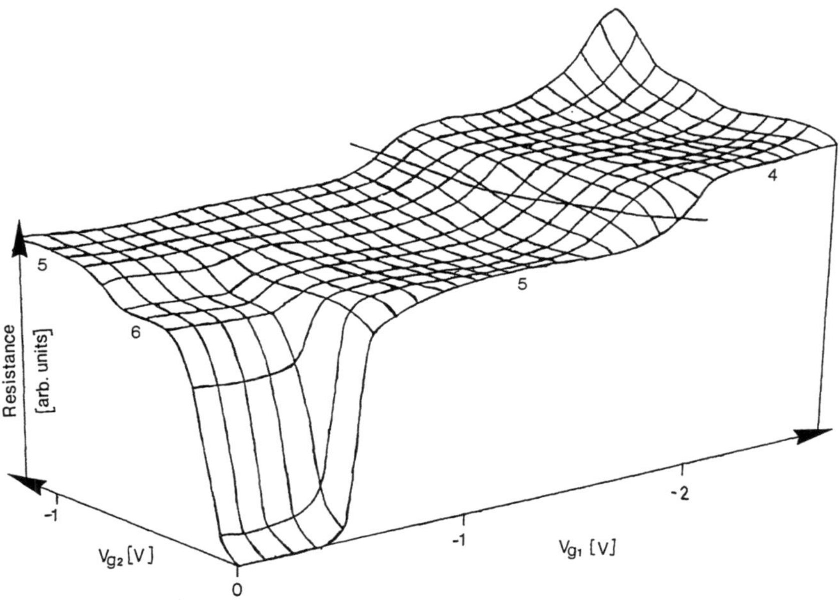

Figure 3. The channel resistance is plotted in a three-dimensional representation as a function of the two applied gate voltages [13]. The quantised nature of the resistance is clearly demonstrated and plateaux indices are marked for clarity. The line connecting the points of inflection between plateaux four and five illustrates the perturbation of the confinement due to the proximity of the gates as discussed in the text.

References

1 B.J. van Wees, H. van Houten, C.W.J. Beenakker, J.G.Williamson, L.P.Kouwenhoven, D. van der Marel, and C.T. Foxon, Phys. Rev. Lett. 60, 848 (1988)

2 D.A. Wharam, T.J. Thornton, R. Newbury, M. Pepper, H. Ahmed, J.E.F. Frost, D.G. Hasko, D.C. Peacock, D.A. Ritchie, and G.A.C. Jones, J. Phys. C 21, L209 (1988)

3 R. Johnston and L. Schweitzer, J. Phys. C Lett. (1988) in print

4 A. Widom and R.Tao, Preprint (1988)

5 Y. Isawa, Preprint (1988)

6 E.G. Haanappel and D. van der Marel, Preprint (1988)

7 R. Landauer, Localization Interactions and Transport Phenomena ed. B. Kramer, G. Bergmann and Y. Bruynserade (Heidelberg: Springer) p.38 (1985)

8 K-F. Berggren, T.J. Thonton, D.J. Newson, and M. Pepper, Phys. Rev. Lett. 57, 1769 (1986)
9 K.-F. Berggren and D.J. Newson, Semicond. Sci. Technol. 1, 327 (1986)
10 T.P. Smith, J.A. Brum, J.M. Hong, C.M. Knoedler, H.Arnot, L.Esaki, Preprint (1988)
11 D.A. Wharam, U.A.S. Ekenberg, M. Pepper, H. Ahmed, J.E.F. Frost, D.G. Hasko, D.C. Peacock, D.A. Ritchie, and G.A.C. Jones, Preprint (1988)
12 B.J. van Wees, L.P.Kouwenhoven, H. van Houten, C.W.J. Beenakker, J.E. Mooij, C.T. Foxon and J.J. Harris, Preprint (1988)
13 D.A. Wharam, M. Pepper, H. Ahmed, J.E.F. Frost, D.G. Hasko, D.C. Peacock, D.A. Ritchie, and G.A.C. Jones, to be published J. Phys. C (1988)

Influence of Magnetic Fields on Ballistic Transport in Narrow Constrictions

B. Huckestein, R. Johnston, and L. Schweitzer

Physikalisch-Technische Bundesanstalt, Bundesallee 100,
D-3300 Braunschweig, Fed. Rep. of Germany

1. INTRODUCTION

Progress in the preparation of high mobility narrow two-dimensional systems has recently led to a number of remarkable observations. The most surprising of these was the discovery of a quantised resistance, $R_c = h/(e^2 s(B)N)$, by van Wees et al. [1] and Wharam et al. [2], where N is an integer and $s(B) = 1$ for magnetic induction $B \neq 0$ and $s(0) = 2$. This empirical relation is strongly reminiscent of the integer Quantum-Hall-Effect found in MOS-FETs and heterostructures at low temperatures and high magnetic fields. From their experimental measuring conditions it is concluded [1,2] that the quantised resistance R_c is associated with ballistic transport through a narrow two-dimensional constriction, so that localisation effects due to disorder are not important.

A model has been proposed recently [3] to explain the quantised resistance R_c in terms of dissipationless currents flowing round a cylinder with finite width W. It has been shown that, within linear response theory, $R_c = U_c/I$ is (independent of W) always quantised at $T = 0$. For finite temperatures, however, the width of the system has to be small so that the quantisation can be resolved in the experiments. A peculiar plateau-width dependence as a function of particle density was predicted which may be observable in experiments.

In this paper the influence of a magnetic field on these results will be considered.

2. THEORY

The Hamiltonian describing the dynamics of non-interacting electrons in a two-dimensional system subject to a perpendicular magnetic field B (z-direction) is

$$H(t) = \frac{\hbar^2}{2m}\left[-\frac{\partial^2}{\partial y^2} + \left(i\frac{\partial}{\partial x} + \frac{eBy}{\hbar} + \frac{e}{\hbar}F(t)\right)^2\right] + g\mu_B BS_z, \tag{1}$$

where m is the mass and e the charge of the electron, and $E_x = -\partial F(t)/\partial t$ is the time dependent electric field around the cylinder which creates the current. The last term is the Zeeman splitting with Landé factor g, Bohr magneton μ_B and spin operator S. Dirichlet boundary conditions are imposed to account for the finite width W of the system. The time dependent function F(t) changes smoothly from zero to $2\pi\lambda$ in a finite time and then stays constant. This corresponds to switching on and off an electric field in the x-direction, which creates a current flowing around the cylinder. Linear response gives for the current in the x-direction, in the limit that the cylinder diameter tends to infinity, see [4], equ. (4.8)

Springer Series in Solid-State Sciences, Vol. 87
High Magnetic Fields in Semiconductor Physics II Editor: G. Landwehr

$$I_x = -\lambda \left(\frac{e}{\hbar}\right)^2 \sum_n \int_{-\infty}^{\infty} dk \frac{\partial f_n(k)}{\partial k} \frac{\varepsilon_n(k)}{\partial k} , \tag{2}$$

where $\{\varepsilon_n(k)\}$ is the set of eigenvalues of the Hamiltonian at $t = 0$ and $f_n(k) \equiv f(\varepsilon_n(k))$ is the Fermi function.The voltage was defined previously (equ. (11)of [3]) which gives for the two point dissipationless conductance $S_c = 1/R_c$ at temperature zero

$$S_c = e^2/h\, N_e , \tag{3}$$

where N_e is the number of spectral lines which are cut by the equilibrium Fermi energy; for details see [3]. It is established that the conductance is quantised and it remains to see how it depends on the width or the electron density of the constriction in the presence of a magnetic field.

2.1 Weak magnetic fields

For weak magnetic fields the energy spectrum is obtained from standard perturbation theory, keeping terms to linear order in B

$$E_{n,\sigma}(k) = \frac{\hbar^2}{2m} \left[\left(\frac{n\pi}{W}\right)^2 + k^2\right] + g\mu_B B\sigma , \tag{4}$$

where the spin quantum number for electrons is $\sigma = \pm 1/2$. With $k_F^2 = 2m\, E_F/\hbar^2$ held constant, the (2N – 1)th and (2N)th steps of S_c (W,B) are at system widths W

$$W_{(2N-1)} = \pi N/\left(k_F^2 + \frac{m}{\hbar^2} g\mu_B\, B\right)^{1/2}, \tag{5}$$

$$W_{(2N)} = \pi N/\left(k_F^2 - \frac{m}{\hbar^2} g\mu_B B\right)^{1/2}, \tag{6}$$

with N = 1, 2, ...

The width P of the plateau between steps (2N) and (2N+1) is given by

$$P_{(2N,\, 2N+1)} = \pi(N+1)/\left(k_F^2 + \frac{m}{\hbar^2} g\mu_B\, B\right)^{1/2} - \pi N/\left(k_F^2 - \frac{m}{\hbar^2} g\mu_B\, B\right)^{1/2}, \tag{7}$$

which in the limit $B \rightarrow 0$ gives π/k_F^2, while between steps (2N–1) and (2N) the plateau width is

$$P_{(2N-1,2N)} = \pi/\left(k_F^2 - \frac{m}{\hbar^2} g\mu_B B\right)^{1/2} - \pi/\left(k_F^2 + \frac{m}{\hbar^2} g\mu_B\, B\right)^{1/2}, \tag{8}$$

which disappears in the limit $B \rightarrow 0$.

Since k_F is known from the plateau width at $B = 0$, the product of electron mass and g-value may be obtained from the low field resistance plateaus.

If the sample width is kept constant, the quantised conductance S_c, as a function of electron density ρ, makes the (2N–1)th step at density

$$\rho_{(2N-1)}(B) = \frac{2}{W^2} \sum_n^N \left(N^2 - n^2\right)^{1/2} - \frac{mg\mu_B B}{\pi\hbar^2} \sum_n^{N-1} \left(N^2 - n^2\right)^{-1/2} \tag{9}$$

and the (2N)th step at

$$\rho_{(2N)}(B) = \frac{2}{W^2} \sum_n^N \left(N^2 - n^2\right)^{1/2} + \frac{mg\mu_B B}{\pi\hbar^2} \sum_n^{N-1} \left(N^2 - n^2\right)^{-1/2} \tag{10}$$

for $N > 1$. The first plateau (e^2/h) starts at density zero and the second jump occurs at density $(2mg\mu_B B/\hbar^2)^{1/2}/\pi W$.

2.2 Strong magnetic fields

In the case of strong magnetic fields it is possible to express the eigenvalue spectrum as a power series in terms of the system parameters [5]. The coefficients of this expansion have to be calculated numerically. In Fig. 1 the dimensionless energy spectrum $\varepsilon^0_{n,\sigma} = 2mW^2/\hbar^2 \cdot E_{n,\sigma}$ (k=0) is plotted as a function of the parameter $b = W^2/\ell^2$, where the magnetic length $\ell = [\hbar/(eB)]^{1/2}$. For fixed Fermi energy E_F and magnetic field B the quantised conductance $S_c(W)$ can easily be obtained by looking for those b_N values for which the straight line $\varepsilon = 2m/\hbar^2 \cdot \ell^2 E_F b$ cuts the spectral lines $\varepsilon^0_{n,\sigma}$ The system width, where the Nth step in $S_c(W)$ occurs, is then giv-

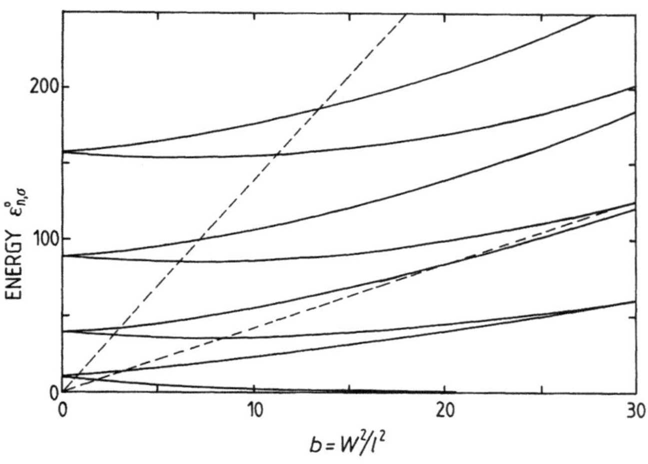

Fig. 1:The four lowest spin-split energy bands ($\varepsilon^0_{n,\sigma} = 2mW^2/\hbar^2 \cdot E_{n,\sigma}(k = 0)$) with $g = 2$ versus $b = W^2/\ell^2$. The dashed lines represent two possible positions of the Fermi energy ($\varepsilon = 2m/\hbar^2\, \ell^2 E_F b$) for fixed ℓ, or two values of the magnetic field for a given E_F

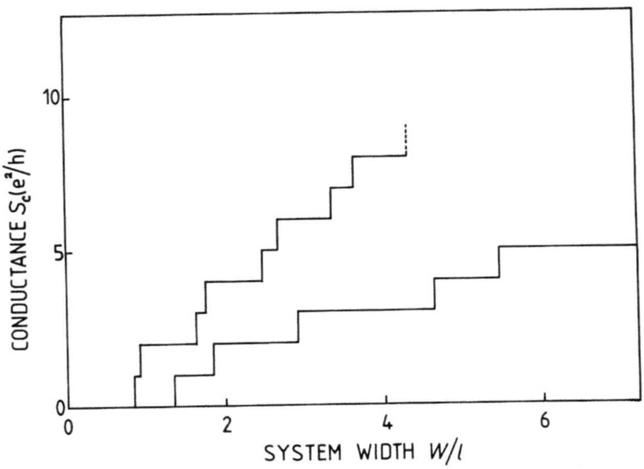

Fig. 2: The quantised conductance S_C as a function of the system width. The curves correspond to the two dashed lines in Fig. 1

en by $W_N = \ell\, b_N^{1/2}$ and the plateau width is $P_{(N,N+1)} = \ell\,(b_{N+1}{}^{1/2} - b_N{}^{1/2})$. In Fig. 2, $S_c(W)$ is plotted for two different ε lines (dashed lines in Fig. 1) and the results agree well with the experimental observations [6].

Since in the limit $b \rightarrow \infty$ the Landau spectrum $\varepsilon^L_{n,\sigma} = (2n-1+g\sigma)b$ is obtained, it is possible to calculate the total number of plateaus N_p for a given $\varepsilon = 2m/\hbar^2 \cdot \ell^2 E_F b$. If $\varepsilon^0_{N,\sigma}$ is the last energy level which in the limit $b \rightarrow \infty$ is smaller than ε and $g<2$, then the total number of plateaus N_p is $2N-1/2(1-2\sigma)$. For $g = 2$, N_p is $2N-1$. In the case $g>2$, due to level crossing, the situation is more complicated.

3. CONCLUSIONS

It has been shown that the conductance S_c is quantised at zero temperature independently of the strength of the magnetic field in a model where the total current and voltage are calculated.

If the magnetic length is much smaller than the width of the system, then the model is the same as that considered for the Quantum Hall Effect [7]. In this regime S_c becomes independent of the width but shows plateaus as a function of the electron density. For this model the conductance S_c is exactly quantised in contrast to the Hall conductance.

REFERENCES

1. B.J. van Wees, H. van Houten, C.W.J. Beenakker, J.G. Williamson, L.P. Kouwenhoven, D. van der Marel and C.T. Foxon, Phys. Rev. Lett. **60**, 848 (1988)
2. D.A. Wharam, T.J. Thornton, R. Newbury, M. Pepper, H. Ahmed, J.E.F. Frost, D.G. Hasko, D.C. Peacock, D.A. Ritchie and G.A.C. Jones, J. Phys. **C21**, L209 (1988)
3. R. Johnston and L. Schweitzer, J. Phys. **C21**, L861 (1988)
4. R. Johnston and L. Schweitzer, Z. Phys. **B69**, 77 (1987)
5. B. Huckestein and R. Kümmel, Z. Phys. **B66**, 475 (1987)
6. B.J. van Wees, L.P. Kouwenhoven, H. van Houten, C.W.J. Beenakker, J.E. Mooij, C.T. Foxon and J.J. Harris, preprint (1988)
7. R. Johnston and L. Schweitzer, Z. Phys. **B72**, 217 (1988)

Part VII

Spin Effects, Cyclotron Resonance in 2D and 3D Systems

Spin-Splitting in Structured Semiconductors

U. Rössler, F. Malcher, and G. Lommer

Institut für Theoretische Physik, Universität Regensburg,
Postfach 397, D-8400 Regensburg, Fed. Rep. of Germany

Spin-splitting of electron states results from a lifting of time reversal or spatial inversion symmetry. In semiconductor heterostructures this can be accomplished by applying a magnetic field or by an inversion asymmetry of the crystal structure or of the built-in potential, respectively. We give a review of these effects for electron states in n-inversion layers of AlGaAs/GaAs heterostructures and MIS-structures on InSb and HgCdTe. The material-specific aspects, in particular the dependence on the energy gap of the terms that give rise to spin-splitting, are discussed.

1. THE ORIGIN OF SPIN-SPLITTING

Spin-degeneracy of electron states results as a combined effect of inversion symmetry in space and time of the system under consideration. (It has to be distinguished from Kramers degeneracy, which is present due to time reversal invariance even in systems without spatial inversion symmetry.) Thus spin-splitting can be caused either by lifting the time reversal symmetry by applying a magnetic field, or by lack of spatial inversion symmetry. In a 3-dimensional solid the latter can be a genuine property of the crystal structure, as e.g. in a zinc blence type semiconductor. But it can be produced also artificially in single heterostructures or asymmetric quantum wells. Inversion asymmetry induced spin-splitting occurs only for electrons with finite wave vector, i.e. for electrons in motion; in semiconductor structures it can be also a combined effect of the electron motion and the built-in electric fields, i.e. a spin-orbit effect.

For bulk semiconductors the spin-splitting of electrons has found considerable interest in recent years in studies of spin-relaxation and optical spin-orientation [1], spin-polarization of photo-emitted electrons from a GaAs (110) surface [2], anisotropic intensity of electric-dipole allowed spin-resonance in InSb [3], but also in detailed investigations of the energy bands [4,5]. In all of these studies the spin-splitting of electron states in the Γ_6 conduction band is ascribed to the inversion-asymmetry induced k^3 term in the dispersion of this band. It results from a contribution $\sim \vec{\sigma} \cdot \vec{\kappa}$ to the conduction band Hamiltonian, where $\vec{\sigma}$ is the vector of Pauli spin-matrices and $\vec{\kappa} = \left(k_x(k_y^2 - k_z^2),\ k_y(k_z^2 - k_x^2),\ k_z(k_x^2 - k_y^2)\right)$. This term, originally introduced by RASHBA and SHEKA [6], acts like a Zeeman term with a "magnetic field" being composed of the electron wave vector $\vec{k} = (k_x, k_y, k_z)$. Actually, this term is only one in a series expansion of the conduction band Hamiltonian in powers of k_x, k_y, k_z including an external magnetic field $\vec{B}$ as shown by OGG [7]. But it is the lowest order spin-splitting term which exists for $\vec{B} = 0$.

Springer Series in Solid-State Sciences, Vol. 87
High Magnetic Fields in Semiconductor Physics II Editor: G. Landwehr

Electron-spin resonance in structured semiconductors has been investigated first by DÄRR et al. [8] for the quasi-two-dimensional electron system in an inversion layer on InSb. The observed reduction of the spin-splitting as compared to the bulk material and its dependence on the surface electron density N_S have been ascribed to the spin-orbit term $\sim \vec{\sigma} \cdot (\vec{k} \times \vec{E})$ where $\vec{E}$ is the built-in electric field of the inversion layer. Originally this term has been introduced for conduction band states in wurtzite type semiconductors [9]. A removal of spin-degeneracy and a reversal of the ordering of Landau levels in an inversion layer on $Hg_{0.79}Cd_{0.21}Te$ as found in Shubnikov–de Haas measurements has been ascribed also to this spin-orbit term [10]. In inversion layers on InSb the calculated spin-splitting of the subband dispersion [11], which is due to the spin-orbit term, has been used to interpret intersubband spectroscopic data [12]. The electric field occurring in the spin-orbit term is an expectation value of the derivative of the interface potential with the subband wave function. Since it is proportional to the force on an electron in this state, it has been argued that this expectation value vanishes, because the subband state is a bound state [8]. However, because of the mass discontinuity at the interface, there is a contribution to the force also from the kinetic energy [13].

Electron-spin resonance in semiconductor heterostructures has been studied by STEIN et al. [14] and gave rise to speculations about a spin-splitting at zero magnetic field because of a linear extrapolation of data obtained at finite magnetic field for $N = 1$ and $N = 2$ Landau levels. An attempt has been made to explain this zero field splitting by the spin-orbit (or RASHBA) term [15]. However, a detailed investigation of Landau levels in AlGaAs/GaAs heterostructures has clearly ruled out a zero-field spin-splitting for Landau levels with small N [16]. This has been confirmed also by more recent experimental investigations of electron-spin resonance in AlGaAs/GaAs single and double heterostructures [17]. A theoretical study has demonstrated that the spin-splitting at the Fermi energy is dominated by Zeeman terms for finite magnetic field but evolves into that of the k^3 term for $\vec{B} \to 0$ [18].

The different aspects of spin-splitting, those connected with the bulk band structure and those caused by interface potentials in structured semiconductors, can be represented in a unified way (Sect. 2), which also allows one to discuss the material specific aspects of the various mechanisms responsible for spin-splitting (Sect. 3).

2. THE k·p APPROACH TO SUBBAND STATES

There are essentially two different concepts to describe electron states in structured semiconductors. The first one starts from the multiband **k·p** Hamiltonian of the Kane model and has been applied first to n-inversion layers on narrow-gap semiconductors [10,19]. Actually the simplest version of the Kane model is used as kinetic energy operator for the subband problem. It considers the conduction band Γ_6 and the spin-orbit split valence bands Γ_8 and Γ_7 as quasi-degenerate and contains besides the energy separations between these states the momentum matrix element P between conduction and valence bands as the single parameter. Diagonalization of this 8×8 Hamiltonian yields isotropic and spin-degenerate but nonparabolic energy bands for the bulk problem. By replacing $k_z \to \frac{1}{i}\partial_z$ and adding the interface potential $U(z)$ the subband problem appears as a set of coupled differential equations:

$$\left(H_{8\times8}\left(k_z \rightarrow \frac{1}{i}\partial_z\right) + (U(z)-E)\,\mathbf{1}_{8\times8}\right)\Psi(z) = 0\,. \tag{1}$$

The simplicity of the spherical Kane model has been essential for its applications to the subband problem [10,11,19]. It allows e.g. to reduce the eight coupled equations (1) to a set of only two for the conduction band envelope functions, which exhibits a spin-splitting of the subband states at finite wave vector $\mathbf{k}_{\|}$, parallel to the interface, due to the spin-orbit term [10]. (The same procedure is possible also for the spherical version of an extended Kane model, which includes also the coupling to the p-antibonding states with a matrix element P' [20].)

While the use of the spherical 8×8 Kane model is justified for inversion layers on narrow-gap semiconductors like $Hg_{0.79}Cd_{0.21}Te$, it becomes questionable for structured semiconductors with larger gap, whose energy bands are increasingly influenced also by remote band contributions. As an example, the k^3 term which causes the spin-splitting of the bulk conduction band, is not contained in the spherical Kane model [21]. This problem can be overcome by taking into account the full $\mathbf{k}\cdot\mathbf{p}$ Hamiltonian of the Kane model or by starting from an extended 14×14 Kane model, which includes fully the p-antibonding conduction bands Γ_7' and Γ_8' [4]. Unfortunately, the complexity of this model prohibits a rigorous reduction of the coupled set of differential equations for the subband problem to only two equations for the conduction band envelope functions, and the numerical treatment of the full problem — even without magnetic field and for rectangular quantum wells — is quite difficult [22].

The other concept to formulate the subband Hamiltonian for conduction band electrons is based on group-theory and is known as invariant expansion [23]. It makes use of the invariance properties of the Hamiltonian by an expansion in products of irreducible tensor components

$$H_{2\times2} = \sum_{\kappa\lambda} a_{\kappa\lambda} \sum_{L} X_L^{(\kappa\lambda)} \mathcal{K}_L^{(\kappa\lambda)^*}(k,E,B) \tag{2}$$

which are constructed from a set of independent 2×2 matrices $(X_L^{(\kappa\lambda)})$ and from components of the electron wave vector, the magnetic field or the built-in electric field $(\mathcal{K}_L^{(\kappa\lambda)^*}(k,E,B))$ and transform according to the irreducible representation Γ_κ. For the bulk conduction band of InSb this concept has been used already by OGG [7]; for n-inversion channels of semiconductor heterostructures it is outlined in Ref. 13 and 16. Additional information is required, however, to arrive at a well-defined Hamiltonian, because group-theory does not provide the material specific prefactors $a_{\kappa\lambda}$ of the invariants. It can be obtained from the first concept by reducing the full $\mathbf{k}\cdot\mathbf{p}$ Hamiltonian of the extended Kane model [4] by higher order perturbation theory to a 2×2 Hamiltonian for the conduction band. By comparison with the invariant expansion (2), the coefficients $a_{\kappa\lambda}$ can be expressed by band parameters [24], which are well known for standard semiconductors [25]. The subband problem is now formulated as

$$\left\{H_{2\times2}\left(k_z \rightarrow \frac{1}{i}\partial_z\right) + (U(z)-E)\,\mathbf{1}_{2\times2}\right\} \quad \begin{Bmatrix} \xi_\uparrow(z) \\ \xi_\downarrow(z) \end{Bmatrix} = 0\,. \tag{3}$$

The 2×2 kinetic energy operator now contains higher order terms in components of the electron wave vector or magnetic and electric field terms. Moreover it depends on z, because the $a_{\kappa\lambda}$ change their values at the interface.

While (3) was obtained by first reducing the bulk $\mathbf{k}\cdot\mathbf{p}$ Hamiltonian by partitioning to a 2×2 conduction band Hamiltonian and then replacing $k_z \rightarrow \frac{1}{i}\partial_z$ and adding the interface potential $U(z)$, the two steps can be performed also in reversed order. In this case, however, the operators arising from $k_z \rightarrow \frac{1}{i}\partial_z$ and the z-dependent potential $U(z)$ appear in energy denominators and can be treated only in an approximate way [20]. On the other hand, these terms lead in a natural way to invariants like the spin-orbit term or allow one to introduce z-dependent masses and g-factors, because the energy separation e.g. between a subband state and the valence band varies with z.

3. SPIN-SPLITTING OF ELECTRONS IN n-INVERSION LAYERS

In this section we present some details of our calculations of Landau levels of electrons in n-inversion layers, give explicit expressions for the relevant terms responsible for spin-splitting and present some results for AlGaAs/GaAs heterostructures and MIS-structures on InSb and $Hg_{0.8}Cd_{0.2}Te$.

For this purpose we consider structures with z as the growth-direction and split the subband Hamiltonian into

$$H_0 = -\frac{\hbar^2}{2}\partial_z \frac{1}{m^*}\partial_z + U(z) + \frac{\hbar^2}{2m^*}k_\parallel^2 \tag{4}$$

the hermitian form of the parabolic approximation, and H_1 which contains the remaining terms of (2). For a magnetic field parallel to the growth direction $\left(\vec{B} = (0,0,B)\right)$ we consider the invariants formed with the irreducible tensor components of Table 1. $\mathcal{K}^{(1,3)^*}$ and $\mathcal{K}^{(1,4)^*}$ transform according to the irreducible representation Γ_1 of T_d and have to be combined with the 2×2 unit matrix to form an invariant.

Table 1: Irreducible tensor components $\mathcal{K}_L^{(\kappa,\lambda)^*}$ considered in the Hamiltonian H_1. $\vec{B} = (0,0,B)$ is the applied magnetic field and $\vec{E} = (0,0,E)$ is the built-in electric field of the heterostructure; $\{a,b\} = \frac{1}{2}(ab+ba)$, $[a,b] = ab - ba$.

$\mathcal{K}_1^{(1,3)^*}$	k^4
$\mathcal{K}_1^{(1,4)^*}$	$\{k_x,k_y\}^2 + \{k_y,k_z\}^2 + \{k_z,k_x\}^2$
$\mathcal{K}_z^{(4,1)^*}$	$[k_x,k_y] = \frac{e}{\hbar c}B$
$\mathcal{K}_x^{(4,2)^*}$	$\{(k_y^2 - k_z^2), k_x\}$
$\mathcal{K}_y^{(4,2)^*}$	$\{(k_z^2 - k_x^2), k_y\}$
$\mathcal{K}_z^{(4,2)^*}$	$\{(k_x^2 - k_y^2), k_z\}$
$\mathcal{K}_z^{(4,3)^*}$	$k^2[k_x,k_y] = k^2\frac{e}{\hbar c}B$
$\mathcal{K}_x^{(4,6)^*}$	$k_y E$
$\mathcal{K}_y^{(4,6)^*}$	$-k_x E$

These terms represent an isotropic and an anisotropic nonparabolicity. The remaining invariants $\mathcal{K}_L^{(4,\lambda)^*}$ transform like Γ_4 and give invariants if combined with σ_L $(L = x, y, z)$, the Pauli spin-matrices. These are responsible for a spin-splitting. Some of these tensor components contain commutators of the components of $\vec{k}$, which are different from zero only in the presence of a magnetic field. They cause the spin-splitting due to lifting the time reversal symmetry. The other spin-splitting terms can be recognized as those due to lack of spatial inversion symmetry and are responsible for a spin-splitting even for vanishing magnetic field.

The eigenvalue problem with H_0 is solved self-consistently for a given heterostructure; the terms of H_1 (Table 1) are considered by a perturbation calculation using the basis $|\nu, N, s>$, where the quantum numbers denote subband (ν), Landau level (N) and spin (s). The expressions of Table 1 have to be converted by $k_x = \frac{1}{\sqrt{2}\lambda_c}(a^+ + a)$, $k_y = \frac{-i}{\sqrt{2}\lambda_c}(a^+ - a)$, where a^+ and a are Landau oscillator operators, and $k_z = \frac{1}{i}\partial_z = \frac{1}{\lambda_c}\zeta$, λ_c being the cyclotron radius.

A complete list of the weighting factors $a_{\kappa\lambda}$ expressed in terms of the band parameters is given in Ref. 24. Here we repeat only the explicit expressions for a_{42} and a_{46}:

$$a_{42} = \frac{C_k P^2}{\sqrt{3} E_g^2} - \frac{4iPQP'}{3}\left[\frac{1}{(E_g + \Delta_0)(E_g' + \Delta_0' - E_g)} - \frac{1}{E_g(E_g' - E_g)}\right] \quad (5)$$

$$a_{46} = \frac{2}{3}\left[\frac{eP^2}{E_0}\left(\frac{1}{E_g} - \frac{1}{E_g + \Delta_0}\right) - \frac{eP'^2}{E_0'}\left(\frac{1}{E_g' + \Delta_0' - E_g} - \frac{1}{E_g' - E_g}\right)\right] \quad (6)$$

where $E_0 = E_g + \Delta_0/3$ and $E_0' = E_g' + 2\Delta_0'/3$. E_g, $E_g + \Delta_0$ and $E_g' + \Delta_0' - E_g$, $E_g' - E_g$ are the energy separations of the bulk conduction band minimum (Γ_6) from the spin-orbit-split valence band ($\Gamma_8 + \Gamma_7$) and the p-antibonding conduction band ($\Gamma_7' + \Gamma_8'$), respectively. P and P' are the corresponding matrix elements of $\frac{\hbar}{m}\mathbf{p}$, while Q is the matrix element between $\Gamma_8 + \Gamma_7$ and $\Gamma_7' + \Gamma_8'$. C_k is the inversion asymmetry induced k-linear term of the valence band [26]. (5) and (6) exhibit as a common feature, that these expressions vanish if the spin-orbit splitting of the bulk bands, Δ_0 and Δ_0', would be zero. Moreover, these weighting factors will be large in systems with small gap E_g and large spin-orbit splitting Δ_0. However, in contrast to a_{46},a_{42} vanishes in a material with spatial inversion symmetry, for which $P' = C_k = 0$ due to parity.

3.1 Results for AlGaAs/GaAs heterostructures

Electron spin-resonance in modulation-doped AlGaAs/GaAs heterostructures revealed the important result, that the spin-splitting depends on the Landau quantum number N, on the electron concentration N_S in the inversion channel, and on the background doping N_D of the GaAs, which accommodates these electrons [14]. All these aspects have been explained quantitatively for single heterostructures in calculations, based on the model outlined before [16]. It has been shown, that the observed reduction of the spin-splitting with respect to that expected from the bulk g-value and its dependence on N can be mainly ascribed to the isotropic k-dependent contribution to the g-factor, $\mathcal{K}_z^{(4,3)^*}$, and to a smaller part to an admixture of the g-factor of AlGaAs due to the penetration of the subband wave function into the large-gap material [16]. Here we

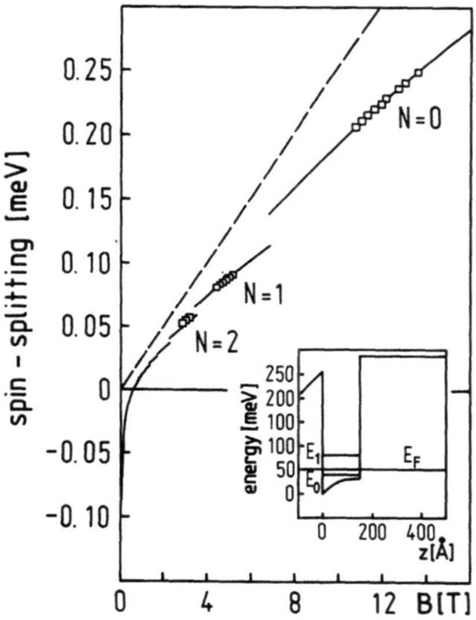

Fig. 1: Calculated spin-splitting of Landau levels at the Fermi energy vs. magnetic field for an $Al_{0.35}Ga_{0.65}As/GaAs$ double heterostructure with $N_S = 3.3 \cdot 10^{11}$ cm^{-2} and $L_z =$ 150 Å and experimental data of Ref. [17]. The dashed line indicates the spin-splitting corresponding to the bulk g-factor. The insert shows the profile of the double heterostructure.

present similar results for a $Al_{0.35}Ga_{0.65}As/GaAs$ double heterostructure in comparison with the experimental results of DOBERS et al. [17] (Fig. 1). The piecewise continuous line represents the calculated spin-splitting of the $N = 0, 1, 2\ldots$ Landau levels for the magnetic-field windows, i.e. at the Fermi energy E_F. As for the single heterostructure our parameter-free calculation is in perfect agreement with the experimental data for $N = 0, 1$, and 2.

For increasing N and $B \to 0$ the product $NB \sim E_F$ remains constant. In this case our calculation, including all terms of H_1, yields a vanishing spin-splitting at a finite magnetic field B_0 and an increase of the spin-splitting with opposite sign, ending for $B = 0$ with the value known from the subband dispersion. This behaviour was found also for a single AlGaAs/GaAs heterostructure in Ref. 18 (Fig. 2). For $N \gg 1$ the secular problem of $H_0 + H_1$, falls approximately into 2×2 blocks with the eigenvalues

$$E_{N\pm} = N\hbar\omega_c^* \pm \frac{1}{2}\left\{(\hbar\omega_c^* + \Delta_z)^2 + 4\Delta_{42}^2\right\}^{1/2} \tag{7}$$

where Δ_z is the Zeeman splitting (from $\mathcal{K}_z^{(4,1)^*}$, $\mathcal{K}_z^{(4,3)^*}$) and

$$\Delta_{42} = \sqrt{2}a_{42}\left[\left(\frac{e}{\hbar c}BN\right)^{3/2} + 2 < \partial_z^2 > \left(\frac{e}{\hbar c}BN\right)^{1/2}\right] \tag{8}$$

the off-diagonal k^3 coupling $\left(\mathcal{K}_L^{(4,2)^*}\right)$. From (7) we obtain for the spin-splitting of the Landau level N

$$2\Delta = E_{N+1,-} - E_{N,+} = \hbar\omega_c^* - \left\{(\hbar\omega_c^* + \Delta_z)^2 + 4\Delta_{42}^2\right\}^{1/2}. \tag{9}$$

For $B \to 0$ ($N \to \infty$) the cyclotron energy $\hbar\omega_c^*$ and the Zeeman splitting Δ_z vanish, but Δ_{42} remains constant and becomes the subband splitting $k_{\parallel F}$ (dotted line of Fig. 2).

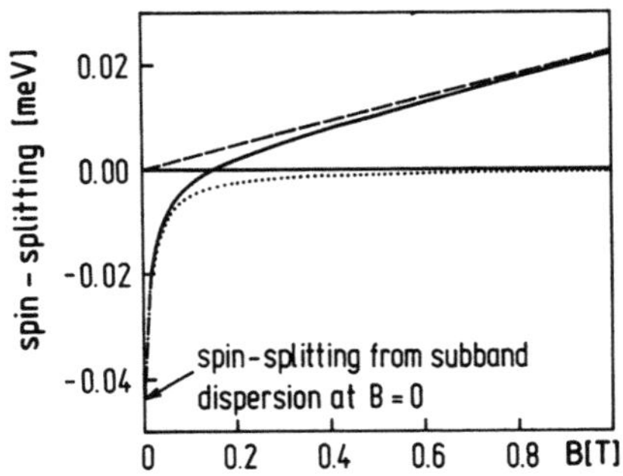

Fig. 2: Calculated spin-splitting of Landau levels at the Fermi energy vs. magnetic field for an $Al_{0.3}Ga_{0.7}As/GaAs$ single heterostructure with $N_S = 1.0 \cdot 10^{11}$ cm^{-2} and $N_D = 0.5 \cdot 10^{11}$ cm^{-2} (solid line). The broken and dotted lines indicate the spin-splitting if only the Zeeman terms or the k^3 term, respectively, are considered [18].

For increasing B, however, $\hbar\omega_c^*$ and Δ_z become much larger than Δ_{42} and dominate the spin-splitting (dashed line of Fig. 2).

3.2 Material-specific aspects

As mentioned in connection with (5) and (6), the weighting factors $a_{\kappa\lambda}$ depend on the momentum matrix elements and band separations. In order to demonstrate this material dependence we give the values of a_{42} and a_{46} for GaAs, InSb and $Hg_{0.8}Cd_{0.2}Te$ together with E_g and Δ_0 in Table 2. It is seen, that the $a_{\kappa\lambda}$ vary by several orders of magnitude. Table 2 contains also the total zero-field spin-splitting 2Δ at $k_{\|F}$ and the separate contributions of the k^3 ($2\Delta_{42}$) and spin-orbit terms ($2\Delta_{SO}$) for inversion layers with the same electron and depletion concentration. These results show, that for AlGaAs/GaAs the zero-field spin-splitting at $k_{\|F}$ is clearly dominated by the k^3 term and the contribution of the spin-orbit term is negligible. However, for systems with smaller gap and larger spin-orbit coupling (Δ_0) the influence of the spin-orbit term, $\mathcal{K}_L^{(4,6)^*}$, increases and becomes even stronger than that of the k^3 term.

Table 2: Upper part: Bulk values of the gap energy E_g and of the spin-orbit splitting Δ_0. Lower part: Spin-splitting of the lowest subband at the Fermi energy for inversion layers with $N_S = 5 \cdot 10^{11}$ cm^{-2} and $N_D = 0.8 \cdot 10^{11}$ cm^{-2} in three different materials.

	GaAs	InSb	$Hg_{0.8}Cd_{0.2}Te$
E_g [eV]	1.519	0.237	0.058
Δ_0 [eV]	0.341	0.810	0.900
	AlGaAs/GaAs	InSb	$Hg_{0.8}Cd_{0.2}Te$
2Δ [meV]	0.111	2.48	9.53
$2\Delta_{42}$ [meV]	0.109	1.61	4.58
$2\Delta_{46}$ [meV]	0.024	1.84	8.36

For AlGaAs/GaAs heterostructures we have shown that the expectation value of the electric field is well defined in terms of the self-consistent Hartree-potential on both sides of the interface and the conduction band offset. Moreover, we found a vanishing expectation value of the force on an electron in the inversion layer due to compensation of two terms, coming from the z-dependence of the kinetic energy (mass discontinuity) and from the electric field [13]. In contrast for MIS-structures, neither the band-offset nor the mass-discontinuity at the semiconductor-insulator interface is known. In our calculations we have modelled the insulator as a large gap system, containing two parameters: the conduction band-offset ΔV_c and the electron mass. Using the condition of vanishing force on a bound state, we eliminated the mass and found for $\Delta V_c \simeq 2$ eV (which we regard as realistic value) the results given in Table 2. In spite of this result we have to confess that all calculations of spin-splittings in MIS-structures suffer from our lack of knowing how to properly describe the insulator.

3.3 Landau levels of inversion layers on $Hg_{0.8}Cd_{0.2}Te$

The results of Table 2 for $B = 0$ show clearly, that for decreasing gap energy the spin-orbit term becomes increasingly important, whereas for the large gap system AlGaAs/GaAs the spin-splitting is dominated by the k^3 term. In this section, we study the effect of both terms on the fan chart of Landau levels, which may serve as fingerprint to detect these terms. Since the striking features occur at relatively low magnetic fields a simplified, but analytical, calculation becomes possible. The spin-orbit term

$$H_{46} = a_{46}\frac{\sqrt{2}}{\lambda_c}E\begin{pmatrix} 0 & -ia^{+} \\ ia & 0 \end{pmatrix} \tag{10}$$

couples Landau levels $|N+1,+\rangle$ and $|N,-\rangle$, whereas the k^3 term (which is simplified by neglecting contributions $\sim a^3$, a^{+3})

$$H_{42} = a_{42}\frac{1}{\sqrt{2}\lambda_c^3}\begin{pmatrix} 0 & a(a^{+}a - 2\zeta^2) \\ a^{+}(a^{+}a - 2\zeta^2) & 0 \end{pmatrix} \tag{11}$$

couples levels $|N-1,+\rangle$ and $|N,-\rangle$. The unperturbed Landau levels are calculated from

$$E^0_{N\pm} = -\frac{E_g^{eff}}{2} + \left\{\left(\frac{E_g^{eff}}{2}\right)^2 + E_g^{eff}\left(\hbar\omega_c^*\left(N+\frac{1}{2}\right) \pm \frac{1}{2}g^*\mu_B B\right)\right\}^{1/2} \tag{12}$$

to account for nonparabolicity. E_g^{eff} is an effective gap, which considers the z-dependent separation of the subband levels from the valence band.

In Fig. 3 we show the drastic modifications of the unperturbed Landau levels $E^0_{N\pm}$ (dashed lines) caused by the spin-orbit term (Fig. 3a) and by the k^3 term (Fig. 3b). The main feature is the crossing of Landau levels, which occurs between different pairs and at different magnetic fields if the coupling is caused by H_{46} or by H_{42}. Only a single Landau level remains unchanged, which is $|0,+\rangle$ in Fig. 3a and $|0,-\rangle$ in Fig. 3b. The crossing of Landau levels due to the spin-orbit term has been found already by OHKAWA and UEMURA [10] and can be detected by merging SdH oscillations. In fact, the experimental results of Ref. 27 seem to be consistent with the fan chart of

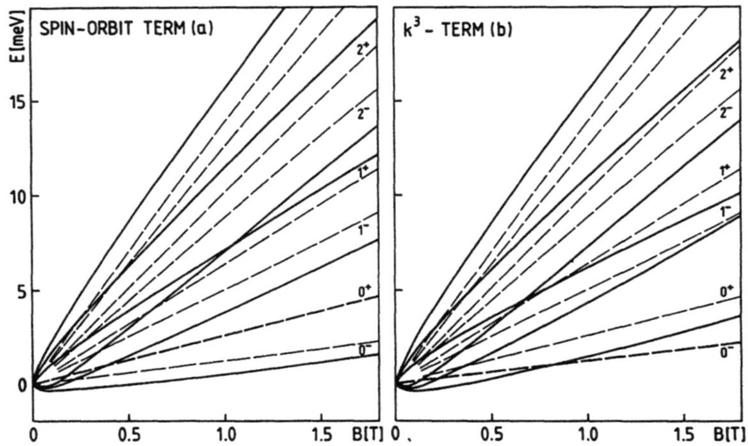

Fig. 3: Energy of Landau levels vs. magnetic field for an inversion layer on HgCdTe showing the influence of (a) the spin-orbit term and (b) of the k^3 term (heavy lines). For comparison the unperturbed energies of (12) are shown (broken lines).

Fig. 3a, indicating a dominance of the spin-orbit term in the $Hg_{0.8}Cd_{0.2}Te$ system. The lacking knowledge of the interface electric field, however, prohibits a confirmation of this interpretation by quantitative calculations.

4. CONCLUSIONS

Spin-splitting in semiconductors and semiconductor structures can be caused by lifting the symmetry of time reversal or of spatial inversion. The former leads to Zeeman terms, which vanish for zero applied magnetic field. The latter arise from inversion-asymmetry of the bulk semiconductor or of the built-in electric field. The material-specific aspects of these terms are discussed and their influence on spin-splitting in several structured semiconductors is studied by quantitative calculations. For AlGaAs/GaAs heterostructures we find perfect agreement with experimental data. The quantitative aspects of spin-splitting in inversion-layers on MIS-structures remain unclear, because the insulator and the interface are not well-defined in our models.

5. ACKNOWLEDGEMENTS

Our work was stimulated by discussions with K. von Klitzing, F. Koch, J. Kotthaus and their co-workers; it was supported by the Deutsche Forschungsgemeinschaft.

6. REFERENCES

1. For a review see: G.E. Pikus and A.N. Titkov: In *Optical Orientation*, ed. by F. Meier and B.P. Zakharchenya (Elsevier Science, New York 1984) p. 73

2. H. Riechert, S.F. Alvarado, A.N. Titkov, and V.I. Safarov: Phys. Rev. Lett. **52**, 2297 (1984)
3. M. Dobrowolska, Y.-F. Chen, J.K. Furdyna, and S. Rodriguez: Phys. Rev. Lett. **51**, 134 (1983)
4. U. Rössler: Solid State Commun. **49**, 943 (1984)
5. N.E. Christensen, M. Cardona: Solid State Commun. **51**, 491 (1984);
   M. Cardona, N.E. Christensen, M. Dobrowolska, J.K. Furdyna, S. Rodriguez: Solid State Commun. **60**, 17 (1986)
6. E.I. Rashba, V.I. Sheka: Fiz. Tverd. Tela **3**, 1735 (1961) [Sov. Phys. – Solid State **3**, 1257 (1961)]
7. N.R. Ogg: Proc. Phys. Soc. (London) **89**, 431 (1966)
8. A. Därr, J.P. Kotthaus, and T. Ando: In *Proc. 13th Int. Conf. Phys. Semicond.*, Rom 1976, ed. by F.G. Fumi (Tipografia marves, Rome 1976) p. 774
9. E.I. Rashba: Fiz. Tverd. Tela **2**, 1224 (1960) [Sov. Phys. – Solid State **2**, 1109 (1960)]
10. F.J. Ohkawa and Y. Uemura: J. Phys. Soc. Jpn. **37**, 1325 (1974)
11. G.E. Marques and L.J. Sham: Surf. Sci. **113**, 131 (1982)
12. K. Wiesinger, W. Beinvogl, and J.F. Koch: In *Proc. 14th Int. Conf. Physics Semicond.*, Edinburgh 1978, ed. by B.L.H. Wilson (Inst. of Physics, Bristol 1978) p. 1215
13. F. Malcher, G. Lommer, and U. Rössler: Superlattices and Microstructures **2**, 267 (1986)
14. D. Stein, K. von Klitzing, and G. Weimann: Phys. Rev. Lett. **51**, 130 (1983);
    D. Stein, G. Ebert, K. von Klitzing, and G. Weimann: Surf. Sci. **142**, 406 (1984)
15. Yu.A. Bychkov and E.I. Rashba: Pisma Zh. Eksp. Teor. Fiz. **39**, 66 (1984) [JETP Letters **39**, 78 (1984)]; J. Phys. C**17**, 6039 (1984); In *Proc. 17th Int. Conf. Phys. Semicond.*, San Francisco 1984, ed. by J.D. Chadi and W.A. Harrison (Springer, New York 1985) p. 321
16. G. Lommer, F. Malcher, and U. Rössler: Phys. Rev. B**32**, 6965 (1985); Superlattices and Microstructures **2**, 273 (1986)
17. M. Dobers, K. von Klitzing, and G. Weimann: Phys. Rev. B**38**, 5453 (1988)
18. G. Lommer, F. Malcher, and U. Rössler: Phys. Rev. Lett. **60**, 728 (1988)
19. Y. Takada, K. Arai, N. Uchimura, and Y. Uemura: J. Phys. Soc. Jpn. **49**, 1851 (1980);
    Y. Takada, K. Arai, and Y. Uemura: In *Physics of Narrow Gap Semiconductors*, Proc. Int. Conf. Linz 1981, ed. by E. Gornik, H. Heinrich, and L. Palmetshofer (Springer, Berlin 1982) p. 101
20. R. Lassnig: Phys. Rev. B**31**, 8076 (1985)
21. The k^3 term has tetrahedral symmetry, see [4] and [5].
22. See e.g. R. Eppenga, M.F.H. Schuurmans, and S. Colak: Phys. Rev. B**36**, 1554 (1987)
23. G.L. Bir and G.E. Pikus: *Symmetry and Strain-induced Effects in Semiconductors* (Wiley, New York 1974)
24. M. Braun and U. Rössler: J. Phys. C**18**, 3365 (1985)
25. Landolt-Börnstein, Series III, Vol. 17a: Semiconductors, ed. by O. Madelung (Springer, New York 1982) and Vol. 22
26. M. Cardona, N.E. Christensen, and G. Fasol: Phys. Rev. Lett. **56**, 2831 (1986)
27. R. Wollrab, R. Sizmann, private communications

Electron Spin Resonance in the Two-Dimensional Electron Gas of GaAs-AlGaAs Heterostructures

M. Dobers[1], *F. Malcher*[2], *G. Lommer*[2], *K. v. Klitzing*[1], *U. Rössler*[1], *K. Ploog*[1], *and G. Weimann*[3]

[1]Max-Planck-Institut für Festkörperforschung, Heisenbergstr. 1 D-7000 Stuttgart 80, Fed. Rep. of Germany

[2]Universität Regensburg, Institut für Physik I, Theoretische Physik, Postfach 397, D-8400 Regensburg, Fed. Rep. of Germany

[3]Walter-Schottky-Institut, TUM, D-8046 Garching, Fed. Rep. of Germany

1. Introduction

The investigation of the magnetooptical excitations in a two-dimensional electron gas (2DEG) has up to now almost always been restricted to the frequency range between far infrared and visible light involving intersubband resonance, cyclotron resonance etc. [1,2]. However, the neighbouring range, covering millimeter and centimeter waves, has been neglected. This is not surprising, since the photon energies are smaller than the thermal energies even at liquid Helium temperatures. Accordingly, if there are two arbitrary energy levels a centimeter wave photon energy apart from each other, a resonant transition should hardly be observable, because these levels are equally thermally populated. In the case of electron spin resonance (ESR) in a 2DEG, however, the conditions are different. Strong exchange interaction ("g-factor-enhancement") [3] can lead to a drastically different population of the spin levels even in GaAs-AlGaAs heterostructures, where the spin-splitting of a Landau level is in the order of hundred μeV (at $B \approx 10T$). Hence, ESR can also be done in a 2DEG [4,5].

The energy spectrum of a 2DEG in a perpendicular magnetic field for a parabolic band is the following [1]:

$$E_{N,m_S} = E_j + (N + \frac{1}{2})\frac{e\hbar}{m^*}B + m_s g \mu_B B, \qquad (1)$$

with E_j being the subband energy, e the electronic charge, $\hbar$ Planck's constant and μ_B the Bohr magneton. $N = 0, 1, 2\ldots$ is the Landau level index and m_s the magnetic spin quantum number, which takes the two values $m_s = +1/2$ ("spin up") and $m_s = -1/2$ ("spin down"). The splitting between two Landau levels depends upon the effective mass m^*, whereas the spin-splitting is determined by the effective g-factor g. Both these parameters are a consequence of the actual band structure so that appropriate experiments can be compared with bandstructure calculations. In this paper we concentrate on the experimental investigation of the Landau level *spin*-splittings in the 2DEG of GaAs-AlGaAs heterostructures and compare the findings with theoretical results.

2. Experimental Method

Conventional ESR-spectrometers in principle detect the resonant absorption of microwave power when the spin-splitting equals the microwave photon energy. This technique [6] is very sensitive as it requires only of the order of 10^{12} flipping spins. For a 2DEG, however, this is not sensitive enough, since with sheet densities of the order of $N_s \approx 10^{11} cm^{-2}$ electrons and areas of the order of square millimeters, one is far below

Springer Series in Solid-State Sciences, Vol. 87
High Magnetic Fields in Semiconductor Physics II Editor: G. Landwehr

detectability. In addition, the presence of free conducting electrons complicates conventional ESR, so that such experiments even on large-area- and multi-layer-samples do not seem to be promising.

The magnetoresistivity ρ_{xx}, however, is affected by ESR. Because the sample itself acts as a detector for ESR, via a sort of photoconductivity, the sensitivity is often high enough to study systematically the Landau level spin splittings. Figure 1 shows the magnetoresistivity of a heterostructure under microwave irradiation. Beside the Shubnikov-de Haas oscillations (SdH-), revealing broad regions of vanishing ρ_{xx} at higher magnetic fields, there is a sharp resonance structure at $B \approx 5.5$ T which is due to ESR. The microwave frequency is chosen so as to produce ESR exactly at $i = 3$ ($i = hN_s/(eB)$ being the filling factor). In this case the resonant change of ρ_{xx} can be so strong that it is even visible on the scale of the SdH-oscillations.

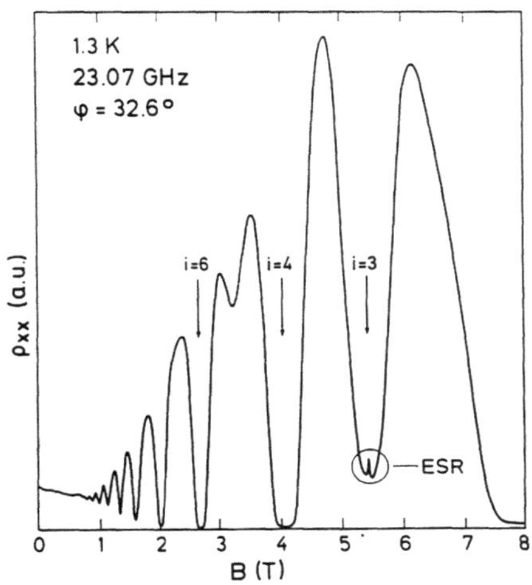

Figure 1: Magnetoresistivity ρ_{xx} of sample 2 under microwave radiation. The filling factors i of some of the minima are given. The sharp peak at $B \approx 5.5$ T is due to ESR. The sample normal has been tilted to the magnetic field by $\varphi = 32.6°$

Depending upon the microwave power and the filling factor, the resonant change of ρ_{xx} may be much weaker [5]. Then, ESR can be detected more sensitively by measuring the *change* of ρ_{xx} due to microwave irradiation $\Delta\rho_{xx}$. Such a spectrum is shown in Fig. 2 (solid line) in comparison with ρ_{xx} itself (dashed line). $\Delta\rho_{xx}$ can be divided into two constituents: a broad spectrum, which is due to non-resonant heating of the electrons and closely related to the temperature-derivative of ρ_{xx}, and a resonant peak (here at $B \approx 10.7$ T). By changing the microwave frequency, the broad spectrum is only varied because of different microwave intensities. The resonant structure, on the other hand, shifts with the applied frequency on top of the broad spectrum.

By this, the magnetic field dependence of the spin-splitting can be studied. Out of many different samples showing in principle similar results, we concentrate on four as examples. The samples are heterostructures grown by molecular beam epitaxy. They consist of a semi-insulating GaAs substrate, followed by a 2.2μm-4μm thick undoped GaAs buffer layer, an undoped $Al_xGa_{1-x}As$ spacer layer (15.6 nm - 39 nm), a Si-doped $Al_xGa_{1-x}As$ layer (33 nm - 49 nm) and an undoped GaAs cap layer (6.6 nm - 25 nm). Whereas the samples 1,3 and 4 have this conventional layer sequence, sample 2 has got an additional short-period superlattice (6×10.4 nm $Al_xGa_{1-x}As$, 5×3.4 nm GaAs) separated by 15 nm from the spacer, so that sample 2 is in fact a single-side doped quantum well of width 15 nm [7]. For the electron concentrations, Aluminum contents and mobilities see the table in section 4.

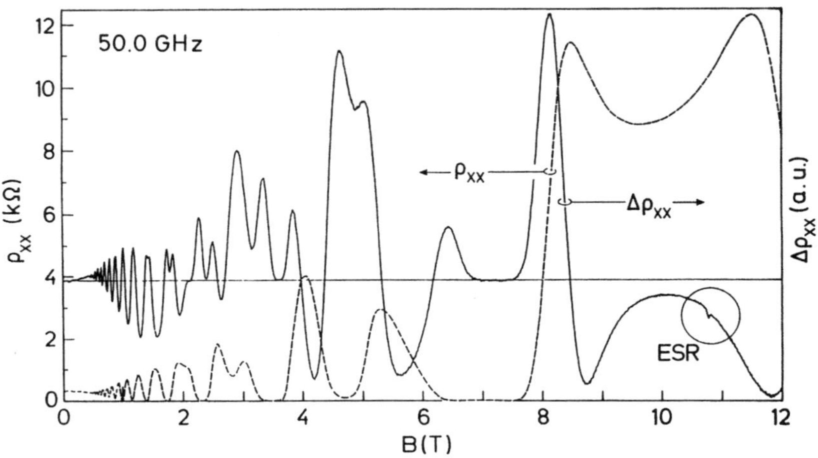

Figure 2: Magnetoresistivity ρ_{xx} of sample 2 (dashed line) and its change due to microwave irradiation ($\nu = 50.0$ GHz) $\Delta\rho_{xx}$ (solid line). ESR occurs at $B \approx 10.7$ T

The samples had standard Hall bar geometry and were located inside an oversized waveguide in front of a moveable short, immersed in liquid Helium. The magnetoresistivity was measured by standard lock-in detection typically utilizing frequencies of the order of 10 Hz. Changes of the sample resistivity due to the chopped microwave radiation (1 kHz) were detected using a second lock-in amplifier. In magnetic fields up to 14.5 T various microwave sources (klystrons and backward wave oscillators) were used to cover the frequency range up to 70 GHz with microwave powers between a few and several hundred mW.

3. ESR-Lineshape

The ESR-line in $\Delta\rho_{xx}$ is not a pure absorption line, but reflects also the conductivity mechanism, which is not understood in detail. Therefore, contrary to the case of conventional ESR [6], a detailed analysis of the ESR-line in microwave photoconductivity is not possible. However, from the approximate width of the lines of the order of $\Delta B = 50$ mT, the homogeneous width of the density of states of the spin levels can be estimated. It is derived to be less than $\Gamma_{ESR} = \frac{1}{2} g \mu_B \Delta B = 1 \mu eV$. This is by orders of magnitude narrower compared with the level-width derived from the dc mobility and, in the samples we investigated, also at least two orders of magnitude narrower than the level-width derived from the cyclotron resonance linewidth [8] (although the cyclotron frequencies are a hundred times larger than the ESR frequencies). Obviously, the scattering mechanisms, leading to level broadening in dc transport and cyclotron resonance, do not affect the spin of the scattered electrons. The strength of spin-flip scattering is hence another characteristic quantity of a sample beside the non-spin-flip scattering, which - as long as it predominates - determines the dc mobility.

The ESR structure in $\Delta\rho_{xx}$ can consist either of an increase or a decrease of ρ_{xx}. As can be seen in Fig. 3, this behaviour depends systematically on the filling factor: in the vicinity of an odd filling factor (i.e. minimum-region of ρ_{xx}), ESR results in an *increase* of ρ_{xx}. Approaching the neighbouring even filling factors, ESR results in a *decrease* of ρ_{xx} (in the maximum regions of ρ_{xx}). Continuing towards the even filling factors, ESR is not measurable at all, because there the spin levels are equally populated, either completely full or empty. These effects give rise to possibilities for further investigations.

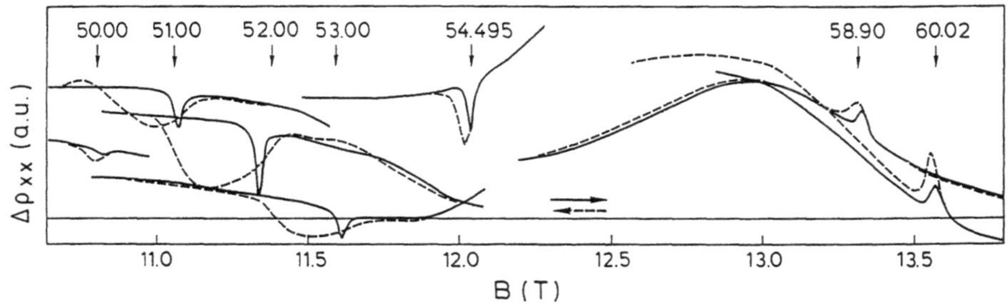

Figure 3: Typical ESR-structures in the change of magnetoresistivity due to microwave radiation $\Delta\rho_{xx}$ for different microwave frequencies (noted in GHz above the respective peaks). The solid lines refer to measurements while sweeping the magnetic field up ($\dot{B} > 0$), the dashed to $\dot{B} < 0$

Another, totally different effect also influences the ESR lineshape. In Fig. 3 every ESR line is given for a magnetic field up sweep (solid lines) and also for a down sweep (dashed lines). The hysteretic behaviour, the so-called Overhauser shift [9], is due to hyperfine interaction between the electronic and the nuclear spins [10]. Sweeping down, the resonance line is shifted, or at least broadened to lower magnetic fields. Sweeping up, on the other hand, the line can be narrowed. These effects are well understood [11]. Here we only want to note that they complicate a possible lineshape analysis.

4. Magnetic field dependence of the spin splitting

The magnetic field dependence of the Landau level spin-splittings ΔE is derived from a series of experiments with different microwave frequencies just as in Fig. 3. Results of sample 1 for $N = 0, 1, 2$ are given in Fig. 4. The spin-splittings of a certain Landau level are restricted to a specific magnetic field range, because ESR in a certain Landau level can only be done, as long as the Fermi level is lying within this Landau level. The measured spin-splittings are well below the dotted line $\Delta E = 0.44\mu_B B$, which is calculated with the bulk GaAs conduction band edge g-factor $|g| = 0.44$ [12]. For a given sample, the g-factor depends on the magnetic field and the Landau level index in accordance with

$$g(B,N) = g_0 - c \cdot (N + \frac{1}{2})B \quad . \tag{2}$$

The experimentally determined g-factors of the four samples are given in Fig. 5. As in Fig. 4, the solid lines are least square fits following Eq. 2.

We have performed self-consistent subband calculations for a subband Hamiltonian obtained from a 14×14 $\underline{k} \cdot \underline{p}$-model for heterostructures and single-sided doped quantum wells with given 2d-electron density N_s, depletion charge density N_{depl} and Aluminum content x of the barrier material. For details we refer to Refs. 13-15. These calculations take into account the conduction band nonparabolicities as well as spin-orbit coupling in the electric field at the interface. The theory yields the following results:

- The dependence of the calculated spin-splittings on the magnetic field and Landau level index agrees with the experimental results.
- For GaAs-AlGaAs heterostructures the reduction of the spin-splitting compared to the bulk-GaAs is caused by the nonparabolicity of the bulk-GaAs band structure. The theoretical results [15] justify the least square fits of the experimental data with Eq. 2. The spin-orbit coupling turns out to be negligbly small [13].

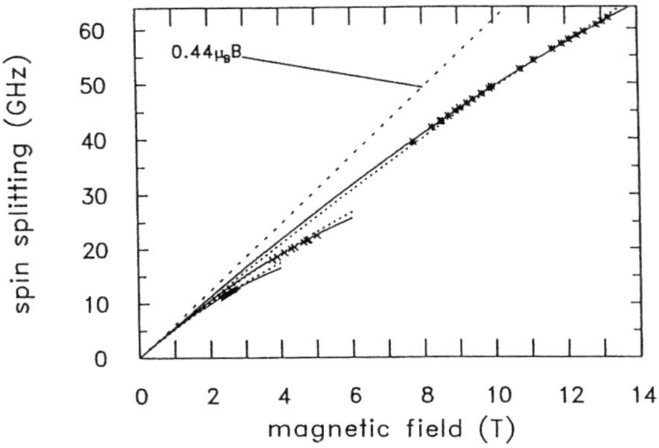

Figure 4: Measured spin-splittings of the three lowest Landau levels of sample 1. The solid lines are least square fits to the experimental data in accordance with Eq. 2. The dashed lines are results of bandstructure calculations

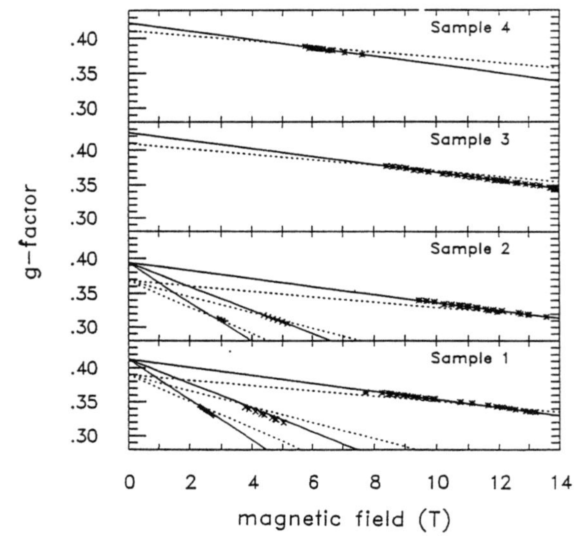

Figure 5: Experimentally determined g-factors of the four samples. The solid lines are least square fits to the experimental data in accordance with Eq. 2. The dashed lines are results of bandstructure calculations

Table Electron densities N_s, mobilities μ at 4.2 K and Aluminum contents x in the barrier materials of the four samples and their respective spin splitting parameters g_0 and c (results of least square fits in accordance to Eq. 2)

	$N_s[cm^{-2}]$	$\mu[\frac{cm^2}{Vs}]$	$x[\%]$	g_0	$c[\frac{1}{T}]$
sample 1	$3.0 \cdot 10^{11}$	335,000	39	0.414	0.0122
sample 2	$3.3 \cdot 10^{11}$	360,000	35	0.394	0.0115
sample 3	$2.3 \cdot 10^{11}$	1,000,000	37	0.425	0.0114
sample 4	$1.3 \cdot 10^{11}$	265,000	47	0.410	0.0083

- With increasing N_s or N_{depl}, the reduction of the spin-splitting increases due to higher values of the Fermi energy and changes in the subband wavefunction, which result in a stronger influence of the nonparabolicity.
- For low Landau level quantum numbers N, the spin-splitting vanishes for $B \to 0$ [13,14]. However, the *observable* spin-splitting (i.e. that at E_F, with $N \to \infty$ for $B \to 0$) evolves into the spin-splitting of the subband dispersion, caused by the inversion asymmetry induced k^3-nonparabolicity [16].

As already mentioned, the calculated spin-splittings depend on the depletion charge density N_{depl}. Unfortunately, this value is not well-defined from the sample preparation and estimated values have to be used in the calculation. In spite of this uncertainty, we achieved good agreement between theory and experiment.

5. Variation of N_s by illumination with red light

One possibility to change N_s is by illumination with the light of a red LED [17-20]. The increase of N_s can be divided into a decaying ("transient photoconductivity") [21] and a stable part ("persistent photoconductivity, PPC") [18,19]. Two possible mechanisms have been discussed for PPC:

- Photons with an energy larger than the GaAs-bandgap create electron-hole pairs in the GaAs. These are separated by the electric field at the interface: the electron moves to the 2DEG and increases N_s, whereas the hole moves in the opposite direction and recombines with an ionized acceptor in the GaAs and thus decreases N_{depl}. This mechanism increases N_s at the cost of N_{depl}.
- The second mechanism is the ionization of DX-centers in the doped AlGaAs and the subsequent tunneling of the electrons into the 2d-layer. This mechanism does not alter N_{depl}.

Successive illumination with short (≈ 1 ms) light pulses increases at first preferentially N_s via the first mechanism until the depletion charge is neutralized. Then the second mechanism becomes dominant.

Figure 6 shows the g-factors of sample 2 as a function of N_s, which was determined from the periodicity of the SdH-oscillations. Results for two runs are presented. After the first run, the sample was warmed up to room temperature and then cooled down again for the next run. If N_{depl} were constant, increasing N_s would lead to a higher Fermi level and hence, due to a stronger influence of nonparabolicity, to a *decreasing*

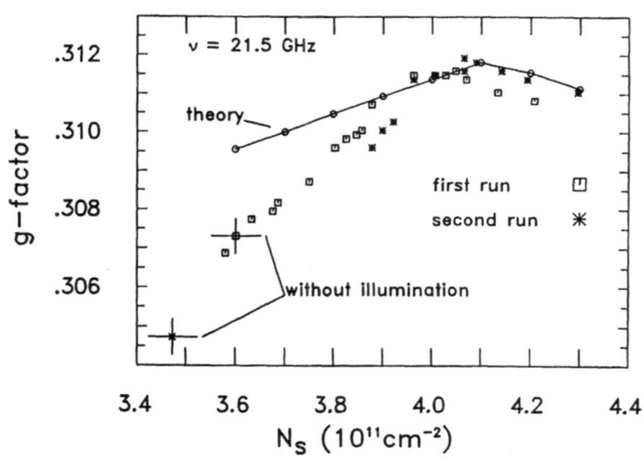

Figure 6: g-factors as a function of the sheet-electron density N_s, varied by illumination with a red LED. Theoretical results (see text) and experimental results for two different cooling cycles of sample 2

g-factor. The experiment, however, initially shows an opposite tendency. Only at $N_s \approx 4.1 \cdot 10^{11} cm^{-2}$ and above does the expected decrease occur.

The first increase of g with N_s can be explained as due to the simultaneously decreasing depletion charge. Obviously, here, the influence of a variation in N_{depl} is stronger than that of N_s. The bending of $g(N_s)$ at $N_s \approx 4.1 \cdot 10^{11} cm^{-2}$ can be explained by the assumption that the depletion charges are totally neutralized and N_{depl} can therefore not further be diminished. The increase of N_s is thus only a result of the ionization of DX-centers. This is confirmed by the fact that a certain increase of N_s requires a much longer illumination time in the region $N_s > 4.1 \cdot 10^{11} cm^{-2}$. Also a current bypass occurs.

The theoretical results are obtained using the model that the change of N_s arises completely from the change of N_{depl}: $\Delta N_s = -\Delta N_{depl}$ until $N_{depl}=0$. Beginning with $N_s = 3.6 \cdot 10^{11} cm^{-2}$ and $N_{depl} = 0.5 \cdot 10^{11} cm^{-2}$ this point is reached at $N_s = 4.1 \cdot 10^{11} cm^{-2}$. The further increase of N_s is not connected with a change of N_{depl} and thus results in the in the expected decrease of g.

The qualitative agreement between theory and experiment shows that the theoretical predictions for the N_s- and N_{depl}-dependence are valid and corroborates the explanation of the PPC-effect.

The first results after cooling down the samples (without any illumination) are indicated by crosses. The sheet carrier concentrations differ slightly, which is a well-known effect. Our measurements indicate that these slight differences in N_s correspond to different depletion charges. From the difference between N_s without illumination and N_s at the bending point of $g(N_s)$, one can estimate the depletion charge, under the assumption $\Delta N_s = -\Delta N_{depl}$, to be $N_{depl} \approx 0.6 \cdot 10^{11} cm^{-2}$ (for the second run).

6. Variation of N_s by the application of a backgate voltage

A further possibility for a variation of N_s is the application of a backgate voltage U_{BG} between the substrate and the 2DEG [22,23]. Using the analogy of a parallel plate capacitor

$$\Delta N_s = \frac{\epsilon_0 \epsilon_{rel}}{e} \cdot \frac{U_{BG}}{d} \quad , \tag{3}$$

the change of the 2d-concentration ΔN_s is proportional to U_{BG}, with ϵ_0 and ϵ_{rel} being the absolute and relative dielectric constants. With a substrate thickness of $d \approx 0.5$ mm, U_{BG} can be of the order of some hundred Volts. As in the case of illumination, the variation of N_s is not the only effect of a substrate bias. In addition, the confining potential at the interface is varied. Again there are two antagonistic effects: $U_{BG} < 0$ decreases N_s and therefore the chemical potential $E_F - E_0$. On the other hand, the confining potential becomes narrower, which makes the subband energy E_0 and consequently also the Fermi energy E_F rise. Detailed calculations are required, to show which of the two influences on E_F, and hence on the g-factor, predominates.

Figure 7 shows $\Delta\rho_{xx}$ with a microwave frequency $\nu = 55.0$ GHz. The different traces refer to different backgate voltages leading to different carrier concentrations (noted in Fig. 7 at the respective curves). With changes in N_s, the broad nonresonant spectrum, which is closely related to the SdH-oscillations, is also varied. In contrast, the resonance position remains nearly constant. The slight changes of the resonance position give rise to the g-factor variation, depicted in the insert of Fig. 7. The increase of g with increasing N_s appears to indicate a decrease of E_F. Apparently, the effect of the variation of the confining potential on E_F is stronger than the increase of the chemical potential. This is also confirmed by the theoretical results. With the simple assumption that the electric field $E_{BG} = U_{BG}/d$, arising from the backgate voltage, is linearly superposed on the confining potential, the confining potential has been evaluated self-consistently.

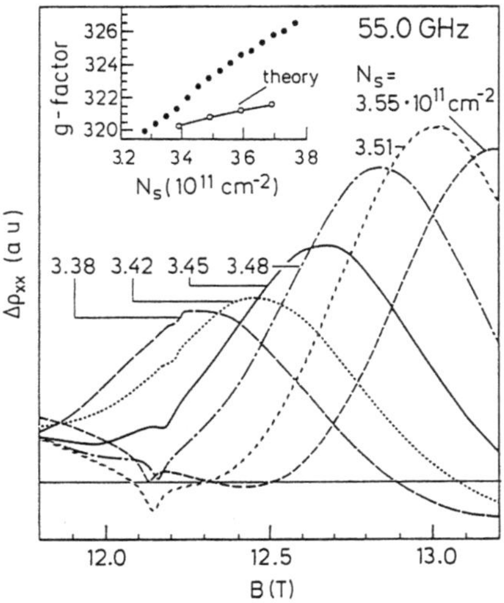

Figure 7: The change of ρ_{xx} due to microwave irradiation (ν = 55.0 GHz) for sample 2. Different traces refer to different sheet electron densities achieved by the application of a backgate voltage. The insert compares the corresponding g-factors with theoretical results

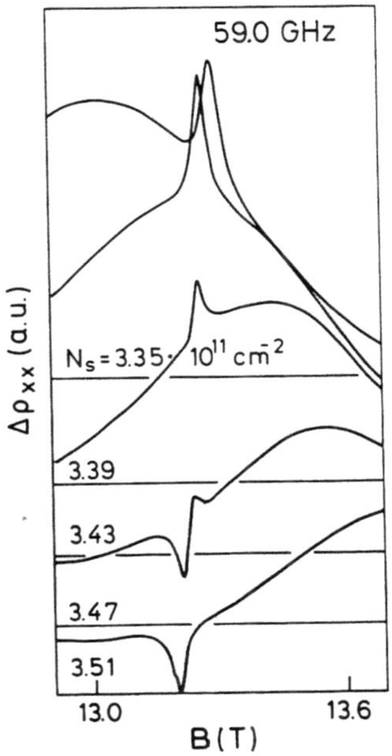

Figure 8: As in Fig. 7, but for a different microwave frequency (ν = 59.0 GHz). For the corresponding g-factors see Fig. 9

If one keeps in mind that the y-axes of the insert shows a very small range of g-factors on a large scale, one can again state good qualitative agreement between theory and experiment. Nevertheless, further effects may play a role here:

Figure 8 shows traces of $\Delta\rho_{xx}$ similar to those of Fig. 7, but in this case for a higher frequency ν = 59.0 GHz. ESR therefore occurs at higher magnetic fields, or lower filling factors. The transition between ESR producing an increasing in ρ_{xx} and a decreasing in ρ_{xx} is clearly visible. The second lowest curve reveals a dispersion-like ESR structure in ρ_{xx} (in this case at $i = 1.07$), bringing the determination of the actual position of the resonance into doubt. It is obviously no longer possible to take the touching point of the resonance structure with the background-parallel tangent. In Fig. 9a the corresponding g-factors are given, the dispersion-like ESR structure being analyzed with respect to *both* minimum *and* maximum. There are clear deviations from the monotonic behaviour of $g(N_s)$, found elsewhere, and this not only at the doubly analyzed measurement ($N_s = 3.47 \cdot 10^{11} cm^{-2}$), but also already in its vicinity.

Here this effect is still small, and one could trace it solely to difficulties in the ESR-lineshape analysis. But the results for ν = 50.0 GHz (Fig. 9) reveal more drastic effects: there is a certain extended range of N_s with two ESR structures and these are at a very large distance from each other. It almost appears to be more appropriate to think of two resonances rather than of one resonance with an untypically broad structure in $\Delta\rho_{xx}$. Believing in two resonances, Fig. 9b shows "dispersion" branches in $g(N_s)$, the first being nearly independent of N_s, the second linearly dependent on N_s. One could speculate about an interaction between these two branches, similar, for example, to the polaron effect in cyclotron resonance [24,25].

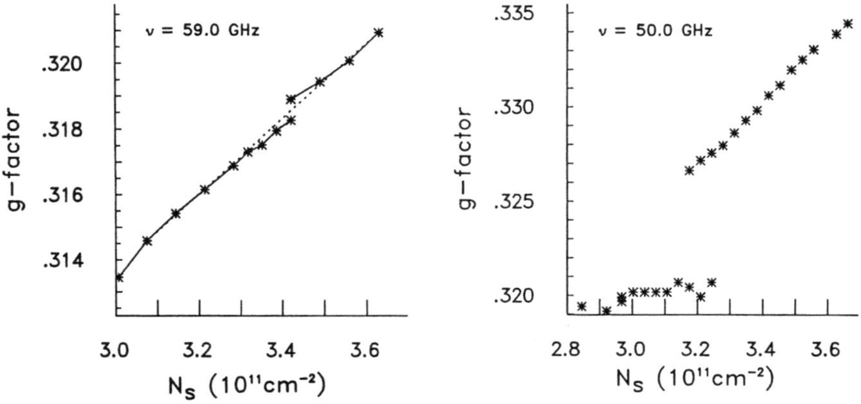

Figure 9: Experimentally determined g-factors in dependence of N_s, varied by a backgate voltage. **a)** (on the left) $\nu = 59.0$ GHz, the solid and the dashed traces are only guides for the eye **b)** (on the right) $\nu = 50.0$ GHz

This all remains speculation. We cannot explain this "double-resonance" behaviour of the ESR. At some filling factors (i.e. frequencies) these effects are very pronounced, at others not. Again we have no logical explanation. For example, we have not seen any peculiarities at simple fractional filling factors, comparable to the case of cyclotron resonance [26]. There are therefore still more experimental data to be collected.

In any case, three aspects have come to light indicating an immediate filling factor dependence of ESR:

- the existence of certain restricted magnetic field ranges for the detection of ESR (cf. section 2),
- the transition between an ESR-produced increase and decrease of ρ_{xx} (cf. also Fig. 3) and
- the "double-resonance" behaviour.

If there is a possibility to investigate the conductivity mechanism by the aid of ESR, these three ought to be appropriate points of attack.

7. Conclusions

In GaAs-AlGaAs heterostructures we find a magnetic-field- and Landau-level-dependent reduction of the spin-splitting with respect to the bulk-GaAs value, which can be well explained by self-consistent bandstructure calculations, considering the nonparabolicity of the bulk band structures. Although the measurements have been done in a type of transport experiment, all spin-splittings agree well with the single particle energy levels determined theoretically, and do not seem to be influenced by many particle interaction and scattering. In addition to a first approximation there is no inherent filling factor dependence of the spin-splittings. Nevertheless, a more rigorous investigation reveals details, which could lead to a study of the conductivity mechanism and the density of states in strong magnetic fields by use of ESR.

8. Acknowledgement

This work has been supported by the Stiftung Volkswagenwerk, the Deutsche Forschungsgemeinschaft and the Bundesministerium für Forschung und Technologie.

9. References

[1] T. Ando, A.B. Fowler, F. Stern, *Rev. Mod. Phys.* **54**, 437 (1982)

[2] U. Merkt *Festkörperprobleme* **27**, 109 (1987)

[3] T. Ando, Y. Uemura, *J. Phys. Soc. Jpn.* **37**, 1044 (1974)

[4] D. Stein, K. v. Klitzing, G. Weimann *Phys. Rev. Lett.* **51**, 130 (1983)

[5] D. Stein, G. Ebert, K. v. Klitzing, G. Weimann, *Surf. Sci.* **142**, 406 (1984)

[6] C.P. Poole Jr. *Electron Spin Resonance*, Interscience Publishers, New York, London, Sydney 1967

[7] G. Weimann, *Festkörperprobleme* **26**, 231 (1986)

[8] H. Sigg, private communication

[9] A.W. Overhauser, *Phys. Rev.* **92**, 411 (1953)

[10] For an overview see: A. Abragram, *The Principles of Nuclear Magnetism*, Clarendon Press, Oxford, 1961

[11] M. Dobers et al., to be published

[12] C. Weisbuch, C. Herrmann, *Phys. Rev.* **B15**, 816 (1977), D.J. Chadi, A.H. Clark, R.D. Burnham, *Phys. Rev.* **B13**, 4466 (1976)

[13] F. Malcher, G. Lommer, U. Rössler, *Superlattices and Microstructures* **2**, 267 (1986)

[14] G. Lommer, F. Malcher, U. Rössler, *Superlattices and Microstructures* **2**, 273 (1986)

[15] G. Lommer, F. Malcher, U. Rössler, *Phys. Rev.* **B32**, 6965 (1985)

[16] G. Lommer, F. Malcher, U. Rössler, *Phys. Rev. Lett.* **60**, 728 (1988)

[17] H.L. Störmer, R. Dingle, A.C. Gossard, W. Wiegmann, M.D. Sturge, *Solid State Commun.* **29**, 705 (1979)

[18] A. Kastalsky, J.C.M. Hwang, *Solid State Commun.* **51**, 317 (1984)

[19] E.F. Schubert, J. Knecht, K. Ploog, *J. Phys. C: Solid State Phys.* **18**, L215, (1985)

[20] J.J. Harris, D.E. Lacklison, C.T. Foxon, F.M. Selten, A.M. Suckling, R.J. Nicholas, K.W.J. Barnham, *Semicond. Sci. Technol.* **2**, 783 (1987)

[21] J.X. He, K.P. Martin, R.J. Higgins, *Phys. Rev.* **B36**, 6508 (1987)

[22] H.L. Störmer, A.C. Gossard, W. Wiegmann, *Appl. Phys. Lett.* **39**, 493 (1981)

[23] K. Hirakawa, H. Sakaki, J. Yoshino, *Phys. Rev. Lett.* **54**, 1279 (1985)

[24] H. Sigg, H.J.A. Bluyssen, P. Wyder, *Solid State Commun.* **48**, 897 (1983)

[25] H. Sigg, P. Wyder, J.A.A.J. Perenboom, *Phys. Rev.* **B31**, 5253 (1985)

[26] G.L.J.A. Rikken, H.W. Myron, C.J.G.M. Langerak, H. Sigg, *Surf. Sci.* **170**, 160 (1986)

Overhauser-Shift of the ESR in the Two-Dimensional Electron Gas of GaAs-AlGaAs Heterostructures

M. Dobers[1], *K. v. Klitzing*[1], *J. Schneider*[2], *G. Weimann*[3], *and K. Ploog*[1]

[1]Max-Planck-Institut für Festkörperforschung, Heisenbergstr. 1
D-7000 Stuttgart 80, Fed. Rep. of Germany
[2]Fraunhofer Institut für Angewandte Festkörperphysik,
Eckerstr. 4, D-7800 Freiburg, Fed. Rep. of Germany
[3]Walter-Schottky-Institut, TUM, D-8046 Garching

1. Introduction

The study of electron spin resonance (ESR) of the two-dimensional conduction electrons in GaAs-AlGaAs-heterostructures [1,2] revealed a strong coupling between electronic and nuclear spins. This coupling results in a dynamical nuclear spin polarization [3,4] and hence reciprocally in a shift of the ESR-line, the Overhauser-shift [5,6,3]. These effects together with very long nuclear spin relaxation times give rise to hysteresis and instabilities in the ESR. By applying radio frequencies in order to perform nuclear magnetic resonance (NMR) we have been able to detect, as well as ESR, also NMR in the magnetoresistivity ρ_{xx} of GaAs-AlGaAs-heterostructures.

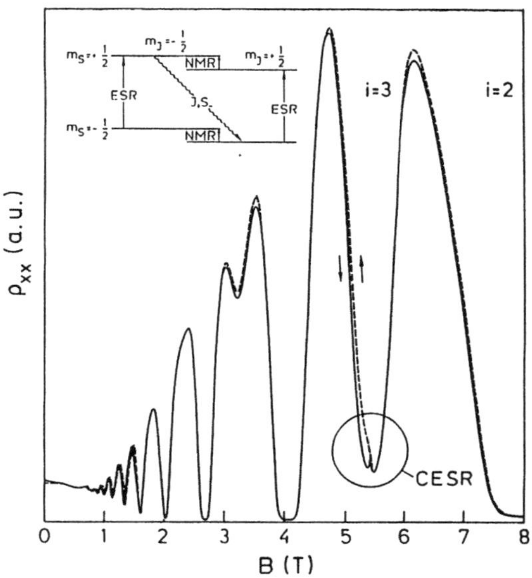

Figure 1: Magnetoresistivity ρ_{xx} under microwave radiation with a frequency $\nu = 23.1$GHz as a function of the magnetic field. Solid line: $\dot{B} = +2.35$T/min, dashed line: $\dot{B} = -2.35$T/min. The filling factor i for some of the minima is indicated. The sharp structure at $B = 5.5$T is due to conduction electron spin resonance. The sample normal is tilted by an angle of the order of 30° with respect to the magnetic field. The insert shows the energy level scheme of a system of a nuclear and an electronic spin, in order of their respective magnetic spin quantum numbers

2. Theoretical Background

The Hamiltonian for the electronic spin $\underline{S}$ does not only consist of a Zeeman term but includes also hyperfine coupling to nuclear spins $\underline{I}$:

Springer Series in Solid-State Sciences, Vol. 87
High Magnetic Fields in Semiconductor Physics II Editor: G. Landwehr

$$H = g\mu_B \underline{B} \cdot \underline{S} + A\underline{I} \cdot \underline{S} \quad . \tag{1}$$

Here, g is the magnetic-field- and Landau-level-dependent g-factor [1,2], μ_B Bohr's magneton and A the hyperfine interaction constant. An ensemble average over all nuclear spins leads to the electronic spin splitting

$$\Delta E = +g\mu_B B + A < I > \quad , \tag{2}$$

which is no longer only a function of the magnetic field B, but depends also upon the nuclear spin polarization $< I >$. In thermodynamic equilibrium $< I >$ is even at liquid helium temperatures very small (for ^{75}As at $B = 6T$: $< I >< 10^{-3}$), so that in high magnetic fields the hyperfine term is negligible compared to the Zeeman term. Therefore, in the energy spectrum of a two-dimensional electron gas (2DEG)

$$E_{N,m_s} = E_0 + (N + \frac{1}{2})\frac{\hbar e}{m^*}B + g\mu_B B m_s + A < I > m_s \tag{3}$$

the last term may be omitted, as usual ($N = 0, 1, 2, \ldots$ the Landau level quantum number; $m_s = +1/2, -1/2$ the magnetic spin quantum number of the electrons). With a non-vanishing $< I >$ the electronic spin splitting is increased and the resulting Overhauser-shift B_N of the ESR-line is given by

$$\Delta E = g\mu_B(B + B_N) = g\mu_B(B + \frac{A}{g\mu_B} < I >) \tag{4}$$

as is illustrated in the insert of Fig. 2. The Overhauser-shift can easily be very large if the nuclear spins are dynamically polarized. It is again the hyperfine interaction that allows a mutual flip of electronic and nuclear spins

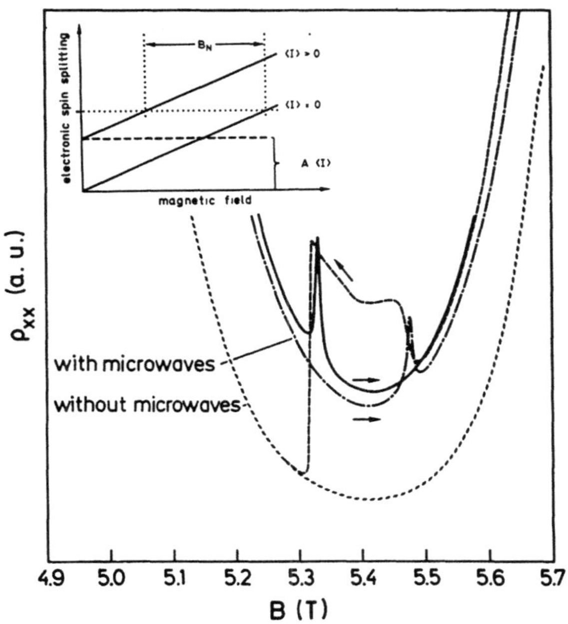

Figure 2: Magnetoresistivity ρ_{xx} in the vicinity of filling factor $i = 3$ without (lower trace) and with microwave radiation; $\dot{B} = \pm 2.35T/min$, the sign of $\dot{B}$ being indicated by the arrows. The background increase of ρ_{xx} by microwaves is due to a non-resonant heating [1,2]. The insert illustrates the different contributions to the electronic spin splitting (cf. text)

$$A\underline{I}\cdot\underline{S} = A[\frac{1}{2}(I_+S_- + I_-S_+) + I_zS_z] \quad . \tag{5}$$

As is illustrated in the insert of Fig. 1, besides the nuclear-spin-conserving electronic spin relaxation, there is another relaxation channel, that is mediated by one of the so-called flipflop-terms AI_+S_-. Due to the extremely long nuclear relaxation times, as long as the electronic spins are relaxing to their ground state (for example during ESR), the nuclear spin polarization steadily increases.

3. Experimental Results

We detected ESR by the change of ρ_{xx} due to microwave illumination as is explained in more detail in Ref. 2. For simplicity, we show only results for sample 2 in Ref. 2. Its magnetoresistivity under microwave illumination is shown in Fig. 1. The solid line, taken while sweeping the magnetic field up ($\dot{B} > 0$) reveals a sharp ESR-structure at $B \approx 5.5T$. On the other hand, while sweeping the magnetic field down ($\dot{B} < 0$), the resonance is broadly smeared out towards lower magnetic fields. During ESR the nuclei are dynamically polarized, which results in an Overhauser-shift towards lower fields. *Increasing* the magnetic field leads only to a dynamical narrowing of the ESR-line, because $\dot{B}$ and the rate of change of the resonance position $\dot{B}_{res}$ are of opposite sign. *Decreasing* the magnetic field, by contrast, leads to a dynamical broadening of the ESR-line towards lower magnetic fields, because both $\dot{B} < 0$ and $\dot{B}_{res} < 0$. With sufficient microwave power and, consequently, strong enough dynamic nuclear spin polarization, even the condition $\dot{B} = \dot{B}_{res}$ can be reached, resulting in a locking of the ESR-position to the external applied magnetic field. Fig. 2 illustrates the Overhauser-shift more strikingly. The dash-dotted line was taken with the nuclear spins being in thermal equilibrium, $< I > \approx 0$. The resonance position is therefore to a good approximation only determined by the Zeeman term. With $\dot{B} < 0$, the resonance is shifted ($\dot{B}_{res} \approx \dot{B}$), yielding an ESR-enhanced magnetoresistivity. When the microwaves are switched off (here at $B \approx 5.3T$), $< I >$ and therefore also the Overhauser-shift B_N is no longer increased. Sweeping up a moment later with microwaves switched on again (solid line) gives ESR at the position, where the microwaves were previously switched off in the preceeding $\dot{B} < 0$-run. This indicates that the nuclear spin polarization remains nearly constant on the time scale of an ESR-experiment. From the Overhauser-shift of, in this case $B_N \approx 0.2T$, using the values for bulk-GaAs ([7], cf. the table), the nuclear spin polarization can be evaluated to be $< I > \approx 0.06$. This is at least two orders of magnitude larger as compared to $< I >$ at thermal equilibrium.

For a proper determination of electronic g-factors it is essential that $< I >$ is vanishingly small. This can be achieved by saturating the NMR. By NMR, the nuclear spin polarization and thereby the Overhauser-shift is abruptly decreased, resulting in a sudden flip-back of the shifted ESR-line towards its unshifted position. In Fig. 3, this becomes visible by a sudden decrease of the ESR-enhanced magnetoresistivity to its unenhanced value. Using a small NMR-coil, different radio-frequencies have been applied to the sample, during $\dot{B} < 0$. The position of the sharp decreases of ρ_{xx} as a

Table : A list of the three isotopes in GaAs, with their respective spins I, abundances a and NMR-frequencies ν_{NMR}. The last row gives the contribution of each isotope to the Overhauser shift in *bulk*-GaAs (after Ref. [7])

	I	a (%)	ν_{NMR} at $B = 5.5T$	B_N (T)
^{69}Ga	3/2	60.4	56.21 MHz	$-0.91\cdot < I^{^{69}Ga} >$
^{71}Ga	3/2	39.6	71.41 MHz	$-0.78\cdot < I^{^{71}Ga} >$
^{75}As	3/2	100	40.12 Mhz	$-1.84\cdot < I^{^{75}As} >$

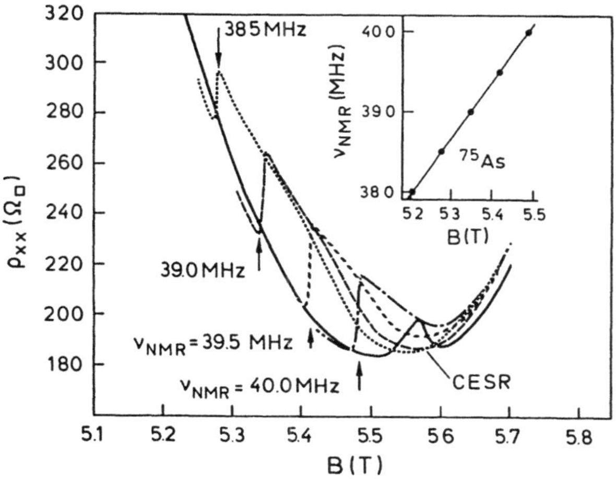

Figure 3: Magnetoresistivity ρ_{xx} in the vicinity of filling factor $i = 3$ under continuous microwave ($\nu_{ESR} = 23.4$GHz) and radio frequency radiation. Solid trace: $\dot{B} > 0$, other traces: $\dot{B} < 0$. The NMR of the ^{75}As nuclei results in a sharp decrease of ρ_{xx} (cf. text). Measurements for different NMR frequencies ν_{NMR} are shown. The insert gives the magnetic field dependence of ν_{NMR}

function of the applied frequency (cf. the insert in Fig. 3) unambiguously identifies the NMR as that of the ^{75}As-nuclei.

Another way to achieve $< I >\approx 0$ is simply to wait for a complete nuclear spin relaxation. The relaxation of a dynamically created $< I >$ is, due to $< I >\propto B_N$, connected with a relaxation of the shifted ESR-line back to its eqilibrium position. By observing the time dependence of the back-shift of the ESR-line, the nuclear spin relaxation can be determined. We observed, depending upon the temperature and the filling factor, nuclear relaxation times of the order of $T_1 = 10min \ldots 20min$. Obviously, T_1 is much shorter than reported for high-purity bulk-GaAs [8], but in the presence of free electrons, the relaxation via $A/2(I_+S_- + I_-S_+)$ (Korringa relaxation) reduces T_1 [3,9].

The Korringa-relaxation depends upon the probability of an electronic spin-flip at the Fermi-level (without changing the electronic energy). A study of the magnetic field and temperature dependence of T_1 therefore promises to give information on the electronic density of states [3,9].

4. Conclusions

Studying the Overhauser-shift of the ESR in the 2DEG of GaAs-AlGaAs-heterostructures, we have been able to prove the nuclear origin of the observed hysteresis effects. In addition we observed NMR of the lattice nuclei exclusively in the direct vicinity of the 2DEG and presented a method of determining the spin relaxation time of these nuclei.

5. Acknowledgement

This work has been supported by the Stiftung Volkswagenwerk and the Bundesministerium für Forschung und Technologie.

6. References

[1] M. Dobers, K. v. Klitzing, G. Weimann to be published in Phys. Rev. B

[2] M. Dobers, F. Malcher, G. Lommer, K. v. Klitzing, U. Rössler, K. Ploog, G. Weimann *Proc. Int. Conf. on the Application of High Magnetic Fields in Semiconductor Physics*, Würzburg, 1988 (this volume)

[3] A. Abragam *The Principles of Nuclear Magnetism* Clarendon Press, Oxford, 1961

[4] G. E. Pake, T. L. Estle *The Physical Principles of Electron Paramagnetic Resonance* W. A. Benjamin Inc., Reading, Massachussets, 1973

[5] A. W. Overhauser *Phys. Rev.* **92**, 411 (1953)

[6] For a review see: D. Paget, V. L. Berkovits *Optical Orientation* ed. by F. Meier and B. P. Zakharchenya in *Modern Problems in Condensed Matter Sciences*, Vol. 8, Elsevier Science Publ. B.V., North Holland, Amsterdam, New York, 1984, p.381

[7] D. Paget, G. Lampel, B. Sapoval, V. S. Safarov *Phys. Rev.* **B15**, 5780 (1977)

[8] J. A. McNeil, W. G. Clark *Phys. Rev.* **B13**, 4705 (1976)

[9] I. D. Vagner, T. Maniv, E. Ehrenfreund *Solid State Commun.* **44**, 635 (1982)

Combined Resonance in Systems of Different Dimensionality

E.I. Rashba

L.D. Landau Institute for Theoretical Physics, Academy of Sciences of the USSR, SU-Moscow 117334, USSR

1. Introduction

In this paper we concentrate on two aspects of the physics of combined resonances (COR). i)Let us begin with pecularities of COR which are specific for systems with different dimensionalities. Experimental findings of recent years have provided us with highly interesting data on 3D-systems (bulk), 2D - heterojunctions, 1D - dislocations and zero-D - impurity centers. It is highly instructive to compare the mechanisms controlling the intensity of COR in these systems. From the extensive bulk of the available experimental data we shall select and discuss only those which are especially convincing and most closely connected with the basic ideas of this paper. ii) There exist two basic types of COR excitation mechanisms. We term them here as intrinsic and extrinsic mechanisms. The first one is due to specific spin-orbit (SO) terms in the effective mass approximation (EMA) Hamiltonian of band carriers. Exactly this mechanism has been originally predicted by RASHBA [1], and later was investigated in numerous experimental works. The intensity of COR caused by this mechanism is especially large for band carriers. For electrons bound to defects, e.g., to donors, it is usually considerably lower [2]. There are also extrinsic mechanisms of COR, which seems to have been discovered originally by BELL in his experiments on spin resonance (SR) in strongly doped n-InSb[3]. However, the theoretical recognition and the formulation of the problem came later [4]. The extrinsic COR mechanisms are due to the existence of impurities and defects. The number of papers, in which extrinsic COR was observed, is rather small so far. However, we feel that the field is promising and that very interesting results can be expected. It is reasonable to emphasize from the very beginning that both problems, i) and ii), are not totally independent as can be seen from what follows.

There is some difference in the use of the term COR between the Soviet and the Western scientific literature. In Soviet papers the meaning of this term is the same as in the original paper [1] where the notion of COR has been introduced, and in the review paper [5] as well. In these papers the term COR is used for every electric-dipole resonance transition changing the electron spin state. The resonance may be a pure spin resonance, in this case it occurs at the Zeeman frequency ω_s. To be specific, here we use the term elecric-dipole spin resonance (EDSR). However, in many cases the COR frequency is a combinational frequency. For example, for band ele-

Springer Series in Solid-State Sciences, Vol. 87
High Magnetic Fields in Semiconductor Physics II Editor: G. Landwehr

ctrons usually such a frequency is a combination of the cyclotron frequency ω_c and the spin frequency ω_s (like $n\omega_c \pm \omega_s$). We term such a COR transition as a combinational frequency resonance (CFR) or electric-dipole CFR (EDCFR). In what follows we shall use listed terms as defined here. In particular, we shall use the term COR in its wider meaning, and not in the restricted sence (equivalent to EDCFR). Such a restriction is often introduced in the papers of some Western authors.

2. General

The interaction of an electronic spin with an a.c. electric field **E** may be written as

$$e(\hat{\mathbf{v}}\mathbf{A})/c \qquad \text{or} \qquad e(\hat{\mathbf{r}}\mathbf{E}) \; . \tag{1}$$

Here **A** is its vector potential, and $\hat{\mathbf{r}}$ and $\hat{\mathbf{v}}$ are the operators of coordinate and velocity in the effective mass representation. As usual, $\hat{\mathbf{v}}$ is a polynomial in quasimomentum $\hat{\mathbf{k}}$, and $\hat{\mathbf{r}}=i\partial/\partial\hat{\mathbf{k}}+\mathbf{X}(\hat{\mathbf{k}})$, where $\mathbf{X}$ is a polynomial in $\hat{\mathbf{k}}$ [6]. As applied to COR, only the terms including both the Pauli matrices σ and also $\hat{\mathbf{r}}$ or $\hat{\mathbf{v}}$ are of importance. Of course, these terms arise due to spin orbit (SO) interaction. Since

$$\hat{\mathbf{v}} = i[H,\hat{\mathbf{r}}] \; , \tag{2}$$

where H is a Hamiltonian (we put $\hbar=1$), the well-known equation

$$< f \parallel \hat{\mathbf{v}} \parallel i > = i\omega_{fi} < f \parallel \hat{\mathbf{r}} \parallel i > \tag{3}$$

holds for an arbitrary initial state i and a finite state f, ω_{fi} is the energy difference between them. Hence, the calculation of the matrix elements of $\hat{\mathbf{v}}$ or $\hat{\mathbf{r}}$ is a matter of convenience (for detail see [7]). The most important SO terms for crystals having different symmetry will be given in the next sections.

The form of the SO terms responsible for COR depends on the symmetry of the system. When the symmetry is lowered, new SO terms enter H and $\hat{\mathbf{r}}$. When this occurs due to some geometrical restrictions, e.g., near an interface, the COR patterns for bulk carriers and for the ones confined near the interface may be quite different and their comparison is very instructive. In particular, one should expect that COR for 2D carriers, confined near the interface, will be much stronger than for 3D carriers in the bulk. However, some connection may exist between the parameters responsible for COR in 2D and in 3D. So, the comparative investigation of COR with carriers having different dimensionality may be very useful.

3. Combined resonance in 3D

In this case let us restrict ourselves only to electrons in InSb, since for them the most complete experimental data are available. As a result, it is possible to investigate independently the contribution of different mechanisms to COR.

Let us begin with COR arising from the term in the EMA Hamiltonian which is of the lowest order in **k**. It may be written as [8]:

$$H_{so} = \delta_0(\sigma\hat{\varkappa}), \quad \hat{\varkappa}_x = \hat{k}_y\hat{k}_x\hat{k}_y - \hat{k}_z\hat{k}_x\hat{k}_z \quad . \tag{4}$$

This term exists due to the inversion asymmetry of InSb. It follows from (2) and (4) that the transitions at the frequencies ω_s, $\omega_c \pm \omega_s$, $2\omega_c \pm \omega_s$ are allowed. Since (4) is a cubic (and not a spheric) invariant, the angular indicatrices of the COR intensity must be highly anisotropic [8]. The EDSR caused by this mechanism has been discovered only recently by DOBROWOLSKA et al. [9] in the longitudinal Voigt geometry $\mathbf{E}\|\mathbf{B}\perp\mathbf{q}$, where $\mathbf{q}$ is a wave vector, and $\mathbf{B}$ a d.c. magnetic field. In this paper the shape of the angular indicatrices, predicted in [8], has been confirmed and the interference of EDSR and EPR has been discovered. This interference is observable since δ_0 has a moderate magnitude. Its observation has allowed the authors of a number of subsequent papers [10-12] to determine the value of $\delta_0 \simeq 225$ eVxÅ^3 and even its sign (which depends on the specific choice of the reference system).

In the spheric approximation the theory of COR has been developed by SHEKA [13] in terms of the Kane Hamiltonian. Since in this theory the lowest term in k in H_{so} is of the order of k^4, the corresponding COR mechanism is known as a nonparabolicity mechanism. In the lowest approximation the theory can be formulated in the simplest way in terms of the SO contribution to $\hat{\mathbf{r}}$ [14]

$$\hat{\mathbf{r}}_{so} = (g/4m_0)\ (\ E_G^{-1} + (E_G + \Delta)^{-1}\ )\ [\ \sigma \times \hat{\mathbf{k}}\ ]\ . \tag{5}$$

Here m_0 and g are the free electron mass and the g-factor of a band electron, E_G is a gap width and Δ is the SO splitting. Since $\hat{\mathbf{r}}_{so}$ is a spheric vector (and not a cubic one) the COR is isotropic. In the longitudinal polarization $\mathbf{E}\|\mathbf{B}$ COR is allowed only at the combinational frequency $\omega_c + \omega_s$. This EDCFR has been observed by McCOMBE et al. [15]. Due to the large factor entering (5), EDCFR is considerably stronger than EDSR caused by the k^3-mechanism.

Under the conditions of uniaxial strain T the k-linear terms arise in the electron dispersion law [16]. In this case COR is especially strong [1]. When ν is a unit vector directed along one of the high symmetry axes and $T\|\nu$, then H_{so} is

$$H_{so} = \alpha \, [\, \sigma \times \hat{\mathbf{k}} \,]\nu \, . \tag{6}$$

Equation (6) is the simplest Hamiltonian of SO interaction, well-known from the theory of energy spectra of wurtzite type crystals. It has been shown experimentally in [17] that the intensity of EDSR in the region of T about one kbar increases as T^2, i.e., the mechanism of EDSR caused by (6) is more effective than the one caused by (4) even at a very small α. A similar result has been obtained in [18] by a somewhat different method.

Therefore, all the basic COR mechanisms have been observed at present in n-InSb. The terms proportional to k and k^3 are important for us especially as applied to 2D systems, which we shall consider below.

4. Combined Resonance in 2D

SR of electrons confined in 2D layers in GaAs/AlGaAs heterojunctions have been reported by STEIN et al. [19]. The observed resonance showing two pecularities: i) Its frequency $\omega_s(B)$ was nearly linear in B for large B. However, the extrapolation of the data into the low B region resulted in a finite offset $\omega_0>0$. ii) The intensity of the resonance was anomalously high.

First, both pecularities have been ascribed to the effect of SO interaction chosen in the form (6), where ν is a unit vector perpendicular to the confining layer [20]. This model describes qualitatively both observed regularities, and the value $\alpha \simeq 2.5\times10^{-10}$ eVxcm has been found from the dependence $\omega_s(B)$ [20]. With this value of α the ratio of the intensities $I_{EDSR}/I_{EPR} \simeq 10^7$, it is proportional to α^2 [1]. This estimate shows that the electric-dipole mechanism of the SR dominates strongly over the magnetic-dipole one, and explains observation ii). Such a description seemed highly tempting, since only one unknown parameter α entered the theory, and the dependence $\omega_s(B)$ made it possible to find it. However, it has been shown [21] that in GaAs the behaviour of $\omega_s(B)$ in the high B region is dominated by the dependence $g(B)$ which comes from the bulk k^4- nonparabolicity. Hence, α cannot be found from the dependence $\omega_s(B)$ and must be estimated independently. Without describing details we only mention here that several mechanisms contribute to α. One of them is the bulk k^3-nonparabolicity (4) [22]. The contribution to α due to it may be estimated as

$$\alpha \simeq \delta_0 (\pi/a)^2 \, , \tag{7}$$

here a characterizes the width of the confining layer. According to [23] it is this contribution to α that dominates in GaAs/AlGaAs heterojunctions. Using the value $\delta_0 \simeq 35$ eVxÅ^3 [24] we obtain the result that the lower bound for α is 10^{-10}eVxcm . The same estimate holds for typical symmetrical quantum wells [25].

So, we may conclude that for electrons in GaAs/AlGaAs heterojuntions the EDSR must be a dominating mechanism of the SR. There exists direct experimental confirmation of this statement [26]. For holes the Hamiltonian (6) is applicable only in a very narrow region of k values [25]. However, no doubt the EDSR is strong as well. The COR has been observed previously in other 2D systems.

5. Combined resonance in 1D

Let us consider the EDSR with carriers bound to dislocations. The EDSR band (*Ch*-band) has been discovered in Si on [1$\bar{1}$0] dislocations after annealing of crystals. It is a remarkable property of the *Ch*-band that it is only observed when $\mathbf{E}\|\mathbf{l}$, where $\mathbf{l}$ is the dislocation direction [27]. If one admits that such a dislocation has an invariant vector $\mathbf{b}\perp\mathbf{l}$, the Hamiltonian may be again written in the form (6) with $\nu\|\mathbf{b}$. One should bear in mind that $\mathbf{k}\|\mathbf{l}$. The angular indicatrix of the EDSR intensity is

$$I(\theta, \theta_{\mathbf{E}}, \Phi) = I_0 \cos^2\theta_{\mathbf{E}} (1 - \sin^2\theta \sin^2\Phi) . \tag{8}$$

Here $\theta_{\mathbf{E}}$ and θ are the angles between the polar axis $\mathbf{l}$ and vectors $\mathbf{E}$ and $\mathbf{B}$, and Φ is the azimuth angle of $\mathbf{B}$ in the $\mathbf{b},[\mathbf{b}\times\mathbf{l}]$-plane. Experimental data are described by this dependence so well that the applicability of the Hamiltonian (6) is without any doubt [27].

The results of KVEDER et al. [27] seem to be the first convincing evidence of the band (not hopping) mechanism of the transport of carriers along dislocations.

Let us mention in conclusion, that k-linear terms responsible for EDSR appeared in the systems, described in sects. 4 and 5, due to existence of extended defects: interfaces and dislocations.

6. Combined Resonance in zero-D

Impurity centers can be considered as zero-D systems. The theory of EDSR for donors has been developed as applied to SO interaction of the type (6) in [2]. The most striking difference between the zero-D case and all others is manifested in nonmagnetic crystals for donors with an orbitally nondegenerate ground state in the weak magnetic field region. For EDSR in this case the initial and finite states, i and f, are components of the same Kramers doublet. As a result $\langle i \| \hat{\mathbf{r}} \| f \rangle \to 0$ when $\mathbf{B} \to 0$. For band carriers, i.e., in higher dimensionalities, i and f cannot be the Kramers conjugate states (for a general position of $\mathbf{k}$ in the Brillouin zone). Hence, this restriction is absent for them. As a result, in zero-D the EDSR intensity acquires a small factor $(\omega_s/\bar{E})^2 \ll 1$, $\bar{E}$ is a typical separation between energy levels. However, when B is so large that $\omega_s \gg \bar{E}$, the intensity of EDSR with electrons bound to donors is nearly the same as with band electrons. A very interesting situation arises for intermediate fields B, when the Zeeman splitting ω_s becomes close to the donor excitation energy, e.g., to $E_{2p} - E_{1s}$. Under these conditions the intensity of EDSR must increase drastically. This gigantic resonance in the intensity of EDSR, predicted in [2], has been discovered experimentally [28] on donors in a semimagnitic semiconductor $Cd_{1-x}Mn_xSe$. For this system the condition $\omega_s \simeq E_{2p} - E_{1s}$ is fulfilled in moderate fields B due to a very large magnitude of the g-factor ($g \simeq 100$). The treatment of the experimental data by theories [2] and [29] leads to nearly the same value of $\alpha \simeq 2.3\times10^{-10}$eVxcm (cf.(6)).

7. Combined Resonance Induced by Lattice Imperfections

There is experimental evidence for the fact that lattice imperfections produce new mechanisms of COR. i) There is a strong and narrow band of EDSR in highly doped n-InSb which remains metallic at T=0 [3]. A high intensity of this band cannot be provided by the k^3 and k^4- mechanisms [4]. ii) In specimens of Ge doped by As strong EDSR has been observed, but only in highly compensated samples. These observations seemed puzzling since in Ge the EDSR was not observed with free carriers, and EDSR with electrons bound to donors was expected to be even weaker (sect.6).

The results of paper [3] have not been confirmed yet, and a theoretical interpretation of them is missing (cf. [4]). The observations of GERSHENSON et al.[30] may be explained [4] when one takes into account i) the multivalley structure of the energy spectrum of Ge, ii) the strong anisotropy of the valleys and iii) that the random electric field of the impurities (charged due to compensation) mixes the wave functions of different valleys. The effective SO coupling constant is $(g_\parallel - g_\perp)(P_\parallel - P_\perp)$, where $g_\parallel$, $g_\perp$ and $P_\parallel$, $P_\perp$ are the principal components of the tensors of the electronic g-factor and the electric polarizability of donors (found in the single valley approximation). The basic conclusions of the theory by MEL'NIKOV and RASHBA [4] are: i)uniaxial strain suppresses EDSR, ii) the EDSR band has a considerably larger width than the EPR band, iii) it has broad, slowly decreasing wings, iv) its width increases with ω_s (i.e., with B) and v) its intensity is highly anisotropic. These statements are in agreement with experimental data [31].

As has been mentioned [32], when the exchange field in certain magnetic material or an external magnetic field depend on $\mathbf{r}$, this results in a new mechanism of COR since the Zeeman term in the Hamiltonian $H_z \propto (\sigma\mathbf{B}(\mathbf{r}))$ includes both σ and $\mathbf{r}$. A situation like this occurs for shallow impurity centers in semimagnetic semiconductors, where the extrinsic mechanism of COR is acting side by side with the intrinsic one (sect. 6) [33]. The former results in EDCFR arising from exchange interaction of the electron with magnetic ions $H=(J/2)(\mathbf{S}\sigma)$, here $\mathbf{S}$ is spin of an ion. The resonance frequency is $\omega=(g\mu_B - 2\mu)B$, here μ_B and μ are the Bohr magneton and the magnetic moment of an ion, respectively. The intensity shows resonance enhancement when $\omega \simeq E_q - E_{1s}$, here the symmetry of the excited state q may differ from the symmetry of p-states. The intensity is also large for low ω, this unusual behaviour is caused by the fact that the existence of the ion spin **S** removes the restriction on the matrix element which follows from Kramers' theorem (sect. 6). The intensities for extrinsic and intrinsic mechanisms may compete. However, they may be separated due to different angular dependence of the intensities. It seems plausible that EDCFR observed in [34] arises due to extrinsic mechanism.

8. Conclusion

In conclusion we mention the following. There are specific peculiarities of COR inherent in every dimensionality. The lowering of dimensionality occurs by inserting specific defects into a crystal. For instance, the inserting of one or two interfaces results in the appearence of 2D carriers in a heterojunction or a quantum well. Inserting of regular extended defects lowers the symmetry of the system, but does not abolish the band nature of the energy spectrum. As a result, COR is usually stronger in 2D than in 3D. In some cases it is possible to bring into correspondence those terms in the SO Hamiltonian of a 2D system which are responsible for COR with definite terms in the SO Hamiltonian of a "parent" 3D system (of course, we have in mind the EMA Hamiltonians). A similar statement holds for 1D systems. The situation is quite opposite for zero-D systems, i.e., for impurity centers. When COR is caused in this case by any intrinsic mechanism, it is weaker than in 3D. Study of COR existing due to intrinsic mechanisms is a powerful tool to investigate those fine details of the dispersion law which are caused by SO interaction.

There are new trends in the physics of COR connected with the investigation of extrinsic mechanisms. When extended defects (interfaces or dislocations) are responsible for COR, their investigation can give unique information on the geometry of defects (e.g., vector **b** for dislocations, sect. 5). In the case when an extrinsic mechanism is inherent in point defects, aggregate centers and random fields or strains, COR can give information on the geometry of defects and statistical properties of random systems.

References

1. E. I. Rashba, Sov. Phys. - Solid State 2, 1109 (1960)
2. E. I. Rashba and V.I.Sheka, Sov. Phys. - Solid State 6, 114, 451 (1964)
3. R. L. Bell, Phys. Rev. Lett. 9, 52 (1962)
4. V. I. Mel'nikov and E.I.Rashba, Sov. Phys. - JETP 34, 1353 (1972)
5. E. I. Rashba, Sov. Phys. - Uspekhi 7, 823 (1965)
6. E. I. Blount: in *Solid State Physics*, Vol.13, 305 (1962)
7. E. I. Rashba and V. I. Sheka: in *Landau Level Spectroscopy*, ed. by G.Landwehr and E.I.Rashba (North Holland, in press)
8. E. I. Rashba and V. I. Sheka, Sov. Phys. - Solid State 3, 1257, 1357 (1961)
9. M. Dobrowolska, Y. Chen, J. K. Furdyna, and S. Rodriguez, Phys. Rev. Lett. 51, 134 (1983)

10. Y.-F. Chen, M. Dobrowolska, J. K. Furdyna, and S. Rodriguez, Phys. Rev. B32, 890 (1985)
11. V. I. Sheka and L. S. Khasan, Sov. Phys. - JETP Letters 41, 72 (1985)
12. M. Cardona, N. E. Christensen, M. Dobrowolska, J. K. Furdyna, and S. Rodriguez, Solid State Commun. 60, 17 (1986)
13. V. I. Sheka, Sov. Phys. - Solid State 10, 2470 (1964)
14. Y. Yafet: in *Solid State Physics*, Vol.14, 1 (1963)
15. B. D. McCombe, S. G. Bishop, R. Kaplan, Phys. Rev. Lett. 18, 748 (1967)
16. G. L. Bir and G. E. Pikus, *Symmetry and Strain-Induced Effects in Semiconductors* (Halsted Press, 1974)
17. M. Kriechbaum, R. Meisels, F. Kuchar, and F. Fantner: in *Proc. 16th Int. Conf. Phys. Semiconductors*, ed. by M.Averous (North Holland, 1983), p.444
18. C. Jagannath and R. L. Aggarwal, Phys. Rev. B32, 2243 (1985)
19. D. Stein, K. von Klitzing, and G. Weimann, Phys. Rev. Lett. 51, 130 (1983)
20. Yu. A. Bychkov and E. I. Rashba, Sov. Phys. - JETP Lett. 39, 78 (1984)
21. G. Lommer, F. Malcher, and U. Rossler, Phys. Rev. B32, 6965 (1985)
22. Yu. A. Bychkov and E. I. Rashba: in *Proc. 17th Int. Conf. Phys. Semiconductors*, ed. by J.D.Chadi and W.A.Harrison (Springer, 1985), p.321
23. F. Malcher, G. Lommer, and U. Rossler, Superlattices and Microstructures 2, 267 (1986)
24. G. E. Pikus, V. A. Maruschak, and A. N. Titkov, Fiz. Tekh. Poluprovodn. 22, 185 (1988)
25. E. I. Rashba and E. Ya. Sherman, Phys. Lett. A129, 175 (1988)
26. H. L. Stormer, private communication
27. V. V. Kveder, V. Ya. Kravchenko, T. R. Mchedlidze, Yu. A. Osip'yan, D. E. Khmel'nitskii, and A. I. Shalynin, Sov.Phys. - JETP Lett. 43, 255 (1986)
28. M. Dobrowolska, H. D. Drew, J. K. Furdyna, T. Ichiguchi, A. Witowski, and P. A. Wolff, Phys. Rev. Lett. 49, 845 (1982)
29. S.Gopolan, J. K. Furdyna, and S. Rodriguez, Phys. Rev. B32, 903, 1985
30. E. M. Gershenson, N. M. Pevin, and M. S. Fogel'son, Sov. Phys. - JETP Lett. 12, 139 (1970)
31. E. M. Gershenson, N. M. Pevin, I. T. Semenov, and M. S. Fogel'son, Sov. Phys. Semicond. 10, 104 (1976)
32. S. I. Pekar and E. I. Rashba, Sov.Phys.- JETP 20, 1295 (1965)
33. Yu. G. Rubo, L. S. Khasan, V. I. Sheka, and A. S. Ioselevich, Pis'ma Zh. Eksp. Teor. Fiz. 47, 555 (1988)
34. R. Stepniewski, Solid State Commun. 58, 19 (1986)

Recent Cyclotron Resonance Work

T. Ohyama

Department of Physics, College of General Education, Osaka University, Toyonaka, Osaka 560, Japan

Far-infrared as well as microwave cyclotron and impurity resonance are described for materials as semi-insulating GaAs, GaInAs bulk- and hetero-junctions, and the II-VI semiconductor ZnSe which have recently attracted much attention. By means of the experiments, various new features are clarified under thermal equilibrium and nonequilibrium conditions.

1. INTRODUCTION

Since the first observation of cyclotron resonance in germanium single crystals by DRESSELHAUS, KIP and KITTEL [1] and LAX, ZEIGER, DEXTER and ROSENBLUM [2], cyclotron resonance experiments became one of the most powerful and direct tools to obtain information about the band structure of semiconductors, semimetals and metals and about the transport behaviour of these materials.

The success of the far-infrared laser cyclotron resonance experiments achieved in 1966 by BUTTON, GEBBIE and LAX [3] helped to overcome a limitation experienced by microwave spectroscopy in studying the cyclotron resonance in materials with low mobility. In addition, the application of time-resolved techniques, as well as strong pulsed photoexcitation enables one to investigate the dynamical behaviour of photoexcited semiconductors [4, 5].

This report presents very recent results on the cyclotron resonance in semiconductors as semi-insulating GaAs, ternary alloy GaInAs as well as GaInAs-AlInAs heterojunctions, and the II-VI compound semiconductor ZnSe.

2. EXPERIMENTAL PROCEDURES

Wavelengths of 119 and 220 μm are obtained from a discharge-type H_2O laser, and 84 and 172 μm from a discharge-type D_2O laser, all of which are operated in the pulsed mode at the repetition of 30 Hz and, if needed, in synchronized combination with the photoexcitation light pulses and / or electric field pulses at 15 Hz. The excitation light is provided by a xenon flash lamp with a pulse width of 1 μs. The transmitted FIR laser beam is detected by an n-type InSb Putley detector or an n-type germanium photoconductive detector. The absorption spectra are obtained as ln (I_0 / I_E) by a two-channel boxcar averager with an aperture of 0.5 μs. Here I_E and I_0 are the transmitted beam intensities of the pulsed FIR laser with and without the excitation. In the event of an experiment without any excitation, I_E should be taken as the transmitted laser-beam intensity, I_0 being kept constant.

The temperature is varied between 4.2 and 150 K and measured with a calibrated carbon-glass sensor pressed onto the sample holder and regulated with a temperature controller.

The microwave setup is a nonresonant reflection-type wave-guide system working at 35 GHz. The FIR experiments are carried out in the Faraday configuration, and a magnetic field up to 10 T is applied perpendicular to the sample face except for ZnSe samples with twin crystal boundaries.

3. RESULTS AND DISCUSSIONS

3.1 Semi-Insulating GaAs

The origin and nature of the EL2 center which is the dominant mid-gap state in LEC-grown semi-insulating GaAs (S.I.-GaAs) has been subject of controversy since its first observation. Besides its importance for semiconductor technology, EL2 itself is a basically interesting object for basic research. A large variety of models for the origin of EL2 have already been proposed, but there has been no satisfactory explanation until now. The main probe for investigating the characteristics of that deep center has so far been restricted to photoluminescence, photocapacitance measurements and Deep Level Transient Spectroscopy (DLTS) [5]. Both the FIR magneto-absorption and microwave photoconductivity measurements enable us to get an experimental novel insight into the dynamical behaviour and the electronic properties of the EL2 center.

Experiments were performed on one In-doped (sample A) and two undoped (samples B and C) GaAs crystals. Sample C was annealed in a nitrogen atmosphere at 800°C for thirty minutes.

In order to avoid undesirable excitation, we employed two kinds of filters: one to cut off wavelengths shorter than 0.9 μm (1.38 eV), hereafter called IR filter, and the other to cut off wavelengths longer than 0.75 μm (1.65 eV), IRC filter. By making use of the IR filter, we could suppress the band-gap excitation, so that electrons could be excited only from the mid-gap states. With the aid of the IRC filter, on the other hand, electrons could be excited mainly from the valence band. We are thus able to make experiments under the condition of selective excitation.

A fully compensated sample of S.I.-GaAs contains very few neutral shallow impurities and a great deal of ionized impurities at low temperature equilibrium. It is practically impossible to get proper information of S.I.-GaAs which contains very few neutral shallow impurities and conduction electrons without photoexcitaiton, because only shallow neutral donors and conduction electrons can absorb photons in the FIR region employed in this experiment.

Suppose that the sample is illuminated by intrinsic or extrinsic light. Most of electrons excited from the valence band or from the mid-gap states are instantly captured by ionized shallow impurities which are neutralized as a consequence. In this way we can investigate even the character of S.I.-GaAs through the FIR magneto-optical absorption of photo-neutralized impurities and short-lived conduction electrons.

A series of time resolved traces under photoexcitation with the IRC filter is shown in Fig. 1. The signal showing up at 1.9 T arises from the so-called impurity-cyclotron resonance (ICR), or the 1s - $2p_{+1}$ Zeeman

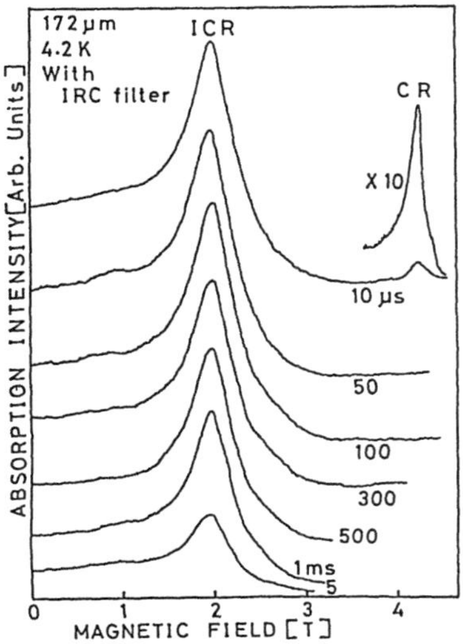

Fig. 1. Time-resolved CR and ICR with IRC filter for sample C at the wavelength of 172 μm. The trace of the CR signal at the top is taken with an enlarged scale.

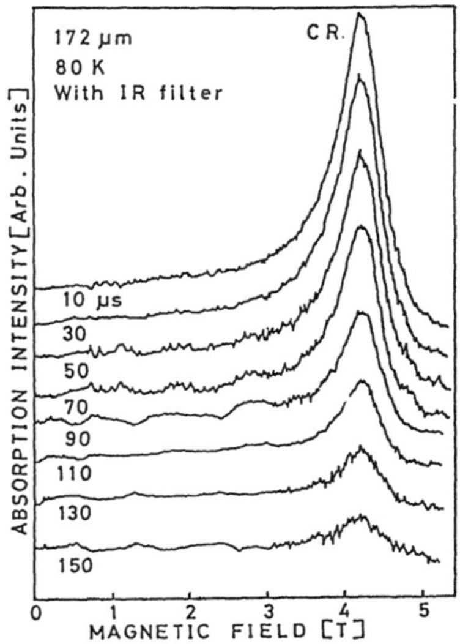

Fig. 2. Time-resolved CR absorption with IR filter for sample C at the wavelength of 172 μ m.

transition, of typical shallow donor electrons. The weak cyclotron resonance (CR) of conduction electrons, on the other hand, is observed at 4.3 T. In this case, electrons involved in both resonances are mainly excited from the valence band and not from the mid-gap states. The ICR decays very slowly. Though the absorption signal decays after each photopulse, we can obtain the same signal intensity for any length of time so long as the gate position of the boxcar is fixed. Almost the same signals are observed right after starting the experiment employing the IR filter for the IRC filter. Both ICR and CR signals, however, decrease rapidly in intensity as the time proceeds. The diminished signal never recovers its original strength below 130 K. This feature must be connected with the photo-quenching effect, which is a characteristic behaviour of EL2 centers in S.I.-GaAs [6].

Figure 2 shows typical traces of CR absorption at 80 K obtained and with an IR filter. The data are taken at various delay-times after photoexcitation. At this temperature most of the electrons have left the donors, so that no ICR absorption is observed. The CR absorption taken at such a high temperature shows very little photo-quenching, even for an extrinsic excitation with an IR filter. The lifetime of electrons photoexcited from EL2 centers shortens rapidly as the temperature is raised. This fact is also confirmed by the microwave photoconductivity measurement at various temperatures.

For understanding all the observations, we propose a model of the metastable state for the EL2 center after photo-quenching. The ground state of the EL2 center will be composed of an As_{Ga} antisite and a shallow impurity that is in the vicinity and coupled very weakly. Provided that the stable configuration of the defect is strongly charge-state-dependent, successive release and capture of an electron by the center through photoexcitation will result in a migration or in a transition to a metastable atomic configuration. In consequence a complex center consisting of an As_{Ga} antisite strongly coupled with a shallow impurity is built up, and the EL2 center is raised to its metastable state. Exactly speaking, both the shallow donor involved in the ICR absorption and the As_{Ga} antisite are raised to their metastable states. If the EL2 center turns into its metastable state by photoexcitation with the IR filter without forming a complex center, the ionized shallow donor is still neutralized by an electron excited from the valence band by the intrinsic light and should be able to contribute to the ICR. Our experimental result, however, contradicts this model.

At higher temperature ($T > 50$ K), shallow donors can hardly bind electrons, so that no ICR absorption is observed as shown in Fig. 2. In such a situation, the time variation of the photoexcited conduction electron density is given by the simple relation:

$$n_e(t) = n_0 e^{-t/\tau_c} \tag{1}$$

where n_0 is the density of initially created electrons and τ_c the lifetime of electrons given by $1/\tau_c = N_m^+\langle v\sigma_c\rangle$. Here N_m^+ is the density of the ionized EL2 center, $\langle v\rangle$ the mean square thermal velocity of electrons and τ_c the electron capture cross-section. Figure 3 shows the dependence of the electron capture rate $1/\tau_c$, obtained both at 172 μm (1744 GHz) and at 35 GHz microwave measurements, on the reciprocal temperature together with the photoquenching rate $1/\tau_q$. We can deduce for the electron capture cross-section:

$$\sigma_c = \sigma_\infty \exp(-\Delta E_A/k_B T) \tag{2}$$

where ΔE_A is a characteristic activation energy, k_B the Boltzmann constant and σ_∞ the capture cross-section for $T = \infty$. The solid line in Fig. 3 yields ΔE_A = 46 meV. ΔE_A may originate from a barrier in the conduction band related to the Franck-Condon shift of the EL2 center.

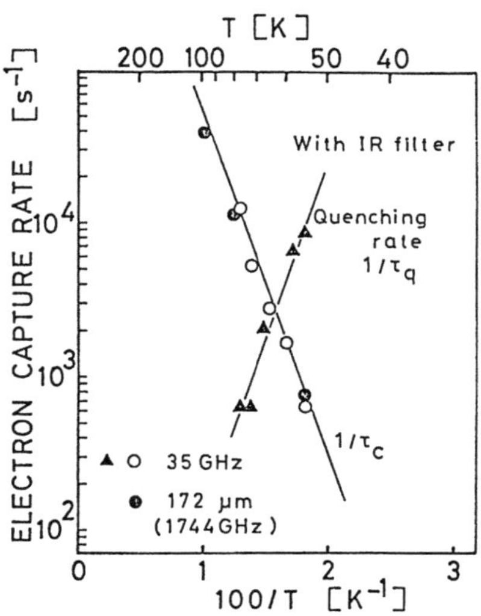

Fig. 3. Temperature dependence of the electron capture rates $1/\tau_c$ by the EL2 center. Open circles represent the results obtained from the microwave photoconductivity, solid circles the results from the FIR CR. The straight line yields the value of the activation energy ΔE_A = 46 meV. The Photo-quenching rates $1/\tau_q$ are also shown.

3.2 GaInAs and GaInAs-AlInAs Hetero-Structures

The ternary III-V alloys are becoming important as a promising device material in long-wavelength optical communication systems and for high-speed field-effect transistor applications.

The evaluation of the scattering probability in a ternary alloy is considerably more complicated than those in element- and binary alloy semiconductors, because the ternary compounds exhibit an additional alloy scattering process which is not found in other materials.

In this paper, we report on the investigation of impurity scattering as well as impact-ionization mechanisms from ICR and CR absorption measurements in GaInAs, to understand the dynamical behaviour of ternary compounds. In addition, we have made both magneto-resistance and CR absorption measurements to clarify the quantum transport features of two-dimensional (2D) electrons in selectively doped GaInAs-AlInAs heterostructured samples.

Four n-type $Ga_{0.47}In_{0.53}As$ samples and one $Ga_{0.47}In_{0.53}As$-$Al_{0.48}In_{0.52}As$ heterojunction were employed for the present experiment. All the samples were grown on InP substrates, which were wedge-shaped to avoid interference effects. Their carrier concentration are as follows: sample A ($n = 3.8 \times 10^{14}\ cm^{-3}$), sample B ($n = 6.7 \times 10^{14}\ cm^{-3}$), sample C ($n = 1.7 \times 10^{15}\ cm^{-3}$) and sample D ($n = 8.3 \times 10^{15}\ cm^{-3}$).

Typical traces of the cyclotron resonance absorption at 4.2 K of sample A are shown in Fig. 4 for a wavelength of 119 μm and at various delay-times after photoexcitation. We have no absorption signal for the CR of conduction electrons at 4.2 K without photoexcitation, since very few electrons exist in the conduction band, and only the ICR is observed. Through the time-dependence of the absorption intensity and the linewidth analysis, we obtain the scattering rate of photoexcited electrons. The inverse relaxation time $1/\tau$ derived from the simple relation,

$$1/\tau = \omega \Delta B / 2B_r \quad , \tag{3}$$

where ΔB is the linewidth measured between the half-maxima, B_r the resonance field and ω the angular frequency of the FIR-laser used, is shown in Fig. 5 as a function of the electron-density n_e. We assume that the electron density variation of $1/\tau$ can be written

$$1/\tau = a + b\, n_e^{\eta} \quad , \tag{4}$$

where "a" is a density-independent limiting value of $1/\tau$, mainly arising from impurity as well as alloy scatterings, and "b" and "η" are constants. The linear dependence ($\eta = 1$) on n_e appears to fit the data reasonably well over a wide range of n_e.

Donor electrons are easily excited into the conduction band by the electric field. As stated above, we employ a differential method. In that case, the absorption spectra are obtained as $\ln(I_0/I_E)$, that is related to the change in the absorption coefficient, $\Delta\alpha$, induced by the electric field excitation. As the absorption coefficient for the ICR and the CR is in general proportional to the density of neutral impurities and conduction electrons, respectively, $\Delta\alpha$ must be proportional to the density difference $N_E - N_0$. Here N_E is the density of the absorbers with excitation and N_0 that without excitation. In the case $N_E < N_0$, the resultant

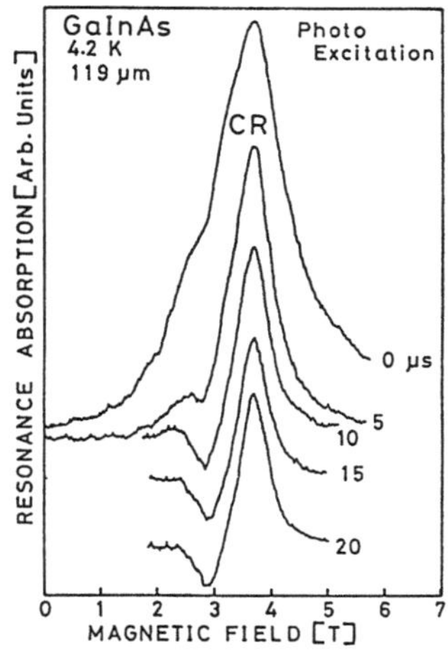

Fig. 4. Time-resolved CR signals for sample A, taken under pulsed photoexcitation.

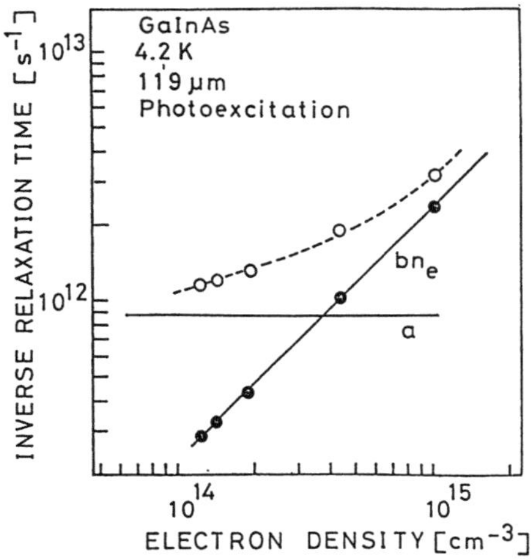

Fig. 5. Inverse relaxation time (open circles) is plotted against electron density. The variation of $1/\tau$ is derived from the Eq. (4). Solid circles show a linear dependence on n_e.

signal thus takes a negative from. Figure 6 shows typical traces of FIR magneto-absorption obtained at 4.2 K and for the wavelength of 119 μm at various values of the electric field. Decrease of the absorption arising from the ICR and increase of the CR are obvious, as the electric field is increased beyond 10 V/cm. The inset shows signals obtained with the electric field below 15 V/cm. At a lower electric field, $\Delta\alpha_{ICR}$ exceeds $\Delta\alpha_{CR}$, but the situation has reversed above 20 V/cm. On further increase of the electric field, the intensity ratio $\gamma = \Delta\alpha_{CR} / \Delta\alpha_{ICR}$ saturates above 50 V/cm. This behaviour is understood as follows: at a lower electric field, electrons excited from donors into the conduction band remain at the band-tail where the electron mobility is extremely small, so that they hardly contribute to the sharp CR absorption.

Linewidth measurements of the CR absorption have been carried out for various samples. It is found that $1/\tau$ is practically independent of the electric field. By means of the electric field, donor electrons are excited into the conduction band. Thereby the densities of neutral donors, ionized donors and conduction electrons might change remarkably. The experimental results, however, indicate that the dominant cause of the line-broadening at 4.2 K is obstacles other than the above-mentioned scatterers. We should keep it in our mind that the ternary compounds exhibit an additional alloy scattering process. HARRISON and HAUSER [7] have derived an expression for alloy scattering, based on the initial treatment by BROODS. They obtained the energy-dependent scattering rate as

$$1/\tau_{AL} = \frac{3\sqrt{2}\pi (k_B T)^{1/2} m^{*3/2} \; x\,(1-x)\,(\Delta E)^{1/2} \varepsilon^{1/2}}{8\hbar^4 N_a} \, , \tag{5}$$

where $\varepsilon = E/k_BT$ is the electronic kinetic energy reduced by k_BT, N_a is the density of alloying sites, x and 1-x are the mole fractions of the binary end compounds and ΔE is a scattering potential.

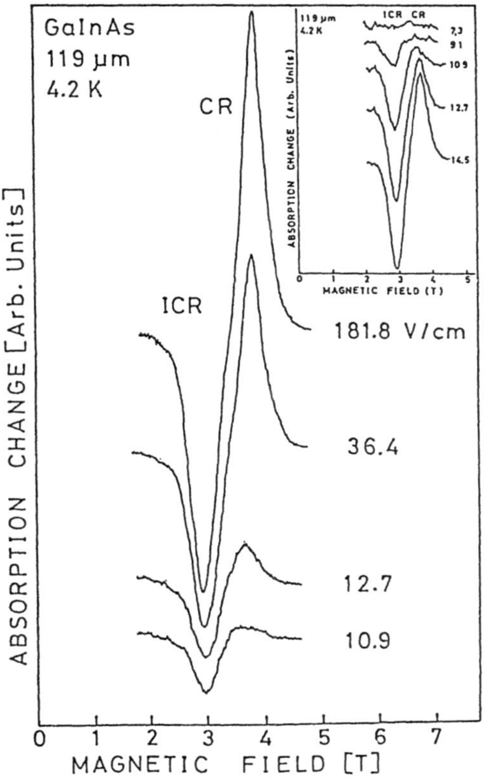

Fig. 6. Dependence of the CR as well as the ICR absorption signals on the electric field with the pulsed electric field modulation technique.

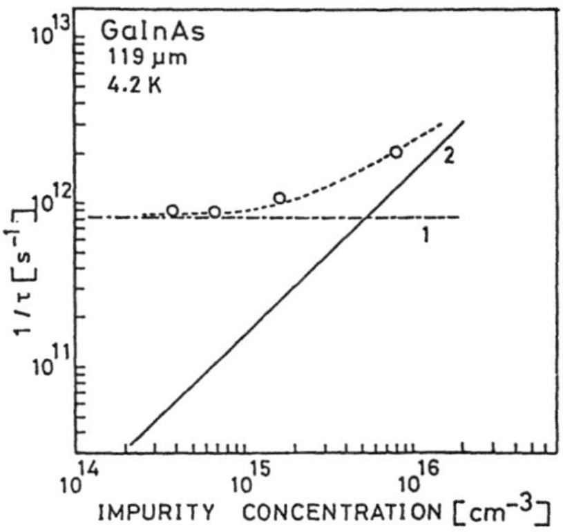

Fig. 7. Dependence on the neutral-donor concentration of the inverse relaxation time obtained at 4.2 K. Curve 2 shows the contribution of neutral impurity scattering. To fit the experimental data, an additional contribution shown by 1, which is independent of impurity concentration, is necessary.

A definite donor concentration dependence is shown in Fig. 7. We assume that the impurity concentration variation of $1/\tau$ can be written

$$1/\tau = 1/\tau_{AL} + CN_n \quad , \tag{6}$$

where N_n is the concentration of neutral impurities and C is a constant. At 4.2 K, the contribution from phonon scattering can safely be neglected on account of a much smaller scattering compared with the other mechanisms. Comparison of experimental results with the calculated scattering rate is performed assuming a Maxwell-Boltzman distribution function and using the numerical values appropriate to $Ga_{0.47}In_{0.53}As$; namely, $m^* = 0.041\ m_0$, $x = 0.47$ and $N_a = 9.5 \times 10^{21}\ cm^{-3}$. In this way the alloy scattering potential is estimated to be $\Delta E = 1.1$ eV for GaInAs.

Both magnetoresistance and the CR absorption measurements were made on a rectangular shaped GaInAs-AlInAs heterojunction at 4.2 K. Very pronounced Shubnikov-de Haas oscillations appear in the magnetoresistance. The oscillations with two obviously different periods reveal the occupation of two electronic subbands. We may thus deduce the 2D electron density of two subbands to be 1.4×10^{12} cm^{-2} and 3.3×10^{11} cm^{-2}. We have also observed CR absorption in the same sample for the wavelength of 119 μm. The effective mass is found to be $m^* = 0.053\ m_0 \pm 0.002\ m_0$ and a typical experimental trace is shown in

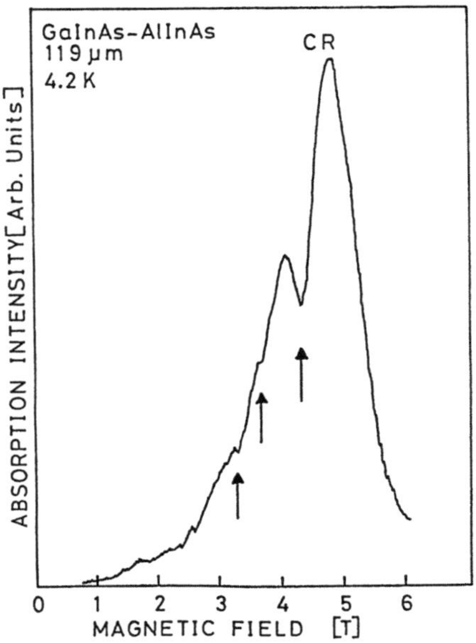

Fig. 8. The CR absorption for GaInAs-AlInAs heterojunction at the wavelength of 119 μm. Several structures in the CR absorption due to a quantum effect are observed.

Fig. 8. This effective mass value is somewhat larger than the effective mass obtained by the CR for the bulk $Ga_{0.47}In_{0.53}As$ ($m^* = 0.041\ m_0$) and for a heterostructured sample ($m^* = 0.048\ m_0$) with lower 2D electron density [8]. The variation is probably caused by a nonparabolicity of the conduction band. A characteristic feature of the CR of 2D electrons are the quantum oscillations of the absorption amplitude which appear at low temperatures and sufficiently high 2D electron densities. This type of oscillation, which is predicted by Ando [9], was observed in inversion layers on Si surface [10]. For the present system we find quantum oscillations in the resonance amplitude for the first time (arrows in Fig. 8). Employing the above-mentioned modulation detection under photoexcitation, quantum oscillations can clearly be recognized.

3.3 Cyclotron Resonance of 2D Electrons in Twin Crystal Boundaries

ZnSe is expected to be one of the most promising II-VI materials for fabrication of efficient blue-light-emitting devices. An intricate effect, however, called self-compensation, imposes difficulties on growing high quality ZnSe single crystals, and consequently on investigating transport phenomena in detail. Recently we have for the first time succeeded in observing both electron [11] and hole cyclotron resonance [12], and thus have been able to discuss some sophisticated physics of ZnSe. In this paper we report some characteristic behaviour of 2D electrons in twin crystal boundaries of ZnSe.

Crystals of ZnSe were grown by a kind of transportation method of the vapor phase. We employed two samples. One sample was cut from an as-grown crystal (sample A) and the other from an ingot after "zinc-dip" treatment (sample B). After "zinc-dip" treatment, the material showed a strong n-type conduction.

Through X-ray analysis and microscopic observation with optical polarizers, it was found that both samples contain some twin crystal boundaries.

At 4.2 K an electron CR peak for sample A is observed at B = 0.2 T in a 35 GHz measurement, only right after an excitation pulse. The peak lies on a broad plateau. At higher temperatures the plateau turns into a secondary resonance peak as shown in Fig. 9. The resonance field is remarkably sensitive to the magnetic field direction. This is characteristic for 2D-behaviour.

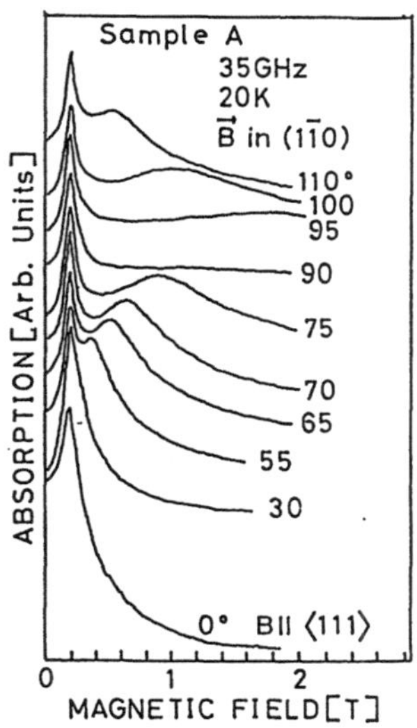

Fig. 9. The CR absorptions obtained with 35 GHz microwave set up at 20 K for ZnSe sample A are shown at different angles between the magnetic field and the crystal growth direction ⟨111⟩.

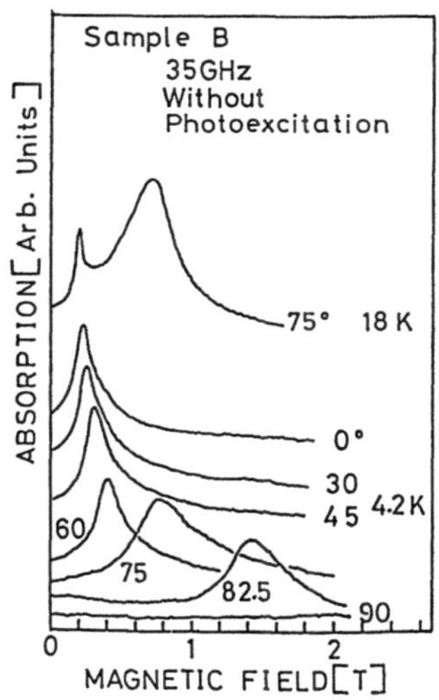

Fig. 10. Traces of the 35 GHz CR absorption at 4.2 K from ZnSe sample B for various angles. Data are taken without photoexcitation. The trace on the top is taken at 18 K and the 3D electron signal together with that of 2D electron is observable.

Figure 10 shows the CR traces from sample B at 4.2 K for various angles between the magnetic field direction and a crystal growth-direction ⟨111⟩. It should be noted that these signals are observed without photoexcitation. The resonance field is again sensitive to the magnetic field direction. The above observations can be well understood by considering that the carriers giving rise to the plateau at 4.2 K for sample A and to the 2D resonance for both samples are most likely bound, loosely or tightly depending on the position of Fermi level, in potential wells of twin crystal boundaries. The localized 2D electrons can thermally be excited into the conduction band to become three dimensional.

In summary, various new features of electronic and defect properties in GaAs, GaInAs and ZnSe have been observed by infrared laser- as well as microwave cyclotron resonance (CR) and impurity cyclotron resonance (ICR). Especially, the "photo-quenching effect" of cyclotron resonance absorption in semi-insulating GaAs, and the 2D nature of electrons confined in a GaInAs-AlInAs interface and in twin crystal boundaries in ZnSe have been demonstrated. In addition, the electron-alloy scattering under quantum limit conditions has been discussed.

The author is greatly indebted to the Sony Central Research Laboratory and NTT Basic Research Laboratories for their generous offer of GaInAs and GaInAs-AlInAs heterostructured samples. The author also wishes to thank Prof. E. Otsuka for his guidance and careful attention throughout. Special thanks are due to Dr. H. Nakata, Dr. K. Fujii, Mr. H. Kobori, Mr. T. Tomaru and Mr. K. Sakakibara for their valuable help in the analysis and experiments. He further expresses his thanks to Dr. M. Isshiki and Prof. K. Igaki for preparing the ZnSe samples.

This work has been partially supported by the Scientific Research Grant-in-Aid #61114008 for Special Project Research on "Alloy Semiconductor Physics and Electronics", from the Ministry of Education, Science and Culture.

References

1. G. Dresselhaus, A.F. Kip and C. Kittel: Phys. Rev. 92, 827 (1953).
2. H. Lax, H.J. Zeiger, R.N. Dexter and E.S. Rosenblum: Phys. Rev. 93, 1418 (1954).
3. K.J. Button, H.A. Gebbie and B. Lax: IEEE J. Quantum Electron. 9E2, 206 (1966).
4. T. Ohyama, T. Yoshihara, T. Sanada, K. Murase and E. Otsuka: Phys. Rev. Lett. 27, 33 (1971)
5. T. Ohyama and E. Otsuka: In Infrared and Millimeter Waves, ed. by K.J. Button, vol. 8 (Academic Press, New York, London 1983) p. 213.
6. See, for example, Proc. the 4th Conf. on Semi-Insulating III-V Materials, Hakone 1986,ed. by H. Kukimoto and S. Miyazawa (OHM, North-Holland, 1986)
7. J.W. Harrison and J.R. Hauser: J. Appl. Phys. 47 292 (1976).
8. J.C. Portal, R.J. Nicholas, M.A. Brummell, A.Y. Cho, K.Y. Cheng and T.P. Pearsall: Solid State Commun. 43, 907 (1982).
9. T. Ando: J. Phys. Soc. Jpn. 38, 989 (1975).
10. Abstreiter, G.J.P. Kotthaus, J.F. Koch and G. Dorda: Phys. Rev. B14, 2480 (1976).
11. T. Ohyama, K. Sakakibara, E. Otsuka, M. Isshiki and K. Igaki: Phys. Rev. B37, 6153 (1988).
12. T. Ohyama, K. Sakakibara and E. Otsuka: Jpn. J. Appl. Phys. 26, L136 (1987).

Influence of Repulsive Scatterers on the Cyclotron Resonance in Two-Dimensional Electron Systems with Controlled Acceptor Impurity Concentration

H. Sigg, J. Richter, K. v. Klitzing, and K. Ploog

Max-Planck-Institut für Festkörperforschung, Heisenbergstr. 1
D-7000 Stuttgart 80, Fed. Rep. of Germany

We review experimental results of the influence of impurities on the cyclotron resonance (CR) in two-dimensional electron systems (2DES). Using AlGaAs/GaAs samples with well-defined species and concentration of acceptor impurities we can analyze in detail several effects on the CR, e.g., filling factor dependent linewidth, strong linewidth narrowing, impurity concentration dependent shifts of the CR line position etc. We show that these effects are determined by collective properties of the random distribution of repulsive scatterers, leading to a reinterpretation of the harmonic potential model by Mikeska and Schmidt. In the light of our experiments on samples with well-defined impurities we can show, that much of the experimental CR data obtained so far even in so-called high mobility samples, are still essentially governed by the influence of impurities.

1. Introduction

The investigation of the cyclotron resonance (CR) of the two-dimensional electron gas (2DEG) goes back to the mid-seventies when such 2DEG systems were realized in Si-MOS inversion layers for the first time. The earliest CR investigations reported by Abstreiter et al. [1] and Allen et al. [2] already gave an indication of the complexity of the CR in 2D electron systems. It was found that the effective mass $m_c^* = \frac{eB}{\omega_c}$ and linewidth of the CR depend on the electron density n_e and on the CR frequency ω_c. Also, the quality of the samples was found to be important: sometimes a decrease of m_c^* followed by a spectacular linewidth narrowing at low densities was observed [3] and sometimes an increase of m_c^* for increasing wavelength was found [4]. These have been attributed to a localization of the 2D electron in a random potential [3]; alternatively a condensation into a Wigner glass has also been discussed [5]. In spite of the numerous and extensive studies in this area [6] these effects are still controversial and there is still some ambiguity about their origin, although - applying Kohn's theorem - the necessity of some extrinsic imperfections was anticipated in all models proposed.

More recently, the CR of the 2DEG has mainly been studied in the much purer AlGaAs/GaAs heterostructures. In these systems the CR was believed to show much more clearly the 2D single particle energy spectrum; hence, the 2D polaron effect [7], the nonparabolicity of the band [8], and the coupling to higher subbands [9,10,11] have been studied and successfully been described using such single particle models. But additionally, the complexity of the CR known from the Si inversion layer system has been recovered. A splitting of the CR has been observed in very high mobility samples [12], linewidth broadening connected to the filling of the Landau levels [13] and spin levels [14] has been found, and asymmetric CR lineprofiles have been reported [15]. An illustrative example of the various effects on linewidth and lineposition of the CR in AlGaAs/GaAs heterostructures was given by Ensslin et al. using samples with a front gate [16]. At very low density of the 2DEG a decrease of m_c^*, similar to what is observed in Si inversion

Springer Series in Solid-State Sciences, Vol. 87
High Magnetic Fields in Semiconductor Physics II Editor: G. Landwehr

layers mentioned above, was found by Chou et al. [17]. However, these recent CR investigations on the AlGaAs/GaAs system have still not solved the older problems, although the presence of imperfection is again assumed to be important.

We review in the following, extended studies on the CR of the 2DEG in heterostructures with intentionally built-in impurities of a given species and of well defined and characterized concentrations.

In a first series of measurements we investigated e-beam irradiated samples where the concentration could be controlled in a defined and reversible way via the irradiation doses and subsequential annealing process [18]. e-beam irradiation creates acceptor-like defects which are homogeneously distributed over the whole sample thickness. We will refer to this in the following as "volume" doping. In addition to these structures we have also investigated so-called $\delta-$sheet doped samples, where, during the MBE growth, Be-acceptors were built-in at a well defined distance from the interface [19].

Sample preparation and experimental set up are described in Section 2. The experimental results are presented in Section 3 and are discussed in Section 4. A model description is given, that allows a quantitative comparison with earlier CR results obtained in Si inversion layer systems and a recent investigation of high mobility AlGaAs/GaAs heterostructures [20]. The results are summarized in Section 5.

2. Preparation and Experimental Techniques

In the first series of measurements we studied high mobility modulation doped AlGaAs-GaAs heterostructures with carrier mobilities μ_e and electron sheet densities n_e of typically $200,000\ cm^2/Vs < \mu_e < 800,000\ cm^2/Vs$ and $2 \cdot 10^{11} cm^{-2} < n_e < 5 \cdot 10^{11} cm^{-2}$, respectively. The samples were cooled down to 77 K and were irradiated by 1 MeV electrons. It is known that the defects induced by this irradiation are mainly vacancy-interstitial pairs in the As sublattice which recombine at temperatures above $200^0 C$ [21]. These defects give rise to a number of electron traps, with energy states which are both, for the GaAs and the AlGaAs material, located about 50 meV or more below the conduction band (see Fig. 1a). Traps with energies below the Fermi energy in the band diagram of Fig. 1a bind an electron and become negatively charged. Since these electrons mainly come from the 2DEG there is the effect that irradiation always leads to a reduction of the interface electron density n_e. This is important since we then can estimate the sheet density of the ionized scatterers $n_{sc} \cong \Delta n_e$. E.g., for an irradiation of about $2 \cdot 10^{17}$ electrons/cm^2 we found $n_{sc} \cong \Delta n_e = 1 \cdot 10^{11} cm^{-2}$. In a second series of samples, the acceptor-doping was achieved by δ-sheet doping techniques [22]. The built-in dopant was Be and has been located within the GaAs material at 2.5 nm from the AlGaAs-GaAs interface. In Fig. 1, we have visualized schematically - i.e., without screening effects included - the effective potential which arises from the randomly distributed ionized acceptors. In the following we refer to this effective potential for the electrons of the 2DEG as "scattering potential". The essential difference between the two types of samples is that the strength of the potential for each scatterer is constant for the δ-sheet doped sample, whereas it varies for the electron irradiated samples due to the randomly varying distances of the impurities from the interface.

We have performed transport measurements and far infrared (FIR) transmission experiments in magnetic fields up to 15 T. Magnetotransport was used to characterize the dc mobility and carrier concentration of the samples. Both far infrared lasers and a fast scanning Fourier transform spectrometer were used in the experiments. In all experiments the temperature was 2.2 K if not otherwise specified.

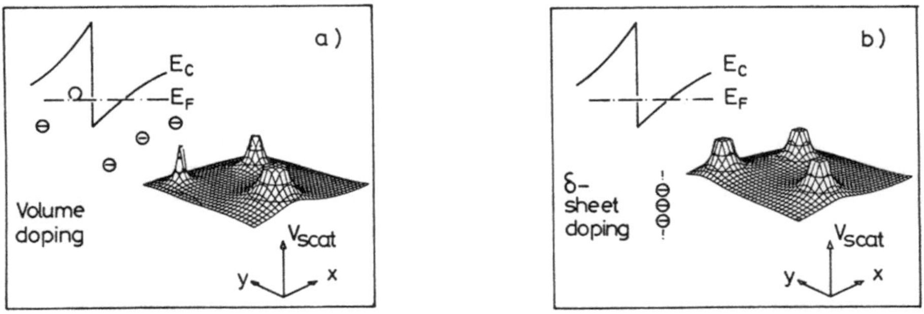

Figure 1: Schematic band diagram of the AlGaAs-GaAs interface conduction band with indicated energy position and charge of the acceptor states for (a) - the case of a volume doping as obtained by 1 MeV electron irradiation and for (b) - the case of δ-sheet doping with Be. In the figures (a) and (b) to the right, the scattering potential V_{scat} is sketched for the two kinds of acceptor distributions

3. Experimental Results

Figure 2a shows the plots of the CR spectra obtained at various magnetic fields on an irradiated sample from the first series of measurements presented in Ref. [18]. A very similar series of CR spectra is shown in Fig. 2b which has been obtained from a Be δ-sheet doped AlGaAs-GaAs heterostructure [19]. For both samples, the CR linewidth - defined by the fullwidth at half maximum - is found to be small at high magnetic field but increases with decreasing magnetic fields. The spectra taken at low fields - i.e. for $B_2 = 4.06(5.40)T$ in Fig. 2a (2b) - exhibit a split resonance. It will become obvious, that this split resonance consists of two CR's - one which exists only at low magnetic fields, i.e., $0 < B \leq B_2$ and one at high magnetic field $B \geq B_2$. In the following we will refer to the former resonance as "low field CR", and we will call the latter "high field CR." From the Shubnikov-de Haas experiments it is found that B_2 is the magnetic field which corresponds to filling factor 2 of the spin resolved LL, i.e., $\nu = \frac{n_e h}{eB} = 2$.

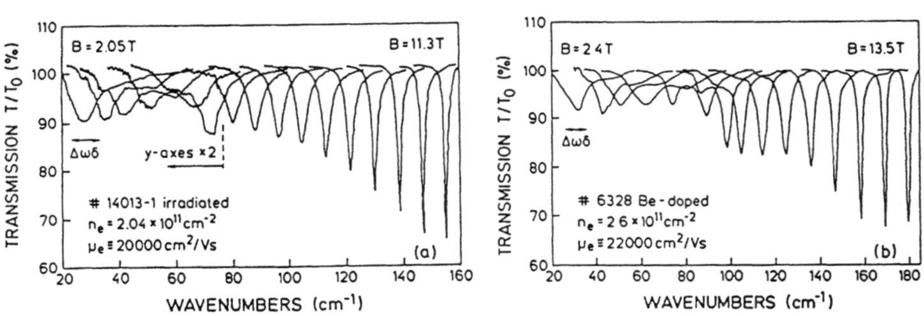

Figure 2: Normalized CR transmission spectra - taken at T= 2.2 K and various magnetic fields of constant interval within the indicated range - for (a) an irradiated and (b) a Be δ- sheet doped AlGaAs/GaAs heterostructure. Indicated is the CR linewidth $\Delta\omega_\delta$ which is calculated from the dc-mobility μ_e

The linewidth of the low field CR ($\nu > 2$) is found to be consistent with the measured low dc mobilities in these samples. This, and the observation of an extremely narrow high field CR is a common feature for all the investigated e-beam irradiated and Be-doped samples.

A detailed analysis of the CR is shown in Fig. 3 in a plot of the linewidth $\Delta\omega$ versus the magnetic field B. The CR data are obtained from an irradiated sample. At $B \simeq B_2$ a linewidth of about 12 cm^{-1} for both the low field (+) and high field (o) CR is observed. $\Delta\omega$ is nearly constant for $B < B_2$, however - as already shown in Fig. 2 - $\Delta\omega$ strongly decreases with increasing magnetic fields for $B > B_2$. The strong decrease in $\Delta\omega$ flattens out at $B \approx 7T$. The distinction made in a low field and high field CR is evident also from the analysis of the magnetic field dependence of the CR frequency ω_c. This is shown in the inset of Fig. 3 for the irradiated sample of Fig. 3 (o) and for the same sample previous to its irradiation (∇), respectively. We find that the split CR shown in Fig. 2 is the consequence of a frequency up-shift of the high field CR. The line position of the low field CR, however, is not much influenced by the irradiation and shows the usual linear dependence on the magnetic field $\omega_c^o = \frac{eB}{m_c}$, where m_c is the effective mass, $m_c \simeq 0.07m_e$. The magnetic field dependence of ω_c for the high field CR is described best by the expression $\omega_c^2 \cong (\omega_c^o)^2 + (\omega_s)^2$, where ω_s determines the frequency shift. This is shown in Fig. 4 for the case of a Be doped sample.

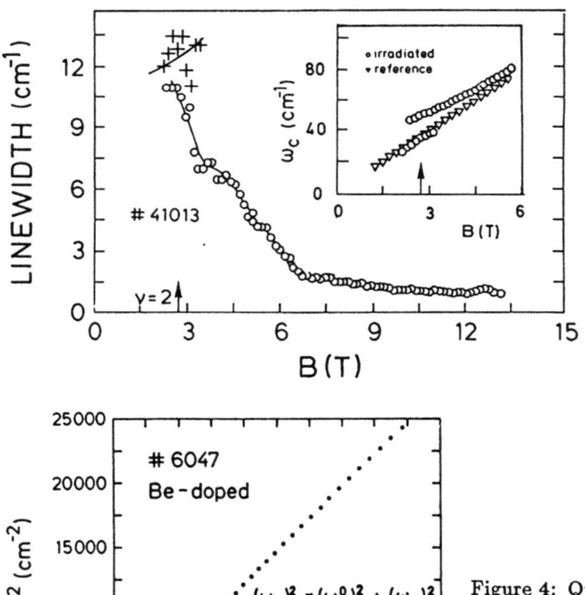

Figure 3: Magnetic field dependence of the CR linewidth, and, in the inset, of the line position obtained from an irradiated sample. The linewidth data of the split resonance are indicated separately. (+) corresponds to the low frequency line, which occurs at ω_c in coincidence with the non-irradiated reference sample (see inset), and the linewidth data indicated by (o) corresponds to the upshifted CR. The arrows indicate filling factor $\nu = 2$

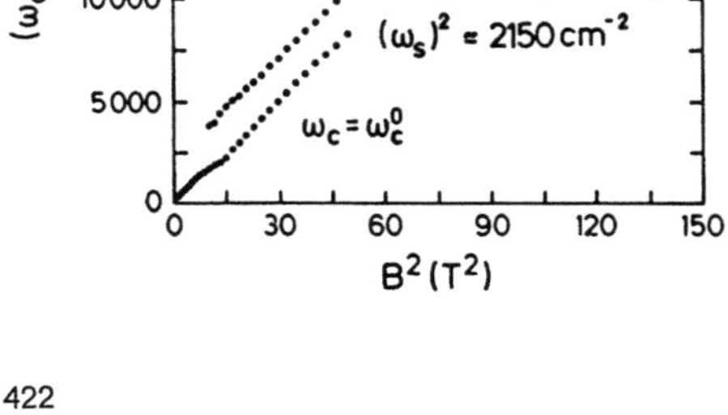

Figure 4: Quadratic plot of the CR frequency ω_c versus magnetic field B for a Be-doped sample. The data is well approximated by $\omega_c^2 \cong (\omega_c^o)^2 + (\omega_s)^2$, where ω_c^o is the CR frequency known for GaAs, with $\omega_c^0 = \frac{e}{m_c^*}B$ and $m_c \approx 0.07m_e$ and ω_s is the upshift in frequency observed at high magnetic fields

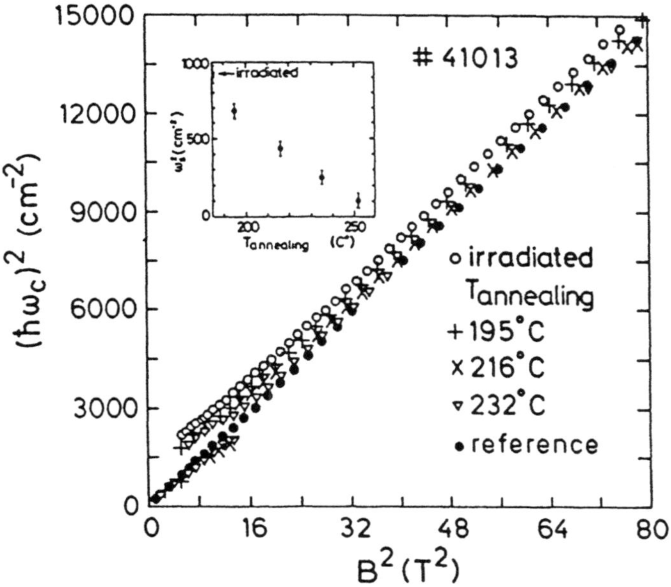

Figure 5: Plot of ω_c^2 versus B^2, showing the result of isochronal annealing processes on an irradiated sample. In the inset, the experimental shift $(\omega_s)^2$ is depicted as a function of the annealing temperature $T_{annealing}$

From the quadratic plot, ω_c^2 versus B^2, in Fig. 4 we obtain that ω_s is zero for the low field CR ($\nu > 2$) and $\omega_s^2 \cong 2150\ cm^{-2}$ for the high field CR ($\nu < 2$). Around filling factor 2 - i.e., close to B_2 - both the shifted and unshifted CR are resolved (c.f. Fig. 4 at $15T^2 < B^2 < 50T^2$). In the final figure of this section we would like to emphasize once more the main point of this shifted CR; namely ω_s, the shift in frequency depends on the scatterer density. This is demonstrated in Fig. 5 which shows the CR line position analysis obtained from the successively annealed sample of Fig. 3. Annealing temperatures are indicated in the figure. The inset of Fig. 5 shows ω_s^2 versus the temperature of the corresponding annealing cycle.

4. Discussion

We first discuss our experimental results emphasizing the obvious similarities with some of the earlier results obtained in electron inversion layers in Si-MOS structures. The most striking similarity is the occurrence of a narrow CR linewidth at high magnetic field and corresponding low filling factors. For the Si-electron inversion layers, CR linewidths as narrow as $\Delta\omega = 5\ cm^{-1}$ have been reported [6]. For the case of the heterostructures investigated here, $\Delta\omega$ smaller than $1\ cm^{-1}$ has been observed. In both systems, the narrow linewidth goes together with a line shift of the CR to higher energies. The magnetic field dependence of ω_c in both systems can then be described by $\omega_c^2 \cong (\omega_c^o)^2 + (\omega_s)^2$. Line shifts up to $\omega_s = 20\ cm^{-1}$ had been found in Si-MOS systems [23]. In our study, the largest shift was about $\omega_s \simeq 45\ cm^{-1}$ and has been observed in the Be-doped sample shown in Fig. 4. These qualitative similarities between the CR in Si-MOS systems and the two differently prepared acceptor-doped AlGaAs/GaAs heterostructures suggests a common mechanism to be the origin of the CR line narrowing and line shifts. The experiment on the irradiated samples shows that the line shift ω_s depends on the annealing temperature (c.f. Fig. 5). Since the annealing temperature determines the recombination rate of the Frenkel-like defects [21], we know that the number of defects, and therefore acceptor states, are subsequently reduced after each of the annealing cycles applied for the experiment shown in Fig. 5. Hence, the line shift dependence on the annealing temperature shown in Fig. 5 means, that ω_s depends on the scatterer density. This is a very spectacular finding, because it says that the CR is governed by a "collective" property of the scatterer distribution. It should be noticed that annealing of the samples does not influence the random type of scattering potential shown in Fig. 1a, and does not change the strength of the individual repulsive scatterer. Additional support for the finding that collective properties of the scattering potential are determining the observed CR behaviour at high magnetic field, is obtained from the experimental observation that the CR behaves qualitatively the same in both the AlGaAs/GaAs systems with respectively a δ-sheet and "volume" acceptor doping. From that we can learn that the shifted and narrow CR is related to some average value of scattering strength; i.e., the CR does not distinguish between the two scatterer distribution shown in Fig. 1a and 1b, respectively. We do not know exactly the distribution and species of the scatterers in the Si-MOS systems, however, from the analogy with the comparison made above we have strong evidence that the observed CR line shift in these systems are due to a similar kind of scatterer distributions, i.e., it is due to ionized acceptors from the space charge layer - this would correspond to a volume doping with scatterers of random strength (c.f. Fig. 1a) - or/and due to interface imperfections with scattering potential similar to the case of Fig. 1b.

We will return to this in more details below. Firstly, we will discuss the CR behaviour at low fields. In chapter 3 it is mentioned that the CR measured at low fields is not shifted and its linewidth $\Delta\omega$ is found to be consistent with the Hall mobility μ_e: i.e., $(\Delta\omega_\delta)^2 = \frac{2}{\pi}\cdot\omega_c^2(B\cdot\mu_e)^{-1}$. This expression has been derived by Ando [24] for the case of δ-scatterers within the so-called self-consistent Born approximation (SCBA). In subsequent studies by Lassnig and Gornik [25], and by Ando and Murayama [26] it has been shown that the same expression also applies for our case of scattering on Coulomb centers located at the interface. The SCBA result for the irradiated sample with $\mu_e \cong 1.5\cdot 10^4 cm^2/Vs$ therefore is $\Delta\omega = \Delta\omega_\delta \cong 10\ cm^{-1}\sqrt{B}$ (B in Tesla). This value is in quite good agreement with the low field data shown in Fig. 3. However, the SCBA model cannot explain the high field data of a narrow and line shifted CR. The good agreement of the SCBA calculation with the low field data - if not accidental - and its disagreement with the high field data suggests that the 2DEG undergoes a drastic change when the quantum limit of the single LL occupation is reached. Considering the arguments given above, that the random distribution of repulsive scatterers is responsible for the narrow line shifted CR, we may assume that at low field, i.e., $\nu > 2$, the scattering potential for the higher LL states can no longer be represented by such a simple distribution of only repulsive scattering centers. The scattering potential is smoothed due to screening of the scatterer by electrons in the lower lying LL energy states. In Ref. [5], it has been mentioned that the high field properties of the CR also in Si-MOS systems are connected to the quantum limit of single LL occupation.

It is worthwhile noting that in a large number of publications (c.f. Ref. [6]), the narrow CR line has been discussed in terms of an anomalous narrowing. However, our experiment suggests a rather different point of view. The CR at high magnetic field is as narrow as if the repulsive scatterers were not present. This different viewpoint is evidenced by the CR linewidth analysis of the successively annealed sample shown in Fig. 6a.

There, $\Delta\omega$ is plotted versus the sheet carrier density n_e, which has been determined after each annealing cycle. As mentioned in section 2, we can estimate the scatterer density from the 2DEG sheet density, because each scatterer corresponds to a trap state which had captured one electron.

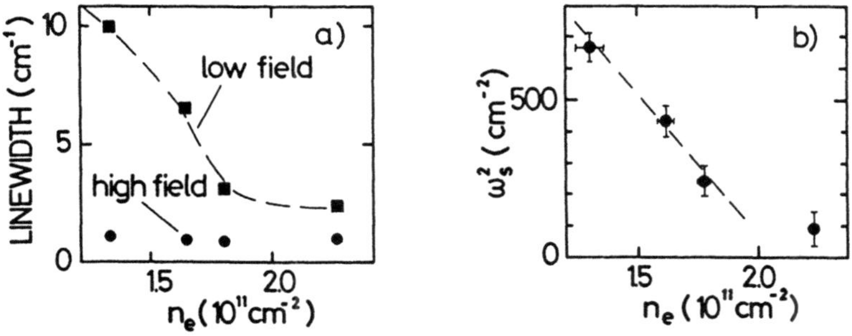

Figure 6: (a) The linewidth for the high field (•) and low field (■) and (b) the measured upshift in frequency ω_s^2, obtained from a series of measurements on a subsequently annealed AlGaAs/GaAs heterostructure are plotted versus the 2DEG carrier density n_e. Since an increase in n_e is correlated to the number of recombined defects the x-scale can be related to the density of repulsive scatterers $n_{sc} = (N_0 - n_e)\gamma$ (see text)

Therefore, we obtain $n_{sc} = (N_0 - n_e)\gamma$ where γ is a constant of order 1 that gives the probability that the captured electron "belongs" to the 2DEG and where N_0 corresponds to the electron density for which $n_{sc} = 0$. This value for the experiment of Fig. 6 is $N_0 = 2.0 \cdot 10^{11} cm^{-2}$ and is obtained from interpolating the scatterer dependent effects ($e.g. \omega_s^2$) to zero. The CR linewidth measured at low fields (■, Fig. 6a) strongly decreases with increasing n_e and correspondingly decreasing n_{sc}. The strong decrease flattens out at $n_e = N_0 \cong 2 \cdot 10^{11} cm^{-2}$. $\Delta\omega$ of the high field CR (•, Fig. 6a) is extremely small and shows no significant dependence on n_{sc}. This suggests, that at high magnetic fields the electrons are avoiding the scatterers so perfectly as if no scatterers are present. The influence of the scatterers is reflected only in the shift of the line position which shows a characteristic dependence on n_{sc}. From Fig. 6b we find that the shift ω_s^2 (obtained from the analysis of Fig. 5) depends linearly on n_e and interpolates to zero at $N_0 = 2.0 \cdot 10^{11} cm^{-2}$. Assuming $\gamma = 1$ we obtain that ω_s^2 is proportional to n_{sc} with a constant of proportionality equal to $8.5 \cdot 10^{-9}$ determined from Fig. 6b.

We now give a phenomenological model which allows a microscopic interpretation of the observed shift ω_s^2, and in particular its linear dependence on the scatterer density. A quadratically shifted CR can be described phenomenologically by considering an additional harmonic potential $V_{osc}(r) = \frac{m^*}{2}\omega_s^2 r^2$. The CR spectrum of the bound electron is to a good approximation $\omega_c^2 \cong (\omega_c^o)^2 + (\omega_s)^2$ for $\omega_s^2 \ll \omega_c^0$. A similar energy spectrum also applies to a harmonic potential well with finite extension radius r_{max}, as has been shown by Mikeska and Schmidt [27]. We adapt their model to our case, however, in a different way. In the Mikeska and Schmidt model the parameters of the harmonic oscillator potential are determined by the individual impurity, corresponding to a shifted CR with frequency shift ω_s that is related to the individual properties of the scatterers. We propose a model which explains the experimentally observed "collective" behaviour of the scatterer induced effects, i.e., the fact that the shift ω_s is determined by the average scatterer density and not their individual properties. In our interpretation we relate the two model parameter of Mikeska and Schmidt, r_{max} and $V_0 = V_{osc}(r_{max})$, to relevant system parameters of the randomly distributed repulsive scatterers. Whereas ω_s^2 can be determined from the experimentally observed energy shifts, it is not directly possible to extract V_0 and r_{max} since the model provides us with the ratio $\frac{V_0}{r_{max}^2} = \frac{m^*}{2}\omega_s^2$. However, if we make the most simple and obvious assumption that (a) - V_0 does not change after annealing processes and (b) - that $r_{max} = l_{sc}/2$ ($l_{sc} = (n_{sc})^{-\frac{1}{2}}$ is the mean scatterer distance) we find immediately the experimentally observed dependence

$$\omega_s^2 = \frac{8}{m^*} V_0 \cdot n_{sc} \quad . \tag{1}$$

This finding strongly supports our assumption and the proposed model. Within this assumption we also can estimate the potential V_0 and find for the irradiated sample $V_0 = 1.5\ meV$.

It is now also interesting to consider the Si-MOS system in the light of our model. Wilson et al. [23] have found for their sample, e.g., at zero substrate bias V_{SD} a quadratic shift of $\omega_s^2 = (15\ cm^{-1})^2$. However, the scatterer density is not known. Thus, for an estimation, we scale our value of V_0, assuming a Coulomb potential type of binding energy, i.e. V_0 is proportional to m^*/ϵ^2, and get for the Si-MOS system $V_0 = 7.5 \cdot 1.5\ meV$. With this value we calculate from (1) $n_{sc} = 3.5 \cdot 10^9\ cm^{-2}$. The corresponding volume density of the space charge acceptor doping N_A can be estimated from the extension of the electron wavefunction in the Si layer which is approximately 50 Å. Thus, the obtained space charge doping is $N_A = 7 \cdot 10^{15} cm^{-3}$, much larger than what was specified in Ref. [23], i.e., $N_A = 1 - 2\ 10^{15} cm^{-3}$. This discrepancy does not necessarily mean that the assumption of the Coulomb type of scaling for V_0 is not appropriate. In fact, the observed dependence of ω_s on V_{SD} suggests the presence of interface imperfections. This would correspond not to a "volume" doping, rather to a sheet doping which exhibits a much stronger interaction with the inversion layer electrons.

Recently, Nicholas et al. [20] have investigated the CR in samples with some of the highest mobilities that are available with today techniques. They found, concerning the energy shift, a very similar behaviour as discussed here. In their samples, ω_s^2 is typically $(1.85\ cm^{-1})^2$. Using our experimentally determined relation, $\omega_s^2 = 8.5 \cdot 10^{-9} n_{sc}$, we obtain for wavefunction extension of ~ 200Å a volume acceptor density $N_A \cong 2 \cdot 10^{14} cm^{-3}$. This value is consistent with the known residual p-type doping concentration in MBE grown heterostructures. Thus, our model predicts very well essential features of the high field CR. It describes quantitatively the energy shifts and explains qualitatively the observation of small linewidths. However, the model so far is too simple to explain the latter quantitatively, e.g., the continuous decrease of $\Delta\omega$ in the range $2 > \nu > 1$ (cf. Fig. 3). The behaviour of the CR at the transition from the high field to the low field regime, i.e., $\nu < 2$ to $\nu > 2$, seems very complex, in particular in high mobility samples when - in contrast to Fig. 4, - only one CR is observed in the vicinity of $\nu = 2$, indicating the presence of interaction effects [16,20]. The occupation of higher LL has also a dramatic effect on the effective scattering potential due to the changed screening properties of the 2DEG.

5. Summary and Conclusions

The CR of the 2DEG in two different kinds of acceptor impurity doped AlGaAs/GaAs heterostructures has been investigated. The CR has been found to be very similar for both systems. The common origin of these similarities has been attributed to a random distribution of negatively charged impurities. The main features of the CR are: at low fields - a rather broad non-shifted CR line, and at high magnetic field a CR line which becomes increasingly smaller with increasing magnetic field and which is up-shifted in frequency as $\omega_c^2 \cong (\omega_c^o)^2 + (\omega_s)^2$. From the experiment, a phenomenological model is deduced of an electron bound in a harmonic potential which is determined by the collective properties of the random scatterer distribution: the width of the potential is determined by the mean distance between the repulsive scatterers and the constant potential energy V_0 has been related to a characteristic (anti) bound energy of the Coulomb scatterer. This evident relationship, that collective properties of the random scatterer distribution causes the experimentally observed shift of the CR frequency, might be very useful for a further and more rigorous understanding of the CR of the 2DEG.

Acknowledgement

We would like to acknowledge financial support from the Bundesministerium für Forschung und Technologie (BMFT).

6. References

[1] G. Abstreiter, P. Kneschaurek, J.P. Kotthaus, J.F. Koch, *Phys. Rev. Lett.* **32**, 104 (1974)
[2] S.J. Allen Jr., D.C. Tsui, J.V. Dalton, *Phys. Rev. Lett.* **32**, 107 (1974)
[3] J.P. Kotthaus, G. Abstreiter, J.F. Koch, R. Ranvaud, *Phys. Rev. Lett.* **34**, 151 (1975)
[4] T.A. Kennedy, R.J. Wagner, B.D. McCombe, D.C. Tsui, *Phys. Rev. Lett.* **35**, 1031 (1975)
[5] B.A. Wilson, S.J. Allen, Jr., D.C. Tsui, *Phys. Rev. Lett.* **44**, 479 (1980)
[6] For reviews see: J.P. Kotthaus, *J. Magn. Magn. Mater.* **11**, 20; B.A. Wilson, S.J. Allen, Jr., D.C. Tsui, *Phys. Rev.* **B24**, 5887 (1981); T. Ando, A.B. Fowler, F. Stern, *Rev. Mod. Phys.* **54**, 432 (1982)
[7] H. Sigg, P. Wyder, J.A.A.J. Perenboom, *Phys. Rev.* **B31**, 5253 (1985)
[8] F. Thiele, U. Merkt, J.P. Kotthaus, G. Lommer, F. Malcher, U. Rössler, G. Weimann, *Solid State Commun.* **62**, 841 (1987)
[9] Z. Schlesinger, J.C.M. Hwang, S.J. Allen Jr., Phys. Rev. Lett.**50**, 2098 (1983)
[10] G.L.J.A. Rikken, H. Sigg, C.J.G.M. Langerak, H.W. Myron, J.A.A.J. Perenboom, *Phys. Rev.* **B34**, 5590 (1986)
[11] S. Oelting, A.D. Wieck, E. Batke, U. Merkt, *Surf. Sci.* **196**, 273 (1988)
[12] Z. Schlesinger, S.J. Allen, J.C.M. Huang, P.M. Platzman, N. Tzoar, *Phys. Rev.* **B30**, 435 (1984)
[13] Th. Englert, J.C. Maan, Ch. Uihlein, D.C. Tsui, A.C. Gossard, *Solid State Commun.* **46**, 545 (1983)
[14] Z. Schlesinger, W.I. Wang, A.H. MacDonald, *Phys. Rev. Lett.* **58**, 73 (1987)
[15] E. Batke, H.L. Störmer, A.C. Gossard, J.H. English, *Phys. Rev.* **B37**, 3093 (1988)
[16] K. Ensslin, D. Heitmann, H. Sigg, K. Ploog, *Phys. Rev.* **B36**, 8177 (1987)
[17] M.J. Chou, T.C. Tsui, G. Weimann, *Phys. Rev.* **B37**, 848 (1988)
[18] H. Sigg, D. Weiss, K. v. Klitzing, *Surf. Sci.* **196**, 293 (1988)
[19] J. Richter, H. Sigg, K. v. Klitzing, K. Ploog, *to be published*
[20] R.J. Nicholas, M.A. Hopkins, D.J. Barnes, M.A. Brummell, H. Sigg, D. Heitmann, K. Ensslin, J.J. Harris, C.T. Foxon, G. Weimann, *to be published*
[21] D. Pons, J.C. Bourgoin, *J. Phys. C: Solid State Phys.* **18**, 3839 (1985)
[22] K. Ploog, *J. Cryst. Growth* **81**, 304 (1987)
[23] B.A. Wilson, S.J. Allen, Jr., D.C. Tsui, *Phys. Rev.* **B24**, 5887 (1981)
[24] T. Ando, *J. Phys. Soc. Japan* **38**, 989 (1975)
[25] R. Lassnig, E. Gornik, *Solid State Commun.* **47**, 959 (1983)
[26] T. Ando, Y. Murayama, *J. Phys. Soc. Japan* **54**, 1519 (1985)
[27] H.J. Mikeska and H. Schmidt, *Z. Physik* **B20**, 43 (1975)

Cyclotron Resonance in GaAs/AlGaAs Heterojunctions

G.Y. Hu and R.F. O'Connell*

Department of Physics and Astronomy, Louisiana State University, Baton Rouge, LA 70803, USA

The Generalized Quantum Langevin Equation approach is extended to study the cyclotron resonance (CR) of interacting two-dimensional electronic systems. Analytical forms for the memory functions, which are free from the divergences encountered in conventional theories, are obtained from first principles. The elimination of the divergence is achieved by the inclusion of the effects of the fluctuation of the center of mass. In our theory there exists two resonance peaks in the CR absorption spectrum: the conventional magnetoplasmon absorption peak at the frequency $\omega > \omega_c$, and another peak at $\omega < \omega_c$ due to the fluctuation effect. The positions, amplitudes, and the relative spacing of these two peaks depend mainly on the magnetic field B, the electron density n_s, the mobility of the sample and the temperature. Our results for the CR spectrum are consistent with the experiments of Schlesinger et al. In addition, using only one parameter (diffusion constant), our theory gives a very good fit to the experimental line shape results of Muro et al. on GaAs heterojunctions, including its dependence on B and n_s.

1. Introduction

It is well-known that in the presence of a perpendicular magnetic field, each energy band of an ideal two-dimensional (2D) electron gas is quantized into a series of discrete Landau levels [1], and the corresponding free electron polarization function has a divergence problem. Even though a lot of theoretical efforts in dealing with this problem were made in the literature [1-5], a rigorous cyclotron resonance (CR) theory, which also agrees with the experimental results is still pending.

Recently, we have shown that there is a physical effect which can eliminate the divergence viz fluctuations, and that then a CR theory can be formulated [6]. In Ref. 6 we applied our theory to the short range impurity case, and obtained a good fit to the heretofore unexplained CR experimental data of Si(001) for the cyclotron effective mass and the scattering time for a range of electron densities. In this paper, we apply our theory to an interacting 2D electron long-range impurity system, and we study the CR phenomenon for the low temperature and high magnetic field (quantum limit) case, in particular with application to the GaAs/AlGaAs heterojunctions. The formulation is based on the use of the generalized quantum Langevin equation (GLE), which we introduced recently to study electronic quantum transport [7]. In our approach, the dynamical conductivity $\sigma(\omega)$ is obtained in terms of the memory function $M(\omega)$, which

---

* On leave from Department of Physics, Fudan University, Shanghai, China.

Springer Series in Solid-State Sciences, Vol. 87
High Magnetic Fields in Semiconductor Physics II Editor: G. Landwehr

is essentially the Fourier transform of the memory function appearing in the GLE. When the fluctuation effect of the center of mass is included, a divergence-free calculation for the CR can be performed. Our theoretical results will be compared to some recent experimental data [8,9] on the CR spectrum of the GaAs/AlGaAs heterojunctions.

2. Formulation

In our formalism, the a.c. conductivity is expressed in the familiar form [3-6]

$$\sigma_{\pm}(\omega) = \frac{i n_s e^2/m}{\omega \mp \omega_c + M(\omega)} , \tag{1}$$

where m, -e, and n_s are the electron effective mass, charge, and surface density respectively, and $\omega_c = eB/mc$ is the CR frequency. In addition, the memory function $M(\omega)$ in the random phase approximation (RPA) and in the case of long-range impurity scattering, may be written as

$$M(\omega) = \omega_c \frac{n_i}{n_s} \frac{e^2}{\kappa_{sc} \ell_o \omega_c} \int dx \, \frac{x^2 e^{-ax}}{y \, H(x)} \left\{ \frac{1}{\varepsilon(x,o)} - \frac{1}{\varepsilon(x,y)} \right\} , \tag{2}$$

where $x = q\ell_o$, $y = \omega/\omega_c$, $\ell_o = (\hbar/m\omega_c)^{1/2}$, $a = 2\alpha/\ell_o$, α is the setback distance between the doped layer and the 2D electron gas in the GaAs heterojunction, and n_i is the impurity density. κ_{sc} is the dielectric constant of the static lattice and $\varepsilon(\vec{q},\omega)$ is the RPA dielectric function of wave vector $\vec{q}$, which contains the Coulomb potential treated by the standard Fang-Howard-Stern (FHS) variational scheme for the 2D electron gas [1]. In addition, H(x) is the standard result for the structure factor resulting from the FHS variational treatment of the electron-electron and electron-impurity interactions.

The evaluation of the memory function $M(\omega)$ of (2) remains a controversial question, arising from the proper handling of the well-known divergence of the free electron polarization function associated with the RPA dielectric function $\varepsilon(\vec{q},\omega)$. In a recent paper [6], we have demonstrated that there is a physical effect which can eliminate the divergences viz fluctuations, and that the electron polarization function associated with the dielectric function $\varepsilon(\vec{q},\omega)$ is

$$\pi_{nn'}(q,\omega) = \frac{f(\varepsilon_n) - f(\varepsilon_{n'})}{\hbar\omega - (\varepsilon_{n'} - \varepsilon_n) + i\, Dq^2} , \tag{3}$$

where $\varepsilon_n = (n + 1/2)\,\hbar\,\omega_c - \varepsilon_F$, ε_F is the Fermi energy of the electron gas, f(x) is the Fermi distribution function, and D is the diffusion constant defined in terms of the ensemble average of the coordinate fluctuations $\delta R^2(t)$ of the center of mass

$$D = \lim_{t\to\infty} \frac{1}{2t} \overline{\delta R^2(t)} . \tag{4}$$

As we have shown elsewhere, D is proportional to the inverse of the number of particles N and thus the fluctuation effects are most pronounced for small systems with low-electron concentrations. Since D is difficult to evaluate numerically, we will treat it as a parameter (while retaining the property of being inversely proportional to N) in the following calculation.

3. Numerical Results and Discussion

The memory function (2) is the key quantity in our theory. Our numerical calculation reveals that the real part $M_1(y)$ is an oscillating function of ω, which is positive at the low frequency side and has its first zero at $\omega \gtrsim \omega_c$, while the imaginary part is a positive function. On the other hand, from (1), the absorption peak position is mainly determined by the solution of $M_1(y)/\omega_c = 1-y$. Thus, due to the oscillating behavior of $M_1(y)$, there may exist two resonance peaks in the CR absorption spectrum in our theory. One is the conventional magnetoplasmon absorption peak at the frequency $\omega > \omega_c$ (where $M_1(y) < 0$), and the other peak is at $\omega \leqslant \omega_c$ due to fluctuation effects. The positions, magnitudes, half-widths, and the separation of these two peaks depends on the B, n_s and the mobility of the sample. Our results are very consistent with the experiments of Schlesinger et al. [8] and Muro et al. [9]. Here we provide a detailed comparison between our numerical results and the experimental result of Muro et al., which was performed for a GaAs/AlGaAs FET system at incident wave lengths $\lambda = 118.8$ µm and 170.6µm, with changing B and n_s. In our calculation, for the physical quantities contained in (2) - (4), we take the standard values: $m = 0.07m_e$, $\kappa_{sc} = 13.1$, $\alpha = 50$Å, $n_i = n_s$, and the only parameter $D^* = mD/\hbar$ is determined by fitting to the experimental data. We have chosen the $n_s = 2.63\times10^{11}\text{cm}^{-2}$, $\lambda = 170.6$ µm case as our best fitting point to perform the calculation, and the results are presented in Fig. 1a, where $D^* = 0.2$ is used and where the corresponding calculation for $\lambda = 118.8$ µm has also been shown. For the other values of n_s, we use the property $D^* \sim n_s^{-1}$ to get $D^* = 0.526\times10^{11}\text{cm}^{-2}/n_s$. The numerical results thus obtained are presented in Fig. 1b together with the experimental data of Muro et al. [9] in Fig. 1c.

As can be seen from the figure, the agreement between our theory and the experimental results is quite good in all respects such as the line shape, the peak shift, the line width and their dependence on the magnetic field and the electron density. Furthermore, this is achieved using only one parameter and a relatively simple calculation. In a previous study [6] of the Si(001)-MOS system, the present method was used to obtain good agreement with heretofore unexplained experiments [10]. In the present calculation for GaAs/AlGaAs, the D* chosen is approximately one order larger than that of the Si case. This is expected as GaAs has a smaller effective mass and larger transport time.

Acknowledgements

This research was supported in part by the U. S. Office of Naval Research under contract No. N00014-86-K-0002.

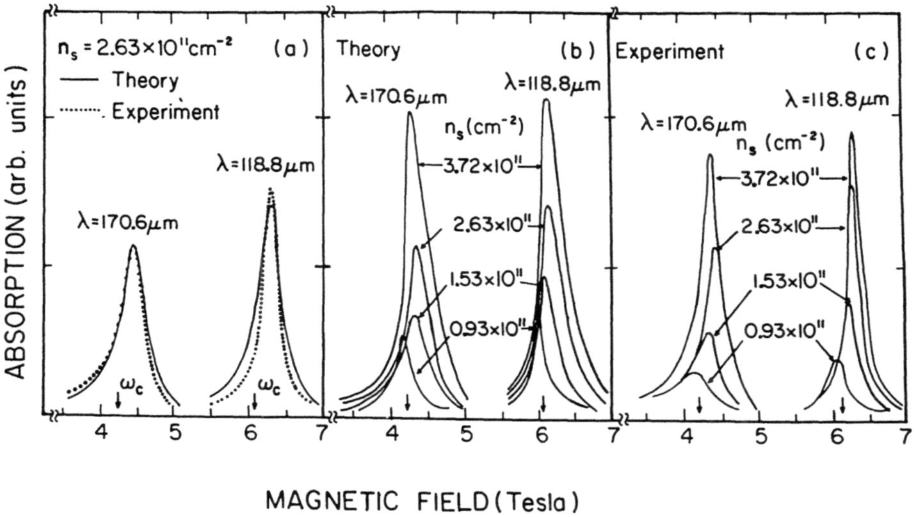

Fig. 1. Theoretical and experimental [9] CR spectrum against the magnetic field at two infrared frequencies for the GaAs/AlGaAs heterojunction. (a) A direct comparison between the theory and experiment at $n_s = 2.63\times10^{11}cm^{-2}$. $D^* = 0.2$ is used to obtain the best fit for the theoretic curves. (b) Theoretical CR spectrum at different electron densities n_s. (c) Experimental CR spectrum taken from Ref. 9.

References

1. T. Ando, A. B. Fowler, and F. Stern, Rev. Mod. Phys. 54, 437 (1982).
2. T. Ando and Y. Uemura, J. Phys. Soc. Jpn. 36, 959 (1974).
3. C. S. Ting, S. C. Ying, and J. J. Quinn, Phys. Rev. Lett. 37, 215 (1976); ibid, Phys. Rev. B16, 5394 (1977).
4. H. Fukuyama, Y. Kuramoto, and P. M. Platzman, Phys. Rev. B18 4980 (1979).
5. C. Kallin and B. I. Halperin, Phys. Rev. B30 5655 (1984); ibid B31, 3635 (1985).
6. G. Y. Hu and R. F. O'Connell, Phys. Rev. B 37, 10391 (1988).
7. G. Y. Hu and R. F. O'Connell, Phys. Rev. B36 5798 (1987); ibid, Physica A, 149, 1 (1988).
8. Z. Schlesinger, S. J. Allen, J. C. M. Hwang, and P. M. Platzman, Phys. Rev. B30 435 (1984).
9. K. Muro, S. Mori, S. Narita, S. Hiyamizu, and K. Nanbu, Surf. Sci., 142 394 (1984).
10. R. J. Wagner, T. A. Kennedy, B. D. McCombe, and D. C. Tsui, Phys. Rev. B22 945 (1980).

Non-parabolicity as a Cause of Oscillations in 2D Cyclotron Resonance

E.B. Hansen and O.P. Hansen

Physics Laboratory, H.C. Ørsted Institute, University of Copenhagen, DK-2100 Copenhagen Ø, Denmark

1. INTRODUCTION

Observation of oscillatory linewidth in the cyclotron resonance of GaAlAs heterostructures was first reported by ENGLERT et al. /1/. The oscillations were found to be correlated with the filling factor of the Landau levels, and the authors /1/ suggested screening of ionized impurities to be the cause. However, shortly after SCHLESINGER et al. /2/ arrived at the opposite conclusion from their measurements: that the Landau level filling factor does not play a major role. Conflicting evidence is still being presented on this issue.

In a recent investigation /3/ we found that non-parabolicity gives rise to an oscillatory behaviour determined by the filling factor.

In the present communication we shall use the method from /3/ to analyse the results of SCHLESINGER et al. /2/, and to comment on the role of saturation effects recently discussed by CHOU et al. /4/.

2. THEORY

2.1 Basic Features of the Conductivity Calculation

We use the Kane two-band model with proper account of the way in which non-parabolicity changes the velocity operator. The conductivity tensor is calculated using the density matrix treatment of galvanomagnetic phenomena. Damping is introduced phenomenologically in terms of a constant relaxation time τ.

To obtain 2D-versions of the 3D-expressions we introduce a sublevel energy and treat it as an adjustable parameter.

2.2 Filling Factors Giving Maxima in Linewidth

Let ω denote the frequency of the applied radiation, then, if $\omega\tau>>1$, it follows from eq. (11) of /3/ that at a resonance the conductivity is given by

$$\sigma^{r}_{xx,res} = \frac{e^2}{2h}\omega\tau\sum_{n,s}\,[f(E_{n,s}) - f(E_{n+1,s})]\,[n+1]\,, \qquad (1)$$

where the superscript r and the subscript xx designates "real part" and "diagonal element".

Figure 1(a) shows a situation where the last, partially filled Landau level has the number n+1. The arrows designate the transitions which contribute to the sum in eq. (1). The relevant Fermi functions are:

Springer Series in Solid-State Sciences, Vol. 87
High Magnetic Fields in Semiconductor Physics II Editor: G. Landwehr

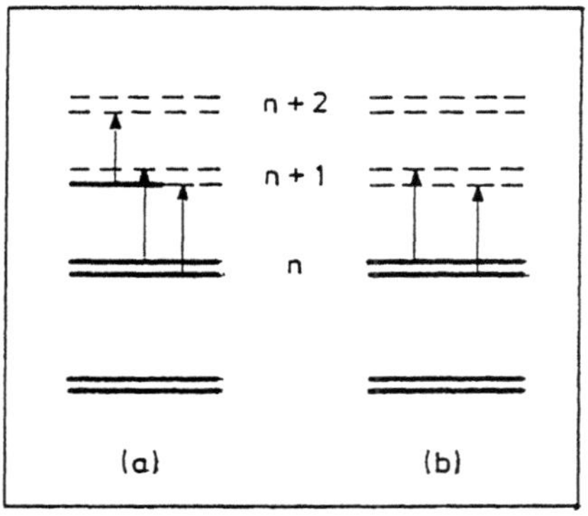

Fig. 1 Filling patterns around the n'th Landau level. Full length, heavy line: filled level. Full length, broken line: empty level. The doubling of lines is to indicate spin-splitting

$$f_{n,\pm\frac{1}{2}} = 1, \quad f_{n+1,-\frac{1}{2}} = \nu - 2(n+1), \quad f_{n+1,+\frac{1}{2}} = f_{n+2,\pm\frac{1}{2}} = 0 , \qquad (2)$$

where $f_{n,s}$ is short for $f(E_{n,s})$, and where ν is the filling factor $N_s h/eB$. The two transitions from n to n+1 contribute by

$$[1 - [\nu - 2(n+1)] + 1-0][n+1]$$

while the transition from n+1 to n+2 gives

$$[\nu - 2(n+1) - 0 + 0 - 0][n+2].$$

Equating the two contributions we find

$$\nu_{m.l} = \frac{4(n+1)(n+2)}{2n+3} , \qquad (3)$$

where m.l means "maximum linewidth". The smallest value of $\nu_{m.l}$ is seen to be 8/3.

The reason why maxima in linewidth occur at filling factors $\nu_{m.l}$ can be understood by comparing figure 1(a) with figure 1(b): in 1(b), where ν is even, there is just one single resonance energy (spin-splitting is indicated but neglected), while in 1(a), due to non-parabolicity, there are two. This also explains the way in which non-parabolicity gives rise to oscillatory behaviour of the linewidth: as the filling is varied, the system will alternate between the two filling patterns (a) and (b).

2.3 Splitting of CR Line by Non-parabolicity

The maximal linewidth at $\nu_{m.l}$ will eventually result in a splitting of the line if the scattering width is sufficiently small. The condition for splitting is

$$(\Delta E_{n,s} - \Delta E_{n+1,s})/\hbar\omega \lesssim 2/\mu B , \qquad (4)$$

where μ is the mobility $(e\tau/m)$.

2.4 Electrodynamics. Saturation effects

The transmission amplitude for the electric field of circulary polarized radiation falling on a thin sheet with conductivity σ can be expressed by (see/3/ for references):

$$t_{\pm} = \frac{2}{1+n+\sigma_{\pm}} , \qquad (5)$$

where n is the refractive index of the semiconductor substrate. From this we have calculated the transmittance of linearly polarized light:

$$T = \tfrac{1}{2}[\,|t_+|^2 + |t_-|^2]. \qquad (6)$$

The so-called saturation effects appear in high conductivity samples, $\sigma >> 1+n$, where the cyclotron active part of the polarized radiation is completely absorbed. This leaves a transmittance of 50%, independent of σ.

The results presented in this communication, as well as in /3/, include the saturation effect,i.e.: no small-signal, $\sigma << 1+n$, approximation has been made.

3. NUMERICAL RESULTS

3.1 Heterojunction of GaAlAs

In SCHLESINGER et al. /2/ a splitting of the CR line was observed in an applied field of 4.63T for a sample with $N_S = 3.9 \times 10^{15}\,m^{-2}$. As read from the figures of this publication the position and magnitude of the splitting are 62 cm^{-1} and 3 cm^{-1} respectively.

Attempting an analysis of these measurements in terms of non-parabolicity it is noted, first that the smallest $\nu_{m,1}$ value is 8/3 for which a field of 4.63T will require a density of $3.0 \times 10^{15}\,m^{-2}$, i.e. substantially smaller than the one quoted. Second, in order to see a splitting, the mobility should be 50 T^{-1} (cf. eq.(4)), which may not contradict the information given in /2/. The remaining input data for the calculations are identical to those in table 1 of reference /3/, i.e.: conventional, bulk GaAs-values. The results are displayed in figure 2, and the following features are noted:

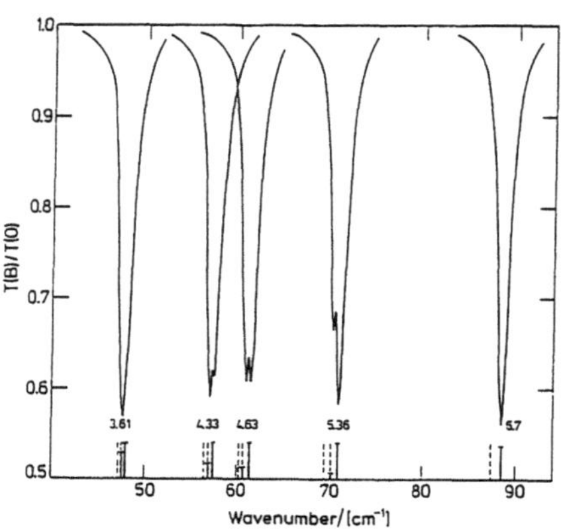

Fig.2 Normalized transmittance for a GaAlAs heterojunction ($N_S = 3.0 \times 10^{15}\,m^{-2}$, $\mu = 50T^{-1}$) versus wavenumber. Field values are given at each curve. Heavy line vertical bars on the wavenumber scale indicate the filling of levels involved in the resonance (full length: completely filled level). Spin-splittings are not shown.

1) the calculated splitting is only 0.6 cm^{-1}
2) the "shoulders" which appear in the lineshapes for 4.33T and 5.36T are in qualitative agreement with the observed ones.

3.2 Quantum Well of InAs

In figure 3 we display the results for an InAs quantum well. These calculations were performed to make a comparision with the measurements of HEITMANN et al. /5/. Carrier density and mobility was chosen in accordance with the values stated in /5/: $8.8 \times 10^{15} m^{-2}$ and $8T^{-1}$. Each dot in figure 3 represents the extremal absorption as found from the calculated resonance line. The extremal absorption is seen to oscillate, periodically in B^{-1}, with maxima at even filling factors. This oscillatory pattern resembles the one found experimentally, however with important differences: the experimental amplitude is considerably larger, and the maxima are found at odd filling factors.

The measurements in /5/ demonstrate that a high carrier density does not necessarily prevent the observation of oscillations, and the calculations displayed in figure 3 show that oscillations result from non-parabolicity, even when saturation effects are taken into account. Thus, we can not agree with the recent statement of CHOU et al. /4/ as regards high carrier density and saturation effects.

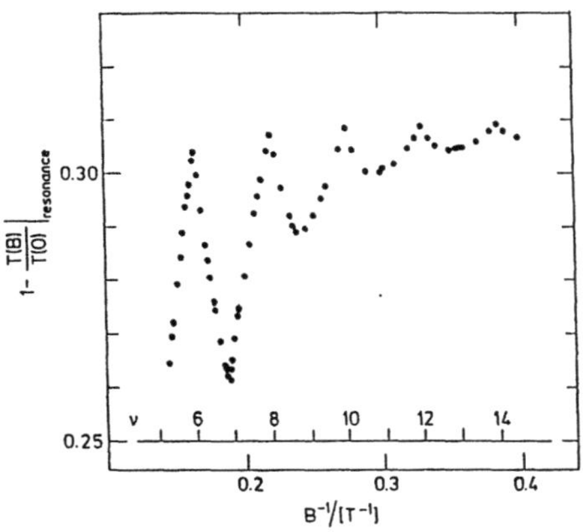

Fig.3 Extremal absorption for an InAs quantum well ($N_s = 8.8 \times 10^{15} m^{-2}$, $\mu = 8T^{-1}$) versus filling factor and reciprocal field.

4. CONCLUSION

Calculations which take non-parabolicity into account result in a number of effects (oscillations, line splitting, shoulders) resembling those found experimentally. However, the calculated effects are less pronounced, and problems with carrier densities (location of oscillation extrema) are encountered.

REFERENCES

1. Th. Englert, J.C. Maan, Ch. Uihlein, D.C. Tsui, and A.C. Gossard : Solid State Commun. 46, 545 (1983)
2. Z. Schlesinger, S.J. Allen, J.C.M. Hwang, P.M. Platzman, and N. Tzoar: Phys. Rev. B30, 435 (1984)
3. E.B. Hansen and O.P. Hansen: Solid State Commun. 66 1181 (1988)
4. M.J.Chou, D.C. Tsui, and G. Weimann: Phys. Rev. B 37 848 (1988)
5. D. Heitmann, M. Ziesmann, and L.L. Chang: Phys. Rev. B 34, 7463 (1986)

Magnetoconductivity of n-GaAs/$Ga_{1-x}Al_xAs$ Heterojunctions in Strong Transverse Electric Fields

M. Kroeker[1], *E. Batke*[1], *U. Merkt*[1], *J.P. Kotthaus*[1], *G. Weimann*[2], *and W. Schlapp*[2]

[1]Institut für Angewandte Physik, Universität Hamburg, D-2000 Hamburg 36, Fed. Rep. of Germany

[2]Forschungsinstitut der Deutschen Bundespost, D-6100 Darmstadt, Fed. Rep. of Germany

Cyclotron resonance absorption and static magnetoconductance in GaAs-AlGaAs heterojunctions are investigated at low temperatures ($\leq$4.2K) in strong magnetic fields (B$\leq$10T) under pulsed source-drain fields (E$\leq$200 V/cm). Transverse E-field induced changes of cyclotron resonance are shown to primarily reflect changes in the inversion electron density, whereas additional electron heating effects can only be deduced in a few cases.

In semiconductor systems with a non-parabolic conduction band cyclotron resonance can be used to study electron redistribution phenomena caused by a strong electric field E transverse to the magnetic field B. This has been extensively investigated in bulk semiconductors such as InSb [1] and has helped to understand hot electron distributions in such $E\perp B$ geometries. For two-dimensional electron systems confined at a heterojunction interface similar experiments have recenctly been reported [2,3] and the E-field induced changes of cyclotron resonance have been linked to electron heating. Here we report on studies of E-field induced changes in the cyclotron resonance of high mobility GaAs-AlGaAs heterojunctions measured in transmission at fixed infrared laser energies $\hbar\omega$ on various sample geometries. Such high frequency experiments are complemented by measurements of the quasi-static magnetoconductance in strong transverse E-fields. Our experiments demonstrate that field induced changes of the electron density dominate electron heating effects in all samples studied here.

Transmission of far-infrared laser radiation of fixed frequency $\hbar\omega$ incident perpendicularly to the surface of GaAs-AlGaAs heterojunctions is measured at temperatures T$\leq$4.2K with magnetic fields normal to the surface and pulsed source-drain fields E. We employ fast photoconductive Ge:Ga and InSb detectors with response times of typically 1μsec and measure the transmission signal at preset times during and after the source-drain excitation pulse using box-car integration. To avoid thermal heating and to be able to separate slow from fast processes short source-drain pulses are applied with duty cycles of less than 10% and typical repetition rates of 1kHz. In addition to the fast response captured by a box-car integrator we record the average transmission with a lock-in amplifier. The samples are modulation-doped GaAs-AlGaAs heterojunctions with a high mobility ($\gtrsim$300000 cm^2/Vsec at 4.2K) two-dimensional electron system and are grown by molecular beam epitaxy [4]. Ohmic source-drain contacts in various geometries are prepared by In-diffusion.

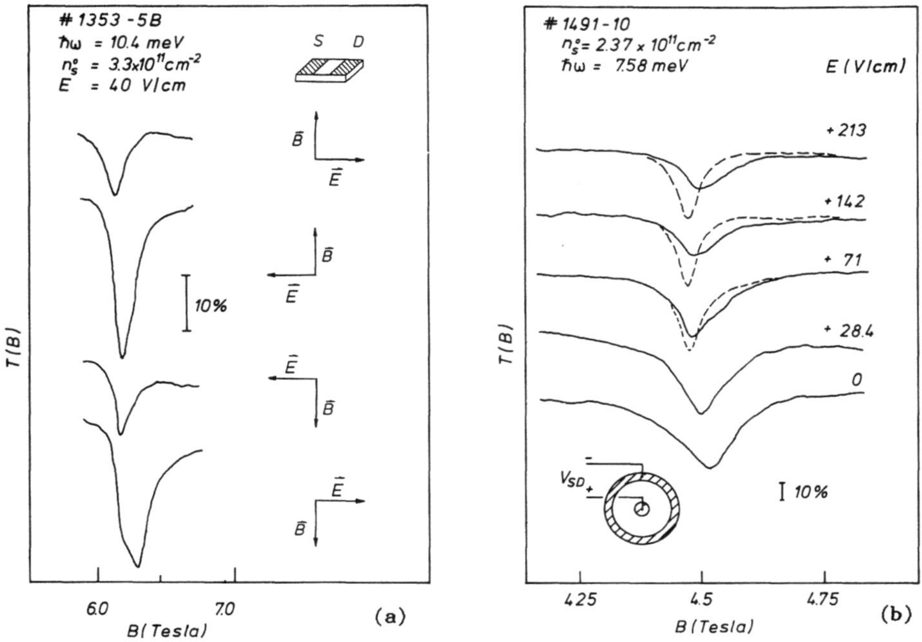

Fig. 1: Cyclotron resonance transmission traces at fixed laser energy ħω in transverse electric fields (a) for a rectangular geometry (b) for a Corbino geometry with an inside diameter of 0.8 mm and a channel length of 1.4 mm. The solid lines are traces measured during a short E-field pulse, the dashed traces are averaged over the pulse repetition cycle.

Typical cyclotron resonance (CR) transmission traces recorded during source-drain excitation (solid traces) are displayed in Fig.1 for rectangular (a) and Corbino geometry (b), respectively. In Fig.1a we demonstrate that CR strength and line shapes in rectangular samples depend on the directions of both, the source-drain field E and the magnetic field B. In further experiments we find that position and lineshape of the CR in transverse E-fields also depend on the optically tested part of the rectangular sample geometry. These effects we attribute to an inhomogeneous field distribution and electron depletion caused by the Hall-effect. To avoid the inhomogeneous current flow in rectangular samples we have used in the following Corbino geometries as in Fig.1b. As expected we find that in Corbino samples the CR signal no longer depends on the direction of B. Surprisingly, we still find CR to depend on the direction of the source- drain field E, which we specify in the following by the polarity of the center contact. In Fig.1b both the CR signal measured during the source- drain pulse with the box-car (solid trace) as well as the average signal measured during a period of the pulse repetition with a lock-in (dashed traces) are shown for various values of E. The equilibrium density of the sample measured from Shubnikov de-Haas oscillations (SdH) at very low source-drain fields is entered as n_s^o. Two effects are immediately visible. With increasing E both CR traces, the fast (solid line) and the averaged signal (dashed line) decrease in strength and shift to lower magnetic fields. The latter effect is stronger for the lock-in signal which also narrows with increasing E. The change in resonance strength as quantitatively derived from fits to the CR traces we attribute to a transverse E-field induced change in inversion electron density n_s that persists at least partially after the source-drain pulse is switched off.

To establish such source-drain field induced changes of the electron density in the 2D channel by a second experiment we have performed time-resolved studies of the quasi-static conductance. In measurements during the source-drain pulse we find that the SdH-oscillations shift from the equilibrium value at pulsed fields as low as 1V/cm. In most cases the shift corresponds to a decrease of n_s and also depends on the polarity of E. At higher transverse fields ($E \gtrsim 10$V/cm) the SdH oscillations vanish but the magnetoconductance depends on size and polarity of E. After the E- field is switched off the return to the equilibrium density n_s^0 has both fast and slow time components. The slow part of return to equilibrium is illustrated in Fig.2 where we show the time-dependence of the magnetoconductance at a low source-drain field ($E \simeq -0.7$ V/cm) after a high field pulse ($E \simeq -71$ V/cm) has been switched off at t=0. The approach of the magnetoconductance to the equilibrium value exhibits SdH oscillations and thus demonstrates that there is a slow mechanism with which the density n_s^0 gets restored.

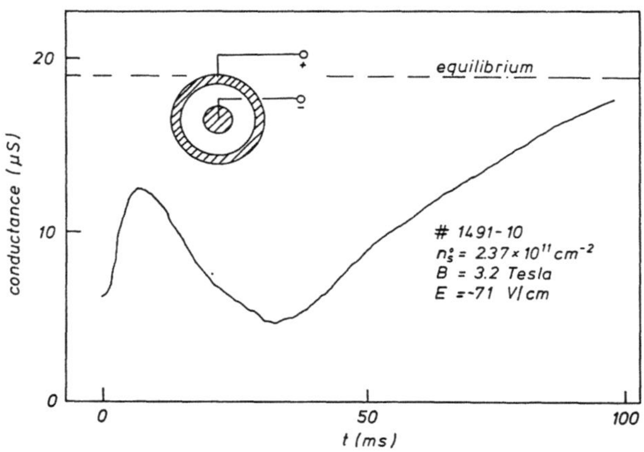

Fig. 2: Time dependence of the magnetoconductance after pulsed source-drain exitation with $E \simeq -71$ V/cm

To evaluate CR traces as in Fig.1 we fit the measured traces with classical CR using a single Lorentzian line wherever possible. Fit parameters are m^*, n_s and τ to describe resonance position, strength and halfwidth, respectively. Fig.3 summarizes results thus obtained on a Corbino geometry sample at two laser energies, i.e, in two regions of resonance magnetic field B. In Fig.3a we plot for $\hbar\omega \simeq 7.58$ meV n_s vs. source-drain field E. Both the box-car and the averaged lock-in data are shown. Generally, one finds n_s to decrease with increasing E but for a given E the values depend on polarity and time scale. In Fig.3b we display the CR mass m^*/m_e vs. the density n_s. Except for the case $\hbar\omega \simeq 7.58$ meV and positive E all data fall on straight lines that reflect the expected nonparabolicity of the cyclotron mass in the magnetic quantum limit /5/. This implies that most of the observed changes of CR with transverse fields result from transverse field induced depletion of carriers in the inversion layer.

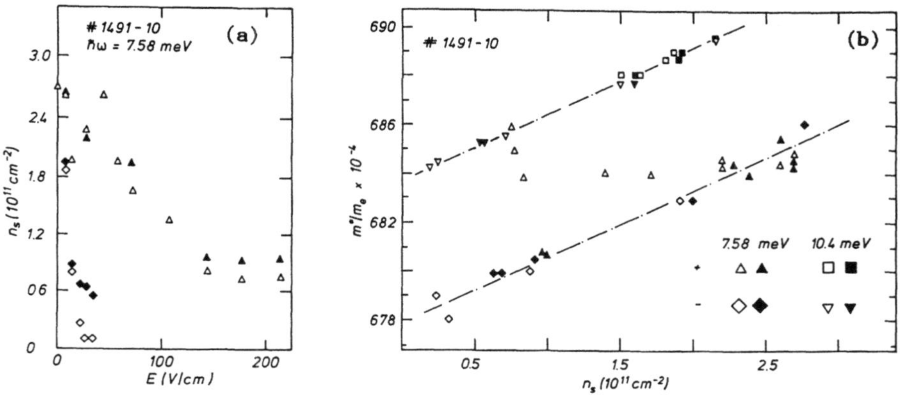

Fig. 3: (a) E-field dependence of electron density n_s extracted from fits to CR (b) CR masses vs. density n_s for two laser energies. As in (a) the different n_s-values result from pulsed transverse fields. The open symbols present data taken during the E-pulse, the filled symbols are time-averaged data. Different symbols indicates different polarities and laser energies. The dashed lines in (b) connect most data points and are consistent with the non-parabolicity of m^*/m_e expected for the transition between the lowest Landau levels.

Only for the case of sufficiently high positive E at the lower laser energy $\hbar\omega$=7.58 meV, we observe that during the pulse the fit mass increases with increasing transverse field and deviates from the straight lines in Fig.3b. This result is what is expected for electron heating.

In conclusion we find that transverse field-induced depletion of the inversion channel is a dominant effect in dynamic and static magnetoconductance studies on GaAs-AlGaAs heterojunctions. At present the detailed mechanisms for this depletion are not understood. Experiments with other geometries indicate that the high transverse field rather than the large source-drain potential differences induce these depletion phenomena.

Financial support of the Deutsche Forschungsgemeinschaft is gratefully acknowledged.

REFERENCES

1. O. Matsuda and E. Otsuka: J. Phys. Chem. Solids 40, 819 (1979)
2. W. Seidenbusch, E. Gornik, and G. Weimann, Physica 134B, 314 (1985), W. Seidenbusch: Phys. Rev. B 36, 1877 (1987)
3. S. Yamada, T. Fukui, K. Tsubaki, and A. Sugimura: Phys. Rev. B 32, 8078 (1985)
4. G. Weimann: In Festkörperprobleme, Advances in Solid State Physics, Vol.28, ed. P. Grosse (Vieweg, Braunschweig, 1986), 231.
5. F. Thiele, U. Merkt, J.P. Kotthaus, G. Lommer, F. Malcher, U. Rössler and G. Weimann: Solid State Commun. 62, 841 (1987)

On the Halfwidth of the Cyclotron Resonance Line in Semiconductors

K. Pastor[1,2], *J. Oberti*[2], *M.L. Sadowski*[1,3], *M. Goiran*[2], *and J. Leotin*[2]

[1]Institute of Experimental Physics, Warsaw University, PL-00-681 Warsaw, Poland
[2]I.N.S.A. Laboratoire de Physique des Solides, Toulouse, France
[3]High Pressure Research Center, Polish Academy of Sciences, PL-01-142 Warsaw, Poland

The cyclotron resonance linewidth is measured in a broad range of temperatures and magnetic field in CdTe. The width due to scattering on ionized impurities and on acoustic phonons is then calculated numerically as a function of temperature, magnetic field and concentration of impurities. Two ways of adding the two contributions are investigated, and the often used Matthiessen rule approximation is found to be valid in the range of parameters in question. The calculated results are compared also with experimental data for InSb.

The evaluation and understanding of the cyclotron resonance (CR) halfwidth has been of interest for quite a long time, since CR data offer a convenient (no contacts) and reliable criterion for the quality of a semiconductor (mobility). The most important result is that the CR linewidth is generated by various scattering mechanisms of the absorbing carriers and the linewidth (ΔB) should be interpreted as the inverse of the relaxation time (τ) of the electric current[1] :

$$\Delta B/B_{res} = 2\,\tau^{-1}/\omega_c \ .$$

Assuming low electron concentration and a non-degenerate semiconductor (hence a Boltzmann distribution) the power absorbed by a sample is[2] :

$$P(\omega) = 4e^2 n_e \hbar/(2m^{*3}T)^{1/2}\int_0^{\infty} dk_z \, \exp(-\hbar^2 k_z^2/2m^* kT) \left\{ \Gamma(k_z) / \left[(\omega-\omega_c)^2 + \Gamma^2\right] \right\} , \quad (1)$$

where n_e is the electron concentration and $\Gamma(k_z)$ is a function describing the width, dependent on the scattering mechanism.

A review of most of the existing theories dealing with the ionized impurity scattering, as well as the presentation of a very advanced technique used to calculate the CR width, was provided by Prasad[2]. The author derived a general expression for the CR width, which in the ultra-

Springer Series in Solid-State Sciences, Vol. 87
High Magnetic Fields in Semiconductor Physics II Editor: G. Landwehr

quantum limit, i.e. considering only cyclotron resonance transitions between the Landau levels N = 0 and N = 1 (or assuming $\hbar\omega_c >> kT$), is as follows :

$$\Gamma(k_z) = (2\pi)^3 N_1 \int d^3q \mid V(q) \mid ^2 t^2 e^{-t} \left\{ [\Gamma(q_z)/[E(k_z) - E(k_z - q_z)]^2 + \hbar^2 \Gamma^2 \right\}, \qquad (2)$$

where V(q) is the Fourier tranform of the scattering potential. $t = 1/2r_o^2q^2$, and $E(k_z)$ is the electron energy. The scattering potential seen by a moving electron is approximated by an unscreened Coulomb potential, since measurements are usually performed in very pure samples, where screening effects may safely be ignored.

The second major contribution to the CR width comes from scattering on acoustic phonons, and the interaction operator is taken in a usual form :

$$\gamma(q) = C_q \exp(iqr),$$

where for the scattering through a deformation potential $C_q = iD(\hbar q/2\rho_m S_o)^{1/2}$, D is the deformation potential, ρ_m the mass dentsity, and S_o the sound speed. For the piezo-electric coupling in cubic crystals $C_q = Pe\langle K(\theta)\rangle(\hbar/2\rho_m S_o q)^{1/2}/\varepsilon_o \kappa$, where P/2 is the only nonvanishing component of the piezo-electric tensor, $K(\theta)$ is a factor reflecting the anisotropy of the interaction and κ is the dielectric constant of the host crystal.

A formula for the function Γ was given by Srinivas et al[3] :

$$\Gamma_{ph}(k_z) = \Gamma_{em} + \Gamma_{ab}, \qquad (3)$$

$$\Gamma_{em} = 2\pi \int d^3q\, [1 + N(q)]\, |C_q|^2 t^2 e^{-t}\, \Gamma \Big/ \left\{ [E(k_z - q_z) - E(k_z) + \hbar\omega_q]^2 + \hbar^2\Gamma^2 \right\},$$

$$\Gamma_{ab} = 2\pi \int d^3q\, N(q)\, |C_q|^2 t^2 e^{-t}\, \Gamma \Big/ \left\{ [E(k_z - q_z) - E(k_z) - \hbar\omega_q]^2 + \hbar^2\Gamma^2 \right\},$$

N(q) is the phonon occupation number $N(q) \approx [\exp(\hbar S_o q/KT) - 1]^{-1}$.

After replacing the Lorentzians by appropriate delta functions, eq. 3 can be solved analytically[3].

In practice it is almost never possible to limit considerations to a single scattering mechanism. It is generally accepted that for various scattering modes (except those due to electron-electron interaction) the transition probabilities are additive. In this case the so-called "Matthiessen rule" holds :

$$\Gamma(k_z) = \sum \Gamma_n(k_z) \ . \tag{4}$$

The sum of the average halfwidths associated with various mechanisms should be introduced into eq. 1 to obtain the total width. The validity of the Matthiessen rule has recently been questioned[3] ; however, since the phonon contibution is independent of the impurity concentration, there appear to be no physical arguments to exclude it.

Using computers, we solved eq. 2 numerically to obtain for the first time the full (not approximate) function $\Gamma(k_z)$ for scattering on ionized impurities. The dependence of the calculated CR widths on the concentration of ionized impurities was found to be mostly linear[4].

We then calculated the CR width due to two mechanisms : ionized impurity and acoustic phonon scattering (via both deformation potential and piezo-electric coupling), first in accordance with Matthiessen's rule and then rigorously, taking both mechanisms into account in eq. 2. The two methods were found to be equivalent with reasonable accuracy.

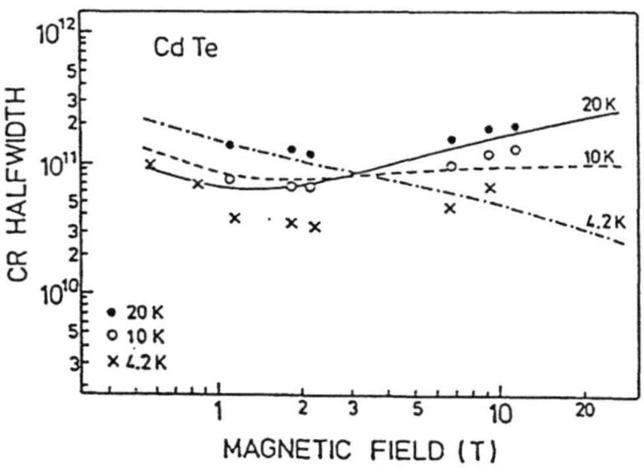

Fig.1.
CR width calculated for CdTe as a function of the magnetic field for three temperatures : 4.2K, 10K and 20K, together with experimental points. The concentration of ionized impurities was 5.10^{13} cm^{-3}, the deformation potential was 6 eV.

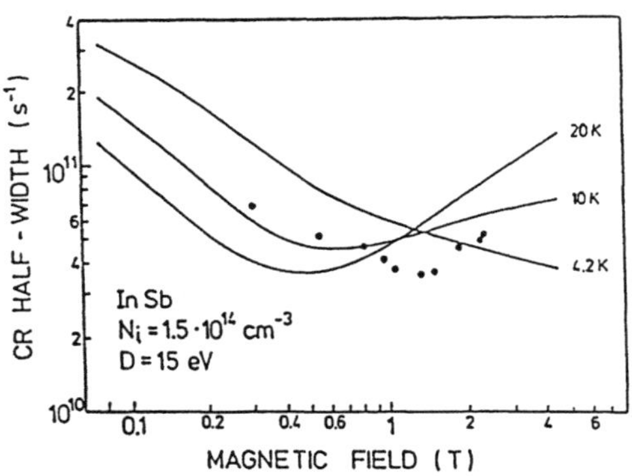

Fig.2.
CR width calculated for InSb as a function of the magnetic field for the same temperatures. The experimental points are from McCombe[6]. The concentration of impurities and deformation potential are indicated.

The results were compared with experimental data for CdTe and InSb. The CdTe samples were grown in a vertical Bridgeman furnace and then zone refined. The concentration of electrically active donors and acceptors was below 10^{14} cm^{-3} with the compensation ratio 2-4. The samples displayed "persistent cyclotron resonance", but its presence did not affect the observed width of the resonance[5]. All the measurements were performed using non-polarized light form an aoptically pumped far-infrared laser operating on several wavelengths longer than 100 m (to avoid the influence of polaron effects); the magnetic field was provided by a pulsed magnet capable of giving 35 T. A fast, helium cooled Ge photoconductor was used as a detector. A flowing-gas type cryostat, equipped with a calibrated Si diode was used to change and control the temperature of the sample in the range 4.2 - 70K with an accuracy of 0.5 - 5K, depending on the temperature. The results were recorded and analyzed on a computerized data acquisition system. The data concerning InSb was taken from McCombe et al[6]. The effect of the piezo-electric coupling in CdTe was found to be comparable with that of the deformation potential scattering, although small enough not to influence the overall behaviour of the CR width.

The following material constants, used in our calculations, were taken from the literature; the values of the deformation potential D, cited in a large range of values, were chosen to obtain the best fit.

	m^*	κ	ρ_m	S_o	D	f (ref.7)
CdTe	0. 10	9. 6	5855	3110	6	$6.4.10^{-5}$
InSb	0.0139	17. 6	5780	5775	15	———

(m^*, κ, f, are dimensionless, D is given in eV, ρ_m and S_o in SI units)

It must be stressed here that we were able to fit the data with virtually no adjustable parameters. The fit is by no means an ideal one, possibly due to the fact that the dependence of ionized impurity concentration on the temperature and magnetic field has not been included.

REFERENCES

1. A. Kawabata, J. Phys. Soc. Jap. 23. 999 (1967)
2. M. Prasad, Physica Status Solidi (b) 109, 11 (1982).
3. TK Srinivas, S. Chaudhury, S. Fujita, J. Phys. Chem. Sol. 44, 417 (1983) and J. Phys. Chem. Sol. 44, 931 (1983).

4. The width was previously considered to vary with the square root of the concentration of ionized impurities.
5. K. Pastor, J. Oberti, M. Goiran, J. Leotin, M. Grynberg, L.C. Brunel, Phys. Rev. B 36, 7737 (1987).
6. B.D. Mc Combe, R. Kaplan, R.J. Wagner, E. Gornik, and W. Mueller, Phys. Rev. B13, 2536 (1976).
7. G.D. Mahan in "Polarons in Ionic Crystals and Polar Semiconductors" p.553, ed. J.T. Devreese, North Holland, 1972.

Magneto-transport and Magneto-optical Studies in a Quasi-Three-Dimensional Modulation-Doped Semiconductor Structure

M. Shayegan [1], *M. Santos* [1], *T. Sajoto* [1], *K. Karrai* [2], *M.-W. Lee* [2], *and H.D. Drew* [2]

[1] Princeton University, Princeton, NJ 08544, USA
[2] University of Maryland, College Park, MD 20742, USA

In modulation-doped semiconductor heterostructures, the carriers are spatially separated from the dopant atoms to reduce the ionized impurity scattering. Because the disorder and the electron-impurity interaction in such structures are reduced, many-body effects can play a significant role in quantum transport properties of the electron system. An example is the fractional quantum Hall effect observed in the high-mobility two-dimensional electron system (2DES) at the $GaAs/Al_xGa_{1-x}As$ heterojunction interface in the presence of a high magnetic field [1]. There is great interest in extending studies of electron-electron interaction effects to three-dimensional (3D) systems [2]. In the case of uniformly doped semiconductors, however, the electron-impurity interaction is strong and usually leads to the magnetic freeze-out of the carriers.

We report here on the first realization of a modulation-doped quasi-3D electron system (3DES) and our characterization of this system using magnetotransport and magnetooptical properties. The structure, grown by molecular beam epitaxy, is a 2000 Å wide undoped $Al_xGa_{1-x}As$ well bounded by $Al_yGa_{1-y}As$ ($y > x$) barriers consisting of undoped (spacer) and doped layers on both sides of the well. The alloy composition in the well (x) is graded so that the combined potentials due to the $Al_xGa_{1-x}As$ and the electric charge in the well produce a square potential well with a nearly uniform electron density [3]. This is schematically shown in Fig. 1. If x is not graded and there is no space charge, then the bottom of the potential is flat (Fig. 1a). However, once the electrons are transferred to such a well, the electrostatic repulsion between the electrons forces them into a stable configuration in which two 2DES are formed at the two heterointerfaces between the $Al_yGa_{1-y}As$ and $Al_xGa_{1-x}As$ (Fig. 1b). If the well-width is more than a few hundred Å wide, then the two 2DES are isolated. On the other hand, if x is graded so that a parabolic potential well (Fig. 1c) is formed, then when electrons of correct density are placed in this well, a system of nearly uniformly distributed electrons in a flat potential (Fig. 1d) may be expected. To see this in another way, let us assume that the electron system in Fig. 1d has been realized. A uniform distribution of charge in the well gives rise to an electrostatic potential of the form $\phi = -e^2nz^2/2\epsilon_s$, where e is the electron charge, n is the 3D electron density in the well, ϵ_s is the dielectric constant of the well, and z is the distance measured from the center of the well. The grading for x in the well therefore has to be chosen correctly to cancel the above quadratic potential. Note that in order to justify the assumption of uniform distribution of electrons in the well, it is necessary that several subbands of the well are occupied (so that electron wavefunctions for different subbands average to a nearly uniform spatial distribution).

The structure (M77) consists of a 2000 Å wide $Al_xGa_{1-x}As$ well in which x is quadratically varied between $x = 0$ (at the center of the well) and $x = 0.19$ (at the edges of the well). The barriers on each side of the well consist of undoped (165 Å thick) and doped (few hundred Å thick) $Al_{0.27}Ga_{0.73}As$ [4]. Assuming that the conduction-band offset (ΔE_c) for $GaAs/Al_xGa_{1-x}As$ heterostructure is equal to 900 x (in units of meV) [5], the electron density in the well required to cancel the quadratic ΔE_c in Fig. 1c is expected

Springer Series in Solid-State Sciences, Vol. 87
High Magnetic Fields in Semiconductor Physics II Editor: G. Landwehr

to be $\simeq 2\times10^{16}\mathrm{cm}^{-3}$ [using $\phi = -\,e^2nz^2/2\epsilon_s$ with $\epsilon_s = 13\ \epsilon_o$ where ϵ_o is the permitivity of free space]. Also, the expected height of the final square potential well (Fig. 1d) is $\sim$ 70 meV. Note that if the areal density of electrons in the well is less than the design value, we still expect (ignoring the edge effects) a nearly uniform density in a well whose effective width is smaller than 2000 Å, and whose barriers, however, are graded.

We measured the transverse (ρ_{xx}) and Hall (ρ_{xy}) resistivities in the conventional manner (magnetic field $\vec{B}$ in the z direction perpendicular to the plane of the sample), as well as the longitudinal resistivity (ρ_{zz}) (with $\vec{B}$ in the plane of the sample along the direction of the current). The low-field Hall data indicate an electron areal density $n_s = 2.5\times10^{11}\mathrm{cm}^{-2}$ and a mobility $\mu = 1.2\times10^5\mathrm{cm}^2/\mathrm{Vs}$. The ρ_{xx} and ρ_{xy} data also exhibit the integral quantum Hall effect expected for a high-mobility 2DES [4]. The positions of ρ_{xx} minima for B $\gtrsim$ 1.5 T are consistent with $n_s = 2.5 \times 10^{11}\mathrm{cm}^{-2}$. In the field range B $\lesssim$ 1.5 T, however, the positions (in B) of the maxima/minima in ρ_{xx} are *not* consistent with only one electric subband being occupied [4]. The ρ_{zz} data (Fig. 2) can be used to determine the number of occupied electric subbands. When the magnetic field is applied parallel to the quantum well, the potential due to the magnetic-field confinement as well as that due to the well have to be considered. The result of such a hybrid (electric-magnetic) potential is a non-linear eigen energy vs. B fan-diagram in which the energy

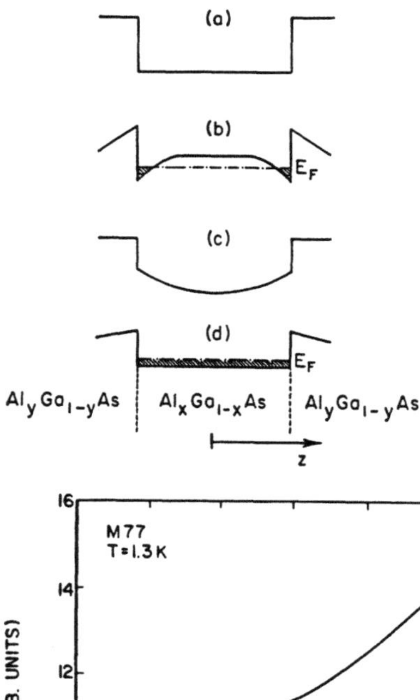

Fig. 1 The conduction-band edge (E_c) of the structure is shown schematically as a function of distance (z). In (a) and (b), E_c in the $Al_xGa_{1-x}As$ well is shown without and with electrons respectively. In (c) and (d), E_c is shown for a *graded* $Al_xGa_{1-x}As$ well without and with electrons in the well respectively. In (b), two isolated 2DES are formed while in (d) a nearly uniform carrier density may be expected. The structure is modulation-doped outside the well.

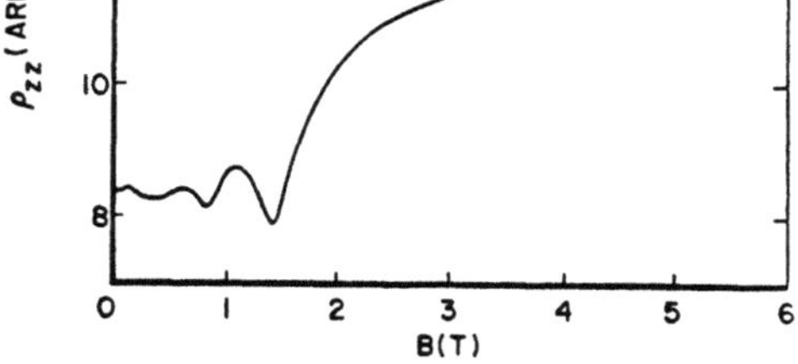

Fig. 2
Magnetotransport coefficient ρ_{zz} for sample M77.

levels at B = 0 are determined by the well-width, while as B $\rightarrow\infty$, they approach the bulk Landau-levels [6]. The ρ_{zz} maxima in Fig. 2 are expected to occur as the Fermi level E_F intersects these energy levels [4,6]. The fact that three maxima are observed in ρ_{zz} is direct evidence that four electric subbands are occupied at B = 0. The positions of the observed maxima are also qualitatively consistent with the non-linear fan-diagram calculated for a 2000 Å wide well in a parallel magnetic field [4].

Far infrared transmission measurements were performed in the Faraday geometry with unpolarized light. The sample was wedged at 4° in order to avoid line-shape distortions due to interference fringes [7]. The experimental spectra for B applied along the growth direction are shown in Fig. 3. In the range B > 2 T, the cyclotron resonance (CR) line-shape shows a striking asymmetry which appears to be due to a satellite absorption peak on the low frequency side. This satellite is well resolved only in the vicinity of 6 T and only in the highest mobility sample. The oscillator strengths of the CR peaks are consistent with the electron density determined from the magnetotransport measurements. In order to check the dimensionality of the electron gas, the sample was mounted in such a way that the growth direction was tilted 32° ±2° with respect to the B axis. Above 7 T, the CR frequency does not depend on the tilt angle. This suggests quasi-3D character of the electron gas. Below 7 T, however, the CR is shifted to ***higher*** frequencies when the sample is tilted from 0° to 32°. At 5 T this shift is about 10% of the untilted CR frequency. It is important to note that this shift is in the opposite direction than that expected for a strictly 2DES. Although the line-width of the CR in the tilt angle experiment is 50% larger, the line-shape displays the same asymmetry as observed in the untilted measurement. This indicates that the satellite peak is not due to an uncontrolled 2DES at some interface of the heterostructure. The experiments were repeated at T = 5, 10, 15, and 50K. Although the lines broaden with temperature, their shapes remain nearly unchanged showing that the intensity of the satellite peak is not increased at the cost of the main peak.

Since the Al concentration varies quadratically across the quantum well, the effective mass (m^*) is also expected to vary along the growth axis [8]. From the growth parameters we expect $m^*(z)$ to vary from $0.067m_0$ in the center of the quantum well to $0.082m_0$ at the edges of the well, where the Al concentration is 19%. In Fig. 4 the expected CR frequencies for bulk GaAs and $Al_{0.19}Ga_{0.81}As$ versus B are represented as solid and dashed lines respectively. Also plotted in Fig. 4 are our experimental points representing the

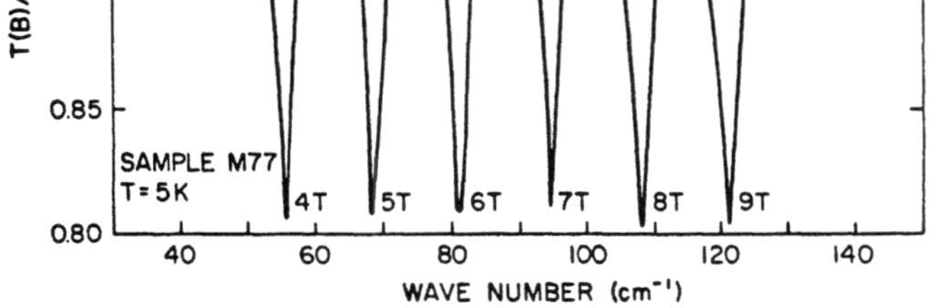

Fig. 3 Transmission spectra for sample M77 for different magnetic field strengths.

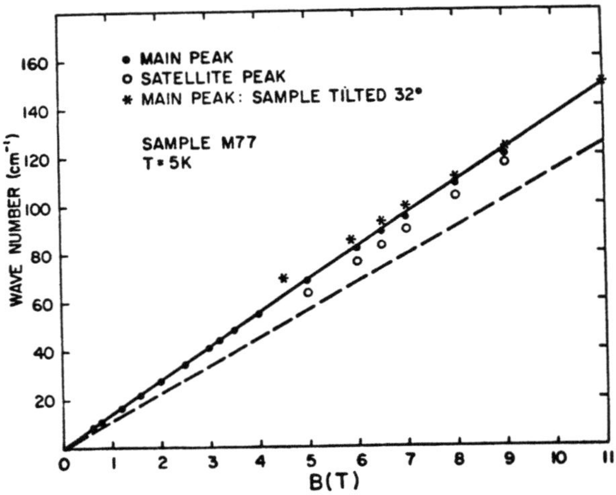

Fig. 4 Positions of the CR and satellite peaks. The solid (dashed) line is the expected CR position for bulk n-GaAs ($Al_{0.19}Ga_{0.81}As$) which should correspond to the center (edges) of the well.

positions of both the main and the shoulder peaks in our CR spectra. Although we do not yet understand the line-shape quantitatively, we qualitatively interpret these results in terms of the different m^* that electrons in different regions of the quantum well would experience.

In summary, a modulation-doped quasi-3DES and its preliminary characterization using magnetotransport and magnetooptical measurements are reported. Growth of similar structures with wider wells and more closely-spaced energy levels, which are expected to better approximate a 3D system, is planned.

We thank V.J. Goldman, J. Jo, C.T. Liu, and D.C. Tsui for useful discussions, and B. Brandt, M-W. Lee, D. Romero, L. Rubin, C. Silvestre and H.P. Wei for technical assistance. Support of this work by NSF grants No. ECS-8553110, DMR-8705002, and DMR-8704670, and grants by the IBM Corporation, GTE Laboratories, Inc., and the New Jersey Commission on Science and Technology is acknowledged.

1. For a review, see D.C. Tsui and H.L. Störmer, IEEE J. Quantum Electron. **QE-22**, 1711 (1986).
2. B.I. Halperin, Japan. J. of Appl. Phys. **26**, Suppl. 26-3, 1913 (1987).
3. The idea of using a well with graded composition was suggested in Ref. 2.
4. For details, see M. Shayegan et al., Appl. Phys. Lett. (in press).
5. See, e.g., J. Menendez et al., Phys. Rev. B **33**, 8863 (1986).
6. J. Yoshino, H. Sakaki, and T. Hotta, Surf. Sci. **142**, 326 (1984).
7. T.A. Kennedy et al., Sol. State Commun. **18**, 275 (1976).
8. M. Zachau, F. Koch, G. Weimann, W. Schlapp, Phys. Rev. B **33**, 8564 (1986).

N_S-Dependent Polaron Effects in GaAs-(Ga,Al)As Heterojunctions*

C.J.G.M. Langerak[1], *J. Singleton*[1], *D.J. Barnes*[2], *P.J. van der Wel*[1], *R.J. Nicholas*[2], *M.A. Hopkins*[2], *T.J.B.M. Janssen*[1], *J.A.A.J. Perenboom*[1], *and C.T.B. Foxon*[3]

[1]High Field Magnet Laboratory and Research Institute for Materials, Toernooiveld, NL-6525 ED Nijmegen, The Netherlands
[2]The Clarendon Laboratory, Parks Road, Oxford OX1 3PU, UK
[3]Philips Research Laboratory, Cross Oak Lane, Redhill, Surrey RH1 5HA, UK

The cyclotron resonance (CR) of the two-dimensional electron gas (2DEG) in GaAs-(Ga,Al)As heterojunctions has been studied for 2D carrier densities N_s between 0.8×10^{11} cm^{-2} and 5.4×10^{11} cm^{-2}. The use of reflectivity allowed the CR to be recorded up to energies only 1.1 meV below the LO phonon. The results show that the resonant polaron effect occurs at the LO phonon energy, and that it becomes strongly suppressed due to occupation effects as N_s increases, explaining previous conflicting reports of "enhanced" or "reduced" polaron effects.

The apparently conflicting reports of "enhanced" [1,2] and "reduced" [3,4,5] resonant polaron effects in the 2DEG have generated much interest. We report a study of the resonant polaron cyclotron resonance in GaAs-(Ga,Al)As heterojunctions as a function of 2D carrier density N_s, which resolves this controversy. The heterojunctions were grown by MBE at Philips Research Laboratories, and values of N_s between 0.8×10^{11} cm^{-2} and 5.4×10^{11} cm^{-2} were obtained by using persistent photoconductivity in three samples (G156, G29, G278). The experiments were carried out at 1.5 K, and N_s was determined from Shubnikov-de Haas oscillations.

As the GaAs substrates of the samples do not transmit between the TO and LO phonon energies, the CR was measured in reflectivity. Figure 1 shows typical reflectivity data for two different values of N_s: outside the reststrahlen band the CR is observed as a peak whilst within it the CR appears as a minimum. The latter result is due to the superposition of CR absorption upon the high reflectivity of the GaAs substrate [6]. Note that the CR moves to a lower field as N_s is increased, and that this shift is much larger close to the LO phonon energy: this is a manifestation of the suppression of the polaron effect.

In order to deduce the effective mass m^* reliably from traces like those in Fig. 1, a calculation of the full dielectric response of the heterojunction was fitted to the experimental resonances [6]. The m^* thus extracted are shown as a function

* Work supported by FOM and NWO (Netherlands) and SERC (UK).

Springer Series in Solid-State Sciences, Vol. 87
High Magnetic Fields in Semiconductor Physics II Editor: G. Landwehr

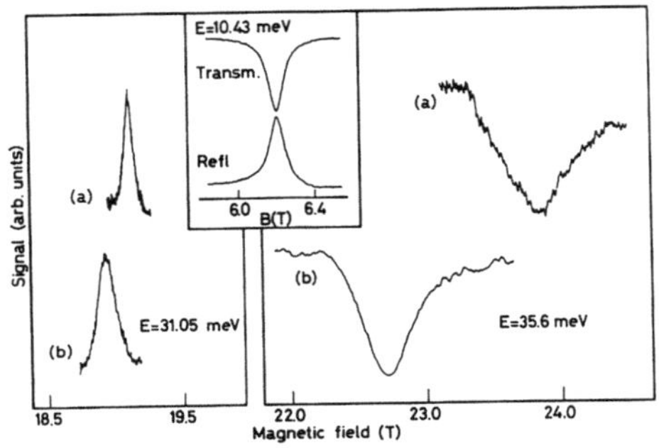

Fig. 1. Normalised reflectivity for G278 at CR energies below and within the reststrahlen band. Data are shown for $N_s=3.4\times10^{11}$ cm^{-2} (a) and $N_s=5.4\times10^{11}$ cm^{-2} (b). The inset shows that the reflectivity peak corresponds to the transmission minimum outside the reststrahlen band.

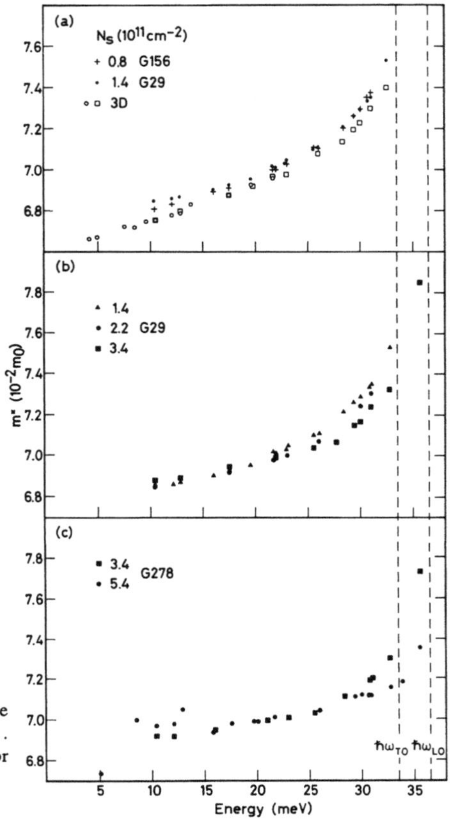

Fig. 2. Effective masses obtained from the measurements *vs.* energy at several values of N_s. Data from bulk GaAs (3D) are shown for comparison with the low N_s heterojunctions in (a).

of energy in Fig. 2 for several values of N_S; also shown are data from bulk GaAs [3,7,8]. The heterojunction m^* are almost all taken within the quantum limit, and so show a background linear increase with energy, due to the band non-parabolicity [6], on which the polaron contribution is superimposed. At low N_S the resonant polaron effect appears to be comparable to that in the bulk, but on increasing N_S, the mass enhancement is much reduced. Note that for all N_S, m^* continues to increase with energy as the LO phonon energy is approached, indicating electron-LO-phonon coupling: there is no evidence to support earlier suggestions that coupling to phonons with frequencies below the LO phonon may occur in some 2D systems, particularly at high N_S [9,10].

The true strength of the polaron effect in Fig. 2 is masked by other factors depending on either N_S or energy E. In order to remove these, we represent the effective mass in the absence of polaron effects by the following expression [11]:

$$\frac{1}{m_b^*(E)} = \frac{1}{m_0^*}\left(1+\left(\frac{2K_2}{E_g}\right)(E+\langle T\rangle_z)\right)$$

where m_0^* is the bare bulk GaAs band-edge effective mass ($0.0652\, m_e$ [7]) and $<T>_z$ represents the kinetic energy due to confinement. K_2 represents the band non-parabolicity: its value is ~ -1.4 [11]. In the quantum limit, CR occurs between the first and second Landau levels, and so E is set equal to their average energy, *i.e.* $E = \hbar\omega_c$. The equation is used to normalise the experimental m^* to reveal the polaron contribution, and the results for four different values of N_s are shown in Fig. 3, along with data for bulk GaAs. At low energies the 2DEG-m^* lie below the bulk data, but approach them as N_s is reduced. This agrees with one-polaron theory, derived in the limit of low N_s, but including the finite width of the 2DEG, which predicts a non-resonant part of the polaron effect which is approximately the same in 2D and 3D [12]. By 33 meV, the polaron contribution to m^* for the two lowest values of N_s appears to be enhanced over that in the bulk data, and to be increasing more rapidly with energy. Thus for low N_s, the resonant polaron effect in GaAs-(Ga,Al)As heterojunctions appears to be enhanced over that in bulk GaAs, in qualitative agreement with one-polaron theory [12], although the theory differs in form from the data here.

In order to assess the factors contributing to the N_s -dependence of the polaron effect, we turn to the results of Memory-Function calculations including occupation effects, dynamical screening and the finite thickness of the 2DEG [13]. The theoretical curve derived for $N_s = 4\times10^{11}\ \mathrm{cm}^{-2}$ is shown as a solid line in Fig. 3. There is reasonable agreement between experiment and theory, suggesting that most of the observed N_s -dependence of the polaron effect is due to occupation effects [6,13]. In fact the suppression of the polaron effect seems to be even stronger than predicted by the theory, with the data for $N_s = 5.4\times10^{11}\ \mathrm{cm}^{-2}$ showing almost no polaron enhancement except close to the LO energy. This may be due to the neglect of the spin-splitting in the Memory-Function model, which, if included, would further increase the occupancy of the levels.

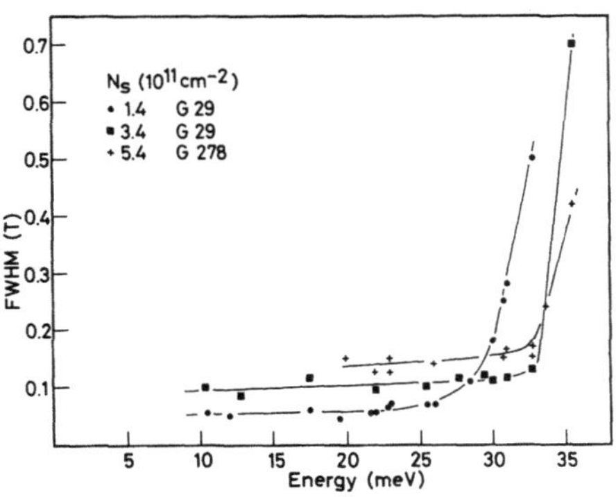

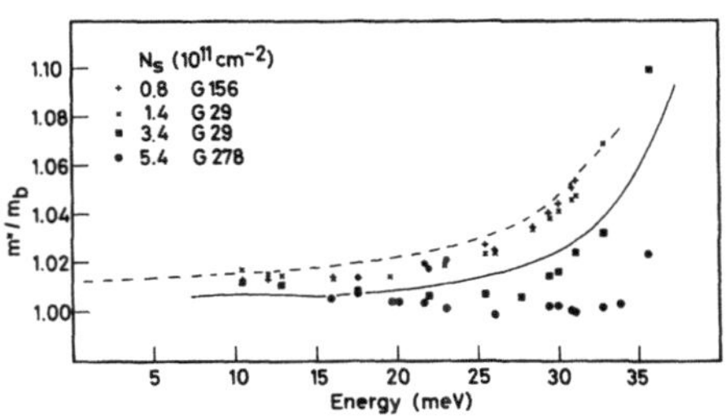

Fig. 3. Normalised effective mass for various N_s *vs.* energy compared with bulk results (dash-dotted line) and results of the Memory-Function model for $N_s = 4\times10^{11}\ \mathrm{cm}^{-2}$ (solid line).

Fig. 4. Linewidth (full width at half maximum, FWHM) *vs.* energy for various N_s.

In Fig. 4 the CR linewidths are plotted for various N_s as a function of energy. A strong increase in the CR linewidth is seen as the LO phonon is approached: this cannot be attributed to the dielectric response alone, and is due to the emergence of optic phonon emission as another scattering mechanism [6]. Increasing N_s suppresses this, so that the start of the increase in linewidth is moved to higher energies, and the size of the effect is reduced. Thus the linewidth exhibits the N_s-dependence of the polaron effect in the same way that m^* does.

In summary, we have observed that the strength of the non-resonant polaron coupling in GaAs-(Ga,Al)As heterojunctions approaches that in bulk GaAs as N_s tends to zero, and that the resonant polaron effect for low N_s is enhanced over that in 3D. Increasing N_s suppresses both non-resonant and resonant coupling, due chiefly to occupation effects, and this can be used to explain the reports of "enhanced" and "reduced" polaron effects in the 2DEG [6]. Our experiments indicate coupling to the LO phonon in GaAs-(Ga,Al)As heterojunctions, in contrast to magnetophonon resonance (MPR) measurements, which indicated a lower energy phonon [14]. Possible reasons for the difference are discussed elsewhere [6], but we remark that CR is a very direct way of observing the phonon involved in the polaron effect: the derivation of a phonon energy from MPR is much more involved.

References

1. M. Horst, U. Merkt, and J.P. Kotthaus: Phys. Rev. Lett. **50**, 754 (1983).
2. J. Singleton, R.J. Nicholas, and F. Nasir: Solid State Commun. **58**, 833 (1986).
3. H. Sigg, P. Wyder, and J.A.A.J. Perenboom: Phys. Rev. B **31**, 5253 (1985).
4. M. Horst, U. Merkt, W. Zawadzki, J.C. Maan, and K. Ploog: Solid State Commun. **53**, 403 (1985).
5. M. Ziesmann, D. Heitmann, and L.L. Chang: Phys. Rev. B **35**, 4541 (1987).
6. C.J.G.M. Langerak, J. Singleton, P.J. van der Wel, J.A.A.J. Perenboom, D.J. Barnes, R.J. Nicholas, M.A. Hopkins, and C.T.B. Foxon: to be published.
7. M.A. Hopkins, R.J. Nicholas, W. Zawadzki, P. Pfeffer, D. Gauthier, J.C. Portal, and M.A. DiForte-Poisson: Semicond. Sci. Technol. **2**, 568 (1987).
8. H. Sigg, J.A.A.J. Perenboom, P. Pfeffer, and W. Zawadzki: Solid State Commun. **61**, 685 (1987).
9. R.J. Nicholas, L.C. Brunel, S. Huant, K. Karraï, J.C. Portal, M.A. Brummell, M. Razeghi, K.Y. Cheng, and A.Y. Cho: Phys. Rev. Lett. **55**, 883 (1985).
10. K. Karraï, S. Huant, G. Martinez, and L.C. Brunel: Solid State Commun. **66**, 355 (1988).
11. M.A. Hopkins, R.J. Nicholas, M.A. Brummell, J.J. Harris, and C.T. Foxon: Phys. Rev. B **36**, 4789 (1987)
12. F.M. Peeters, Wu Xiaoguang, and J.T. Devreese: Phys. Scripta T **13**, 282 (1986).
13. F.M. Peeters, Wu Xiaoguang, and J.T. Devreese: Surf. Sci. **196**, 437 (1988).
14. M.A. Brummell, R.J. Nicholas, M.A. Hopkins, J.J. Harris, and C.T. Foxon: Phys. Rev. Lett. **58**, 77 (1987).

Polarons in 2D-Systems Subjected to a Magnetic Field

*J.T. Devreese** and *F.M. Peeters***

University of Antwerp (U.I.A.), Department of Physics,
Universiteitsplein 1, B-2610 Antwertp, Belgium

A theoretical survey will be given of the properties of *quasi-two-dimensional polarons* in a magnetic field. One polaron and many-polaron effects will be discussed. The cyclotron resonance absorption spectrum of a gas of quasi-two-dimensional polarons is calculated using linear response theory. The calculation is based on a memory function formalism. Band non-parabolicity and the many-body character of the system with inclusion of the full *dynamical screening* of the electron-phonon interaction is incorporated. Comparison is made with experimental results on the cyclotron mass in $GaAs/Al_xGa_{1-x}As$ heterostructures. We found that for electron densities larger than $1.4 \times 10^{11} cm^{-2}$ the polaron mass renormalization is smaller than for the corresponding bulk GaAs system.

1. Introduction

Many of the present day 2D electronic systems are made out of III-V compound semiconducting materials which are weakly polar in nature. Examples of such systems are $GaAs$-$Al_xGa_{1-x}As$ heterostructures and superlattices, p-InSb inversion layers, $Ga_xIn_{1-x}As$ heterostructures, InAs-GaSb systems, ... In those weakly polar semiconductors electron-LO-phonon interaction is important and polaron effects may be observable.

Spatial confinement of electrons to a quasi-two-dimensional(Q2D) layer[1] will influence the electron-phonon interaction. Transport measurements[2] (mobility, hot electron relaxation,...) gives information on *dissipation* as due to the electron-phonon interaction and is theoretically related to the imaginary part of the self-energy. On the other hand cyclotron resonance experiments, in particular the position of the cyclotron resonance peak, gives information on *non-dissipative* aspects of this interaction like polaron mass renormalization and coupling between Landau levels. This aspect of the interaction is often described by the real part of the self-energy. In this paper we will discuss the latter aspect of the electron-phonon interaction when a magnetic field is present and when the electrons are confined to a two-dimensional(2D) layer with a thickness d.

One polaron theories[3,4] predict that polaron effects should be larger in confined systems than in the equivalent bulk material. This is evident from the approximate *scaling relation*[5,6] for the polaron ground state energy

* Also at R.U.C.A., B-2020 Antwerp and University of Technology, NL-5600 MB Eindhoven.

** Senior Research Associate of the National Fund for Scientific Research (Belgium)

Springer Series in Solid-State Sciences, Vol. 87
High Magnetic Fields in Semiconductor Physics II Editor: G. Landwehr

$$E_{2D}(\alpha) = \frac{2}{3} E_{3D}\left(\frac{3\pi}{4}\alpha\right) \quad , \tag{1}$$

and the polaron mass[7]

$$m^*_{2D}(\alpha) = m^*_{3D}\left(\frac{3\pi}{4}\alpha\right) \quad . \tag{2}$$

From these relations the 2D-result can be obtained from the corresponding 3D-result. The above scaling relations are exact within second-order perturbation theory and are approximately valid for arbitrary electron-phonon coupling strength α. The Feynman polaron approximation satisfies exactly Eqs. (1) and (2) (note that these equations are derived in the limit of zero magnetic field). In recent studies[8–13] on the cyclotron resonance of 2D polarons in *GaAs*-heterostructures it was found that many-particle aspects of these systems are very important and reduce the polaron effects considerably.

This is different from 3D systems where the one particle approximation in e.g. bulk GaAs[14] is a very good approximation. The electron density is typically of the order of $n_e \sim 10^{13} cm^{-3}$ which leads to a Fermi energy $E_F \ll k_b T$ and consequently Boltzmann statistics is valid. In typical GaAs-heterostructures the electron density is $n_e \sim 10^{11} - 10^{12} cm^{-2}$ which for a typical electron layer thickness of $100\AA$ results in an effective 3D electron density which is 4-5 orders of magnitude larger than for typical 3D systems. The resulting Fermi energy satisfies $E_F \gg k_b T$ and consequently the electrons will obey Fermi-Dirac statistics. The occupation probabilities of the Landau states is important and screening of the electron-phonon interaction may influence the results.

Often[15–20] one calculates the cyclotron resonance mass from the position of the Landau levels(E_n) in the following way: $m^* = m_b \omega_c / \omega_c^*$ (m_b is the electron band mass and $\omega_c = eH/m_b c$ is the cyclotron resonance frequency for a non-interacting electron) where the cyclotron resonance frequency is determined by $\omega_c^* = (E_1 - E_0)/\hbar = \omega_c - (\Delta E_1 - \Delta E_0)/\hbar$ with ΔE_n the electron-phonon correction to the position of the nth Landau level.

In this paper we will advocate a different approach and calculate the *magneto-optical absorption spectrum* directly. This is the experimental measured quantity in a cyclotron resonance experiment. A full *dynamical screening* theory will be presented and applied to the GaAs-heterostructure system. No phenomenological parameters are introduced in the calculation.

2. The energy levels of a 2D polaron in a magnetic field

First we will study the system consisting of *one* electron moving in a plane and the electron in subjected to a magnetic field which is directed perpendicular to this plane. The energy of such an electron will be quantized into Landau levels $E_n = \hbar\omega_c(n + \frac{1}{2})$ with $n = 0, 1, 2, ...$ and $\omega_c = eB/m_b c$ is the cyclotron resonance frequency of a band electron with mass m_b. Due to the weak interaction of the electron with the LO phonons the landau levels will be shifted by ΔE_n. Within second-order perturbation theory the energy shift of the nth Landau level is given by

$$\Delta E_n = -\sum_{m=0}^{\infty} \sum_{\mathbf{q}} \frac{|M_{n,m}(\mathbf{q})|^2}{D_{n,m}} \quad . \tag{3}$$

where

$$M_{n,m}(\mathbf{q}) = < m, \mathbf{q}|H_I|0, n > \quad , \tag{4}$$

is the matrix element of the electron-phonon interaction operator H_I. The electron-phonon interaction term in the Hamiltonian is

$$H_I = \sum_{\mathbf{q}} \left(V_q a_q e^{i\mathbf{q}\cdot\mathbf{r}} + V_q^* a_q^+ e^{-i\mathbf{q}\cdot\mathbf{r}} \right) \quad , \tag{5}$$

with $a_q(a_q^+)$ the annihilation (creation) operator of an LO phonon with momentum $\hbar\mathbf{q}$ and energy $\hbar\omega_{LO}$. The interaction coefficient is

$$|V_q|^2 = \frac{1}{q}\frac{\sqrt{2}\pi\alpha}{A} \quad , \tag{6}$$

in 2D where we used units such that $\hbar = m_b = \omega_{LO} = 1$ and A is the surface area of the system. The energy dominator in Eq.(3) is given by

$$D_{n,m} = \hbar\omega_{LO} - \Delta_n + \hbar\omega_c(m - n) \quad , \tag{7}$$

where the choice: (1) $\Delta_n = 0$ leads to Rayleigh-Schrödinger perturbation theory (RSPT) which gives accurate results for ΔE_n if $\omega_c \ll \omega_{LO}$; (2) $\Delta_n = \Delta E_n$ results in Wigner-Brillouin perturbation theory (WBPT) which can account for the *splitting* of the degenerate energy levels; and (3) $\Delta_n = \Delta E_n - \Delta E_n^{RSPT}$ gives an "improved" Wigner-Brillouin theory (IWBPT) as was developed in Ref.21 and further discussed in Ref.18. IWBPT gives the correct *pinning* behavior for small α.

For a parabolic conduction band and in the ideal 2D case it is possible to obtain a closed analytic expression for the energy shift ΔE_n due to the electron-phonon interaction[17]:

$$\Delta E_n = -\frac{\pi}{2}\alpha\frac{1}{\sqrt{\omega_c}}\sum_{m=0}^{n}\binom{n}{m}\left(\frac{(2m-1)!!}{2^m}\right)^2\frac{1}{m!}\cdot\frac{\Gamma((1-\Delta_n)/\omega_c - m)}{\Gamma((1-\Delta_n)/\omega_c + 1/2)} \quad , \tag{8}$$

where $\binom{n}{m}$ is the well-known combinatorics, $(2m-1)!! = (2m-1)(2m-3)...3\cdot 1$ with $(2m-1)!! = 1$ if $m = 0$ and $\Gamma(x)$ is the gamma function. For $\Delta_n = 0$ Eq.(9) was also obtained by Larsen[16] for $n = 0, 1, 2$. Within the IWBPT approximation Eq.(9) has to be solved self-consistently which was done in Ref. 17. The results of such a calculation are that the polaron energy levels are: (1) shifted to lower energies, (2) at small magnetic fields they are bended downward due to a mass renormalization, (3) at $n\omega_c \simeq \omega_{LO}$ the nth Landau levels do not cross the LO-phonon+(E_0) level, and (4) for $\omega_c \to \infty$ all the Landau levels become pinned to $\hbar\omega_{LO} + \Delta E_0^{RSPT}$. When one compares the 2D result with the equivalent 3D result one notes that in 2D the energy shift, the mass renormalization, and the level splitting at $n\omega_c \simeq \omega_{LO}$ are more pronounced than for the 3D polaron.

We will consider the limiting behaviour of the polaron correction for certain values of the magnetic field. First for small magnetic fields (i.e. $n\omega_c \ll \omega_{LO}$) the energy difference between two successive Landau levels is

$$\Delta E_{n+1} - \Delta E_n = \alpha\frac{\pi}{8}\omega_c\left(1 + \frac{9}{8}(n+1)\omega_c + \frac{5}{128}(30n^2 + 60n + 29)\omega_c^2 + ...\right) \quad , \tag{9}$$

which results in the cyclotron mass

$$\frac{m^*}{m_b} = \frac{1}{1 - \alpha\frac{\pi}{8}(1 + \frac{9}{8}\omega_c + \frac{145}{128}\omega_c^2 + ...)} \quad . \tag{10}$$

For $n\omega_c = \omega_{LO}$ two unperturbed levels will cross but because of the presence of the electron-phonon interaction they will split. The splitting is found to be[17] twice

$$|\Delta E_n| = \sqrt{\alpha}\sqrt{\sqrt{\frac{\pi}{n}}\frac{(2n-1)!!}{2^{n+1}n!}} \quad , \tag{11}$$

which has the characteristic $\alpha^{1/2}$-dependence which is different from the $\alpha^{2/3}$-dependence as found for the 3D polaron.

In the limit of $\omega_c \rightarrow \infty$ the Landau levels become pinned. We found[17] the following limiting behaviour

$$E_n = E_0^{RSPT} + 1 + \frac{1}{2}\omega_c - \frac{\alpha}{\sqrt{\omega_c}}\frac{\sqrt{\pi}(2n-1)!!}{2^{n+1}n!} \quad . \tag{12}$$

3. The magneto-optical absorption spectrum

Theoretically there are two different viewpoints in calculating the cyclotron resonance frequency. First, and the most often used in polaron physics because of its apparent simplicity, relies on a calculation of the position of the Landau levels E_n from which one obtains: $\omega_c^* = (E_1 - E_0)/\hbar$. This approach was elaborated on in the forgoing section. An alternative approach which we want to advocate is by calculating the *cyclotron resonance absorption spectrum* itself. It is this quantity which is measured experimentally. The position of the cyclotron resonance peak then gives the cyclotron resonance frequency.

Within linear response theory the cyclotron resonance absorption spectrum in the Faraday active mode configuration is proportional to the real part of the dynamical conductivity tensor

$$Re\sigma_+(\omega) \sim \frac{Im\Sigma(\omega)}{(\omega - \omega_c - Re\Sigma(\omega))^2 + (Im\Sigma(\omega))^2} \quad , \tag{13}$$

where $\Sigma(\omega)$ is the memory function which is a function of the unperturbed cyclotron resonance frequency ω_c, of the electron-phonon coupling constant α and also of the width of the 2D electron layer. The memory function is in essence a force-force correlation function where the time evolution is restricted to the Liouville space which is perpendicular to the velocity operator $\dot{x}+i\dot{y}$. In order to calculate the memory function we use second-order perturbation theory and i) decouple the electron-phonon system, ii) the phonons are taken at thermal equilibrium and iii) the electron system can be calculated within different approximations. To first order in the electron-phonon coupling constant the memory function takes the form

$$\Sigma(\omega) = \frac{1}{\omega}\int_0^\infty dt(1 - e^{i\omega t})ImF(t) \quad , \tag{14}$$

with

$$F(t) = -\sum_{\mathbf{k}} \frac{k_\parallel^2}{n_e m_b \hbar} |V_{\mathbf{k}}|^2 \left[i\,D(k_\parallel, t) + i\,n(\omega_{LO})D^R(k_\parallel, t)\right] e^{-i\omega_{LO} t} \quad , \tag{15}$$

where $n(\omega) = (e^{\beta\hbar\omega}-1)^{-1}$ is the occupation number. $D(k_{\|}, t)$ $(D^R(k_{\|}, t))$ is the electron (retarded) density-density correlation function[22]. $V_{\mathbf{k}}$ is the interaction coefficient of the electron-phonon interaction and is given by

$$V_{\mathbf{k}} = i\omega_{LO}\left(\frac{4\pi\alpha}{Vk^2}\right)^{1/2}\left(\frac{\hbar}{2m_b\omega_{LO}}\right)^{1/4}\langle\psi_0|e^{ik_z z}|\psi_0\rangle \quad . \tag{16}$$

if we include the non-zero thickness of the 2D electron layer. $\psi_0(z) = (b^3/2)^{1/2}ze^{-bz/2}$ is the variational wave function of the electron in the direction normal to the 2D electron layer[1]. The variational parameter b is taken as $b = (48\pi Nm_b e^2/\hbar^2\epsilon_0)^{1/3}$, where $N = n_d + (11/32)n_e$ and n_d and n_e are the depletion and carrier charge densities respectively. In Eq.(17) only the lowest subband is included and all higher electric subbands are neglected. This approximation should be valid for systems with not too high electron density ($n_e < 10^{12}cm^{-2}$) and low temperature.

At low temperatures one has $Im\Sigma(\omega) \simeq 0$ and consequently the cyclotron resonance frequency is given by the non-linear equation

$$\omega - \omega_c - Re\Sigma(\omega) = 0 \quad . \tag{17}$$

From the solution $\omega = \omega_c^*$ the cyclotron resonance mass is defined as $m^* = m_b\omega_c/\omega_c^*$. The physical meaning of Eq.(17) is apparent: $Re\Sigma(\omega_c^*)$ is the shift in the cyclotron resonance peak due to the electron-phonon interaction. Note also that theoretically the cyclotron mass is always referred to the electron band mass which is not a directly meausurable quantity.

The one polaron theory is able to account[23,24] for the basic *qualitative* behaviour of the cyclotron resonance peak. In a next step we will investigate how *many-particle* aspects of the electron system will influence the absorption spectrum and more specifically the influence on the position of the cyclotron resonance frequency will be investigated in detail.

As a first step the *occupation* effect (i.e. the Pauli-blocking effect) and the influence of *static screening* was investigated in Refs. 25 and 26. The occupation effect is taken into account if in Eq.(15) the polarization function $\Pi = \hbar D$ is approximated by Π_0 which is the polarization of a non-interacting 2D electron gas[27,28]. To include the effect of the *static screening* into the memory function calculation $|V_{\mathbf{k}}|^2$ in Eq.(15) is substituted by $|V_{\mathbf{k}}|^2/\epsilon^2(k_{\|})$ where $\epsilon(k)$ is the static dielectric function which we took within the RPA approximation[27,28] but where we retained the magnetic field dependence.

We have extended the above results recently[29] and have included a full *dynamical screening* of the electron-phonon interaction. Within such a calculation the real part of the memory function can be obtained by substituting Eq.(15) into Eq.(14). In the zero temperature limit we obtain

$$Re\Sigma(\omega) = \sum_{\mathbf{k}} \frac{k^2 f(k,b)}{n_e m_b \omega}\frac{|V_k|^2}{v(k)}\frac{\omega^2}{\pi}\int_{-\infty}^{+\infty} dx \frac{1}{[(x+\omega_{LO})^2-\omega^2](x+\omega_{LO})} Im\frac{1}{\epsilon(k,x)} \quad , \tag{18}$$

where $\mathbf{k}$ is a 2D wave vector. The sum over k_z has been performed and leads to the well-known form factor $f(k,b)$ where the width of the 2D electron layer is given by $\sqrt{<(z-<z>)^2>} = \sqrt{3}/b$. In Eq.(18) $v(k)$ is the Fourier transform of the unscreened 2D Coulomb interaction and $\epsilon(k,\omega)$ is the *frequency* dependent dielectric function of the 2D electron gas which is considered here in the random-phase-approximation(RPA) with the inclusion of a magnetic field.

In deriving the above expressions for the cyclotron resonance spectrum the following assumptions are made: (1)the interaction of the electrons is with the bulk LO-phonons of GaAs and the LO-phonons are assumed to be dispersionless; (2)perpendicular to the electron layer the electron wave function is chosen to be of the Fang-Howard type. Only the occupation of the lowest electric subband is incorporated; (3)the calculation is done for the zero temperature limit which is reasonable in view of the low experimental temperature ($T \sim 4.2K$); (4)the electron conduction band is taken to be parabolic.

4. Comparison with experiment

In order to make a detailed quantitative comparison between the experimental measured cyclotron resonance mass and the theoretical calculated values it is necessary to incorporate the electron band non-parabolicity which is very important for the GaAs-heterostructures under investigation here. The band non-parabolicity will be taken into account via the assumption of a *local parabolic band* approximation. Larsen[30] has shown that this is valid for GaAs. In this approach the Landau levels in the absence of polaron coupling are determined from the two-level Kane theory as elaborated by Lassnig and Zawadzki[31]. They found

$$\epsilon_{np} = -\frac{E_g}{2} + < U > + \frac{E_g}{2}\sqrt{1 + \frac{4}{E_g}(< T > + \epsilon_{\|})} \quad , \tag{19}$$

with $< U >$ the average electron potential energy in the z-direction $< U > = (12\pi e^2/\epsilon_0 b)(n_d + \frac{11}{16}n_e)$ and $< T > = \hbar^2 b^{*2}/8m_b$ the average electron kinetic energy in the z-direction and $\epsilon_{\|} = \hbar\omega_c(n+1/2)$. $b^* = b\sqrt{x}$ is determined from the equation $x^3 - px - q = 0$ with $p = \hbar^2 b^2/2m_b E_g$ and $q = (1 + 4\epsilon_{\|}/E_g)$.

Due to the band non-parabolicity the cyclotron resonance frequency will be different from ω_c even in the absence of any polaron effects. The shifted cyclotron resonance frequency is given by $\hbar(\omega_c)_{np} = \epsilon_{np}(n = 1) - \epsilon_{np}(n = 0)$. In order to incorporate polaron effects together with the band non-parabolicity, we insert $(\omega_c)_{np}$ and b^*, as obtained from above, into the memory function. This results in the non linear equation

$$\omega - (\omega_c)_{np} - Re\Sigma(\alpha, (\omega_c)_{np}, b^*; \omega) = 0 \quad , \tag{20}$$

where the solution $\omega = \omega_c^*$ equals the theoretical cyclotron mass which is affected both by band non-parabolicity and by polaron effects.

Note that in the present approximation the correct behaviour for either vanishing electron-phonon coupling, or parabolic energy band limit, is obtained. Also in the limit of zero magnetic field the correct behaviour is obtained because we included *all* the intermediate states. Further the polaron effect and the band non-parabolicity are not considered to be additive.

A quantitative comparison of the polaron mass renormalization was made[26] for GaAs - heterostructures with a typical electron density of $n_e \sim 4 \times 10^{11} cm^{-2}$. Earlier we found[22] that the *non-zero width* of the 2D electron layer considerably reduces the 2D *one* polaron results for the polaron mass(by a factor of 2 for $n_e \sim 4 \times 10^{11} cm^{-2}$ and $\omega_c/\omega_{LO} \sim 0.8$). Inclusion of the occupation probabilities for the Landau levels leads to a further reduction[25,26] (a factor of 1.7 for $n_e \sim 4 \times 10^{11} cm^{-2}$ and $\omega_c/\omega_{LO} \sim 0.8$). We found[26] that screening of the electron-phonon interaction further weakens the

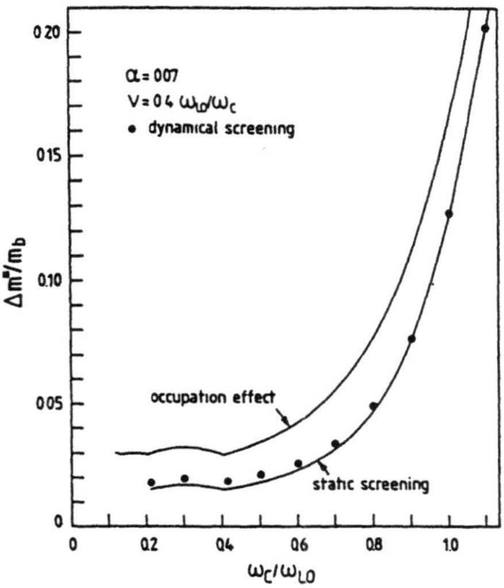

Fig. 1: The polaron correction to the cyclotron mass. The many-body effects are treated within different approximations: occupation effect, static screening and dynamical screening. The electron density is $4 \times 10^{11} cm^{-2}$ and a parabolic conduction band is assumed.

effect of the electron-phonon interaction but that treating the screening dynamically[29] does not alter the results significantly as compared to the static screening results (see Fig.1). Static screening overestimates the reduction of the mass renormalization. This is most apparent in the weak field case (i.e. $\omega_c/\omega_{LO} < 0.5$) where the polaron mass renormalization is small. In the magnetic field range where polaron effects become observable (i.e. $\omega_c/\omega_{LO} > 0.5$) a static screeening theory gives, within 5%, the same results as a full dynamical screening theory. Thus for all practical purposes the numerical much simpler static screening approach will be sufficiently reliable when we compare theory with experiment.

In Fig. 2 the experimental result for the cyclotron mass as obtained by Hopkins *et al*[11] (solid dots) is compared with the present theoretical results. A very close agreement between theory and experiment could be obtained over the whole magnetic field range if the band mass is taken to be $m_b/m_e = 0.0661$. Polaron effects are evident because band non-parabolicity alone would give only a linear(roughly) increase of the cyclotron mass with magnetic field strength(dashed curve).

Next we want to address the question whether or not the polaron shift in the cyclotron resonance peak position is larger or smaller in real 2D systems than in 3D systems. We know that, due to the occupation effect and due to screening, the polaron effect decreases with increasing electron density. This was recently also confirmed experimentally by Langerak *et al*[13]. Therefore it is sufficient to analyse only the low electron density sample of Hopkins *et al*[11]. The electron-phonon contribution to the shift in the experimental cyclotron resonance frequency(ω^*) is extracted by subtracting the shift due to band non-parabolicity $\omega_c^0 = (\omega_c)_{np}$. The result is shown in Fig. 3 where, for comparison, we also show the electron-phonon contribution to the polaron cyclotron mass for an ideal 2D system (one polaron) and for the 3D system. From Fig. 3 we may conclude that in $GaAs/Al_xGa_{1-x}As$ heterostructures polaron effects below the resonance condition are *smaller* than in bulk GaAs if $n_e \geq 1.4 \times 10^{11} cm^{-2}$.

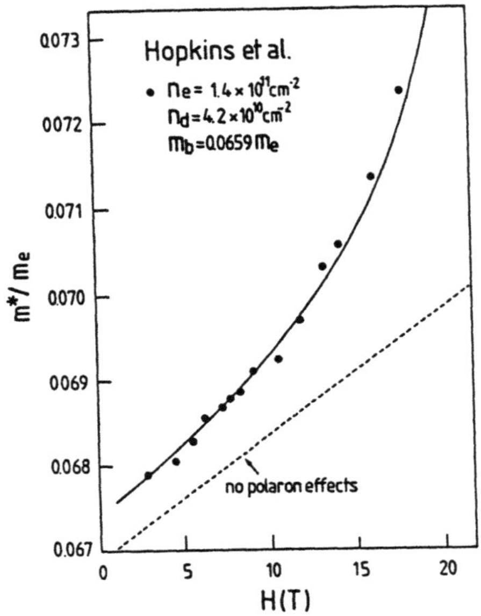

Fig. 2: Cyclotron mass as function of the magnetic field. The solid dots are the experimental results, the full curve is the present theoretical result and the dashed curve corresponds to the theoretical results without polaron effects.

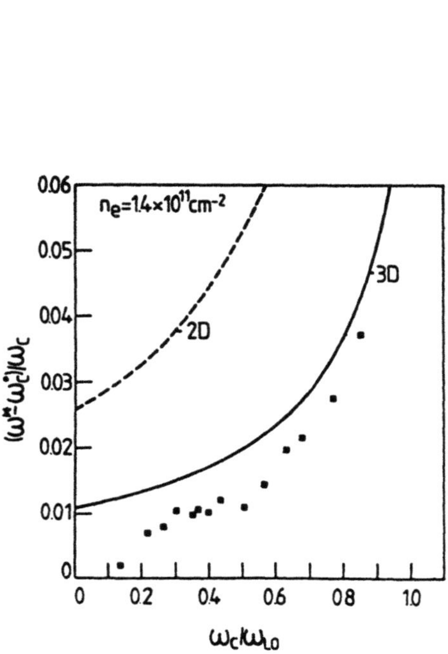

Fig. 3: The polaron contribution to the effective electron mass within a one polaron picture for an ideal 2D (dashed curve) and a 3D (full curve) system of GaAs. The experimental results for $n_e = 1.4\times10^{11}cm^{-2}$ are given by the full squares.

Up to now all cyclotron resonance measurements in GaAs-heterostructures are for $\omega_c^* < \omega_{LO}$. Measurements are needed in the $\omega_c^* > \omega_{LO}$ region. The discontinuity at the resonance condition $\omega_c^* = \omega_{LO}$ is a direct consequence of the electron-phonon interaction and is not obscured by effects from band non-parabolicity. Splitting of the cyclotron resonance peak around $\omega_c^* \sim \omega_{LO}$ is a consequence of the removal of the degeneracy of the energy level $E_1 = \frac{3}{2}\hbar\omega_c$ and $E_0 + \hbar\omega_{LO} = \frac{1}{2}\hbar\omega_c + \hbar\omega_{LO}$. This splitting has been observed in polar 3D semiconductors, in InSb-inversion layers[8,23] and in $Ga_xIn_{1-x}As$-heterostructures[32] but not in GaAs-heterostructures (because magnetic fields such that $H > 22T$ are needed).

Acknowledgements

This work is sponsored by F.K.F.O.(Fonds voor Kollektief Fundamenteel Onderzoek, Belgium), project No. 2.0072.80. We would like to thank Wu Xiaoguang for a stimulating collaboration in a large part of the research presented here and C. Langerak and J. Singleton for sending us their experimental results prior to publication.

References

[1] T. Ando, A.B. Fowler and F. Stern, Rev. Mod. Phys. **54**, 437 (1982).
[2] See e.g. J. Shah, in *The Physics of the Two-Dimensional Electron Gas*, edited by J.T. Devreese and F.M. Peeters (Plenum Press, New York, 1987).
[3] S. Das Sarma and A. Madhukar, Phys. Rev. **B22**, 2823 (1980).
[4] S. Das Sarma and B.A. Mason, Ann. Phys. (N.Y.) **163**, 78 (1985).
[5] Wu Xiaoguang, F.M. Peeters and J.T. Devreese, Phys. Rev. **B31**, 3420 (1985).
[6] F.M. Peeters and J.T. Devreese, Phys. Rev. **B36**, 4442 (1987).
[7] F.M. Peeters, Xiaoguang Wu and J.T. Devreese, Phys. Rev. **B37**, 933 (1988).
[8] U. Merkt, M. Horst and J.P. Kotthaus, Physica Scripta **T13**, 272 (1986).
[9] H. Sigg, P. Wyder and J.A.A.J. Perenboom, Phys. Rev. **B31**, 5253 (1985).
[10] M. Horst, U. Merkt, W. Zawadzki, J.C. Maan and K. Ploog, Solid State Commun. **53**, 403 (1985).
[11] M.A. Hopkins, R.J. Nicholas, M.A. Brummell, J.J. Harris and C.T. Foxon, Superlatt. and Microstr. **2**, 319 (1986); Phys. Rev. **B36**, 4789 (1987).
[12] F. Thiele, U. Merkt, J.P. Kotthaus, G. Lommer, F. Malcher, U. Rössler and G. Weimann, Solid Stat. Commun. **62**, 841 (1987).
[13] C.J.G.M. Langerak, J. Singleton, P.J. van der Wel, J.A.A.J. Perenboom, D.J. Barnes, R.J. Nicholas, M.A. Hopkins and C.T.B. Foxon (to be published)
[14] M. Ziesmann, D. Heitmann and L.L Chang, Phys. Rev. **B35**, 4541 (1987).
[15] S. Das Sarma, Phys. Rev. Lett. **52**, 859 (1984).
[16] D.M. Larsen, Phys. Rev. **B30**, 4807 (1984).
[17] F.M. Peeters and J.T. Devreese, Phys. Rev. **B31**, 3689 (1985).
[18] G. Lindemann, R. Lassnig, W. Seidenbusch and E. Gornik, Phys. Rev. **B28**, 4693 (1983).
[19] D.M. Larsen, Phys. Rev. **B30**, 4595 (1984).
[20] R. Lassnig, Surface Sci. **170**, 549 (1986).
[21] D.M. Larsen and E.J. Johnson, J. Phys. Soc. Jpn., Suppl. **21**, 443 (1966).
[22] Wu Xiaoguang, F.M. Peeters and J.T. Devreese, Phys. Rev. **B34**, 8800 (1986); F.M. Peeters, Wu Xiaoguang and J.T. Devreese, Physica Scripta **T13**, 282 (1986).
[23] M. Horst, U. Merkt and J.P. Kotthaus, Phys. Rev. Lett. **50**, 754 (1983).
[24] J. Singleton, R.J. Nicholas, R.C. Rogers and C.T.B. Foxon, Surf. Sci. **196** (1988).
[25] Wu Xiaoguang, F.M. Peeters and J.T. Devreese, in *the Proc. of the 18th Int. Conf. on the Phys. of Semicond.*, Ed. O. Engström (World Scientific, Singapore, 1987) 589.
[26] Wu Xiaoguang, F. M. Peeters, and J. T. Devreese, Phys. Stat. Sol. (b) **143**, 581 (1987).
[27] N. J. Horing and M. M. Yildiz, Ann. Phys. (N.Y.) **97**, 216 (1976).
[28] M. L. Glasser, Phys. Rev. B **28**, 4387 (1983).
[29] Wu Xiaoguang, F. M. Peeters, and J. T. Devreese, Phys. Rev. **B36**, 9760 (1987).
[30] D.M. Larsen, Phys. Rev. B **36**, 3304 (1987).
[31] R. Lassnig and W. Zawadzki, Surface Sci. **142**, 388 (1984).
[32] R.J. Nicholas, L.C. Brunel, S. Huant, K. Karrai, J.C. Portal, M.A. Brummell, M. Razeghi, K.Y. Cheng and A.Y. Cho, Phys. Rev. Let. **55**, 883 (1985).

Part VIII

Semimagnetic Semiconductors, 2D and 3D

Magneto-optic Phenomena in Diluted Magnetic Semiconductors

A.K. Ramdas

Department of Physics, Purdue University, West Lafayette,
IN 47907, USA

1. Introduction

Diluted magnetic semiconductors (DMS) [1] – typically, the tetrahedrally coordinated II-VI compound semiconductors in which the cations have been replaced by transition metal ions (e.g., $Cd_{1-x}Mn_xTe$ or $Cd_{1-x}Fe_xSe$) – exhibit striking magnetic phenomena owing to the large spin-spin exchange interaction between the localized magnetic moments of the d-electrons of the magnetic transition metal ion (e.g., Mn^{2+}) and the band electrons (the 'sp-d' exchange interaction) as well as the anti-ferromagnetic coupling between the magnetic ions. The latter comes into play more significantly as the temperature is lowered and the concentration of the magnetic ions (x) is increased. The unique properties of DMS's are brought out in a particularly spectacular fashion in magneto-optic phenomena like the Faraday effect, the Zeeman effect of the exciton and Raman scattering from magnetic excitations. The magnetic features which uniquely distinguish DMS's are illustrated in this review with examples from the research program of the author and his collaborators.

2. Zeeman Effect: Free Exciton

The reflectivity spectrum of the zinc blende DMS, $Zn_{0.72}Mn_{0.28}Te$ measured at T = 1.3 K, shows a sharp signature at E_o = 2.592 eV corresponding to the free exciton. An additional feature at 2.156 eV is associated with Mn^{2+} which occurs in all the Mn-based DMS's, provided the band gap is larger than this energy. In Fig. 1 the magnetic field dependence of the Zeeman components of the free exciton, recorded in the Faraday geometry, is displayed [2]. The four components labeled "a", "b", "c" and "d" in Fig. 1 correspond to the four transitions in Fig. 2; "a" and "b" are observed in $\hat{\sigma}_+ = (1/\sqrt{2})(\hat{x}+i\hat{y})$ and "c" and "d" in $\hat{\sigma}_- = (1/\sqrt{2})(\hat{x}-i\hat{y})$ polarization of the incident light propagating along the external magnetic field $\vec{H} \parallel \hat{z}$. The enormous magnitude of the Zeeman splitting and its saturation at high fields are striking features very clearly evident. The energies of the four Zeeman components are: $\hbar\omega_a = E_o + 3B - 3A$; $\hbar\omega_b = E_o + B + 3A$; $\hbar\omega_c = E_o - B - 3A$; and $\hbar\omega_d = E_o - 3B + 3A$. Here $A = (1/6)(\alpha M/g_{Mn}\mu_B)$; $B = (1/6)(\beta M/g_{Mn}\mu_B)$; α and β are the 's-d' and the 'p-d' exchange integrals for the conduction and valence band, respectively; M = magnetization per unit volume; g_{Mn} = Landé g factor of the Mn^{2+} spins = 2 corresponding to the $^6S_{5/2}$ ground state of Mn^{2+} and μ_B = the Bohr magneton. The magnetization for dilute system of Mn^{2+} is, to a good approximation for small x, given by $M = -g_{Mn}\mu_B x N_o \langle S_z^{Mn}\rangle$ and $\langle S_z^{Mn}\rangle = -5/2\, B_{5/2}(\eta)$, where N_o is the number of unit cells per unit volume, $\langle S_z^{Mn}\rangle$ is the thermal average of the Mn^{2+} spins along the direction of the magnetic field, $\eta = (g_{Mn}\mu_B H)/(k_B T)$, k_B = the Boltzmann constant and $B_{5/2}$ is the Brillouin function B_J for J = 5/2. The large excitonic Zeeman splittings are completely dominated by the spin splittings of the conduction and valence bands produced by the large 's-d' and 'p-d' exchange interaction. For $Zn_{1-x}Mn_xTe$, $N_o\alpha$ = 190 meV and $N_o\beta$ =

Springer Series in Solid-State Sciences, Vol. 87
High Magnetic Fields in Semiconductor Physics II Editor: G. Landwehr

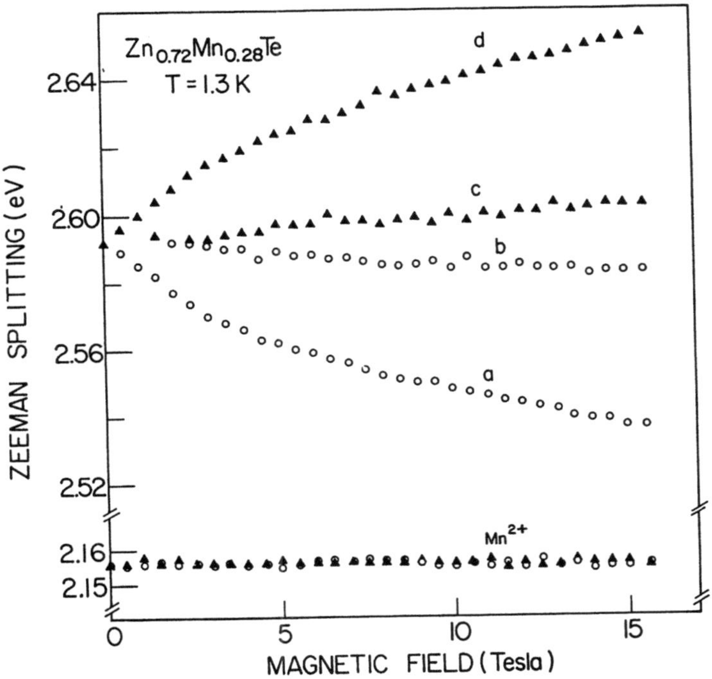

Fig. 1. Magnetic field dependence of the energies of the Zeeman components of the free excitons (a, b, c, and d) and the internal Mn^{2+} transition in $Zn_{0.72}Mn_{0.28}Te$ at 1.3 K. The 'c' and 'd' components are observed in the $\hat{\sigma}_-$ and the 'a' and 'b' transitions in the $\hat{\sigma}_+$ polarization, respectively

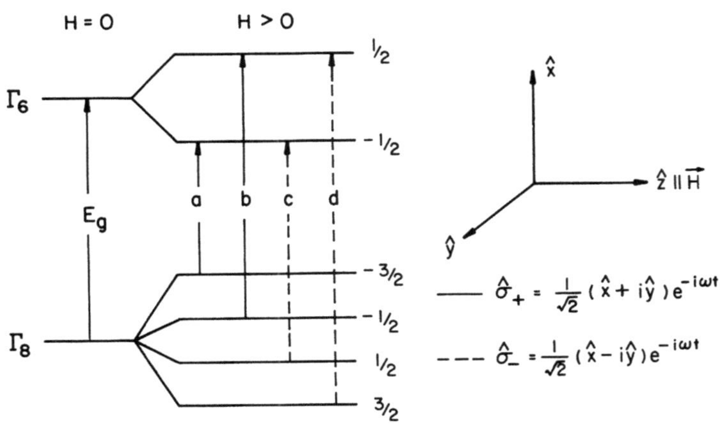

Fig. 2. Left: schematic illustration of the $\Gamma_8 \rightarrow \Gamma_6$ interband transition in a zinc blende diluted magnetic semiconductor in the presence of an external magnetic field allowed in the Faraday geometry. Right: Faraday geometry and the circular polarization of light

-1090 meV [3]. In contrast to the Zeeman effect of the free exciton, the energy of the Mn^{2+} feature is unaffected by the magnetic field. The significance of this observation has been discussed in Ref. [2].

The large Zeeman effect of the free exciton and its initmate relationship with the magnetization in DMS's is related to the phenomenological parameters $N_o\alpha$ and $N_o\beta$. The microscopic mechanisms underlying their large values and their signs are discussed in [4] and [5]. We note here that the Zeeman components of the free exciton can also be observed in photoluminescence and their large Zeeman shifts can often bring the low energy components in the region of the Raman lines associated with the zone center optical phonons [6].

3. Faraday Effect

In contrast to the II-VI semiconductors, the Mn-based II-VI's exhibit enormous Faraday rotations [3],[7]. This spectacular enhancement of the Faraday rotation (Θ_F), the so called *giant Faraday rotation*, is again a consequence of the large sp-d exchange interaction.

Consider white light propagating along the optic axis ($\hat{c}$) of a uniaxial, wurtzite DMS like $Cd_{1-x}Mn_xSe$. Let the DMS be placed between two polarizers set at 45° with respect to each other, the sample being placed in a variable-temperature optical cryostat equipped with a superconducting coil and $\vec{H} \parallel \hat{c}$, and the transmitted light be spectrally analyzed with a double monochromator. In Fig. 3 we present such a transmission curve for zero magnetic field and at 10 kG for $Cd_{0.75}Mn_{0.25}Se$. At zero field, the onset of transmission for $\hbar\omega \leq E_g$ is clearly noticed. When the magnetic field is turned on, pronounced oscillations are observed in the transmission spectrum —"Faraday in-

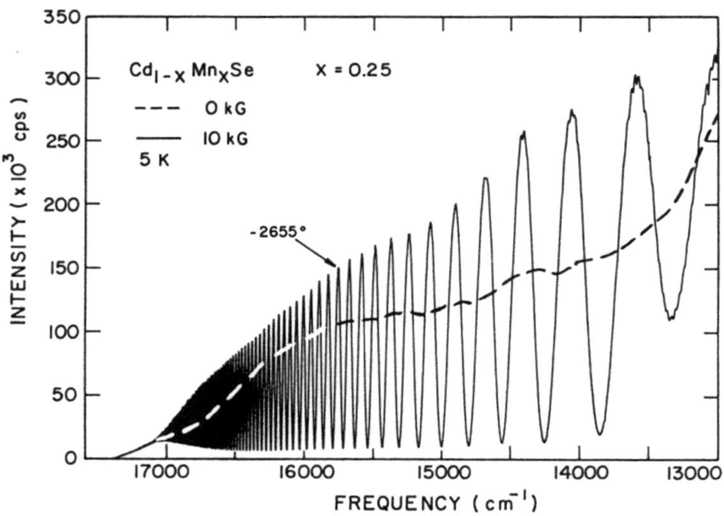

Fig. 3. Faraday effect in $Cd_{1-x}Mn_xSe$ (x = 0.25), T = 5 K. The dashed line represents the transmission of light as a function of its frequency for a 3 mm thick sample located between two polaroids with their axes at 45° and at zero magnetic field. The light propagates along the optic axis, $\hat{c}$. The oscillations occur when a 10 kG magnetic field is applied along $\hat{c}$ [8]

tensity"; each successive peak represents an additional Faraday rotation of 180°. The absolute Faraday rotation as a function of frequency is determined by monitoring the transmission at several fixed photon energies as the magnetic field is swept. By setting the polarizer and the analyzer at angles other than 45° with respect to each other, the entire Faraday-rotation dispersion can be determined. Figure 3 demonstrates the giant negative Faraday rotation in DMS's and its large dispersion as the photon energy approaches the absorption edge [8].

As is well known, Faraday rotation is the consequence of the circular birefringence caused by the circularly polarized Zeeman components of a given optical transition. The dispersion of the Verdet constant (V) — Faraday rotation per unit magnetic field per unit length — in DMS's is, in the main, associated with the Zeeman effect of the free exciton [7]. The dispersion of V in $Cd_{1-x}Mn_xTe$ is shown in Fig. 4(a); the free exciton energy deduced from these data is in excellent agreement with that obtained from piezomodulated reflectivity [9]. The Brillouin-function-like behavior of Θ_F *vs* H is shown in Fig. 4(b). The sign of Θ_F is negative in contrast to that in the nonmagnetic II-VI compounds. The large magnitude of Θ_F; its enhancement for photon energies approaching those of the Zeeman components of the exciton; the (H/T) dependence of Θ_F — all of these once again are related to the large 'sp-d' exchange interaction and the magnetization (M). In fact, for small x and large T, Θ_F is linearly proportional to $(\beta-\alpha)$ and M [3]. In Fig. 5 we show the transmission of $Cd_{0.75}Mn_{0.25}Se$ at a photon energy (E) at a *fixed* fraction away from the energy gap (E_o) recorded as a function of T^{-1}, using the same arrangement employed in Fig. 3. Again, each successive minimum represents an additional 180° Faraday rotation. The increase in Θ_F and its trend towards saturation with decreasing T are clearly evident. In Fig. 6, we display the "Faraday" oscillations in the transmitted light as a function

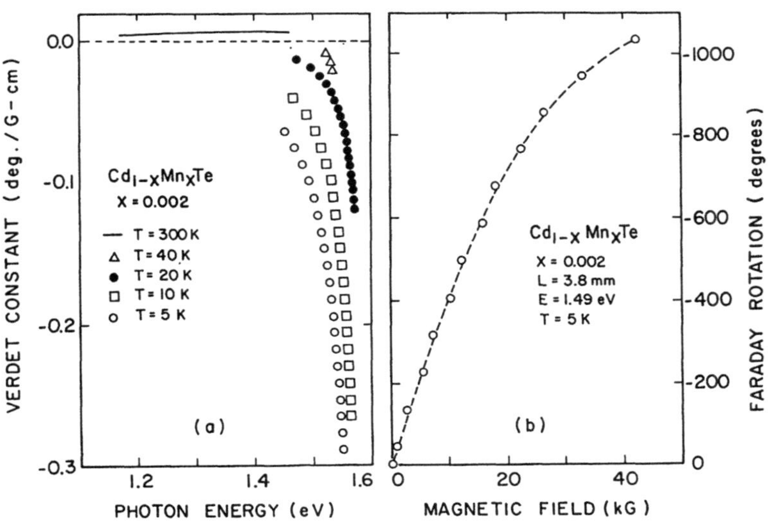

Fig. 4. (a) Verdet constant dispersion in $Cd_{1-x}Mn_xTe$ (x = 0.002). (b) Faraday rotation Θ_F as a function of magnetic field at a fixed photon energy E = 1.49 eV. The dashed line is a fit to the data of the form $\Theta_F = \Theta_{Sat}B_{5/2}(\eta)$, $\Theta_{Sat}B_{5/2} = -1227°$ and T = 4.5 K [3]

of H for $Cd_{0.99}Mn_{0.01}Se$ at T = 1.8 K and E = 1.71 eV. The increase in Θ_F with $\vec{H}$ and the larger spacing between successive maxima (minima) as $\vec{H}$ increases are features which illustrate the origin of the Faraday rotation in the large 'sp-d' exchange interaction with its saturation at high fields. The decrease in the amplitude of the oscillations with $\vec{H}$ illustrates the in-

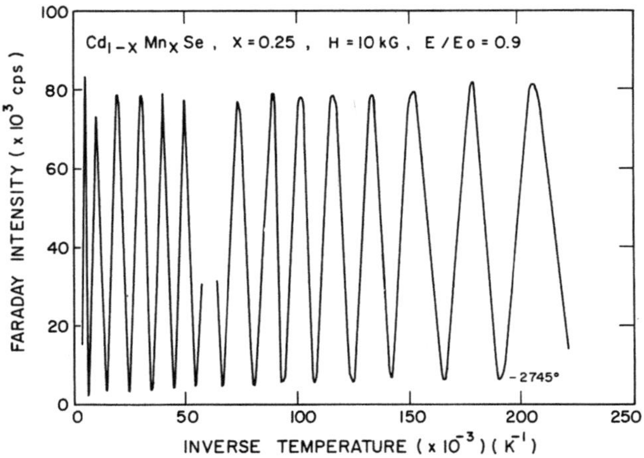

Fig. 5. Oscillations in the transmitted intensity ("Faraday" intensity) of light through $Cd_{1-x}Mn_xSe$ (x = 0.25) using the arrangement in Fig. 3 as a function of T^{-1} and with the photon energy E = 0.9 E_o, E_o = energy gap [8]

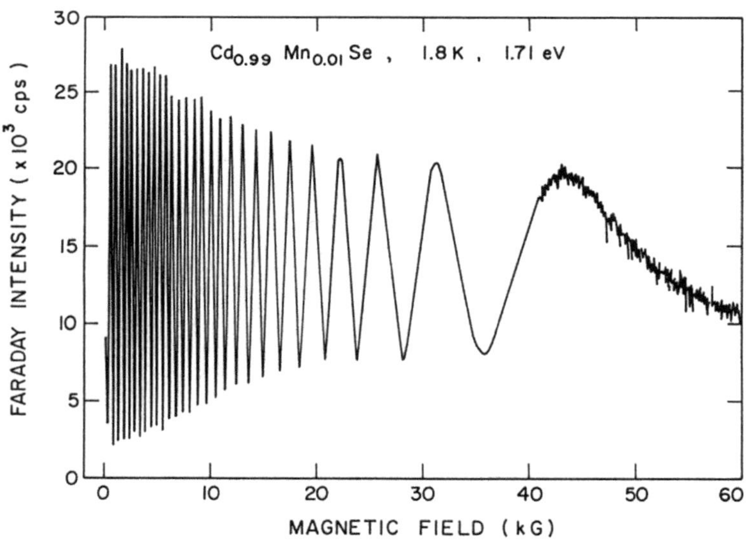

Fig. 6. "Faraday" oscillations in the intensity of light at a fixed photon energy = 1.71 eV transmitted through $Cd_{1-x}Mn_xSe$ (x = 0.01) as a function of H. Thickness = 2.9 mm and T = 1.8 K

creasing absorption which sets in as the low energy circularly polarized Zeeman component approaches the fixed photon energy of the incident light, *viz*, the circular dichroism (Faraday ellipticity).

4. Spin-flip Raman Scattering from Donor Electrons

Figure 7 shows the Raman spectrum of a gallium doped $Cd_{1-x}Mn_xTe$ (x = 0.03) in the presence of an external magnetic field. The features labeled 'SF' and 'PM' are observed in the presence of an external magnetic field. The experiment was performed with $\vec{H} \parallel \hat{z}$, the incident laser light along $\hat{z}$, scattered at right angles and analyzed along $\hat{z}$. The observation of the Stokes Raman lines in $(\hat{\sigma}_+,\hat{z})$ only and that of the anti-Stokes in $(\hat{\sigma}_-,\hat{z})$ is characteristic of magnetic excitations. In this section we discuss 'SF'.

The g-factor characterizing 'SF' is significantly larger than 2. For small x, it increases with x, and follows the $B_{5/2}(\eta)$ function in its dependence on H and T. All the characteristics can be fully accounted for in terms of the spin-flip of an electron bound to a gallium donor [10]. The donor electron in an external magnetic field experiences large 's-d' interaction or, equivalently, an 'amplified' magnetic field; αN_o of 220 meV deduced from the data for $Cd_{1-x}Mn_xTe$ compares very well with values obtained from other magneto-optic phenomena. The Raman line at ω_{SF} has been observed even at H = 0. This phenomenon has been described in the literature as the "Bound Magnetic Polaron" and has been shown to arise from the polarization of the Mn^{2+} ions by the donor electron — again through 's-d' interaction — as well as from the thermodynamic fluctuations around $\langle\vec{M}\rangle = 0$ at $\vec{H} = 0$ [10],[11].

Spin-flip Raman scattering from donors in $Cd_{1-x}Co_xSe$ and $Cd_{1-x}Fe_xSe$ has also been observed [12]. The contrast between Mn^{2+}, Co^{2+}, and Fe^{2+} with respect to the nature of their ground state is worth noting. The atomic

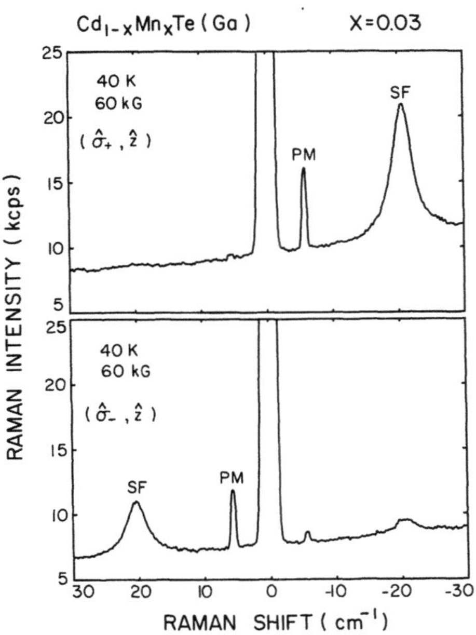

Fig. 7. Raman spectra of $Cd_{1-x}Mn_xTe$:Ga (x = 0.03) showing the $\Delta M_S = \pm 1$ transitions within the Zeeman multiplet of Mn^{2+} (PM) and the spin-flip of electrons bound to Ga donors (SF) 1 kcps ≡ 10^3 counts/sec [10]

ground state of Mn^{2+}, $^6S_{5/2}$, in a tetrahedral environment characteristic of, say, a zinc blende structure undergoes a crystal field splitting into a Γ_8 quadruplet and a Γ_7 doublet. Experimentally this crystal field splitting is very small and can be ignored; as expected, the $g_{Mn^{2+}} = 2$ is consistent with atomic $^6S_{5/2}$ like ground state. Co^{2+} has an unfilled $3d^7$ electron configuration. The atomic ground state of Co^{2+}, $^4F_{9/2}$, splits into an orbital singlet (Γ_2) and two higher lying triplets (Γ_5,Γ_4). The crystal field splitting separating the lowest Γ_2 ground state from the Γ_5 and Γ_4 states is so large that all the magnetic phenomena are controlled by Γ_2 along with its total spin of 3/2. Thus, the magnetic behavior of Co^{2+} is qualitatively analogous to that of Mn^{2+}. It is very different from that of Fe^{2+} which has complex level structure and a non-magnetic ground state; the Fe based DMS's thus exhibit van Vleck paramagnetism [13].

Raman scattering associated with the spin-flip of electrons bound to donors in $Cd_{1-x}Co_xSe$ and $Cd_{1-x}Fe_xSe$ has been observed by Bartholomew et al. and Suh et al. [12]. The magnetic field dependence of ω_{SF} in $Cd_{1-x}Co_xSe$ at various temperatures shown in Fig. 8(a) indicates a large s-d exchange interaction (αN_o = 320 meV) and is clear evidence of a bound magnetic polaron. The data for the spin-flip of donor electrons in $Cd_{1-x}Fe_xSe$ are displayed in Fig. 8(b). The magnetic field and temperature dependence of ω_{SF} in $Cd_{1-x}Fe_xSe$ are qualitatively different from that in $Cd_{1-x}Mn_xSe$ and $Cd_{1-x}Co_xSe$. The non-magnetic nature of Fe^{2+} is clearly indicated by the zero Raman shift at zero field, i.e., by the absence of the bound magnetic polaron [12]. Heiman et al. [14] have reported and discussed the lack of bound magnetic polaron in $Cd_{1-x}Fe_xSe$.

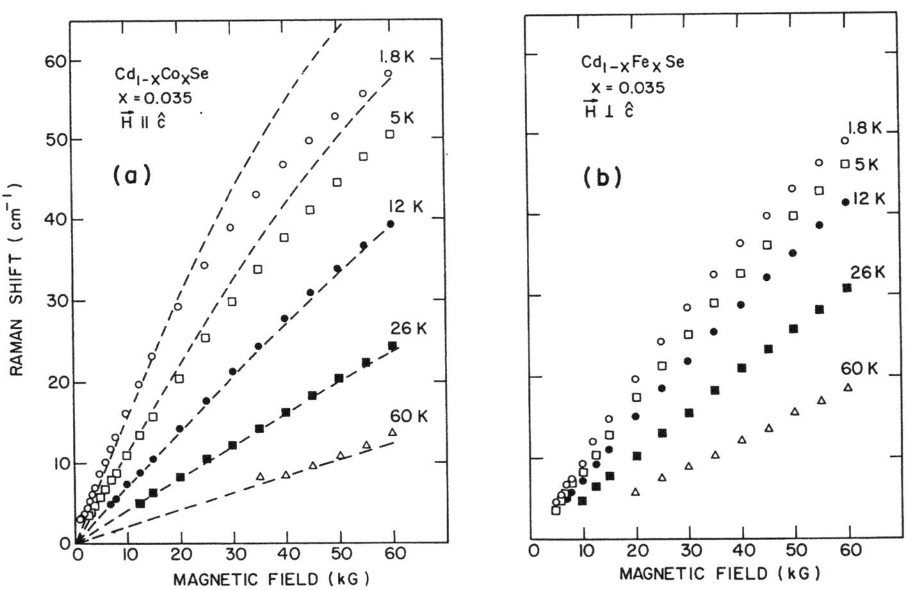

Fig. 8. Spin-flip Raman shift (ω_{SF}) from donor-bound electrons in (a) $Cd_{1-x}Co_xSe$ and (b) $Cd_{1-x}Fe_xSe$ as a function of applied magnetic field [12]

5. Raman-Electron Paramagnetic Resonance of Magnetic Ions in DMS's

In the paramagnetic phase, the application of an external magnetic field lifts the sixfold degeneracy of the $^6S_{5/2}$ ground state of Mn^{2+}. The Zeeman multiplet is then characterized by $E(M_S) = g\mu_B H M_S$, with $g = 2$, μ_B = the Bohr magneton, and M_S the projection of the spin along H. The line labeled 'PM' in Fig. 7 originates from the $\Delta M_S = +1$ (Stokes) and -1 (anti-Stokes) spin-flip transitions between the adjacent sublevels of the Zeeman multiplet yielding a Raman shift of $\hbar\omega_{PM} = g\mu_B H$ [15]. Indeed the magnetic field dependence of the 'PM' line yields g = 2 and what we have here is the Raman-electron paramagnetic resonance of the $3d^5$ electrons of Mn^{2+}. The details of the polarization effects are consistent with this interpretation. On the basis of the resonance studies, a microscopic picture involving virtual inter-band transitions between the valence-band maximum and the conduction band minimum accompanied by a spin-flip of Mn^{2+} through exchange interaction has been invoked to explain the Raman-EPR. Being a polar crystal, the Raman-EPR can also appear in conjunction with the excitation of LO phonons, i.e., additional Raman lines appear with frequency shifts $\omega_{LO} \pm \omega_{PM}$.

In Fig. 9 the resonance enhancement of the Raman line at $\omega_{LO} - \omega_{PM}$ is demonstrated in $Cd_{0.95}Mn_{0.05}Te$. The resonance condition is fulfilled when the large magnetic-field-induced exciton splitting allows one of its lower frequency components to be brought into coincidence with the Raman line. Selection rules and the scattering configuration used favor the $\omega_{LO} - \omega_{PM}$ for the resonance enhancement where LO is the CdTe-like zone center longitudinal optical phonon. The figure shows the temperature tuning of this resonance enhancement, the magnetic field being kept fixed. As the temperature is increased from 5 K to 14 K the low frequency excitonic component

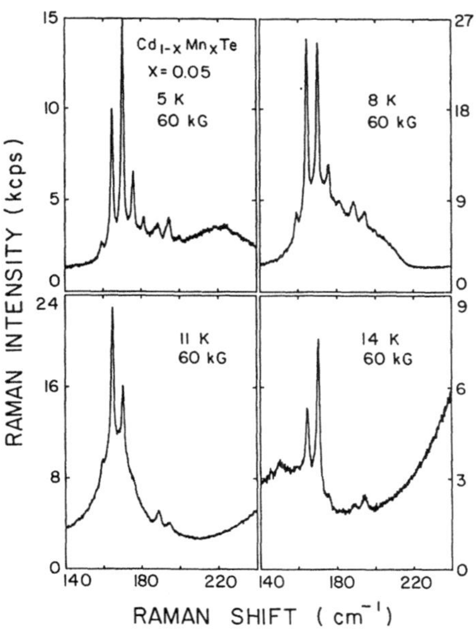

Fig. 9. Temperature variation of the resonance Raman scattering in $Cd_{1-x}Mn_xTe$ (x=0.05) in the region of the LO phonons with H = 60 kG, excited with the λ_L = 7525Å radiation of a Kr^+ laser. The most prominent line at T=5 K is the "CdTe" like zone center LO phonon while that at T = 11 K is the $\omega_{LO}-\omega_{PM}$ line [6]

moves to higher frequencies towards the zero field position as the magnetization decreases. The resonance enhancement reaches a peak for T = 11 K and the $\omega_{LO} - \omega_{PM}$ line is the most prominent line in the spectrum. This is a case of "out-resonance," i.e., resonance associated with the scattered photon energy approaching an electronic transition energy, the excitonic Zeeman component in the present case [6].

The ω_{PM} associated with $Cd_{1-x}Co_xSe$ (x = 0.035) corresponding to the transition between the Zeeman split Γ_2 ground state levels of Co^{2+} has been observed [12]. The linear magnetic field dependence of ω_{PM} of Co^{2+} yields g = 2.32. Below 20 K, this Raman peak shifts to higher frequencies signaling the onset of magnetic ordering. Similar behavior was already noted in $Cd_{1-x}Mn_xTe$. Since the ground state of Fe^{2+} is nonmagnetic in the DMS's studied, it is not surprising that no ω_{PM} was observed in $Cd_{1-x}Fe_xSe$.

Acknowledgements

The work reported in this review received support from the U.S. National Science Foundation in the following manner: Faraday Effect and Zeeman Effect of the Free Exciton (Grant No. DMR 86-16787); Spin-flip Raman Scattering in $Cd_{1-x}Co_xFe$ and $Cd_{1-x}Fe_xSe$ (Grant No. DMR 85-20866). The author thanks his collaborators Professor S. Rodriguez, Professor J.K. Furdyna, Dr. D.U. Bartholomew, Dr. Y.R. Lee, Dr. E.-K. Suh and Ms. E.-S. Oh for many stimulating discussions.

References

1. For a general survey and an overview the reader is referred to *Diluted Magnetic Semiconductors* edited by J.K. Furdyna and J. Kossut, volume 25 of *Semiconductors and Semimetals* edited by R.K. Willardson and A.C. Beer (Academic, New York, 1988).
2. Y.R. Lee, A.K. Ramdas and R.L. Aggarwal, Phys. Rev. B33, 7383 (1986).
3. D.U. Bartholomew, J.K. Furdyna and A.K. Ramdas, Phys. Rev. B34, 6943 (1986).
4. K.C. Hass and H. Ehrenreich, Journal of Crystal Growth 86, 8 (1988).
5. A.K. Bhattacharjee, G. Fishman and B. Coqblin, Physica B117 & 118, 449 (1983).
6. D.L. Peterson, D.U. Bartholomew, A.K. Ramdas and S. Rodriguez, Phys. Rev. B31, 7932 (1985).
7. J.A. Gaj, R.R. Galazka and M. Nawrocki, Solid State Commun. 25, 193 (1978).
8. E.-S. Oh, D.U. Bartholomew, A.K. Ramdas and J.K. Furdyna (unpublished).
9. Y.R. Lee, A.K. Ramdas and R.L. Aggarwal (unpublished).
10. D.L. Peterson, D.U. Bartholomew, U. Debska, A.K. Ramdas and S. Rodriguez, Phys. Rev. B32, 323 (1985).
11. See P.A. Wolff in Ref. [1] pp. 413-454.
12. D.U. Bartholomew, E.-K. Suh, S. Rodriguez, A.K. Ramdas, U. Debska and J.K. Furdyna (unpublished). See also E.-K. Suh, D.U. Bartholomew, J.K. Furdyna, U. Debska, A.K. Ramdas and S. Rodriguez, Bull. Am. Phys. Soc. 32, 802 (1987).
13. H. Serre, G. Bastard, C. Rigaux, J. Mycielski and J.K. Furdyna in the *4th International Conference on the Physics of Narrow-gap Semiconductors*, Linz, 1982, edited by E. Gornik, H. Heinrich and L. Palmetshofer (Springer, New York) p. 321.
14. D. Heiman, A. Petrou, S.H. Bloom, Y. Shapira, E.D. Isaacs and W. Giriat, Phys. Rev. Lett. 60, 1876 (1988).
15. A. Petrou, D.L. Peterson, S. Venugopalan, R.R. Galazka, A.K. Ramdas and S. Rodriguez, Phys. Rev. B27, 3471 (1983).

Magnetooptics at Γ and L Points of the Brillouin Zone and Magnetization Studies of Semimagnetic Semiconductors $Cd_{1-x}Mn_xTe$ and $Zn_{1-x}Mn_xTe$ with $0.01 < x < 0.73$

D. Coquillat[1], *J.P. Lascaray*[1], *A. Benhida*[1], *J. Deportes*[2], *A.K. Bhattacharjee*[3], *and R. Triboulet*[4]

[1]Groupe d'Etudes des Semiconducteurs (UA 357), Université des Sciences et Technique du Languedoc, Place E. Bataillon, F-34060 Montpellier Cedex, France

[2]Centre National de la Recherche Scientifique, Laboratoire Louis Néel, 166X, F-38042 Grenoble, France

[3]Laboratoire de Physique des Solides, Université Paris-Sud, F-91405 Orsay, France

[4]Laboratoire de Physique des Solides, CNRS, 1, Place A. Briand, F-92190 Meudon, France

Abstract: Magnetization and magnetoreflectivity measurements at Γ point of B.Z. were performed in magnetic field up to 5.5T i) at 4.2K for $0.02 < x < 0.73$. An anomalous behavior for high concentration is explained in terms of a second order perturbation calculation, ii) in temperature range 1.8K - 60K for $0.01 < x < 0.30$. An important variation of values of $\Delta E/|x<S_z>|$ with temperature is observed . ΔE is the Zeeman splitting and $x<S_z>$ the average spin component . At L point of B.Z. the splitting of E_1 structure compared to splitting at Γ point is very small. A variation of exchange integrals with wavevector has been found necessary to explain the data.

$Cd_{1-x}Mn_xTe$ and $Zn_{1-x}Mn_xTe$ belong to a diluted antiferromagnetic system called semimagnetic semiconductors (SMSC) with extremely interesting physical properties such as large Faraday rotations , giant Zeeman splittings of the band-carriers levels ,the occurrence of bound magnetic polarons , and "step-like" character of high field magnetization . It is the purpose of this paper to review the effect of exchange interaction between a band electron and the d electrons localized on Mn ions i) at the Γ point at 4.2K for x as high as 0.73, ii) in the temperature range $1.8K < T < 300K$ for $0.01 < x < 0.30$ and iii) at the L point of the Brillouin zone by means of magnetooptical experiments.

The exciton in II-VI SMSC with Zinc-Blende structure splits into four Zeeman components in Faraday configuration. For crystals with low Mn concentration, magnetoreflectivity spectra show a strong and a weak component in each circular polarization σ^+, σ^- (1,2). Combined magneto-reflectivity and magnetization in $Cd_{1-x}Mn_xTe$ for $x < 0.30$ at 1.5K revealed a proportionality of Zeeman splitting and magnetization. This was interpreted in terms of a simple exchange interaction model (3) and can be written as :

$$\Delta E = - x < S_z > N_o (\alpha-\beta)$$

Springer Series in Solid-State Sciences, Vol. 87
High Magnetic Fields in Semiconductor Physics II Editor: G. Landwehr

where ΔE is the energy difference between the $\sigma+$ and $\sigma-$ strong component, N_o is the number of cations per unit volume, α and β being the exchange integrals for conduction and valence bands. The average spin component $x < S_z >$ is given by :

$$x < S_z > = M_m\ m(x)/(g\ \mu_B\ N_a)$$

where N_a is Avogadro's number, m(x) is the molar mass of compound, μ_B is the bohr magneton and g = 2. According to this model ΔE as a function of $|x<S_z>|$ is a straight line of slope $N_o\ (\alpha-\beta)$. For x < 0.3, both $Cd_{1-x}Mn_xTe$ (2) and $Zn_{1-x}Mn_xTe$ (4) show only a small variation of the slope with x. For concentrated alloys, only strong components are visible due to the broadening of the structures. Recently we have reported magnetization and magnetoreflectivity in $Cd_{1-x}Mn_xTe$ and $Zn_{1-x}Mn_xTe$ with 0.02 < x < 0.73 at 4.5K and magnetic fields up to 5.5T (5,6). Using a standard magnetoreflectivity technique, ΔE remains measurable up to x = 0.73 in contrast with reference 4. ΔE was thus plotted against $|x < S_z >|$ for the given values of x. The results for $Cd_{1-x}Mn_xTe$ are shown in Fig. 1.

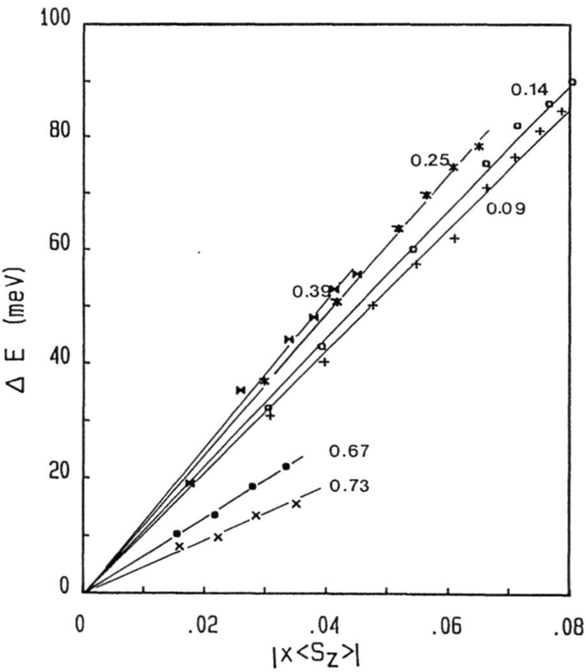

Fig.1 : Zeeman splitting versus average spin component per cation in $Cd_{1-x}Mn_xTe$ for different values of x.

It can be seen that the plots are roughly linear. In $Cd_{1-x}Mn_xTe$ the slope increases slowly from 0.09 to 0.39 and decreases drastically for x = 0.67 and 0.73. In $Zn_{1-x}Mn_xTe$, the slope remains unchanged for x up to 0.25 and then decreases by about 65% as x passes from 0.25 to 0.71.

The decrease in the concentrated alloys is too large to be attributed to a variation of the exchange parameter. A second order perturbation calculation (7) provides an explanation. The expression of splitting between strong component is given by :

$$\Delta E = [N_o\beta\ (1 - (3/2)\ r_v) - N_o\alpha\ (1 - (3/2)\ r_c)]\ \mu$$
$$- (\pi/4) x\ S(S+1)\ [\,|N_o\alpha|\ r_c^{\ 3/2} + |N_o\beta|\ r_v^{\ 3/2}]\ \sqrt{|\mu|}\ ,$$

where $\mu = x < S_z >$, $r_c = |N_o\alpha / W_c|$ and $r_v = |N_o\beta / (3 W_v)|$

W_c and W_v being the width of the conduction and valence bands respectively assumed parabolic. The correction is found to increase with the Mn concentration and the ratio of the exchange parameters to the band width.

While many investigations have been devoted to Zeeman splitting at pumped liquid Helium temperature , the available data as a function of temperature are rather scare . Here, we describe new results of magnetoreflectivity and magnetization obtained for $Cd_{1-x}Mn_xTe$ having Mn mole fraction $0.01 < x < 0.30$ for the temperature range $1.8 < T < 60K$ and B = 5.5T. The magnetic field was applied on zero field cooled sample in both experiments. For this range of composition the comparaison between magnetoreflectivity and magnetization data is possible according to the mean-field model (3). The ratio $\Delta E /|< S_z >|$ in such a case corresponds to $N_o(\alpha-\beta)$. As a result , the plot of values $\Delta E/|x< S_z >|$ versus temperature (Fig. 2) shows an important temperature dependence.

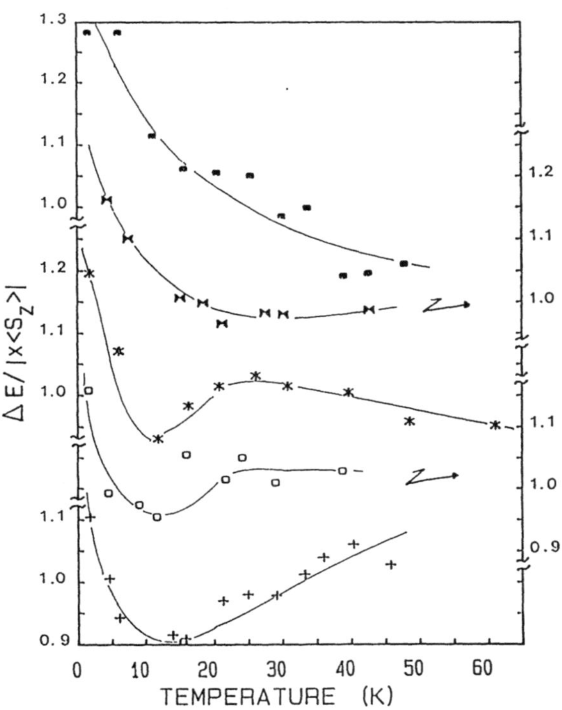

Fig.2 : $\Delta E/|x<S_z>|$ versus temperature in $Cd_{1-x}Mn_xTe$ for x=0.01 ( ), 0.09 ( ), 0.14 ( ), 0.25 ( ),0.30 ( ).For T>60K strong and weak components are mixed. Continuous lines added for clarity.

Values of $\Delta E/|x<S_z>|$ decrease strongly by about 20% as T increases from 1.8K up to about 15K for x = 0.01, 0.09, 0.14. Then the ratio increases by about 10% and seems to remain constant. Points are reported only when the weak and strong components are not mixed. For higher composition, $\Delta E/|x< S_z >|$ exhibits a rather monotonic decreasing up to 25K for x = 0.25 and up to 40K for x = 0.30, and then remains constant. The inversion of slope observed for x = 0.01, x = 0.09 and x = 0.14 is attributed to the different conditions of experiment for magnetoreflectivity and magnetization. During magnetoreflectivity measurement, the sample is illuminated with a tungsten filament lamp, in consequence the temperature of the surface of the sample is higher than the measured temperature. Finally it seems that $\Delta E/|x< S_z >|$ decreases strongly at low temperature for all studied compositions and then remains constant up to 50K.

First experimental observations of Zeeman splittings at L point of the Brillouin zone by a Magneto Circular Dichroïsm Technique (8,9) have shown that for electron states far from the center of the Brillouin zone, the exchange effects are much less spectacular than at Γ point. Quantitative data by Ginter et al (9) produce a ratio of the splitting value at L point to that at Γ point $\Delta E_L/\Delta E = 1/26$. In order to relate exchange splittings at different points of the Brillouin zone, they proposed a simple tight binding model and obtained $\Delta E_L/\Delta E = 1/4$ much higher than the experimental values. D. Coquillat et al have carried out systematic magnetooptical measurements of interband transitions in the vicinity of E_1 reflectivity structure in $Cd_{1-x}Mn_xTe$ (10) and $Zn_{1-x}Mn_xTe$ (11) in the range of alloys composition $0.01 < x < 0.39$, temperature $4.5K < T < 100K$ and magnetic field up to 5.5T. These measurements allow one to determine absolute values of circular polarization rate $P = (I_{\sigma+} - I_{\sigma-}) / (I_{\sigma+} + I_{\sigma-})$ and reflectivity R. Similarly to (9), in a certain spectral range, P(E) report quite well the form of logarithmic derivative of reflectivity dlnR(E)/dE. In such a case P can be written :

$$P = (\Delta E/2)\, d\ln R(E)/dE .$$

The magnetic field dependence of the E_1 structure splitting measured at 4.5K resembles strikingly those known for the Γ point. The proportionality of ΔE_L and ΔE for $Cd_{1-x}Mn_xTe$ and $Zn_{1-x}Mn_xTe$ is visible in Fig. 3.

The coefficient of proportionality of the two splittings remains the same. The results give a ratio $\Delta E_L/\Delta E \simeq 1/16$ for $Cd_{1-x}Mn_xTe$ and 1/20 for $Zn_{1-x}Mn_xTe$ for varying magnetic field and temperature in the whole range of studied compositions. These proportionality coefficients are much smaller than those calculated in the model of Ginter et al based on wave-vector independent exchange integrals α and β. That difference cannot be justified only by the approximation of the tight binding model. Most probably it is due to a reduction of the ion-carrier exchange constant at the L point compared to the Γ point. This idea is supported by some other results (12,13) which explain the temperature and composition variation of the energy gap in SMSC by the fluctuations of the magnetic ion system. By fitting experimental data, the authors find that the ion-carrier exchange parameter for the valence band decreases with the wavevector. A qualitative agreement of both results is quite clear.

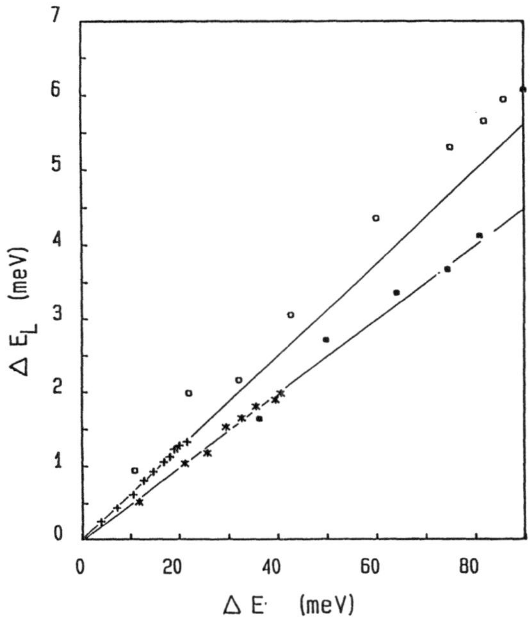

Fig.3 : Splitting ΔE_L plotted versus ΔE for x at T=4.5K, B varies from 0.5T to 5.5T for $Cd_{1-x}Mn_xTe$: x=0.01 ( ), x=0.14 ( ) and for $Zn_{1-x}Mn_xTe$: x=0.02 ( ), x=0.17 ( ).

REFERENCES :

1) J.A. Gaj, P. Byszewski, M.Z. Cieplak, G. Fishman, R.R. Galazka, J. Ginter, M. Nawrocki, N. Guyen the Khoi, R. Planel, R. Ranvaud, A. Twardowski, Inst. Phys. Conf. Ser. n° 43, chap. 29, Edimbourg, 1979.
2) J.A. Gaj, R. Planel, G. Fishamn, Solid State Comm. 29, 435, 1979.
3) J.A. Gaj, J. Ginter, R.R. Galazka ,Phys. Stat. Sol.(b) ,89 ,655,1978
4) G. Barilero, C. Rigaux, M. Menant, Nguyen Hy Hau, W. Giriat Phys. Rev. B.32,5144,1985.
5) J.P. Lascaray, M.C.D. Deruelle, D. Coquillat,Phys. Rev. B.35,675,1987.
6) J.P. Lascaray, D. Coquillat, J. Deportes, A.K. Bhattacharjee To be published in Phys. Rev. B.
7) A.K. Bhattacharjee, Solid State Com. 65,275,1988.
8) E. Dudziak, J. Brzezinski, L. Jedral, Proc. XI. Conf. Phys. SC Compounds,Jaszowiec,Poland, Pol. Acad. Sci. Publ. Warsaw, 1982, p 166.
9) J. Ginter, J.A. Gaj, Le Si Dang, Solid State. Comm. 48,849,1983.
10) D. Coquillat, J.P. Lascaray, M.C. Desjardins-Deruelle, J.A. Gaj, R. Triboulet, Solid State Comm. 59,25,1986.
11) D. Coquillat, J.P. Lascaray, J.A. Gaj, J. Deportes, J.K. Furdyna To be published.
12) J.A. Gaj, A. Golnik, Acta. Phys. Pol. A71,197,1987.
13) J.A. Gaj, A. Golnik, J.P. Lascaray, D. Coquillat, M.C. Desjardins-Deruelle, Mat. Res. Soc. Symposia Proceedings, Vol. 89,59,1987.

Magnetic Polarons and Other Spin Effects in II-VI Semimagnetic Semiconductors and Their Superlattices

S.A. Jackson

AT&T Bell Laboratories, Murray Hill, NJ 07974, USA

We review experimental and theoretical aspects of bound magnetic polarons (BMP) in II-VI semimagnetic semiconductors, using the $CdTe/Cd_{1-x}Mn_xTe$ system as a prototype. We show how the semiclassical approach of Heiman, Wolff and Warnock [2] can be generalized for magnetic polarons in a quantum well (QW) system. In particular, we calculate the variation of the exciton binding energy in a (111) QW as a function of magnetic field applied perpendicular to the epilayers. Separately, we show how spin fluctuations in this system lead to an enhanced exciton binding via a spin flip induced spin wave exchange between electron and hole, and give the size of the extra binding for x = .25 at T = 2° K.

1. Introduction

There has been much recent work on the phenomenon and properties of magnetic polarons in semimagnetic semiconductors [1]. These materials are mixed semiconducting crystals (usually II-VI in many cases studied to date), whose lattice is made up, in part, of substitutional magnetic ions. Among the systems studied are those listed below.

$Hg_{1-x}Mn_xTe$	$Zn_{1-x}Mn_xTe$
$Hg_{1-x}Mn_xSe$	$Hg_{1-x}Fe_xSe$
$Cd_{1-x}Mn_xTe$	$Cd_{1-x}Fe_xTe$
$Cd_{1-x}Mn_xSe$	$Pb_{1-x}Mn_xTe$
$Cd_{1-x}Mn_xS$	$Pb_{1-x}Mn_xSe$
$Zn_{1-x}MnS$	Stannites

The semiconductors $Cd_{1-x}Mn_xTe$ and $Cd_{1-x}Mn_xSe$ have been particularly well studied. The discussion in this paper will focus on $Cd_{1-x}Mn_xTe$ as a prototypical wide-gap semi-magnetic semiconductor. This semiconductor crystallizes in the zincblende structure for Mn concentrations less than 70% and is a direct bandgap system. The lattice constant, bandgap and effective mass are tunable functions of Mn concentration, x. The bandgap varies approximately as $\Delta E = 1.5\ \mathrm{eV} \cdot x$. The spin orbit (S-O) splitting in the II-VI semiconductors is generally large (~ 1 eV) making them potentially good candidates for non-linear optics since the S-O splitting and bandgap can be comparable.

What makes the semi-magnetic semiconductors interesting is the presence of the magnetic ions, Mn^{2+}. At high concentrations of Mn, the exchange interaction between the local moments is important and leads to $Cd_{1-x}Mn_xTe$ exhibiting paramagnetic, possibly spin glass, and antiferromagnetic phases as a function of x and temperature. The Hamiltonian describing the local moment interaction is

$$H = -\sum_{i \neq j} J_{ij}\, S_i \cdot S_j \quad . \tag{1}$$

There is a second, important, exchange interaction in this system, which is a Kondo-like interaction between a band electron or hole and the local moments of the Mn ions,

Springer Series in Solid-State Sciences, Vol. 87
High Magnetic Fields in Semiconductor Physics II Editor: G. Landwehr

$$H = -\sum_j J(\mathbf{r} - \mathbf{R}_j)\mathbf{s} \cdot \mathbf{S}_j \tag{2}$$

where s and S_j are spin operators of a band electron or hole and a magnetic ion, respectively, and J is the exchange constant. This interaction, at low Mn concentrations, gives rise to magnetic polaron effects, leading to the interesting magneto-optical effects in these materials: giant Faraday rotations, giant Zeeman splittings and zero-field Stokes shifts in spin-flip Raman scattering. Our problem then is, given the existence of bulk bound magnetic polarons (BMP) and the solution of the BMP problem, are there free magnetic polaron (FMP) effects: i) in a quantum well (QW) system and ii) in the bulk due to spin fluctuations. We are concerned with the extension to a QW system of the semi-classical BMP approach of Heiman, Wolff and Warnock, [2] and with how to go beyond the semi-classical treatment to handle spin fluctuations.

In this paper we will focus on two aspects of the magnetic couplings in $Cd_{1-x}Mn_xTe$. First, after a review of bound magnetic polarons in bulk semiconductors, we discuss a calculation of magnetic polaron effects in a quantum well system. Secondly, at higher concentrations of Mn we discuss how spin fluctuations in the vicinity of a transition from the paramagnetic to an ordered phase get reflected in the luminescence spectra of $Cd_{1-x}Mn_xTe$ as an enhanced exciton binding energy, which we relate to the magnetic susceptibility of the Mn ion subsystem and calculate within a spin wave approximation.

2. Magnetic Polarons in a Quantum Well System

Evidence for bound magnetic polarons appears in both the luminescence and spin flip Raman scattering spectra of $Cd_{1-x}Mn_xTe$. Luminescence data of Golnik et al [3]. showed a very rapid variation with temperature of the acceptor bound exciton ($\mathring{A}X$) binding energy, which becomes greater as x increases and is unlike the usual temperature dependence of $\mathring{A}X$ in pure CdTe. The spin flip Raman frequency is finite for B=0 (and independent of T) [1]. Flipping the spin thus takes energy, even for B=0, implying a net spin polarization.

These phenomena now are known to result from the sp-d exchange interaction [4] between band electrons/holes and Mn local moments given in equation (2). A mean field approach consists of looking for band electron or hole states in an external magnetic field with a given spin configuration of magnetic ions. In finite magnetic field, H, the electron energy levels become (for spin up and down)

$$E_{\ell\pm} = (\ell + 1/2)\hbar\omega_c \pm 1/2\,(g^* \mu_B H + N_s \alpha <S_z>) \tag{3}$$

where $\alpha = \dfrac{<S|J|S>}{\Omega_0}$ is the exchange constant and g^* the electron spin g-factor;

$$<S_z> = S_0\, B_{5/2}\left[\frac{5\mu_B H}{k(T+T_0)}\right] \tag{4}$$

where $S_0 = 5/2$ and $B_{5/2}$ is a Brillouin function. Equation (3) may be written

$$E_{\ell\pm} = (\ell + 1/2)\hbar\omega_c \pm 1/2\, g_{eff}\, \mu_B H \tag{5}$$

where

$$g_{eff} = g^* + \frac{N_s\, \alpha <S_Z>}{\mu_B H} \; . \tag{6}$$

In the presence of the sp-d exchange interaction, the electron behaves as if it has an effective g-factor which is extremely large (e.g. $g_{eff} \approx 100$ for $H \sim 10$ kG).

Heiman, Wolff and Warnock [2] calculated the spin-flip Raman scattering spectrum of (Cd, Mn)Se. Their treatment is based on the effective mass Hamiltonian

$$H = \frac{p^2}{2m^*} - \frac{e^2}{\epsilon_0 r} - \alpha \sum_j (\mathbf{s}_e \cdot \mathbf{S}_j)\, \delta(\mathbf{r} - \mathbf{R}_j) \quad . \tag{7}$$

Assuming no mixing of orbital and spin degrees of freedom and using a trial wave function, they performed an ensemble average over low lying spin states (treating the Mn spin as classical variables) to calculate the partition function and free energy. This was then minimized with respect to the range parameter in the trial wavefunction to determine the enhanced binding energy of the electron in the presence of the sp-d exchange interaction and to calculate the spin-flip Raman spectrum. The partition function in this treatment is

$$Z = \underset{\{s, S_j\}}{\mathrm{Tr}} \left\{ e^{-\beta E_0}\, e^{\beta\alpha \sum_j |(s \cdot S_j)\, |\phi(R_j)|^2|} \right\} \tag{8}$$

and

$$G = -kT \ln Z \quad . \tag{9}$$

Recent photoluminescence studies by Nurmikko et al. [5] in CdTe/Cd(Mn)Te strained layer superlattices (SLS) have revealed the existence of two-dimensional (2D) excitons resulting from the binding at the interface of heavy holes, due to a combination of alloy fluctuation [6,7] and magnetic polaron effects [8,9]. There is an anisotropy in their luminescence spectra in a finite magnetic field, depending upon whether the applied magnetic field is parallel ($B_{\|}$) or perpendicular ($B_{\perp}$) to the epilayers (See Fig. (1)). There is a clear red shift of the luminescence peak with increasing magnetic field, for $B_{\perp}$. The SLS studied by Nurmikko were grown on a Cd(Mn)Te buffer layer, lattice matched to the Cd(Mn)Te epilayers, in which case the Cd(Mn)Te layers are unstrained while the CdTe layers are under in-plane biaxial compression with uniaxial tension in the z-direction.

The lattice mismatch strain lifts the valence band degeneracy and splits the $J = 3/2$, $M_J = \pm 3/2$ state and the $J = 3/2$, $M_J = \pm 1/2$ state near $\mathbf{k} = 0$. The sp-d coupling gives a further spin splitting separately of the $M_J = \pm 3/2$ and the $M_J = \pm 1/2$ states. For the holes, we

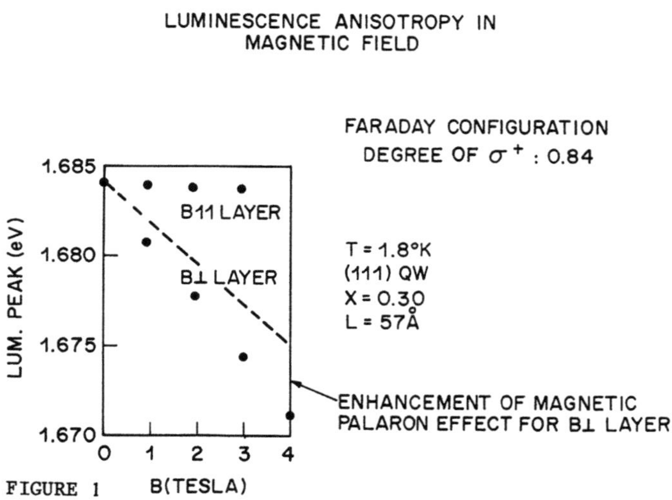

FIGURE 1

have a different kinetic energy operator in the Hamiltonian for different spin projectors because of strain induced mass anisotropy. Without Coulomb and spin effects, the Hamiltonians H_h and H_e for the hole and electron may be written

$$H_h = (K.E.)_{J_z} \pm \frac{d}{\sqrt{3}} (J_{\parallel}^2 - \frac{1}{3} J^2) \epsilon_{xy} \tag{10}$$

(in the strained region for (111) QWs) and

$$H_e = K.E. + V_0 ; \tag{11}$$

V_0 is the band offset (= bandgap difference between CdTe and $Cd_{1-x}Mn_xTe$) for the conduction band. The strain splitting then provides the effective z-potential for the holes. The spin splittings are such that the luminescence is mainly due to the conduction to valence band transition: $|\frac{1}{2}, -\frac{1}{2}>_{C.B.} \rightarrow |\frac{3}{2}, -\frac{3}{2}>_{V.B.}$ giving rise to σ^+ recombination with electric vector $\frac{1}{\sqrt{2}}(x + iy) e^{-i\omega t}$ (Faraday geometry). In this case, for magnetic field perpendicular to the epilayers, we expect strong coupling to the σ^+ recombination, while for magnetic field parallel to the epilayers, we expect the coupling to be not very strong.

The combined effect of strain and sp-d exchange coupling means that we have different effective mass Hamiltonians in the CdTe and CdMnTe layers. The Hamiltonians are as given below, H =

$(k = \nabla)$

Strain : CdTe	Spin Exchange: CdMnTe	
Holes		
$(M_J = \mp 3/2)$		
$-\frac{\hbar^2}{2m_z^{(1)}} \nabla_z^2 - \frac{\hbar^2}{2m_\perp^{(1)}} \nabla_\perp^2 + \frac{d\epsilon_{xy}}{\sqrt{3}}$	$-\frac{\hbar^2}{2m_{hh}} \nabla_z^2 - \frac{\hbar^2}{2m_\perp^{(1)*}} \nabla_\perp^2 - \sum_j J_h s_h \cdot S_j \delta(\mathbf{r}_h - \mathbf{R}_j)$	(12)
$(M_J = \mp 1/2)$		
$-\frac{\hbar^2}{2m_z^{(2)}} \nabla_z^2 - \frac{\hbar^2}{2m_\perp^{(2)}} \nabla_\perp^2 - \frac{d\epsilon_{xy}}{\sqrt{3}}$	$-\frac{\hbar^2}{2m_{\ell h}} \nabla_z^2 - \frac{\hbar^2}{2m_\perp^{(2)*}} \nabla_\perp^2 - \sum_j J_h s_h \cdot S_j \delta(\mathbf{r}_h - \mathbf{R}_j)$	(13)
Electrons		
$(m_s = \pm 1/2)$		
$\frac{\hbar^2}{2m^*} (\nabla_z^2 + \nabla_\perp^2) + E_G$	$\frac{\hbar^2}{2m^*} (\nabla_z^2 + \nabla_\perp^2) + V_0(z) - \sum_j J_e s_e \cdot S_j \delta(\mathbf{r}_e - \mathbf{R}_j)$	(14)
	Excitons	
$-\frac{\hbar^2}{2\mu_\parallel^{(1)}} \nabla_\parallel^2 - \frac{\hbar^2}{2\mu_\perp^{(1)}} \nabla_\perp^2 - \frac{e^2}{\epsilon r} + E_G - \frac{d\epsilon_{xy}}{\sqrt{3}}$	$-\frac{\hbar^2}{2\mu_\parallel^{(2)}} \nabla_\parallel^2 - \frac{\hbar^2}{2\mu_\perp^{(2)}} \nabla_\perp^2 - \frac{e^2}{\epsilon r} + E_G + V_0(z) - J_e \sum_j s_e \cdot S_j \delta(\mathbf{r}_e - \mathbf{R}_j) + J_h \sum_j s_h \cdot S_j \delta(\mathbf{r}_h - \mathbf{R}_j)$.	(15)

We now calculate the partition function in order to obtain the free energy. The luminescence line is an $\ell = 1$, $\Delta m_\ell = -1$ transition. Therefore we use the Wigner-Eckhart theorem to get the appropriate weighting of the $|1/2, -1/2>_{C.B.}$ and $|3/2, -3/2>_{V.B.}$ states in the $\ell = 1$, $\Delta m_\ell = -1$ matrix element, (taking proper account of the fact that the hole is a time-reversed electron state, i.e.)

$$Z = \underset{\{S_j\}}{\mathrm{Tr}} \; <1,-1| e^{-\beta<H>_\psi} |1,-1> , \tag{16}$$

where $<H>_\psi = \int d^3r_e \, d^3r_h \, \psi^*(r_e, r_h) \, H \, \psi(r_e, r_h)$ is the Hamiltonian weighted by the exciton envelope function and the spin ensemble average is over Mn spins.

The QW separates the electron and hole motion in z. We assume that the motion perpendicular to the superlattice growth direction (in the plane of the layers) is still correlated (due to Coulomb and spin effects). Thus, we pick a trial wave function of the form

$$\psi(r_e, r_h) = \phi_e(z_e)\phi_h(z_h)\psi(R_\perp)\chi(r_\perp) \quad . \tag{17}$$

The form of the trial z- and ⊥ wavefunctions are

$$\phi(z) = \frac{\nu^{1/2}}{\pi^{1/4}} e^{-\frac{\nu^2}{2}(z-z_0)^2} \quad \text{(for e and h)} , \tag{18}$$

$$\chi(r_\perp) = \frac{\eta}{\sqrt{\pi}} e^{-\eta^2 r_\perp^2} \quad \text{(relative notion)} , \tag{19}$$

$$\psi(R_\perp) = \frac{\alpha}{\sqrt{\pi}} e^{-\alpha^2 R_\perp^2} \quad \text{(C.M. motion)} , \tag{20}$$

where ν, z_0, η are variational parameters and α is fixed by estimates of exciton localization distances in the plane of the interface ($\alpha^{-1} \approx 20$ Å). Previous studies have shown that ν_e and z_0^e for the electron wavefunction are determined by the conduction band edge potential which for II-VI systems of this type is essentially the band gap difference. We evaluate the spin trace in eqn. (16) to calculate Z and $G = -kT \, \ell n \, Z$, in the presence of a magnetic field prependicular to the layers ($B_\perp$). In varying ν_h and η, we use two values of z_o^h: $z_o^h = 0$ and $z_o^h = L/2$ (L = QW width), i.e. we place the hole at the QW center and the edge to see which yields the greatest shift in the excitron binding energy. We find $z_o^h = L/2$ gives the lowest energy. For $B_\perp = 0$, L = 50Å, x = 0.25 and T = 2°K, ΔE = 35 meV for a (111) QW and ΔE = 9 meV for an (001) QW. These numbers reflect the combined effect of Coulomb interaction and magnetic polaron (sp-d) coupling. The variation of the luminescence peak energy with magnetic field ($B_\perp$) for the Nurmikko et al. [5] data for a 57Å (111) QW with x = 0.30 is indicated by the solid line in Fig. 1; our results are given by the dashed line. The results indicate two things: first, Coulomb effects are greater than magnetic polaron effects in these QW systems for $B_\perp = 0$. The greater shift of the exciton luminescence peak for a (111) QW than for an (001) QW reflects this. For (111) QWs the strain generated valence band offsets are greater, leading to greater confinement of holes in the non-magnetic region of the QW system, even for holes in the vicinity of the interface. This enhances the exciton binding due to the Coulomb term in the Hamiltonian. However, the very rapid variation of the luminescence peak energy with magnetic field is a reflection of magnetic polaron coupling. Our results give for the electron and hole an effective g-factor, $g_{eff} = 20g$, compared with $g_{eff} = 27g$ for the experimental data shown (note the difference in our x and L values vs. the experiment). We thus expect to observe similarly strong magneto-optical effects in QW systems as in the bulk, but with particular signatures of the two-dimensional nature of he excitons.

3. Spin Fluctuations and Spin Waves in (Cd, Mn)Te

Unaccounted for in the previous discussion were spin fluctuation effects in luminescence due to dissipative terms in the Hamiltonian (i.e. off-diagonal terms: $s^+S_j^- + s^-S_j^+$). Photoluminescence experiments of Golnik et al. [13] as well as more recent neutron scattering results of Giebultowicz et al. [10] show that, at higher Mn concentrations ($x \gtrsim 0.20$), magnetic ordering of the Mn subsystem occurs, and the spin fluctuations and correlations associated with it play a large role in determining the luminescence spectra of (Cd, Mn)Te. Earlier magnetization and susceptibility measurements [11] of (Cd, Mn)Te indicated the presence of a possible spin glass transition at low temperature which became more pronounced as x increased, with strong antiferromagnetic effects. Indeed, more recent frequency dependent susceptibility measurements by Geschwind et al. [12] suggest a low temperature transition to a dynamically inhibited antiferromagnetic state. The neutron scattering work of Giebultowicz et al. [10] clearly shows that the system orders antiferromagnetically (AF) at low temperature in a type III AF structure, with a finite, anisotropic correlation length.

The variation of the exciton peak in the data of Golnik et al. [3] shows a rapid temperature dependence with a peak developing at a low temperature. The temperature of the peak is essentially that identified from susceptibility measurements as the spin glass or AF transition temperature in (Cd, Mn)Te. This behavior was confirmed in a later, more complete, set of luminescence experiments in (Cd,Mn)Te [13]. In these experiments, as a magnetic field was applied and increased, the luminescence feature exhibiting the strong temperature dependence merged into the free exciton live and disappeared. This suggests the behavior to be the result of spin fluctuation effects which diminish in a magnetic field. In fact, when the difference in peak positions, ΔE_o, between the free exciton and the temperature dependent feature is plotted, it can be fit to a similar form as the susceptibility, i.e.

$$\Delta E_o = \frac{\tilde{E}\Lambda(T)}{T + |\theta|\,\Lambda(T)} , \tag{21}$$

where $\tilde{E}$ is a constant; $\Lambda(T) = 1 - q(T)$ and $q(T)$ is the Edwards-Anderson spin glass order parameter, and $\theta =$ Curie temperature $= -45^\circ K$, indicating strong AF effects. Ching and Huber [14] performed a numerical simulation of the ground state and dynamical structure factor, $S(q,\omega)$, of a dilute Heisenberg FCC system with AF nearest neighbor and next nearest neighbor interactions. They found that the ground state is like that of a type III AF with short range order. The inelastic excitation spectrum of Giebultowicz [10] can be well-fit by a cubic average of the magnon dispersion relation of a type III AF with a dilution factor (for x) put in (with damping). This corroborates the idea that there is a strong AF content to the low temperature phase of (Cd, Mn)Te.

We now show how these results can be brought together into a consistent picture which explains the luminescence results. As the transition temperature is approached, short range correlations between spins develop in the Mn subsystem. These are measured by $S(q,\omega)$, which, by the Fluctuation-Dissipation Theorem, is related to the dynamical susceptibility i.e.

$$\frac{2}{\omega}\,\chi''(q,\omega) \approx S(q,\omega) \ . \tag{22}$$

Fluctuations and correlations in (Cd, Mn)Te require a consideration of the off-diagonal terms in the sp-d coupling in the Hamiltonian. When electrons and holes are optically generated, they perturb (excite) the Mn subsystem through the sp-d coupling. If the excitation is small, and *it is* for $x > 0.20$, compared with the Mn-Mn spin coupling, then the response of the system is, to a good approximation, given by the linear susceptibility. We might expect nonlinear effects to be important for higher power absorption than is generally the case in photoluminescence experiments.

Beginning with spin terms in H

$$\begin{aligned} H_{Spin} = & -\sum_j J_e \delta(\mathbf{r}_e - \mathbf{R}_j)\mathbf{s}_e \cdot \mathbf{S}_j \\ & - \sum_j J_h\, \delta(\mathbf{r}_h - \mathbf{R}_j)\mathbf{s}_h \cdot \mathbf{S}_j \\ & - \sum_{i \neq j} J_{ij}\, \mathbf{S}_i \cdot \mathbf{S}_j \ , \end{aligned} \tag{23}$$

the first two terms act as perturbing potentials. Linear response theory lends to an interaction between an electron and a hole due to spin flip,

$$V_{eff} = -x^2(\alpha N_o)(JN_o)\int d^3q\, e^{iq\cdot(r_e - r_h)} \cdot \left[s_e^+ \chi_\pm(q) s_h^- + s_e^- \chi_\mp(q) s_h^+\right], \tag{24}$$

where $\alpha(\alpha N_o = .22$ eV) is the electron exchange constant, and $J(JN_o = .88$ meV) is the hole exchange coupling constant. We have an effective sp-d exchange induced spin flip pair interaction between an electron and a hole due to each being coupled to the Mn spin fluctuations. $\chi(q)$ can be related to S(q) through the fluctuation dissipation theorem and we can either calculate S(q) or extract it from neutron scattering. For purposes of illustration, and to get an estimate of the enhanced exciton binding energy, we make a spin wave approximation to $\chi(q)$, following the method of Holstein and Primakoff, and use the concentration and cubic averaged dispersion relation for the type III AF

$$\omega(q) = 4\,x\,JA\cdot B\;, \tag{25}$$

where

$$A = \left[\cos\left(\frac{q_x a}{2}\right)\cos\left(\frac{q_y a}{2}\right) + \cos\left(\frac{q_x a}{2}\right)\cos\left(\frac{q_z a}{2}\right) + \cos\left(\frac{q_y a}{2}\right)\cos\left(\frac{q_z a}{2}\right)\right]^{1/2}, \tag{25a}$$

$$B = \left\{1 - \frac{1}{2}\left[\cos\left(\frac{q_x a}{2}\right)\cos\left(\frac{q_y a}{2}\right) + \cos\left(\frac{q_x a}{2}\right)\cos\left(\frac{q_z a}{2}\right) + \cos\left(\frac{q_y a}{2}\right)\cos\left(\frac{q_z a}{2}\right)\right]\right\}^{1/2}, \tag{25b}$$

where a = lattice const. Using a trial wavefunction for the exciton envelope function of the form

$$\phi(r) = (\pi a^3)^{-1/2}\, e^{-r/a}\;, \tag{26}$$

where a is a variational parameter, we find for the enhanced excitonic binding energy from V_{eff}: for x = 0.4, ΔE_0 = 42 meV. Golnik et al. [13] find, just above the peak in luminescence energy, ΔE = 60 meV. These results indicate that AF ordering and spin fluctuations play a critical role in exciton dynamics in semi-magnetic semiconductor systems. Light scattering is a probe, with great sensitivity, of magnetic phase transitions in these systems, which together with susceptibility and neutron scattering measurements can further enhance our understanding of magnetic effects in these systems, particularly in QWs and strained-layer superlattices.

References

1. P. A. Wolff: In *Diluted Magnetic Semiconductors*, ed. by J. K. Furdyna and J. Kossut, Semiconductors and Semimetals, Vol. 25 (Academic Press, San Diego, 1988) p. 413.
2. D. Heiman, P. Wolff and J. Warnock, Phys. Rev. B*27* 4848 (1983).
3. A. Golnik, J. Gaj, M. Nowrocki, R. Planel and C. Benoit à la Guillaume, J. Phys. Soc. Japan, Suppl. A*49*, 819 (1980).
4. J. A. Gaj: In *Diluted Magnetic Semiconductors*, ed. by J. K. Furdyna and J. Kossut, Semiconductors and Semimetals, Vol. 25 (Academic Press, Dan Diego, 1988) p. 275.
5. A. V. Nurmikko, L. A. Kolodziejski and R. L. Gunshor: In *Dilute Magnetic Semiconductors*, Vol. 89, Proc. of Mat. Res. Soc. (Pittsburgh, 1987).

6. S.-K. Chang, A. V. Nurmikko, L. A. Kolodziejski and R. L. Gunshor, Phys. Rev. B*33*, 2589 (1986) (and references cited therein).

7. S. A. Jackson, Journal de Physique, Coll. C*5*, 349 (1987).

8. J. W. Wu, A. V. Nurmikko and J. J. Quinn, Phys. Rev. B*34*, 1080 (1986).

9. S. A. Jackson and C. R. McIntyre: In *Diluted Magnetic Magnetic Semiconductors*, Vol. 89, Proc. of Mat. Res. Soc. (Pittsburgh, 1987).

10. T. M. Giebultowicz and T. M. Holden: *In Diluted Magnetic Semiconductors*, ed. by J. K. Furdyna and J. Kossut, Semiconductors and Semimetals, Vol. 25 (Academic Press, San Diego, 1988) p. 125.

11. S. Nagata, R. Galazka, D. Mullin, H. Akbarzadeh, G. Khattak, J. Furdyna and P. Keesom, Phys. Rev. B*22*, 3331 (1980).

12. S. Geschwind, A. Ogielski, G. E. Devlin, J. Hegarty and P. Bridenbaugh, J. Appl. Phys. to be published.

13. A. Golnik, J. Ginter, J. A. Gaj, J. Phys. C*16*, 6073 (1983).

14. W. Y. Ching and D. L. Huber, Phys. Rev. B*30*, 179 (1984).

High-Field Investigations on Semimagnetic Semiconductors

M. von Ortenberg

The Institute for Solid State Physics, The University of Tokyo,
Roppongi, Minato-ku, Tokyo 106, Japan
Permanent address: Physikalisches Institut der Universität
Würzburg, D-8700 Würzburg, Fed. Rep. of Germany

We present DC-magneto-transport and CR-measurements on semimagnetic semiconductors in the conventional, submegagauss, and megagauss-range. Especially the Fe-doped Hg-compounds exhibit a particulary high mobility so that the internal quantum structure can be probed directly.

1. INTRODUCTION

Semimagnetic Semiconductors (Diluted Magnetic Semiconductors) represent a relatively new class of mixed crystals combining classical semiconductor properties with those of magnetic materials. This hybridization leads to a tunability of the energy-band structure of the charge carriers, which can now be controlled via the external parameters of temperature and magnetic field. The extension of this concept of initially ternary compounds to quaternary or even higher complexity gives one the possibility to "tailor" any kind of band structure necessary for technical application. By means of Fe-doping the mobility of these compounds can be increased to the order of 10^6 $cm^6/Vsec$. The following discussion will present mostly new results, which have been obtained in the joint effort of an international cooperation and will be published in more detail separately.

2. Fe-DOPING IN SEMIMAGNETIC SEMICONDUCTORS

2.1 HgSe:Fe

Using Fe as paramagnetic dopant in the host material HgSe a series of interesting phenomena can be observed due to the energy degeneracy of the localized Fe^{++}-state with the conduction-band states as visualized in Fig. 1. For $n_{Fe++} > 5*10^{18}$ cm^{-3} the Fermi level gets pinned to the Fe^{++}-state and an isoenergetic exchange between localized and delocalized electrons becomes possible. In the experiment this fact manifests itself in a drastic improvement of the mobility and

Springer Series in Solid-State Sciences, Vol. 87
High Magnetic Fields in Semiconductor Physics II Editor: G. Landwehr

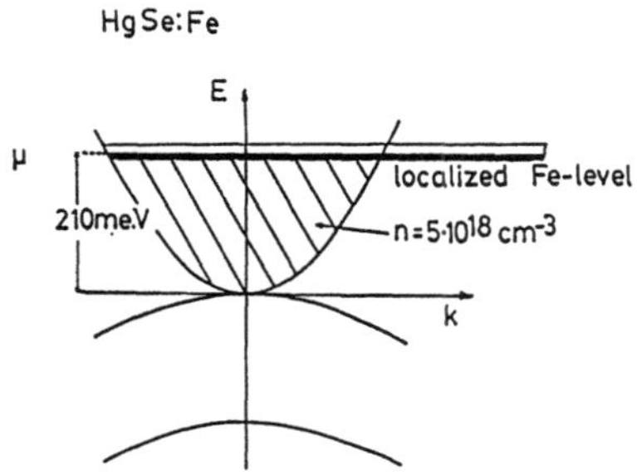

Fig. 1 In HgSe:Fe the level of the localized Fe^{++}(3d6)-state is degenerate with the conduction band and lies about 210 meV above the conduction band edge. For Fe^{++}-concentrations higher than $5*10^{18}$ cm^{-3} the Fermi energy is pinned to the Fe^{++}-level.

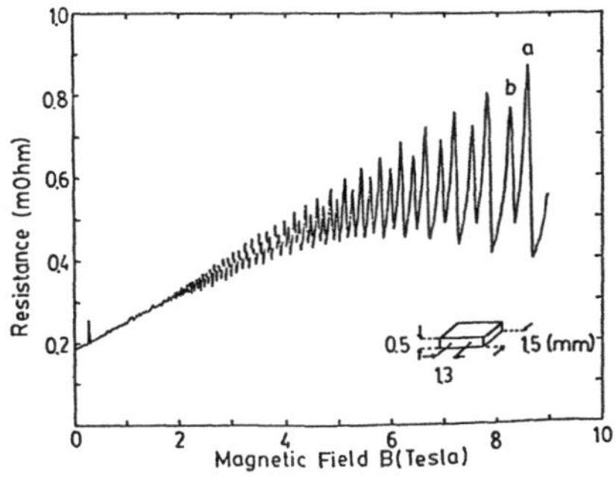

Fig. 2 The Shubnikov-de Haas oscillations in the transverse magneto resistance reflect directly the spin-splitted density of states and indicate that the quantized system has only small damping.

Dingle temperature. The formation of a space-charge superlattice by Fe^{+++}-ions as proposed by MYCIELSLKI [1] is only possible for very high Fe-concentration and seems to be very unlikely for the present concentrations of some 10^{19} cm^{-3}. The excellent mobility of the $5*10^{18}$ cm^{-3} Fe-doped, Fermi-level pinned HgSe sample is visualized in the Shubnikov-de Haas effect in the transverse magneto resistance at a temperature of T = 0.4 K in Fig. 2 [2]. The data represent directly the one-dimensional density of states of the motion parallel to the magnetic field atached to each Landau level multiplied with the scattering probability [3]. In contrast to Mn as paramagnetic dopant Fe does not produce a pronounced enhancement of the spin splitting. It should be noticed that a- and b-set show different scattering probability due to the superposition of the change in the core potential and the spin-dependent exchange interaction of the paramagnetic scatterers. Spin-orbit coupling as discussed by NARITA and SUIZU [4] produces a weighting of a- and b-set opposite to the experimental result.

Due to the pinning of the Fermi-level a fluctuation of electrons between localized and delocalized states is induced by any change in the free-density of states, hence by an external magnetic field. This effect can be detected directly in the Hall-resistance of HgSe:Fe as shown in Fig. 3 and leads to the "Three Dimensional Analogue of the Quantum Hall Effect" [5]. There is an excellent agreement of data, in which the linear part of the Hall effect has been suppressed, and simulation. Please note the equal weighting of a- and b-set, because the Hall effect is not affected by the scattering

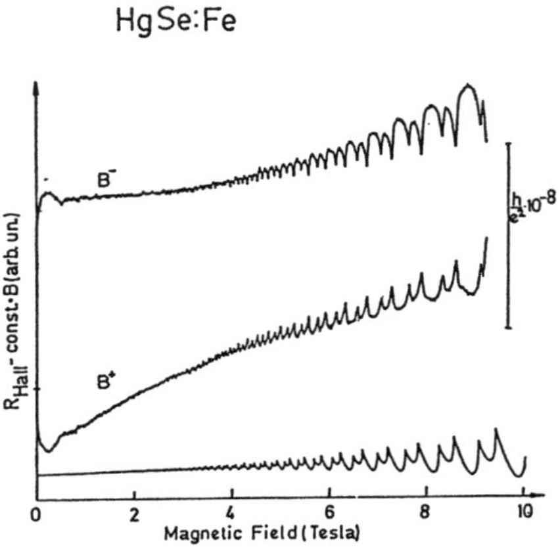

Fig. 3 Experimental data of the "Three-Dimensional Analogue of the Quantum Hall Effect" in comparison with the simulation. A term linear in B has been substracted.

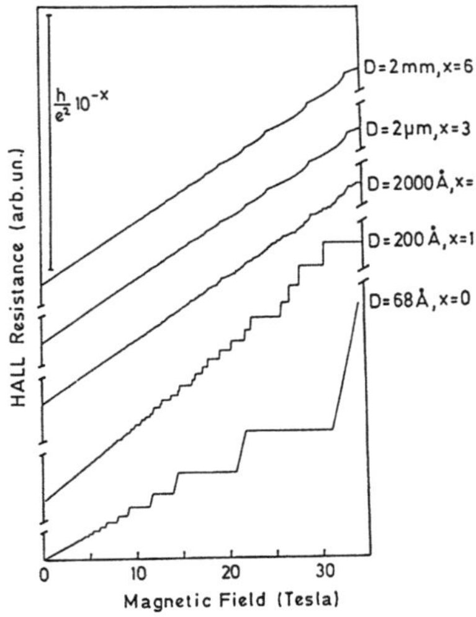

Fig. 4 The simulation of the HALL-resistance shows the manifestation of the 2D-quantum structure with decreasing sample thickness D.

probability. By reducing successively the sample thickness and thus going smoothly from 3D to 2D the whole transition range of the Quantum Hall Effect should be accessible to experimental investigation as shown by the simulation in Fig. 4 [5]. In the 2D-case the slope between the different plateaus will provide detailed information about the width of the Fe^{++}-level.

The internal dynamics between the localized and delocalized systems leads also to a "Spin Fluctuation" if the free-density of states is modulated. In Fig. 5 we have plotted the second derivative of the magnetization with respect to the magnetic field as recorded in the experiment by a 2f-modulation technique and the corresponding simulation [6]. We have corroborated by change of different parameters that the recorded data are not affected by eddy-current effects in the highly conductive sample but reflect directly the "Spin Fluctuation" of the system.

While the above data were all detected in DC-magnetic fields up to 10 Tesla, the extension of the experiments into the sub-megagauss and even megagauss range seemed to be desirable. Those fields, however, can be produced only in pulsed operation involving inevitably the presence of a large dB/dt. For highly conductive materials as are high-mobility semiconductors and superconductors the usual DC-technique applied for transport-properties in pulsed magnetic fields, however, leads to disastrous experimental results, since every field-noise produces eddy-currents and affects directly the probe-signal. This essential disadvantage of pulsed magnetic fields with respect to DC-fields could be overcome by introducing high-frequency, phase-sensitive modulation technique in pulsed-magnetic field experiments. Another advantage of the lock-in technique is, that no tedious compensation of the inevitable pick-up voltage due to finite contact loops is necessary. Also there is no reaction of the contact leads of the sample to the Lorentz force, since the modulation frequency of the current is so high, that even for currents of some hundred mA no motion of the wires is detected. The schematics of the experimental arrangement is reproduced in Fig. 6 [7]. Due to the frequency limitation to 1 MHz of the presently available components this technology could be used only in the submegagauss range of nondestructive coils for pulse duration of some msec. The extension of the measuring technology to higher frequencies and thus for use in the megagauss regime of destructive coils with pulse duration of some μsec is in progress. We like to emphasize that by the combination of "lock-in"-technique and pulsed high-field technology the submegagauss range has now the same measuring sensitivity which was so far a privilege of DC-fields. In Fig. 7 we have reproduced the unprocessed transient-recorder traces of the transverse magneto resistance in HgSe:Fe for fields up to 40 Tesla [7]. The application of high magnetic fields allows now to detect unambigously the "Three Dimensional Analogue of the Quantum Hall Effect" directly without linear compensation as shown in Fig. 8 [2]. The data are symmetric for field up and down.

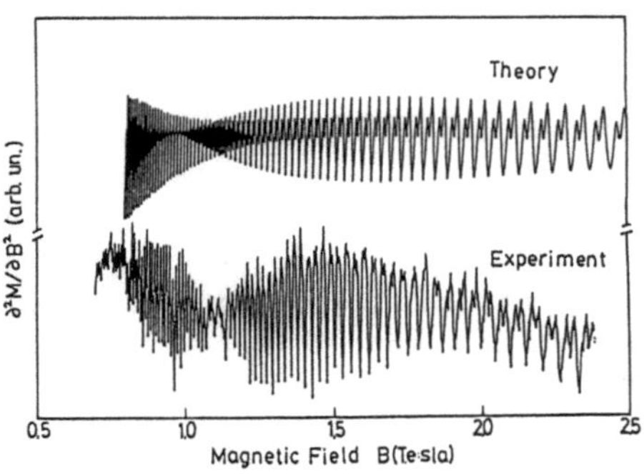

Fig. 5 Experimental results and theoretical simulation of the magnetization data on HgSe:Fe are in excellent agreement. The beating effect at B=1 Tesla is due to the inversion asymmetry splitting and modifies the spin splitting.

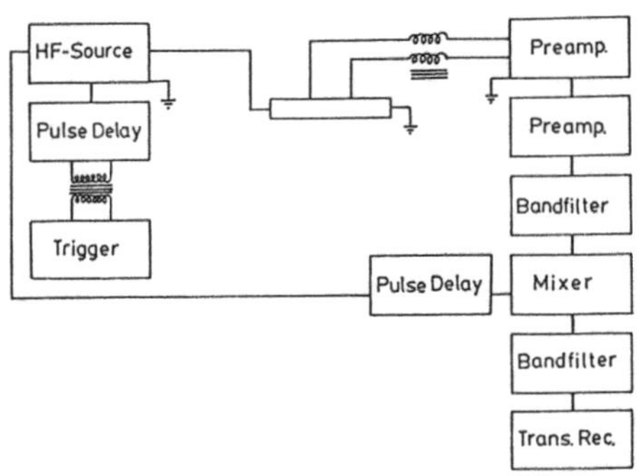

Fig. 6 The schematic of the high-frequency (→1 MHz) phase sensitive detector system for output bandwidth > 100 kHz.

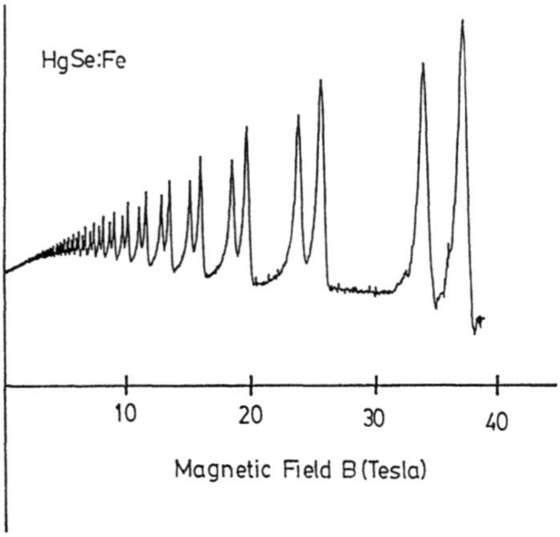

Fig. 7 The SdH-structure of the transverse magneto-resistance of HgSe:Fe images directly the density of states. Data recorded by phase-sensitive detection in pulsed field.

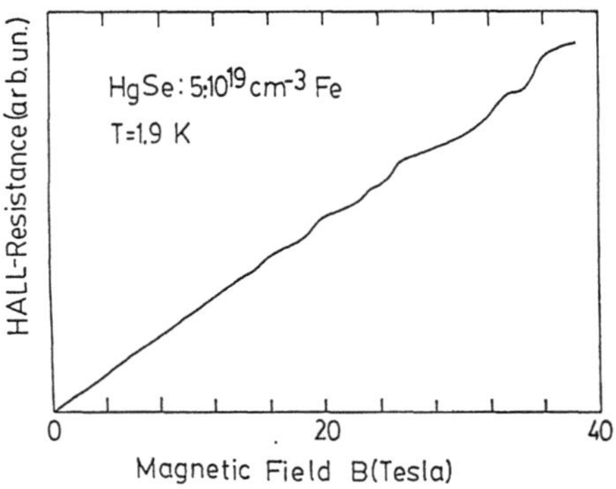

Fig. 8 The "Three Dimensional Analogue of the Quantum Hall Effect" of HgSe:Fe recorded by phase-sensitive detection in pulsed field.

We tried also a RF-transmission technique in the submegagauss range with HgSe:Fe, since this method is also applicable in the megagauss range [8]. The pronounced development of helicon-waves, however, for the applied frequencies of some 10 MHz obscured the Shubnikov-de Haas structure of the data. As a matter of fact a refractive index to the order 10^4 was thus observed in HgSe:Fe [2]. To overcome the particular problems of megagauss-transport imvestigations we introduced a new kind of measuring technique, namely "Eddy-Current Spectroscopy" [9]. Due to the inevitable presence of a large dB/dt in the megagauss range extremely strong currents are induced even in minute samples of HgSe:Fe and thus producing a strong magnetic moment, which can be detected by megagauss magnetization equipment. The reliability of this method has been tested in the submegagauss range using a high-frequency magnetic-field modulation technique with phase-sensitive detection [9]. As a

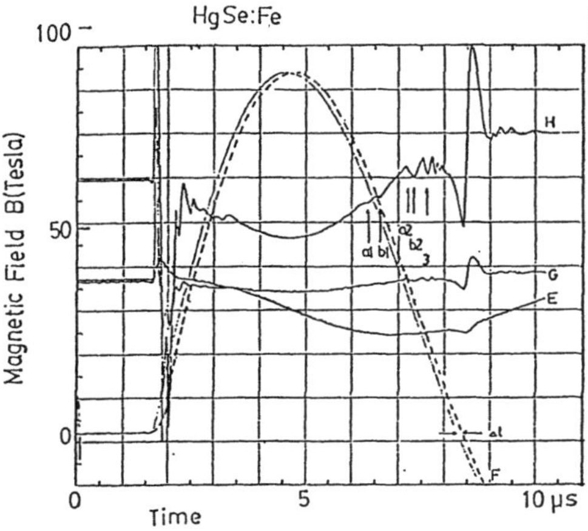

Fig. 9 Eddy-Current Spectroscopy on HgSe:Fe resolves clearly the quantum structure of the system in the megagauss range.

matter of fact in sample grains of only 100 µm diameter of HgSe:Fe with a resistivity of $0.8*10^{-5}$ Ohm*cm a current of the order 10^4 A is induced in the surface and thus changing the effective internal field considerably. Assuming a dB/dt = 100Tesla/3µsec a field shift of 15 Tesla is produced. The corresponding data are plotted in Fig. 9 [9]. The oscillations due to the a1 and b1 Landau level are clearly visible, whereas the b0-peak is still out of range. The 1/B position of the levels follows exactly the extrapolation of the low-field data.

2.2 $Hg_{1-x}Mn_xSe$:Fe

To study the pronounced enhancement of the g-factor using Mn as "spin dopant" also in the Fermi-level pinned HgSe:Fe system we extended the measurements of DOBROWOLSKI et al [10] also to the submega- and megagauss range. In Fig. 10 we give a direct comparison of the transverse magneto-resistance up to 40 Tesla for HgSe:Fe (n_{Fe} = $5*10^{19}$ cm^{-3}) and $Hg_{1-x}Mn_xSe$:Fe (n_{Fe} = $1*10^{19}$ cm^{-3}, x = 0.01) in the upper and lower part respectively [11]. The enhancement of the spin splitting due to the sd-exchange interaction of the quasi-free carriers and the localized 3d-electrons of the Mn^{++} is clearly visible. Notice the change in the phase of the Shubnikov-de Haas structures due to the decrease of the cyclotron mass for the Mn-doped material for smaller interaction gap. It should be noted that for the low temperature the essential part of the data is related to complete polarization of the Mn^{++}-spin to S_z = -2.5. To confirm the suggested indication of the spin levels we investigated the temperature dependence of the spin-splitting up to 40 Tesla as represented by the data in Fig. 11 [11]. The Shubnikov-de Haas oscillations are clearly visible even at a temperature of 92 K, where the spin splitting is considerably decreased because of to the depolarization of the Mn^{++}-spins. Due to the smearing out of the Fermi function (kT = 7.5 meV) and the increased scattering by acoustic phonons the line width, however, is essentially increased. The low-temperature data for the submega- and megagauss range are summarized in Fig. 12.

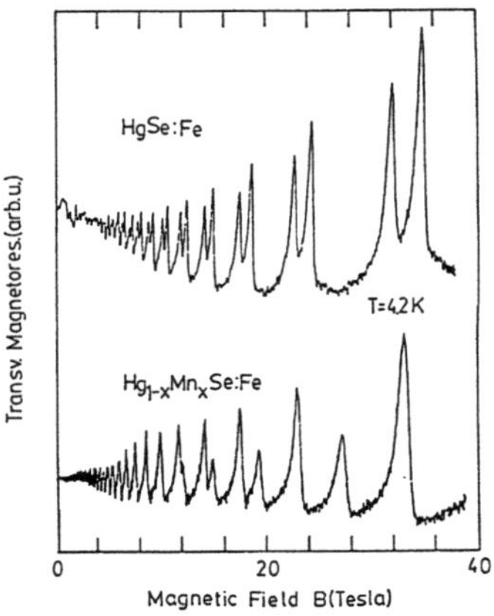

Fig. 10 The direct comparison of the SdH-structures in HgSe:($5*10^{19}$ cm^{-3} Fe) and $Hg_{0.99}Mn_{0.01}Se$:($1*10^{19}$ cm^{-3} Fe) demonstrates the strong influence of the Mn^{++}(sd)-exchange inter-action on the spin splitting.

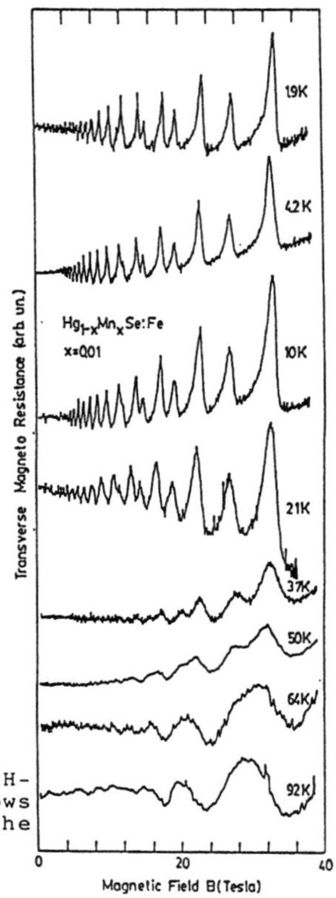

Fig. 11 The temperature dependence of the SdH-structures in $Hg_{0.99}Mn_{0.01}Se$:($1*10^{19}$ cm^{-3} Fe) shows clearly that with increasing temperature the Mn^{++}(sd)-exchange interaction is turned off.

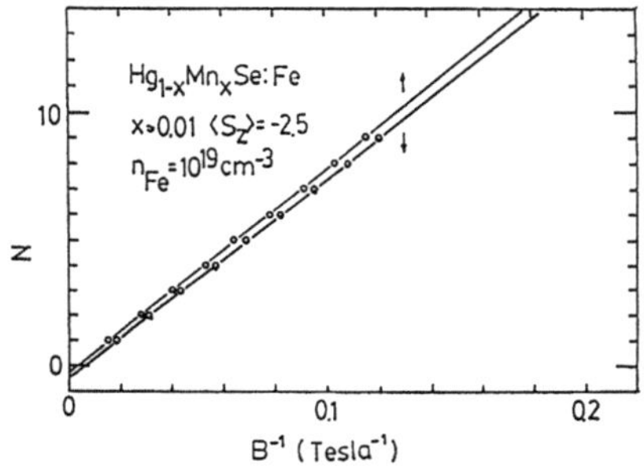

Fig. 12 From the different slopes of the N(1/B)-plots we derive for the exchange-cpoupling constant $J*m_c$ = 46.3 meV$*m_0$.

Neglecting inversion-asymmetry splitting we derive for the product of the effective exchange-coupling constant and the cyclotron mass $J^* m_c$ = 46.3 meV*m_0. To determine the cyclotron mass experimentally in a material with carrier concentration of $5*10^{18}$ cm^{-3} extremely thin samples and short wavelength are advisable. The corresponding experiments using CO_2-laser radiation in the megagauss range are in progress.

3. p-type $Pb_{1-x}Mn_xTe$

Whereas the most pronounced semimagnetic properties are observed with materials belonging to the II-VI-group semimagnetic behaviour is also found in IV-VI-systems. Especially $Pb_{1-x}Mn_xTe$ exhibits a series of particularities in the transport and optical properties [12]. Whereas the conduction band in $Pb_{1-x}Mn_xTe$ seems to be not so strongly affected by the exchange interaction, the valence band is more strongly modified. The disadvantage of p-type $Pb_{1-x}Mn_xTe$, however, is the relatively low carrier mobility. Applying the above discussed experimental techniques we investigated hot-wall grown, epitaxial layers [13]. The samples were [111]-oriented on BaF_2-substrates and carrier concentration of some 10^{17} cm^{-3}. The transverse magneto resistance is plotted in Fig. 13 and shows only broad oscillations of the ($7.2*10^{17}$ cm^{-3}, x = 0.016)-carrier system with the quantum-limit in the high-field region. The relatively low carrier mobility is responsible for the broadening of the oscillations. No spin splitting is resolved. The CO_2-laser cyclotron-resonance transmission spectra of the 1.3 µm thick layer are reproduced in Fig. 14

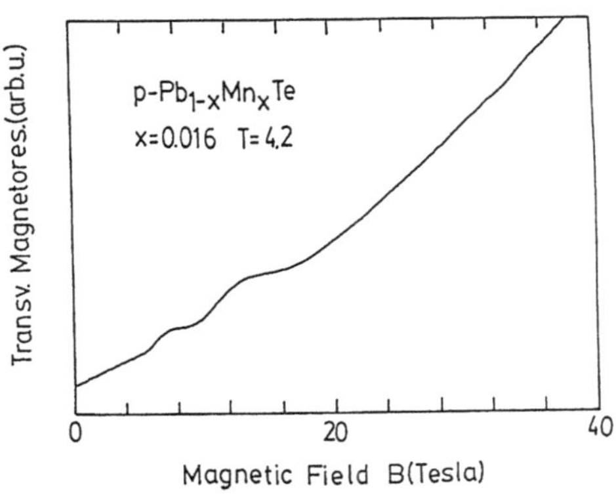

Fig. 13 Due to the low mobility the SdH-strucure of p-$Pb_{1-x}Mn_xTe$ (x=0.016) is rather smeared out. p = $7.2*10^{17}$ cm^{-3}.

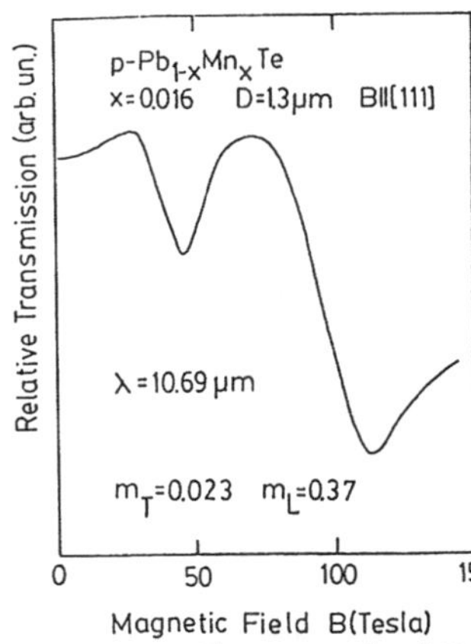

Fig. 14 The magneto-transmission of $Pb_{1-x}Mn_x$ with two cyclotron resonances.

showing clearly the resonances of the two sets of ellipsoids for B//[111] [14]. No spin-splitting of the cyclotron resonance due to nonparabolicity nor any spin-flip, combined-resonance nor impurity-transition is observed. The evaluation of the data results in the ellipsoidal mass parameters $m_T = 0.023*m_0$ and $m_L = 0.37*m_0$. No special indication of semimagnetic behaviour was observed in the present high-field data.

4. SUMMARY

High magnetic field spectroscopy in the submega- and megagauss ranges demonstrate clearly the semimagnetic properties of the ternary and quaternary Fe-doped systems. Due to high- frequency modulation techniques in pulsed magnetic fields the measuring sensitivity is comparable to those in DC-magnetic fields.

5. ACKNOWLEDGMENT

The author would like to express his sincere gratitude for the successful and joyful international exchange interaction in the quaternary Japanese-German-Polish-Austrian compound. Special thanks are expressed to his Japanese hosts Professor Noboru Miura and his coworkers for a lively and impressive introduction to the Japanese way of life.

6. REFERENCES

[1] J. Mycielski
Solid State Commun., **60**, 168 (1986)

[2] M. von Ortenberg, W. Staguhn, S. Takeyama, T.Sakakibara, Y. Iye, N. Miura, W. Dobrowolski, A. Mycielski, R.R. Galazka
to be published

[3] R. Kubo, S.J. Miyake, N. Hashitsume
Solid State Physics **17**, 269 (1965)

[4] S. Narita and K. Suizu
Suppl. of Progress of Theoret. Physics, 57, 187 (1975)

[5] M. von Ortenberg, O. Portugall, W. Dobrowolski, A. Mycielski, R.R. Galazka, F. Herlach
J. Phys C: Solid State Physics **21**, 5393 (1988)

[6] M. von Ortenberg, O. Portugall, W. Dobrowolski, A. Mycielski, R.R. Galazka
Proc. "19th Intnl. Conf. on the Physics of Semicond." Warsaw 1988

[7] M. von Ortenberg, W. Staguhn, F. Boebel, S. Takeyama, T. Sakakibara, N. Miura
J. Phys. E: Scientific Instrumentaions, in press

[8] T. Sakakibara, T. Goto, N. Miura
Proc. "2nd Intnl. Symposium of High Field Magnetism" Leuven , Belgium 1986, to be published in Physica B+C

[9] M. von Ortenberg, S. Takeyama, K. Amaya, N. Miura
to be published

[10] W. Dobrowolski, B. Witkowska, R.R. Galazka
this volume

[11] M. von Ortenberg, W.Staguhn, S.Takeyama, T. Sakakibara, K. Amaya, N. Miura
to be published

[12] G. Bauer, H. Pascher, M. von Ortenberg
to be published

[13] M. von Ortenberg,W. Staguhn, S. Takeyama, T. Sakakibara, N. Miura, G. Bauer
to be published

[14] S. Takeyama, H. Yokoi, M. von Ortenberg, N. Miura, G. Bauer
to be published

Shubnikov–de Haas Effect in $Hg_{1-x}Mn_xSe$:Fe

W. Dobrowolski, J. Kossut, B. Witkowska, and R.R. Gałązka

Institute of Physics, Polish Academy of Sciences,
Al. Lotników 32/46, PL-02-668 Warsaw, Poland

1. INTRODUCTION

Fe donor level in HgSe is degenerate with the conduction band and lies about 210 meV above its edge [1] (see, lower part of Fig. 1a). This fact gives rise to several unusual properties of Fe doped HgSe [see, e.g., 2]. For low iron concentration ($N_{Fe} < N^*_{Fe} \approx 4.5 \times 10^{18} cm^{-3}$) all resonant donors are ionized to Fe^{3+} state (as shown schematically in upper part of Fig. 1a). The properties of the crystal at low temperatures are then similar to those of HgSe doped with, e.g., gallium [3,4]. When the iron concentration exceeds N^*_{Fe} the neutral and ionized donors coexist (mixed valence regime – see, Fig. 1b). In such circumstances the system is able to lower its energy by redistribution of electrons between ionized and neutral donors, which leads to spatial correlations of positions of ionized donors [5]. Ultimately a "charge superlattice" of ionized donors can be formed. In consequence, the low momentum transfer scattering processes, important in the case of the scattering by charged centers (dominant mechanism in HgSe at LHe temperatures), are excluded. The electron mobility and the Dingle temperature T_D achieve then anomalous values (respectively, high and low). On the other hand, the resonant character of the donors could open an additional very effective scattering channel – the resonant scattering. However, as pointed out in [6], formation of the Coulomb gap [7] in one electron d–band density of states (caused by inter – site repulsion) suppresses this mode of scattering.

The present paper attempts to answer the question whether a similar picture is applicable to semimagnetic (diluted magnetic) semiconductor HgMnSe:Fe.

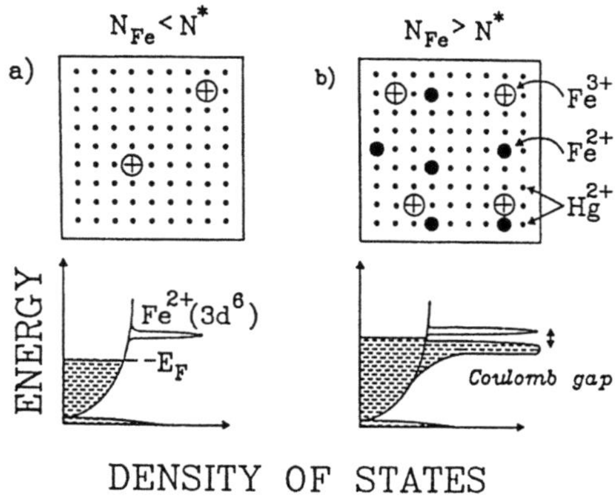

Fig. 1 Upper part: schematic representation of HgSe cation sublattice (dots) with substituting Fe donors in two charge states (pluses - ionized donors, full symbols - neutral donors) in two regions of Fe concentrations: (a) $N_{Fe} < N^*_{Fe}$ (all donors ionized); and (b) $N_{Fe} > N^*_{Fe}$ (mixed valence region). Lower part shows one-particle density of states showing the Coulomb gap E_C for $N_{Fe} > N^*_{Fe}$. The hatched area represents occupied states

Springer Series in Solid-State Sciences, Vol. 87
High Magnetic Fields in Semiconductor Physics II Editor: G. Landwehr

2. EXPERIMENTAL RESULTS AND DISCUSSION

Samples of $Hg_{1-x}Mn_xSe$:Fe with x $\approx$ 0.003 and doped with iron up to 5 $\times$ 10^{19} cm^{-3} were grown by the modified Bridgman method. The molar fractions x were determined by the density measurements. Prior to mounting in the holder the unoriented samples were etched in bromine. Gold wire contacts were soldered with indium. Table I shows the basic characteristics of the investigated samples.

Table I

sample number	iron doping level $[cm^{-3}]$	electron concentration $[cm^{-3}]$	electron mobility $[cm^2/Vs]$	Dingle temperature [K]
2784-3	2×10^{18}	4.2×10^{18}	2.1×10^4	10.0±4
2785-3	6×10^{18}	4.5×10^{18}	2.1×10^4	4.3±1.8
2786-3	1×10^{19}	4.5×10^{18}	3.3×10^4	2.3±1.
2825-2a	2×10^{19}	4.7×10^{18}	2.4×10^4	—
2826-4a	5×10^{19}	5.0×10^{18}	3.0×10^4	3.1±0.8

The Shubnikov – de Haas (SdH) oscillations were observed in magnetic fields up to 6T and in temperatures 1.6K – 30K. Figure 2 shows the oscillatory part of the transverse magnetoresistance for two samples: $Hg_{0.997}Mn_{0.003}Se$: 2×10^{18} Fe cm^{-3} and 1×10^{19} Fe cm^{-3}. The sample with higher doping level (and higher electron concentration!) exhibits more pronounced oscillations which suggest that the Dingle temperature in this sample is substantially lower (in contrast with expected increase of T_D with growing electron concentration).

The least-squares fitting technique [8] was used to determine the amplitudes and phases of the harmonic content of the oscillatory transverse magnetoresistance. A typical result is shown in Fig 3. The upper part presents the ratio of fundamental to second harmonics calculated for two samples: $Hg_{0.997}Mn_{0.003}Se$ doped with 2×10^{18} cm^{-3} and 5×10^{19} cm^{-3} of Fe. In the latter the second harmonic is less damped ($\approx$ 5 times) than in the first sample. In the first sample $T_D \approx$ 10K which is comparable to the value in HgSe:Ga. Further analysis of the amplitude in the second sample yields the value of the Dingle temperature differing by a factor of 3 (i.e., $T_D \approx$ 3K). The anomalously low values of the Dingle temperature found in all investigated samples with $N_{Fe} \geq 6 \times 10^{18}$ cm^{-3} strongly suggest that, similarly to HgSe:Fe, also in $Hg_{1-x}Mn_xSe$:Fe the scattering rate is reduced. As in HgSe:Fe this anomaly is presumably due to the decreased effectiveness of the ionized impurity scattering because of the partial spacial ordering of the ionized donors. Basing ourselves on the Dingle temperature and electron mobility data we may conclude that samples with $N_{Fe} \geq 6 \times 10^{18}$ cm^{-3} belong to mixed valence regime.

Beside similarities we observed also differences between HgSe:Fe and $Hg_{1-x}Mn_xSe$:Fe, namely, different values of the reduction factor of the scattering rate. It is evident that some additional scattering mechanism contributes significantly in the case of the mixed crystals. The fact that the Dingle temperature is less influenced by this mechanism suggests a short range nature of an associated scattering potential. Likely candidates in this context are the alloy scattering and the resonant scattering which becomes effective again when the Coulomb gap is diffused.

We would like to point out that the observed reduction of the scattering rate may be of advantage in the investigation of the magnetic nature of $Hg_{1-x}Mn_xSe$. Figure 4

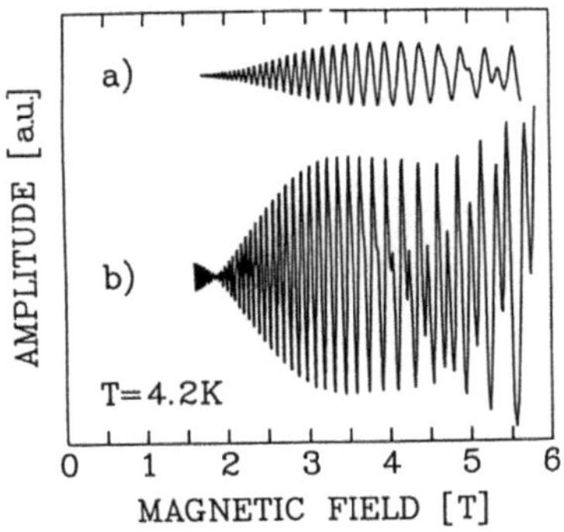

Fig. 2 Shubnikov – de Haas oscillation in (a) $Hg_{0.997}Mn_{0.003}Se$: 2×10^{18} Fe cm^{-3} and (b) $Hg_{0.997}Mn_{0.003}Se$: 1×10^{19} Fe cm^{-3}

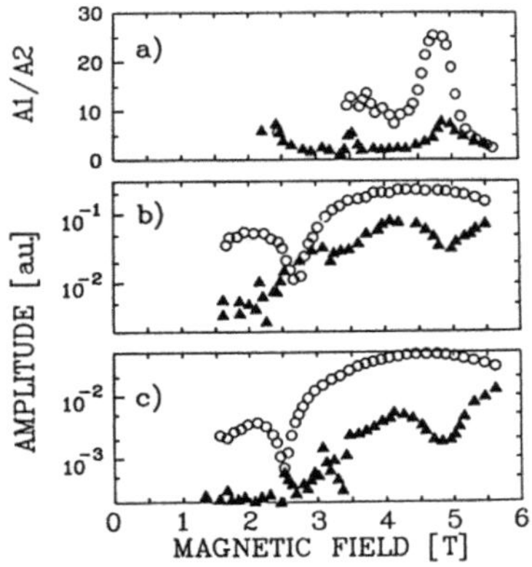

Fig. 3 (a) The ratio of the fundamental to second harmonics (circles: sample $Hg_{0.997}Mn_{0.003}Se$: 2×10^{18} Fe cm^{-3}), triangles $Hg_{0.997}Mn_{0.003}Se$: 1×10^{19} Fe cm^{-3}) (b) and (c): fundamental (circles) and second harmonics (triangles) in samples $Hg_{0.997}Mn_{0.003}Se$: 2×10^{18} Fe cm^{-3} and $Hg_{0.997}Mn_{0.003}Se$: 5×10^{19} Fe cm^{-3}, respectively

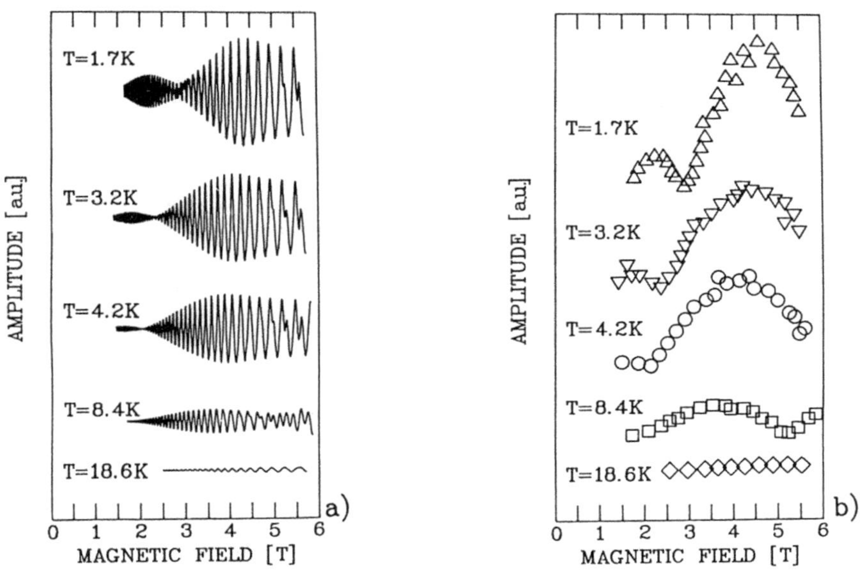

Fig. 4 (a) The Shubnikov – de Haas oscillations in $Hg_{0.997}Mn_{0.003}Se$: 5×10^{19} Fe cm^{-3} at several temperatures; (b) corresponding values ot the fundamental amplitudes as a function of the magnetic field. Note that the amplitude in the vicinity of 3T increases when the temperature grows from 1.7K to 3.2K

shows the Shubnikov – de Haas oscillation recorded at several temperatures. Because of very low Dingle temperatures it is possible to observe an anomalous dependence of the amplitude vs. temperature or a strong variation of the spin splitting with the temperature and magnetic field, in samples with very high electron concentration (5×10^{18} cm^{-3}). Previously, such effects were observable only in samples with substantially lower electron concentration [see, e.g., 9]. An investigation of these features together with a study of magnetic properties are underway.

Acknowledgments: The authors are indebted to Dr. A.Mycielski for many valuable discussions.

REFERENCES

1. A.Mycielski, P.Dzwonkowski, B.Kowalski, B.A.Orłowski, M.Dobrowolska, M.Arciszewska, W.Dobrowolski and J.M.Baranowski, J. Phys. C: Solid State Phys. **19**, 3605 (1986)
2. A.Mycielski, J.Appl. Phys. **63**, 3279 (1988); W.Dobrowolski, K.Dybko, C.Skierbiszewski, T.Suski, E.Litwin–Staszewska, A.Miotkowska, J.Kossut and A.Mycielski, in *Proc. 19th Int. Conf. on Phys. of Semicond.*, Warsaw 1988, in press
3. W.Dobrowolski, K.Dybko, A.Mycielski, J.Mycielski, S.Piechota, M.Palczewska, H.Szymczak and Z.Wilamowski, in *Proc. 18th Int. Conf. on Phys. of Semicond.*, Stockholm 1986, p. 1743, ed. by O.Engström (World Scientific, Singapore 1987)
4. F.S.Pool, J.Kossut, U.Debska and R.Reifenberger, Phys. Rev. **B35**, 3900 (1987)
5. J.Mycielski, Solid State Commun., 60, 165 (1986)
6. T.Dietl, Jpn. J. Appl. Phys. **26**, Suppl. 26-3 (1987)
7. B.I.Shklovskii and A.L.Efros, in *Electronic Properties of Doped Semiconductors*,ed. by M.Cardona, P.Fulde and H.J.Queisser, Springer Series in Solid State Sci., Vol. **45** (Springer, Berlin, Heidelberg 1984) and references therein
8. R.Reifenberger and D.A.Schwarzkopf , Phys. Rev. Lett. **50**, 907 (1983); M.Vaziri and R.Reifenberger, Phys.Rev. **B32**, 3921 (1985)
9. S.Takeyama and R.R.Gałązka, Phys. Stat. Sol b **96**, 413 (1979); P.Byszewski, M.Z.Cieplak and A.Mongird–Gorska, J. Phys. C: Solid St. Phys.,**13**,5383 (1980); H.M.A.Schleijpen and F.A.P.Blom, Phys. Stat. Sol b **135**, 605 (1986)

Zeeman Studies of MBE Grown CdTe in High Magnetic Fields

R.N. Bicknell-Tassius, T.L. Kuhn, W. Ossau, and G. Landwehr

Physikalisches Institut der Universität Würzburg,
D-8700 Würzburg, Fed. Rep. of Germany

I. Introduction

The II-VI semiconductors feature bandgaps ranging form the near UV to th far IR as shown schmatically in Fig. 1. As a consequence, these materials could be used in a wide variety of electronic and optoelectronic devices [1,2]. However, their use in active device applications have been severely limited because of problems that have been encountered in attempts to control their electrical properties through substitutional doping [3]. In particular, the interest in CdTe stems not only from its interesting optical and electronic properties, but also from its close link to the important infrared material HgCdTe. In recent years, molecular beam epitaxy and organometallic chemical vapor deposition have been employed to prepare CdTe thin films of high structural perfection. In order for CdTe thin films to have significant commercial application, however, systematic control of their electrical properties through substitutional doping will be essential.

Recently a new technique, photoassisted molecular beam epitaxy (PAMBE), has been developed to aid in the substitutional doping of II-VI compound semiconductors [4,5]. The PAMBE technique has been employed to not only prepare highly conducting indium doped CdTe, but also p-type CdTe, n-type CdMnTe and n-type CdTe-CdMnTe superlattices [6]. High structural perfection HgCdTe has also been grown using the PAMBE technique [7]. Preliminary

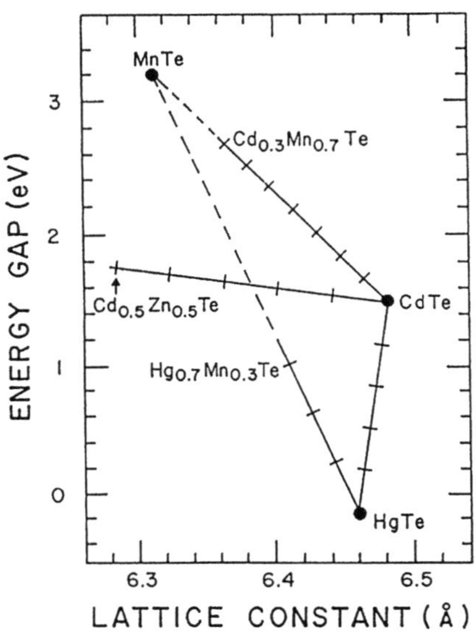

Figure 1: Lattice constant and energy gaps for several II-VI Semiconductors

Springer Series in Solid-State Sciences, Vol. 87
High Magnetic Fields in Semiconductor Physics II Editor: G. Landwehr

device fabrication, including p-n homojunctions and MESFETS have also been recently reported [8]. The photoassisted MBE process differs from conventional MBE in that the substrate is illuminated during the MBE growth. The incident light provides high energy low momentum photons at the growth surface during the deposition process. The photon energies associated with visible light are considerably larger than the thermal energies that are available at typical MBE growth temperatures. As a result, new effects may be observed from the introduction of light during film growth, including enhacement of surface migration of constituent atoms, modification of surface bonding, conversion of surface molecules into atoms, and modification of surface potentials through generation of photoexcited electron-hole pairs.

Substitutional doping of CdTe and other II-VI semiconductors has been a longstanding problem. Self-compensation effects are believed to be the major reason for the generally poor activation of dopants in these materials. In CdTe the introduction of an n-type dopant, such as indium, is generally accompanied by the formation of donor-neutral vacancy complexes which provide the compensation mechanism. Nonequilibrium thin film growth techniques may provide a means to reduce the formation of the compensating centers. In particular, the photoassisted MBE technique appears to greatly reduce the density of the self-compensating centers.

In the present work, the effect of the illumination power density on the PAMBE technique is investigated. The PAMBE technique gives rise to large changes in the point defect nature of the growing samples. The power density employed also has a dramatic effect on the point defect density. This can be clearly seen through substantial changes in the observed low temperature photoluminescence spectrum for samples grown using the PAMBE technique.

II. Experimental Details

The epilayer growths were carried out in a four chamber RIBER 2300 epitaxy system shown in Fig. 2. This system has been used to prepare, by conventional MBE, high quality CdTe and CdMnTe as well as superlattices of CdMnTe-CdTe. Illumination during the PAMBE process was provided by an argon-ion laser operating on the 514.5 nm line. Power densities up to 350 mW/cm^2 have been investigated. Thus, for the growth rates of 1-3 Åemployed, the photon flux

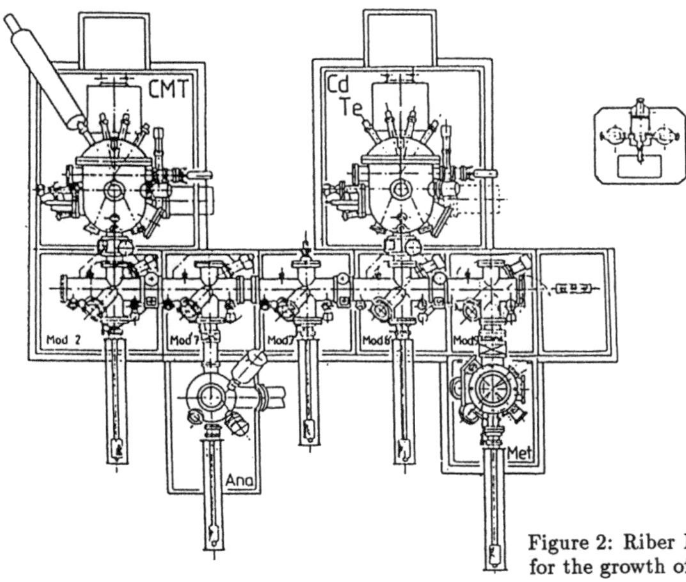

Figure 2: Riber MBE 2300 system used for the growth of II-VI semiconductors

was several orders of magnitude greater that the cadmium or tellurium impingement rate at the growth surface. Chemimechanically polished (100) CdTe wafers were used as substrates. Prior to insertion into the MBE system the substrates were first degreased using standard solvents. Next, they were etched in a bromine-in-methanol solution. Finally, they were rinsed in methanol and blown with dry nitrogen before insertion into the MBE load-lock. Immediately prior to the film growth the substrates were annealed in the MBE system at a temperature of 300-400 °C. During the film growth, the substrates were maintained at temperatures of 160-320 °C. Substrate temperatures were calibrated using the melting point on In and Sn mounted on the from of the molybdenum transfer block. The optical properties of the pambe samples were investigated using low temperature photoluminescence (PL) at liquid helium temperatures The samples were mounted in a helium bath cryostat. The Pl signal was excited using a Coherent INNOVA 100-20 argon-ion laser (514.5 nm). The luminescence was detected using a Jobin-Yvon HR-1000 grating monochromater with a cooled GaAs photocathode photomultiplier and standard photon-counting electronics. The magnetic field was produced using a 9.5 T split coil superconducting magnet.

III. Results and Discussion

A. Conventional MBE CdTe.

The first step in investigating the effect of light on the MBE growth process was to grow a series of CdTe epilayers by conventional MBE. This allowed an optimization of the growth parameters and also served as a basis of comparison when illumination effects were investigated. These samples were grown under identical conditions except that substrate temperature was varied between 230-320 °. All four spectrum are similar to those observed in bulk single crystal CdTe [9]. The luminescence spectra occurs in three regions, the Excitonic (1.58-1.60 eV.), the electron to acceptor and donor to acceptor (1.50-1.57 eV.) and the defect band (1.42-1.50 eV.). At the lowest substrate temperatures employed the spectra are dominated by the lower energy transitions with the excitonic transitions being quite small in comparision. As the substrate temperature is increased the spectra become dominated by recombination in the excitonic region.

An expanded view of the excitonic region in these four samples is shown in Figure 3. In all four samples the main excitonic recombination is the primary bound exciton which occurs

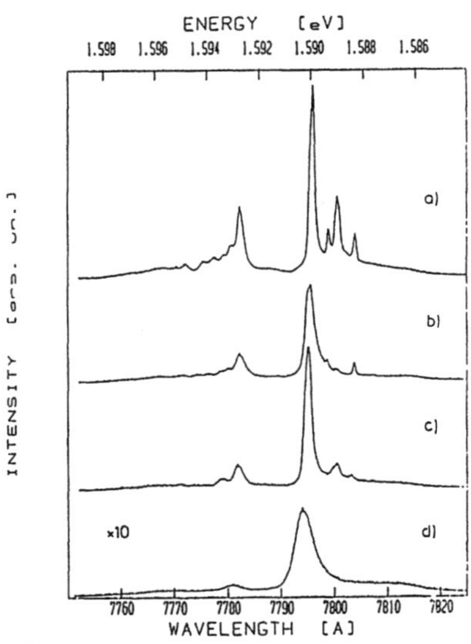

Figure 3: Photoluminescence spectrum for conventional MBE epilayers using substrate temperatures of a) 320 C, b) 290 C, c) 260 C and d) 230 C.

at 1.5902 eV. One sees the general improvement in the optical quality of the samples as the substrate temperature employed during MBE growth is increased. A number of other sharp excitonic transitions can be identified in the samples grown at higher temperatures. These include acceptor-bound excitonic transitions which have been associated with copper (1.5896 ev), silver (1.5885 eV), or nitrogen, sodium, and lithium (1.5892 eV) [10,11]. In the donor-bound region the main feature occurs at 1.5929 eV and has been associated with chlorine. It should be noted that the transitions are quite narrow (0.35 meV) and show no shift at different substrate temperatures.

The above results demonstrate that high quality CdTe epilayers can be grown by conventional MBE and demonstrate the effect of substrate temperature on the luminescence of the epilayers.

B. Photoassisted MBE of CdTe

In order to investigate the effect of the power density employed in the photoassisted MBE process a series of samples using power densities between 80 and 320 mW/cm^2 and a substrate temperature of 290 °C was grown. The photoluminescence spectra for these samples are shown in Figure 4. The spectra for the samples grown using a power density of 80 and 320 mW/cm^2 are similar to the samples grown by conventional MBE, though a general improvement in quality is observed.

The luminescence spectra for the sample grown using a power density of 160 mW/cm^2 is dramatically different from any of the samples grown by conventional MBE. The spectrum is completely dominated by a line which we interpret as an acceptor bound excitonic transition at 1.5898 eV and its phonon replica.

An expanded view of the excitonic region of the PAMBE grown epilayers is shown in Figure 4. These spectra again demonstrate the dramatic changes that occur in the point defects of the samples grown by the photoassisted MBE technique.

In particular, the sample grown using 160 mW/cm^2 is quite remarkable. The only observed transition is the primary acceptor bound exciton which is now shifted by 0.3 meV and split into two components at 1.5898 eV and 1.5994 eV. This transition is very bright, more than an order of magnitude brighter than the other PAMBE grown samples, and very narrow (0.18 meV).

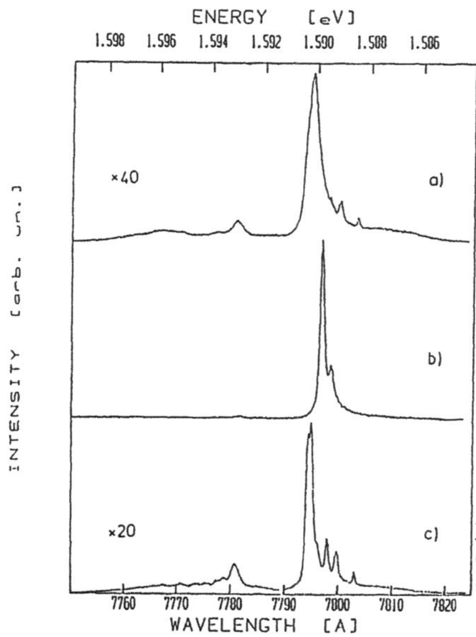

Figure 4: Photoluminescence for PAMBE grown epilayers using a) 320 mW/cm^2, b) 160 mW/cm^2, c) 80 mW/cm^2.

C. Zeeman Studies

In addition we have studied linear and quadratic Zeeman effects of the acceptor-bound-exciton lines. Figure 5 shows the photoluminescence spectra at zero magnetic field for the sample grown using 80 mW/cm^2 (spectra a), which is similar to the spectra of samples grown using conventional MBE. Spectra b) is observed for the sample grown with 160 mW/cm^2. Again the dramatic change in the luminescence caused by the photoassisted MBE process is demonstrated. Line **A** , which is the dominant luminescence feature in spectra a) is tentatively attributed to exciton recombination at an unidentified acceptor (vacancy-impurity complex) [14]. This line is not observed in spectra b), however, we observe a new luminescence line **B** with a smaller halfwidth and a strong increase in intensity. Line **B** is also present in spectra a) as a small shoulder on the low energy side of line **A**. This unambigously shows that line **B** is a new line and not identical with line **A** shifted to lower energies only. In both samples we detect luminescence of excitons bound to the residual impurities copper and silver. Due to the huge intensity of line **B** in spectra b) the other line are not visible on this scale.

Further evidence that line **B** is a new luminescence feature comes from the Zeeman pattern recorded in the Voigt and Faraday configurations. In addition to the splittings of the line we also observe an overall diamagnetic shift to higher energies with increasing field strength for all four acceptor bound excitons (**A**, **B**, (Cu^o,X), (Ag^o,X)). The energy shift and the splitting pattern obtained for both samples discussed here are shown in Fig. 6. At high magnetic fields the components interfere and careful analysis of the spectra is required to deduce the correct splitting scheme. Polarization effects aid in the analysis of the splitting. In Fig. 6 the circles σ- and the crosses π- polarization. The line are drawn as an aid to the eye. We observe two drastically different splitting behaviors. On one side there is a symmetrical splitting behavior in Voigt configuration with four σ- and two π- subcomponents for the copper and silver bound excitons as well as for line **A** . For discussion of the results we choose the line of (Ag^o,X) as

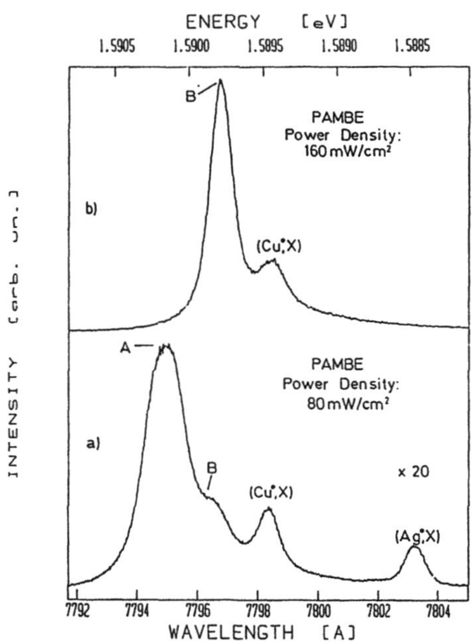

Figure 5: Photoluminescence spectra of the acceptor-bound-exciton at zero magnetic field.

representative. On the other side there is an asymmetric splitting for the luminescence line **B** , which will be discussed later.

In fig 7. we have plotted the energy shift of the center of gravity for the (Ag^o,X) line. This diamagnetic shift has been fit to the theory of Cabib et. al. [15], using a dielectric constant of $\epsilon = 9.8$ [12]. The best fit is obtained with a reduced mass of $\mu = (0.0875 \pm 0.005)m_o$. This value is slightly smaller that that obtained by Molva et. al. [12], who have fit their data with an effective mass of $\mu = 0.093\ m_o$. There authors applied magnetic fields up to 4.5 T and used a simple hydrogenic model for the diamagnetic shift. At field strengths of 4T their model based on first order perturbation theory is already in error by more than 5 %. Therefore out data of the diamagnetic shift obtained for field strengths up to 9.5 T and analyzed with the more

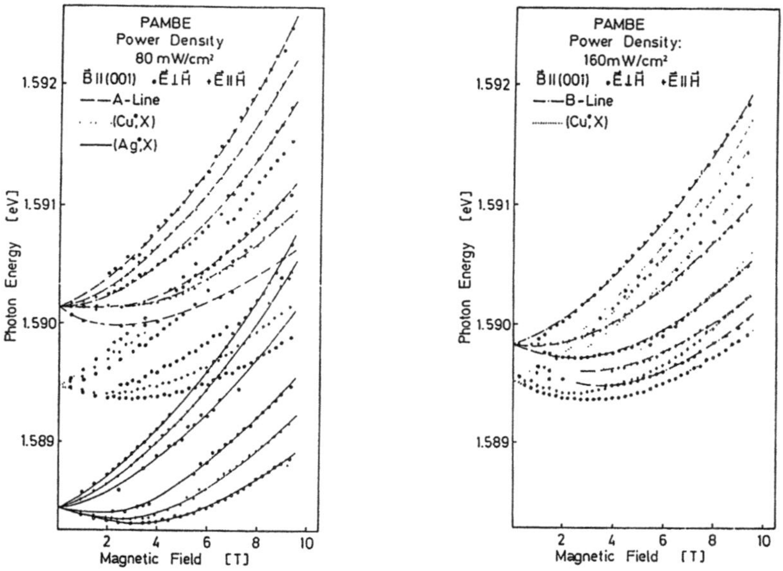

Figure 6: Magnetic field dependence of the Zeeman components of the acceptor- bound-exciton lines, circles represent σ- and crosses π- polarizations. An asymmetric splitting of the **B** line can be clearly seen.

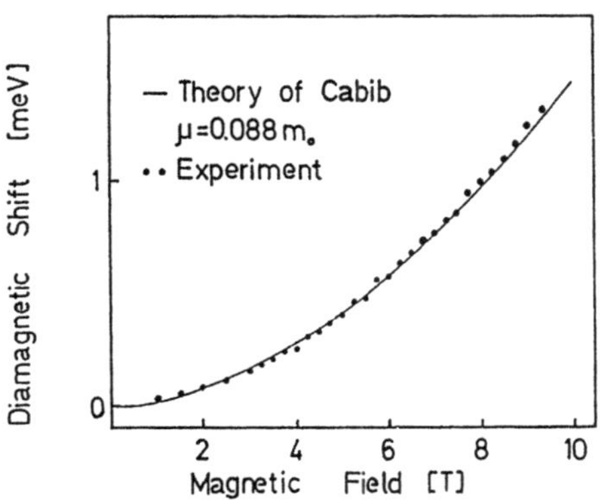

Figure 7: Diamagnetic shift of the center of gravity of the (Ag^o,X) line. the solid line is a fit to the theory of Cabib et. al. [15].

precise theory of Cabib et al. are more reliable. The mass value obtained for the reduced mass of the exciton is in close agreement with the effective mass ov the conduction band electron m = (0.0963±0.0008) m_o [16,17]. This strongly supports the pseudonor model for the (Ag^o,X) complex suggested by Rühle and Bimberg [18]. In this model the two holes of the complex are strongly bound be the short range potential of the acceptor, thus producing a long range Coulomb potential which binds the electron in donor-like states.

The linear Zeeman effect obtained by subtracting the overall diamagnetic effects from the energy shifts are consistent with the model of exciton recombination at neutral acceptors with T_d symmetry, indicating that the acceptors are indeed on lattice sites. The ground state is a J = 1/2 state for both Ag and Cu bound excitons. For the **A**-line the observed polarization of the subcomponents is not as pronounced as for the copper and silver bound excitons. Therefore, the origin of the line is not as certain as that of the (Cu^o,X) and (Ag^o,X). These bound exciton complexes are made up of two holes with j = 3/2 and one electron with j = 1/2. In the j-j coupling scheme, this will result in bound exciton states described by the total angular momentum 1/2, 3/2, and 5/2. The fact that only tow σ subcomponents are observed in the Zeeman spectrum (Fig. 8a) indicates that the ground state is a J = 1/2 state. The same situation has been observed for neutral acceptor. exciton complexes, including Sn [19,20] and Cu [21], in GaAs. The final state selection rules used in this work for the copper and silver bound excitons are shown in Fig. 8a as an insert.

A quantative analysis of the linear Zeeman effects have been made by using the effective spin Hamiltonian:

$$H_{\frac{1}{2}} = g_e \mu_B \vec{J} \cdot \vec{K}$$

for the initial bound-exciton state J = 1/2 and

$$H_{\frac{3}{2}} = \mu_B (K \vec{J} \cdot \vec{H} + L(J_x^3 H_x + J_y^3 H_y J_z^3 H_z))$$

for the final acceptor state J = 3/2 in cubic symmetry [22,23]. The best fist to the data for field orientation B||(001) has been obtained with the following set of parameters:

$g_e = -1.77 \pm 0.02$
$K = 0.56 \pm 0.05$
$L = -0.04 \pm 0.03$

These values are only slightly different from those obtained by Molva et. al. with magnetic fields up to 4.5 T. For a (A^o,X) complex one would expect a diamagnetic splitting effect of the acceptor ground state J = 3/2, which splits apart the center of gravity of the m_j =| 1/2 | Zeeman components from the m_j =| 3/2 | components. This effect is proportional to $(a_o\ H)^2$, where a_o is the effective Bohr radius of the acceptor of interest. Due to the H dependence we expected to detect a diamagnetic splitting for the largest applied field strength of 9.5 T. Within our experimental error the Zeeman spectra are symmetrical showing that the diamagnetic splitting is negligibly small, indicating that the bound hole is strongly localized and the diamagnetic shift effects (Fig. 7) are due to the extended states of the bound exciton. This again confirms the pseudodonor model of an electron bound by a long range Coulomb potential produced by holes strongly bound to the acceptor.

The Zeeman pattern of the luminescence line **B** observed in samples grown under illumination of 160 mW/cm^2 is completely different. We obtain an asymmetric splitting as can be seen in Fig. 7b, and in addition the line seems to have two components. Because of the asymmetric splitting the overall diamagnetic shift of the center of gravity is not easily deduced for the **B**-line. As the diamagnetic shift for copper and silver bound excitons as well as the **A**-line are identical in this spectral region, we assumed the same overall diamagnetic shift for the **B**-line. The linear

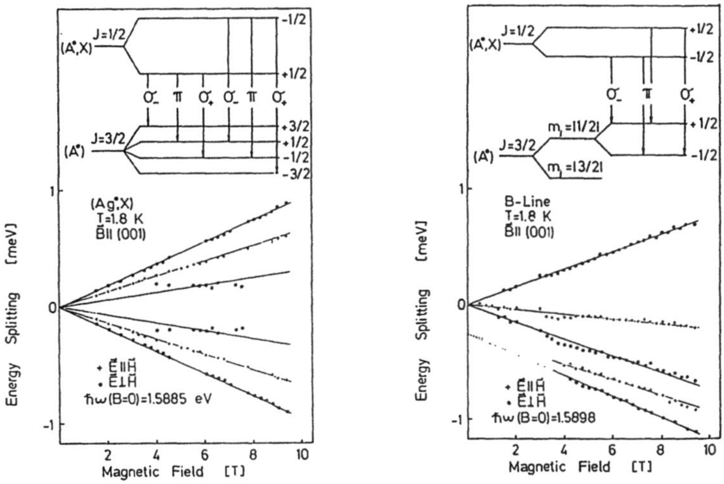

Figure 8: Linear Zeeman splitting scheme of the (Ag^o,X) line and of the B-line.

Zeeman pattern obtained with this assumption is show in Fig. 8b. We obviously see a linear but asymmetric splitting of the **B**-line into five subcomponents. The two components observable at high magnetic fields only have an intersection point with the energy axis of about 0.25 meV lower than that of the other three components. The photoassisted MBE process does not change the intensity and splitting behavior of th excitons bound to substitutional impurities. Therefore, it does not produce any stress during the film growth. As we observe an influence on line **A** only, we assume that line **A** and **B** have similar origins. The luminescence of line **A** is attributed to a vacancy-impurity complex [14]. We suggest that this complex is changed be the photoassisted MBE growth. As the splitting behavior for line **B** shows no T_d symmetry one possible explanation may be that the impurity or vacancy undergoes a site change producing a crystal field. This crystal field splits the J = 3/2 state into $M_j = \mid 1/2 \mid$ and $M_j = \mid 3/2 \mid$ states. It is known that uniaxial stress shifts the $M_j = \mid 1/2 \mid$ state to higher energy, while the $M_j = \mid 3/2 \mid$ state to lower energy as shown in the insert of Fig. 8b. Applying this recombination model it is possible to explain the observed splittings for the **B**-line. We observe only the three high energy components. These lines are due to transitions between the exciton J = 1/2 state and the $M_j = \mid 1/2 \mid$ states of the acceptor resulting in two σ -subcomponents and two unresolved π components. The two components at lower energies may be transitions between the J = 1/2 exciton and the $m_j = +3/2,+1/2$ hole states of another substitutional acceptor like Sb, because the linear Zeeman pattern of these two components are nearly the same as those obtained for Ag and Cu. To prove if this recombination scheme for the **B**-line is correct further magnetic field studies are under way to investigate the symmetry of the crystal field producing the splittings of the **B**-line luminescence.

IV. Summary

The above results show conclusively that illumination of the substrate during MBE growth has profound effects on the photoluminescence properties of the epilayers. It has also been demonstrated that the choice of illumination power density is critical. The PAMBE technique has been employed in the growth of high quality undoped CdTe. The photoluminescence spectrum is dominated by a single intense acceptor bound excitonic transition that is extremely narrow (0.18 meV). Zeeman studies performed on CdTE samples grown by PAMBE suggest a site change within an impurity-vacancy complex. The rate of site changes seems to be correlated to the intensity of laser light used during growth. Furthermore, we have studied linear and quadratic Zeeman effects for impurity bound exciton complexes (Ag^o,X) and (Cu^o,X). The relevant splitting factors are g_e = -1.77 for electrons and K = +0.56 and L = -0.04 for the hole parameters describing the isotropic and anisotropic Zeeman effects, respectively.

V. Acknowledgements

The work for this project was supported by the Bundesministerium für Forshung und Technologie (Bonn). We wish to acknowledge the assistance of A. Waag and M. Möller during the course of this work.

References

1. F.V. Wald, Rev. Phys. Appl. **12**, 277 (1977).
2. R.O. Bell, Rev. Phys. Appl. **12**, 391 (1977).
3. Y. Marfaing, Rev. Phys. Appl. **12**, 211 (1977).
4. R.N. Bicknell, N.C. Giles, J.F. Schetzina, Appl. Phys. Lett. **45**, 1095 (1986).
5. R.N. Bicknell, N.C. Giles, J.F. Schetzina, Appl. Phys. Lett. **49**, 1735 (1986).
6. R.N. Bicknell, N.C. Giles, J.F. Schetzina, Appl. Phys. Lett. **50**, 691 (1987).
7. J.W. Han, S. Hwang, Y. Lansari, J.W. Cook, J.F. Schetzina, Bull. Am. Phy. Soc. **33** , 586 (1988).
8. D.L. Dreifus, R.M. Kolbas, K.A. Harris, R.N. Bicknell, R.L. Harper, and J.F. Schetzina, Appl. Phys. Lett. **51**, 931 (1987).
9. K.Zanio, **Cadmium Telluride**,in **Semiconductors and** Semimetals, vol. **13**, eds. R.K. Willardson and A.C. Beer (Academic Press, New York, 1978).
10. J.L. Pautrat, J.M. Francou, N. Magnea, E.Molva, and K. Saminadayar, J. Cryst. Growth **72**, 194 (1985).
11. E. Mplva, J.L. Pautrat, K. Saminadayar, G. Milchberg and N. Magnea, Phys. Rev. **30**, 3344 (1984).
12. E. Molva, Le Si Dang, Phys. Rev. **27**, 6222 (1983).
13. T. Taguchi, J. Yamamoto, J. Shirafuji, and Y. Inuishi, Sol. Stat. Comm. **19**, 1037 (1976).
14. R.O. Bell, Sol. Stat. Comm. **7**, 913 (1975).
15. D. Cabib, E. Fabri, and G. Fiorio, Nuovo Ciemento **19**, 181 (1972).
16. A.L.Mears, and R.A. Stradling, Sol. Stat. Comm. **7**, 1267 (1972).
17. Le Si Dang, G. Neu, and R. Romestain, Sol. Stat. Comm. **44**, 1187 (1982).
18. W. Rü hle and D. Bimberg, Phys. Rev. **12**, 2382 (1975).
19. W. Schairer, D. Bimberg, W. Kottler, K. Cho, and M Schmidt, Phys. Rev. **13**, 3452 (1972).
20. D. Bimberg, K. Cho, and W. Kottler, **Colloque Physique sous** Champs Magnetiques Intense, p. 211 (1975).
21. F. William, D. Bimberg, M. Blätte, Phys. Rev. **7**, 2473 (1973).
22. Y. Yafet, and D.J. Thomas, Phys. Rev. **31**, 2405 (1963).
23. A. Abragam, and B. Bleaney, **Electron Paramagnetic Resonance of** Transition Ions (Claredon Press, Oxford 1970), p. 721.

Analysis of Exchange Interactions in Semimagnetic Semiconductors from High Field Magnetization

A. Bruno and J.P. Lascaray

Groupe d'Etudes des Semiconducteurs, Université des Sciences et Techniques du Languedoc, Place E. Bataillon, F-34060 Montpellier Cedex, France

Abstract: Magnetic properties of $II_{1-x}Mn_xVI$ and $Pb_{1-x}A_xVI$ (A = Eu, Gd and Mn) semimagnetic semiconductors (SMSC) have been investigated using high magnetic field magnetization (up to 42T). The results compared to available data in SMSC and magnetic semiconductors (MnVI ,EuVI,GdVI...),are analysed in order to study the exchange interactions between magnetic ions.

Recently, many efforts have been made to determine the exchange interaction values in semimagnetic semiconductors (SMSC) materials with accuracy. These experiments have employed various measurement techniques, as susceptibility, high magnetic field magnetization (HMFM), specific heat, neutron scattering, and Raman scattering ... Among these techniques, HMFM is in our opinion the method which gives the most information on magnetic properties /1 to 8/ (determination of the exchange interactions between nearest neighbors (J_1) from the step-like magnetization, analysis of concentration and distribution of magnetic ions ...). In this paper, we analyse by exchange mechanism the J_1 values determined by HMFM and collected in /9/ for II-VI SMSCs and in /10, 11, 12/ for lead salt SMSCs .

Figure 1 represents the exchange constant versus the cation-cation distance for various magnetic and semimagnetic semiconductors containing manganese and for diatomic Mn_2 molecules. The exchange constant and the cation-cation distance measured for Mn_2 in which the direct exchange is dominant correspond approximatively to the values observed in MnO, MnS, MnSe, $Mg_{1-x}Mn_xO$ and $Ca_{1-x}Mn_xO$ with NaCl crystallographic structure (curve 2). The comparison between the molecule Mn_2 and Mn compounds suggests that direct exchange may be possible in Mn compounds in agreement with KONDO's assumption /13/. However, T.N. CASSELMAN et al. /14/ contradict this hypothesis, based on the value of the ionic radius of Mn^{++} (0.8 Å) and on the fact that optical experiments on manganese salts indicate that the anions shield the manganese ions from one another. The last authors confirm the existence of a right angle superexchange in NaCl structure compounds which decrease with the direct distance as evidenced from curve 2. Consequently, the exchange constant in the molecule Mn_2 and in the solid, due to a different exchange mechanism, should not be compared although these values are in accord. As a matter of fact, in the case of Mn_2, the s-electrons of the Mn atoms allow the binding and also, through exchange, the interaction between the localised spins on the Mn2+ sites.And in the $II_{1-x}Mn_xVI$ SMSCs, the binding is obtained from Mn s-anion p hybridization and magnetic ions interact through the sp-electrons by superexchange.

The superexchange mechanism which depends on the interionic anion should be related to the cation-anion-cation (c.a.c.) angles and distances/9/. Figure 2 represents the II-VI SMSC exchange constant versus

Springer Series in Solid-State Sciences, Vol. 87
High Magnetic Fields in Semiconductor Physics II Editor: G. Landwehr

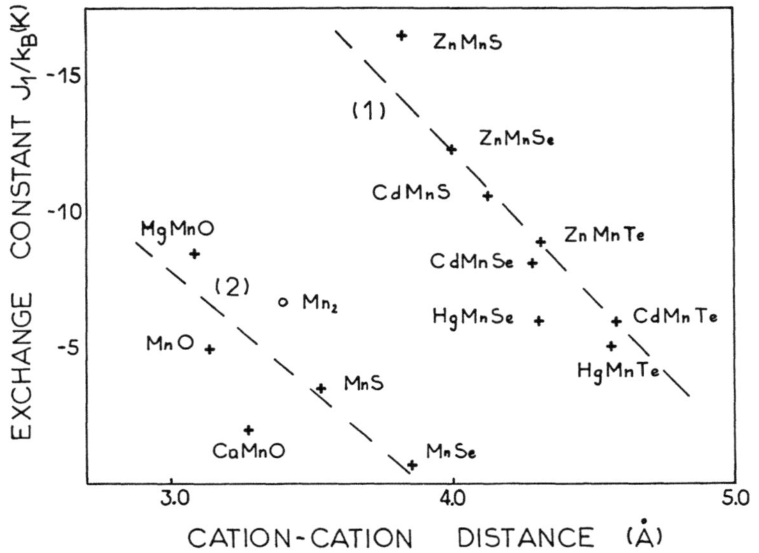

Fig.1 Nearest neighbour exchange constants, J_1, as a function of cation-cation distances for $II_{1-x}\ Mn_x\ VI$ SMSCs (curve 1) and other manganese compounds (curve 2). The circle (o) represents the values of the manganese diatomic molecule /21/. The MnS, MnSe values are given in /22/, MnO in /23/, $Mg_{1-x}Mn_xO$ and $Ca_{1-x}Mn_xO$ in /25/

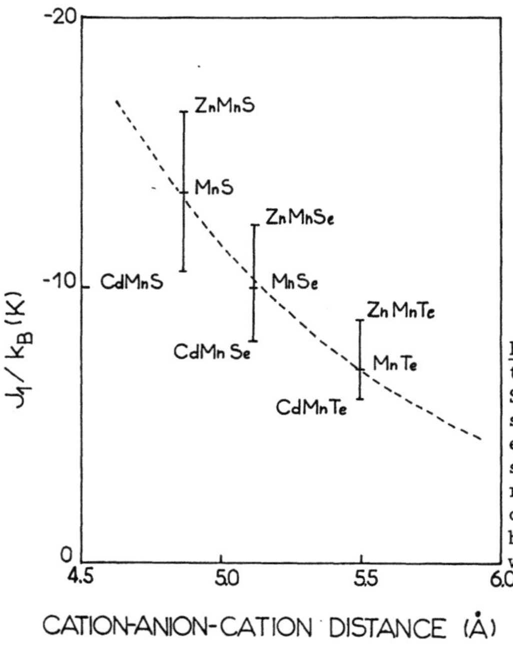

Fig. 2 : Exchange constants versus the cation-anion-cation distances. SMSC with the same anion have the same distance d_{cac} and their exchange constants are related by solid lines. The dashed line represents the J_1 variation, determined from equation /1/, of hypothetical manganese calcogenides without distorsion.

the Mn^{2+}-anion - Mn^{2+} distances. The ternary compounds with the same anion are represented with the same c.a.c. distance (d_{cac}) in agreement with the results of A.BALZAROTTI et al. /15/ and of /9/. MnS, MnSe and MnTe represent hypothetical compounds (Mn in tetrahedral sites) without distorsion for which the exchange constants depend on only c.a.c. distance. For these compounds, the c.a.c. angle is the same (θ_o = 109.6°). For zinc or cadmium SMSCs, the c.a.c. angle is respectively weaker or higher than θ_o. Across the series MnS, MnSe and MnTe, we observe a monotonically exchange constant decrease versus c.a.c. distance. This behavior may be explained from a p-d overlap decrease. The dashed line in fig. 2 is phenomenologically described by an exponential law given by : $J_1 = J_o \exp(-d_{cac})$, where d_{cac} is in Angström and J_o is a fitting parameter. Consequently, we confirm from this analysis that the superexchange is dominant in SMSCs, as evidenced theoretically by B.E.LARSON et al /16/. However , it is not the only mechanism. In mercury compounds, a ferromagnetic exchange has shown up /9/. This exchange is related to virtual interband transitions enhanced in narrow gap semiconductors.

Figure 3 represents the exchange constant between rare earth magnetic ions in the NaCl structure versus c.a.c. distance. We compare our results obtained on lead salt SMSCs ($Pb_{1-x}Gd_xTe$ /10,11/, $Pb_{1-x}Gd_xSe$ and $Pb_{1-x}Eu_xTe$ /12/) with the magnetic semiconductor exchange constants (europium /17/ and gadolinium /18/ chalcogenides). For lead salt compounds, distorsions may exist as in II-VI SMSCs /15/. Consequently, for $Pb_{1-x}R_xVI$, the possible distorsions are taken in account by the large error bars of the c.a.c. distance included between the Pb-VI and R-VI distances.

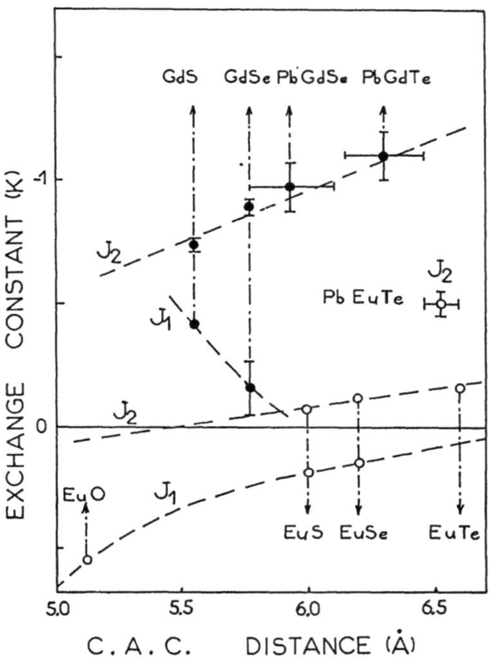

Fig 3 : Exchange constants versus c.a.c. distance between Gadolinium ions (•) and Europium ions (o) for compounds with NaCl structure.

The same exchange mechanism is expected in Gd semimagnetic and magnetic semiconductors owing to the J_2 continuity versus the c.a.c. distance. Thus, the extrapolation of J_1 values for higher c.a.c. distances ($Pb_{1-x}Gd_xSe$ and $Pb_{1-x}Gd_xTe$) confirms the J_2 predominance (J_1 may be neglected) determined from an analysis of the magnetic properties /11/. The J_1 values are of opposite signs for Eu and Gd compounds. The reason is that, for Eu chalcogenides, J_1 is mainly due to the Goodenough mechanism. The magnitude of this exchange is inversely proportional to the energy difference of d and f levels (E(d,f)) /17/. E(d,f) is very important for Gd compounds due to a large value of the Gd $4f^7$ energy below the top of the valence band /19/. Consequently, the Goodenough exchange is negligible in Gd compounds and the antiferromagnetic J_1 value may be explained by a right angle superexchange previously observed in Mn compounds with NaCl structure. This exchange decreases with the c.a.c. distance. This mechanism also exists in Eu chalcogenides, but the antiferromagnetic contribution is compensated by the prevailing ferromagnetic Goodenough exchange. We note that, up to now, the right angle superexchange has not been considered in exchange studies of Eu chalcogenides. Another interesting point, in fig.3, is the $Pb_{1-x}Eu_xTe$ exchange value which is three times more important than those in EuTe with the same lattice constant. We expect a supplementary mechanism due to interband transitions, as suggested in /20/, which is enhanced in narrow gap materials. The same interaction probably exists in Gd compounds (conductor (C) or semiconductor (SC)) and may explain the large J_2 values in comparison with those of Eu chalcogenides. This mechanism depends surely on the presence of s-electrons of SCs and Cs which do not exist in insulating .And the fact that RKKY interaction is negligible /12/ and that J_2 is predominant shows that it is also related to the p-electrons .

Consequently, this exchange interaction may be a superexchange enhanced through the s-electrons polarised by exchange with ion spins. A careful study of these mechanisms is in progress.

We have collected the exchange interaction values determined from HMFM and analysed them in terms of magnetic ion - anion - magnetic ion distance. We have shown that the exchange interaction in SMSCs depends both on the crystallography and on the band structure of the host materials. Thus, the weak values of the exchange constants in lead compounds in comparison with II-VI materials should be related mainly to the large c.a.c. distance between nearest neighbor ions which decreases considerably the superexchange (4.3 Å < d_{cac} < 5.5 Å for magnetic ions in tetrahedral sites (fig. 2) and 5.5 Å < d_{cac} < 6.5 Å in octahedral sites (fig. 3)).

References

1. R.L. Aggarwal, S.N. Jasperson, Y. Shapira, S. Foner, T. Sakakibara, T.Goto, N.Miura, K. Dwight and A. Wold, in Proceedings of the 17th International Conference on the Physics of Semiconductors, San Francisco 1984, edited by J.D. Chadi and W.A. Harrison (Springer, N.Y. 1985)p.1419.
2. R.L.Aggarwal, S.N.Jasperson, P.Becla, R.R.Galazka, Phys. Rev.B32, 5132, (1985).
3. Y. Shapira, S. Foner, P. Becla and D.N. Domingues, M.J. Naughton and J.S. Brooks, Phys. Rev. B33, 356 (1986).
4. N.Yamada, S.Takeyama, T.Sakakibara, T.Goto and N.Miura, Phys. Rev. B34, 4121 (1986).
5. G. Barilero, C. Rigaux, Nguyen Hy Hau, J.P. Picoche, N. Giriat, Solid State Commun. 62, 345 (1987).

6. J.P. Lascaray, M. Nawrocki, J.M. Broto, M. Rakoto and M. Demianiuk, Solid State Commun. 61, 401 (1987).
7. J.P. Lascaray, A. Bruno, M. Nawrocki, J.M. Broto, J.C. Ousset, S.Askenazi and R. Triboulet, Phys.Rev. B35, 6860 (1987).
8. Y. Shapira, N.F. Oliveira, Jr. Phys. Rev. B35, 6888 (1987).
9. J.P. Lascaray and A. Bruno, International Conference on Magnetism, Paris, 25-29 July 1988. To be published in J. de Physique.
10. A. Bruno, J.P. Lascaray, M. Averous, J.M. Broto, J.C. Ousset and J.F. Dumas. Phys.Rev. B35,2068 (1987).
11. A.Bruno, J.P.Lascaray, M.Averous, G.Fillion, J.F.Dumas, Phys.Rev. B37, 1186, 1988-II.
12. A. Bruno, J.P. Lascaray and M. Averous, unpublished.
13. J. Kondo, Progr. Theoret. Phys. (Kyoto) 22, 41 (1959).
14. T.N. Casselman, F. Keffer, Phys. Rev. B4, 498 (1960).
15. A. Balzarotti, N. Motta, A. Kissiel, M. Zimnal - Starnawska, M.T.Czyzyk and M. Podgorny, Phys. Rev. B31, 7526 (1985).
16. B.E. Larson, K.E. Hass, H. Ehrenreich, A.E. Carlsson - Solid State Commun., Vol 56, p 347-350 (1985).
17. P.Wachter, Handbook on the physics and chemistry of rare earths edited by K.A.Gschneidner and L.Eyring, North Holland Publishing Company (1979).
18. R. Hauger, E. Kaldis, G. Von Schultess, P. Wachter and Ch. Zürcher, Journal of Magnetism and Magnetic Materials 3 (1976) 103-108.
19. E. Kaldis, A. Schlegel, P. Wachter and Ch. Zürcher - Journal of Magnetism and Magnetic Materials 3 (1976) 1-3.
20. L. Liu, Phys. Rev. B26, 975 (1982).
21. C.A. Baumann, R.J. Vanzee, S.V. Bhat, W. Weltner, Jr., J. Chem. Phys. 78 (1) (1983).
22. J.J. Banewicz, R.G. Heidelberg, Allan H. Luxem.,J. Phys. Chem.65, 615 (1961).
23. M.E. Lines and E.D. Jones, Phys. Rev. 139, A1313 (1965).
24. E.A. Harris, J. Phys. C : Solid State Phys. Vol 5, 338 (1972).

Critical Behavior of the Hall Coefficient and Dielectric Susceptibility near the Anderson-Mott Transition in p-$Hg_{1-x}Mn_xTe$

J. Jaroszyński, T. Dietl, M. Sawicki, T. Wojtowicz, and W. Plesiewicz

Institute of Physics, Polish Academy of Sciences,
Al. Lotników 32/46, PL-02-688 Warszawa, Poland

Results of millikelvin measurements of magnetoresistance, Hall effect, and dielectric susceptibility near the magnetic field-induced non-metal-metal transition are presented for $Hg_{0.92}Mn_{0.08}Te$. The critical behavior is discussed in terms of the Anderson-Mott transition, driven by electron-electron interactions in a spin-polarized band.

1. INTRODUCTION

It is well-known that doped semiconductors undergo a transition from insulating to metallic phase when the impurity concentration is increased. Under suitable circumstances, this transition can be tuned by an external perturbation, such as stress or a magnetic field [1,2]. It now seems that the transition is driven by disorder-modified electron-electron interactions, at least in the situation when long living spin fluctuations are destroyed by spin-flip scattering or spin-splitting [3]. The model predicts σ near the metal-insulator transition (MIT) to be continuous, and characterized by a critical exponent $\nu = 1$, in the expression $\sigma(y) = \sigma_c(y/y_c - 1)^\nu$, where y is a parameter that controls the disorder. Such an exponent is indeed observed in most materials [1,2]. Moreover, a universal relation is expected [3] between the exponent α describing the temperature dependence of the conductivity at $y = y_c$, $\sigma(T) \sim T^\alpha$, and that which characterizes the divergence of the static dielectric susceptibility χ, $\zeta = 1 - 1/\alpha$, where $\chi \sim (y_c - y)^\zeta$.

In the case of narrow-gap p-$Hg_{1-x}Mn_xTe$, the insulator-to-metal transition can be tuned by a magnetic field [4]. The dominant effect that leads to the transition is redistribution of holes between spin-split Γ_8 subbands, produced by a giant p-d exchange spin-splitting. This splitting is known [5] to be connected with a change of the transverse hole mass from a heavy mass to a light one. In consequence, in classically weak magnetic fields σ exhibits rather unusual anisotropy, $\sigma_{xx} > \sigma_{zz}$, where z is the magnetic field direction. Furthermore, the redistribution results in strong spin-polarization of the hole liquid, and thus in significant reduction of its spin fluctuations.

Our previous millikelvin studies of σ at criticality in p-$Hg_{0.915}Mn_{0.085}Te$ have demonstrated that $\nu = 1^{+0.2}_{-0.1}$ for both transverse and longitudinal configurations [4]. Moreover, by fitting the dependence of σ on T, the value of α was estimated to be $0.65^{+0.4}_{-0.2}$ [4]. Besides, the behavior of phonon-assisted hopping conductivity was analyzed [6] and found to be consistent with the presence of a soft Coulomb gap in the one-particle density of states.

This paper presents results of the most recent studies, in which – in addition to the resistance measurements – the dielectric susceptibility χ and the Hall coefficient R_H have been examined near the MIT in a p-$Hg_{0.92}Mn_{0.08}Te$ sample.

Springer Series in Solid-State Sciences, Vol. 87
High Magnetic Fields in Semiconductor Physics II Editor: G. Landwehr

2. SAMPLES AND EXPERIMENT

A crystal of $Hg_{0.92}Mn_{0.08}Te$ was grown by the Bridgman method and doped with compensating In impurities to the level of $7 \times 10^{15}cm^{-3}$. The samples were cut from carefully selected single-crystalline grains. The net acceptor concentration, $p \equiv N_A - N_D$, estimated from the plateau values of the Hall coefficient in the range of $50 - 70$ K, was found to be $1.5 \times 10^{17}cm^{-3}$. Mercury vacancies are thought to form acceptor centers in HgTe alloys. The samples used for dielectric susceptibility and transport measurements were cut from the same slice. After etching, thin indium layers were evaporated on opposite faces of one of the samples, forming a capacitor with an area $\sim$ 5 mm^2 and thickness $\sim$ 300 μm.

The measurements were performed in a dilution refrigerator, down to 30 mK and up to 60 kOe. To determine the dielectric susceptibility, the complex conductance $Y = G + i\omega C$ was measured by using a three probe a.c. bridge, operating at 93 kHz. Small voltages ($\leq 10^{-4}$V) and low input powers ($\sim 10^{-12}$ W) were used. Conventional, six probe resistance and Hall effect measurements were made using of a d.c. technique. The input powers were kept below 10^{-11} W. All the results were obtained in the ohmic regime.

3. RESULTS AND DISCUSSION

Figure 1 presents the temperature dependence of $1/R_H$, and of equivalent circuit capacitance. The latter gives the capacitance of the sample in series with that of a Schottky contact barrier. It is seen that, at sufficiently low temperatures, $1/R_H$ approximately obeys a $T^{1/2}$ law, as is commonly observed in the case of conductivity. The capacitance C exhibits a much stronger temperature dependence, $C(T) = C(0) + BT^2$, as noted previously in Si:P [7] and AlGaAs [8]. As may be seen in Fig. 1, our results for the lowest

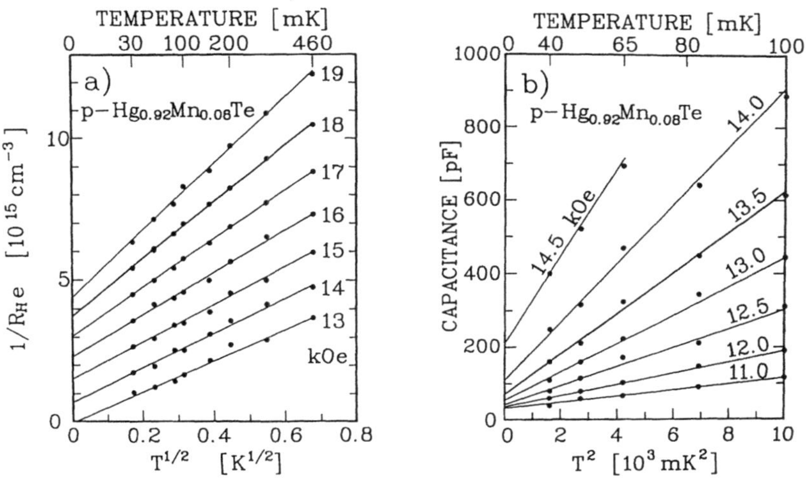

Fig. 1 (a) Inverse Hall coefficient vs. square root of the temperature, and (b) parallel equivalent circuit capacitance vs. the square of the temperature, in various magnetic fields. The straight-line fits indicate the $T^{1/2}$ and T^2 dependences for $1/R_H$ and C, respectively.

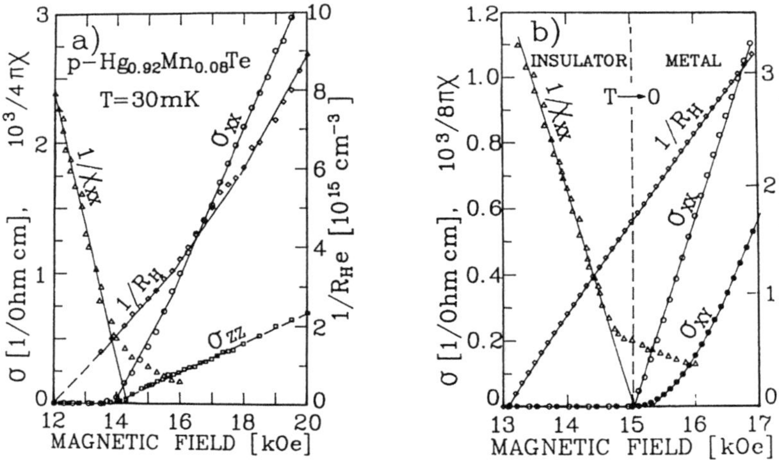

Fig. 2 (a) The conductivity tensor components σ_{xx} and σ_{zz}, as well as the inverse of the Hall coefficient R_H and of the dielectric susceptibility χ, measured near the insulator-to-metal transition at 30 mK. (b) Values of χ, R_H, σ_{xx} and σ_{xy}, extrapolated to $T = 0$. The solid lines represent critical behavior, with parameters displayed in Table 1.

measurement temperature give values of R_H and χ close to those found by extrapolation to $T = 0$.

Figure 2a shows $1/R_H$ and $1/\chi \sim 1/C$ at 30 mK in the critical region, together with the conductivity tensor components σ_{xx} and σ_{zz}, extracted from the Hall effect and resistivity measurements. Values of $1/R_H$, $1/\chi$, σ_{xx} and σ_{xy}, extrapolated to $T = 0$, are depicted in Fig. 2b.

It is seen in Fig. 2 that χ strongly increases when the MIT is approached from the insulating side. However, it does not diverge to infinity, most probably because of the presence of the Schottky barrier capacitance $C_S \approx 1$ nF. If the region where C_S becomes important is disregarded, the critical field deduced from the dependence of χ on H coincides with that given by $\sigma(H)$ measurements, $H_c \simeq 15$ kOe. Furthermore, the dependence of $1/\chi(H)$ is seen to be essentially linear with the magnetic field. This implies that $\zeta = -1 \pm 0.2$, which gives $\alpha = 0.5 \pm 0.1$. The latter is to be compared to the value deduced from the dependence of σ on T, namely, $\alpha = 0.65^{+0.4}_{-0.2}$ [4]. It should be noted, however, that on the one hand, the experimental data are perturbed by a Schottky barrier capacitance (whose field dependence is unknown) and, on the other, the contribution from phonon-assisted processes may not be negligible [9]. By fitting the $T = 0$ values with the formula $X = X_c \mid H/H_c - 1 \mid^{\delta}$, we obtain the values of X_c, H_c and δ given in Table 1.

Similary, in studying the field dependence of the Hall coefficient, we found that R_H does exhibit critical behavior, as shown in Fig. 1. The critical field turns out, however, to be substantially lower that than deduced for the $\chi(H)$ and $\sigma(H)$ dependences. Since MIT occurs in classically weak fields ($\sigma_{xx} > \sigma_{xy}$), the latter implies that the critical exponent of σ_{xy} is twice that for σ_{xx}. This, and the results for other semiconductors [10], call for theoretical studies of R_H near the MIT, with electron-electron interactions included in these considerations [11].

Table 1. The parameters obtained from fits of the experimental data, extrapolated to $T = 0$ with the formula $X = X_c \mid H/H_c - 1 \mid^\delta$ in p-$Hg_{0.92}Mn_{0.08}Te$.

X	X_c	H_c [kOe]	Critical exponent
σ_{xx} [$(\Omega\ \mathrm{cm})^{-1}$]	9.5±0.5	15.1±0.2	1.05±0.1
σ_{xy} [$(\Omega\ \mathrm{cm})^{-1}$]	31.2±5	15.1±0.2	1.89±0.1
σ_{zz} [$(\Omega\ \mathrm{cm})^{-1}$]	1.6±0.2	14.9±0.2	1.06±0.1
R_H [cm^3/C]	670±70	13.1±0.2	−0.9±0.1
$4\pi\chi$	60[a)]	15.0±0.3	−1±0.2

a) the error in the absolute value of χ_c is probably of the order of 100%.

REFERENCES

1. see, e.g., G.A. Thomas: *Phil. Mag.* **52**, 479 (1985)
2. see also, Anderson Localization, eds. T. Ando, H. Fukuyama (Springer, Berlin 1988)
3. A. M. Finkel'stein: *Zh. Eksp. Teor. Fiz.* **86**, 367 (1984) [*Sov. Phys. JETP* **59** 212 (1984)]; C. Castellani, C. Di Castro, P. A. Lee, M. Ma *Phys. Rev.* **B30**, 527 (1984)
4. T. Wojtowicz et al.: *Phys. Rev. Lett.* **56**, 2419 (1986); T. Wojtowicz et al.: In *High Magnetic Fields in Semiconductor Physics*, ed. by G. Landwehr, Springer Ser. Solid-State Sci., Vol. 71, 442 (Springer, Berlin, Heidelberg 1987). T. Dietl et al.: in Ref. 2, p. 58
5. J.A. Gaj, J. Ginter, R.R. Gałązka: *Phys. Status Solidi (b)* **89**, 655 (1978); A. Mycielski, J. Mycielski: *J. Phys. Soc. Jpn.* **49**, Suppl. A:797, (1980)
6. T. Wojtowicz et al.: *Proc. Int. Conf. on High Field Magnetism*, Leuven 1988, *Physica B*, in press
7. M.A. Paalanen, T.F. Rosenbaum, G.A. Thomas, R.N. Bhatt: *Phys. Rev. Lett.* **51**, 1896 (1983)
8. S. Katsumoto, in Ref. 2, p.45
9. R.N. Bhatt, *J. Phys. C* **18** L537 (1985)
10. D.W. Koon, T.G. Castner: *Phys. Rev. Lett.* **60**, 1755 (1988), and the references therein
11. for the theory of R_H in the absence of interactions, see, B. Shapiro, E. Abrahams: *Phys. Rev.* **B24**, 4025 (1981)

Magnetophonon Resonance Recombination Studies of $Hg_{1-x-y}Cd_xMn_yTe$ Using LPE Crystals

T. Uchino, K. Takita, and K. Masuda

Institute of Materials Science, University of Tsukuba,
Tsukuba City 305, Japan

A new type of magnetophonon effect associated with a recombination process with two TA-phonon emissions is studied in diluted magnetic semiconductors $Hg_{1-x-y}Cd_xMn_yTe$ by using high quality LPE crystals. The band parameters are determined for the zero-gap region from the analysis of experimental results of magnetophonon resonance (MPR) recombination. An energy gap dependence of the resonant acceptor states is also discussed based on the MPR recombination measurements.

1. Introduction

The observation of a new type of magnetophonon resonance effect peculiar to zero-gap semiconductors has been reported in HgTe and $Hg_{1-y}Mn_yTe$ ($y \leq 1.0\%$) using liquid phase epitaxy (LPE) crystals [1-3]. This is the MPR recombination of heated electrons and holes with emission of two TA-phonons. This paper mainly reports experimental results of extended studies of this type of magnetophonon effect in diluted magnetic semiconductors $Hg_{1-x-y}Cd_xMn_yTe$. Diluted magnetic semiconductors $Hg_{1-x-y}Cd_xMn_yTe$ have recently attracted much attention in semiconductor physics [4], because their electronic band structure and magnetic properties can be controlled by varying the alloy composition. However, the study of $Hg_{1-x-y}Cd_xMn_yTe$ is not sufficient to understand the electronic structure in comparison with the other narrow gap semiconductors. By carrying out MPR recombination measurements in high magnetic fields, it is possible to determine the accurate value of the band parameters near the band edge, because TA phonons with definite energies in the range of 2∿4meV were used as a probe.

2. Samples

$Hg_{1-x-y}Cd_xMn_yTe$ crystals have been grown on CdTe substrates for the first time by LPE method from Hg-rich solutions. The thickness of the LPE layers was from 40μm to 150μm. The single crystal nature of these crystals is confirmed by X-ray diffraction. The compositions of the epilayers were measured by an X-ray microprobe analyzer (XMA). From XMA analyses, LPE layers show excellent compositional uniformity in the surface and along the growth direction. In the interface region, however, compositional gradients were observed within the width of less than 20μm, which is due to the interdiffusion between source materials and the substrates. By using the LPE crystal growth method, samples were obtained in various alloy compositions in the range: $0 \leq x \leq 0.3$ and $0 \leq y \leq 0.1$. In the early work on $Hg_{1-x}Cd_xTe$ [5], it was reported that epilayers with homogeneous composition were not obtained from Hg-rich solutions. The present result of $Hg_{1-x-y}Cd_xMn_yTe$ is in contrast to the early data. By cutting from a

Springer Series in Solid-State Sciences, Vol. 87
High Magnetic Fields in Semiconductor Physics II Editor: G. Landwehr

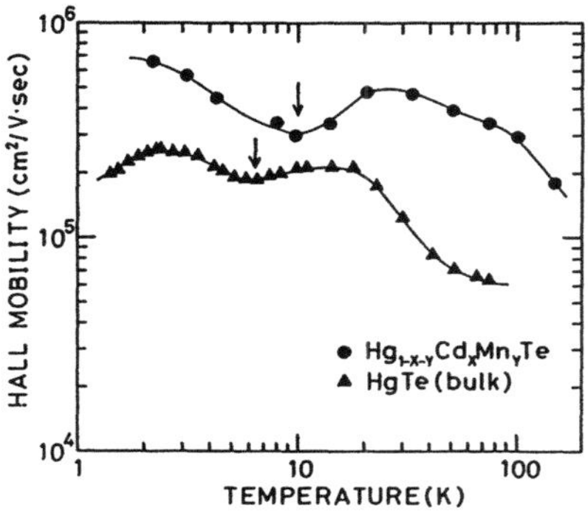

Fig.1 Temperature dependence of electron Hall mobility for LPE-$Hg_{1-x-y}Cd_xMn_yTe$ sample with x=4.1%, y=1.2%. For comparison, the result of bulk-HgTe sample is also shown. This bulk-HgTe sample with high mobility was obtained by annealing in Hg-vapor for more than 5 months at 185°C.

single wafer, three different samples were prepared for longitudinal magnetoresistance measurements with current and magnetic field parallel to high symmetry crystallographic directions, i.e., <100>, <110> or <111>.

An example of the temperature dependence of the electron Hall mobility is shown for LPE-$Hg_{1-x-y}Cd_xMn_yTe$ sample with x=4.1%, y=1.2% (i.e., $Eg \simeq -145meV$) in Fig.1. The result of bulk-HgTe sample is also shown for comparison. This bulk-HgTe sample with high mobility was grown by a modified Bridgman method and annealed for more than 5 months at 185°C in Hg-vapor. It can be seen that the LPE crystal shows high mobility even as-grown. A large dip is observed around 10K in the electron mobility curves as is shown by the arrows in Fig.1. A shift of the dip position to higher temperatures was observed with increasing energy gap. The anomaly in the temperature dependence of the Hall mobility can be interpreted as effect of an acceptor state. This acceptor can also be observed in magneto-optical measurements [6] and transport data [7] in $Hg_{1-x}Cd_xTe$.

3. Magnetophonon Resonance Recombination

Second derivatives of the longitudinal magnetoresistance under heated electron conditions were investigated in $Hg_{1-x-y}Cd_xMn_yTe$ at 4.2K. In this experiment, samples with negative energy gaps were investigated. Fig.2 shows typical examples of second derivative curves and the calculated Landau-level scheme of $Hg_{1-x-y}Cd_xMn_yTe$ with x=4.1%, y=1.2% for the case of H//<100>. Some small peaks were observed in the upper curve of $-\partial^2\rho/\partial H^2$ under heated electron conditions. Significant structure was not observed in the lower curve in comparison with the upper one, because it is still in the ohmic conduction region. The observed peaks are interpreted to be due to recombination of heated electrons and holes with emission of two TA-phonons as is shown schematically in the upper part of Fig.2. A modified Pidgeon-Brown (MPB) model [8] including exchange terms was applied to analyze the experimental data. The calculated energy scheme based on the model and the interpretation of the observed peaks are shown in Fig.2. As the band parameters of the MPB model in the calculation, previously obtained $Hg_{1-y}Mn_yTe$ values are used. In Fig.2, arrows with solid and broken lines show the transitions connected with the emission of two TA-phonons with energies of 5.2meV and 6.2meV, respectively. The energies of the TA-phonons are determined by similar MPR recombination measurements in HgTe. As can be seen in Fig.2, the peaks around 3.5T and 6.5T did not show good agreement with the calculated results and shifted to higher magnetic fields. This disagreement may be explained by assuming that the TA-phonon

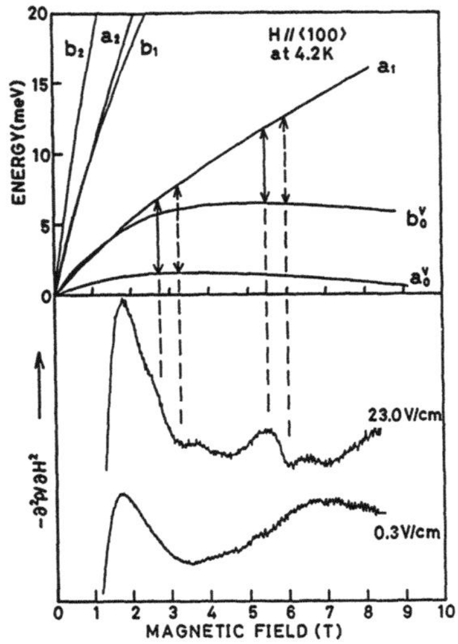

Fig.2 Typical examples of second derivative curves of $Hg_{1-x-y}Cd_xMn_yTe$ (x=4.1%, y=1.2%) and the calculated Landau-level scheme in the case of H//<100>.

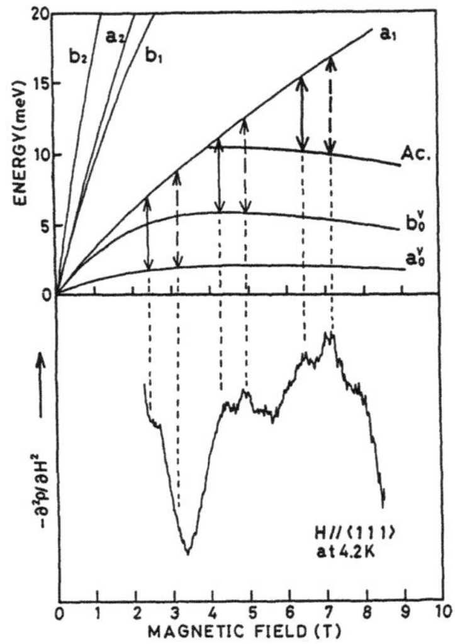

Fig.3 A comparison of the observed oscillation with the calculated Landau-levels in the case of H//<111>, for the sample with x=4.1%, y=1.2%.

density of states has two peaks at 2.6meV and 3.4meV in $Hg_{1-x-y}Cd_xMn_yTe$ in contrast to the values of 2.6meV and 3.1meV in HgTe. Based on this analysis, the exchange integral β has been determined as β=0.62eV. This value of β agrees with our previous results in $Hg_{1-y}Mn_yTe$ [2].

The magnetic field direction dependences of the peak positions were also investigated in relation to the crystallographic axis. In the case of H//<111>, a typical example of the observed oscillations is shown in the lower part of Fig.3. The peak positions are different from the H//<100> case in Fig.2. This means that the Landau-level scheme is different due to the anisotropy of the band. The peak positions in the case of H//<111> were shifted a little bit to higher fields with respect to calculations using the band parameters of HgTe. This fact means that the anisotropy factor is very large provided that the alloy composition in a single wafer of the LPE-crystal can be assumed to be homogeneous. Consequently, the band parameters of the MPB-model for $Hg_{1-x-y}Cd_xMn_yTe$ are deduced as follows; γ_1=3.0, γ_2=-0.5, γ_3=1.7, κ=-1.3, P=8.3x10^{-8}eV•cm, Δ= 1.08eV, β=0.62eV. The set of γ-parameters is that of GROVES et al. [9] but modified in γ_3 to express the observed anisotropy of the band. The value of 2.2 of the anisotropy factor (γ_3-γ_2) is larger than that of GROVES et al. (1.2) and our previous results in HgTe and $Hg_{1-y}Mn_yTe$ for y≦1.0% (1.5). However, this value agrees with the value of MIURA et al. [10] for $Hg_{1-x-y}Cd_xMn_yTe$.

A couple of peaks around 6.5T in Fig.3 are not assigned to transitions between Landau-levels. These peaks are interpreted as transitions associated with an impurity level which is located at E_A above the top of the valence band as illustrated by the bold line denoted by Ac. in the upper part of Fig.3. Here E_A=4.4meV is assumed. This acceptor binding energy is larger than that of $Hg_{1-y}Mn_yTe$ with y=0.5% deduced from the MPR recombination measurements (i.e., E_A=2.0±0.2meV) [3]. This tendency of the acceptor binding energy to increase with energy gap is also observed in the temperature dependence of the Hall mobility as mentioned in the preceding section.

Figure 4 shows an example of the MPR recombination oscillations of $Hg_{1-x-y}Cd_xMn_yTe$ with x=1.1%, y=0.65% (i.e., Eg=-216meV) for H//<100>. The agreement between the calculated energy scheme and the experimental results is good as can be seen in Fig.4. It should be noted that the calculation was carried out using the above mentioned band parameters. Here we also assume that the TA phonon density of states has two peaks at 2.6meV and 3.4meV. As described in this paper, the experimental results of $Hg_{1-x-y}Cd_xMn_yTe$ are explained fairly well by the proposed new set of band parameters.

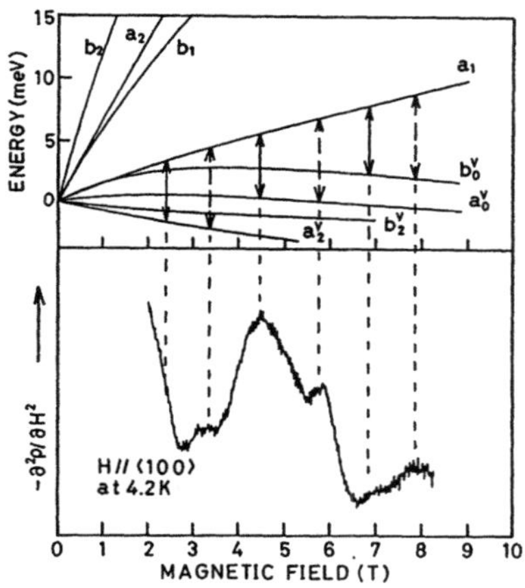

Fig.4 Landau-level scheme of $Hg_{1-x-y}Cd_xMn_yTe$ with x=1.1%, y=0.65% for H//<100> case (upper part). Possible magnetophonon resonance recombination transitions for $2\hbar\omega_{TA}$= 5.2meV and 6.8meV are indicated by arrows with solid lines and broken lines, respectively. The lower part is an example of the observed resonance.

References

1. K. Takita, T. Ippōshi and K. Masuda: Solid State Commun. 52, 1021 (1984).
2. K. Takita, T. Uchino, T. Ippōshi and K. Masuda: Solid State Commun. 56, 603 (1985).
3. K. Takita, T. Uchino, T. Gochou and K. Masuda: Solid State Commun. 61, 535 (1987).
4. R. R. Galazka: Proc. 18th Int. Conf. Physics of Semiconductors, Stockholm, (1986) ed. by O. Engstrom, (World Scientific, Singapore, 1987) p.1727.
5. J. E. Bowers, J. L. Schmit, C. J. Speerschneider and R. B. Maciolek: IEEE Trans. Electron Devices ED-27, 24 (1980).
6. Y. Guldner, C. Rigaux, A. Mycielski and Y. Couder: phys. stat. sol. (b) 81, 615 (1977).
7. A. Mauger and J. Friedel: Phys. Rev. B12, 2412 (1975).
8. M. Jaczynski, J. Kossut and R. R. Galazka: phys. stat. sol. (b) 88, 73 (1978).
9. S. H. Groves, R. N. Brown and C. R. Pidgeon: Phys. Rev. 161, 779 (1967).
10. N. Miura, S. Takeyama, S. Tanaka and N. Yamada: Proc. 18th Int. Conf. Physics of Semiconductors, Stockholm, (1986) ed. by O. Engstrom, (World Scientific, Singapore, 1987) p.1775.

Interband Electron-Phonon Interaction in Magnetooptics of $Hg_{1-x}Mn_xTe$

P. Pfeffer and W. Zawadzki

Institute of Physics, Polish Academy of Sciences,
PL-02-668 Warsaw, Poland

Resonant two-mode polarons due to phonon-induced electron transitions between the conduction and the valence bands are considered and shown to describe the observed effect in the zero-gap semimagnetic material $Hg_{0.95}Mn_{0.05}Te$. The interband electron-phonon interaction allows one to investigate an involved structure of the heavy-hole magnetic subbands in zero-gap semiconductors.

The existing theoretical work on resonant polarons has been concerned with intraband situations in which the electron-phonon interaction (EPI) couples electron states in the same band. However, the smallness of the gap opened by a magnetic field in symmetry-induced zero-gap semiconductors[1] offers a possibility to match the gap to the optic phonon energy $\hbar\omega_L$ and thus to observe interband polarons in which EPI couples electron states across the gap. Such interband resonant polarons have been observed recently by Bauer et al. [2] in $Hg_{1-x}Mn_xTe$ by means of infrared magneto-absorption. In the following we describe the interband polaron behavior, interpret the results of [2] and indicate new possibilities of interband EPI in investigations of the electronic spectrum.

The initial Hamiltonian reads

$$H = P^2/2m_o + V_o + H_{so} + H_{exch} + H_F + H_{phn} , \qquad (1)$$

where $\vec{P} = \vec{p} + e\vec{A}$, V_o is the periodic potential, H_F is EPI, H_{phn} describes the free phonons, and $H_{exch} = J(\vec{r}-\vec{R})\vec{S}\cdot\vec{\sigma}$ is the exchange interaction between Mn ions and mobile electrons. The first four terms describe a semimagnetic semiconductor in the presence of a quantizing magnetic field. The resulting Landau levels are described within the Pidgeon and Brown model [3] completed by the exchange terms [4]. Band parameters and exchange characteristics are given in [2]. The 8x8 Hamiltonian is diagonalized numerically giving nonparabolic electron and heavy-hole energies in absence the of EPI.

$Hg_{1-x}Mn_xTe$ has two longitudinal optic modes. The polar coupling constants have been determined using the procedure of Swierkowski et al. [5].

$$\gamma_1^2 = \frac{1}{q^2}\,\frac{2\pi e^2\hbar\omega_{L1}}{\varepsilon_\infty V}\left(1 - \frac{\omega_{T1}^2}{\omega_{L1}^2}\right)\left(\frac{\omega_{T2}^2 - \omega_{L1}^2}{\omega_{L2}^2 - \omega_{L1}^2}\right) \qquad (2)$$

Springer Series in Solid-State Sciences, Vol. 87
High Magnetic Fields in Semiconductor Physics II Editor: G. Landwehr

$$\gamma_2^2 = \frac{1}{q^2} \frac{2\pi e^2 \hbar\omega_{L2}}{\varepsilon_\infty V} \left(1 - \frac{\omega_{T2}^2}{\omega_{L2}^2}\right)\left(\frac{\omega_{T1}^2 - \omega_{L2}^2}{\omega_{L1}^2 - \omega_{L2}^2}\right) . \tag{3}$$

For x = 0.05 we have used $\hbar\omega_{L1} = 17.13$, $\hbar\omega_{T1} = 14.67$, $\hbar\omega_{L2} = 24.06$, $\hbar\omega_{T2} = 23.82$ (in meV, cf. [6]) and $\varepsilon_\infty = 10.6$. The resulting polar constant for the first mode is almost ten times larger than for the second one.

We are interested in energies of the upper (conduction) state cn (at $k_z = 0$) perturbed by EPI with the lower (heavy hole) state hn'. The electron self-energy is

$$\Sigma_c^n(E) = \sum_q \frac{|\langle q, hn' | H_F | 0, cn\rangle|^2}{E + i\gamma_o - (\varepsilon_h^{n'} + \hbar\omega_L)} - i\gamma_o . \tag{4}$$

Contribution $i\gamma_o$ eliminates a nonphysical divergence. The polaron energies are given by maxima of the spectral function

$$A_c^n(E) = -\frac{1}{\pi} \frac{\Gamma}{(E - \varepsilon_c^n - \Delta)^2 + \Gamma^2} , \tag{5}$$

where $\Sigma(E) = \Delta(E) + i\Gamma(E)$.

We consider first the phonon-induced interband transition $b_c(0) \rightarrow b_{hh}(-1)$. The $b_{hh}(-1)$ state has a simple dispersion: $\varepsilon = -b_o - b_2 k_z^2$. Fig. 1 shows the calculated polaron branches in good agreement with experimental data. The next phonon-induced transition is $a_c(0) \rightarrow a_{hh}(-1)$. The dispersion of $a_{hh}(-1)$ state has a "camel-back" shape (cf. Fig. 2). Since in phonon-induced transitions there is a sizeable momentum transfer, the electron from the bottom of $a_c(0)$ subband can go not only to the local minimum of $a_{hh}(-1)$ subband at $k_z = 0$ but also to the two local maxima at $k_z \neq 0$. Both these

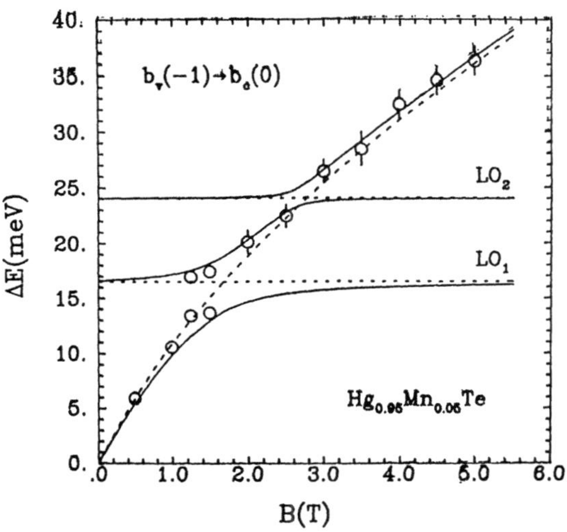

Fig.1 Two-mode interband polaron in $Hg_{0.95}Mn_{0.05}Te$ due to $b_c(0) \rightarrow b_{hh}(-1)$ transition. Solid lines are calculated, open circles experimental [2].

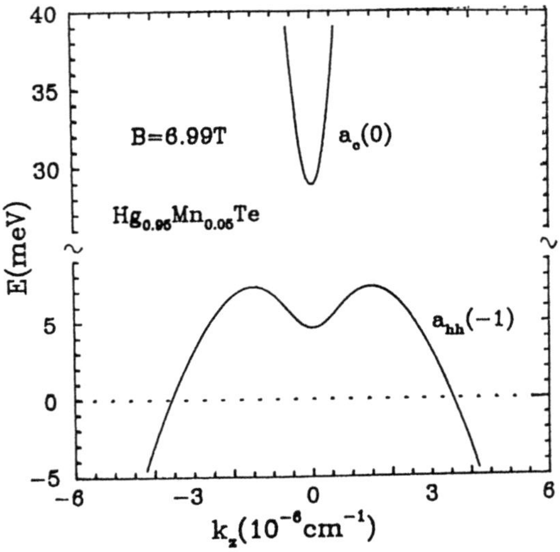

Fig.2 Dispersions of $a_c(0)$ and $a_{hh}(-1)$ Landau-Luttinger subbands in $Hg_{0.95}Mn_{0.05}Te$ at B = 6.99 T.

points have a singular density of states and they cause a resonant polaron behavior acting as two separate levels. This results in a more complicated polaron pattern, shown in Fig.3.

A particularly interesting situation occurs at the resonance with the second phonon mode (around B = 7T), which has not been investigated experimentally. At these fields the calculated spectral function predicts almost equal intensities of all three polaron branches (Fig. 4). An analysis of the resonant polaron behavior in this range gives a possibility to reveal the camel-back shape and to measure directly the energy difference between the local maxima and the local minimum of the k_z-dispersion.

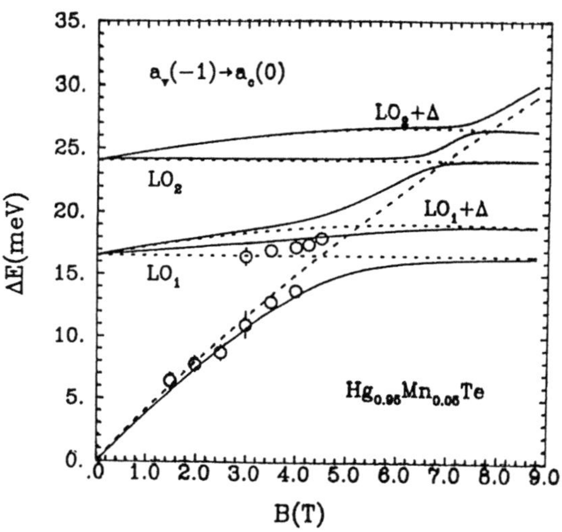

Fig.3 Two-mode interband polaron in $Hg_{0.95}Mn_{0.05}Te$ due to $a_c(0) \rightarrow a_{hh}(-1)$ transition. Solid lines are calculated, open circles experimental [2].

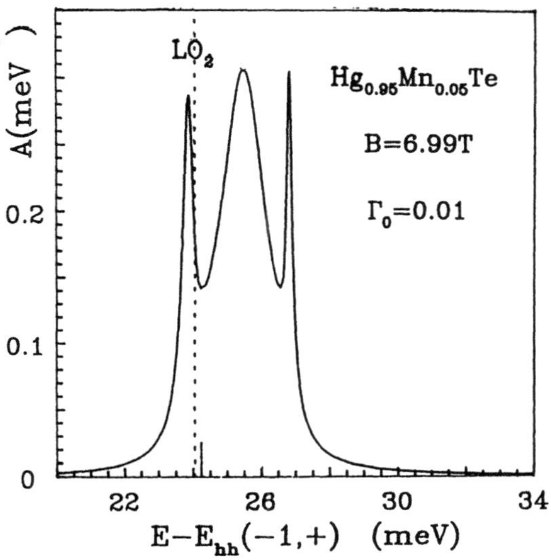

Fig.4 Spectral function of the polaron shown in Fig.2 near resonance with the second optic mode

We have also described resonant two-mode intra-conduction-band polarons in $Hg_{1-x}Mn_xTe$ crystal with x = 0.106, observed in [2]. Again a good account of the experimental data has been obtained with the use of phonon frequencies determined in ref. [6] and ε_∞ = 14.5, confirming our understanding of the band structure and dielectric properties of the semimagnetic zero-gap $Hg_{1-x}Mn_xTe$ semiconductors.

We are pleased to acknowledge informative discussions with Dr J.Kossut.

Literature

1. J. Kowalski and W. Zawadzki, Sol. State Comm. 13, 1433 (1973).
2. G. Bauer et al. Phys. Rev. B31, 2040 (1985).
3. C.R. Pidgeon and R.N. Brown, Phys. Rev. 146, 575 (1966).
4. R.R. Galazka and J. Kossut, Springer Lecture Notes in Physics, Vol. 133 (Ed. W. Zawadzki, Berlin 1980) p. 245.
5. L. Swierkowski et al. Sol. State Comm. 27, 1245 (1978).
6. W. Gebicki and W. Nazarewicz, Phys. Stat. Sol. (b) 80, 307 (1977).

Part IX

Electron-Phonon Interaction, Magneto-phonon Effect

Magnetic Field Dependence of Acoustic Phonon Emission and Scattering in 2D Electron Systems

L.J. Challis, A.J. Kent, and V.W. Rampton

Department of Physics, University of Nottingham,
Nottingham NG7 2RD, UK

Magnetic field-dependent phonon emission and scattering from a two-dimensional electron gas (2DEG) is reviewed.

1. INTRODUCTION

Investigation of phonon emission and scattering from a 2DEG is already a rich field of study. The information it is capable of providing on the electron-phonon interaction is much more detailed than that obtainable in any other way since fewer averages are involved over phonon wave-vector and polarization. It is also clear that the techniques provide a valuable diagnostic tool for studying properties such as hot electron effects, Quantum Hall breakdown and so on. We concentrate here on magnetic field-dependent effects. A more general review of work in this area up to February 1987 is given by CHALLIS et al [1].

2. PHONON EMISSION

At low temperatures ($kT_e \ll \hbar\omega_{opt}$), the predominant energy loss mechanism from a heated 2DEG is by acoustic phonon emission. There are marked differences in the angular distribution of the phonons emitted from that of a 3D system because of the wave-vector selection rule $q_{\parallel} \lesssim 2k_F$. This has the effect of constraining most of the emission to lie within a cone of half-angle determined by $2k_F$, ROTHENFUSSER et al [2]. The frequency distribution may often not be very different from that of a 3D heater, however, since in both cases it is usually dominated by the density of states and occupation numbers of the 3D phonons [3].

Major changes occur in both distributions in quantising magnetic fields. When $kT_e \sim \hbar\omega_c$, a significant proportion of the emission should be at the cyclotron frequency, ω_c, and its harmonics with the rest appearing at lower frequencies as a result of intra-Landau level transitions. Evidently the proportion at ω_c depends on parameters such as the position of the Fermi energy, the ratio $kT_e/\hbar\omega_c$ and the width of the Landau levels. Detailed calculations assuming only cyclotron emission ($kT_e \sim \hbar\omega_c$, $\mu B \gg 1$) have been made by TOOMBS et al [4] and an example of the angular distribution of the TA emission from a Si MOSFET is given in Fig. 1. The Fermi level is midway between the N=0 and N=1 Landau levels. (The calculation neglects phonon focussing effects which are particularly strong for $\vartheta \sim 0°$). The structure at ω_c is due to the different angular distributions of the various $N+1 \rightarrow N$ Landau transitions contributing to the emission, in particular the $1 \rightarrow 0$ and $2 \rightarrow 1$. As T_e falls to $kT_e \ll \hbar\omega_c$, the dominant process becomes low frequency intra-Landau level emission at least when E_F is in or near to one of the levels. When it is mid-gap we

Springer Series in Solid-State Sciences, Vol. 87
High Magnetic Fields in Semiconductor Physics II Editor: G. Landwehr

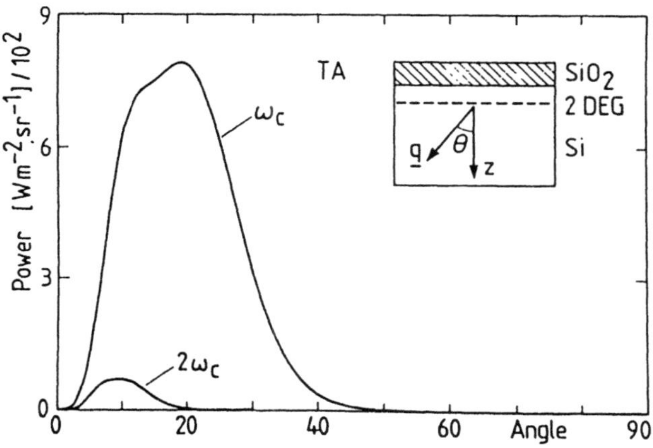

Figure 1 Example of the angular distribution of the TA phonon emission from a hot 2DEG in a quantizing magnetic field [4]

are approaching the Quantum Hall regime and dissipationless current flow, and the phonon emission in this regime seems of particular interest.

The first experiments to demonstrate cyclotron phonon emission were by KENT et al [5,6] using the arrangement shown in Fig. 2. 100ns current pulses were passed through the 2DEG in a (100) Si MOSFET and the resulting TA and LA phonon pulses were detected at the opposite face of the substrate using CdS bolometers. The silicon acts as a low-pass filter to phonons with a roll-off around 800 GHz caused by isotope scattering [7]. So as we tune the cyclotron frequency (ν_c=147B GHz) through the transmission function, while maintaining the input power constant, we expect a fall in signal proportional to the intensity at ν_c.

The results shown in Fig. 3 demonstrate the increasing proportion of cyclotron emission as the power and so electron temperature increases. It was also shown that the emission is re-distributed in favour of smaller angles as the field is increased. This was examined using 4 small bolometers and more detailed information has recently been obtained by HURLEY et al using a phonon-imaging technique in which the detection point is moved by raster scanning a laser beam. Further details and results are shown elsewhere in these proceedings [8].

Phonon emission studies have also been made at much lower powers, 10-500μW mm^{-2}, when $kT_e \ll \hbar\omega_c$. Pulse techniques cannot be used at these levels and these experiments by HEWETT et al [9] were carried out using the steady state technique shown in Fig. 4(a). The Si MOSFET sample is mounted in a vacuum and connected thermally at one end to a 1K heat sink. Since the substrate is transparent to phonons, their angular distribution determines the temperature distribution on the opposite face. Fig. 4(b) shows how the temperature directly opposite the 2DEG varies with gate voltage V_g and so sheet density N_s in a field of 6T and for constant power input. The forward intensity oscillates as the Fermi level moves through the Landau levels with maxima when E_F lies within a level and minima when it is in a gap. In the Quantum Hall limit the source-drain resistance $R_{SD}=R_H$ but the power dissipation, I^2R_H cannot of course occur in most of the 2DEG so the phonon emission must occur near to 2 opposite corners

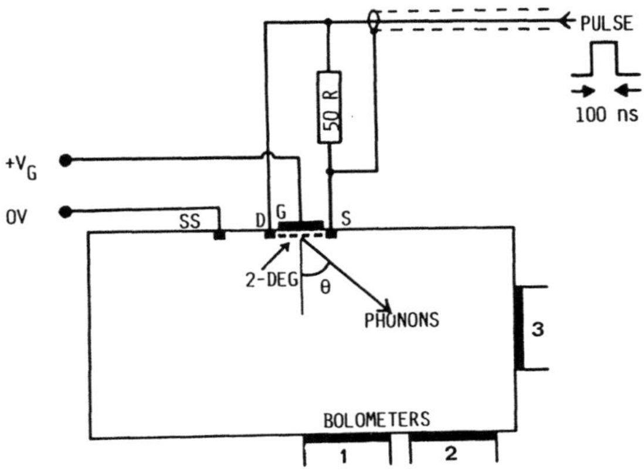

Figure 2 Experimental arrangement for observing cyclotron phonon emission from a hot 2DEG in a quantizing magnetic field [5,6]

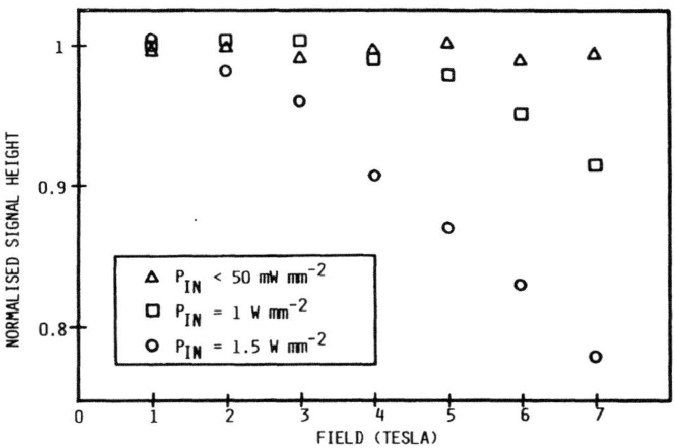

Figure 3 Decrease in intensity as the cyclotron emission is tuned through the roll-off in the substrate transmission [5,6]

where $\underline{E}.\underline{J} \neq 0$. Fig. 4(b) shows that the phonon emission when E_F is mid-gap ($\hbar\omega_c/kT_e \sim 25$), is less strongly directed forwards than from the intra-Landau level processes that occur when E_F is within a level. The power and field dependence of these effects have been examined in detail and further investigations are in progress.

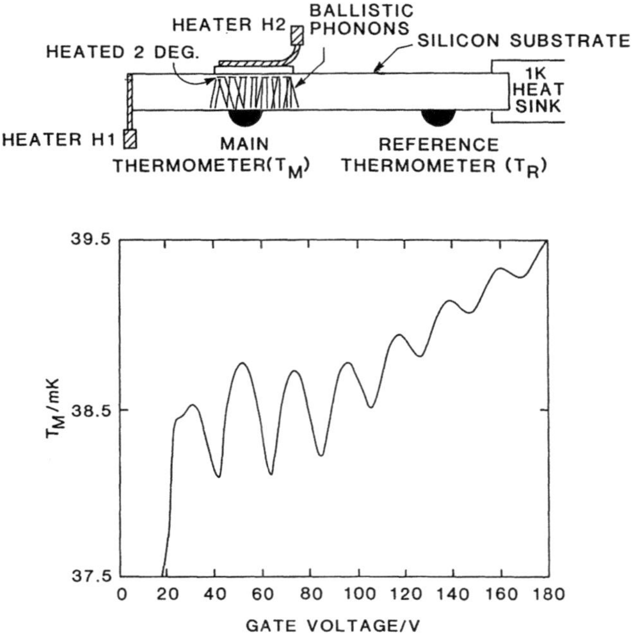

(a)

(b)

Figure 4 (a) Experimental arrangement for observing phonon emission in the Quantum Hall regime, $kT_e \ll \hbar\omega_c$ (b) Oscillations in T_M caused by changes in the angular distribution of the phonon emission as E_F moves through the Landau levels. $B \doteq 6T$, $P = 64\mu Wmm^{-2}$. T=1.8K [9]

3. ABSORPTION AND REFLECTION

A 2DEG is fairly transparent to a beam of phonons although a small fraction appears to be reflected and an even smaller fraction absorbed [10]. The absorption can be measured from the rise in temperature of the 2DEG which is determined from changes in the source-drain resistance $R_{SD}(T)$. The reflection can be determined from the loss in transmission from an incident beam if we assume absorption<<reflection. The first experiment on absorption in fields using thermal phonons is due to EISENSTEIN et al [11]. The 2DEG, in a GaAs/AlGaAs heterostructructure at 0.2K, was connected in Corbino geometry and the energy absorbed from an incident heat pulse was detected from the change produced in σ_{xx}. Since $kT_e \ll \hbar\omega_c$, the absorption was at low frequencies ($\lessdot$20GHz) as a result of intra-Landau level transitions. The phonon intensity in this range was varied by passing the pulse through ^{4}He which down-converts to $\lessdot$20GHz a proportion which decreases with increasing pressure. The resulting decrease in absorbed energy is shown in Fig. 5 which also shows that this increases with N, the level index.

The first experiments showing loss at the cyclotron frequency were carried out by transmission through the 2DEG in a Si MOSFET [12,6]. Fig. 6 shows that, at low fields, there is a substantially increasing loss in

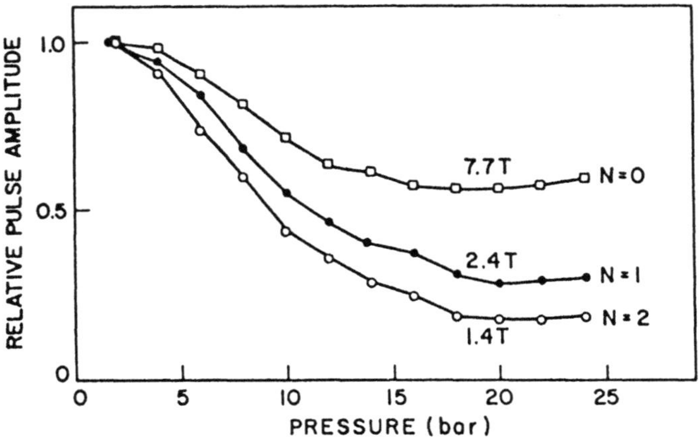

Figure 5 Absorbed energy in a 2DEG (GaAs/AlGaAs) as a function of the pressure on the ^{4}He filter. N is the index of the highest occupied Landau level [11]

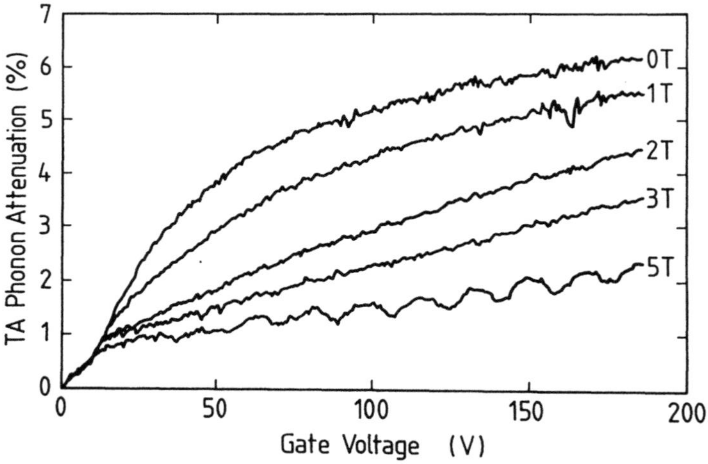

Figure 6 Loss in TA phonon intensity on transmission through a 2DEG (Si MOSFET) [12,6]

intensity as the sheet density is increased by the gate voltage (N_S=2.9x10^{14}V_gm^{-2}). The loss from the broad phonon spectrum is decreased by field quantization and oscillations appear when B>3.5T.

The oscillations are shown in more detail in Fig. 7 which now shows the transmitted signal. Interestingly their phase, relative to the SDH oscillations in source-drain resistance, changes by π between 4 and 5T. At 4T the loss is greatest when E_F lies within a Landau level while at 5T the maximum occurs for E_F in the gap. We attribute this to the existence of two loss mechanisms. Intra-Landau level transitions are most effective when E_F is in a Landau level while inter-Landau level or cyclotron

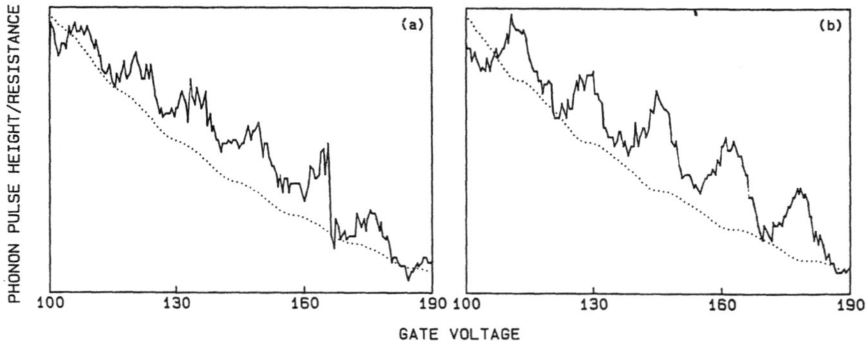

Figure 7 Quantum oscillations in the transmission of a 2DEG as E_F is moved through the Landau levels (a) 4T (b) 5T. The broken lines show SDH oscillations in source-drain resistance [12]

processes peak when E_F is in a gap. The phase change occurs when there is a change in the dominant process and the field at which this happens can be increased by increasing the average frequency of the phonons, i.e. the power to the heater. We recall that since this is a transmission experiment it seems likely that the effects are mostly due to changes in reflection rather than absorption.

The effect of the two processes has also been seen in absorption by HARDY et al [13]. The absorption by a 2DEG in a Si MOSFET was measured using R_{SD} as a bolometer and results are shown in Fig. 8. At the lower powers and so lower average phonon frequencies, the phase of the oscillations relative to SDH oscillations in R_{SD} shows that intra-Landau level transitions are dominant as expected. However as the power and so the frequency is increased, additional peaks in absorption appear in between the others which are attributed to cyclotron absorption. A very useful feature of this system is that it provides us with a tunable detector at the cyclotron frequency.

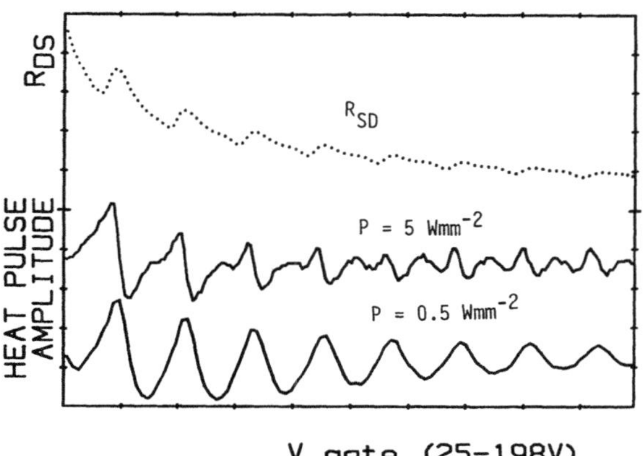

Figure 8 The energy absorbed by a 2DEG (Si MOSFET) as the gate voltage is increased at B=6T. The broken line shows R_{SD} [13]

Field dependent phonon scattering from a 2DEG has also been seen in GaAs/AlGaAs heterostructures. This was first observed by EISENSTEIN et al. [14] in the thermal conductivity of a 50μm GaAs wafer with a multilayer heterostructure on one face. The high conductivity value for T<1K means that the phonon scattering is largely specular and so is very sensitive to relatively weak diffuse scattering processes.

Figure 9 shows the temperature difference ΔT along the sample for a constant heat input. Since the phonons have frequencies $\nu \lesssim$ 100 GHz they are only scattered by intra-Landau level transitions in this field range ($\nu_c \gtrsim$ 1000 GHz). Hence, Fig. 9 represents the field dependence of the diffuse part of this intra-Landau scattering. It is presumed to be due to absorption followed later by re-emission and is greatest (ΔT = maximum) when E_F lies within a Landau level and least when it is in a gap, as expected. The scattering strength can be deduced from the difference.

Related effects have also been seen in transmission. Fig. 10 shows pronounced oscillations in heat pulse transmission as the field is varied: NEWTON et al [15]. Since the heater temperature ~10K, it emits phonons with frequencies up to ~1 THz but isotope scattering limits this to 700 GHz. So for B>2T, the oscillations can again only be due to intra-Landau level transitions and, as expected, the transmission passes through sharp peaks when E_F lies in a Landau gap. The oscillations for B<2T are attributable to both types of transition and analysis is in progress to obtain quantitative information on the reflectivity.

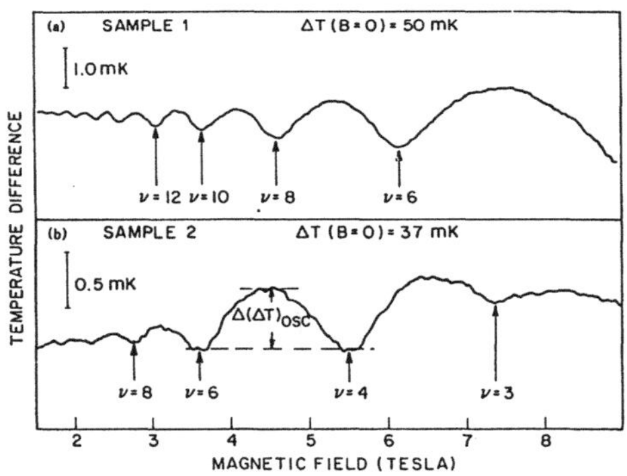

Figure 9 Quantum oscillations in ΔT along a GaAs wafer due to diffuse scattering of the phonon current caused by a set of 2DEGs [14]. (a) T=0.7K (b) T=0.35K. The ν = 3 minimum is a spin gap

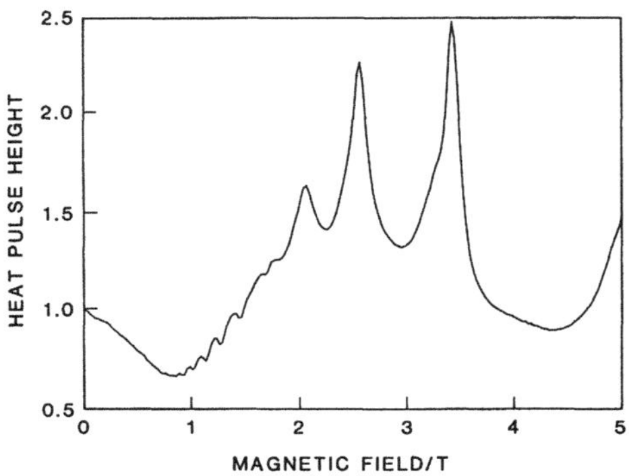

Figure 10 LA phonon transmission through a 2DEG (GaAs/AlGaAs) showing peaks when E_F is in the Landau gaps [15]

4. ULTRASONIC AND SURFACE WAVES (SAW)

The first experiments with coherent phonons were surface acoustic wave studies of a 2DEG (GaAs/AlGaAs) at 70 MHz: WIXFORTH et al [16]. The field dependent attenuation is shown in Fig. 11 and similar results have recently been obtained at 200 MHz by RAMPTON et al [17]. Detailed discussion is given elsewhere in these proceedings [18] and here we note only that the acoustic wavelengths appear to be sufficiently long in comparison with other lengths (k_F^{-1} and the electron mean free path) that the attenuation can be interpreted in terms of macroscopic variables, in particular, $\sigma_{xx}(B)$ [16].

Figure 12 shows data by RAMPTON et al [17] from the first studies using bulk ultrasound. 10 GHz pulses of LA phonons passed through a 2DEG in a GaAs/AlGaAs heterostructure at normal incidence and were detected by a CdS bolometer as a function of magnetic field. The strong interaction for this geometry in which $q_{||} = 0$ is interesting and somewhat surprising and detailed interpretation and further measurements are in progress.

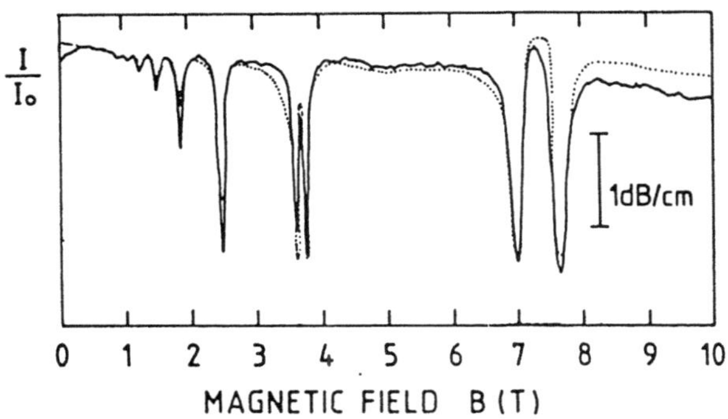

Figure 11 Attenuation of SAW by a 2DEG (GaAs/AlGaAs) [16] (The size of the attenuation is in fact about half that shown here [18])

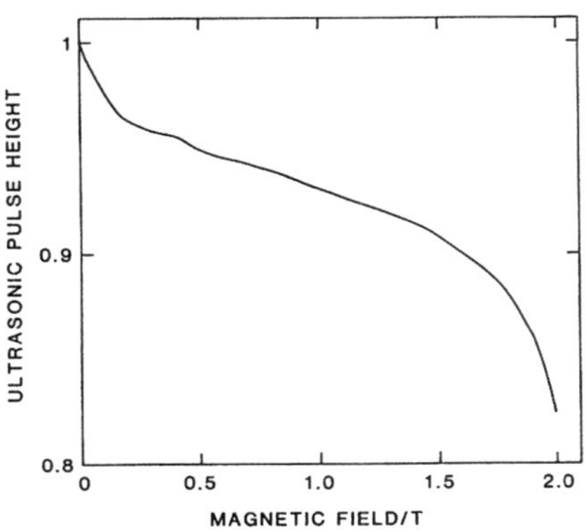

Figure 12 10GHz ultrasonic transmission through a 2DEG (GaAs/AlGaAs) [17]

5 ACKNOWLEDGMENTS

We are grateful for many discussions with our collaborators: P.J.A. Carter, A.G. Every, G.A. Hardy, P. Hawker, N.P. Hewett, D.C. Hurley, K.B. McEnaney, T. Miyasato, D. Neilson, M.I. Newton, F.F. Ouali, P.A.Russell, F.W.Sheard, G.A. Toombs and Y.B. Wahab and to GEC, NATO and SERC for financial support.

6 REFERENCES

1. L.J. Challis, G.A. Toombs and F.W. Sheard: In Physics of Phonons, ed. by T. Paskiewicz, Springer, Lect. Notes in Physics, Vol.285 (Springer, Berlin 1987) p.348
2. M. Rothenfusser, L. Koster and W. Dietsche: Phys. Rev.B34, 5518 (1986)
3. see F.W. Sheard, L.J. Challis and G.A. Toombs, to be published, for a recent discussion
4. G.A. Toombs, F.W. Sheard, D. Neilson and L.J. Challis: Solid State Commun.64, 577 (1987). In practice the emission measured at the opposite face of the substrate to the 2DEG includes phonons thermalised by inelastic reflection from the top face
5. A.J. Kent, M.I. Newton, V.W. Rampton, G.A. Hardy, P. Hawker, P.A. Russell and L.J. Challis: Jap.J.Appl.Phys., 26, supp 26-3, 1757 (1987)
6. A.J. Kent, V.W. Rampton, M.I. Newton, P.J.A. Carter, G.A. Hardy, P.A. Russell and L.J. Challis: Surf.Sci.,196, 410 (1988)
7. The roll-off frequency decreases with t, the thickness of the substrate as $t^{-1/4}$. The value given is for t=5cm
8. D.C. Hurley, G.A. Hardy, A.J. Kent, V.W. Rampton and L.J. Challis: these proceedings
9. N.P. Hewett, P.A. Russell, L.J. Challis, V.W. Rampton and A.J. Kent: these proceedings. Work in zero field is described by L.J. Challis, N.P. Hewett, P.A. Russell, V.W. Rampton and A.J. Kent: In Proc. 19th Int.Conf.Phys.Semiconductors, Warsaw (1988), in press
10. This was first discussed by J.C. Hensel, B.I. Halperin and R.C. Dynes: Phys.Rev.Letts, 51, 2302 (1983) and is supported by our absorption experiments [13]
11. J.P. Eisenstein, V. Narayanamurti, H.L. Stormer, A.Y. Cho and J.C.M. Hwang: In Proc. 5th Int.Conf. on Phonon Scattering in Condensed Matter, Urbana, 1986, ed. by A.C. Anderson and J.P. Wolfe, Springer, Ser.Solid.St.Sci., Vol.68 (Springer, Berlin 1986) p.401
12. V.W. Rampton, M.I. Newton, A.J. Kent, P.J.A. Carter, G.A. Hardy, P.A. Russell and L.J. Challis: In Jap.J.Appl.Phys.,26, Supp. 26-3, 1755 (1987)
13. G.A. Hardy, A.J. Kent, V.W. Rampton and L.J. Challis, these proceedings; A.J. Kent, G.A. Hardy, P. Hawker, V.W. Rampton, M.I. Newton, P.A. Russell and L.J. Challis: Phys.Rev.Letts. 61, 180 (1988)
14. J.P. Eisenstein, A.C.Gossard and, V.Narayanamurti: Phys.Rev.Letts, 59, 1341 (1987); J.P. Eisenstein, A.C. Gossard and V. Narayanamurti; Surf.Sci. 196, 445 (1988)
15. M.I. Newton, P.J.A. Carter, V.W. Rampton, M. Henini and O.H. Hughes: In Proc. 19th Int.Conf.Phys.Semiconductors, Warsaw (1988) in press
16. A.Wixforth, J.P. Kotthaus and G. Weimann: Phys.Rev. Letts., 56, 2104 (1986)
17. V.W. Rampton, M.I. Newton, P.J.A. Carter, L.J. Challis and A.J. Kent: In Proc. 11th Conf. Utilisation of Ultrasonic Methods for studying Properties of Condensed Matter, 1988; Acta Phys. Slovaka, in press
18. A. Wixforth and J.P. Kotthaus: these proceedings

Absorption of Phonons by a Two-Dimensional Electron Gas in the Silicon Inversion Layer in a Quantizing Magnetic Field

G.A. Hardy, A.J. Kent, V.W. Rampton, and L.J. Challis

Department of Physics, University of Nottingham,
Nottingham NG7 2RD, UK

1. Introduction

Acoustic phonon scattering by a two-dimensional electron gas (2DEG) in the silicon inversion layer was first investigated by HENSEL et al.[1] using heat pulse techniques. In the absence of an applied magnetic field interesting effects associated with the reduced dimensionality of the electronic system were observed. The application of a magnetic field normal to the 2DEG quantizes the electron system into a set of discrete Landau levels separated in energy by $\hbar\omega_c$, where the cyclotron frequency ω_c=eB/m*. Under these conditions the scattering of electrons is expected to involve the absorption or emission of cyclotron phonons with energy $\hbar\omega = n\hbar\omega_c$. Evidence of this has been observed in recent experiments [2]; the attenuation of a ballistic heat pulse by a 2DEG in a magnetic field displayed quantum oscillations associated with the position of the Fermi level with respect to the Landau level spectrum. From phase changes in these oscillations it was deduced that both inter and intra-Landau level processes were taking place. The field dependent absorption of phonons by a 2DEG in a GaAs/AlGaAs heterostructure was studied by EISENSTEIN et al.[3], using the device as a detector. A significant absorptivity for low frequency phonons leading to intra-Landau level transitions was observed. In this experiment we use the 2DEG in a Si MOSFET to detect ballistic heat pulses generated by a thin film heater and to study the dependence of the phonon absorption on applied magnetic fields of up to 7T.

2. Experimental Techniques

The experimental arrangement is shown in figure 1. The heat pulses are generated using a 1mm x 1mm, 50Ω constant an heater fabricated on the (100)

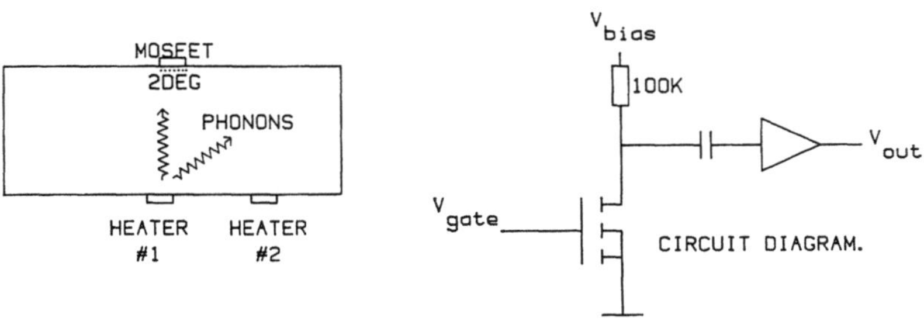

Fig. 1 Experimental arrangement showing electrical connections to device.

Springer Series in Solid-State Sciences, Vol. 87
High Magnetic Fields in Semiconductor Physics II Editor: G. Landwehr

face of the 5mm thick 1000Ωcm Si wafer. The MOSFET is on the opposite side of the wafer, it has a gate area of 1mm x 1mm and a peak mobility of $4500cm^2V^{-1}s^{-1}$. The sample is immersed in liquid helium at 4.2K in the bore of a 7T superconducting solenoid. A small bias current ≈0.1mA is passed between the source and drain contacts of the device and the changes in the channel resistance due to the incident heat pulse show up as changes in the source-drain voltage. This voltage is fed to a high impedance amplifier and then on to a signal averaging system based on a digital storage oscilloscope and microcomputer. An example of the time resolved heat pulse signals is shown in figure 2.

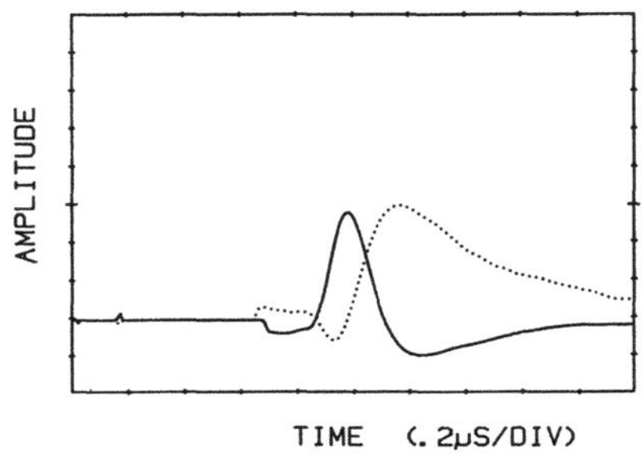

Fig. 2 Heat pulse signals detected by MOSFET at B=7T; heater input power $=5Wmm^{-2}$ and pulse length =100ns.

Solid line: E_F between two Landau sub-bands.

Dotted line: E_F within a Landau level.

(Signals not to same scale)

3. Results and Discussion

The sign of the voltage signal shown in figure 2 indicates that the heat pulse causes an increase in the channel resistance, implying a decrease in electron mobility. The absence of any signal corresponding to longitudinal mode phonons is related to the geometry of the experiment. Normally incident longitudinal phonons do not have the necessary wavevector component in the plane of the 2DEG to interact with the electrons. The transverse mode, however, does possess such a component which arises from the nature of the phonon focussing near to the [100] direction in silicon.

Figure 3 shows the variation of signal amplitude as the gate voltage is swept between 25 and 200 volts in a 7 Tesla magnetic field; the results for two different heater powers are shown along with the Shubnikov-de Haas oscillations in the channel resistance. The results for the lower heater power show the signal height to be a maximum when the Fermi level, E_F, is coincident with a Landau sub-band. At the higher power an intermediate peak is seen which corresponds to E_F lying between two Landau levels; we

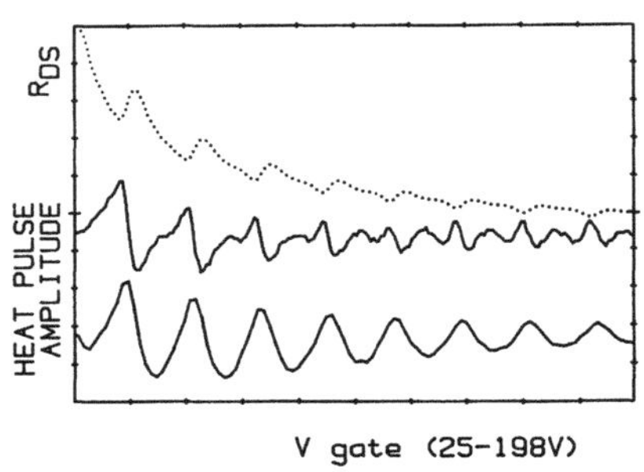

Fig. 3 Variation of signal amplitude with gate voltage at B=7T.

Lower trace: Heater imput power $P_{in}=0.5Wmm^{-2}$.

Upper trace: $P_{in}=5Wmm^{-2}$.

Dotted line: Shubnikov de Haas oscillations in drain-source resistance.

attribute this signal to the absorption of cyclotron frequency phonons. Only at the higher heater power is the proportion of cyclotron phonons in the thermal pulse sufficient to produce this intermediate signal.

The probability of phonon absorption is proportional to $f_1(1-f_2)$, where f_1 and f_2 are the Fermi factors of the initial and final electron states of energy E_1 and E_2 respectively. This function has a maximum at $(E_1+E_2)/2$, so that for cyclotron frequency phonons $\omega_c=(E_2-E_1)/\hbar$ the absorption is a maximum when E_F lies mid-way between two Landau levels, while the probability of absorption of lower frequency phonons associated with intra-Landau level transitions is a maximum when E_F coincides with a Landau level.

The nature of the time resolved signals (figure 2) differs for intra and inter-Landau level processes. The signal observed when E_F coincides with a Landau level is similar to that obtained in zero field [4]. We believe this is due to an increase in electron temperature during the time of arrival of the heat pulse which reduces the mobility through a reduction in the screening of the oxide charge scattering. The absorption of a cyclotron phonon excites an electron from a full to an empty Landau subband. The availability of nearby empty states leads to an increase in the scattering rate for the excited electrons. This causes an immediate change in device mobility, hence the observed signal is sharper and peaks at the phonon time of flight across the sample following the excitation pulse.

Figure 4 shows the variation of the inter-Landau level signal as a function of heater power at B=5T. If this signal is due to the absorption of cyclotron phonons then its amplitude should vary as the proportion of cyclotron frequency phonons in the heat pulse. The phonon power, $P_e(\omega)$, emitted at frequency ω is given by (1).

$$P_e(\omega) = \text{const} \times \omega^3 \left[\frac{1}{\exp(\hbar\omega/kT_h)-1} - \frac{1}{\exp(\hbar\omega/kT_o)-1}\right] \quad (1)$$

T_o is the lattice temperature and T_h is the heater temperature which is related to the heater input power by (2).

$$P_{in}=\sigma(T_h{}^4-T_o{}^4) \quad \text{Wmm}^{-2}, \quad (2)$$

where σ is the acoustic mismatch constant.

The solid line in figure 4 was obtained by evaluating (1) at the cyclotron frequency, $\omega_C=4.6\times10^{12}\,s^{-1}$, and using the acoustic mismatch constant as a

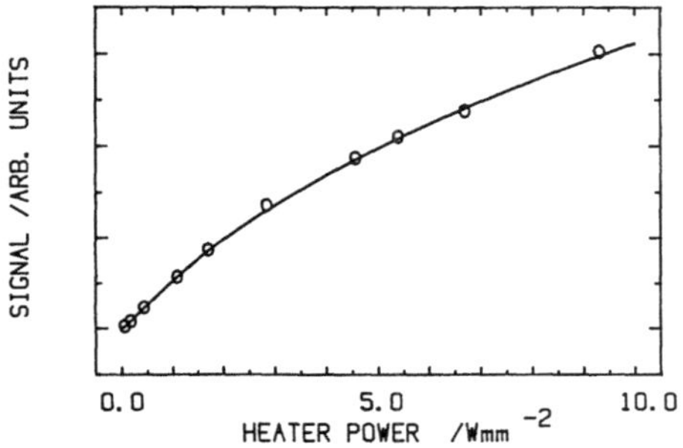

Fig. 4 Amplitude of signal obtained for E_F between Landau levels as a function of heater input power. The solid line is calculated theoretically assuming cyclotron phonon absorption.

parameter to fit it to the data. A value of σ=150 was required, which is in good agreement with the value of 160 from acoustic mismatch theory [5].

Using a heater at 45° to the normal to the plane of 2DEG we were able to observe a signal due to longitudinal mode phonons as well as the transverse. Both modes behaved in the same way in a magnetic field, but the absorption was weaker than in the normal incidence geometry this is probably due to the angular dependence of the 2D electron-phonon interaction [6]. There were some other differences between this and the normal geometry; in particular the signal changed sign when E_F was coincident with the edge of a Landau level. This implies a decrease in the channel resistance and we believe it is due to phonon assisted activation of electrons from the localised states between Landau levels.

4. Conclusions

We have used a Si MOSFET to detect ballistic heat pulses. In a magnetic field when E_F lies between two Landau levels evidence of cyclotron phonon absorption is observed. We believe that this experiment is a useful tool for the study of the electron-phonon interaction in 2D systems, giving more direct information than can be obtained by traditional transport measurements. Furthermore we have shown that this system may be used as a tunable, frequency selective phonon detector.

Acknowledgements

We wish to thank Dr. C.M.K. Starbuck, Department of Electronics and Information Engineering, Southampton University, for preparing the MOSFET samples. We are also grateful to Drs. P.A. Russell and M.I. Newton for their assistance in these experiments. This work was supported by a grant from the Science and Engineering Research Council of the UK.

References

1. J.C. Hensel, R.C. Dynes and D.C. Tsui:Phys. Rev. B28, 1124 (1983)
2. A.J. Kent, V.W. Rampton, M.I. Newton, P.J.A. Carter, G.A. Hardy, P. Hawker, P.A. Russell and L.J. Challis: Surface cience 196, 410 (1988)
3. J.P. Eisenstein, V. Narayanamurti, H. L. Stormer, A.Y.Chu and J.C.M. Hwang: Proc. 5th. Int. Conf. on Phonon Scattering in Condensed Matter, 401 (Springer, Berlin 1986)
4. A.J. Kent, G.A. Hardy, P. Hawker, V.W. Rampton, M.I. Newton, P.A. Russell and L. J. Challis: Phys. Rev. Lett. 28, 180 (1988)
5. W. Kappus and O. Weiss: J.Appl. Phys. 44, 1947 (1973)
6. G.A. Toombs, F. W. Sheard, L.J. Challis and D. Neilson: Solid State Commun. 64, 577 (1987)

Magnetophonon Resonance of a Two-Dimensional Electron Gas in AlGaAs/GaAs Single Heterojunctions

C. Hamaguchi, N. Mori, H. Murata, and K. Taniguchi

Department of Electronics, Osaka University, Suita City, Osaka 565, Japan

The theory of magnetophonon effect in two-dimensional electron gas in AlGaAs/GaAs heterostructures is developed and the derived formula of the oscillatory magnetoresistance $\Delta\rho \sim \sum_r \exp(-2\pi r\gamma)\cos(2\pi r\omega_0/\omega_c)$ shows a good agreement with the experimental observations. The Landau level broadening is calculated for various scattering processes. The electron concentration dependence of the magnetophonon oscillation amplitude observed recently by BRUMMELL et al. is shown to be well explained in terms of the Landau level broadening due to the remote-impurity scattering. Magnetophonon effective mass is estimated from the theory based on the k·p perturbation method to take into account the conduction band nonparabolicity.

1. INTRODUCTION

Magnetophonon resonance (MPR) experiments are known to be very useful to investigate electron-optic phonon interactions in two-dimensional electron gas systems of heterostructures. Experiments on MPR have been carried out by several groups /1/ and revealed that the magnetophonon mass is larger than the effective mass in bulk GaAs, reflecting the conduction band nonparabolicity. BRUMMELL et al./2/ found that the phonon frequencies determined from the cyclotron mass and MPR extrema in AlGaAs/GaAs are significantly below the bulk GaAs LO phonon values. The present authors /3/ carried out MPR experiments in AlGaAs/GaAs/AlGaAs single quantum well structures and theoretical analysis of the MPR effect by taking into account the confinement of the LO phonons in GaAs layer, and found that the MPR arises from the inter-subband transition in addition to the intra-subband transition by absorbing or emitting the confined LO phonons. Recently BRUMMELL et al./4/ found that the amplitude of MPR oscillations decreases rapidly with increasing carrier concentration in AlGaAs/GaAs. They attributed this decrease to screening. Although numerical analyses have been carried out by LASSNIG and ZAWADZKI /5/ and MORI and HAMAGUCHI /6/, no one succeeded in deriving an analytical formula. In this paper we report a very useful formula for analyzing the MPR effect in two-dimensional electron system which is derived by using the Kubo formula /7/ and the Fang-Howard function /8/. Using this formula and Landau level broadening due to the remote-impurity scattering we will show that the carrier concentration dependence observed by BRUMMELL et al./4/ is quantitatively explained.

2. THEORY OF MAGNETOPHONON RESONANCE

Our theory is based on perturbation expansion of the Kubo formula and harmonic analysis used first by BARKER /9/ and extended to the case of crossed high electric and high magnetic fields by MORI et al./10/, where

Springer Series in Solid-State Sciences, Vol. 87
High Magnetic Fields in Semiconductor Physics II Editor: G. Landwehr

the magnetoconductivity is evaluated by including Landau level damping. We extend the treatment to the case of two-dimensional electron system, where the Fang-Howard trial function is used for the electronic state and the divergences due to the singularity of the density of states are removed by introducing collision broadening. When the spectral density is given by Lorentzian form, harmonic analysis leads to the purely oscillatory part of the transverse magnetoresistivity $\Delta\rho$ given by

$$\frac{\Delta\rho}{\rho_0} = 2\overline{\Delta\rho}\sum_{r=1}^{\infty}\exp(-2\pi r\gamma)\cos(2\pi r\omega_0/\omega_c) \tag{1}$$

where γ is the damping parameter given by $2\Gamma/\hbar\omega_c$, Γ is the Landau level broadening, $\omega_c = eB/m^*$ and ω_0 are the cyclotron and optic phonon frequencies, B the magnetic field, and m^* the effective mass (magnetophonon mass). The preexponent $\overline{\Delta\rho}$ is given by

$$\overline{\Delta\rho} = \frac{3}{2^7\pi}\left(1 + \frac{1}{3}\frac{\hbar^2 b^2}{2m^*}\frac{1}{\hbar\omega_0}\right)^{-1} N_0(N_0 + 1)\left(\frac{\omega_0^2\tau_0 m^*}{N_s\hbar}\right)\left(\frac{e^2 b}{k_B T\varepsilon_0}\right)\left(\frac{1}{\kappa_\infty} - \frac{1}{\kappa_0}\right)\Theta \tag{2}$$

where N_0 is the Bose-Einstein distribution of optic phonons, τ_0 is the relaxation time of the electrons at vanishing magnetic field, N_s is the sheet electron density, b is the variational parameter in the Fang-Howard function, $\varepsilon_0\kappa_0$ and $\varepsilon_0\kappa_\infty$ are the static and high frequency dielectric constants, and Θ is given by $\Theta = (\omega_c/\omega_0)\sum_n\{f(E_n) - f(E_n+\hbar\omega_0)\}$ which is almost independent of the magnetic field with f(E) the Fermi-Dirac distribution function and E_n the energy of n-th Landau level. The details of the derivations of these equations and the broadening energies are given elsewhere /11/. The non-oscillatory part is given by $2\overline{\Delta\rho}(1/\pi)\tan^{-1}(1/\gamma)$. Oscillatory features of the experimental observations are known to be well expressed by the formula given by Eq. (1) from the analogy of the Barker's theory derived for bulk materials.

The broadening of Landau levels is caused by various scatterings such as remote-impurities in AlGaAs layer, background impurities in GaAs layer, polar optic phonons, acoustic phonons, and interface roughness. We evaluated the broadening by using the one electron retarded Green function method and calculating the self-energy, and the results will be published elsewhere /11/. We present here the Landau level broadening due to remote-impurity scattering because as shown later this process dominates the broadening in AlGaAs/GaAs heterostructures which we are interested in. The broadening due to the remote impurity scattering is given by

$$\Gamma_{RIS} = \frac{1}{4\pi}\left(\frac{e^2}{2\varepsilon_0\kappa_0}\right)^2 N_{is}\int_0^\infty d\zeta\, \frac{e^{-2L_s(\sqrt{2\zeta}/l_B)}}{\varepsilon^2(\sqrt{2\zeta}/l_B)(1 + \sqrt{2\zeta}/bl_B)^6}|J_{NN}(\zeta)|^2\}^{1/2} \tag{3}$$

where $\varepsilon(\sqrt{2\zeta}/l_B)$ is the dielectric function with screening effect, $l_B = \sqrt{\hbar/m^*\omega_c}$ is the classical radius of the ground (N = 0) Landau level, L_s is the spacer layer thickness, $N_{is} = N_i L_i$ is the sheet impurity density in AlGaAs with doping width L_i, and $J_{NN}(z)$ is defined by BARKER /9/ and also given by Eq. (15) of reference 10.

In Fig. 1 we show the broadening as a function of sheet electron density for the remote-impurity scattering at T = 180K and for $N_s = 3.0\times10^{11}$ cm^{-2}, where we assumed a constant value of the layer thickness 100 A. Also presented in Fig. 1 are the Landau level broadening due to optic phonon scattering, acoustic phonon scattering and interface roughness scattering,

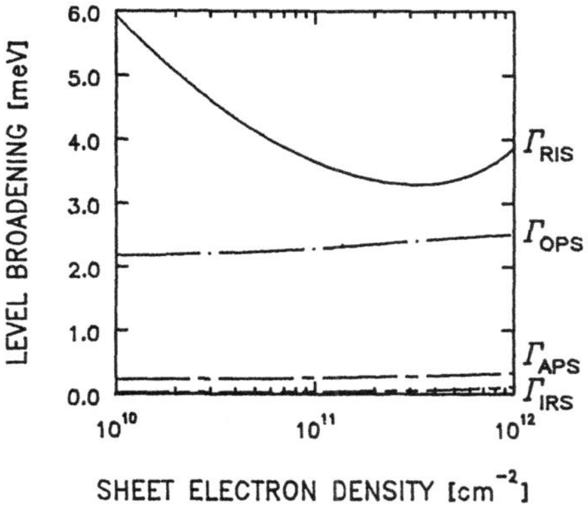

Fig. 1 Calculated Landau level broadening as a function of sheet electron density for remote-impurity Γ_{RIS}, polar optic phonon Γ_{OPS}, acoustic phonon Γ_{APS}, and interface roughness Γ_{IRS} scatterings in AlGaAs/GaAs heterostructures at 180 K, where the spacer layer thickness is assumed to be 100 A.

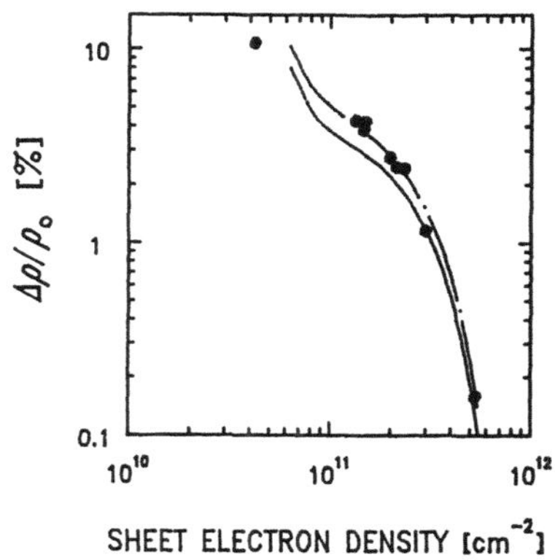

Fig. 2 Sheet electron density dependence of oscillation amplitude in AlGaAs/GaAs heterostructures at 180 K, for N = 3 resonance; solid circles (experimental /7/), and dash-dotted and solid curves are calculated from Eq. (1) with for the sheet impurity density N_{depl} = $1X10^{10}$ and $5X10^{10}$ cm^{-2}, respectively.

where we find that the remote-impurity scattering dominates the broadening effect. Since we assumed a constant spacer layer thickness, the broadening decreases with increasing the sheet electron density, which reflects the fact that the screening effect plays an important role in the region of the low electron density. The increase in Γ_{RIS} in the region of high N_s arises from the fact that the term N_{is} dominates in determining Γ_{RIS} in Eq. (3).

The amplitude of the MPR oscillation reported by BRUMMELL et al./4/ is shown in Fig. 2 as a function of the sheet electron density at T=180K and for N=3 ($N\hbar\omega_{c^*} = \hbar\omega_0$), where the sheet electron density is changed by changing the spacer layer thickness and keeping the doping density constant. In that case the electron sheet density is evaluated from the following relation of the charge neutrality condition,

$$N_{is} = N_s + N_{depl} \tag{4}$$

where N_{depl} is the depletion charge density in GaAs layer which is assumed to be acceptor. Since the GaAs layer is not doped intentionally, we can assume that the depletion charge density is constant in the series of samples used by BRUMMELL et al./4/. Therefore the sheet electron density is controlled by changing the spacer layer thickness and keeping the doping density in AlGaAs layer constant. Their sample characteristics are reported by FOXON et al./12/. It is evident from these considerations that the Landau level broadening due to the remote-impurity scattering is dominated by two factors in eq. (3), the sheet impurity density N_{is} in AlGaAs layer and the spacer layer thickness L_s. Using the calculated broadening energies from Eq. (3) we evaluated the oscillation amplitudes as a function of the sheet electron density from Eq. (1) at T = 180 K for

N = 3, which are shown in Fig. 2 by the solid curve for N_{depl} = $5X10^{10}$ cm^{-2} and by the dash-dotted curve for N_{depl} = $1X10^{10}$ cm^{-2}. We find a very good agreement between the experimental and calculated results. Therefore we can conclude that the damping of the magnetophonon oscillations is caused by the remote-impurity scattering.

3. MAGNETOPHONON MASS

As stated in 1. the effective mass determined from the MPR effect is much higher than the mass obtained in bulk materials, which is explained in terms of conduction band nonparabolicity. The Landau levels are calculated by using the five level k·p perturbation theory developed by LASSNIG /13/. A typical example is shown in Fig. 3, where the magnetophonon masses for N = 3 in the ground subband (solid curve), the first excited (dash-dotted) and the second excited subband (dash-double dotted) are plotted as a function of the sheet electron density in AlGaAs/GaAs single heterojunctions. The large effective mass around 0.072m is in good agreement with the experimental observations.

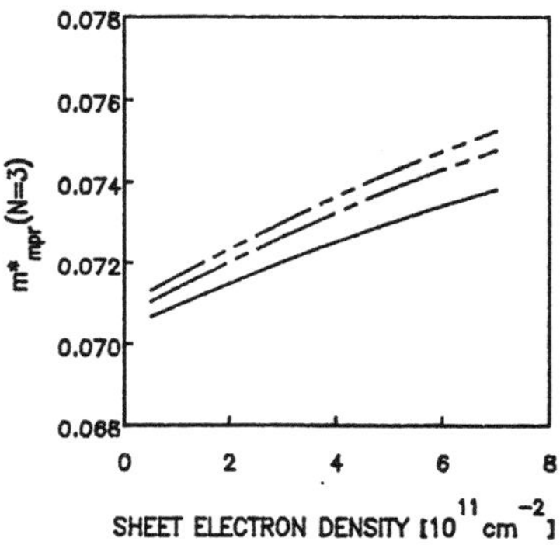

Fig. 3 Magnetophonon effective masses for N = 3 in the ground (solid curve), first excited (dash-dotted curve) and second excited (dash-double dotted curve) subbands as a function of sheet electron density in AlGaAs/GaAs single heterojunctions at 180 K.

References

1. For review see R.J. Nicholas, Prog. Quantum Electron. 10, 1 (1985).
2. M.A. Brummell, R.J. Nicholas, M.A. Hopkins, J.J. Harris and C.T. Foxon, Phys. Rev. Lett. 58, 77 (1987).
3. N. Mori, K. Taniguchi, C. Hamaguchi, S. Sasa, and S. Hiyamizu, J. Phys. C 21, 1791 (1988).
4. M.A. Brummell, D.R. Leadley, R.J. Nicholas, J.J. Harris and C.T. Foxon, Surf. Sci. 196, 451 (1988).
5. R. Lassnig and W. Zawadzki, J. Phys. C 16, 5453 (1983).
6. N. Mori and C. Hamaguchi, Technol. Repts. Osaka Univ. 37, 127 (1987).
7. R.Kubo, S.J.Miyake, and N.Hashitsume, Solid State Phys. 16, 269 (1965).
8. F. F. Fang and W. E. Howard, Phys. Rev. Lett. 16, 797 (1966).
9. J. R. Barker, J. Phys. C 5, 1657 (1972).
10. N. Mori, N. Nakamura, K. Taniguchi, and C. Hamaguchi, J. Phys. Soc. Jpn. 57, 205 (1988).
11. N.Mori, H.Murata, K.Taniguchi, and C.Hamaguchi, Phys. Rev. B. in press.
12. C. T. Foxon, J. J. Harris, R. G. Wheeler, and D. E. Lacklison, J. Vac. Sci. Technol. B 4, 511 (1986).
13. R. Lassnig, Phys. Rev. B 31, 8076 (1985).

Carrier Concentration Dependent Phonon Frequencies Deduced from Magnetophonon Resonance in GaInAs/InP Quantum Wells

D.R. Leadley[1], *R.J. Nicholas*[1], *L.L. Taylor*[2], *S.J. Bass*[2], *and M.S. Skolnick*[2]

[1]Clarendon Laboratory, Parks Road, Oxford OX1 3PU, UK

[2]RSRE, St. Andrews Road, Malvern, Worcester WR14 3PS, UK

Magnetophonon resonance has been studied in GaInAs/InP quantum wells and heterojunctions. The frequency of the optic phonon coupling to the electrons of the 2DEG decreased from $272cm^{-1}$ to $229cm^{-1}$, as the carrier density was increased. This is in stark contrast with the expected, fixed value of $273cm^{-1}$ for the LO phonon, but is in qualitative agreement with carrier-concentration-dependent Raman measurements. Possible explainations for this striking change are explored, including interaction with coupled phonon-2D plasmon modes.

1. Introduction

This paper reports magnetoresistance studies of the 2-D electron gas (2DEG) in GaInAs-InP heterostructures, between 300K and 50K. In this temperature region the electron mobility is limited by optic phonon scattering, thus the nature of phonons coupling to the 2DEG is very important. That the electrons are confined to the ternary material, where there will be both GaAs- and InAs-like optic phonon modes present [1,2], makes electron-phonon interactions in GaInAs-InP complicated. Magnetophonon resonance (MPR) measurements have previously been reported in heterojunctions [3] and multi-quantum wells [4], where phonons with lower energy than the GaAs-like LO mode were found. Phonon energies in GaInAs-InP have also been studied in photoluminescence [5] and Raman scattering [6,7], where recent measurements on single quantum wells [8] have shown frequncies varying with electron concentration, and again lower than the GaAs-like LO mode.

2. Experiments

The samples studied were grown by MOCVD at RSRE [9] and consist of three heterojunctions and a series of single quantum wells of well widths from 100Å to 800Å. A range of carrier concentrations (N_s) from 0.8 to $7.6\times10^{11}cm^{-2}$ was achieved, by using modulation doped and undoped wells, and could be varied in a given sample by photoexcitation with a red LED.

Resistances were measured in a magnetic field of up to 9T over the above temperature range. For samples of high N_s the magnetoresistance was dominated by Shubnikov-de Haas oscillations at low temperatures. However, at 120K, where a systematic study was performed, the MPR amplitude was much greater.

Oscillations in the magnetoresistance occur whenever an optic phonon energy, ω_o, is equal to the separation between an integral number of Landau levels [10], giving an increased scattering rate and hence maxima in the resistivity. This leads to a resonance condition:-

$$\hbar\omega_o = N\hbar\omega_c = N\hbar eB_N/m^*$$

where the product $NB_N = B_F$ is the fundamental field of the series and m^* is a suitably defined effective mass. These oscillations are typically ~5% of the zero

Springer Series in Solid-State Sciences, Vol. 87
High Magnetic Fields in Semiconductor Physics II Editor: G. Landwehr

field resistivity and are extracted by subtracting a voltage proportional to the magnetic field, or taking the second derivative of the signal.

The effective mass relevent to the required transitions, m^*_{MPR}, is obtained from cyclotron resonance measurements at the same temperature and correcting the measured mass, m^*_{CR}, for non-parabolicity and polaron effects using [10] :–

$$m^*_{MPR} = m^*_{CR} \left[1 + \frac{(N-1)}{N} K_2 \frac{\hbar\omega_0}{E_g} \right]^{-1} .$$

It was found that the experimental effective mass varied between samples in the range $0.048m_e$ to $0.055m_e$, but was constant as N_s was changed in a given sample.

3. Results

Figure 1 shows the second derivative of the magnetoresistance for the 100Å wells with varying carrier densities. The peaks can clearly be seen moving to lower magnetic field as N_s increases. The labelled peaks form a series periodic in 1/B, with fundemental fields changing from 16.8T to 13.0T corresponding to a range of phonon frequencies from $272cm^{-1}$ to $234cm^{-1}$ as shown in Fig.2. This figure shows data from the other samples studied and includes the multiquantum wells of ref.4 as boxes. Note the smooth curve from one sample to the next demonstrating that

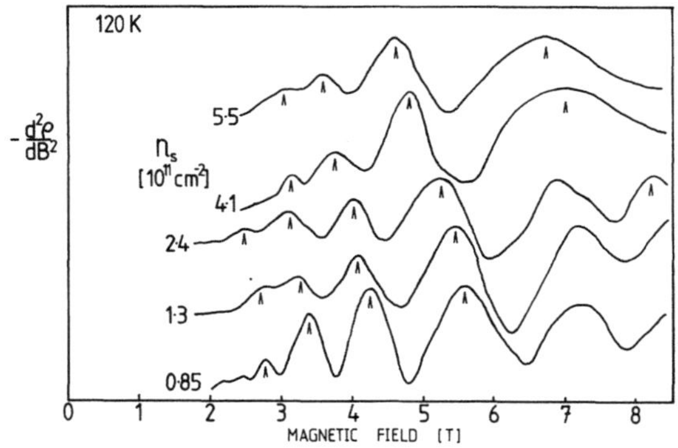

Fig.1 Second derivative recordings of magneto-phonon resonances in 100Å QWs shown for various values of N_s. The marked peaks form series periodic in 1/B.

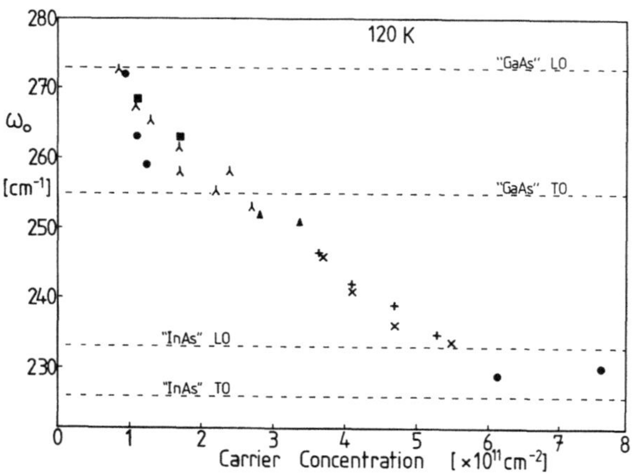

Fig.2 N_s dependence of phonon frequencies deduced from MPR.
× + ⅄ –100Å QWs
Δ – 300Å QW
o – heterojunction
□ – QWs of Ref.4.

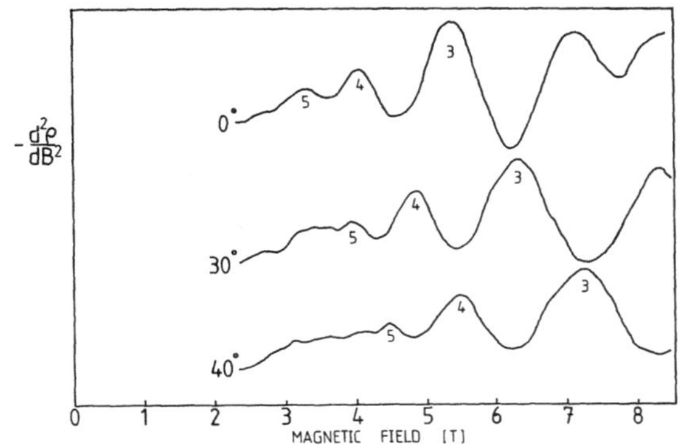

Fig.3 Second derivative recordings of MPR in a 100Å QW for various angles between the magnetic field and the normal to the 2DEG.

population of the well by doping or photoexcitation is equivalent. These results are probably related to recent measurements on GaAs–GaAlAs [11], where a frequency below the LO phonon was also deduced from MPR in samples with $N_s \sim 1\times10^{11}cm^{-1}$.

There is an additional peak seen in low N_s samples (see Fig.1) at around 7T which forms part of a second series corresponding to a phonon of frequency $\sim350cm^{-1}$. This is close to the InP LO mode and is evidence for scattering by a remote phonon, as reported by PORTAL et al.[4] and discussed theoretically by several authors [12,13]. Alternative explanations include penetration of the electron wavefunction into the barrier material [4] or alloying at the interface [9].

On rotation of the samples in the magnetic field the resonance positions move to higher field as Bcosθ, showing that the resonances come from two-dimensional electrons (Fig.3). However, a striking difference is now seen between the quantum wells and heterojunctions: the oscillation amplitudes remain constant for the QWs, but fall off rapidly with rotation in the case of the heterojunctions. This latter effect has previously been seen in GaAs–GaAlAs [11] and in other GaInAs–InP heterojunctions [14], with the rate of decrease apparently depending on the carrier density and subband separations.

4. Discussion

It has previously always been assumed that the electrons of the 2DEG couple to LO phonons. In GaInAs this should be the GaAs-like LO mode at $273cm^{-1}$, as its oscillator strength is three times that of the InAs-like mode [2]. So why does the phonon frequency change with N_s, as suggested by these results?

The Raman scattering measurements of MOWBRAY et al [8] on similar samples show strong coupling to a mode which decreases in frequency from $272cm^{-1}$ to $261cm^{-1}$ as their incident laser power increased. (Higher laser power means more carriers excited, although it is difficult to actually measure N_s in these experiments.) This may be related to the magnetophonon results if intersubband scattering can couple to the collective excitations of the electron–phonon plasma.

In contrast results on heavily doped bulk PbTe [10,15] show coupled mode behavior in optical measurements, but the magnetophonon oscillations are firmly pinned to the unscreened LO phonon frequency. This was explained by the much larger wavevector associated with MPR than optical processes.

There have been several calculations of the dispersion of coupled plasmon–LO–phonon modes as a function of N_s [16–18], some including coupling to subbands. The coupled mode frequencies depend critically on the relative values of the inter–subband separations, $\hbar\omega_{01}$, and the optic phonon energies. When ω_{01} is small there is conventional plasmon coupling with modes above the LO and below the TO phonons. This situation corresponds to the wide QWs and heterojunctions. For $\omega_{01} > \omega_{LO}$ an intermediate mode exists which decreases in frequency from the LO to the TO value as N_s increases. This would be the situation for the narrow wells (*e.g.* for 100Å a finite square well calculation gives $\omega_{01} \sim 740cm^{-1}$), and may suggest a qualitative explanation for the behaviour shown in Fig.2.

The situation is complicated by the indeterminate nature of the optic phonons in GaInAs, which is often assumed to show full two mode behaviour. If this were the case the coupled mode frequencies would not fall monotonically with increasing N_s. PEARSALL et al.[7] have in fact shown that only the modes at $273cm^{-1}$ and $226cm^{-1}$ have well defined LO and TO character, and intermediate phonons, usually assigned to the two-mode behaviour, may be localised.

However, quantitative calculations predict a much slower fall in the coupled mode frequency than observed in our experiments. Furthermore, one would expect both quantitive and qualitative differences between samples of different well width and between QWs and heterojunctions.

5. Conclusion

The results presented here demonstrate unequivocally a strong change in optic phonon scattering as N_s is increased. This may be related to screening of the optic phonon electric field, or the increased degeneracy of the system, as has recently been found to strongly reduce polaron coupling in GaAs-GaAlAs [19]. A proper description of the problem may require the use of dynamic, q-dependant screening, and possibly also a full treatment of the alloy phonon modes including damping.

6. References

1. M. Brodsky and G. Lucovsky : *Phys. Rev. Lett.* 21 990 (1968)
2. K.J. Nash, M.S. Skolnick & S.J. Bass : *Semicond. Sci. Technol.* 2 329 (1987)
3. D. Gauthier, L. Dmowski, J.C. Portal, D.R. Leadley, M.A. Hopkins, M.A. Brummell, R.J. Nicholas, M. Razeghi & P. Maurel : *Superlatt. and Microstruct.* 4 201 (1988)
4. J.C. Portal, J. Cisowski, R.J. Nicholas, M.A. Brummell, M. Razeghi and M.A. Poisson : *J.Phys.C* 16 L573 (1983)
5. M.S. Skolnick, K.J. Nash, P.R. Tapster, D.J. Mowbray, S.J. Bass and A.D. Pitt : *Phys. Rev.* B35 5925 (1987)
6. C.J. Pickering : *Electron. Mater.* 10 901 (1981)
7. T.P. Pearsall, R. Carles and J.C. Portal : *Appl. Phys. Lett.* 42 436 (1983)
8. D.J. Mowbray, W. Hayes, J.A.C. Bland, M.S. Skolnik & S.J. Bass, *Semicond. Sci. Technol.* 2 822 (1987)
9. M.S. Skolnick, P.R. Tapster, S.J. Bass, A.D. Pitt, N. Apsley & S.P. Aldred : *Semicond. Sci. Technol.* 1 29 (1986)
10. R.J. Nicholas : *Prog. Quant. Electron.* 10 1 (1985)
11. M.A. Brummell, D.R. Leadley, R.J. Nicholas, J.J. Harris & C.T. Foxon : *Surf. Sci.* 196 451 (1988)
12. R. Lassnig and W. Zawadzki : *Surf. Sci.* 142 361 (1984)
13. N. Sawaki : *Surf. Sci.* 170 537 (1986)
14. D.R. Leadley, R.J. Nicholas, D. Gauthier, J.C. Portal, M. Razeghi, L.L. Taylor, S.J. Bass and M.S. Skolnick : *to be published.*
15. J.C.Ramage, R.J.Nicholas, R.A.Stradling & F.Kuchar : *J.Phys.C* 10 L611 (1977)
16. G. Abstreiter, R. Merlin and A. Pinczuk : *J. Quant. Electron.* 22 1771 (1986)
17. F.M. Peters, Xiaoguang Wu and J.T. Devreese : *Phys. Rev.* B36 7518 (1987)
18. T. Yuasa and M. Ishii : *Phys. Rev.* B37 7001 (1988)
19. C.J. Langerak, J. Singleton, P.J. van der Wel, J.A. Perenboom, D.J. Barnes, R.J. Nicholas, M.A. Hopkins & C.T. Foxon : *Submitted to Phys. Rev. B.*

Part X

Magneto-optics in 3D Systems

Inversion Asymmetry and Magneto-optics in Semiconductors

S. Rodriguez

Department of Physics, Purdue University, West Lafayette,
IN 47907, USA

By virtue of Kramers' theorem, in the absence of an externally applied magnetic field, the energy levels of a system containing an odd number of electrons are, at least, doubly degenerate. In the band theory of solids the electrons are supposed to move in an effective periodic potential, its states being characterized by a band index ν, a wave vector $\vec{k}$ confined within the fundamental Brillouin zone (BZ) and a spin quantum number. For each set of values of ν and $\vec{k}$ there are two orthogonal spin states which we call s and s'. Denoting the energy eigenvectors by $|\nu,\vec{k},s\rangle$ we have $\langle\nu,\vec{k},s'|\nu,\vec{k},s\rangle=\delta_{ss'}$. The states $|\vec{k},s\rangle$ and $|-\vec{k},s'\rangle$ (we omit the band index for simplicity) are degenerate. If the crystal possesses a center of inversion $|\vec{k},s\rangle$ and $|-\vec{k},s\rangle$ are also degenerate. We conclude therefore that, in crystals with a center of inversion, there are, at least, two degenerate states at each point $\vec{k}$ in the BZ. Figure 1 shows schematically the energy as a function of $\vec{k}$ near $\vec{k}=0$ for the top of the valence band and the local minimum of the lowest conduction band in semiconductors having the diamond structure. Figure 2 shows the similar structure for zinc-blende semiconductors, in which the spin splittings have been grossly exaggerated. The symbols Γ stand for the usual irreducible representations generated by the Bloch states at k=0 using the notation of Koster et al. [1]. They correspond to the double valued representations of O_h for diamond and T_d for zinc-blende.

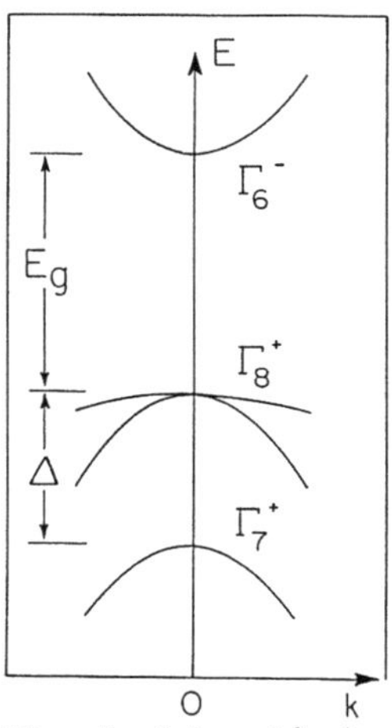

Fig. 1. Schematic band structure of semiconductors of octahedral symmetry (O_h)

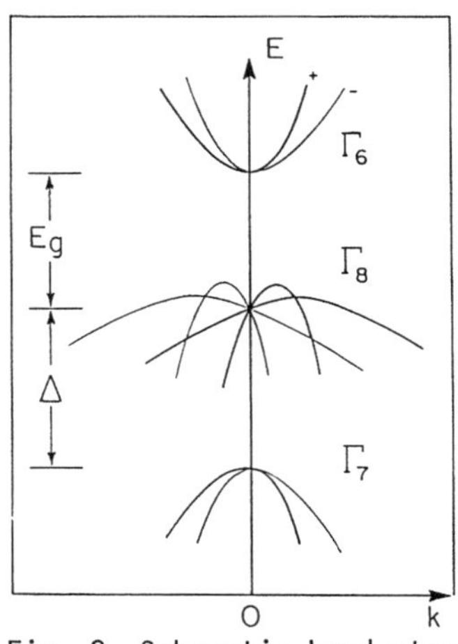

Fig. 2. Schematic band structure of zinc-blende semiconductors (T_d)

Springer Series in Solid-State Sciences, Vol. 87
High Magnetic Fields in Semiconductor Physics II Editor: G. Landwehr

The spin-splitting of the Γ_6 levels in Fig. 2 is proportional to the third power of the wave vector [2] and is anisotropic. For $\vec{k}$ parallel to [110] the energy difference between the orthogonal spin states is $\Delta E=\delta_0 k^3$, the axis of quantization being $[1\bar{1}0]$, the lower energy of the doublet corresponding to the spin pointing along $[\bar{1}10]$ for $\delta_0>0$. The sign of δ_0 is ambiguous unless the polarity of the positive [111] axis is defined. In what follows, the convention is made of taking this direction parallel to the electric dipole moment of the primitive cell, i.e., along the line pointing from the anion to the cation. This polarity can be determined experimentally owing to the substantial physico-chemical difference between (111) surfaces terminating in positively and negatively charged ions [3].

No direct measurement of ΔE is available for any material. However, a number of indirect determinations of δ_0 have recently been made. They are based on measurements of the electron spin relaxation time [4], on spin-polarized photoemission [5] and on far infrared spin resonance [6].

The experiments of Dobrowolska et al. [6] and of Chen et al. [6] in n-type InSb revealed a strong anisotropy in the intensity of the spin resonance of electrons in the Γ_6 conduction band. In particular, reversal of the applied magnetic field $\vec{B}_0$ or of the direction of incidence, $\hat{n}$, of the radiation along the [110] axis in the parallel Voigt geometry (OV, electric field $\vec{E}$ of the incident wave parallel to $\vec{B}_0$) resulted in a change of the spin resonance intensity by a factor of about 2. The effect was attributed to interference between magnetic-dipole and inversion-asymmetry-induced electric-dipole transition amplitudes. The relative intensities at the spin resonance as the orientation of the magnetic field is varied are shown in Fig. 3 together with the result of the theoretically calculated transition probabilities after adjusting the parameter δ_0 when $\vec{B}_0$ forms angles of 90° and 270° with the [001] direction. Agreement between theory and experiment is obtained setting [7] $\delta_0 = 2.2 \times 10^{-22}$ eV cm^3.

Similar phenomena in p-type zinc-blende materials are the subject of theoretical investigations by Bassani et al. [8] and by La Rocca et al. [9]. The energies $E(\vec{k})$ as a function of $\vec{k}$ for hole states near the top of the valence band are obtained using the 4×4 effective mass matrix

$$H = H_0 + H_A \tag{1}$$

where

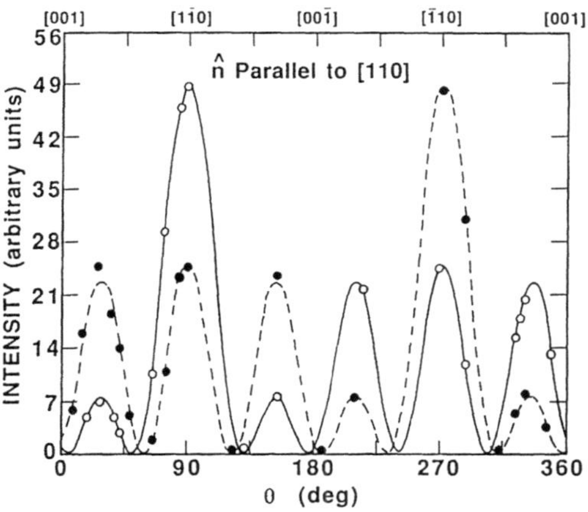

Fig. 3. Intensity of the conduction electron spin resonance of InSb in the ordinary Voigt (OV) geometry as a function of the angle θ between $\vec{B}_0$ and [001]. Full and open circles are the data [6]. The dashed line is identical to the continuous line but is shifted by 180° (see text)

$$H_o = -\frac{\hbar^2}{m}[\frac{1}{2}(\gamma_1 + \frac{5}{2}\gamma_2)k^2 - \gamma_2 \sum_i k_i^2 J_i^2 - \gamma_3 \sum_{i<j} k_i k_j \{J_i, J_j\}] \quad (2)$$

and

$$H_A = (C/\sqrt{3})[\{J_x, J_y^2 - J_z^2\}k_x + \{J_y, J_z^2 - J_x^2\}k_y + \{J_z, J_x^2 - J_y^2\}k_z] . \quad (3)$$

Here J_x, J_y and J_z are the components of the angular momentum operator for $j = 3/2$ along the cubic axes, γ_1, γ_2 and γ_3 are the Luttinger [10] parameters and C is the strength of the inversion asymmetry Hamiltonian introduced by Dresselhaus [11] and by Kane [12]. The sums over i and j extend over x,y,z and the curly brackets in (2) and (3) indicate the commutator of the operators within.

The energy levels of electrons near the top of the valence band in the presence of a magnetic field $\vec{B}_o = B_o\hat{\zeta} = \nabla\times\vec{A}_o$ are obtained replacing $\vec{k}$ by the operator

$$\vec{k} = -i\nabla + (e\vec{A}_o/\hbar c) \quad (4)$$

and adding the Zeeman terms

$$H_Z = -(e\hbar/mc)[K\vec{B}_o\cdot\vec{J} + Q\sum_i J_i^3 B_{oi}] \quad (5)$$

to (2). We observe that while H_o is even with respect to inversion, H_A is odd. In the presence of $\vec{B}_o$ we describe the hole eigenstates using the even parity Hamiltonian H_o+H_Z. We find [10] four sequences of Landau levels in the axial approximation. Figure 4 shows the eigenvalues for $k_\zeta = \vec{k}\cdot\hat{\zeta} = 0$ for the first few levels (n=0,1,2). Within this approximation, electric-dipole optical transitions occur between states of different parity. One-photon parity-conserving transitions can occur within the electric-dipole approximation because the actual eigenstates do not have definite parities owing to mixing of even and odd levels by the odd part, H_A, of the Hamiltonian.

The interaction of the carriers with the incident photons is obtained adding the term

$$(e\vec{A}/\hbar c) = \mathrm{Re}[(eA\hat{\varepsilon}/\hbar c)e^{i\vec{q}\cdot\vec{r} - i\omega t}]$$

to the operator $\vec{k}$ and $\vec{B} = \nabla\times\vec{A}$ to the magnetic induction $\vec{B}_o$. Considering only one-photon absorption or emission, the electron-photon interaction is

$$H' = (eA/2c)V\,e^{-i\omega t} + (eA^*/2c)V^\dagger\,e^{i\omega t} , \quad (6)$$

where

$$V = \hat{\varepsilon}\cdot\vec{v} + \frac{i}{2}\{\hat{\varepsilon}\cdot\vec{v}, \vec{q}\cdot\vec{r}\} - (i\omega/e)\sqrt{\varepsilon}(\hat{n}\times\hat{\varepsilon})\cdot\vec{\mu} . \quad (7)$$

Here $\hat{\varepsilon}$ is a unit vector parallel to the electric field of the incident radiation, $\vec{v} = \hbar^{-1}\partial H/\partial\vec{k}$ is the velocity operator, $\vec{q} = (\omega/c)\sqrt{\varepsilon}\,\hat{n}$ is the wave vector of the photon of angular frequency ω inside the material of dielectric function $\varepsilon(\omega)$ and $\vec{\mu} = -\partial H_Z/\partial B_o$ is the intrinsic magnetic moment.

To obtain the intensity of a particular absorption line we need the matrix element $\langle\psi_f|V|\psi_i\rangle$ between initial (ψ_i) and final (ψ_f) states. These states do not have a definite parity because of the admixture already discussed. We take this into account by expressing $\psi_{i,f}$ as $\exp(iS)|i,f\rangle$ where $|i\rangle$ and $|f\rangle$ are the eigenstates of $H_S = H_o+H_Z$ in which H_o, properly symmetrized to

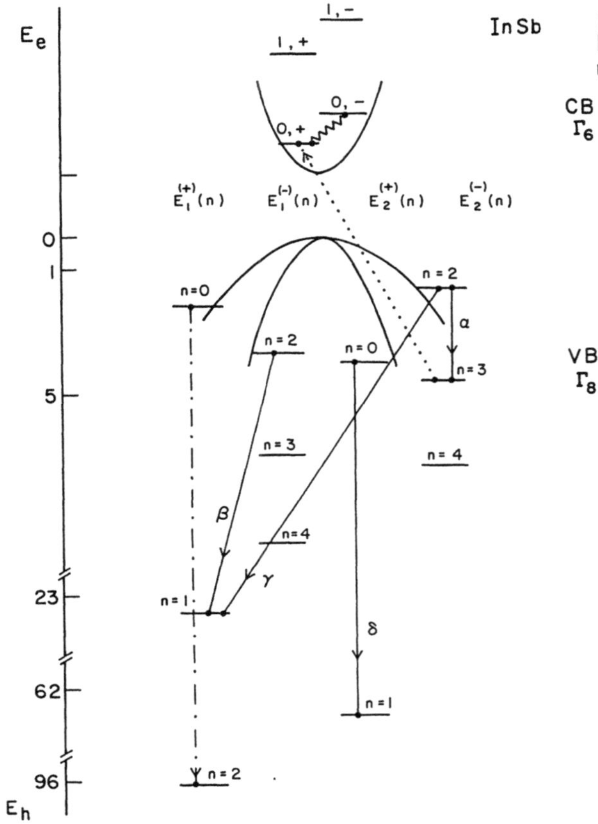

Fig. 4. Landau levels for holes in InSb for $k_\zeta = 0$ in units of $(-\hbar eB_0/mc)$

ensure Hermiticity, contains $\vec{A}_0$. The operator S is selected so that $H_A + i[H_S, S] = 0$. Then $\langle\psi_f|V|\psi_i\rangle = \langle f|\exp(-iS)V\exp(iS)|i\rangle = \langle f|V_{eff}|i\rangle$, where

$$V_{eff} = \hat{\varepsilon}\cdot\vec{v} + i[\hat{\varepsilon}\cdot\vec{v}_S, S] + \frac{i}{2}\{\hat{\varepsilon}\cdot\vec{v}_S, \vec{q}\cdot\vec{r}\} - (i\omega/e)\sqrt{\varepsilon}(\hat{n}\times\hat{\varepsilon})\cdot\vec{\mu} + \ldots \quad . \quad (8)$$

The first term in (8) gives the electric dipole interaction of which only the even part, namely $\hat{\varepsilon}\cdot\vec{v}_A = \hbar^{-1}\,\hat{\varepsilon}\cdot\partial H_A/\partial\vec{k}$, is of interest in the present context; the second term, involving $\vec{v}_S = \hbar^{-1}\,\partial H_0/\partial\vec{k}$ gives the effect of the admixture to first order in S. The third and fourth contributions correspond to the electric quadrupole and the magnetic dipole interactions. In calculating the relative intensities of particular transitions it was necessary to keep both these terms because they gave comparable contributions to the matrix element of V_{eff}.

The interference effect responsible for the change in the intensity of an individual absorption line upon reversal of either $\vec{B}_0$ or of $\vec{q}$ results from the relative signs of the first two terms in V_{eff} compared to those of the last two as functions of the direction of $\vec{B}_0$ (or of $\vec{q}$).

Several parity conserving transitions have been studied. As an example we describe results for the n=0 to n=2 line in InSb shown as a dashed-dotted line in Fig. 4. A measure of the anisotropy is the parameter $\Delta = (I_+ - I_-)/(I_+ + I_-)$ where I_+ and I_- are absorption intensities for opposite senses of $\vec{B}_0$. In Fig. 5 we show $|\Delta|$ as a function of B_0 at the resonance frequency. The numerical results were obtained using $C = -9.2$ meV Å for InSb with $\vec{B}_0 \parallel [111]$ and $\vec{q} \parallel [\bar{1}10]$ in the EV geometry ($\vec{E} \perp \vec{B}_0$). We note that at $B_0 \approx 50$ kG reversal of $\vec{B}_0$ leads to a complete suppression of the resonance. The anisotropy of this resonance for a magnetic field of 50 kG lying in the (112) plane is shown in Fig. 6. A similar result for InSb for a magnetic field of 70 kG in the (110) plane is exhibited in Fig. 7. The dashed line is identical to the solid line but is shifted by 180°.

Fig. 5

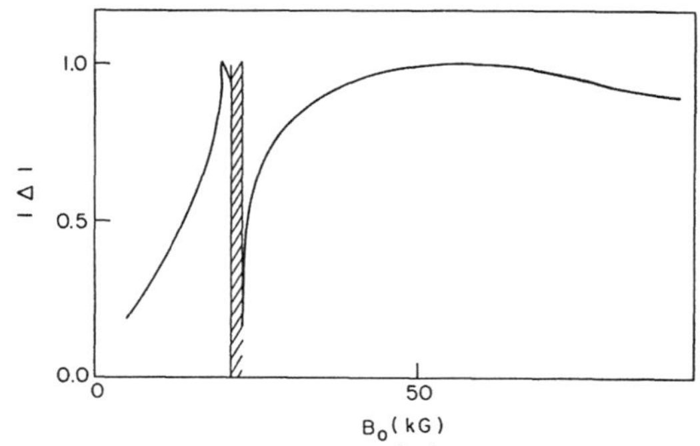

Fig. 6

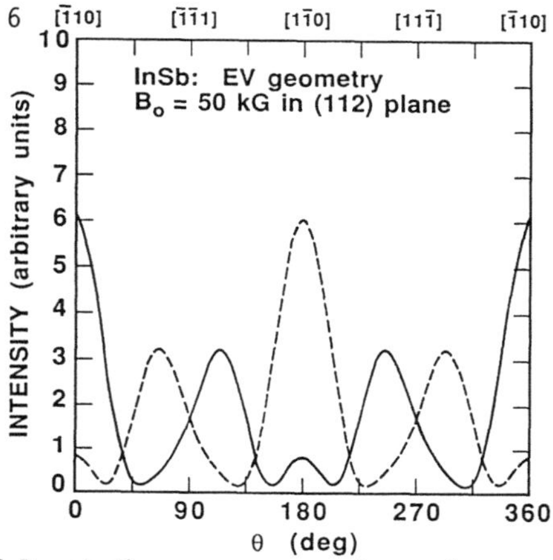

Fig. 5. The parameter $|\Delta|$ as a function of B_0 at the resonance n=0 → n=2 indicated by a dashed-dotted line in Fig. 4 in the extraordinary Voigt (EV) geometry: $\vec{B}_0 \parallel [111]$, $\vec{q} \parallel [\bar{1}10]$. The shaded region is that for which the incident wave is reflected.

Fig. 6. Anisotropy of the line in Fig. 5 for a magnetic field of 50 kG in the (112) plane; θ is the angle between $\vec{B}_0$ and $[\bar{1}10]$

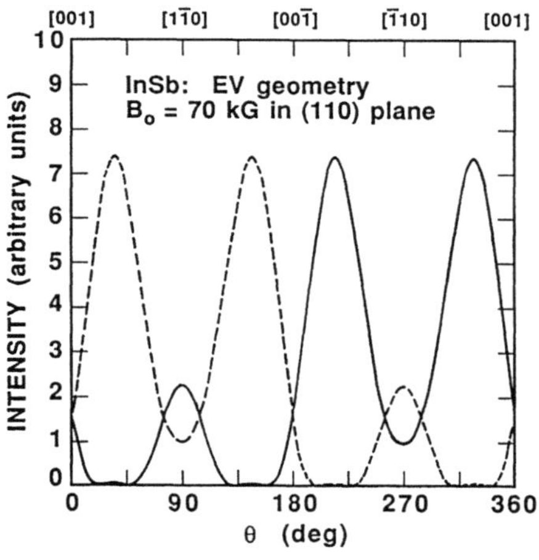

Fig. 7. Anisotropy of the line in Fig. 5 for a magnetic field of 70 kG in the (110) plane; θ is the angle between $\vec{B}_0$ and [001]

The transition discussed above has not been observed yet. However, an order of magnitude estimate shows that its intensity is comparable to that of electric-dipole-induced spin resonance observed in n-type material [6]. It is therefore suggested that a study of the $\Delta n=2$ transition mentioned here will yield the value (and sign) of the parameter C in zinc-blende semiconductors.

In addition to the results mentioned above, we mention the effect of reducing the symmetry of the crystal by application of uniaxial stress. We limit our discussion to the interpretation of experiments by Kuchar et al. [13] and Kriechbaum et al. [14] who carried out far infrared spin-resonance experiments in InSb in the Faraday geometry and in the presence of a uniaxial stress. The Γ_6 conduction electrons experience an additional spin-orbit interaction of the form

$$H_\varepsilon = a[\sigma_x(\varepsilon_{zx}k_z-\varepsilon_{xy}k_y)+c.p.] + b[\sigma_x k_x(\varepsilon_{yy}-\varepsilon_{zz})+c.p.] , \qquad (9)$$

where a and b are constants and $\varepsilon_{ij}(i,j=x,y,z)$ are the components of the strain tensor.

Within $\vec{k}\cdot\vec{p}$ perturbation theory involving the Γ_6 conduction band and the Γ_8 and Γ_7 states of the spin-orbit split valence band

$$a = \frac{2}{3} PC_2 \left(\frac{1}{E_g} - \frac{1}{E_g+\Delta}\right) \qquad (10)$$

and b = 0. P is proportional to the matrix element of the momentum between Γ_8 and Γ_6 states, E_g the fundamental energy gap, Δ the spin-orbit splitting of the Γ_8 and Γ_7 states [15] and C_2 is a deformation potential constant defined by Trebin et al. [16]. We take the electron Landau levels, $|ns\rangle$ $(n=0,1,2,\ldots;s=\pm\frac{1}{2})$ as the zero-order states. Taking account of the mixing caused by H_ε is equivalent to replacing $|ns\rangle$ by

$$\chi_{ns} = |ns\rangle + \sum_{n's'} |n's'\rangle\langle n's'|H_\varepsilon|ns\rangle(E_n^{(s)}-E_{n'}^{(s')})^{-1} + \ldots , \qquad (11)$$

where $E_n^{(s)} = (n+\frac{1}{2})(\hbar eB_0/m^*c)+g\mu_B B_0 s$ is the energy of the state $|ns\rangle$ prior to introducing H_ε (and $H_A = \delta_0[\sigma_x k_x(k_y^2-k_z^2)+c.p.]$), B_0 is the dc magnetic field and g the Landé splitting factor. Strictly speaking g is a tensor but we need not be concerned with these details here. We require the matrix elements of $\vec{v}$ between the states $\chi_{0,\frac{1}{2}}$ and $\chi_{0,-\frac{1}{2}}$. To first order in H_ε this quantity consists of two parts: (i) the term originating from the zero-order velocity operator $\vec{v}_0 = \hbar\vec{k}/m^*$ and the first-order admixture of $|1,-\frac{1}{2}\rangle$ with $|0,\frac{1}{2}\rangle$ and of $|1,\frac{1}{2}\rangle$ with $|0,-\frac{1}{2}\rangle$ and (ii) the element of $\vec{v}_\varepsilon = \hbar^{-1}\partial H_\varepsilon/\partial\vec{k}$ between the zero order states $|0,\frac{1}{2}\rangle$ and $|0,-\frac{1}{2}\rangle$. These contributions turn out to be of comparable magnitudes. The authors of Refs. [13] and [14] calculated (i) correctly but omitted (ii). The similar contributions arising from H_A are calculated in the same way.

In the experimental work of Refs. [13] and [14] the external magnetic field was parallel to $[1\bar{1}2]$ and the stress, T, was applied along [110]. In the Faraday configuration, used by these authors, the matrix elements of $\vec{v}$ in the cyclotron resonance active (CRA), $\hat{\varepsilon}_+$, and in the cyclotron resonance inactive (CRI), $\hat{\varepsilon}_-$, configurations are

$$\langle \chi_{0,-\frac{1}{2}} | v_{CRA} | \chi_{0,\frac{1}{2}} \rangle = (i\cos\beta/\hbar\sqrt{2}) L_-[bT(s_{11}-s_{12}) + (\delta_0 eB_0/2\hbar c)(3\cos^2\beta-1)] \tag{12}$$

and

$$\langle \chi_{0,-\frac{1}{2}} | v_{CRI} | \chi_{0,\frac{1}{2}} \rangle = (i\cos\beta/2\sqrt{2}\hbar) L_+[aTs_{44}-(6\delta_0 eB_0/\hbar c)\sin^2\beta] \ . \tag{13}$$

Here

$$L_+ = 1 + (\hbar eB_0/m^* c)(E_0^{(+)}-E_1^{(-)})^{-1} \ , \tag{14}$$

$$L_- = 1 + (\hbar eB_0/m^* c)(E_0^{(-)}-E_1^{(+)})^{-1} \ , \tag{15}$$

β is the angle $\vec{B}_0$ makes with the [100] axis ($\cos\beta=\sqrt{2/3}$ in the experiments mentioned), s_{ij} are the elastic compliance coefficients and T the force per unit area. In (13) we have omitted the magnetic dipole term. We have estimated its value as it pertains to the experiment of Refs. [13] and [14]. We find that it is in quadrature with the terms displayed in (13) and that it contributes less than 3% of the intensity of the spin-flip transition. We remark that, omitting the terms proportional to δ_0 and to b, our results are the same as in Refs. [13] and [14] except that they use $(\hbar eB_0/m^*c)(E_0^{(+)}-E_1^{(-)})^{-1} \approx -0.76$ for L_+ instead of the correct value deduced from (14), i.e., $L_+ \approx 0.24$. This leads to a correction of the value of C_2 by a factor of 3.2, the corrected value being $|C_2| \approx 5$ eV.

Jagannath and Aggarwal [17] measured the intensity of the stress-induced generation at the spin resonance frequency in InSb. They applied the stress and $\vec{B}_0$ along [111]. The directions of the incident and emitted beams were [1$\bar{1}$0]. In the ordinary Voigt configuration (OV: radiation linearly polarized parallel to $\vec{B}_0$) the spin-flip transition, in this case, is entirely due to magnetic-dipole radiation. In the extraordinary Voigt configuration (EV: radiation linearly polarized perpendicular to $\vec{B}_0$), the matrix element of the [$\bar{1}\bar{1}$2] component of $\vec{v}$ is

$$6^{-\frac{1}{2}} \langle \chi_{0,-\frac{1}{2}} | \vec{v} | \chi_{0,\frac{1}{2}} \rangle \cdot [\bar{1}\bar{1}2] = (iL_+/\hbar\sqrt{3})[\tfrac{1}{2}aTs_{44} + \delta_0 eB_0/\hbar c] \ . \tag{16}$$

Jagannath and Aggarwal used the same expression for L_+ as the authors of Refs. [13] and [14]. Thus, their value of C_2 must also be corrected. For their experiment this adjustment leads to $|C_2| \approx 3.8$ eV.

Calculations by Cardona et al. [18] using pseudopotential and linear combination of atomic orbitals (LCAO) for InSb yield C_2 = 2.44 eV and 4.5 eV, respectively. The agreement between these results and the corrected experimental values reported in the present work is gratifying.

Acknowledgements

This work was supported by the U.S. National Science Foundation (Grant No. DMR 86-16787). The author wishes to thank F. Bassani, N. Kim, G. C. La Rocca, and J. K. Furdyna with whom he has studied the subjects on which this paper is based.

References

1. G.F. Koster, J.O. Dimmock, R.G. Wheeler, and H. Statz, Properties of the Thirty-two Point Groups (MIT Press, Cambridge, Massachusetts, 1963).
2. E.I. Rashba and V.I. Sheka, Fiz. Tverd. Tela (Leningrad) 3, 1735; 3, 1836 (1961)[Sov. Phys.-Solid State 3, 1257 (1961); 3, 1357 (1961)].
3. H.C. Gatos and M.C. Lavine, in Progress in Semiconductors, vol. 9, edited by F. Gibson and R.E. Burgess (Heywood, London, 1965) pp. 1-45; J. Phys. Chem. Solids 14, 169 (1960).
4. A.T. Gorelenok, V.A. Marushchak, and A.N. Titkov, Izv. Akad. Nauk., SSSR Ser. Fiz. 50, 290 (1986).
5. H. Riechert, S.F. Alvarado, A.N. Titkov, and V.I. Safarov, Phys. Rev. Lett. 52, 2297 (1984). See also H.-J. Drouhin, C. Hermann, and G. Lampel, Festkörperprobleme 25, 254 (1985); H. Riechert, H.-J. Drouhin, and C. Hermann, Phys. Rev. (to be puslished).
6. M. Dobrowolska, Y.-F. Chen, J.K. Furdyna, and S. Rodriguez, Phys. Rev. Lett. 51, 134 (1983); Y.-F. Chen, M. Dobrowolska, J.K. Furdyna, and S. Rodriguez, Phys. Rev. B 32, 890 (1985); S. Gopalan, J.K. Furdyna, and S. Rodriguez, Phys. Rev. B 32, 903 (1985).
7. With the convention mentioned above regarding the polarity of the positive [111] direction $\delta_0 > 0$. See M. Cardona, N.E. Christensen, M. Dobrowolska, J.K. Furdyna, and S. Rodriguez, Solid State Commun. 60, 17 (1986).
8. F. Bassani, G.C. La Rocca, and S. Rodriguez, Phys. Rev. B 37, 6857 (1988).
9. G.C. La Rocca, S. Rodriguez, and F. Bassani, Phys. Rev. (to be published).
10. J.M. Luttinger, Phys. Rev. 102, 1030 (1956).
11. G. Dresselhaus, Phys. Rev. 100, 580 (1955).
12. E.O. Kane, J. Phys. Chem. Solids 1, 249 (1957).
13. F. Kuchar, R. Meisels, and M. Kriechbaum, in Proceedings of the Fourth International Conference on the Physics of Narrow Gap Semiconductors, Linz, Austria, edited by E. Gornik, H. Heinrich, and L. Palmetshofer (Springer, Berlin, 1982) p. 197.
14. M. Kriechbaum, R. Meisels, F. Kuchar, and E. Fantner, Physica 117B and 118B, 444 (1983).
15. For InSb we use the numerical parameters in Refs. [13] and [14] augmented by other quantities from other references. We take $P=9.4\times10^{-8}$eVcm, $E_g=0.2352$ eV, $\Delta=0.803$ eV, $m^*=0.015$ m, $g=-41.6$ at 50 kG, $\delta_0=2.2\times10^{-22}$eVcm^3 [6], $s_{44}=3.18\times10^{-3}(\mathrm{kbar})^{-1}$.
16. H.-R. Trebin, U. Rössler, and R. Ranvaud, Phys. Rev. B 20, 686 (1979).
17. C. Jagannath and R.L. Aggarwal, Phys. Rev. B 32, 2243 (1985).
18. M. Cardona, N.E. Christensen, and G. Fasol, Phys. Rev. (in press).

Magnetic Field Dependence of Carrier and Exciton Diffusion in Photoexcited Ge

K. Fujii, T. Tomaru, T. Ohyama, and E. Otsuka

Department of Physics, College of General Education,
Osaka University, Toyonaka, Osaka 560, Japan

To investigate the magnetic field dependence of the carrier diffusion, the nonequilibrium electron-exciton system in photoexcited Ge is studied. Three types of experiments are performed, that is, a time resolved magneto-optical absorption measurement employing a far infrared laser and the same employing microwaves, and a space-resolved luminescence measurement without applying a magnetic field. It is found that 1) a quasi-equilibrium relation between free carriers and excitons exists. 2) The diffusion coefficient clearly decreases with increasing magnetic field. 3) This magnetic field dependence can be explained qualitatively by the random walk model of electrons.

1. INTRODUCTION

Diffusion of nonequilibrium carriers is one of the most important subjects of transport in semiconductors. The diffusion of the carriers is expected to depend strongly on magnetic field, because the motion of carriers is quantized by the magnetic field. The diffusion phenomena of nonequilibrium carriers have been investigated in connection with the diffusion of electron-hole drops (EHD) [1]. But there are some ambiguities in the diffusion coefficient of carriers, because of the existence of EHDs. There have been a few studies on the diffusion of free carriers themselves [2-4]. The aim of this study is to obtain the diffusion coefficients of the photoexcited carriers and excitons and to investigate the influence of the magnetic field on them.

Three types of experiments were performed to investigate the dependence of the diffusion coefficient D on magnetic fields for wide range. The diffusion coefficient D under strong, weak and zero magnetic field can be obtained by means of magneto-optical absorption measurements employing a far-infrared (FIR) laser or microwaves and by space-resolved photoluminescence measurement respectively.

2. EXPERIMENTAL PROCEDURE

In this experiment, high purity germanium samples, in which the residual impurity concentration was less than 10^{12} cm^{-3}, were employed. In the magneto-optical absorption measurements, samples were always set in the Faraday configuration and the magnetic field was up to 10 T. Samples with rectangular shape were employed, the dimension of which was typically $4 \times 7 \times 1 \times mm^3$. A xenon flash lamp or Ar^+ laser were employed as an excitation light for the magneto-optical absorption measurements and photoluminescence measurements, respectively. The excitation arrangement for the magneto-optical absorption measurement is shown in Fig. 1 . The shaded area on the sample is covered by black polyethylene and a metal plate to shield the excitation light and the FIR laser beam. The absorption

Springer Series in Solid-State Sciences, Vol. 87
High Magnetic Fields in Semiconductor Physics II Editor: G. Landwehr

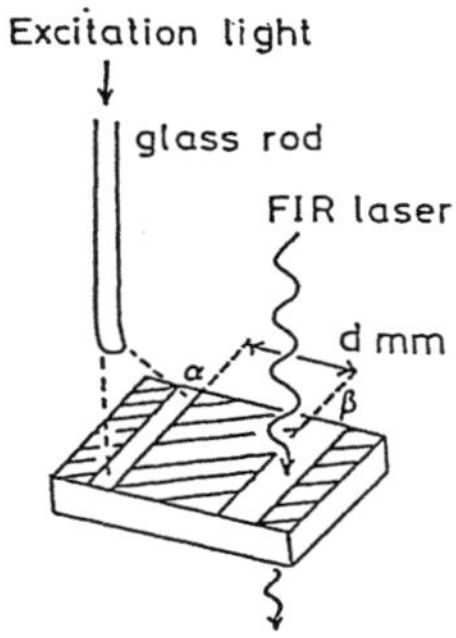

Fig. 1 Sample arrangement.

signal through the strip region β, separated from the photoexcited strip region α by the distance d mm, was observed. In microwave magneto-optical absorption measurements, the arrangement employed was similar to that of SANADA et al. [2]. A water filter was employed to attain surface excitation. An experimental set-up of the photoluminescence measurement is similar to that of CULBERTSON et al. [3] to obtain space-resolved signals. The spatial resolution is ~0.3 mm. The direction of the excitation light is different from that of CULBERTSON.

3. EXPERIMENTAL RESULTS

3.1 Magneto-optical absorption measurements

Figure 2 shows a time variation of magneto-optical absorption curves at the strip region β separated from the photoexcited region α by d = ~1 mm in the arrangement shown in Fig. 1. The FIR laser wavelength is 119 μm. Two peaks are observed, i.e., the peak due to the Zeeman absorption of excitons and that due to the cyclotron resonance of holes. The numeral on the right of each curve indicates the delay time after pulsed photoexcitation. The maximum absorption intensity of both peaks is obtained at 22 μs of the delay time. This delay is quite large and occurs due to the separation between the excited and observed regions. For signals of electrons and holes, large delays are also observed.

Figure 3 shows the time variation of the absorption intensities of the exciton signal derived from the absorption curves in Fig. 2. To determine the diffusion coefficient, a one-dimensional diffusion process along the direction x will be assumed. The diffusion equation has a solution of the form:

$$n(x,t) = \frac{N_0}{\sqrt{4\pi Dt}} \exp\left[-\frac{x^2}{4Dt} - \frac{t}{\tau_e} \right] \tag{1}$$

where $n(x,t)$, τ_e are density and lifetime of the diffusing particles, respectively. The quantity N_0 is the number of the particles generated with delay time = 0. Considering the widths of both regions and integrating the carrier density, we try to fit these data with Eq. (1), using D and τ_e as fitting parameters. Here the lifetimes of the carriers can be deduced directly from the time variation of the magneto-optical absorption in the arrangement that a whole sample surface is illuminated uniformly (This arrangement will be called as "standard arrangement") [5]. These lifetimes can be chosen as the initial values of the fitting parameter τ_e in Eq. (1). There remains the other fitting parameter, D. The solid and broken lines in Fig. 3 are obtained as results of the fitting based on Eq. (1).

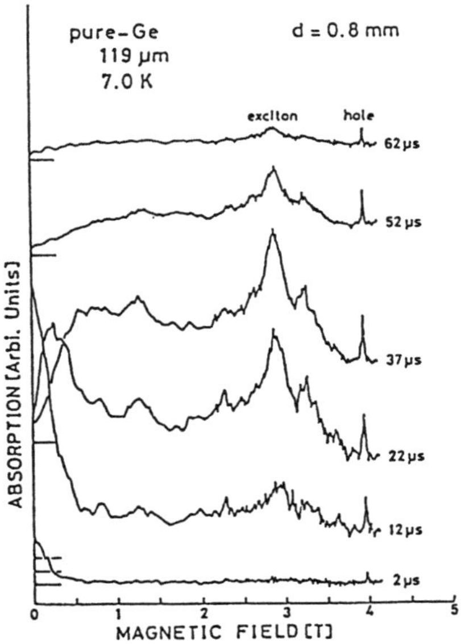

Fig. 2 A series of time resolution traces in the arrangement shown in Fig. 1 at 7.0 K.

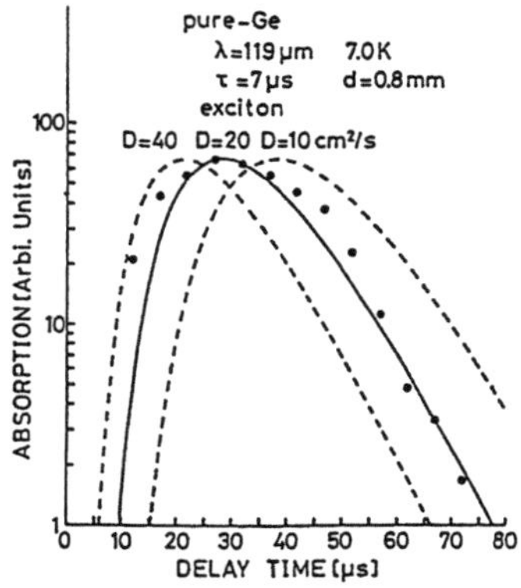

Fig. 3 Time variation of the absorption intensity due to excitons at 3 T. Solid and broken lines shows the results of fitting.

The best fit is obtained at D = 20 cm^2/s for the diffusion coefficient of excitons at 7.0 K. At this temperature, diffusion coefficients of electrons and holes are found to have the same value as that of excitons. This suggests that a quasi-equilibrium relation between electrons, holes and excitons is exists in this condition as well as in the measurement under the standard arrangement [5]. It should be noted that the resonance signals were obtained at 3 T.

In the microwave absorption measurements, electron and hole cyclotron signals are observed at 0.05 T and 0.1 T, respectively. Both absorption intensities attain the maximum values at 17 μs of the delay time. In this case, the electron lifetime deduced from the time variation signals under standard arrangement does not give a good fitting. So the electron lifetime is also treated as a fitting parameter. The best fit is obtained at 500 cm^2/s for D and 3 μs for the lifetime.

3.2 Luminescence measurement

The spatial dependence of the luminescence intensity due to excitons (denoted by EX) under steady state excitation is shown in Fig. 4. The luminescence intensity is expected to decrease exponentially with increasing distance x from the illuminated region. The slope of this curve gives the quantity $(D\tau_e)^{-1/2}$. Using the exciton life time 10 μs obtained under standard arrangement, the diffusion coefficient of exciton is derived to be 2000 cm^2/s at 4.2 K under no magnetic field. At 1.9 K, the exciton diffusion coefficient is 380 cm^2/s, which is smaller than that at 4.2 K. This temperature dependence reflects by the stability of EHD.

4. DISCUSSION

The results are tabulated in Table 1. It can be seen that the diffusion coefficient decreases with magnetic field. The value obtained from the

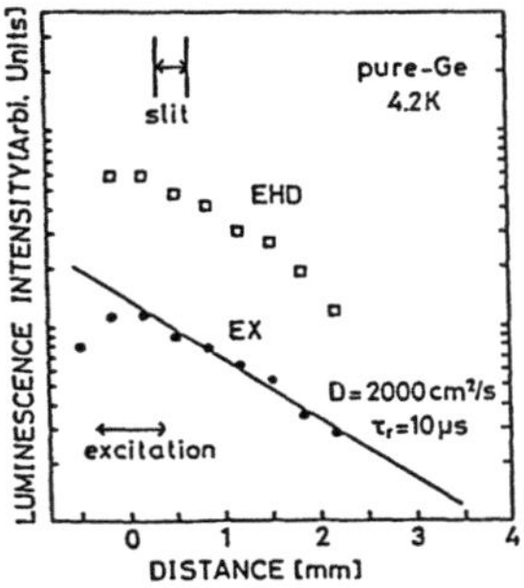

Fig. 4 Spatial variation of the luminescence intensities due to excitons (EX) and due to electron-hole drops(EHD), respectively.

Table 1

magnetic field(T)	ℓ(Å)	τ (s)	$D=\frac{\ell^2}{2\tau}$	D (exp.)	technique
0	17000	10^{-10}	140	2000	photoluminescence (exciton, EHD)
0.1	1700	10^{-10}	1.4	500	microwave CR (electron hole)
3	150	10^{-11}	0.11	50	FIR CR (exciton)

luminescence measurements gives only the upper limit of D at zero magnetic field, since the space resolution is not so good. Our results under weak and no magnetic field agree with the previous results [2,3].

The dependence of D on the magnetic field can be explained qualitatively by the random walk model. In this model, electrons skip by a length ℓ at one scattering event and the diffusion coefficient D is written as follows:

$$D = \ell^2/2\tau \ ,$$

where τ is the momentum relaxation time that is deduced from the width of the cyclotron resonance line for electrons and holes. Under no magnetic field, the length ℓ is given by $v\tau$, where v is a thermal velocity, $(kT/m^*)^{1/2}$. When an external magnetic field is applied normal to the diffusion direction of carriers, electrons are skipping by a cyclotron radius at one scattering event and the length ℓ is of the order of the cyclotron radius. Under strong magnetic field, ℓ is assumed to be a cyclotron radius of the first Landau level, that is $(ch/eB)^{1/2}$. Furthermore, in a weak magnetic field, ℓ is $(2kTm^*/\pi)^{1/2}(c/eB)$ by averaging the electron distribution. Calculated values of D for the three cases are also listed in Table 1. Though there are large differences by more than one order of the magnitude, the magnetic field dependences of D can be explained qualitatively.

References

1. For a review, see J. C. Hensel, T. G. Phillips and G. A. Thomas: Solid State Phys. 32 (1977) eds. H. Ehrenreich, F. Seitz and D. Turnbull (Academic Press, New York).
2. T. Sanada, T. Ohyama, E. Otsuka: Solid State Commun. 17 999 (1975).
3. J. C. Culbertson, R. M. Westervelt and E. E. Haller: Phys. Rev 34 6980 (1986).
4. Ya. E. Pokrovskii and K. I. Svistunova: Sov. Phys. Solid state 13 1241 (1971).
5. K. Fujii, T. Tomaru, T. Ohyama, E. Otsuka: J. Magn. Soc. Jpn. 11 Suppl. 125 (1987).

Neutral Bound Excitons at Intermediate to High Magnetic Fields

F. Dujardin[1], *B. Stébé*[1], *and G. Munschy*[2]

[1]Centre Lorrain d'Optique et Electronique des Solides, Université de Metz et Ecole Supérieure d'Electricité, 2, Rue Edouard Belin, F-57078 Metz Cedex 3, France

[2]Groupe d'Optique Nonlinéaire et d'Optoélectronique, Institut de Physique et Chimie des Matériaux de Strasbourg, Unité mixte 380046, CNRS, ULP, EHICS, 5, Rue de l'Université, F-67084 Strasbourg Cedex F, France

The influence of an external constant magnetic field on the ground state energy of an exciton bound to a neutral shallow acceptor impurity is studied in the effective mass approximation using a variational non-adiabatic method well-adapted for intermediate to high magnetic fields. The results permit to discuss the validity of those previously obtained in the low and high field limits using respectively perturbation and adiabatic variational methods.

1. INTRODUCTION

The recent availability of high quality samples stimulates new research on the optical properties of bound exciton states [1]. Nevertheless, apart from the analysis of the Zeeman splitting, we have hardly any theoretical information predicting their behaviour in a magnetic field. We have therefore undertaken a general study of the properties of these complexes in a magnetic field. In a first approach [2], we have computed the diamagnetic shift of an exciton-neutral acceptor complex (A°,X) for small values of the reduced field parameter γ. We have shown that the diamagnetic shift of a bound exciton differs from that of an exciton and how it depends on the effective mass ratio $\sigma = m_e^*/m_h^*$. This study is no longer valid when the magnetic and Coulomb energies are of the same order of magnitude. This restriction can occur for relatively low fields when the hole mass is small. On the other hand the opportunity to use now higher magnetic field facilities gives a new impetus to high field studies. The very high field limit has been studied [3] in the adiabatic approximation, which is justified if the Coulomb correlations are negligible compared to the magnetic energy. In this work we present a non-adiabatic method for the computation of the ground state energy of an exciton bound to a neutral acceptor, suitable in the case of intermediate fields.

2. HAMILTONIAN

We study explicity the case of the (A°,X) complex, consisting of two holes (1,2) and one electron (e) bound to an ionized acceptor centre in the presence of an external uniform magnetic field B directed along the z-axis. By interchanging the electrons and the holes, this complex is quite analogous to the neutral donor bound exciton (D°,X). The effective envelope Hamiltonian may be simplified by using the usual atomic units $a_A = \varepsilon\hbar^2/m_h^* e^2$ for length, $2|E_A| = m_h^* e^4/\varepsilon^2\hbar^2$ for energy and the dimensionless effective magnetic field parameter $\gamma = \hbar\omega_c/2|E_A|$, where $\omega_c = eB/m_h^* c$ is the effective hole

Springer Series in Solid-State Sciences, Vol. 87
High Magnetic Fields in Semiconductor Physics II Editor: G. Landwehr

cyclotron frequency related to the magnetic field directed along the z-axis. ε is an appropriate dielectric constant taking into account the polarization effects. By using the Lorentz gauge and neglecting the spins as well as the electron-hole exchange interaction [4], the Hamiltonian of the system is given by

$$H = H_0 + H_1' + H_2', \qquad (1)$$

where H_0 is the zero-field Hamiltonian

$$H_0 = T_h + \sigma^{-1} T_e + V \qquad (2)$$

with

$$T_h = -(\nabla_1^2 + \nabla_2^2)/2, \qquad T_e = -\nabla_e^2/2, \qquad (3)$$

$$V = -\frac{1}{r_1} - \frac{1}{r_2} + \frac{1}{r_e} + \frac{1}{r_{12}} + \frac{1}{r_{1e}} - \frac{1}{r_{2e}} \qquad (4)$$

whereas H_1' and H_2' denote the orbital magnetic terms

$$H_1' = \gamma(-L_{1z} - L_{2z} + \sigma^{-1} L_{ez})/2 \qquad (5)$$

$$H_2' = \gamma^2 (\rho_1^2 + \rho_2^2 + \sigma^{-1} \rho_e^2)/8 \qquad (6)$$

with $\rho_i^2 = x_i^2 + y_i^2$, and where L_{iz} are the projections of the angular momentum operators of the three mobile particles. The linear magnetic term H_1' gives rise to the orbital Zeeman splitting, whereas the two terms H_1' and H_2' are in general responsible for the diamagnetic shift. Their influence depends on the value of the effective magnetic field γ, which for acceptor impurities may often be one order of magnitude smaller than for donor impurities.

3. LOW AND HIGH FIELD LIMITS

At zero magnetic field, the best values of the binding energies of the (A^o,X)-complex have been obtained [5] using a Page and Fraser type wave function. For small enough values of the effective magnetic field γ, the magnetic terms H_1' and H_2' may be considered as perturbations with respect to the zero-field Hamiltonian H_0. In this limit, we have variationally determined [2] the eigenvalues of the Halmiltonian (1) using the same zero-field function but by varying its linear variational parameters. The results so obtained are displayed for $\sigma = 1$ in Fig.1. This low field approximation is expected to hold for γ-values up to 0.5; despite the spherical symmetry of the zero-field wave function it does not reflect the actual symmetry of the (A^o,X)- complex when subjected to an external magnetic field.

For very high effective magnetic fields $\gamma \gg 1$, the Coulomb energies are negligible compared to the magnetic energy $\hbar\omega_c$. In this limit, the system behaves like three quasi-free particles interacting independently with the field. The energies of the carriers become quantized in the plane perpendicular to the magnetic field but remain quasi-continuous in its direction. The Schrödinger equation of the transverse motion separates into three one-particle equations giving rise to the well-known Landau levels [6,7].

Strictly speaking, the longitudinal and transverse motions cannot be separated when the Coulomb interaction does not vanish. However, for high enough magnetic fields, we can generalize the procedure introduced in the

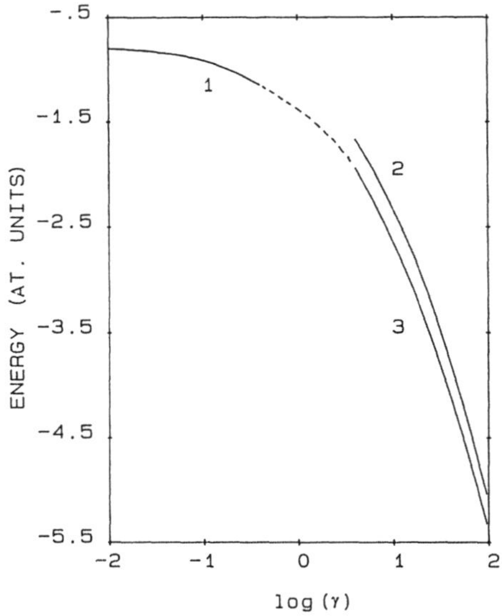

Fig.1 : Energy of the (A°,X) complex versus the reduced field parameter γ
1) variational-perturbation method
2) adiabatic method
3) non-adiabatic method

case of hydrogenic problems [8]. In this high field limit, the Coulomb potential in the Hamiltonian may be considered as a small perturbation and the two motions may be tentatively separated in the sense of the Born and Oppenheimer approximation.

Finally the total ground state energy E has been determined [3] by solving variationaly the equation corresponding to the motion in the direction of the magnetic field (Fig.1). Our results appear to be more adequate for small σ-values because our wave function does not take into account the correlations between the mobile particles.

4. NON-ADIABATIC METHOD FOR INTERMEDIATE FIELDS

When the adiabatic method breaks down, the transverse and longitudinal motion can no longer be separated. In this case we solve directly the Schrödinger equation by means of the variation method. We generalise the cylindrical Gaussian basis used by Aldrich and Greene [9] to study the effect of an arbitrary magnetic field on an hydrogen-like system and we choose the following trial basis :

$$\Psi_i = \Psi_{1i}\Psi_{2i}\Psi_{\bullet i}, \tag{7}$$

$$\Psi_{1i} = \rho_1^{|M_{1i}|} e^{iM_{1i}\varphi_1} \exp(-\alpha_{1i}\rho_1^2 - \beta_{1i} z_1^2) \tag{8}$$

and similar expressions for Ψ_{2i} and $\Psi_{\bullet i}$. Our basis wave function generalises that used in the high field limit, where the parameters α_{1i}, α_{2i} and $\alpha_{\bullet i}$ are all equal to $\gamma/4$ and we have restricted ourselves to the case $M = 0$ which was expected to correspond to the ground state energy as proved in a particular case. The parameters α_i and β_i have been determined variationally using a ten term basis (7).

Figure 2 shows the variations of the ratio of the parameter α_i of the first function $(M_1 = M_2 = M_\bullet = 0)$ by its value in the adiabatic limit (i.e.

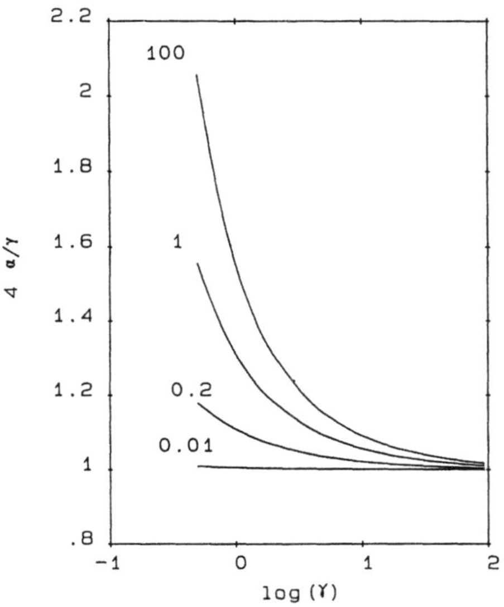

Fig.2 : Ratio of the coefficient α_1 by its value obtained by the adiabatic method (i.e. $\gamma/4$) for several values of the mass ratio σ.

$\gamma/4$), as a function of γ. We notice that when γ increases, this ratio tends to unity. The adiabatic method is then well justified. For a given value of γ, when the mass ratio increases, the parameter α_1 increases also and takes values rather different from $\gamma/4$. This confirms that the adiabatic approximation becomes less accurate for large values of the mass ratio where the present method gives better results.

5. DISCUSSION

As a result, for a given value of the reduced field parameter, it appears that the present results are better than those obtained by our previous adiabatic method for large values of the mass ratio because the Coulomb correlations cannot be neglected (Fig.1). On the other hand, for small values of the mass ratio, the adiabatic method remains better. In this case, the above correlations are less important.

Our results allow to estimate qualitatively the transition energy between a neutral acceptor and the (A°,X)-complex in a magnetic field :

$$h\nu = h\nu_X + E - E_A - E_X \qquad (9)$$

knowing the exciton transition energy $h\nu_X$ in the same field. The exciton and acceptor binding energies E_X and E_A may be obtained from the results for excitons and hydrogenic systems [9,10]. The same results hold for the analog (D°,X)-complex by interchanging the electron and hole effective masses. Nevertheless, in this study we have restricted ourselves to isotropic spherical and non-degenerate electron and hole bands. This approximation becomes questionable for materials with j=3/2 type hole band structure. In this case the best results would probably arise from the use of an experimental "mean" hole mass deduced for instance from the observed exciton spectra.

REFERENCES

1. W.M. CHEN and B. MONEMAR, Phys. Rev. B 36, 7948 (1987)
2. F. DUJARDIN, B. STEBE and G. MUNSCHY, phys.stat.sol.(b) 126, 329 (1984)
3. F. DUJARDIN, B. STEBE and G. MUNSCHY, phys.stat.sol.(b) 141, 559 (1987)
4. B. STEBE and G. MUNSCHY, Solid State Commun. 40, 663 (1981)
5. B. STEBE and G. MUNSCHY, Solid State Commun. 35, 557 (1980)
6. L. LANDAU, Z. Phys. 64, 629 (1930)
7. M.H. JOHNSON and B.A. LIPPMANN, Phys. Rev. 76, 828 (1949)
8. Y. YAFET, R.W. KEYES and E.N. ADAMS, J. Phys. Chem. Solids 1, 737 (1956)
9. C. ALDRICH and R.L. GREENE, phys. stat. sol.(b) 93, 343 (1979)
10. D. CABIB, E. FABRI and G. FIORIO, Solid State Commun. 9, 1517 (1971)

Zeeman Effect of the Carbon Acceptor in GaAs*

J. Schubert, M. Dahl, and E. Bangert

Physikalisches Institut der Universität Würzburg,
Am Hubland, D-8700 Würzburg, Fed. Rep. of Germany

1. Introduction

Experimental as well as theoretical studies of the Zeeman splitting of shallow acceptors in semiconductors with degenerate valence bands have been reported for more than 20 years /1-11/. Although far-infrared spectroscopy provides a large number of lines the interpretation suffers by the fact that a unique assignment to initial and final Zeeman sublevels is rather difficult. Therefore additional information from applied external stress/11/, different orientations of an external magnetic field, temperature dependence or theoretical calculations /2/ is necessary to extract g-values from the experimental data. The group theoretical considerations of the acceptor problem of Bhattacharjee and Rodriguez /8/ and the formulation in terms of tensor operators /4-7/ supplemented each other in an ideal way, such that a transparent calculation of the g-values became possible.

The knowledge of the acceptor states in GaAs was found to be rather qualitative. In the first experimental studies /9/ the determination of the g-values, especially for the lower acceptor levels, was hampered by the incomplete resolution of the absorption lines. Bimberg /10/ observed the electron-acceptor luminescence, therefore only g-values of the ground states were obtained.

2. Experimental Details

Our sample was GaAs bulk material grown by the LEC technique. An unusual property of the sample was its persistent p-type Drude absorption after bleaching the EL2 absorption /12/. This persistent hole concentration ($p = 9\ 10^{15} cm^{-3}$) allowed a convenient investigation of the neutral carbon acceptor.

The transmission spectra were taken by an FIR- Fourier spectrometer. The polarized light was focussed onto the sample positioned in the bore of a superconducting split-coil optical magnet. Behind the sample a light pipe system was mounted to collect the light on the He-cooled Si-bolometer used as detector. The sample was immersed in superfluid He at a temperature of T=1.2K.

We measured in Voigt-geometry the splitting of the 'G- line' ($\tilde{\nu} = 122.5 cm^{-1}$) for the three orientations $B \parallel \langle 100 \rangle$, $B \parallel \langle 111 \rangle$ and $B \parallel \langle 110 \rangle$; typical spectra are shown in fig.1. The position of the absorption lines were determined by fitting

* This work was supported by the Deutsche Forschungsgemeinschaft

Lorentz-profiles to the data. The results for the three different orientations of the B-field are presented in fig.2.

Bhattacharjee and Rodriguez gave a complete group theoretical solution of the acceptor problem. They calculated in terms of two parameters g_1' and g_2' analytical

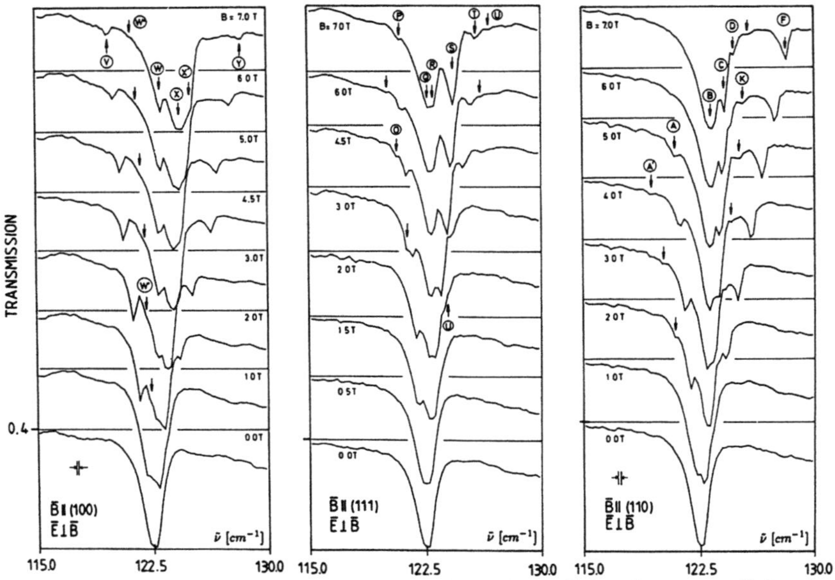

Fig. 1: Transmission spectra for $B \parallel \langle 100\rangle$, $B \parallel \langle 111\rangle$ and $B \parallel \langle 110\rangle$; $E \perp B$; $T = 1.2K$

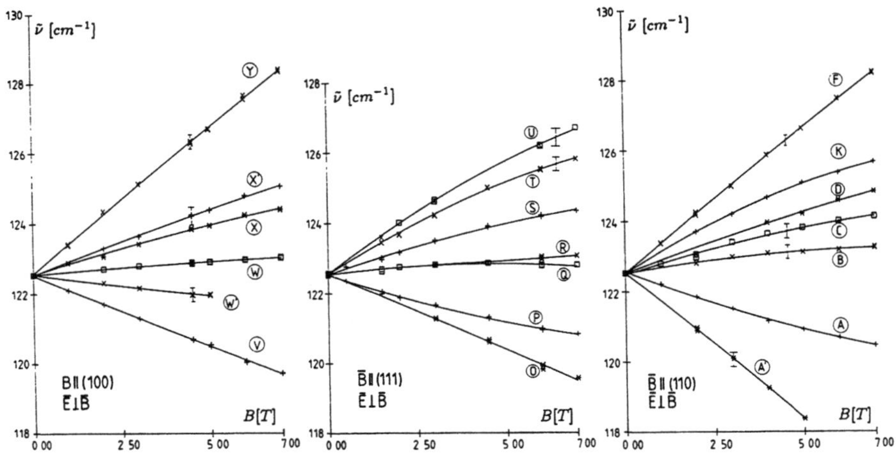

Fig. 2: Field dependence of the splitting of the G-line for $B \parallel \langle 100\rangle$, $B \parallel \langle 111\rangle$ and $B \parallel \langle 110\rangle$; $E \perp B$; T=1.2K

Tab. 1: Experimental and theoretical g-values for the ground and the first excited state of the carbon acceptor

	Experiment		Theory	
	g_1'	g_2'	g_1'	g_2'
$\Gamma_8(1S_{3/2})$	$\mp 0.22 \pm 0.01$	$\mp 0.15 \pm 0.01$	0.527	−0.142
	g_1^G	g_2^G	g_1^G	g_2^G
$\Gamma_8(2P_{3/2})$	$\pm 2.24 \pm 0.06$	$\mp 0.98 \pm 0.03$	2.201	−0.871

expressions for the linear splitting of the acceptor levels for the three orientations of the magnetic field. In addition they deduced the selection rules and expressions for the oscillator strength. Based on these results we extracted the g-values given in table 1 from the experimental data. A unique set of g-values could only be determined using the experimental information (i.e. energy splitting, oscillator strength and thermalisation effects) from all the three orientations.

3. Theory

Since 1970 Baldereschi, Lipari and Altarelli /4-7/ published a series of papers which demonstrate a method to solve the shallow acceptor and exciton problem considering the degenerate valence bandstructure.

This method is based on the translation of Luttingers kp-matrix into spherical tensor operators such that the powerful technique of reduced matrix elements can be applied to calculate the angular dependent parts of the wavefunctions. In our problem, the Zeeman splitting of the acceptor states, this method works as follows:

The acceptor Hamiltonian H, including the magnetic field ($B \parallel [001]$), is divided into a field independent part H_0, a linear part $\mu_B B\hat{G}$, where $\hat{G}$ is called the Zeeman operator, and a term of second order in the magnetic field which is not necessary in our case:

$$H = H_0 + \mu_B B\hat{G} + (...)B^2 \quad . \tag{1}$$

In terms of spherical tensors the Zeeman operator $\hat{G}$ reads as follows:

$$\begin{aligned}\hat{G} = -2\kappa J_z - \gamma_1 L_z + \frac{6\gamma_3 + 4\gamma_2}{5}\left(\frac{1}{2}\sqrt{\frac{5}{3}}V_0^1 + i\sqrt{\frac{5}{2}}X_0^1\right) \\ + (\gamma_3 - \gamma_2)\left(\sqrt{\frac{2}{5}}V_0^3 - i\sqrt{\frac{2}{5}}X_0^3 - iX_4^4 + iX_{-4}^4\right) \quad ,\end{aligned} \tag{2}$$

$$V^{(p)} = \left(L \times J^{(2)}\right)^{(p)} = +i\sqrt{2}\left((r \times k)^{(1)} \times J^{(2)}\right)^{(p)},$$

$$X^{(p)} = \left((r \times k)^{(2)} \times J^{(2)}\right)^{(p)}; \quad J^{(2)} = (J \times J)^{(2)} \quad ,$$

where J is the 3/2 angular momentum operator acting on the fourfold Γ_8 Bloch set and $k = -i\nabla$, r and L operate on the envelope functions. All composed tensor operators are constructed by Edmonds' rule:

$$(A^{(i)} \times B^{(k)})^p_q = (-1)^{i-k+q}\sqrt{2p+1}\sum_m A^{(i)}_m B^{(k)}_{q-m} \begin{pmatrix} i & k & p \\ m & q-m & -q \end{pmatrix} . \quad (3)$$

The linear Zeeman splitting is now found by calculating the expectation values of $\hat{G}$ with the appropriate magnetic field free acceptor states. For the construction of the acceptor states which are the eigenstates of H_0 we follow again Baldereschi et al. /4-7/ and devide H_0 in a spherical symmetric and a cubic part:

$$H_0 = H_{sph} + H_{cub} \quad (4)$$

Solving H_{sph} one ends with two coupled differential equations for the radial part of the wave function. We solved these equations by combining smoothly a power series ansatz, appropriate for small r values, with an asymptotic expansion, appropriate for large values of r. This method gives exact results for all eigenvalues and eigenfunctions of H_{sph}.

The anisotropic cubic part H_{cub} couples different spherical states. Since $H_{cub} \sim \delta = \gamma_3 - \gamma_2$, and $\delta < \gamma_1, \gamma_2, \gamma_3$ holds, the mixing of different spherical states by H_{cub} is expected to be small. Indeed a calculation shows, that for the $1S_{3/2}$ acceptor ground state, which is well separated energetically from the excited states, the contribution of H_{cub} is negligible. Thus the spherical $1S_{3/2}$ state is used as the acceptor ground state. This is not true for the first excited state $2P_{3/2}$ of H_{sph} because it is close in energy to the $2P_{5/2}$ state. Coupling both states by H_{cub}(symmetry O_h) leads to a $\Gamma_8(2P_{3/2})$-, a $\Gamma_8(2P_{5/2})$- and a $\Gamma_7(2P_{5/2})$-state. The $\Gamma_8(2P_{3/2})$ is the lowest of these states and therefore the first excited state of the acceptor with components $\Gamma_8(2P_{3/2})_M$, where $M = +3/2, +1/2, -1/2, -3/2$:

$$\begin{aligned}
\Gamma_8(P_{3/2})_{+3/2} &= a(2P_{3/2,+3/2}) + b[\frac{1}{\sqrt{6}}(2P_{5/2,+3/2}) + \sqrt{\frac{5}{6}}(2P_{5/2,-5/2})] = \Gamma_{8,-1/2} , \\
\Gamma_8(P_{3/2})_{+1/2} &= a(2P_{3/2,+1/2}) - b(2P_{5/2,+1/2}) \qquad = \Gamma_{8,-3/2} , \\
\Gamma_8(P_{3/2})_{-1/2} &= a(2P_{3/2,-1/2}) + b(2P_{5/2,-1/2}) \qquad = \Gamma_{8,+3/2} , \\
\Gamma_8(P_{3/2})_{-3/2} &= a(2P_{3/2,+3/2}) - b[\frac{1}{\sqrt{6}}(2P_{5/2,-3/2}) + \sqrt{\frac{5}{6}}(2P_{5/2,+5/2})] = \Gamma_{8,+1/2} .
\end{aligned} \quad (5)$$

Here $\Gamma_8(P_{3/2})_M$ (left hand side of (5)) is the rotation group notation while $\Gamma_{8,m}$ (right hand side of (5)) refers to the T_d group notation.

This structure of the states results from the O_h symmetry whereas the values of the mixing coefficients a, b are calculated from the explicit form of H_{cub} together with the spherical radial wavefunctions. The calculation of the expectation values of the $\hat{G}$ operator with respect to the states $1S_{3/2} = \Gamma_8(1S_{3/2})_M$ and $\Gamma_8(2P_{3/2})_M$ are performed with the reduced matrix element technique. For the ground state $\Gamma_8(1S_{3/2})_M$ this is a relatively simple task, an explicit formula is derived:

$$g_1' = -2\kappa + \left(\frac{8}{5}\kappa - \frac{4}{5}\gamma_1 - \gamma_2 + \frac{13}{5}\gamma_3\right)\langle g_0 \mid g_0\rangle + \left(\frac{26}{15}\gamma_2 - \frac{2}{15}\gamma_3\right)\langle g_0 \mid r\frac{d}{dr} \mid f_0\rangle ,$$

$$g_2' = -\frac{14}{15}(\gamma_3 - \gamma_2)\left(3\langle g_0 \mid g_0\rangle - 2\langle g_0 \mid r\frac{d}{dr} \mid f_0\rangle\right) , \tag{6}$$

where f_0 and g_0 are the radial functions as defined in /4/. The complicated structure of the $\Gamma_8(2P_{3/2})_M$ states combined with the 8 operators appearing in $\hat{G}$ leads to a large number of contributions (about 80 for $M = 3/2$) therefore the expectation values for $\hat{G}$ are calculated numerically in this case.

Before comparing the theoretically and experimentally deduced g values we would like to remember that the M labels for the four state functions of $\Gamma_8(P_{3/2})$ are different when they are considered as a $D_{3/2}$ basis of the rotation group (label M) or as a Γ_8 basis of the T_d group (label m). Comparing the appropriate Clebsch Gordon coefficients /14/ - used for the $D_{3/2}$ basis - with the T_d - coupling coefficients the relations given in equation (5) are obtained. Therefore the calculated G, G' and the measured g_1', g_2' values obey the equation /13/:

$$MG + M^3G' = mg_1' + m^3g_2' \text{ or : } g_1' = -\frac{1}{12}(40G + 91G') \;\; g_2' = \frac{1}{3}(4G + 10G').$$

For the $1S_{3/2}$ ground state one gets $M = m$.

4.Discussion

The series of spectra presented in Fig.1 reveals for the first time the Zeeman splitting if the 'G-line' of the shallow carbon acceptor in GaAs. This is remarkable since we used a bulk sample and not epitaxially grown material. A glance at the three spectra shows that they depend very markedly on the orientation of the magnetic field. This anisotropy is clearly demonstrated by the splitting diagrams in fig.2 and fig.3. Thus it seems to us not very surprising that the g-values deduced from the data reflect this property showing a very large value for g_2' . Here our result conflicts with /10/ where a negligible g_2' is reported. On the other hand the g_1' value of /10/ is larger than our value. The g- values of /10/ result in a similar energy splitting as ours but the transition probabilities are in contradiction to the thermalisation effects observed in our experiments.

In the theoretical description the anisotropy, which is of course an effect of H_{cub}, enters our calculation in two ways: Through the last part of the Zeeman operator (1) which is proportional to $\gamma_3 - \gamma_2$ and the mixing of the different Γ_8 states (see equation (5)). With the Luttinger parameters $\gamma_1 = 6.95, \gamma_2 = 2.1, \gamma_3 = 2.9$ and $\kappa = 1.1$ we calculatetd the mixing coefficients $a = 0.9898$ and $b = 0.1422$.

The calculation of the ground state values g_1' and g_2' using equation (6) is easily performed but the results are not very reliable. Equation (6) gives only sums and differences of the Luttinger parameters which compensate to give the relatively small g-values. This causes large errors in g_1' and g_2' even for small uncertainties in the Luttinger parameters. Consequently this difference between experiment and theory should not be taken too seriously.

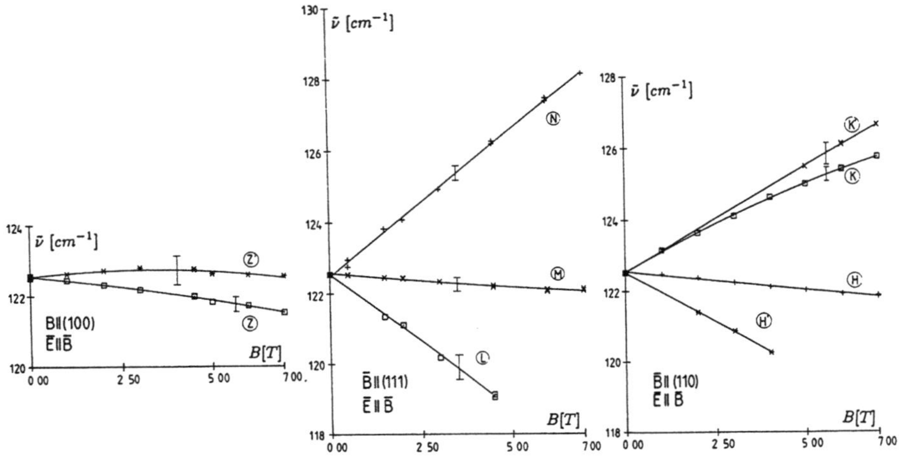

Fig. 3: Field dependence of the splitting of the G-line for $B \parallel \langle 100 \rangle$, $B \parallel \langle 111 \rangle$ and $B \parallel \langle 110 \rangle$; $E \parallel B$; T=1.2K

The situation is much better for the case of the excited state $\Gamma_8(2P_{3/2})$ where the calculation leads to values G_1^G and G_2^G which are not affected very much by small variations of the Luttinger parameters. We found a surprisingly good agreement between the experiment and our values calculated form the shallow acceptor model.

The most prominent result is that the anisotropy of the GaAs valance band has its correspondence in the anisotropy of the shallow acceptor g-values.

Acknowledgement: The authors are indebted to Prof. J. Schneider, Freiburg, for calling this problem to their attention and for supplying the sample. They also want to thank Prof. G. Schaack, Würzburg, for his continuous support.

1. Fisher P., Fan Y.H.: Phys. Rev. Lett. 5, 195 (1960)
2. Soepanagkat H.P., Fisher P.: Phys. Rev. **B8**, 870 (1973)
3. Lin-Chung P.J., Wallis R.F.: J. Phys. Chem. Solids **30**, 1453 (1969)
4. Baldereschi A., Lipari N.O.: Phys. Rev. **B8**, 2697 (1973)
5. Baldereschi A., Lipari N.O.: Phys. Rev. **B9**, 1525 (1973)
6. Lipari N.O., Altarelli M.: Solid State Commun. **32**, 171 (1979)
7. Lipari N.O., Altarelli M.: Solid State Commun. **33**, 47 (1980)
8. Bhattacharjee A.K., Rodriguez S.: Phys. Rev. **B6**, 3836 (1972)
9. Kirkman R.F., Stradling R.A., Lin-Chung P.J.: J. Phys. **C11**, 419 (1978)
10. Bimberg D.: Phys. Rev. **B18**, 1794 (1978)
11. Freeth C.A., Fisher P., Simmonds,P.E.: Solid state Commun. **60**, 175 (1986)
12. Dischler B., Fuchs F., Kaufmann U.: Appl. Phys. Lett. **48**, 1282 (1986)
13. Bleaney B.: Proc. Phys. Soc. Lond. **73**, 937 (1959)
14. Edmonds R.A.: Angular Momentum in Quantum Mechanics, Princton, University Press (1960)

Coherent Anti-Stokes Raman Scattering and Magnetooptical Interband Transitions in $Pb_{1-x}Eu_xSe$

P. Röthlein[1], *G. Meyer*[1], *H. Pascher*[1], *and M. Tacke*[2]

[1]Experimentalphysik, Universität Bayreuth, D-8580 Bayreuth, Fed. Rep. of Germany

[2]Fraunhofer Institut für Physikalische Meßtechnik, D-7800 Freiburg, Fed. Rep. of Germany

Introduction

The ternary compound PbEuSe has attracted considerable interest due to its technical application as semiconductor laser material covering the interesting wavelength region in the mid-infrared [1]. Precise data on the electronic band structure are important for the improvement of the optoelectronic devices. Further it is interesting to investigate the exchange interaction between the free carriers and the paramagnetic Eu-ions in this dilute magnetic IV-VI semiconductor (IV-VI DMS). The question if a molecular field approach for the spin splittings of valence and conduction band is applicable is not answered for all semimagnetic IV-VI compounds. Karczewski and v.Ortenberg [2] observed a zero field splitting of the energy gap in PbMnS from laser emission experiments. Pascher et al. [3] found that such a splitting does not exist in PbMnTe.

From the analysis of magnetooptical interband absorption experiments in strong magnetic fields all bandparameters can be calculated. Precise information on the spin splittings of the Landau states can be obtained by coherent Raman scattering.

Band Model

The energies of the Landau states of PbEuSe in a magnetic field are described within the framework of a molecular field approximation. It is based on the **k**·**p** model for PbSe [4,5,6]. In this model for fixed Φ (denoting the angle between B and [111]) and magnetic field the Landau level energies are obtained by diagonalization of a 4x4 matrix for each Landau quantum number. This procedure neglects the mixture of different Landau states which is very weak in PbSe but can be strong in PbEuSe due to the exchange coupling of the free electrons with the localized Eu^{++}-spins. In order to be able to take such an interaction into account one has to diagonalize larger matrices.

Zawadzki [7] has developed a molecular field approach for this interaction of the mobile carriers with the localized magnetic moments. A modification of his scheme using larger than 4x4 matrices which is described in detail in Refs. 3 and 8 is used to fit the experimental data obtained for PbEuSe. The model contains exchange parameters A, a_1 and B, b_1 for the valence- and conduction band, respectively, which can be determined by fitting the observed dependence of the spin splittings on the magnetic field.

Magnetooptical Interband Transitions

Magnetooptical interband transitions in $Pb_{1-x}Eu_xSe$ ($x \le 0.01$) require radiation wavelengths between 4.5 and 5.5 μm. Coherent radiation in the

Springer Series in Solid-State Sciences, Vol. 87
High Magnetic Fields in Semiconductor Physics II Editor: G. Landwehr

desired range was obtained from frequency doubling of the radiation of a Q-switched CO_2-laser in a phase matched Te-crystal. The samples investigated were p-type epitaxial films with Eu-contents $x < 0.01$ and had carrier concentrations of the order of $10^{17} cm^{-3}$. The transmission spectra were observed in Faraday configuration **B**‖**k**‖[111] with circularly polarized σ^+ and σ^- radiation and in Voigt configuration **B**⊥**k**, **B**‖[1$\bar{1}$0], **E**‖**B**.

Figure 1 shows the interband transmission of a sample with x=0.008 as a function of the magnetic field for the two circular polarizations in Faraday configuration. In comparison to PbTe where a difference between the transmission minima for σ^+ and σ^- polarized radiation is hardly observable, in PbEuSe this difference is a very pronounced effect as it is in PbMnTe. But whereas in PbMnTe the σ^+-resonances occur at higher magnetic fields than σ^-, in PbEuSe it is vice versa. From this fact one can conclude that the spin splitting of the valence band is diminished in PbEuSe and not increased as in PbMnTe. For this conclusion also it is assumed that the main interaction between localized spins and free carriers occurs in the valence band. This assumption follows from a comparison of Faraday- and Voigt data and from the Raman data presented in the following section.

In Fig.2 the results of the transmission data in Faraday configuration are summarized as a fan chart. The insert indicates the selection rules. From these data the gap energy E_g and the matrix elements $P_\perp$ and $P_\parallel$ can be determined by a least squares fit. Because of the small x-values of our samples we assumed the far band parameters to be the same as in PbSe [6]. The exchange parameters are fitted to the CARS data described in the following section. With these parameters theoretical transition energies were calculated by the theory outlined in the previous section. From Fig.2 the reasonable agreement between theory and experimental results can be seen.

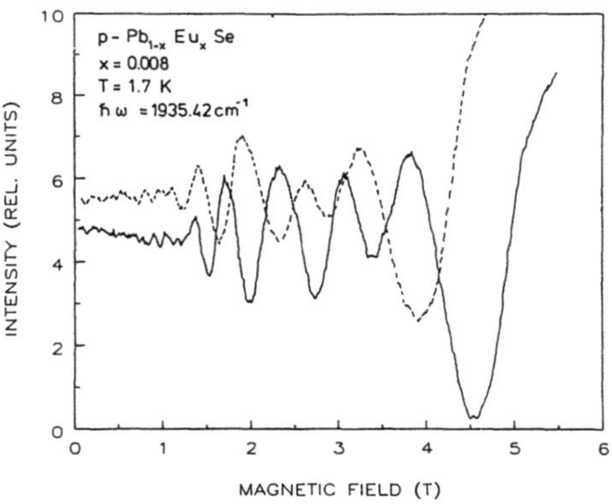

Fig.1: Transmission vs magnetic field in Faraday geometry.
---- σ^+
—— σ^-

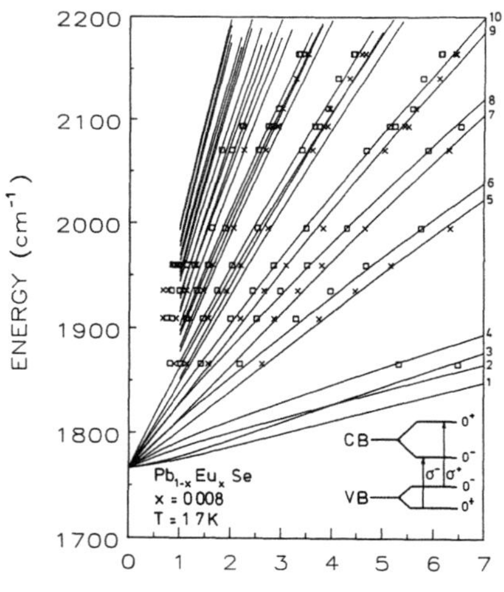

Fig.2: Interband transition energies vs magnetic field in Faraday geometry. x: experiment σ^-; □ experiment σ^+; — calculated
$1: 0^+ \to 0^-$ 70.53°; $2: 0^- \to 0^+$ 70.53°
$3: 0^+ \to 0^-$ 0°; $4: 0^- \to 0^+$ 0°
$5: 1^+ \to 1^-$ 70.53°; $6: 1^- \to 1^+$ 70.53°
$7: 1^+ \to 1^-$ 0°; $8: 1^- \to 1^+$ 0°,
$9: 2^+ \to 2^-$ 70.53°; $10: 2^- \to 2^+$ 70.53°

Optical Four Wave Mixing

If two laser beams with frequencies ω_L and ω_S ($\omega_L-\omega_S=\Delta\omega$) interact in a semiconductor sample, strong resonances of the radiation at $\omega_{AS}=\omega_L+\Delta\omega$ occur if $\hbar\Delta\omega$ is equal to the spin splitting energy of valence or conduction band. The radiation at ω_{AS} is generated by coherent anti-Stokes Raman scattering (CARS). In narrow gap semiconductors the experiment has to be performed with infrared lasers due to the strong interband absorption of the visible radiation used in a conventional CARS apparatus. Our experimental alignment is described in detail in Ref.9. The intensity and photon energy of the lasers is large enough to produce a considerable number of minority carriers in PbEuSe with $x\leq 0.01$ by two photon absorption. Therefore in the p-type material electron and hole resonances can be observed as well.

Figure 3a shows an experimental recording of the CARS- intensity versus magnetic field for fixed $\Delta\omega$. The four resonances observed are due to spin flip transitions within the n=0 Landau states of holes and photoexcited electrons. In Fig 3b the recording for the same $\Delta\omega$ is plotted for a higher temperature. Whereas the spin resonances of the holes shift their resonant magnetic fields strongly with temperature T, the corresponding ones for the electrons are nearly independent of T. At decreasing temperatures the spin resonances of the holes are shifted by the interaction with the localized Eu^{++}-spins to larger resonant magnetic fields, that means the spin splitting is diminished as concluded also from the interband experiments.

Recordings as plotted in Fig.3 are taken for a large number of laser lines with different $\Delta\omega$. From these results the exchange parameters are fitted by a least squares method. In Fig.4 the experimentally determined

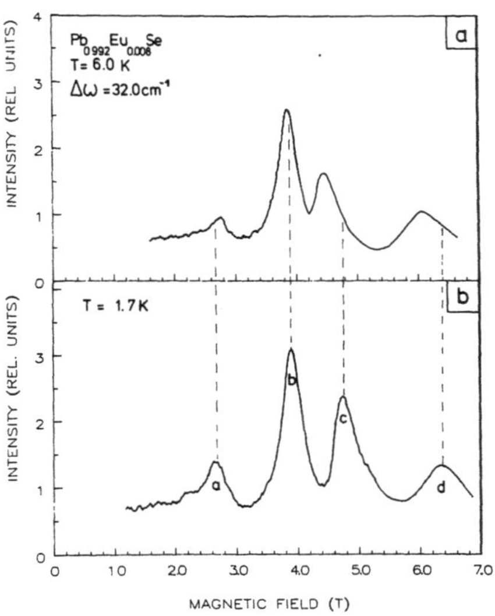

Fig.3: CARS intensity vs magnetic field
(a): T=6.0 K (b): T=1.7 K
a: cb 35.26°; b: cb 90°
c: vb 35.26°; d: vb 90°

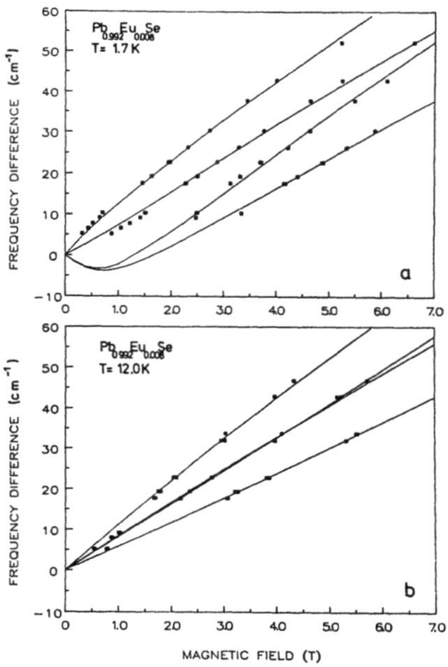

Fig.4: Frequency differences for spin transitions vs magnetic field
(a): T=1.7 K (b): T=12 K
■ experimental results
— calculated transitions

magnetic field dependences of the spin splittings are compared to calculated ones. The agreement shows that the molecular field approach is well suited for the calculation of the spin split Landau states.

Discussion

In Table I the exchange parameters and gap energies of PbEuSe are summarized together with the results of a 1% PbMnTe sample.

Table I: Parameters

Sample	T(K)	E_g(meV)	A(meV)	a_1(meV)	B(meV)	b_1(meV)
PbSe	1.5	146.3±0.3	—	—	—	—
PbEuSe; x=0.0035	1.7	181.1±0.7	—	—	—	—
PbEuSe; x=0.0045	1.7	189.1±0.7	102±10	96±10	21±5	24±5
PbEuSe; x=0.0080	1.7	219.0±0.7	78±10	65±10	19±5	8±5
PbEuSe; x=0.0080	6.0	219.0±0.7	77±10	59±10	25±5	3±5
PbEuSe; x=0.0080	12.0	219.0±0.7	70±10	56±10	31±5	7±5
PbMnTe; x=0.010	1.8	221.9±0.5	-182±15	-288±15	-33±10	27±5

As compared to PbMnTe in PbEuSe the signs of the exchange parameters are inverted (except b_1) and the absolute values are smaller. There is also a weak dependence of the parameters on the Eu^{++}-content x. This dependence shows the necessity of some corrections to the molecular field approach outlined in this paper. In contrast to other DMS the spin splitting of the valence band is reduced and not enhanced by the exchange interaction. As in PbMnTe [3] the experiments give no evidence of a zero field splitting of the energy bands in PbEuSe.

Acknowledgements

Very important discussions with G.Bauer, E.Bangert and W.Zawadzki are gratefully acknowledged. This work was sponsored by the Deutsche Forschungsgemeinschaft, Bonn.

References

1 P.Norton, M.Tacke, J. Crystal Growth 81, 405 (1987).
2 G.Karczewski, M.v.Ortenberg, Proc. 17th Int. Conf. on the Physics of Semiconductors, ed. J.D.Chadi and W.A.Harrison (Springer, 1985)
3 H.Pascher, P.Röthlein, G.Bauer, L.Palmetshofer, Phys.Rev.B36,9395(1987)
4 D.L.Mitchell, R.F.Wallis, Phys. Rev. 151, 581 (1965)
5 M.S.Adler, C.R.Hewes, S.D.Senturia, Phys. Rev. B7 ,186 (1973)
6 H.Pascher, G.Bauer, R.Grisar, to be published in Phys. Rev.B
7 J.Niedwodniczanska- Zawadzka, J.B.Elsinger, L.Palmetshofer, A.Lopez-Otero, E.J.Fantner, G.Bauer, W.Zawadzki, Physica B+C 117&118B,458(1983)
8 H.Pascher, P.Röthlein, I.Roschger, G.Bauer, Proc.19th Int Conf. on the Physics of Semiconductors, ed. W.Zawadzki, (1988)
9 H.Pascher, Appl. Phys B34, 107 (1984)

Part XI

Magneto-transport in 3D Systems

The Shubnikov–de Haas Effect: A Powerful Tool for Characterizing Semiconductors

D.G. Seiler

Center for Applied Quantum Electronics, Department of Physics, University of North Texas, Denton, TX76203, USA

A review is given of the principles involved in using the Shubnikov-de Haas (SdH) effect to characterize three-dimensional (3D) properties of semiconductors. Extensive studies of the SdH effect have been carried out on a wide variety of semiconductors over the last three decades that prove its scientific and technological usefulness. Newly created artificially structured narrow gap materials grown by molecular beam epitaxy (MBE) often have 2D properties which can also be investigated by the SdH effect. Examples are given for MBE grown HgTe single layer films and HgTe-CdTe superlattices.

1. INTRODUCTION

The science and technology of semiconductors include their growth and characterization as well as the study of scientific phenomena and their corresponding technical applications. The new generation of semiconductor applications (e.g., solar cells, solid state lasers, infrared detectors, transistors, etc.) requires new and improved materials. Studies of the Shubnikov-de Haas (SdH) effect have proven to be an extremely valuable tool for characterizing semiconductors in very unique ways. The organization of this article starts with a brief description of the SdH effect in Section 2. In Section 3, a review of the various properties characterized by the SdH effect is given. Numerous SdH studies have been carried out on almost every conceivable semiconductor or alloy system and these studies are categorized in Section 4. Finally, in Section 5 - MBE grown HgTe and HgTe-CdTe superlattices - we report results from our laboratory on artificially structured materials that exhibit two-dimensional behavior. It is hoped that this article will provide a stimulus for increased use of the SdH effect as a routine tool for characterizing semiconductors and for the development of further novel ways to enhance our understanding of semiconductors.

2. THE SHUBNIKOV-DE HAAS (SdH) EFFECT

The SdH effect is an oscillatory variation of magnetoresistance as a function of magnetic field occurring at liquid helium temperatures. These oscillations arise because of the quantization of energy and momentum by the magnetic field (simple descriptions are given by any number of authors, see for example BLAKEMORE [1]). The density of states then exhibits maxima at the Landau level energies $E = (N + 1/2)\hbar\omega_c$, where the integer N is the Landau level number and $\omega_c = eB/m^*$. These maxima in the density of states move to higher energies as B is increased and successive maxima pass through the Fermi energy E_F. Appreciable scattering in a degenerate semiconductor at low temperatures can only take place near E_F, so as successive maxima move past E_F, the scattering rate and resistivity oscillate. These resistivity oscillations, periodic in 1/B, are the SdH oscillations.

For the transverse case, the expression usually used to describe the relationship between the SdH oscillations and the parameters characterizing the material is

$$\frac{\Delta\rho}{\rho_0} = \sum_{R=1}^{\infty} \frac{5}{2}\left[\frac{RP}{2B}\right]^{1/2} \frac{\beta T m' \cos(R\pi\nu)}{\sinh(R\beta T m'/B)}\, e^{-R\beta T_D m'/B} \cos 2\pi(R/PB - 1/8 - R\gamma). \qquad (1)$$

Springer Series in Solid-State Sciences, Vol. 87
High Magnetic Fields in Semiconductor Physics II Editor: G. Landwehr

Equation (1) is a generalization of the ADAMS and HOLSTEIN [2] relation that includes the effects of a nonparabolic band [3] and spin splitting [4]. It is appropriate when $\rho_{yx} >> \rho_{xx}$, when E_F is constant, and when scattering within one Landau level can be neglected. In (1), $\Delta\rho$ is the deviation of ρ_{xx} from the nonoscillatory background resistivity, ρ_0 is the zero-field resistivity, R is the index of summation in the Fourier series, B is the transverse magnetic field (in the z direction), T is the temperature, $\beta = 2\pi^2 k_B m_0/\hbar e = 14.707$ T/K is a collection of constants, $P = \hbar e/E_F m^*$ is the SdH period, $m' = m^*/m_0$ is the ratio of the effective mass to the free electron mass, T_D is the Dingle temperature, γ is the Onsager phase factor, and ν is the spin splitting factor related to the effective g-factor g^* by $\nu = m'g^*/2$. Very good agreement has been found between the theoretical relation (1) and the experimental data for n-InSb [5,6] and for n-InAs [7]. It is widely used to interpret experimental SdH oscillatory data. A generalized expression for the longitudinal case (magnetic field in the x direction, parallel to the sample length) is given by (1) when the factor 5/2 is replaced by 1 [8]. However, longitudinal data are not in agreement with this longitudinal expression [9, 10].

Certain conditions must be satisfied in order to observe the SdH oscillations. The "thermal blurring" of the Fermi energy level must be small compared to the separation of the Landau levels, and the separation of the Landau levels must be less than the Fermi energy:

$$k_B T << \hbar\omega_c < E_F. \quad (2)$$

As the magnetic field is increased, $\hbar\omega_c = \hbar eB/m^*$ increases and successive Landau levels depopulate. For low fields such that $\hbar\omega_c << E_F$, the SdH amplitude may be too small to measure, and for high fields such that $\hbar\omega_c \simeq E_F$, no more Landau levels are left to depopulate.

Another condition that must be satisfied to observe SdH oscillations is that the uncertainty-principle collision broadening of the Landau levels must be much less than the separation of the levels:

$$\Gamma << \hbar\omega_c. \quad (3)$$

This condition can be expressed in terms of the scattering time $\tau = \hbar/\Gamma$ as

$$\omega_c\tau >> 1. \quad (4)$$

Using the approximation that the mobility scattering time τ_μ is approximately equal to the collision broadening scattering time τ, condition (4) can be written in terms of the mobility $\mu = e\tau_\mu/m^*$ as

$$\mu B >> 1. \quad (5)$$

To satisfy condition (2), one needs low temperatures and high enough magnetic fields appropriate to E_F. The condition expressed as (3) through (5) is more easily satisfied if the material has a small effective mass and a high mobility. E_F can be increased by increasing the electron concentration n, but if the concentration is made too high, the mobility may decrease to the point that (5) is not satisfied.

A number of questions should be considered when assembling the equipment for SdH experiments: What materials are to be measured? What material parameters are of interest? What temperatures are needed and how should they be measured, maintained, and controlled? What magnetic fields are needed and available and how should they be measured, maintained and controlled? How should the SdH data be recorded and analyzed to extract the parameters of interest? Each of these questions must be addressed to achieve state-of-the-art measurement capabilities.

In a dc measurement of the SdH oscillations, the sample temperature and current are fixed and the magnetic field is slowly changed while the potential difference at the sample contacts is measured with a sensitive enough voltmeter. A real-time plot is obtained on an xy plotter by driving the X-axis with the output of a Hall probe (or something proportional to the field) and the Y-axis with the analog output of the voltmeter. Although it is possible to analyze the plotter traces manually, it is better to record the data in a digital form for more sophisticated computer analysis.

The ac measurement technique makes use of ac magnetic field modulation and phase sensitive detection. By modulating the magnetic field at a low audio frequency (20-50 Hz) and setting the lock-in amplifier to detect at the modulation frequency (or twice the modulation frequency), an output signal is obtained which is proportional to the first (or second) derivative of the magnetoresistance. The SdH oscillations detected by the ac technique are modified by the multiplicative Bessel function factors $J_1(x)$ {or $J_2(x)$}, where $x = 2\pi RB_m/PB^2$, where Bm is the constant amplitude of the sinusoidal modulation field ($B_m << B$) and R is the SdH harmonic given in (1).

3. CHARACTERIZATION OF PROPERTIES AND PARAMETERS BY THE SdH EFFECT

The SdH effect has provided a wealth of information about semiconductor band structures by allowing the measurement of effective masses, effective g-factors, and Fermi surface cross-sectional areas, from which band parameters and band symmetries can be obtained. It has been used for a variety of purposes, including carrier concentration determinations, measurement of time-resolved hot electron temperatures induced from either dc electric fields or lasers, determining changes in band structures due to hydrostatic and uniaxial stress, and investigating scattering effects such as scattering due to magnetic impurities and spin-flip processes. The effect has been applied in other more practical ways such as determining the amount of substrate-induced strain in epitaxial films, studying the effects of irradiation of fast electrons, and characterizing structural phase transitions. The quality of the material being studied can also be quantitatively determined through measurements of the Dingle temperature.

Table I gives a brief summary of the various properties and parameters able to be characterized by the SdH effect. Once the period P of the SdH oscillations is determined, we also know the Fermi wave vector k_F, the extremal cross-sectional area A normal to the field, the carrier concentration n, and when m^* is known the Fermi energy. Some useful relationships for a simple band are as follows:

$$A = 2\pi e/\hbar P \quad ; \quad A_F = \pi k_F^2 = 2\pi\, m^*\, E_F/\hbar^2 \quad ;$$

$$n = k_F^3/3\pi^2 = 5.66 \times 10^{15}/\, P^{3/2} \quad [\text{where n is in cm}^{-3} \text{ and P in T}^{-1}] \quad ;$$

$$E_F = \frac{\hbar^2}{2m^*}(3\pi^2 n)^{2/3} = \frac{3.69 \times 10^{-15}\, n^{2/3}}{(m^*/m_o)} \quad [\text{where } E_F \text{ is in meV}] \quad .$$

Table I. CHARACTERIZATION OF VARIOUS PROPERTIES BY THE SHUBNIKOV- DE HAAS (SdH) EFFECT

SdH Period or Harmonic Frequency Content Measurement (Fourier Analysis for Multiple Frequencies)

Determines Size and Shape of Fermi Surfaces which are related to E vs k relations

Carrier Concentrations
- Single or multi-populated bands

Carrier Concentration Dependence
- Provides estimates of band parameters (E_g, P, m*, higher band parameters)
- Band symmetries
- Band nonparabolicity

Anisotropies (e.g., warping)
- Band shapes
- Band parameters

Spin-splitting or Harmonic Content
- g-factor
- Exchange interaction parameters
- Spin-flip scattering mechanisms

Hydrostatic and Uniaxial Pressure Dependence
- Deformation potentials
- Band structure changes
- Study of resonant acceptor states

SdH Amplitude Measurement

Temperature Dependence
- Effective mass (concentration dependence gives nonparabolicity of E vs k relations)
- Exchange interactions for semi-magnetic semiconductors

Magnetic Field Dependence
- Dingle temperature (scattering mechanism information)
- Beating effects (e.g., inversion asymmetry splitting)

DC Electric Field Dependence
- Hot electron temperatures

4. SdH STUDIES OF SEMICONDUCTORS AND ALLOYS

Table II lists the various semiconductors and alloys studied along with the frequency of published papers. Most major semiconductor systems are listed.

Table II. FREQUENCY OF SdH STUDIES DONE ON VARIOUS SEMICONDUCTORS AND ALLOY SYSTEMS

1.	$Bi_{1-x}Sb_x$	(>)19	27.	HgSe	29
2.	$(Bi_{1-x}Sb_x)_2Te_3$	3	28.	HgTe	13
3.	Bi_2Se_3	7	29.	$Hg_{1-x}Zn_xSe$	3
4.	$Bi_2(Se_xTe_{1-x})_3$	1	30.	InAs	19
5.	Bi_2Te_3	20	31.	$In_{1-x}Ga_xP_{1-y}As_y$	5
6.	Cd_3As_2	19	32.	$In_{1-x}Ga_xSb$	3
7.	$Cd_3(As_{1-x}P_x)_2$	1	33.	InP	2
8.	$(Cd_{1-x}Mn_x)_3As_2$	8	34.	InSb	74
9.	$Cd_{3-x-y}Mn_xZn_yAs_2$	1	35.	InSe	1
10.	Cd_3P_2	6	36.	$KTaO_3$	3
11.	CdSb	4	37.	$Pb_{1-x}Eu_xTe$	1
12.	$CdSnAs_2$	3	38.	$Pb_{1-x}Ge_xTe$	2
13.	$Cd_{3-x}Zn_xAs_2$	2	39.	$Pb_{1-x}Mn_xTe$	2
14.	$Ga_{1-x}Al_xAs$	2	40.	PbS	2
15.	$Ga_{1-x}Al_xSb$	3	41.	PbSe	5
16.	GaAs	11	42.	$PbSe_{1-x}Te_x$	12
17.	GaSb	25	43.	$Pb_{1-x}Sn_xSe$	27
18.	Ge	8	44.	$Pb_{1-x}Sn_xTe$	27
19.	$Hg_{1-x-y}Cd_xMn_yTe$	4	45.	PbTe	17
20.	$Hg_{1-x}Cd_xSe$	4	46.	$PtSb_2$	2
21.	$Hg_{1-x}Cd_xTe$	27	47.	Sb_2Te_3	1
22.	$Hg_{1-x}Fe_xSe$	6	48.	Sn(Gray)	7
23.	$Hg_{1-x}Fe_xTe$	1	49.	SnTe	6
24.	$Hg_{1-x}Mg_xTe$	1	50.	$SrTiO_3$	1
25.	$Hg_{1-x}Mn_xSe$	8	51.	Te	(>)9
26.	$Hg_{1-x}Mn_xTe$	10			

5. MBE GROWN HgTe AND HgTe-CdTe SUPERLATTICES

New artificially structured II-VI narrowgap materials, created by molecular beam epitaxy (MBE) technology are characterized by unique properties related to their small effective masses and nonparabolic energy bands. These systems are of great practical interest because of the ability to separately "tune" both the band gap and the effective mass to their desired values for use in applications such as for infrared detectors. The SdH effect, as discussed previously, is ideally suited to determining these effective mass values as well as other needed and important information on sample properties. Additionally, the SdH effect allows one to discriminate between three-dimensional (3D) and two-dimensional (2D) behavior because of the characteristic features of 2D behavior as the magnetic field is rotated with respect to the interface. It has been very effectively utilized to characterize III-V MBE grown materials.

The purpose of this part of the paper is to present SdH results that we have obtained on both HgTe single crystal films and HgTe-CdTe superlattices. Our results are summarized in Table III.

We have recently published results on the HgTe films and refer the interested reader to Refs. 11 and 12 for details. Here we just give a brief summary. SdH oscillations in sample 259(001) grown in a (001) direction on a CdZnTe substrate exhibited only bulk or 3D behavior.

TABLE III. SUMMARY OF MBE GROWN MATERIALS AND THEIR PROPERTIES DETERMINED BY THE SdH EFFECT

HgTe Films

Sample	Subband s	Electron Concentration $N_s(10^{11}\ cm^{-2})$ or $n(cm^{-3})$	Effective Mass m_s^*/m_o	Fermi Energy E_F (meV)	Dingle Temperature T_D (K)
HgTe-259(001) (2.4 μm thick)	Bulk	$3.2 \times 10^{15}\ cm^{-3}$	0.025	3.1	6.3
HgTe-259(112) (2 μm thick)	0	1.83	0.032	13.7	5.2
	1	1.17	0.029	9.7	4.5
	2	0.54	0.026	4.9	4.0
	Bulk	$3.0 \times 10^{15}\ cm^{-3}$	0.025	3.1	3.5
HgTe-193-1(112) (2 μm thick)	0	2.44	0.0345	17.0	7.2
	1	1.60	0.0305	12.6	6.1
	1	0.75	0.0265	6.8	5.8
	Bulk	$6.1 \times 10^{15}\ cm^{-3}$	---	4.9	---

HgTe-CdTe Superlattices

Sample	Type	$\mu(cm^2/V.sec)$ (at 10K and 1kG)	$m^*(m_o$ units)	Comments
HgTe-CdTe(99) 70 layers-HgTe(57Å)CdTe(19Å)	p-type	10^4	0.012	Rotation plot shows 2D behavior
HgTe-CdTe(107) 100 layers-HgTe(84Å)CdTe(37Å)	p-type		0.009	
HgTe-CdTe(203) 65 layers-HgTe(200Å)CdTe(35Å)	n-type	6.3×10^4	0.021	Rotation plot shows 2D behavior
HgTe-CdTe(204) 75 layers-HgTe(54Å)CdTe(50Å)	n-type	10^4		No SdH oscillations seen
HgTe-CdTe(133) 75 layers-HgTe(43Å)CdTe(61Å)	p-type	2.4×10^3		No SdH oscillations seen

The electron concentration determined from the SdH period agreed with 1 kG Hall Measurements. The electron effective mass determined, $0.025 \pm 0.004\ m_0$, is consistent with the band edge value of GROVES et.al. [13]. Two-dimensional behavior is found to arise from electrons constrained in an accumulation layer near the CdTe-HgTe interface on samples grown on (112) oriented CdTe substrates. Effective masses and occupation densities of three electric subbands in two samples with different total densities were determined. The masses of lower subbands are higher, their values decreasing with decreasing total electron density, while converging to the bulk value at lower densities. This agrees with predictions of an asymmetric triangular potential well model and a pronounced band nonparabolicity - characteristics of 2D structure in narrow gap semiconductors. Thus for MBE grown HgTe films, the SdH effect has proven capable of providing important, detailed information about the nature of the films and interface region. This information is quantitative and useful for improving the crystal growth process.

Results for SdH studies on several n- and p-type HgTe-CdTe superlattice samples are also given in Table III. Figure 1 shows the angular dependence of the second-derivative-like SdH behavior of a p-type sample #99 as the magnetic field is rotated from the perpendicular to the HgTe-CdTe interfaces ($\theta = 0°$) to the parallel orientation ($\theta = 90°$). The extrema positions are seen to follow the cos θ shift that are typical of 2D behavior. The temperature dependence of these oscillations is shown in Fig. 2. All oscillations at these low fields are damped out at 15K. In order to extract the effective mass value in the usual manner, assuming one carrier dominance, the ratio of the SdH amplitudes $A(T)/A(T_0) = \dfrac{T \sinh (\beta T_0\, m^*/Bm_0)}{T_0 \sinh (\beta T\, m^*/Bm_0)}$ is plotted in Fig. 3. The data fit this theoretical dependence quite well when $m^* = 0.012\ m_0$, a low hole mass. SdH measurements on another p-type sample #107 at these low fields also produced a low mass, $0.009\ m_0$, indicating that these low mass results were not isolated to just one sample. Fits of the temperature dependence of the electrical transport data gave 4 K energy gaps of 30 ±5 meV for #99 and 3 ±2 meV for #107 assuming a -40 mV valence-band offset and a hole mass of $0.4\ m_0$ [14].

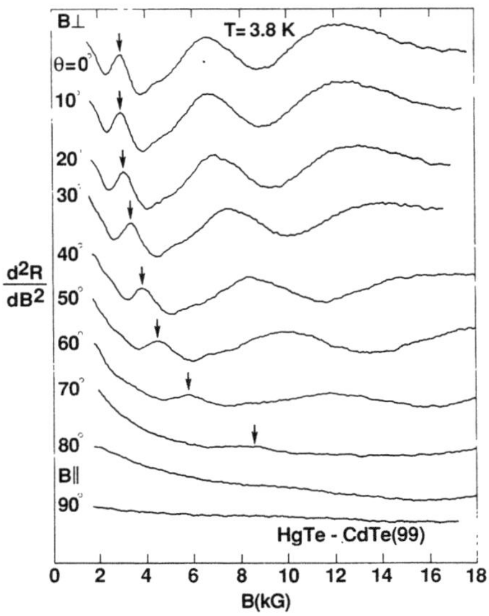

Fig. 1. Angular dependence of SdH oscillations that shows 2D behavior

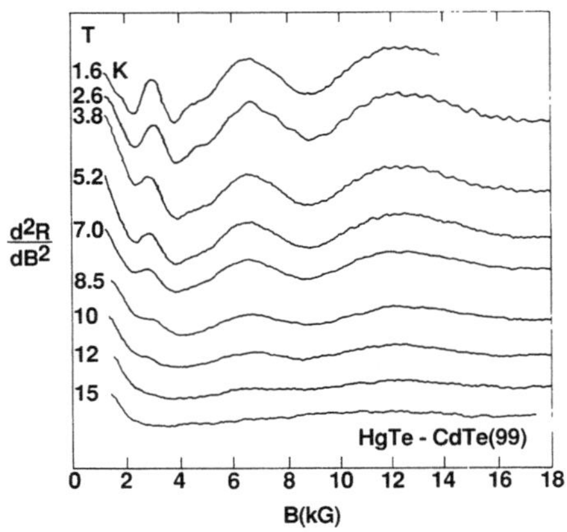

Fig. 2. Temperature dependence of SdH oscillations in low field region

Recent installation of a 120 kG superconducting magnet in our laboratory allowed us to extend the SdH measurements to higher fields as shown in Fig. 4 for sample #99. There is clear evidence for more than one type of carrier. The temperature dependence of the SdH amplitudes at 50 kG gives a mass of ≈ 0.075 m_0. Measurements and analysis of σ_{xy} vs B by the magnetoconductivity tensor component method [15] also shows that there are two holes. We note that evidence for two hole conduction at low temperatures was also given very recently by HOFFMAN et.al. [16] using a similar transport method.

Finally, in Fig. 5, we show SdH oscillations in an n-type superlattice, #203. The 2D nature of these oscillations is demonstrated by the angular rotation data shown in Fig. 6. Temperature dependent amplitude measurements shown in Fig. 7 give an electron effective mass of 0.021 ± 0.004 m_0, which is HgTe like as expected.

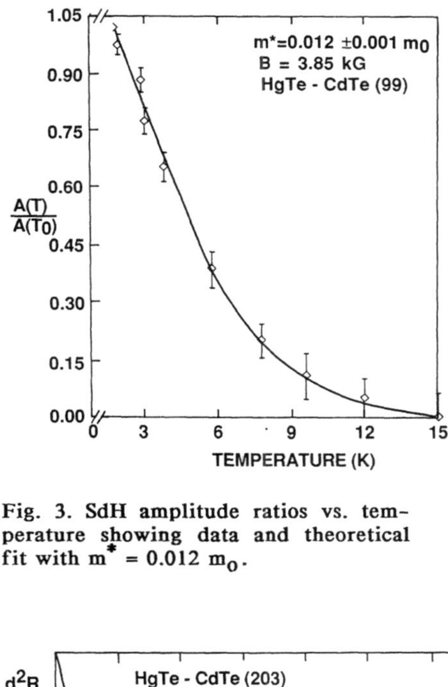

Fig. 3. SdH amplitude ratios vs. temperature showing data and theoretical fit with $m^* = 0.012\ m_0$.

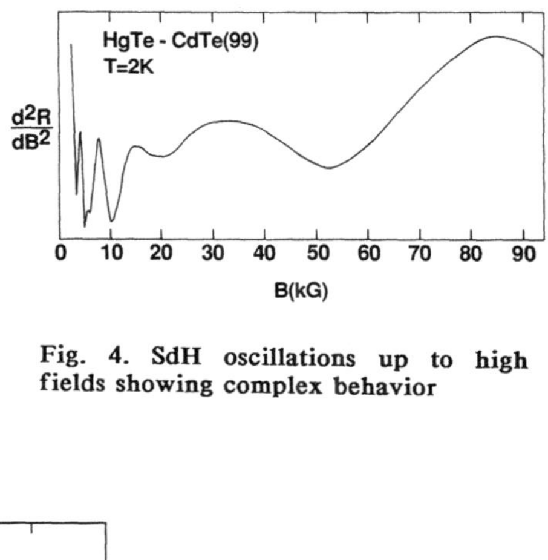

Fig. 4. SdH oscillations up to high fields showing complex behavior

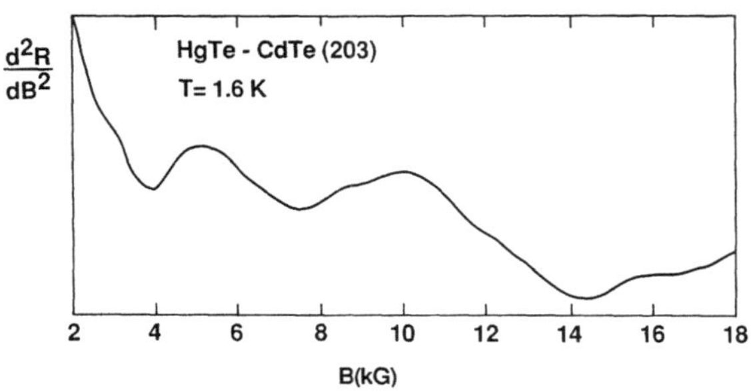

Fig. 5. SdH oscillations in an n-type sample

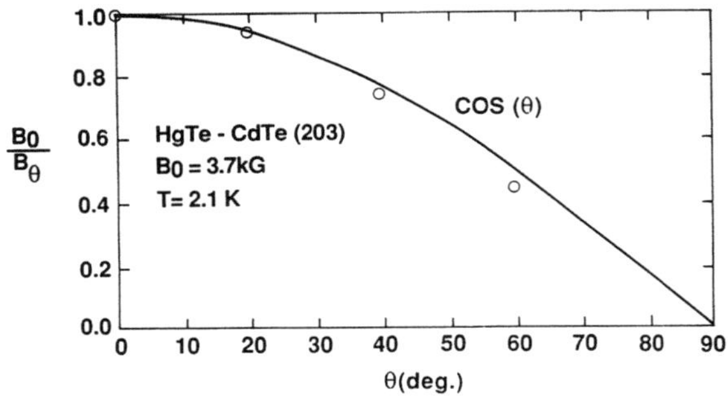

Fig. 6. Shift of the SdH extrema versus field orientation showing the cos θ dependence typical of 2D behavior

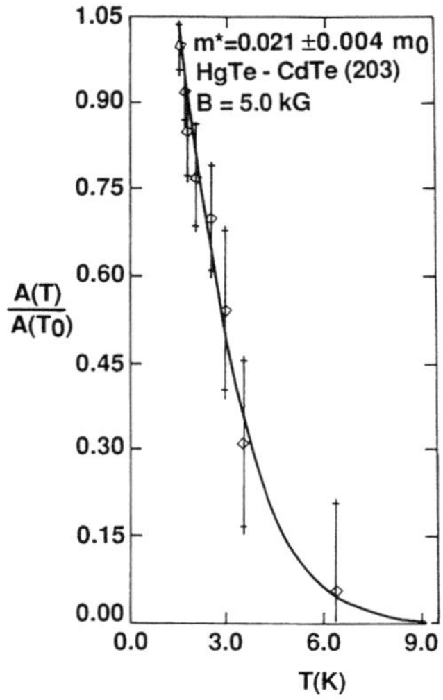

Fig. 7. Temperature dependence of the SdH amplitude showing the theoretical fit for the electron mass $m^* = 0.021\ m_0$.

6. ACKNOWLEDGEMENTS

The two-dimensional results reported here are a result of a collaborative project with R. J. KOESTNER, M. W. GOODWIN, and M. A. KINCH of Texas Instruments. We also acknowledge R. J. JUSTICE for help in taking data and W. ZAWADZKI for insights on 2D physics.

7. REFERENCES

1. J. S. Blakemore, Semiconductor Statistics (Pergamon, New York, 1962), Chap. 1, pp. 48-53.

2. E. N. Adams and T. D. Holstein, J. Phys. Chem. Solids 10, 254 (1959).

3. I. M. Lifshitz and A. M. Kosevich, Sov. Phys.-JETP 2, 636 (1956).

4. M. H. Cohen and E. I. Blount, Philos. Mag. 5, 115 (1960).

5. A. E. Stephens, D. G. Seiler, J. R. Sybert, and H. J. Mackey, Phys. Rev. B 11, 4999 (1975).

6. A. E. Stephens, J. R. Sybert, D. G. Seiler, and H. J. Mackey, in Proc. 14th Int. Conf. Low Temp. Phys., Helsinki, 1975 (American Elsevier, New York 1975), p. 79.

7. A. E. Stephens, R. E. Miler, J. R. Sybert, and D. G. Seiler, Phys. Rev. B 18, 4394 (1978).

8. L. M. Roth and P. N. Argyres, in Semiconductors and Semimetals (Academic Press, New York, 1966), Vol. 1, p. 159.

9. J. Staromlynska, D. M. Finlayson, and R. A. Stradling, J. Phys. C:Solid State Phys. 16, 6373 (1983).

10. J. Hajdu and G. Landwehr, in Strong and Ultrastrong Magnetic Fields and Their Applications (Springer-Verlag, New York, 1985), p. 17.

11. R. J. Justice, D. G. Seiler, W. Zawadzki, R. J. Koestner, and M. W. Goodwin, Appl. Phys. Lett. 52, 1332 (1988).

12. R. J. Justice, D. G. Seiler, W. Zawadzki, R. J. Koestner, M. W. Goodwin, and M. A. Kinch, to be published in J. Vac. Sci. Technol., July/August 1988.

13. S. H. Groves and R. M. Brown, and C. R. Pidgeon Phys. Rev. 161, 779 (1967).

14. M. W. Goodwin, M. A. Kinch, R. J. Koestner, M. C. Chen, D. G. Seiler, and R. J. Justice, J. Vac. Sci. Technol. A5, 3110 (1987).

15. D. L. Leslie-Pelecky, D. G. Seiler, M. R. Loloee, C.L. Littler, Appl. Phys. Lett. 51, 1916 (1987).

16. C. A. Hoffman, J. R. Meyer, E. R. Youngdale, J. R. Lindle, F. J. Bartoli, K. A. Harris, J. W. Cook, Jr., and J. F. Schetzina, 1987 MCT Workshop, to be published in J. Vac. Sci. Technol., July/August 1988.

Percolative Transport in GaAs at 10 T Magnetic Fields: Interpretation via Hydrogen Wavefunctions at Megatesla Fields

G. Bühler[1], *G. Wunner*[1], *R. Buczko*[2], *and J.A. Chroboczek*[3]

[1]Lehrstuhl für Theoretische Astrophysik, Universität Tübingen, D-7400 Tübingen, Fed. Rep. of Germany
[2]Institute of Physics, Polish Academy of Sciences, PL-02-558 Warsaw, Poland
[3]Centre National d'Etudes des Télécommunication, CNS, F-38243 Meylan, France

It is well-known that by the scaling through the dielectric constant and the effective mass hydrogen ground-state wavefunctions at megatesla fields can be used to calculate the properties associated with hydrogenic impurities in direct-gap semiconductors at high laboratory field strengths. In a previous paper[1] we determined percolative transport coefficients employing a WKB approximation for the behaviour of the wavefunction at large distances from the impurity, and demonstrated that these calculations account adequately, within the experimental uncertainties, for magnetoresistance measurements in lightly doped GaAs at low temperatures with magnetic fields up to 10 T. The purpose of the present note is to study to what extent modifications of the theoretical results are expected if the approximations made in ref.[1] with regard to the treatment of the longitudinal part of the wavefunction associated with the ground-state Landau level are dropped. We note that the use of numerically "exact" wavefunctions has the advantage over variational functions used previously in the literature that it provides the correct description of the state in both the vicinity of the donor *and* the asymptotic region which is of importance for percolation.

For a donor in an isotropic semiconductor, the Hamiltonian for an electron under the combined influence of the Coulomb potential of the impurity and a uniform magnetic field assumes, in the effective-mass approximation, the form

$$H = -\Delta - \frac{2}{r} + 2\beta l_z + \beta^2(x^2 + y^2), \tag{1}$$

where energies and lengths are measured in effective Rydberg units and Bohr radii, respectively, and $\beta = B/B_o^*$, with $B_o^* = (a/a^*)^2\ 4.70 \cdot 10^5\ T$. The reference magnetic field B_o^* is chosen in such a way that at B_o^* the cyclotron energy of the electron becomes equal to four times the Rydberg energy, or, equivalently, the Larmor radius equals the Bohr radius. In GaAs $a^* = 200\ a = 99\ \AA$, and thus $B_o^* = 13.4\ T$. In the nonperturbative regime, where Coulomb and Lorentz forces are comparable in magnitude, Schrödinger's equation associated with the Hamiltonian (1) is nonseparable and can be solved only by numerical means. In this region the expansion of the eigenfunctions over the complete orthonormal set of Landau functions $\Phi_{m,n}$ is appropriate:

$$\Psi_{\rm m} = \sum_n \Phi_{m,n}(\rho) g_n(z), \tag{2}$$

where m is the magnetic quantum number. The longitudinal expansion functions, $g_n(z)$, are determined by solving (numerically) the system of coupled-channel equations result-

Springer Series in Solid-State Sciences, Vol. 87
High Magnetic Fields in Semiconductor Physics II Editor: G. Landwehr

ing from the projection of Schrödinger's equation on the different Landau states, their number depending on the desired accuracy of convergence of the eigenvalues (six significant digits for the energies in the present calculations, for tables and details of the computations see ref.[2].)

Determining the percolation exponent first requires evaluating the overlap integral

$$I(\mathbf{R}) = \int \Psi(\mathbf{r})\exp[-i\beta(\hat{\mathbf{z}}\times\mathbf{R})\cdot\mathbf{r}]\Psi(\mathbf{r}-\mathbf{R})\mathrm{d}^3\mathbf{r}, \qquad (3)$$

where $\Psi(\mathbf{r})$ denotes the donor ground state wavefunction. Since we have observed that, with the exception of the immediate vicinity of the impurity nucleus, the main contribution to (2) comes from the $n = 0$ term (cf. Fig. 2 in ref.[2]), in the present analysis we restrict ourselves, in the evaluation of the overlap integral (3), to including only the Landau ground state contribution to the accurate numerical wavefunction. The overlap integral is then found to be given by

$$\begin{aligned} I(\underline{\mathrm{R}}_\perp, Z) &= \mathrm{e}^{-\frac{1}{2}\beta\underline{\mathrm{R}}_\perp^2}\cdot\int g_0(z)g_0(z-Z)\mathrm{d}z \\ &= \mathrm{e}^{-\frac{1}{2}\beta\underline{\mathrm{R}}_\perp^2}\cdot J(Z), \end{aligned} \qquad (4)$$

where $J(Z)$ is the autocorrelation of the longitudinal part of the wavefunction associated with the ($m = 0$) ground state Landau level. From the squared overlap integral, $S = I^2$, the overlap exponent ξ is obtained as $\xi = -\ln S$. The resistance of a random network of impurity centres spanned by resistances $\exp\xi$ is known to be proportional to $\exp\xi_c$, where the latter is determined from the percolation criterion that can be written as

$$Na^{*3}\int_{\xi=\xi_c}\mathrm{d}^3\mathbf{r} = 2.7, \qquad (5)$$

where N is the impurity concentration.

In ref.[1], $g_o(z)$ was approximated by its asymptotic WKB form, viz. $g_o(z) \propto \exp(-E^{1/2}(|z| - E^{-1}\ln|z|))$, and the pertaining autocorrelation integral was also obtained only in an approximate way, the final result being

$$S = I^2 = \mathrm{e}^{-\beta\underline{\mathrm{R}}_\perp^2}\cdot\exp[-2\alpha Z + 2\eta\ln(Z/2)], \qquad (6)$$

with $\alpha = E^{1/2}$ and $\eta = 1 + 2/E^{1/2}$. We have now abandoned these approximations and have evaluated the overlap integral (4) using the *accurate numerical wavefunctions* $g_o(z)$ at the corresponding field strengths. Figure 1 shows a comparison between the surfaces $S = 10^{-6.95}$, at $\beta = 1$, obtained in this way and by the approximation formula (6). For illustrational purposes the shape of the surfaces $S = const.$ at $\beta = 0.7$ obtained by use of the accurate numerical wavefunctions are shown in Fig. 2 for values of the constant decreasing in steps of one decade from 10^{-4} at the innermost contour to 10^{-13} at the outermost contour.

Figure 1 demonstrates that the approximate treatment of ref.[1] overestimates the extension of the surfaces $S = const.$ along the field direction, while the width in the plane $z = 0$ perpendicular to it is underestimated by roughly the same amount. In evaluating the volume integral in (5) these effects approximately cancel, and as a consequence the critical exponents ξ_c following from the condition (5) using the approximations of ref.[1] are not expected to change dramatically when the correct numerical wavefunctions are applied. This enhances the confidence in the theoretical results obtained by the simple analytical approximations.

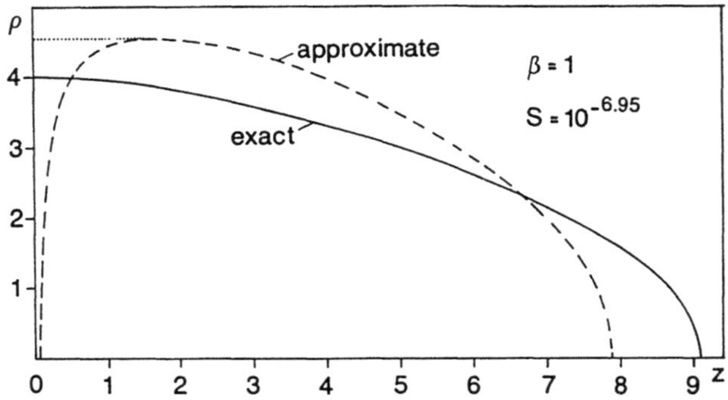

Fig. 1. Comparison (for $\beta = 1$) between the surfaces with constant value of the squared overlap integral $S = 10^{-6.95}$ calculated using the correct numerical longitudinal wavefunction $g_o(z)$ (solid curve) and the WKB approximation of ref.[1] (broken curve). Since this approximation fails in the vicinity of the origin the corresponding surface is continued tangentially at maximum girth. Note that in evaluating the volume in the percolation criterion (5) the deviations of the approximate solution from the correct solution roughly cancel.

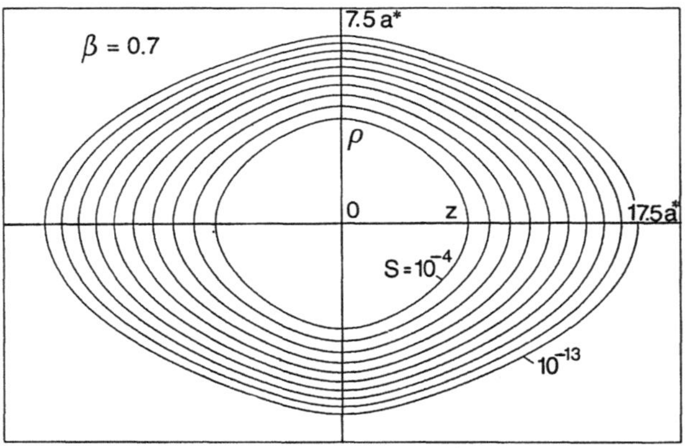

Fig. 2. Surfaces with constant value of the squared overlap integral S calculated using the correct numerical longitudinal wavefunctions $g_o(z)$ at $\beta = 0.7$. The values of S decrease in steps of one decade from 10^{-4} at the innermost contour to 10^{-13} at the outermost contour. The dimensions of the frame in the z and ρ direction are 35 and 15 effective Bohr radii, respectively. Note the growing non-sphericity of the surfaces for decreasing values of S.

As an example, Fig. 3 shows a comparison between magnetoresistance data presented by Taylor[3] in his study of transport in n^+nn^+ GaAs structures, and the results of the approximate analytical evaluation of the condition (5). We find that even this simple treatment accounts quite well for the experimental results. As a next step it will be necessary to study the influence of the inclusion of longitudinal wavefunctions associated with higher Landau levels on the values of the overlap integral (3), and thus on the theoretical values of the magnetoresistance. Work along these lines is in progress.

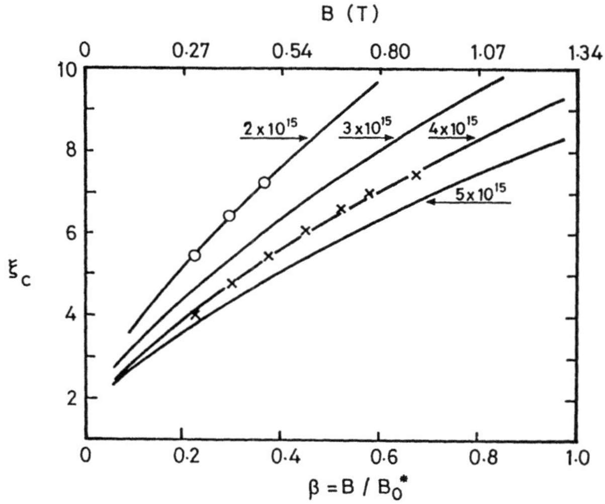

Fig. 3. The critical exponent ξ_c for percolation as a function of magnetic field strength in n-type GaAs ($B_o^* = 13.4\ T$): comparison between theory (solid curves for various donor densities) and experiment. The experimental data (crosses, Taylor[3]) were obtained for $N = 1.5\ 10^{15}\ cm^{-3}$. The first three data points are corrected for the B-dependence of the activation energy for hopping. These corrected data points (circles) follow the dependence calculated for $N = 2\ 10^{15}\ cm^{-3}$, which is close to the actual doping level of the sample.

References

1. R. Buczko, J. A. Chroboczek, G. Wunner: *Phil. Mag. Letts.*, **56**, 251 (1987)
2. W. Rösner, G. Wunner, H. Herold, H. Ruder: *J. Phys. B* **17**, 29 (1984)
3. D. C. Taylor, *Thesis*, University of Nottingham (1987)

Magnetic Field Induced Metal Insulator Transition in PbTe

J. Oswald[1], *B.B. Goldberg*[2], *G. Bauer*[1], *and P.J. Stiles*[3]

[1]Institut für Physik, Montanuniversität Leoben, A-8700 Leoben, Austria
[2]National Magnet Laboratory, MIT, Cambridge, MA 02139, USA
[3]Department of Physics, Brown University, Providence, RI 02912, USA

We have studied the magnetotransport coefficients ρ_{xx}, ρ_{xy}, and ρ_{zz} in epitaxially grown PbTe samples for carrierconcentrations varying from 1.4×10^{16} to $1\times10^{17}cm^{-3}$ in magnetic fields up to 23 T and temperatures between 400 mK and 20 K. The data are discussed in view of several models involving carrier interaction and localization phenomena.

1. Introduction

Among the narrow gap semiconductors PbTe seems to offer several advantages for a study of a metal insulator (MI) transition. In undoped samples the free carriers originate either from Te vacancies (electrons) or Pb vacancies (holes) which form resonant levels within the bands. So far, magnetotransport measurements have not given any evidence for carrier freeze out on localized states within the energy gap [1]. Since the static dielectric constant is of the order of 1000, the hydrogen - like donor binding energies are of the order of 1 μeV. However, Enderlein et al. [2] have shown that dynamical screening effects enhance this energy to about 3μeV, a value which should be used for thermal activation studies. Lattice relaxation effects increase this energy further to about 20μeV. However, since no optical excitation studies are performed but transport measurements the relevant binding energies for the H-like impurities are of the order of 3 μeV a value which will be used for further estimates. Still for samples with carrier concentrations of about $10^{16}cm^{-3}$ and above no bound states will be observable. Such considerations have led Nimtz et al. to consider the many valley semiconductor PbTe as one of possible candidates for a study of a magnetic field induced Wigner condensation [1]. Wigner has predicted the occurrence of a phase transition into a three dimensionally condensed crystalline array for sufficiently low electron concentrations when the mutual Coulomb interaction exceeds the kinetic energy of the electron system. However, for PbTe and other suitable candidates like HgCdTe the accessible experimental parameters make it difficult to extract from magnetotransport experiments unambiguous evidence for a Wigner condensation [3]. The experiments in HgCdTe have also been interpreted as magnetic field induced Mott transition and/or carrier freeze out on impurity centers. Nimtz et al. have found a kink in the magnetic field dependence of the transverse magnetoresistence at sufficiently low temperatures and a decrease of the Hall coefficient R_H with increasing field both in PbTe and PbSnTe. For a better understanding of these phenomena we have performed a study of the magnetotransport coefficient ρ_{xx}, ρ_{zz} and ρ_{xy} up to about 20 T at temperatures down to 450 mK. In order to study either localization phenomena like the Mott or Anderson transition or the Wigner interaction a comparatively low density of

Springer Series in Solid-State Sciences, Vol. 87
High Magnetic Fields in Semiconductor Physics II Editor: G. Landwehr

free carriers is necessary. High magnetic fields reduce the Fermi energy relative to the lowest Landau state. This is necessary as well as low temperatures for all kinds of transitions.

2. Experimental

Two n-PbTe and one p-PbTe sample grown by Hot Wall Epitaxy on BaF_2 (111) substrates were investigated. A Hall bar geometry was lithographically defined (Fig. 1).

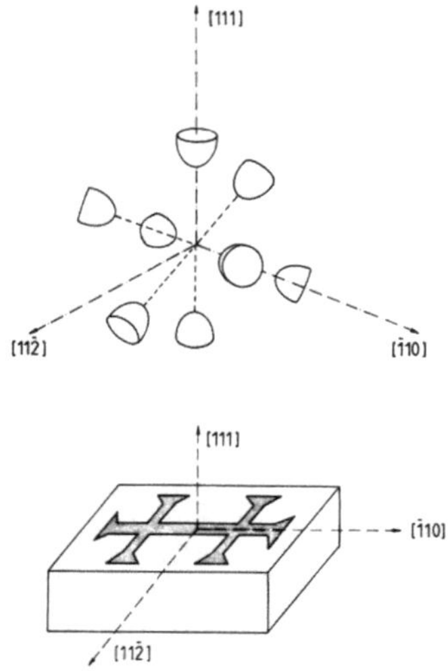

Fig. 1: Surfaces of constant energy in k-space and sample geometry used for magneto transport investigations.

This geometry limits the various orientations of the magnetic field: ρ_{xx} is measured either with B// [111] or B// [11$\bar{2}$], ρ_{xy} with B// [111] and ρ_{zz} with B// [$\bar{1}$10]. Due to the substrate induced strain, the four-fold degeneracy of the L-states at the edge of the Brillouin-zone (Fig. 1) is lifted [9,10]. With respect to the [111] growth direction three oblique valleys both in the conduction band and valence band are shifted upwards in energy by about 5 meV with respect to the [111] valley oriented parallel to it [19]. This shift is of considerable importance for the magnetic fields for which the exterme quantum limit (E.Q.L.) i. e. the occupation of a single Landau level with a given spin state is achieved. The magnetoresistance coefficient was measured with the use of convential lock-in techniques at 14-23 Hz and using currents of the order of 5 μA.

3. Magnetoresistance and Hall Effect

In the lowest concentration sample (n-PbTe A, $n = 1.4 \times 10^{16} cm^{-3}$) the Hall coefficient exhibits a temperature dependence for magnetic fields in excess of about 4 T. For such fields the Fermi energy is less than 0.1 meV above the lowest Landau state. The data shown in Fig. 2 would, naively interpreted, indicate an increase of the carrier concentration at high magnetic fields and low temperatures. The Hall coefficient normalized to its low field value is shown in Fig. 3 for this sample as a function of B with temperature as a parameter. A similar decrease of R_H with B has been

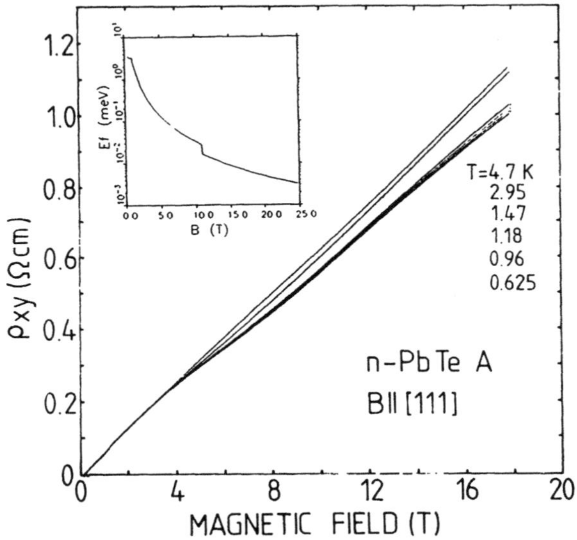

Fig. 2: ρ_{xy} vs B for different temperatures for sample A. Insert: Fermi energy.

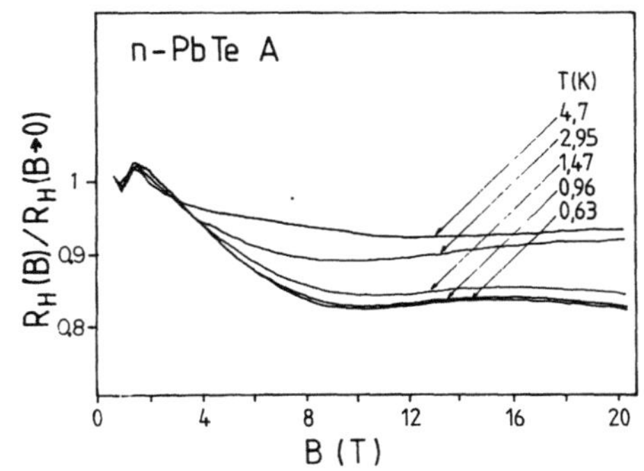

Fig. 3: Hall coefficient ratio vs B at different temperatures.

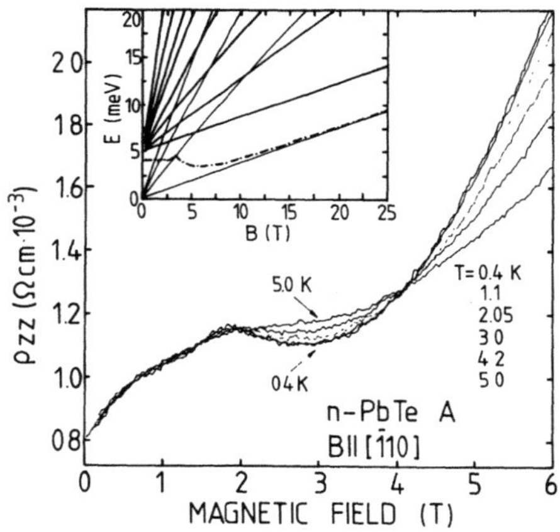

Fig. 4: ρ_{zz} vs B for sample A. Insert: Fermi energy as a function of B for B// [$\bar{1}$10] ; Landau level $\Theta = 35.26^o$ and $\Theta = 90^o$.

observed by Goldman et al. [6] in HgCdTe. However, in HgCdTe with carrier concentrations of about $10^{15}cm^{-3}$ for higher fields, a sharp increase of R_H was observed, interpreted by Goldman et al. as evidence for a MI transition. In Fig.4 and 5 the field dependence of the longitudinal magnetoresistance (B// [$\bar{1}$10]) is shown. The insert shows the dependence of the Fermi energy above the lowest Landau state, which is different from the B//[111] case because of the many valley band structure. The rather strong temperature dependence of ρ_{zz} appears for fields for

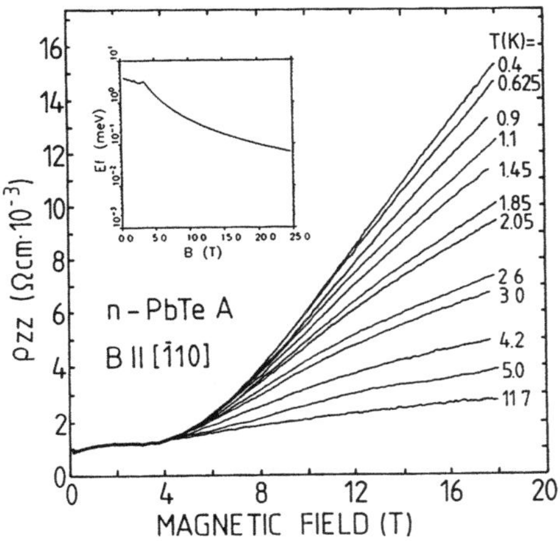

Fig. 5: ρ_{zz} vs B with temperature as a parameter; insert: Fermienergy

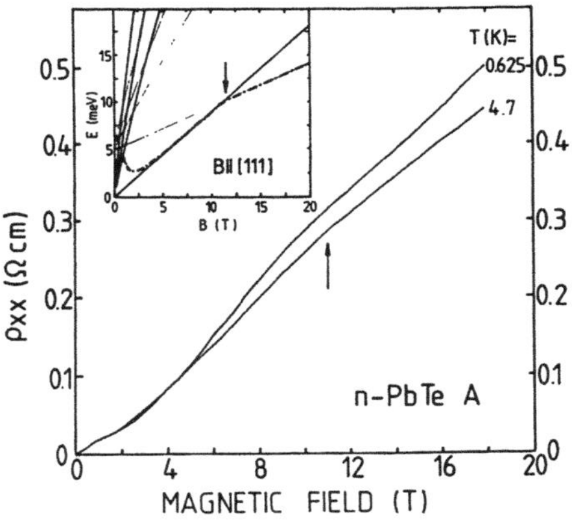

Fig. 6: ρ_{xx} vs B for sample A, B// [111]. Insert: Landaulevels and Fermi energy.

which just one Landau subband, belonging to the $\Theta = 90^{o}$ valley (i.e. the depressed [111] valley) is populated. The transverse magnetoresistance of sample A is shown in Figs. 6 and 7 for two different magnetic field orientations. The apparent temperature dependence is much weaker than in the longitudinal case.
In Fig. 6, the kink in ρ_{xx} at about 11T (arrow) is caused by the transfer of carriers from the [111] to the oblique valleys. It is important to note that sample A is already in the EQL at fields of about 2T. For the [11$\bar{2}$] orientation only the [111] valley is populated and at approximately 3T all carriers are in the O^{-} Landau state.
The temperature dependence of the Hall effect at B = 17.5T is shown in Fig. 8: ln R_{xy} vs 1/T does not indicate an activated behaviour but rather the Hall "dip" already mentioned. The temperature dependence of the transverse magnetoresistance is shown in Fig. 9. At low temperatures ρ_{xx} nearly saturates, a fact which cannot simply be explained by the presence of a surface layer, since ρ_{xx} shows a similar saturation behaviour even at B = 6T and temperatures below 1K.

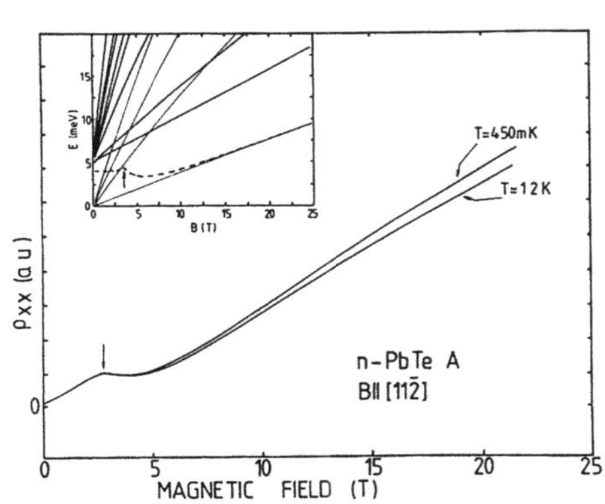

Fig. 7: as Fig. 6 but for B// [11$\bar{2}$]

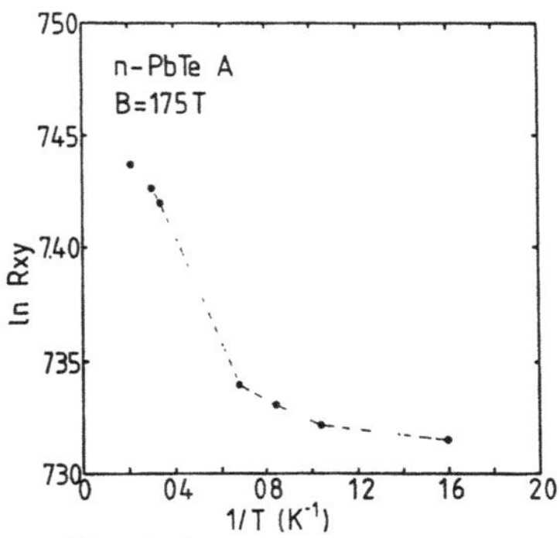

Fig. 8: ln R_{xy} vs 1/T for sample A at B = 5T.

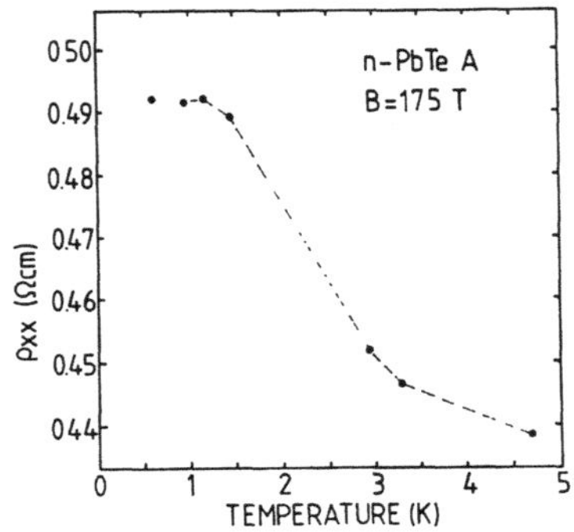

Fig. 9: ρ_{xx} vs T for sample A at B=17T.

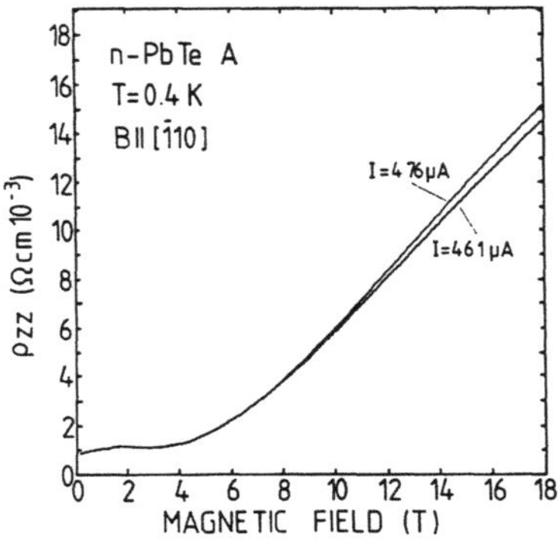

Fig. 10: ρ_{zz} vs B at T=400mK for two different sample currents.

The influence of high electric fields on the transport might well serve to distinguish between the different localisation mechanisms [11]. The dependence of ρ_{zz} on magnetic field and on the current through the sample is shown in Fig. 10. The corresponding electric fields are 17.5 mV/cm and 162 mV/cm, respectively. The traces show a small amount of e^- heating but no strong nolinearities.

Due to the built-in strain, the p-type sample B ($p=6.5x10^{16}cm^{-3}$) is quite different from n - type material. For B// [111] even at the highest field three oblique valleys are always populated. Therefore the orientational anisotropy might influence any localization process. In general the data presented in Figs. 11-13 are quite similar

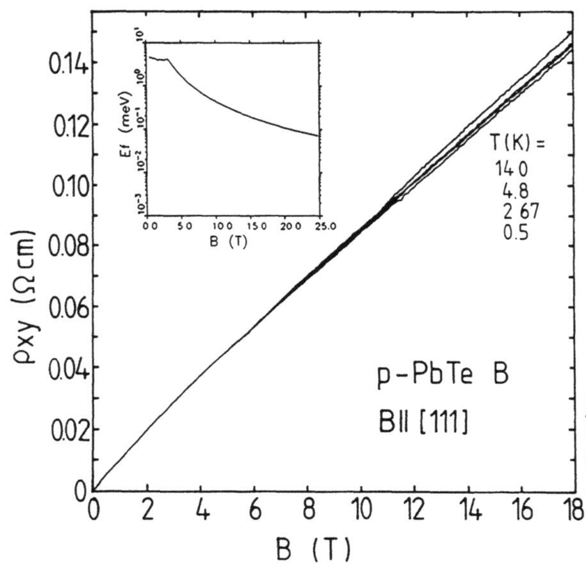

Fig. 11: ρ_{xy} vs B for sample B. Insert: Fermi energy with respect to lowest Landau subband for B// [111].

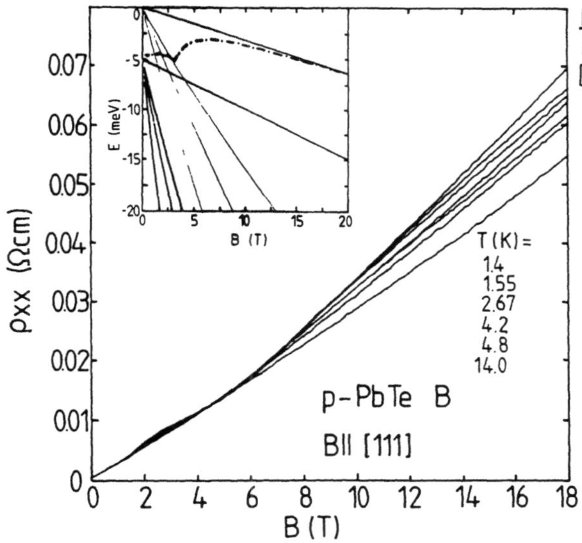

Fig. 12: ρ_{xx} vs B for sample B. Insert: Landau fan chart for B// [111]: $\Theta = 0^{o}$, 70.53^{o}

to n-type sample A despite the fact that in the latter case all carriers are within one pocket. Due to the higher hole concentration of sample B, the lowest value of theFermi energy is about 0.1 meV with respect to the lowest Landau-state. The Hall "dip" is somewhat smaller in sample B in comparison to A. For the longitudinal magnetoresistance (Fig. 13) for B// $[\bar{1}2\bar{1}]$ all carriers are transferred at about 8T into the lowest Landau states of two valleys ($\Theta = 61.81^{o}$). Even so, in this case, a remarkable temperature dependence remains. Thus, for the many valley case, the occupation of two pockets does not prevent the appearence of rather strong temperature dependent magnetotransport phenomena. Similar to the n-type material, the p-type sample also exhibits in ρ_{zz} a dependence of $\ln \rho_{zz} \propto T^{-1/4}$ for temperatures higher than approximately 1.2K (Fig. 14). With increasing carrier concentration the strong dependence of ρ_{xy}, ρ_{xy} and ρ_{zz} on T is diminished and finally disappears. In Fig.15 ρ_{xx} and ρ_{xy} are plotted vs B for sample C ($n = 1.1 \times 10^{17}$ cm^{-3}) and the effect of different temperatures is hardly visible. The extrema in ρ_{xx} (B) are quite useful to establish the correct value of the energy shift of the three oblique valleys with respect to the [111] valley. The two inserts of Fig. 15 show the Landau levels for an unstrained bulk PbTe and for a strained PbTe-layer. For a shift of 5 meV the Shubnikov de Haas like maxima at 12T is caused by carrier transfer from the (b) to the (a) valleys and back.

For low concentrations of both n- and p-type PbTe films, a strong (factor of 2 to 4) temperature dependent increase in ρ_{zz} is observed at high magnetic fields (EQL). This is accompanied by a 5 to 10% decrease in R_H and a similar small increase in ρ_{xx}. The strength of the temperature dependence decreases for increasing carrier concentration, disappearing for n or p greater than $\sim 1.5 \times 10^{17} cm^{-3}$.

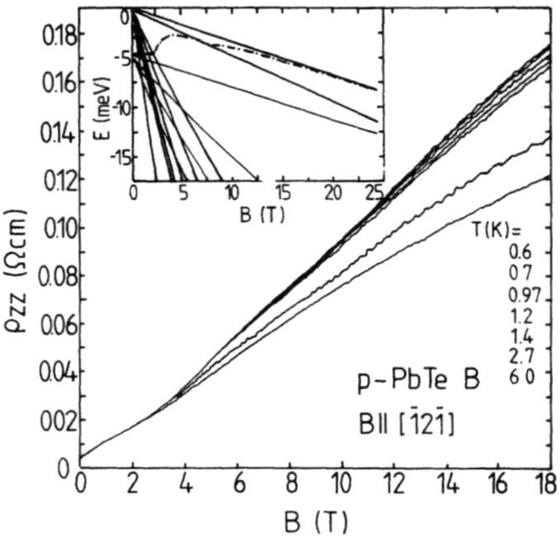

Fig. 13: ρ_{zz} vs B for sample B
Insert: Landau fanchart for
B// [121] Θ = 19.47°, 61.81°, 90°

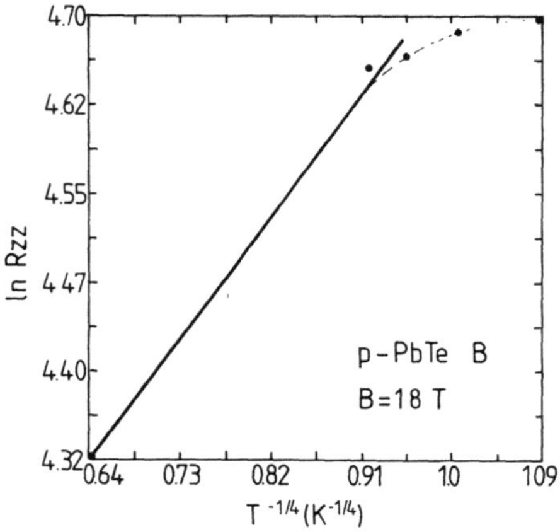

Fig. 14: ln R_{zz} vs $T^{-1/4}$ for sample B.

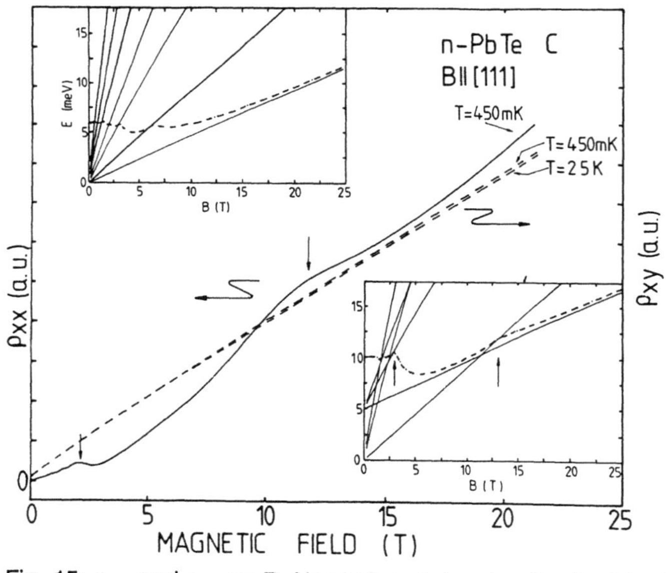

Fig. 15: ρ_{xx} and ρ_{xy} vs B. Upper Insert: Landau fanchart for Θ = 0°, 70.53° without strain induced valley shift; lower insert: shift: $\Delta E = 5$meV

4. Discussion

Historically, the temperature and magnetic field dependence of the transport parameters ρ_{xx}, ρ_{xy} and ρ_{zz} in PbTe and PbSnTe was interpreted by Nimtz et al. [1,8] as evidence for a magnetic field induced Wigner transition. In order to estimate the necessary range of temperarutes and magnetic fields, we use the expressions by Gerhardts [3]. The value of the ratio of the potential energy and the mean kinetic energy $\Gamma = <V>/E_{kin}$ should attain a critical value of $\Gamma \sim 8$. The mean distance between electrons is $r = (4\pi n/3)^{-1/3}$ and the effective Bohr radius $a^* = a_B \varepsilon_s m_0/m^*$. Using for $\varepsilon_s \sim 150$ ([2], instead of $\varepsilon_s = 1500$) we obtain a critical carrier concentration $n < 7 \times 10^{11} cm^{-3}$ for which $r > 7a^*$ [3]. In a magnetic field, the mean carrier energy decreases like the Fermi energy which we have determined quantitatively for different directions of B using the band parameters for PbTe according to [17]. For magnetic fields above 5T, the Fermi energy is less than 200μeV (estimate: $E_F = h^2 (2\pi^2 l^2 n)^2/2m^*$, $l = (h/eB)^{1/2}$) which corresponds to a carrier concentration of about $1.4 \times 10^{16} cm^{-3}$. In order to restrict the kinetic energy, the sample temperature has to be less than the Fermi temperature T_F, i.e. < 2K. Thus this very crude estimate leads to values within the experimentally accessible para-meters. However, a more careful estimate of the phase transition temperature based on the Hartree-Fock approximation to calculate the phase diagram using for $r_s = 0.26$ ($r_s = 4\pi n a^{*3}/3)^{1/3}$) yields $T_c/T^* \sim 0.5$ (T^* denotes the Rydberg temperature which if for PbTe according to Ref. 2 about 0.26K); therefore T_c is about 0.13K (upper limit, $T_c = 17$mK with use of $\varepsilon_s = 1500$). Even magnetic fields higher than 20T would not help considerably to change this situation, since physically one expects a saturation of the melting temperature of the Wigner condensate with B [11,14]. However, these estimates do not rule out the existence of a liquid crystal like highly correlated state above T_c with varying correlation lenghts [14].

The other model which has been put forward in the long lasting controversy on the explanation of high field and low temperature transport phenomena in HgCdTe is a metal insulator transition driven by localization effects. For high magnetic fields the criterion is given by [12,13,14,16]:

$$n\, a_\perp^2\, a_\parallel \sim 0.25^3 , \tag{1}$$

where

$$a_\perp = 2l = 2(h/eB)^{1/2} \text{ and}$$
$$a_\parallel = a^*/ [\ln(a^*/l)^2] . \tag{2}$$

Using the parameters for PbTe, again with $\varepsilon_s \sim 150$ and a magnetic field of 10T we obtain $n = 3 \times 10^{15} cm^{-3}$ ($n = 7 \times 10^{14} cm^{-3}$ for $\varepsilon = 1500$). Due to the anisotropy of the bands the critical density can change by a factor of 3 [18]. Therefore, we nearly approach this limit for sample A under favorable conditions whereas for the investigations on HgCdTe [4] due to the much smaller carrier concentration ($2 \times 10^{14} cm^{-3}$) and the lower value for ε the Mott criterion yields magnetic fields of about 2T. In this context we would like to emphasise that a drastic temperature dependence especially for the ρ_{zz} component is observed for HgCdTe [1,4] for magnetic fields well below the critical field. For HgCdTe, Goldman et al. [6,7] have observed a temperature dependent shift of the resonance field in a cyclotron resonance experiment which was interpreted as evidence for magnetic freeze out of carriers. An observation of impurity cyclotron resonance in p-PbTe in the field and temperature range of interest has been reported as well [19]. Field et al. [14] stated that their observation of a Wigner transition at very low temperatures ($T_W \sim 0.01$K) is not in contradiction with cyclotron resonance (CR) experiments in [6,7]. They argued "that sharp CR lines might be expected even in a perfect crystalline arrangement of electrons because the carriers are also localized" and a gradual increase of transi-

tion energy with field is also possible as the electrons become more tightly bound. All CR experiments reported so far are in a temperature regime above Tw.
The puzzling decrease of the Hall coefficient with increasing magnetic fields is a phenomenon also common to HgCdTe and PbTe (see Fig. 3). Following the model of Goldman et al. [5] the dependence of R_H on B for fields smaller than that corresponding to the Mott transition is a consequence of the presence of infinite metallic donor cluster as well as of shallow isolated donors. With increasing field each isolated donor takes up one electron from the metallic cluster excluding more than the average volume taken before by one electron therefore enhancing the effective density in the conducting cluster. Apparently both disorder and electron correlation effects are physically important in this model. In a magnetic field induced localization regime we expect an exponential dependence of transport coefficients on temperature. For the carrier freeze-out region in n-InSb a dependence $\ln \rho \sim T^{-1}$ is well established. In Fig. 16 we plot ln R_{zz} vs 1/T for various applied fields. It can be seen that a 1/T law if it exists at all is restricted to rather high tem-peratures. Also a plot of ln R_{zz} vs $T^{-1/4}$ (Fig. 17) is inconclusive since the temperature range is too small to yield any strong evidence for a hopping conduction. In Fig. 18 we plot the results of Shayegan et.al. [4], namely ln ρ_{zz} vs $T^{-1/4}$

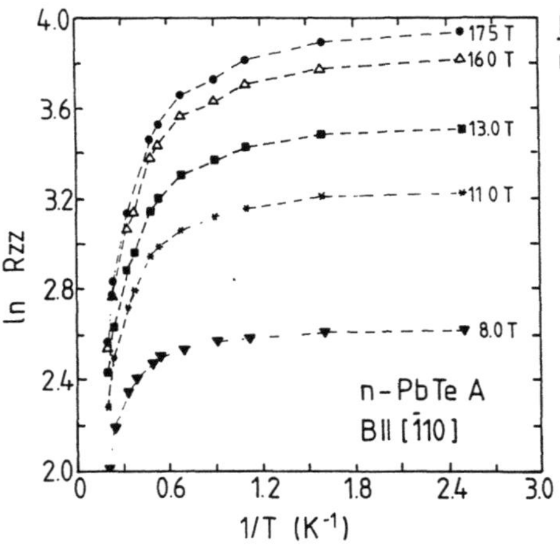

Fig. 16: ln R_{zz} vs 1/T at several applied magnetic fields.

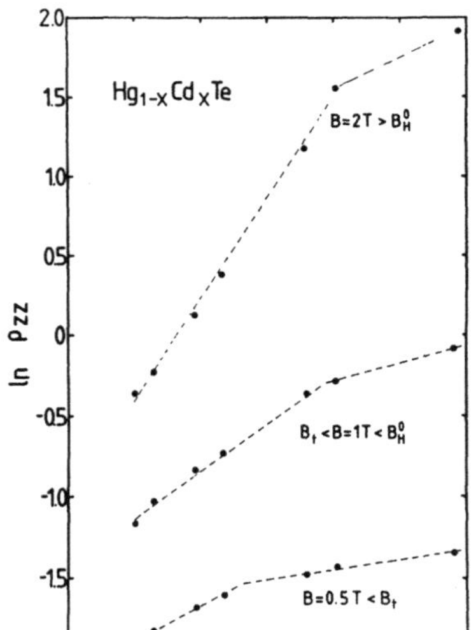

Fig.18 lnρ_{zz} vs T-1/4 at various magnetic fields for HgCdTe. Data taken from Ref[4]

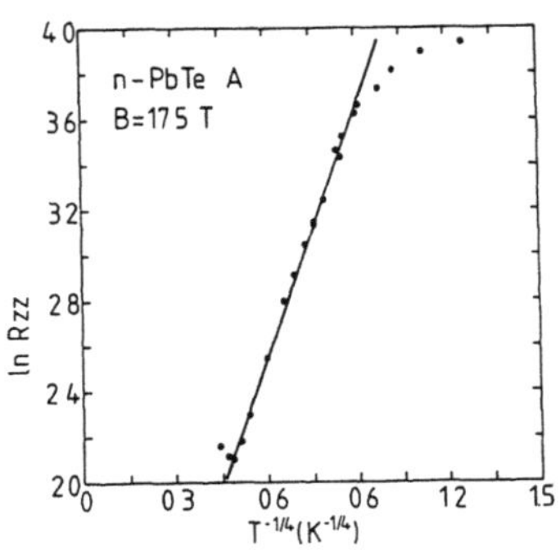

fig. 17: ln R_{zz} vs $T^{-1/4}$ at B = 17.5T

for different applied fields in order to demonstrate that also in their case variable range hopping models are inconclusive even for fields $(B > 2T)$ for which the Mott transition has already occurred.

5. Conclusion

The magnetotransport coefficients of PbTe exhibit striking similarities to those reported by different groups on HgCdTe. The magnetic fields above which a strong temperature and field dependence is observed, especially in the longitudinal magnetoresistance, are only slightly higher (by about a factor of 2) than those reported for HgCdTe with similar energy gap but carrier concentrations of about two orders of magnititude smaller. Applying the criterion given by Gerhardts to our data, the temperatures used in the experiments are more than a factor of 10 too high for Wigner condensation to be observable. In the magnetic field range investigated, the criterion for a metal insulator transition according to the Mott localization is just approached but not reached. The PbTe samples studied contain both Te- and Pb-vacancies. Using the data of Ref [15] we estimate a compensation ratio of about $K = |V_{Te}|/|V_{Pb}| \sim 0.7$ from the low temperature mobility of sample A ($\sim$ 550000 cm^2/Vs) . These vacancies are short range scatterers and represent a source of disorder which will cause local variations of the conduction and valence band edges. It is this disorder which, in our opinion, is the source of the temperature dependence of the magneto transport coefficients and drives rather a Mott-Anderson type of transition. Our study of n- and p-PbTe does not yield conclusive evidence for a transition to condensed phase. Also a Peierls instability or a one-dimensional charge density wave invoked by Bardassis and Das Sarma [20] can be ruled out since there is no indication for a critical temperature in ρ_{zz} and the dependence of the observed anomalies on carrier concentration is opposite to the predictions of the theory. In our opinion both correlation and disorder effects play an important role in determining the magneto transport properties. Some form of a high field, low temperature localization mechanism is likely, though variable range hopping does not explain the entire temperature range measured.

Acknowledgements: Work supported by Fonds zur Förderung der wissenschaftlichen Forschung, Vienna and by National Science Foundation. We thank L. Rubin and B. Brandt of the National Magnet Laboratory for technical assistance, and A. Lopez-Otero for the growth of the samples.

References:

1. G. Nimtz and B. Schlicht: Festkörperprobleme XX, 369 (1980) ed. P. Grosse (Vieweg, Braunschweig).
2. R. Enderlein, F. Bechstedt, W. Hill: in Physics of Narrow Gap Semiconductors, ed. E. Gornik et al. Lecture Notes in Physics, 152, 468 (Springer Berlin, Heidelber) (1982)
3. For a recent review see e. g. R. R. Gerhardts in High Magnetic Fields in Semiconductor Physics, ed. G. Landwehr in Springer Series in Solid State Scienes 71, 482 (1987)

4. M. Shayegan, V. J. Goldman, H. D. Drew, D. A. Nelson, P. M. Tedrow Phys. Rev. B32, 6952 (1985; B31, 6123 (1985).
5. V. J. Goldman, M. Shayegan, H. D. Drew, Phys. Rev. Lett. 67, 1056 (1986).
6. V. J. Goldman, H. D. Drew, M. Shayegan, D. A. Nelson, Phys. Rev. Lett. 56, 968 (1986).
7. J. B. Choi, L. S. Kim, H. D. Drew, D. A. Nelson, Solid State Commun. 65, 547 (1988).
8. G. Nimtz, B. Schlicht, E. Tyssen, R. Dornhaus, L. D. Haas, Solid State Commun. 32, 669 (1979).
9. J. Singleton, E. Kress-Rogers, A. V. Lewis, R. J. Nicholas, E.J. Fantner, G. Bauer, A. Lopez-Otero, J. Phys. C. 19, 77 (1986).
10. G. Bauer, Lecture Notes in Physics 133, 423 (1980).
11. S. B. Field, D. H. Reich, B. S. Shivaram, T. F. Rosenbaum, D. A. Nelson, P. B. Littlewood, Phys. Rev. B33, 5058 (1986).
12. N. F. Mott, M. Kaveh, Advances in Physics 34, 330 (1985).
13. I. M. Tsidilkowskii, Usp. Fiz. Nauk 152, 583 (1987).
14. S. B. Field, D. H. Reich, T. F. Rosenbaum, P. B. Littlewood, D. A. Nelson, submitted to Phys. Rev. B.
15. L. Palmetshofer, K H. Gresslehner, L. Ratschbacher and A. Lopez-Otero, see Ref. 2 p. 391.
16. B. I. Shklovski, A. L. Efros, Electronic Properties of Doped Semiconductors, Springer, Berlin Heidelberg (1984).
17. H. Pascher and G. Bauer, see Ref. 3. p. 400
18. T. Ohtsuki and Y. Ono in Proc. 18th Int. Conf. Phys. Semicond. Vol. 2, p.1223; ed. O. Engstrom (World Scientific Singapure 1986).
19. A. V. Lewis, F. Kuchar, R. J. Nicholas, J. C. Ramage, L. Palmetshofer, Phys. Rev. B28, 2244 (1983).
20. A. Bardassis and S. Das Sarma, Phys. Rev. B29, 780 (1984).

Conductivity and Hall Effect at High Magnetic Fields in Sb-Doped Si near the Metal-Nonmetal Transition

Y. Ochiai [1], *M. Mizuno* [1], *and E. Matsuura* [2]

[1] Institute of Materials Science, University of Tsukuba, Tsukuba, Ibaraki 305, Japan
[2] Institute of Physics, University of Tsukuba, Tsukuba Ibaraki 305, Japan

1. INTRODUCTION

It is well known that an increase of the donor concentration in doped semiconductors leads to a metal-nonmetal transition. The critical concentration of the transition in Sb-doped Si (Si:Sb) has been determined to be 3.0 x 10^{18} cm^{-3} [1] and is slightly lower than that of 3.74 x 10^{18} cm^{-3} in Si:P [2]. The weak localization effect due to quantum interference of scattered electrons near the transition has been widely studied in recent years. In this weak localization regime, a quantum interference correction to the low-temperature conductivity has been calculated from the effects of localization and electron-electron interaction. On the other hand, very near the transition, peculiar behavior has been observed in the low-temperature transport of Si:Sb [3,4]. A large rise in conductivity with decreasing temperatures has been reported in Si:Sb just below the transition [3]. The rise is explained by an electronic phase transition due to a change in the valley degeneracy. However, another explanation of the rise is suggested: it is due to an enhancement of conductivity by destructive interference of the scattered waves in conjunction with spin-orbit interaction. In any way, a certain pseudogap can be assumed in the density of states near the Fermi energy just below the transition.

The scaling nature of the zero-temperature conductivity was also discussed in doped semiconductors. The critical scaling exponent of the conductivity is close to 1/2 in experimental results of Si:P and Si:As [5,6]. However, in theoretical studies [7,8], the absence of scaling for the Hall coefficient R_H has been predicted. It has been recently found that R_H for Si:P and Si:As does not indicate scaling behavior at the transition [6]. On the other hand, it is considered that in the case of a high-Z impurity like Sb, R_H shows scaling because of a contribution of the strong spin-orbit interaction to R_H.

In this paper, we study the critical behavior of the scaling of low-temperature conductivity and R_H of Si:Sb. We will compare the scaling in Si:Sb with that in Si:P and Si:As and also discuss transport anomalies in Si:Sb just below the transition.

2. EXPERIMENTS

The samples used were prepared from uncompensated commercial quality Czochralski-grown Si crystals doped with Sb. A wafer of the Si crystal is cut into bridge-type specimens with dimensions of 0.5 x 0.15 x 0.03 cm^3 with eight side arms of the width 0.15 cm. Two of umpaired Hall contacts were used as voltage electrodes for four-probe resistance measurement. We lapped the surface of the sample mechanically and then chemically etched it, using

Springer Series in Solid-State Sciences, Vol. 87
High Magnetic Fields in Semiconductor Physics II Editor: G. Landwehr

CP4 solution. Thermal annealing was performed in a dry N_2 gas atmosphere during 30 min at 800 C. The lead contact positions of the arms were coated by Sb vacuum evaporation and Au-lead wires were welded to the contact position. The sample was cooled by a dilution refrigerator down to 50 mK. We used a calibrated Ge thermometer and a carbon resister thermometer which was calibrated by a nuclear orientation thermometer consisting of a ^{60}Co single crystal. The sample and thermometers were mounted in a chamber which was made of oxigen-free Cu and was thermally anchored to a mixing chamber. We made low power resistance measurements by using an ac resistance bridge at 25 Hz. Input powers were lowered to the range of 10^{-12} to 10^{-15} W and sample self-heating due to ac excitation was not observed.

We determined the impurity concentration from the room temperature R_H using the concentration dependence of Hall factor in Si:P[7]; supposing that the maximum Hall factor in Si:Sb slightly decreases between the critical concentrations of Si:P and Si:Sb. Actually, the concentration of the maximum Hall factor in Si:As is higher compared to Si:P[9]; the critical concentration of Si:As is 8.5×10^{18} cm^{-3} [6].

We can obtain the zero-temperature conductivity from the $T^{1/2}$ dependence of the conductivities at H=0. Except for the sample very near the transition (3.03×10^{18} cm^{-3}), the low-temperature conduction almost agrees with the $T^{1/2}$ dependence. An enhancement of the low-temperature resistivity has been observed very close to the transition because of surface related defects [4]. Figure 1 shows the zero-temperature conductivity versus n/n_c-1, the critical exponenet s is 0.77 which is larger than that previously reported [3]. In Fig.2 the quantity of $1/n_ceR_H$ is plotted versus n/n_c-1 in the case for data obtained at 1.5 T. R_H values for three lower concentration samples were taken from 0.06 K measurements but others were taken from 1.2 K data. The absence of scaling for R_H has been reported for Si:P and Si:As and the result is drawn by dotted lines in the figure [6]. However, the scaling of $1/R_H$ does not disappear in Si:Sb. The exponenet is close to 1/2 as observed in the conductivity scaling of Si:P and is smaller than that in Ge:Sb as shown in Fig. 2.

For the lowest concentration sample (2.64×10^{18} cm^{-3}) just below the transition, the temperature dependence of the Hall carrier concentration has

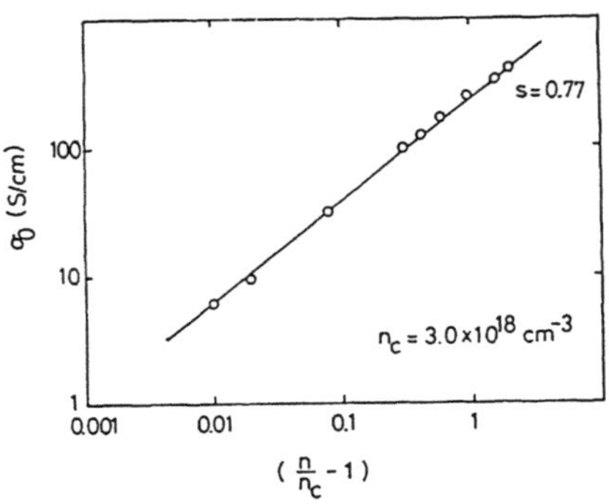

Figure 1. Scaling of the zero-temperature conductivity in Si:Sb

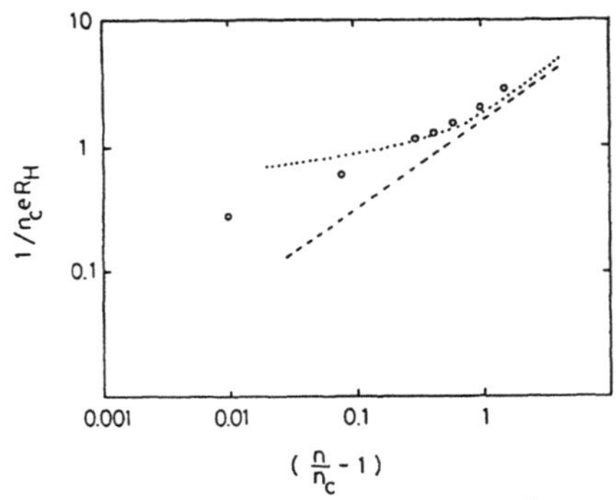

Figure 2. $1/n_ceR_H$ versus n/n_c-1 in Si:Sb (circles), Si:P and Si:As (dotted line; after Ref. 3) and Ge:Sb (dashed line; after Ref. 10)

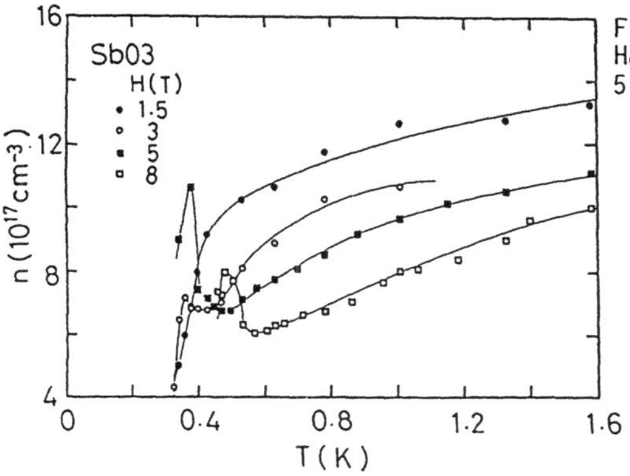

Figure 3. Temperature dependence of Hall concentrations at 1.5, 3, 5 and 8 T

been measured for various magnetic fields as shown in Fig. 3. At higher temperatures than 0.5 K the Hall concentration decreases as the field increases. However, below 0.5 K, the concentrations have a peak at higher fields than 3 T and the peak slightly shifts to high temperatures with increasing fields. Moreover, it seems to be field-independent as the temperature decreases and the concentration drops near 0.3 K. A similar behavior in the temperature dependence has been observed in the low-temperature conductivity for various magnetic fields in just nonmetallic Si:Sb [3].

3. DISCUSSION AND CONCLUDING REMARKS

The absence of scaling for $1/R_H$ has been observed in Si:P and Si:As [6]. However, in Si:Sb, scaling of $1/R_H$ can be seen and the critical exponent is close to 1/2. One possible explanation of the difference in the scaling is due to an effect of compensation. While the critical exponent in conductivity scaling is nearly equal to unity in the case of compensated doped semiconductors, it is also dependent on a relative effect of localization versus electron-electron interaction. Since we use uncompensated samples in this study, compensation is not important in the difference between Si:Sb and others. On the other hand, the localization effect should be an important factor in Si:Sb because the scaling for the conductivity is not close to 1/2.

Another explanation can be considered which is due to an effect of spin-orbit interaction. Although Si:P and Si:As contain impurities with low Z, Sb in Si:Sb has high Z and is expected to have a strong spin-orbit interaction. Since the scaling for $1/R_H$ does not agree with localization predictions [7,8], its appearance in Ge:Sb [10] has been explained by a contribution of the spin-orbit interaction to R_H [6]. An effect of spin-orbit interaction can be discussed in the Hall concentration of just nonmetallic Si:Sb. With lower temperatures the screening due to conduction electrons will gradually become weaker because of the decrease of carrier concentration as shown in Fig. 3. Then the spin-orbit coupling becomes strong at low-temperatures. This means that the strength of the spin-orbit interaction depends on concentration and temperature near the transition. Therefore low-temperature transport in Si:Sb near the transition must be substaintially influenced by spin-orbit interaction. It is suggested that the diffusion length for spin-orbit scattering may be less than the inelastic scattering length at low-temperatures in just nonmetallic Si:Sb [11]. However, a large change in the Hall concentration in Fig. 3 would not be explained by such a short diffusion length only. If the valley degeneracy change by the presence of strong spin-orbit interaction, redistribution of electrons will occur. We consider the above change to be an electronic phase transition predicted by LONG and PEPPER [3]. If this is so, the peaks in Fig. 3 may come from the pseudogap structure of the density of states in just nonmetallic Si:Sb. Futhermore, nonlinear conduction has been observed in nonmetallic Si:Sb [12]. The quasi-high-conductivity transport in this sample may also be explained by the same mechanism as the electronic phase transition in Si:Sb near the transition.

It is found that the critical exponent of the scaling for the conductivity in Si:Sb is larger than that previously observed in doped Si and the scaling for $1/R_H$ does not disappear in Si:Sb. The spin-orbit interaction is very important in Si:Sb near the transition.

ACKNOWLEDGMENT: This work was done at the Crogenics Center in the University of Tsukuba.

REFERENCES

1. T. G. Castner, N. K. Lee, G. S. Cieloszyk, G. L. Salinger; Phys. Rev. Lett. 34, 1627 (1975)
2. T. F. Rosenbaum, K. Andres, G. A. Thomas, R. N. Bhatt; Phys. Rev. Lett. 45, 1723 (1980)
3. A. P. Long, M. Pepper; J. Phys, C: Solod State Phys. 17, L425 (1984)
4. Y. Ochiai, E. Matsuura; Solid State Commun. 49, 441 (1984)
5. G. A. Thomas, T. F. Rosenbaum, R. N. Bhatt; Phys. Rev. Lett. 46, 1435 (1981)
6. D. W. Koon, T. G. Castner; Phys. Rev. Lett. 60, 1755 (1988)
7. H. Fukuyama; J. Phys. Soc. Jpn. 49, 644 (1980)
8. B. Shapiro, E. Abrahams; Phys. Rev. B24, 4025 (1981)
9. P. F. Newman, M. J. Hirsch, D. F. Holcomb; J.Appl.Phys.58,3779(1985)
10. S. B. Field, T. F. Rosenbaum; Phys. Rev. Lett. 55, 522 (1985)
11. N. F. Mott; J. Phys. C: Solid State Phys. 20, 3975 (1987)
12. Y. Ochiai, E. Matsuura; Jpn. J. Appl. Phys. 26, L582 (1987)

Studies of Magnetotransport Measurements of Resonant DX Centres in Heavily Doped GaAs and (AlGa)As Alloys

J.C. Portal[1], *L. Dmowski*[1,*], *D. Lavielle*[1], *A. Celeste*[1], *D.K. Maude*[2,**], *T.J. Foster*[2], *L. Eaves*[2], *P. Basmaji*[3], *P. Gibart*[3], *and R.L. Aulombard*[4]

[1]INSA-CNRS, F-31077 Toulouse-Cedex, SNCI-CNRS, F-38042 Grenoble Cedex, France

[2]Department of Physics, University of Nottingham, Nottingham NG7 2RD, UK

[3]LPSES, CNRS Sophia Antipolis, F-06560 Valbonne-Cedex, France

[4]GES, USTL, F-34060 Montpellier Cedex, France

*On leave from the High Pressure Research Centre, Polish Academy of Science, PL-0-142 Warsaw, Poland

**Current address: Max-Planck-Institut, F-38042 Grenoble Cedex, France

The DX centre in heavily n-doped GaAs and (AlGa)As is investigated using hydrostatic pressure. The DX level is identified with a simple substitutional donor level associated with higher conduction band minima. For the most heavily doped (2.1×10^{19} cm^{-3}) GaAs sample the DX centre is already acting to limit the free electron concentration at atmospheric pressure.

1. INTRODUCTION

The DX centre in n^+(AlGa)As, first reported over a decade ago [1, 2], remains the subject of considerable controversy. The optical, capture and emission cross-sections of DX have been extensively reported [1 - 4]. The large difference between the thermal (0.1 eV) and optical (1.1 eV ionisation energies suggest that the centre undergoes a large lattice relaxation (LLR). These features, together with a relatively large thermal barrier to capture (0.3 eV) give rise to the well-known persistent photoconductivity (PPC) effects observed at low ($T < 120$ K) temperatures.

Several alternative models have been proposed [1, 5 - 7] for the microscopic nature of DX. Historically, the centre was termed "DX" because it was thought to be a donor impurity (D) complexed with an unknown defect (X). More recently it has been suggested [6, 7] that the PPC arises from states associated with the simple substitutional donor. The DX level closely follows the L conduction band minima in (AlGa)As as a function of Al mole fraction (x) and becomes resonant with the Γ conduction band for $x < 0.2$ [8]. This suggests that DX has some of the character of these minima. MORGAN [9] proposed that DX is a triplet state of symmetry T_2 derived from the L-valleys. The T_2 state is driven deeper than the A_1 state associated with the L valley by a Jahn Teller displacement of the atoms of the central cell. Thus the proposed LLR of the donor can be explained without having to invoke a donor defect complex.

Springer Series in Solid-State Sciences, Vol. 87
High Magnetic Fields in Semiconductor Physics II Editor: G. Landwehr

A deep-donor model involving a small lattice relaxation (SLR) has also been proposed [6, 7]. There is much experimental evidence to support both models. A large lattice relaxation is required to explain the large optical ionisation energy (1.1 eV). The pressure dependence of the capture barrier height [10] is also consistent with the LLR model. OSHIYAMA and OHNISHI [11] recently proposed a model in which the donor Si atoms on the Ga site are surrounded by a distorted As tetrahedron. The model is supported by calculations of the energy level structure of clusters. The antibonding state of the donor shifts downward into the band gap upon distortion of the As surroundings. It was suggested that this model could be extended to GaAs under pressure since the local presence of Al is not required to generate the deep level [11]. Recent Mössbauer spectroscopy on Sn-doped (AlGa)As [12] indicate that the electronic structure of DX is considerably different from that of the shallow donor state, with local distortion of cubic symmetry. This is consistent with the donor-displacement model (LLR) proposed by MORGAN [9]. However extended x-ray absorption fine structure (EXAFS) measurement [13] indicates that the lattice relaxation exhibited by the Se DX centre in (AlGa)As is small.

MIZUTA et al [14] have recently shown using Deep Level Transient Spectroscopy (DLTS) measurements that the DX centre becomes occupied in lightly doped n-type GaAs and (AlGa)As with low Al mole fraction (x), at pressures above a critical value, which decreases with increasing x. Since the application of hydrostatic pressure has a similar effect on the band structure to increasing x, it might be expected that the DX centre will move with a similar pressure coefficient to that of the L-minima.

2. EXPERIMENT

In this paper hydrostatic pressure up to 15 kbar and high magnetic fields up to 18 Tesla are used to study the DX centre in a series of heavily doped [(0.2 - 2.1) x 10^{19} cm^{-3}] n-type GaAs and $Al_xGa_{1-x}As$ $(0.25 < x < 0.29)$ samples. The layers were grown by molecular beam epitaxy (MBE) and metal organic vapour phase epitaxy (MOVPE), doped with either Si or Sn and varied in thickness between 0.15 and 12 μm. For the two most heavily doped MBE grown GaAs samples (1.1 and 2.1 x 10^{19} cm^{-3}) enhanced doping was achieved by reducing the susbtrate temperature to 400°C during growth [15]. The carrier concentration was measured at low temperatures (4.2 K) using the Shubnikov-de Haas (SdH) effect. The period Δ (1/B) of the SdH oscillations in the magnetoresistance yields directly the concentration (n) of electrons in the Γ-conduction band according to the relation :

$$\Delta\,(1/B) = \frac{2e}{\hbar}\,(3\pi^2 n)^{-2/3} \quad . \tag{1}$$

Pressure was applied by means of a Polish Cu-Be "liquid clamp" cell. The salient feature of these cells is that the pressure is applied at room temperature. Subsequent cooling to the low temperatures required for the SdH measurements is accompanied by a small decrease in pressure. The pressure is continuously monitored by means of a calibrated n-type InSb resistance manometer mounted inside the cell. All pressures quoted are those measured at 4.2 K. The cooling procedure involved a precool to 77 K of $\sim$20 mins duration, followed by cooling to 4.2 K of $\sim$5 mins duration. The occupation of DX is fixed at some intermediate temperature, probably around 120 K (estimated from the temperature at which the PPC is observed to quench). Below this temperature the capture and emission rates become so slow that the electron population of the trap is no longer in equilibrium with the free-electron density. In order to study PPC, samples could be illuminated under pressure at low temperatures by means of a visible light-emitting diode.

3. RESULTS

Typical SdH measurements on a Si-doped GaAs sample with initial (P = 0) carrier concentration of 1.1×10^{19} cm^{-3} are shown in Fig. 1 for a range of applied pressure.

As the pressure (P) is increased, the period $\Delta(1/B)$ of the SdH oscillations in inverse magnetic field increases, corresponding to a decrease in carrier concentration. Note also that the amplitude of the oscillatory structure increases with increasing pressure (i.e. decreasing n) so that the oscillatory structure becomes observable down to lower magnetic fields. This is a clear indication that the mean collision time τ and mobility μ increase with increasing pressure since the amplitude of the SdH oscillations is controlled by a factor $\omega_c \tau$ where $\omega_c = eB/m^*$ is the cyclotron frequency. Since the period $\Delta(1/B)$ of the SdH oscillations and the zero magnetic field resistance give respectively the carrier concentration n, and the conductivity σ, the variation of μ with P can be determined. Typical results for some of the GaAs and (AlGa)As samples investigated are shown in Fig. 2 and Fig. 3 respectively. For the most heavily doped GaAs sample (Fig. 2 d) and for the (AlGa)As sample with x = 0.29 the carrier concentration falls immediately on increasing the pressure above atmospheric. This indicates that the DX centre in these samples is very close to the Fermi energy and partially occupied even at atmospheric pressure. In more lightly doped samples the carrier concentration remains roughly constant up to a critical value P_c and then falls sharply. The GaAs sample doped at 2×10^{18} cm^{-3} (Fig.2 c) shows no significant change of n up to 15 kbar. The variation of mobility with pressure is shown only for the GaAs samples (Fig. 2). In all samples a decrease in n with P is accompanied by an increase in μ. In lighter doped samples, over the range of pressure for which n remains constant, μ is found to decrease slowly with pressure. This effect is associated with the increase of the electron effective mass of the Γ-conduction band with pressure.

These results indicate that the loss of free carriers with increasing pressure is associated with capture at donor-like localised resonance levels (DX) located between the Γ-and L-conduction band minima. This capture

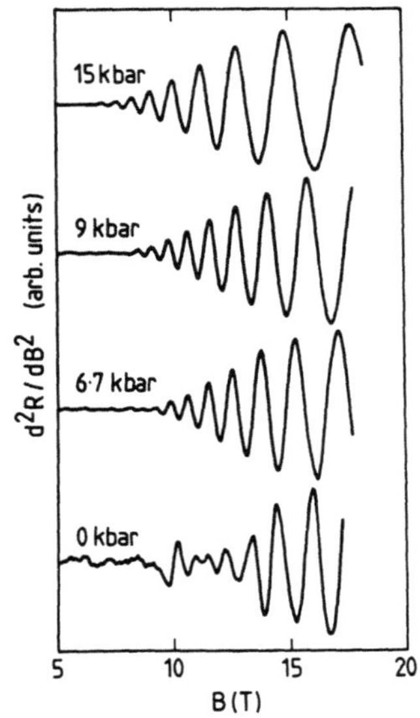

Fig. 1. Typical SdH oscillation in the second derivative of the transverse magnetoresistance of a Si-doped GaAs sample with initial (P = 0) carrier concentration of 1.1×10^{19} cm^{-3}, T = 4 K, J⊥B.

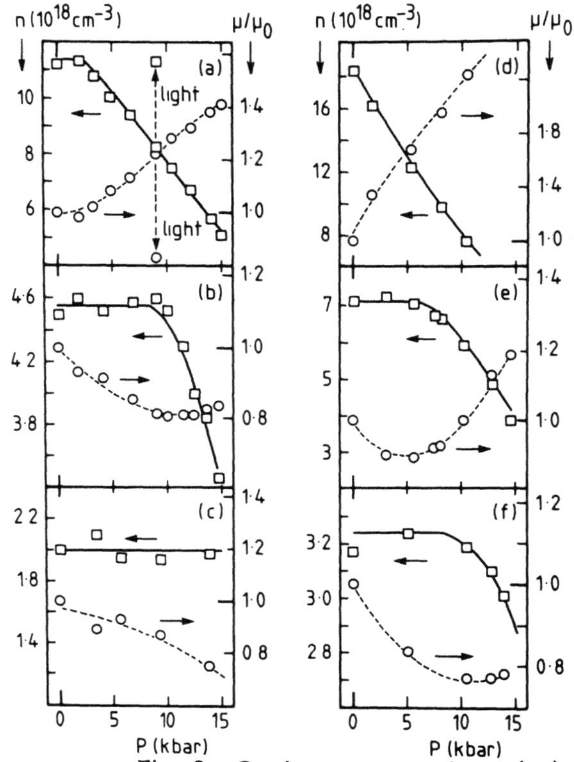

Fig. 2. Carrier concentration n( ) and mobility μ (0) versus pressure P for a range of Si (a, b, c) and Sn (d, e, f) doped GaAs samples. The mobility is normalised to its atmospheric pressure value. The solid lines show the expected variation of n with P calculated using semiconductor statistics as described in the text. The effect of illumination on n and μ at 9 kbar is indicated in (a) for the most heavily Si-doped sample. The broken curves through the mobility data are guides to the eye only.

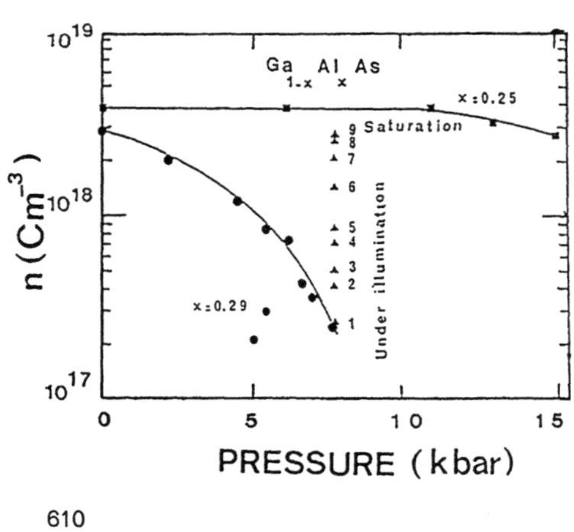

Fig.3 . Carrier concentration n versus pressure P for the Sn-doped (AlGa)As samples. The effect of illumination at 7.8 kbar is shown for the (AlGa)As sample with Al mole fraction x = 0.29.

process is described by the relation $D^+ + e^- \rightarrow D^o$. The neutralisation of ionised donors by electron capture and the corresponding decrease in ionised impurity scattering explains the increase in mobility and in amplitude of the SdH structure. Note, at low temperatures ionised impurity scattering is the dominant mechanism for limiting the mobility. Neutral impurity scattering is only important when the concentration of neutral donors (i.e. DX centres which have captured an electron) is much greater than the free electron concentration [16].

The effect of illumination at low temperature on the most heavily Si-doped GaAs sample at 9 kbar and on the (AlGa)As (x = 0.29) sample at 7.8 kbar is shown in Fig. 2 a and Fig. 3 respectively. Both samples exhibit PPC which characterises the DX centre in (AlGa)As. By illuminating the sample with pulses of light the carrier concentration measured by the SdH effect could be persistently increased in a step-wise manner. Finally the carrier concentration saturates when it reaches its zero pressure value. No further change in n with illumination is observed. The mobility μ, falls to a value somewhat lower than its zero pressure value because of the pressure dependent effective mass. Here it is important to note that the pressure was applied at room temperature and the sample was then cooled slowly and in the dark. Thus a fraction of the DX centres, corresponding to the thermal equilibrium population at $T \sim 120$ K will be occupied after cooling to 4.2 K. Upon illumination the photoionised carriers are prevented from capturing at DX because of the very small capture cross-section at low temperature.

The PPC effects reported here cannot arise from a charge separation model as proposed by QUEISSER and THEODOROU [17]. In our model, the restoration of carriers occurs according to the relation $D^o \rightarrow D^+ + e^-$, so that space charge neutrality is maintained. Using Gauss's law we can estimate roughly the electric field and potential difference that would result if the charge separation model were applicable. For the Si-doped GaAs sample (Fig.2 a), the change in carrier concentration after illumination at 9 kbar is $\Delta n \sim 6 \times 10^{18}$ cm^{-3} over a thickness ($t = 12\,\mu$m) of the conducting layer. The large value at the resulting electric field $E \simeq e\,\Delta nt/\varepsilon\,\varepsilon_o = 5 \times 10^{10}$ V/m and potential difference $\Delta V = e\,\Delta nt^2/2\varepsilon\,\varepsilon_o = 2.8 \times 10^5$ V show that the charge separation model is inapplicable. We stress however, that the model of QUEISSER and THEODOROU is valid for PPC in lighter doped layers.

4. DISCUSSION

The role of hydrostatic pressure in producing the observed trapping out of free electrons from the Γ-conduction band minimum must be related to its effect on the energy of DX and on the conduction band structure. We have previously shown [18], that the variation of n with P can be fitted by assuming that electrons capture at a donor like (DX) level associated with L-conduction band minima.

As the pressure is increased above a critical value, the energy of the DX level moves towards the Fermi energy $\mathcal{E}_F$ of the electrons in the Γ-conduction band, which then remains pinned at the energy of the deep level. The decrease of n with P is then governed by the pressure coefficient of the energy of DX and by the density of states at $\mathcal{E}_F$.

As previously discussed, the occupation of DX is fixed at around T = 120 K (below this temperature the population of DX and the Γ-minimum are no longer in thermal equilibrium). A realistic approach is therefore to calculate the occupancy of DX at T = 120 K. Using standard semiconductor statistics, it is easy to show that, in the absence of compensation, the energy of DX above the Γ-minimum is given by

$$\mathcal{E}_d = \mathcal{E}_F - KT \, Ln \, [\tfrac{1}{2} (\frac{N_{DX}}{n} - 1)] \quad . \tag{2}$$

Here N_{DX} is the number of DX centres which we assume to be equal to the donor doping N_D [19]. The Fermi energy $\mathcal{E}_F$ is calculated taking into account non-parabolicity and the pressure dependence of the effective mass [20, 21]. A value of N_D slightly greater than the zero pressure carrier concentration is used. The calculated values of the energy of the DX level are shown in Fig. 4 for two of the samples. The positions of the X and L minima illustrated in this diagram are the positions for undoped GaAs. Band gap renormalization [22, 23] at the high doping levels used here probably shifts the X and L minima to higher energies relative to the Γ conduction minimum and may also affect the pressure coefficients, even at constant carrier concentration.

As can be seen from Fig. 4 the variation of the energy of DX as a function of hydrostatic pressure is approximately linear. From the slope we can estimate the pressure coefficient for the energy of DX relative to the Γ-minimum. For the GaAs samples with $n \lesssim 5 \times 10^{18}$ cm^{-3} the pressure coefficient of DX is close to that of the L-minima. For a particular dopant, the pressure coefficient of DX increases with increasing doping. For the most heavily doped GaAs samples the pressure coefficient of DX is

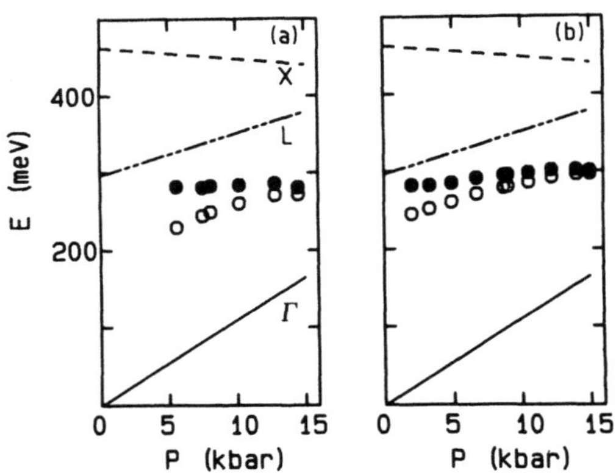

Figure 4. Energy of the DX level (•) and the Fermi energy (0) versus pressure for a) 7.1 x 10^{18} cm^{-3} Sn-doped GaAs sample and b) a 1.1 x 10^{19} cm^{-3} Si-doped GaAs sample. Note that the positions of the X and L minima illustrated are for undoped GaAs. Bandgap renormalisation at these high doping levels probably shifts the X and L minima to higher energies relative to the Γ-conduction band.

close to that of the X-minima. For the (AlGa)As sample the pressure coefficient is close to that of the X-minima even though the doping ($\sim 4 \times 10^{18}$ cm^{-3}) is relatively low. With increasing doping (at constant x), the DX level is shifted to higher energies relative to the Γ-minimum. These results are tabulated in Table 1.

Table 1. The measured carrier concentration n, at P = 0, for the samples studied are indicated in column 2. The Fermi energy ε_F is calculated taking into account non-parabolicity. P_c is the critical pressure at which n starts to fall. N_D is the donor doping level used in the "semiconductor statistics" calculation of the energy of the deep level above the Γ-minimum ε_d. The final column is the pressure coefficient of the energy of the deep level relative to the Γ-minimum used to obtain the fitting curves in Fig. 2 and 3.

Sample	P = 0 carrier conc. $\times 10^{18} cm^{-3}$	ε_F(P = 0) (meV)	P_c (kbar)	N_D $\times 10^{18} cm^{-3}$	ε_d (meV)	$d\varepsilon_d/dP$ meV/kbar
a) GaAs	[Si] 11.3	226	2	12.00	270	-9.4
b) GaAs	[Si] 4.6	126	10	4.60	223	-9.1
c) GaAs	[Si] 2.0	73	Pressure has no effect on n			
d) GaAs	[Sn] 18.0	287	0	21.00	311	-13.3
e) GaAs	[Sn] 7.1	175	7	7.15	251	-10.7
f) GaAs	[Sn] 3.2	109	12	3.25	206	-6.0
g) (AlGa)As (x = 0.29)	[Sn] 2.9	77	0	4.0	91	-11.7

The increase of the energy separation of Γ and DX with increasing doping is consistent with the expected bandgap renormalisation. The direct bandgap is thought to decrease, with increasing doping, by an amount which is of the order of ε_F, the Fermi energy in the Γ-minimum [22, 23]. Hence for the GaAs sample with $n > 5 \times 10^{18}$ cm^{-3}, the increased rate of shift of energy of the DX level relative to the Γ-minimum can be explained by the decreasing bandgap renormalisation as electrons trap out onto DX, hence reducing ε_F. With increasing Al mole fraction x, the X minima decrease in energy and are the lowest conduction band minima for $x > 0.4$. Pressure has a similar effect and GaAs becomes indirect gap for $P > 40$ kbar. There are also donor levels associated with the X-minima, with a binding energy ~ 40 meV [9]. They are known [9] to be of similar symmetry (T_2) to the DX centre. Hence when the X-donor level crosses the DX level a strong mixing and anticrossing effect is expected. This has recently been observed in PL on (AlGa)As samples under pressure [24, 25]. Above a critical composition/pressure the DX centre moves with a similar pressure coefficient to that of the X-minima. This is consistent with our observation that the DX centre in (AlGa)As (x = 0.29) moves with a similar pressure coefficient to that of the X-minima. A similar effect is expected in GaAs under hydrostatic pressure. However, the pressure required (> 20 kbars) is much larger than that employed in the present work.

The zero pressure values of $\mathcal{E}_d$ in GaAs as a function of doping shown in Fig. 5 are calculated by assuming that for pressures below the critical pressure the energy difference between DX and L remains fixed. Values of $\mathcal{E}_d$ reported by THEIS et al. [26] and TACHIKAWA et al. [27] are shown for comparison.

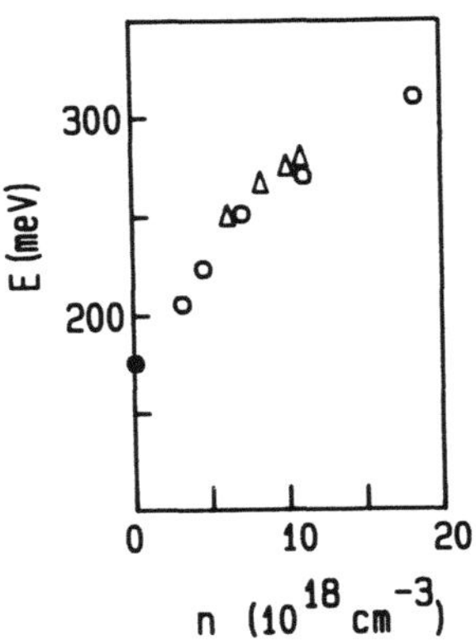

Figure 5. Energy of the DX level ($\mathcal{E}_d$) relative to the Γ -band (0) versus carrier concentration calculated for the GaAs samples using the procedure described in the text. Values reported by THEIS et al. (Δ) and TACHIKAWA et al. (•) for $\mathcal{E}_d$ are shown for comparison.

The solid line in Fig. 2 is the expected variation of n with P, calculated using an iterative method since equation (2) cannot be solved analytically for n. The variation of $\mathcal{E}_d$ with pressure was calculated using the pressure coefficient in Table 1. Non-parabolicity and pressure dependence of effective mass were taken into account when calculating the Fermi energy $\mathcal{E}_F$. For the most heavily Sn-doped GaAs sample ($n = 1.8 \times 10^{19}$ cm^{-3}) a value of $N_D = 2.1 \times 10^{19}$ cm^{-3} considerably greater than the zero pressure carrier concentration was required to fit the experimental data. Subsequent measurements revealed that this sample exhibits PPC at zero pressure. Upon illumination at low temperatures the carrier concentration was persistently increased to 2.1×10^{19} cm^{-3}. This is in excellent agreement with the theoretically predicted value. The effect of the light is to release electrons which have been trapped on DX during the cool-down procedure at atmospheric pressure.

5. MICROSCOPIC NATURE OF DX

It is clear from the data presented above that the DX centre is present at concentrations comparable to the doping level. Far infrared local-vibration mode (LVM) measurements performed on the most heavily Si-doped GaAs samples [28] show that the measured carrier concentration at atmospheric pressure of 1.1×10^{19} cm^{-3} agrees closely with the concentration of Si atoms occuping Ga-sites (1×10^{19} cm^{-3}). The LVM measurements also show that the concentration of Si-Si pairs, Si-X complexes and Si_{As} are at least one order of magnitude lower than the concentration of Si_{Ga}. These observations coupled with the electrical measurements lead us to support the view [6, 7, 14] that the DX centre can be identified with the simple substitutional donor. Such a view would also explain recent results on the hydrogen passivation of Si-donors and DX centres in (AlGa)As [29]. The electrical activity of the shallow donor and DX centre are recovered on annealling with very similar activation energies, suggesting that both have the same chemical origin.

6. CONCLUSION

The pressure induced trapping of electrons and associated PPC in n-type GaAs and (AlGa)As is due to DX levels associated with the higher conduction band minima. The precise location of the DX level relative to the Γ-minimum depends on the free carrier concentration, the DX level moving to higher energy with increasing n. A comparison of our results with the LVM measurements in n^+ GaAs (Si) allows us to identify the DX centre with the simple substitutional donor. For the most heavily doped GaAs sample ($N_D = 2.1 \times 10^{19}$ cm^{-3}) the DX centre is already acting to limit the free carrier concentration at atmospheric pressure.

7. ACKNOWLEDGEMENTS

We are grateful to Dr J.J. Harris, Dr R.B. Beall, Philips Research Laboratories Redhill, Surrey (UK) and to Dr M. Nathan, Dr M. Heiblum, IBM, Yorktown Heights (USA) for providing heavily doped sample GaAs used in these studies. We wish to thank Dr R. Murray and Prof. R.C. Newman for communicating to us their far-infrared absorption measurements and for invaluable discussions, and to Dr J.B. Clegg for performing the SIMS measurements.
This work is supported by SERC, CNRS, CNET, Conseil Regional Midi-Pyrénées and European Communities. One of us (LD) is supported by Ministère des Affaires Etrangères.

8. REFERENCES

1. D.V. Lang and R.A. Logan : Phys. Rev. Lett., 39, 635 (1977)
2. R.J. Nelson : Appl. Phys. Lett., 31, 351 (1977)
3. P.M. Mooney, E. Calleja, S.L. Wirght and M. Heiblum : Proc. Int. Conf. on Defects in Semiconductors, Material Sc. Forum, 10, 417 (1986)
4. R. Legros, P.M. Mooney and S.L. Wright : Phys. Rev. B 35, 7505 (1987)
   D.V. Lang in Deep Levels in Semiconductors, edited by S.T. Pantelides (Gordon and Breach, New-York, 1985), p. 489.

5. A.K. Saxena: Solid State Electron, 25, 127 (1982)
6. H.P. Hjalmarson and T.J. Drummond, Appl. Phys. Lett., 48, 656 (1986). See also H.P. Hjalmarson and T.J. Drummond: Phys. Rev. Lett., 60, 2410 (1986)
7. J.C.M. Henning and J.P.M. Ansems : Semicond. Sc. and Techn., 2, 1 (1987)
8. T.N. Theis and S.L. Wright: Appl. Phys. Lett., 48, 1374 (1986)
9. T.N. Morgan: Phys. Rev. B, 34, 2664 (1986)
10. M.F. Li, P.Y. Yu, E.R. Weber and W. Hansen: Appl. Phys. Lett., 51, 349 (1987)
11. A. Oshiyama and S. Ohnishi: Phys. Rev. B, 33, 4320 (1986)
12. P. Gibart, D.L. Williamson, B. El Jani and P. Basmaji: Int. Symp. GaAs and Related Compounds, Heraklion, Greece (1987). In Inst. Phys. Conf. Ser., n° 91, 379 (IOP Publishing 1988)
13. M. Mizuka and T. Kitano: Appl. Phys. Lett., 52, 126 (1988)
14. M. Mizuta, M. Tachikawa, H. Kukimoto and S. Minomura: Jpn J. Appl. Phys., 24, L143 (1985).
15. J.H. Neave, P.J. Dobson, J.J. Harris, P. Dawson and B.A. Joyce: Appl. Phys. A, 32, 195 (1983).
16. T.J. Drummond and H.P. Hjalmarson: Appl. Phys. Lett., 48, 28 (1986).
17. H.J. Queisser and D.E. Theodorou: Phys. Rev. Lett., 43, 401 (1979).
18. D.K. Maude, J.C. Portal, L. Dmowski, T. Foster, L. Eaves, M. Nathan, M. Heiblum, J.J. Harris and R.B. Beall: Phys. Rev. Lett., 59, 815 (1987).
19. See for example M. Tachikawa, M. Mizuta and H. Kukimoto: Jpn J. Appl. Phys., 23, 1594 (1984), who observe that nearly 100 % of room temperature carriers are generated from DX centres.
20. A. Raymond, J.L. Robert and C. Bernard: J. Phys. C, 12, 2289 (1979).
21. L.G. Shantharama, A.R. Adams, C.N. Ahmad and R.J. Nicholas: J. Phys. C, 17, 4429 (1984).
22. N.W. Ashcroft and N.D. Mermin: Solid State Physics, p. 334 (1976) Saunders College Philadelphia USA by Holt, Rinehart and Winston.
23. R.A. Abram, G.J. Rees and B.L.H. Wilson: Advances in Physics, 27, 799 (1978).
24. M. Chandrasekhar, U. Venkateswaran, H.R. Chandrasekhar, B.A. Vojak, F.A. Chambers and J.M. Meese: 18th Int. Conf. on Physics of Semiconductors, Stockholm, 943, ed. Olof Engström, World Sci. (1986).
25. J.C. Henning and J.P. Ansems: Semicond. Sci. Technol., 3, 361 (1988).
26. T.N. Theis, P.M. Mooney and S.L. Wright: Phys. Rev. Lett., 60, 361 (1988).
27. M. Tackikawa, M. Mizuta, H. Kukimoto and S. Minomura: Jpn J. Appl. Phys., 24, L 821 (1985).
28. L. Eaves, T.J. Foster, D.K. Maude, G.A. Toombs, R. Murray, R.C. Newman, J.C. Portal, L. Dmowski, R.B. Beall, J.J. Harris, M.I. Nathan and M. Heiblum; Int. Symp. GaAs and related compounds, Heraklion, Greece (1987). Inst. Phys. Conf. Ser. n° 91, chap. 4, 355 IOP (1988)..
29. J.C. Nabity, M. Stavola, J. Lopata, W.C. Dautremont-Smith, C.W. Tu and S.J. Pearton: Appl. Phys. Lett., 50, 921 (1987).

Part XII

Reports from High Magnetic Field Laboratories

Recent Topics at the Megagauss Laboratory in Tokyo

N. Miura

Institute for Solid State Physics, University of Tokyo
Roppongi, Minato-ku, Tokyo 106, Japan

A review is given on some recent topics of solid state physics studies in very high magnetic fields up to a few megagauss, which are in progress at the Megagauss Laboratory in Tokyo. The topics include the infrared cyclotron resonance, the magneto-optics of excitons, the critical field determination of high T_c superconductors and magnetism.

1. INTRODUCTION

High magnetic fields provide various new possibilities for semiconductor physics as well as for all other areas of solid state physics. Since the pioneering work of Kapitza in 1924 /1/, many investigations have been made to generate high magnetic fields with a pulsed form beyond the field range which is accessible by steady fields. The so called "megagauss fields" represent the field range above 1MG (100T) which can be produced only by descructive methods. In early days, Fowler et al. reported a production of 1400T by the explosive driven magnetic flux-compression/2/. Because of the destructive nature and the short duration time of the field, experiments in megagauss fields are by no means easy. However, if we properly design the experiment, various interesting experiments become possible in the megagauss range/3/.

For the application in solid state physics, it is required to generate well controllable megagauss fields in a reproducible way. At the Megagauss Laboratory in Tokyo, three different kinds of techniques have been developed to generate very high fields/4-6/. (1) the electromagnetic flux-compression, (2) the single-turn coil technique, and (3) the non-destructive long pulse fields in the sub-megagauss range. The first technique utilizes the electromagnetic force to rapidly squeeze a metal ring (liner) to compress magnetic flux. Using large condenser banks (5MJ+1.5MJ), the maximum field of 350T has so far been obtained. The method is quite destructive because of the large energy consumption but very high fields can be obtained reproducibly. Effort to produce even higher fields is still in progress, in parallel with the applications of the fields. The second technique is a method discharging a very fast pulsed current to a small single-turn coil. Fields up to 150T in the case of a coil with the inside diameter of 10mm and up to 200T in the case of 6mm are produced before the coil is destroyed. The advantage of this technique is that the sample and the sample holder are not destroyed at the field shot, so that we can repeat the experiment on exactly the same sample and measuring system many times. The third method is a use of conventional copper wire wound pulse magnets by which long pulse fields up to 40-50T are produced. Recently, we succeeded in producing a field of 54T without destroying a magnet using a Cu wire strengthened with Nb-Ti alloy multi-fibers inside. The long pulse fields are useful for prelimenary measurements before the megagauss experiments.

Springer Series in Solid-State Sciences, Vol. 87
High Magnetic Fields in Semiconductor Physics II Editor: G. Landwehr

Table I Possible experiments of solid state physics in megagauss fields

	Electromagnetic Flux-compression	Single-turn Coil	Submegagauss long pulse
Maximum field	350T	200T	50T
Duration time	∿5 μs	5 μs	20ms
Cyclotron resonance	A	A	A
Infrared laser Spectroscopy	A	A	A
Magneto-absorption	A	A	A
Magneto-reflection	D	D	A
Luminescence	D	D	A
Faraday rotation	A	A	A
Magnetization		B	A
AC Suscepti-bility	C	C	A
DC Transport			A
AC Transport	C	C	A
Contactless transport	C	B	A
Magneto-striction			B
Microwave absortion		D	D
Ultrasonic Attenuation			D

A:Routinely possible B:Possible depending on conditions
C:Difficult but possible D:To be planned in future

The three methods are conveniently employed for various experiments. Table I summerizes the possible solid state experiments under the fields produced by these methods. In the submegagauss range, most of the solid state experiments are possible. In the megagauss range, however, some of the experiments such as DC transport or magnetization measurements are very difficult, because of the pick up of an enormous voltage induced in the lead wire loop by the large dB/dt. Optical measurements are the most convenient means because they require no lead wires. Recently, various other methods other than optical techniques have also become possible, such as AC transport, AC susceptibility, or magnetization measurements. Particularly, in the single-turn coil method, the capability of repeating the experiment in the same condition has vastly extended the range of possible experiments, since we can subtract the spurious signal which can be obtained by a shot without a sample, as will be described in Section 5. In this paper, we make a review of some highlights among the works recently done using the megagauss facilities.

2. INFRARED CYCLOTRON RESONANCE

In the megagauss range, the energy of the cyclotron motion becomes extremely large. New phenomena such as magnetic field-induced phase transitions take place at some high fields. Alloy compounds PbGeTe or PbSnTe are interesting systems in this respect, since their structural phase transition from the cubic NaCl-type structure to the rhombohedral As-type one is governed by the TO phonon softening in which the virtual inter-band transitions of the valence electrons play an important role. When the density of states near the extrema of the conduction and valence bands are modified by the high field, the transition temperature may be greatly altered/7/. Moreover, even in the end substance of this series PbTe which normally exhibits no such phase transition, the phase transition may be brought about at a finite temperature if the applied magnetic field is sufficiently high. The phase transition is accompanied by the ferroelectric phase transition. Therefore, this can be referred to as the "magnetic field-induced ferroelectric transitions".

The cyclotron resonance was measured in epitaxially grown thin crystals of PeGeTe and PbSnTe in megagauss fields produced by the single-turn coil method, using a CO_2 laser as a light source in the wavelength range 9.1-11 μm. Typical examples of the experimental traces of the cyclotron resonance are shown in Fig.1 for p-type $Pb_{1-x}Ge_xTe$ with x=0.0096 and $p=1.6\times10^{17}cm^{-3}$. The magnetic fields were applied parallel to the <111> axis. Two distinct absorption peaks are observed around 40T(PEAK 1) and 90T(PEAK 2), which are assigned as the absorption corresponding to the two effective masses in different valleys at the L-point for B//<111>. A small peak (PEAK X) is observed at the low field side of the PEAK 1 at low temperatures. This is probably a peak caused by a combined resonance $0^- - 1^+$ which becomes allowed in the low temperature phase with a lower symmetry. The appearance of the PEAK X is an evidence that the crystal is in the As-type phase. The effective masses do not show a dramatic change at the phase transition, but their temperature gradient shows a change at the transition. From

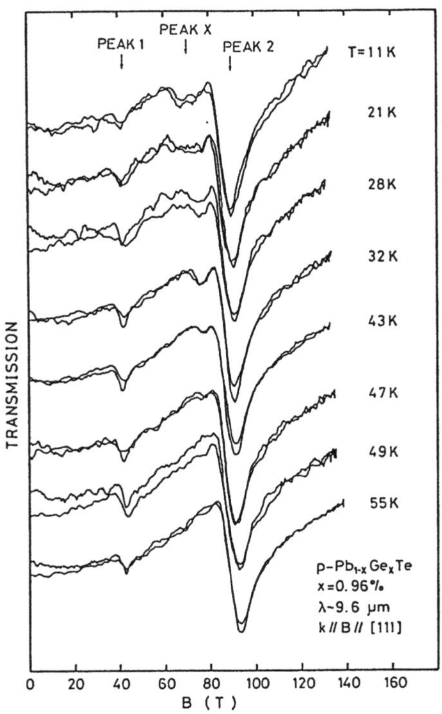

Fig.1 Cyclotron resonance spectra in p-type $Pb_{1-x}Ge_xTe$ at various temperatures.

both the PEAK X and the change of the temperature coefficient of the masses, the transition temperature T_c of this sample is estimated to be 42K at around 42T, 47K at around 78T and 48K at around 91T. The T_c at zero field was estimated to be 35K from the resistance anomaly. Therefore, the transition temperature is found to increase by 13K by the application of the high fields of 91T. In an n-type sample with x=0.0084 and n=8.7 x 10^{16} cm^{-3}, the transition temperature increase of 11K was observed at 78T.

In n-type PbTe, exploration of the field-induced transition was carried out. No sign of such phase transtion was obtained in the cyclotron resonance spectra as shown in Fig.2 within the present field range. On the other hand, a remarkable temperature dependence of the effective mass was observed as shown in Fig.3 which plots the two effective masses as a function of temperature. The heavier mass increases by a factor of 1.60 between 13K and 340K, and the lighter mass by a factor of 1.72. In this substance, it is known that the band gap has large temperature dependence which causes a large effective mass change. However, even if we take into account the total gap change, the observed mass change is much larger than theoretically predicted within the framework of the 2-band model/8/. The effect of the temperature dependence of the remote bands should be small in PbTe. The polaron contribution is much too small to account for the large change. The reason for the large temperature dependence is not clear at this moment, but it may be due to the change in the momentum matrix element. The details will be published elsewhere/9/.

The second example is the cyclotron resonance in n-type GaAs. Although the conduction band structure of GaAs is well known, it shows quite a different feature in very high fields. Figure 4 shows cyclotron resonance spectra in an epitaxial n-GaAs crystal at various temperatures. Four split absorption peaks are resolved at low temperatures, and the relative intensity of the absorption changes with temperature/10/. The four peaks

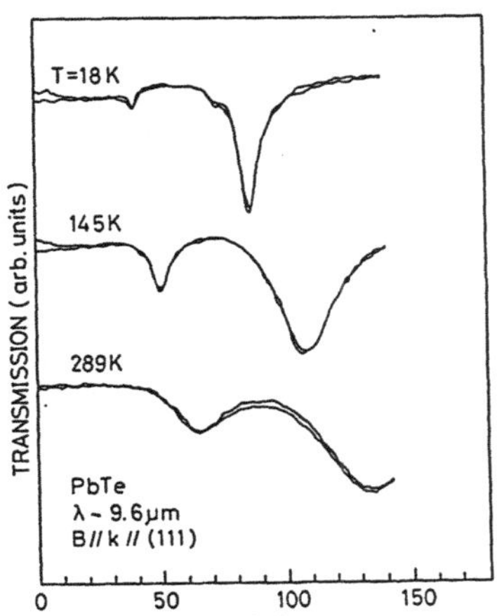

Fig.2 Cyclotron resonance spectra in n-type PbTe at various temperatures.

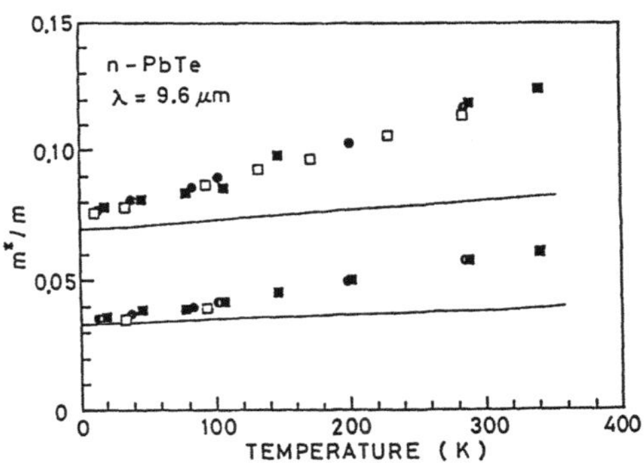

Fig.3 Temperature dependence of the two effective masses of conduction electrons in n-PbTe at $\hbar\omega$ =129meV. B//<111>. The solid lines denote the theoretical lines according to the 2-band model /8/.

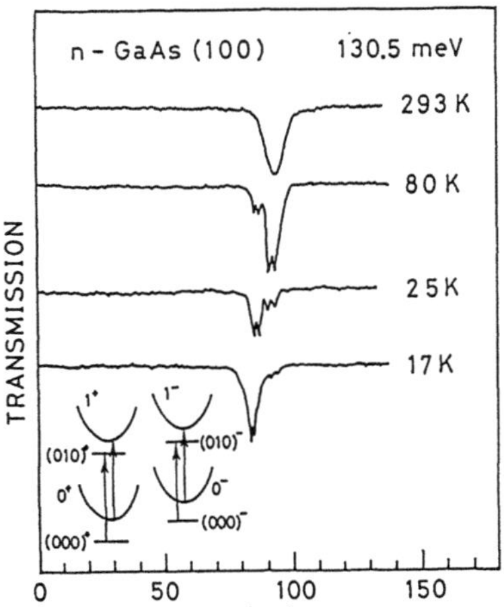

Fig.4 Cyclotron resonance spectra in n-GaAs at various temperatures. $\hbar\omega$ = 130.5 meV. The inset at the left bottom shows the electron transitions between the Landau levels and impurity levels corresponding to the observed four absorption peaks. B//<100>.

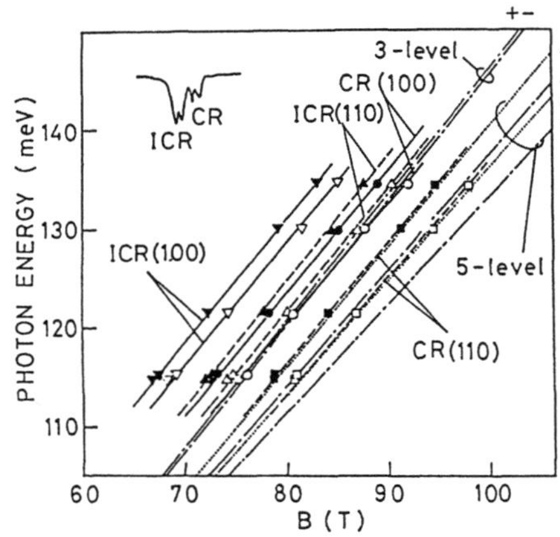

Fig.5 Resonant photon energy versus magnetic field for the four peaks seen in Fig.4. The data for the two magnetic field directions B//<100> and B//<110> are shown. The solid and broken lines were drawn to smoothly connect the data for the guide of eyes. The theoretical lines according the to 3-level model /11/ and the 5-level model /12/. For the latter, the dotted line denotes B//<100>, and the dot-broken line B//<110>.

are assigned as corresponding to the cyclotron resonance and the transition between the donor states (000)-(010) ($1s-2p^+$) as shown in the inset of Fig.4. The fine structure arises from the difference in the transition energy between the two different spin levels because of the band non-parabolicity. The resonant photon energy of the four peaks is plotted in Fig.5 as a function of magnetic field. The cyclotron resonance shows a large non-parabolicity. The effective mass is around $0.08m_0$ in contrast to $0.068m_0$ at the band edge. In addition, there appears a large anisotropy in the conduction band which is usually considered to be spherical in the low field region: i.e. the mass is 5-6% larger for the case B//<110> than for B//<100>. Theoretical lines of the resonant energy are shown in Fig. 5. The 3-level model/11/ explains the overall non-parabolicity, but not the large spin splitting nor the anisotropy. Recently, the spin splitting due to the non-parabolicity was found to increase with increasing field in lower field range below 18T/12/. The increase was ascribed to the contribution of the Γ_7 and Γ_8 bands located above the Γ_6 conduction band. The 5-level model calculation for the Landau levels was carried out including these higher lying bands/10/. As is shown in Fig.5, the 5-level model elucidates the magnitude of the spin-splitting and the anisotropy, but there is a considerable discrepancy in the absolute values of the effective mass, suggesting that the k·p-type theory may be less valid in such high fields. The information was obtained also about the donor binding energy from the impurity peaks.

3. MAGNETO-OPTICS OF EXCITONS

The excitons in very high magnetic fields are of interest from the view point of the fundamental problem concerning a hydrogen atom in high fields, as well as problems concerning the specific properties of each substance. By means of the streak spectrometer using an image-converter camera/14/, magneto-optical spectra of excitons have been investigated in AlGaAs-AlAs quantum wells/15/, anthracene/16/ and so on. Here we present the recent results on Cu_2O and PbI_2.

The so called yellow series excitons in Cu_2O are known to be typical Wannier excitons. The excitons are associated with a direct forbidden transition from the Γ_7^+ valence band to the Γ_6^+ conduction band, so that the exciton series start from N=2. Figure 6 shows an example of the streak photograph of the magneto-optical spectra of N=2 and N=3 excitons in megagauss fields obtained using the single-turn coil technique. The N=2 and N=3 lines are found to shift to the short wavelength side. Moreover, the N=2 line shows a Zeeman splitting. It is for the first time that the Zeeman splitting and the shift of the N=2 line were detected directly. The analysis revealed that the N=2 line shows a quadratic diamagnetic shift against the field up to 40T, and that the amount of the shift and the g-factor are consistently interpreted with lower field measurements/17/. While the N=2 line did not show linear field dependence up to the highest field, the N=3 line tended to approximately the linear dependence above about 50T where the amount of the shift becomes nearly equal to the binding energy. These results suggest that the hydrogen model is applicable to the yellow series excitons in Cu_2O. Details will be published shortly/18/.

Layered crystal PbI_2 has been also known to exhibit prominent exciton absorption peaks. In the case of PbI_2, however, the character of the excitons are less understood because of the relatively small exciton radius(less than 20 A). Moreover, the existence of several polytypes makes the analysis of the data more difficult. In the lower field range below 18T, no diamagnetic shift was observed for N=1-3 exciton lines/19/. In the megagauss fields, a diamagnetic shift of 5.4meV was observed at 100T for an evaporated sample/20/. Recently, we grew good quality epitaxial single crystals by a hot wall technique and a bulk single crystal by a zone refining method/21/. The former has the 4H and the latter has the 2H polytype. In Fig.7, an example of the streak spectra of the magneto-absorption of the ground state excitons in 4H-PbI_2 for B//c /22/. Circularly polarized light was employed to see the dichroism. The sharpness of the exciton lines in these good quality crystals together with the use of the polarized light and an improvement of the measuring system allowed a more accurate measurement than the previous one in the megagauss fields. The data clearly shows the polarization dependent diamagnetic shift due to the Zeeman splitting. The photon energy of the absorption line for both polarizations is plotted as a function of magnetic field in Fig.8. The results on a 2H-type sample at lower fields are also plotted for comparison. From the splitting between the two polarizations, the g-factor was obtained as g=1.1. The diamagnetic shift defined as the midpoint between the two split components shows almost a quadratic dependence on the field, and its magnitude is E_d=6.8 x 10^{-4} B^2meV. The lower field measurement gave the values g=0.89 and E_d=9.7 x10^{-4} B^2meV. The origin of the difference between the data is not clear at this moment. It is probably due to the difference in the polytype. For $B \perp c$ with a Voigt configuration, a smaller g and E_d were observed. Such an anisotropy is of interest from a view point of the magneto-optical effect of the excitons which have an intermediate character between the typical Wannier exciton and the Cationic type small radius exciton as in BiI_3/23/.

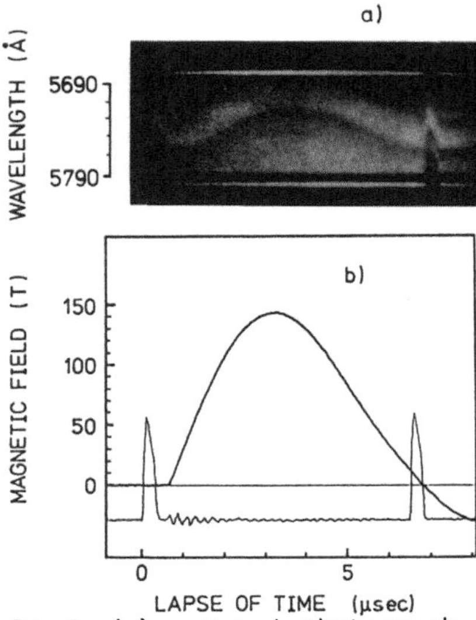

Fig.6 (a) a streak photograph of the magneto-optical spectra of the N=2 excitons in Cu_2O. The N=3 excitons line is also partly seen. T=25.9K. (b) Magnetic field pulse, and the time marker signal which deflects the streak picture at the both edges.

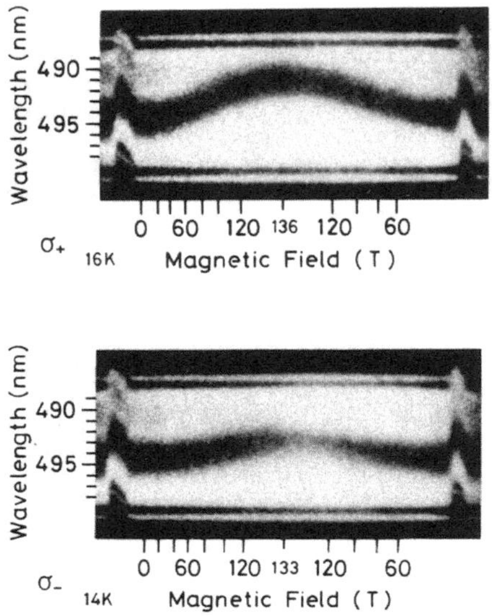

Fig.7 Magneto-optical spectra of the ground state excitons in 4H-PbI_2, which are grown on a CdI_2 substrate by the hot wall epitaxy. B//c. σ_+ and σ_- denote the circular polarization.

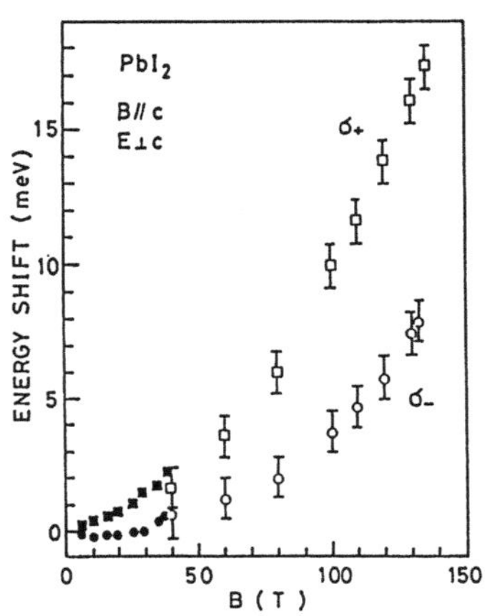

Fig.8 Magnetic field dependence of the energy shift of the ground state exciton in PbI_2 for the two circular polarizations. B//c. The open points stand for the data of 4H-crystal in the megagauss fields and the closed points of 2H-crystal in the long pulse.

4. UPPER CRITICAL FIELD OF HIGH T_c SUPERCONDUCTORS

The recently discovered high T_c oxide super conductors/24,25/ exhibit not only high critical temperatures but also high upper critical field H_{c2}. By extrapolating the data obtained in the non-destructive fields up to 40T/26/, the upper critical field H_{c2} at T=0 is estimated to be 240T for B⊥c and 68T for B//c on the basis of the conventional theory of the dirty superconductors/27/. For constructing the phase diagram of this class of superconductors, the measurements in megagauss fields are needed. The DC magnetoresistance measurement in megagauss fields is extremely difficult because of the large induced voltage by the field. Moreover, it is rather difficult to obtain sufficiently good contacts on the sample for the pulsed measurements. Several different techniques have been employed to measure H_{c2}.

The first method is to measure the derivative of the magnetization. The type II superconductors generally show a hysteresis in the magnetization curve. This is due to the flux-pinning. According to the critical state model/28/, the magnitude of the hysteresis is proportional to the critical current. When the magnetic field sweep changes the sign, a sudden jump occurs because of the hysteresis if the sample is superconducting. As a result, the time derivative signal of the magnetization $\dot{M}$ should show a sharp pulse at the top of the magnetic field. When the sample is in the normal state, such a pulse would not be observed. Therefore, by measuring the pulse signal in $\dot{M}$ at the field top, we can see how the superconductivity approaches to the normal state with increasing field. In sintered samples of $YBa_2Cu_3O_{7-x}$, we have observed the pulse signal at the top of the field up to 101T at 6K /29/. This result provides an evidence that the superconductivity in this material persists in megagauss fields. Figure 9 shows an example of the signal of $\dot{M}$ for a $YBa_2Cu_3O_{7-x}$ single crystal in very high fields up to 87T at the initial temperature of 6K. The fields were produced by the single-turn coil technique. A pronounced pulse can be seen at the top field, indicating that the sample is in the superconducting state. Figure 10 depicts the magnetic field dependence of the magnitude of the hysteresis ΔM thus obtained/30/. In the low field region, ΔM is larger for B//c than for $B_{\perp}c$, because ΔM is proportional to the critical current J_c, and J_c is lager for B//c. With increasing field, however, ΔM decreases more rapidly for B//c than for $B_{\perp}c$ indicating the

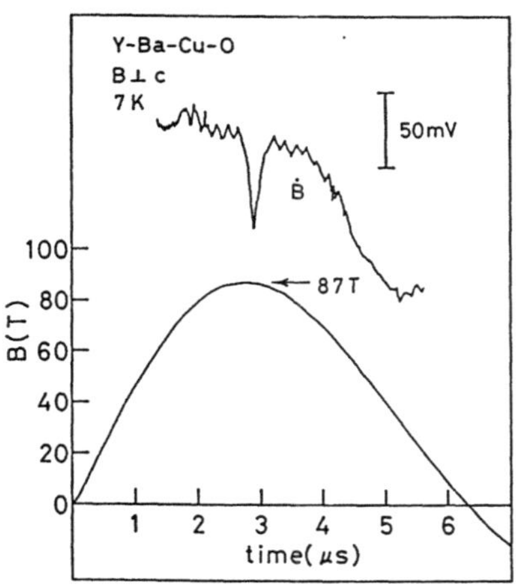

Fig.9 The signal of dB/dt for a single crystal of $YBa_2Cu_3O_{7-x}$ in high pulsed fields by the single-turn coil technique. $B_{\perp}c$. T=7K. A sharp pulse is seen at the top of the field.

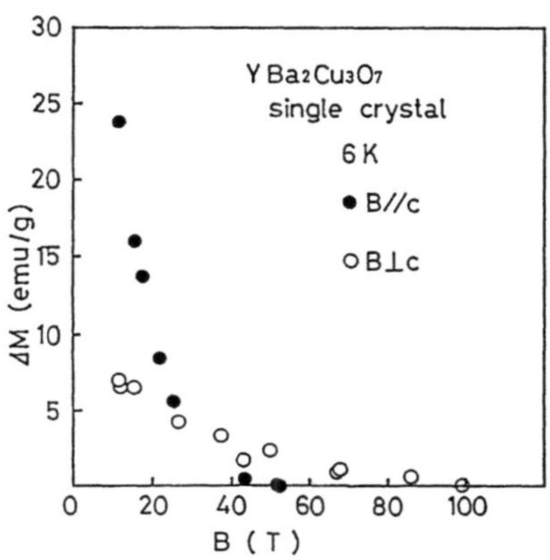

Fig.10 Plot of the magnitude of the magnetization hysteresis of a single crystal $YBa_2Cu_3O_{7-x}$ as a function of magnetic field B. T= 6K.

lower H_{c2}. The highest field at which the hysteresis is still recognizable is 101T $B_{\perp}c$ and 52T for B//c. We are not sure yet whether these fields really correspond to H_{c2}, because of the temperature rise due to the hysteresis and unknown effects of the large dB/dt. However, they provide at least an evidence that the sample is superconducting even under the field above 100T.

Another method for measuring H_{c2} is the contactless AC transmission technique/31/. As shown in the inset of Fig.11, a pair of coils is set on both sides of a thin sample. When an AC current with a frequency of 10-50MHz is sent to one of the coils, the other coil detects a signal whose magnitude depends on the sample resistivity. When the sample is superconducting, almost no signal is detected. When the sample has a finite resistance, the resistivity can be calculated from the observed transmission. Thus this arrangement is a very sensitive means for observing the superconducting transition. Indeed, in the submegagauss range, the transition can be observed with a much better signal to noise ratio than by the DC measurement/31/. Figure 11 shows the transmission of the AC signal for a sintered sample of $YBa_2Cu_3O_{7-x}$ as a function of field at 3 different initial temperatures in the megagauss fields produced by the single-turn coil technique. The signal in the rising field shows an oscillatory noise caused by the large noise at the start of the pulse, but the signal in the falling field shows an excellent magnetoresistance curve. The hysteresis between the sweeps in the rising and falling fields at T_i=78K and 100K is probably due to the temperature rise caused by the Joule heating at the grain boundaries and the magnetization hysteresis. Other methods such as the measurements of the AC magnetoresistance or the infrared transmission would be also useful for the determination of H_{c2}.

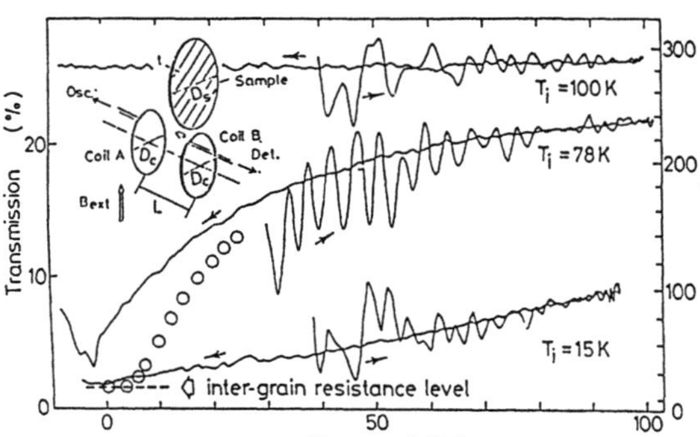

Fig.11 The transmission signal (left scale) and the calculated resisitivity (right scale) of a sintered sample of $YBa_2Cu_3O_{7-x}$ obtained by the AC transmission technique. The frequency of f=22 MHz was used. The sample thickness was 0.16 mm. The circles indicate isothermal change of the resistivity at T=86K obtained by the long pulse field measurement.

5. MAGNETIZATION MEASUREMENTS

The magnetization measurement in megagauss fields is not an easy task because in the measurement using the usual induction method, the external field is much larger than the magnetization, thus dominating the magnetic induction. The Faraday rotation is a convenient alternative method to measure the magnetization, because it reflects the magnetization of the sample. Here we discuss two examples of the study of the magnetization by the Faraday rotation on YIG and $Cd_{1-x}Mn_xTe$.

YIG is a prototype of the family of iron garnet crystals which are typical Neel ferrimagnets. Due to the strong exchange interaction between the 2 sublattices of Fe^{+3} ions which occupy the a- and d-sites, YIG has a high Neel temperature of 559K. When we supply sufficiently high external fields, the a-site sublattice which are normally antiparallel to the d-site sublattice starts to flip towards the direction of the external field. As a result, the spin-cant phase is realized between the ferri-magnetic and paramagnetic phases/32/. This spin-flip transition occurs at a comparable field with the internal field due to the exchange interaction. Therefore, the observation of the spin-flip transition provides the most accurate means to determine the exchange interaction constants over which there has been considerable controversy among the data/32/. Figure 12 shows an experimental trace of the Faraday rotation of YIG at T=158K in megagauss fields produced by the flux-compression. The Faraday rotation signal shows a distinct kink at 250T which corresponds to the transition from the collinear phase to the spin-cant phase. From the position of the kink, the exchange constants are obtained as $J_{ad} = -25.6\ cm^{-1}$ and $J_{aa} = -6.2\ cm^{-1}$/33/.

Diluted magnetic semiconductors $Cd_{1-x}Mn_xTe$ show a paramagnetic behavior when x is small. As x is increased, the antiferromagnetic exchange interaction between the randomly distributed Mn^{2+} ions introduces a complexity in the magnetic properties. For x>0.1, the magnetization is expressed as a sum of the small Brillouin function-type component and a linear component. With increasing x, the linear part becomes dominant and the magnetization never shows a saturation in the lower field measurement below 40T/34/. As a US-Japan Cooperative Project of NSF and JSPS, the spin-alignment experiment in $Cd_{1-x}Mn_xTe$ was carried out by the MIT group using the single-turn coil technique for x=0.1-0.6 /35/. Figure 13 shows the Faraday rotation for a sample with x=0.1. The rotation increases rapidly

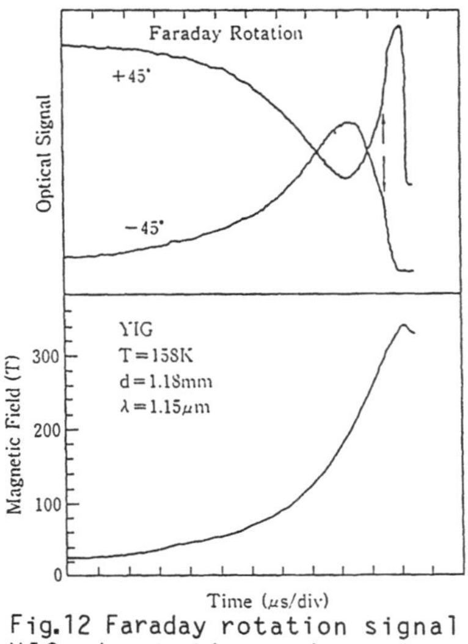

Fig.12 Faraday rotation signal for YIG obtained in the megagauss fields produced by the flux-compression. Two traces for different analyzer angles are shown. At 250T, kinks are observed corresponding to the spin-flip transition. λ =1.15 μm.

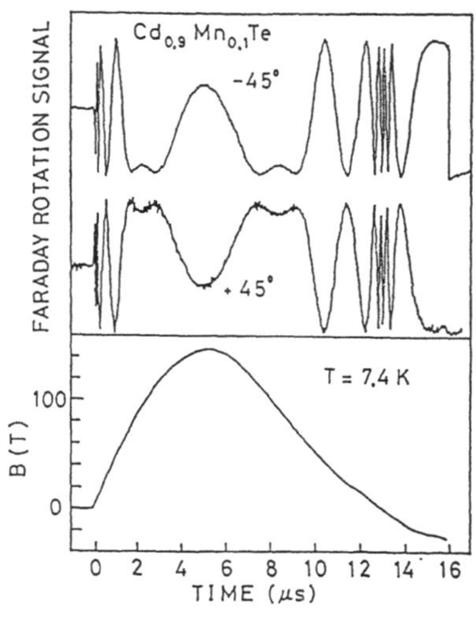

Fig.13 Faraday rotation signal for $Cd_{0.9}Mn_{0.1}Te$. Two traces for different analyzer angles are shown. λ =1.15 μm. T = 7.4 K.

with field in the low field region, but it saturates and starts to decrease at the higher field region. The decrease of the Faraday rotation is due to the ordinary diamagnetic component which CdTe possesses. Compensating the diamagnetic part, it was found that the magnetization almost saturates at 80T for x=0.1 and at 130T for x=0.2. For x>0.3, the magnetization shows no saturation up to 150T, indicating a necessity of higher fields.

In addition to the Faraday rotation, a technique for measuring the magnetization by the direct induction method has also been developed using the single-turn coil method/36,37/. The single-turn coil method has the advantage that we can repeat the experiment with the same pick up system, so that we can make a compensation for the large spurious signal very accurately. The system was successfully employed in the measurement of the metamagnetic magnetization of a 1-dimensional Ising antiferromagnet $CsCoCl_3$/37/.

The author is obliged to the following collaborators for the contributions discussed in this paper: Drs. T. Goto, K. Nakao, S. Takeyama, T. Sakakibara, T. Haruyama, T. Kikuchi, Y. Nagamune, H. Yokoi, K. Tatsuhara, S. P. Najda, W. Zawadzki, G. Bauer, A. Misu, K. Kanisawa, M. Kobayashi, S. Uchida, H. Takei, K. Amaya, D. Heiman and E. Isaacs.

REFERENCES

1. P. Kapitza: Proc. Roy. Soc. A105, 691 (1924), A115, 658 (1927)
2. C. M. Fowler, W.B. Garn and R.S. Caird: J. Appl. Phys. 31, 588 (1960)
3. N. Miura and F. Herlach: In Strong and Ultrastrong Magnetic Fields ed. by F. Herlach (Springer, Berlin, Hidelberg, 1985) p.247
4. N. Miura, T. Goto, K. Nakao, S. Takeyama, T. Sakakibara and F. Herlach: J. Mag. Mag. Mater., 54-57, 1409 (1986)
5. T. Goto, N. Miura, K. Nakao, S. Takeyama and T. Sakakibara: In Megagauss Technology and Pulsed Power Applications ed. by C.M. Fowler, R.S. Caird and D.J. Erickson (Plenum Pub. Co., 1987) p.149
6. N. Miura, T. Goto, K. Nakao, S. Takeyama and T. Sakakibara: To be published in Physica B
7. S. Takaoka and K. Murase: Phys. Rev. B20, 2823 (1979)
8. G. Appold, R. Grisar, G. Bauer, H. Burkhard, R. Ebert, H. Pascher and H.G. Hafele: In Proc. 14th Int. Conf. Phys. Semiconductors ed. by B.L.H. WILSON (The Institute of Physics, 1979) p.1101
9. H. Yokoi et al.: To be published
10. S. P. Najda, S. Takeyama, N. Miura, W. Zawadzki and P. Pfeffer: To be published in Proc. 19th Int. Conf. Phys. Semiconductors (Warsaw, 1988)
11. B. Lax, J.G. Mavroides, H.J. Zeiger and R.J. Keyes: Phys. Rev. 122, 31 (1961)
12. H. Sigg, J.A.A.J. Perenboom, P. Pfeffer and W. Zawadzki: Solid State Commun. 61, 685 (1987)
13. U. Rossler: Solid State Commun. 49, 943 (1984)
14. G. Kido, N. Miura, H. Katayama and S. Chikazumi: J. Phys. E14,349 (1981)
15. N. Miura, S. Takeyama and Y. Iwasa: In Proc. 18th Int. Conf. Phys. Semiconductors, ed. by O. Engstrom (World Scientific Pub. Co., Singapore, 1987) p.715
16. S. Takeyama, M. Kobayashi, A. Matsui, K. Mizuno and N. Miura: In High Magnetic Fields in Semiconductor Physics ed. by G. Landwehr (Springer, Berlin, Heidelberg, 1987) p.536

17. A. Misu, Y. Takada and H. Fukutani: In High Field Magnetism ed. by M. Date (North Holland, Amsterdam, 1983) p.205
18. M. Kobayashi, K. Kanisawa, A. Misu, Y. Nagamune, S. Takeyama and N. Miura: To be published in J. Phys. Soc. Jpn.
19. M. S. Skolnick, L. C. Tanh, F. Levy and G. Harbeke: Physica 89b,143 (1977)
20. N. Miura, G. Kido, H. Katayama and S. Chikazumi: J. Phys. Soc. Jpn. 49 Suppl. A 409 (1986)
21. Y. Nagamune, S. Takeyama, N. Miura, A. Misu and T. Minagawa: Appl. Phys. Lett. 50, 1337 (1987)
22. Y. Nagamune et al.: To be published
23. T. Komatsu, Y. Kaifu, S. Takeyama and N. Miura: Phys. Rev. Lett. 58, 2259 (1987)
24. J. G. Bednorz and K. A. Muller: Z. Phys. B64, 189 (1986)
25. S. Uchida, H. Takagi, K. Kitazawa and S. Tanaka: Jpn. J. Appl. Phys. 26, L1 (1987)
26. T. Sakakibara, T. Goto, Y. Iye, N. Miura, H. Takeya and H. Takei: Jpn. J. Appl. Phys. 11, L1892 (1987)
27. N. R. Werthhamer, E. Helfand and P. C. Hohenberg: Phys. Rev. 147, 295 (1966)
28. C. P. Bean: Rev. Mod. Phys. 36, 31 (1964)
29. K. Nakao, N. Miura, K. Tatsuhara, S. Uchida, H. Takagi, T. Wada and S. Tanaka: Nature 332, 816 (1988)
30. K. Nakao, N. Miura, K. Tatsuhara, S. Uchida, H. Takagi, T. Wada and S. Tanaka, H. Takeya and H. Takei: To be published in Physica B
31. T. Sakakibara, T. Goto and N. Miura: To be published in Physica B
32. N. Miura, I. Oguro and S. Chikazumi: J. Phys. Soc. Jpn. 45, 1534 (1978)
33. K. Nakao, T. Goto and N. Miura: To be published in Proc. ICM 1988
34. D. Heiman, E. D. Isaacs, P. Becla and S. Foner: Phys. Rev. B35, 3307 (1987)
35. E. D. Isaacs, D. Heiman, X. Wang, P. Becla, K. Nakao, S. Takeyama and N. Miura: To be presented at the Int. Conf. Application of High Magnetic Fields in Semiconductor Physics (Wurzburg, 1988)
36. S. Takeyama, K. Amaya, T. Nakagawa, M. Ishizuka, K. Nakao, T. Sakakibara, T. Goto, N. Miura, Y. Ajiro and H. Kikuchi: J. Phys. E 21, 1025 (1988)
37. K. Amaya, S. Takeyama,T. Nakagawa, M. Ishizuka, K. Nakao, T. Sakakibara, T. Goto, N. Miura, Y. Ajiro and H. Kikuchi: To be published in Physica B

Recent Semiconductor Work at the Francis Bitter National Magnet Laboratory

P.A. Wolff

Francis Bitter National Magnet Laboratory, MIT, Cambridge, MA 02139, USA

I. Introduction

For the past five years, the research program of the FBNML semiconductor group has focused on diluted magnetic semiconductors (DMS). Topics of special interest include:

- Bound magnetic polarons (BMP).
- Magnetization steps and Mn^{2+}-Mn^{2+} interactions.
- The metal-insulator transition.
- Ordered DMS.

Work along all of these lines is continuing. The early DMS work at FBNML involved Mn-alloyed II-VI materials. Recently, the program has broadened to include stannites and Fe-alloyed II-VI's. Crystals for this research have been provided by Prof. A. Wold (Brown), Dr. P. Becla (FBNML), and Profs. R. Galazka and A. Mycielski (Warsaw). Their support has been invaluable.

The BMP has been an important theme of FBNML DMS work since its inception. Our goals are to understand BMP energetics and kinetics, to demonstrate the transition from fluctuation-dominated to collective behavior of the BMP, and to develop DMS with stronger polaron interactions. These studies have benefitted from parallel investigations of the magnetic properties of DMS. In particular, the discovery [1] of magnetization "steps" made possible a detailed characterization of the antiferromagnetic Mn^{2+}-Mn^{2+} interactions in materials such as $Cd_{1-x}Mn_xSe$, $Cd_{1-x}Mn_xTe$, etc. It is now clear that these interactions seriously impede BMP formation for $x > 0.1$. The solution to this problem, in our view, is the development of ordered DMS — without nearest neighbor magnetic ions. Work on the stannite crystal, Cu_2MnGeS_4, was motivated by the possibility of testing this idea.

Polish research [2] has shown that Fe-alloyed DMS differ in important ways from their Mn-alloyed counterparts. Our work extends this trend. For example, spin-flip Raman scattering (SFRS) studies of n-CdFeSe show no evidence of the BMP, whereas it is a prominent feature of SFRS spectra of n-CdMnSe. In narrower gap HgCdSe:Fe, we observe striking nonlinear optical effects caused by the Fe^{2+} resonance within the conduction band [3]. Mn^{2+} ions cannot cause such effects because their d-levels are far from the band edges of II-VI semiconductors.

The DMS experiments outlined below were performed in external magnetic fields ranging from $B = 0$ to $B > 100$ T. However, even when $B = 0$, they are essentially high field experiments because the internal (exchange) fields of DMS are in the 10-100 T range.

Springer Series in Solid-State Sciences, Vol. 87
High Magnetic Fields in Semiconductor Physics II Editor: G. Landwehr

II. Collective BMP Regime in CdMnSe

The SFRS experiments of NAWROCKI et al. [4] provided the first evidence for the existence of BMP in DMS. This technique is still one of the best for probing [5-7] their microscopic structure. A striking feature of Nawrocki's experiment was the observation of a zero-field spin splitting of the donor-BMP in n-CdMnSe. By contrast, in conventional semiconductors the spin splitting is strictly linear in applied field. DIETL and SPALEK [8] subsequently demonstrated that this unexpected effect is a result of magnetization fluctuations within the finite Mn^{2+} spin cloud sampled by the donor-bound electron. Although, in the absence of a magnetic field the time-averaged magnetic moment of the Mn^{2+} ions within the electron orbit is zero, it fluctuates about that value with root-mean-square amplitiude $\simeq \mu_B N^{\frac{1}{2}}$, where N ≃ 200 is the number of Mn^{2+} within the orbit. The electron spin, which relaxes rapidly, aligns with this fluctuating internal field to cause the zero field splitting. As a purely statistical effect, the fluctuation-induced splitting was expected to be independent of temperature, as observed in n-CdMnSe for T > 2 K.

At lower temperatures, the BMP theory [5,8] predicts a gradual evolution into a collective regime, wherein the Mn^{2+} spins are fully aligned by the carrier-Mn^{2+} exchange interaction to give a net moment of 50-100 μ_B. This is the true BMP state. There were hints in early SFRS measurements [5] of the beginning of BMP formation at 2 K in n-CdMnSe, but it was then believed that antiferromagnetic (AF) Mn^{2+}-Mn^{2+} interactions would prevent further BMP development at lower temperatures. However, subsequent work [9,10] showed that the magnetic susceptibility of CdMnSe increases by an order of magnitude between T = 2 K and 0.1 K. These observations suggested lower temperature SFRS experiments to observe BMP formation.

Figure 1 shows the zero field splitting of the donor-BMP in $Cd_{0.9}Mn_{0.1}Se$ as measured by ISAACS et al. [11] in the temperature range T = 130 mK to 30 K. The 130 mK experiment is, to our knowledge, the lowest temperature Raman scattering experiment ever performed. These measurements were done with a novel optical fiber system [12] that guides laser light into a dilution refrigerator, suppresses elastically scattered laser radiation, and transmits Raman light to a spectrometer via a second fiber. The experiments were performed with a dye laser tuned just below the band gap of $Cd_{0.9}Mn_{0.1}Se$ to enhance the Raman cross section [5]. Laser powers below 20 μW were required to prevent sample heating at the lowest temperatures. The data clearly demonstrate the transition from fluctuation-dominated to collective behavior at low temperatures. Though continuous — as required

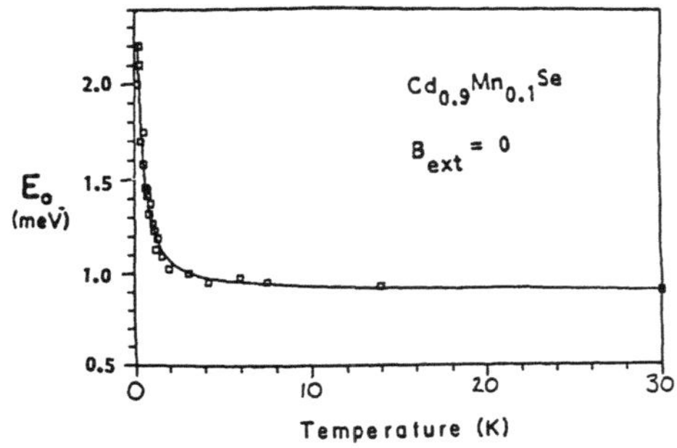

Fig. 1 Spin flip energy vs. temperature for donor-BMP in zero magnetic field

in a finite system — this transition is still surprisingly abrupt. It is well described by the theory [5,8], whose predictions are illustrated by the solid curve in Fig. 1. This fit required knowledge of the Mn^{2+} susceptibility; it was calculated from a cluster model involving both nearest neighbor and next-nearest neighbor Mn^{2+}-Mn^{2+} interactions. ISAACS et al. [11] also observe a substantial decrease of the spin-flip line width — from Γ = 0.75 meV to 0.38 meV — as the BMP enters the collective regime.

III. High Field SFRS

The development of the fiber optic Raman setup has made possible scattering experiments to the highest dc fields, B > 30 T, available at FBNML. Figure 2 shows the magnetic field variation of the spin-flip energy in (CdMn)Se, as measured by ISAACS et al. [13] between T = 0.5 K and 1.5 K. Similar results were obtained in (CdMn)Te. The data clearly demonstrate the magnetization "steps" caused by field-induced breaking of the AF alignment of nearest neighbor Mn^{2+}-Mn^{2+} pairs [1]. At the first, 12.5 T step, the S=1, S_z=1 state of the pair falls below the S=0 state; the second step at 24 T results from a crossing of S=2, S_z=2 and S=1, S_z=1 states. In this field range BMP effects are suppressed, so mean field theory can be used to infer the nearest neighbor Mn^{2+}-Mn^{2+} exchange constants. They are best estimated from the difference of the fields at the two steps via the expression $2|J_{NN}| = g\mu_B(B_2-B_1)$. This procedure cancels perturbations of the step positions caused by next-nearest neighbor Mn^{2+}-Mn^{2+} interactions. The experiments give

$$J_{NN}(\text{CdMnSe})/k_B = -7.5 \pm 0.3\ \text{K}$$

$$J_{NN}(\text{CdMnTe})/k_B = -6.1 \pm 0.2\ \text{K}.$$

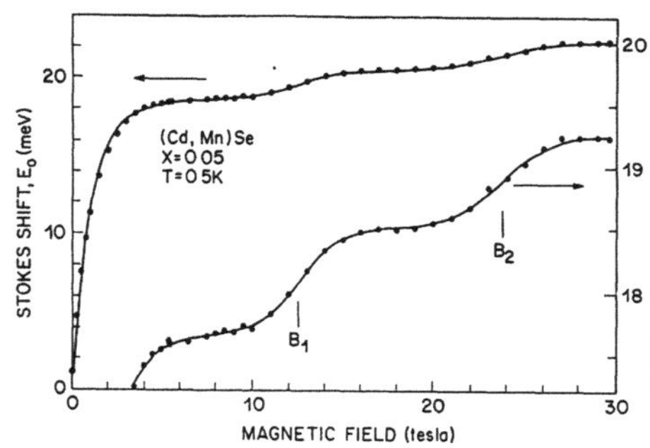

Fig. 2 Stokes shift (Zeeman energy) vs. field in $Cd_{0.95}Mn_{0.05}Se$

IV. BMP in Concentrated Mn-Alloys

Though elegant and intellectually satisfying, the SFRS results outlined above do not give much comfort to an engineer hoping to employ magnetic polarons in devices. To increase BMP energies, and the temperatures at which they form, it is advantageous to use p-type materials and more concentrated Mn-alloys. For example, NHUNG et al. [14] find $E_{BMP} \simeq$ 20 meV for the acceptor-BMP in $Cd_{0.95}Mn_{0.05}Te$; this magnetic energy is twenty times that of the donor-BMP in n-CdMnSe because of the larger exchange constant and more compact acceptor orbit. The simplest theory, which ignores Mn^{2+}-Mn^{2+} interactions, predicts a further linear increase of E_{BMP} with Mn content. Experimentally, however, the polaron energy increases less rapidly with x-value and saturates for $x \approx 0.2$. These effects result from competition between the s-d interaction, which favors ferromagnetic alignment of the Mn^{2+} spins within the BMP, and the antiferromagnetic Mn^{2+}-Mn^{2+} interaction. Two years ago we presented [15] a phenomenological theory of the acceptor-BMP that used experimental magnetization data (to 40 T) to determine the effective exchange potential experienced by the hole. The resulting nonlinear Schrödinger equation was solved numerically to determine BMP energies and wave functions. Two general conclusions emerged from this theoretical work:

i) acceptor-BMP energies saturate at 40 meV for $x \approx 0.2$ in random Mn-alloys.

ii) there is a possibility of achieving much larger polaron energies in ordered DMS, wherein Mn^{2+} ions do not occupy nearest neighbor sites. An acceptor-BMP energy of 240 meV was estimated for $(Cd_3Mn)Te_4$ in the famatinite structure.

To test these ideas, we have extended previous measurements of acceptor-BMP energies in $Cd_{1-x}Mn_xTe$, and performed magnetization studies of the stannite crystal, Cu_2MnGeS_4, that is an ordered, tetrahedrally coordinated analogue of $(Cd_3Mn)Te_4$. Acceptor-BMP energies were measured [16] by photoluminescence studies of intentionally-doped, high purity, p-type $Cd_{1-x}Mn_xTe$ samples. In these crystals the free electron-neutral acceptor (e,A^0) transition dominates DAP luminescence, and provides a sufficiently sharp feature to determine acceptor-BMP energies within ±7 meV. Figure 3 illustrates the temperature variation of E_{BMP} compared to that of the band gap for several samples; the x-value variation of the 5 K binding energy is shown in Fig. 4. The solid curve is the prediction of the phenomenological theory [16]. Magnetization data for this fit were extended to 150 T in a collaborative program [17] with ISSP.

For x-values above $x = 0.15$, acceptor-BMP energies in $Cd_{1-x}Mn_xTe$ are drastically reduced by the Mn^{2+}-Mn^{2+} interaction. The materials are then approaching the spin glass range and become quite difficult to magnetize; for example, $Cd_{0.75}Mn_{0.25}Te$ has a saturation field of 130 T [17]. Since BMP energies scale with the magnetic susceptibility, it is not surprising that they are suppressed. This effect is a result of the large, nearest neighbor Mn^{2+}-Mn^{2+} exchange interaction ($J_{NN}/k_B \approx -6$ K in CdMnTe). It is known, however, that next-nearest neighbor interactions are much weaker. Thus, one anticipates larger susceptibilities — and stronger BMP effects — in crystals without nearest neighbor Mn^{2+}. To test this idea, Prof. Wold's group has begun a program of stannite crystal growth [18]. These materials are ordered, tetrahedrally-coordinated analogues of the Mn-alloyed II-VI's. The particular material studied, $(Cu_2MnGe)S_4$, was selected for ease of growth and the simplicity of its phase diagram. Samples were grown by chemical vapor transport with iodine [18]; excellent, deep red, high optical quality crystals were obtained. Figure 5 illustrates the stannite crystal structure. For our purposes, its crucial feature is the placement of the Mn^{2+} ions. They occupy only one of the tetrahedral sites adjacent to each sulfur ion. Hence, in $(Cu_2MnGe)S_4$ the nearest neighbor Mn^{2+}-Mn^{2+} distance is $\sqrt{2}$ times that in the alloy CdMnTe.

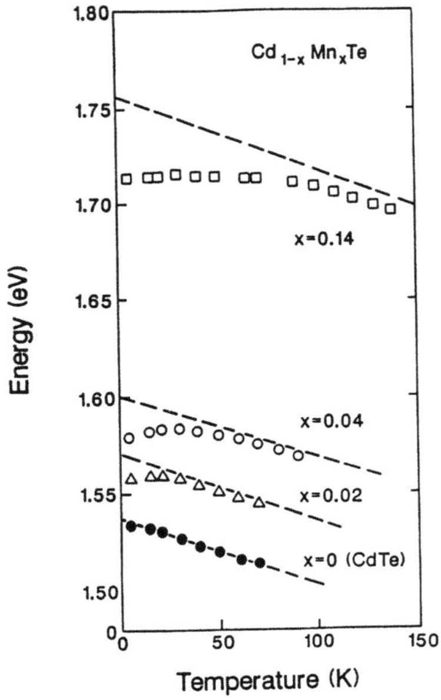

Fig. 3 Energies of acceptor-related (e,A°) transitions vs. temperature

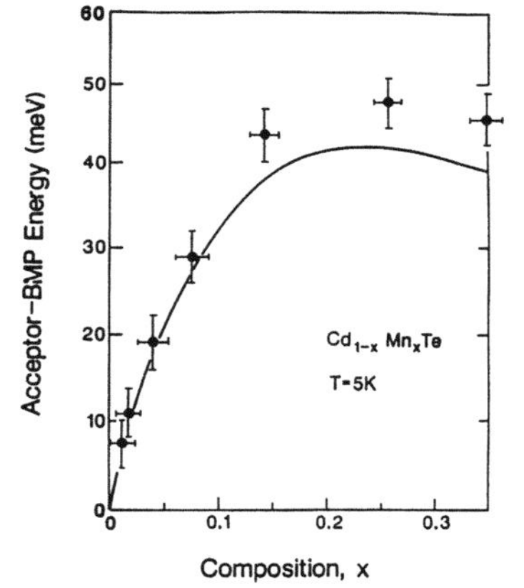

Fig. 4 Composition dependence of acceptor-BMP energy. Solid line is the phenomenological theory

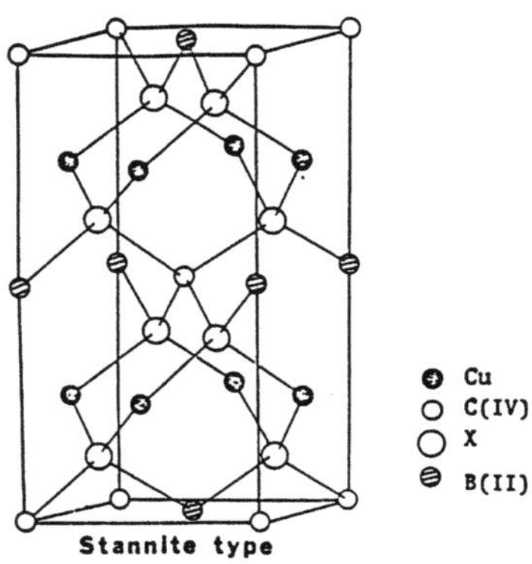

Fig. 5 Stannite crystal structure

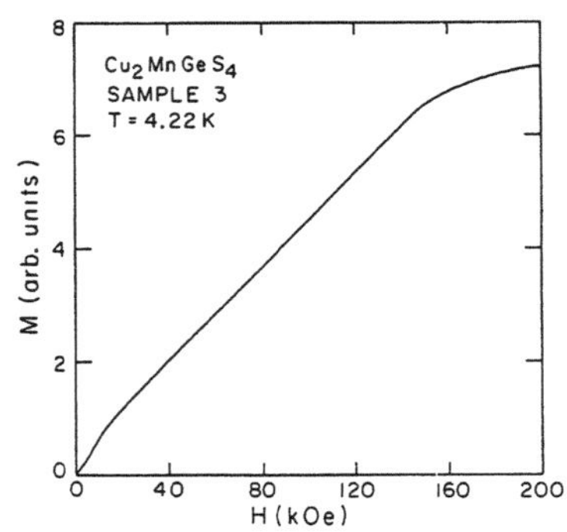

Fig. 6 Magnetization vs. field for Cu_2MnGeS_4

Assuming exchange constants that vary in a comparable way with distance, one thus expects $(Cu_2MnGe)S_4$ to have a susceptibility 5–10 times larger than that of the alloy $Cd_{0.75}Mn_{0.25}Te$ — with the same Mn^{2+} concentration. This hypothesis was confirmed by magnetization measurements of SHAPIRA et al [19], shown in Fig. 6. Below saturation the slope of the 4 K magnetization curve is seven times that of $Cd_{0.75}Mn_{0.25}Te$; the saturation field of $(Cu_2MnGe)S_4$ is 17 T compared to 130 T in $Cd_{0.75}Mn_{0.25}Te$ [17].

In Fig. 6 there is a small, low field region within which the magnetization increases more rapidly than at higher fields. This feature was observed in two samples that are believed to be heavily p-type; it is absent in an insulating sample. A similar effect was previously observed [20] in EuTe and attributed to BMP formation. In Cu_2MnGeS_4 the "polaron" portion of the magnetization curve corresponds to 4% of the saturation magnetization. That is the fraction of the Mn^{2+} ions, within acceptor orbits of radius 10 Å [15], whose density is p = 10^{19} acceptors/cc. The large hole concentrations suggested by these measurements could be a significant advantage of such materials.

Further studies of Cu_2MnGeS_4, including transport and magneto-optics are planned. However, in the long run it may be advantageous to develop MBE techniques for growing $(Cd_3Mn)Te_4$ in the stannite structure [15,21].

V. The Metal-Insulator Transition in p-HgMnTe

Magnetic fields cause a spectacular insulator-to-metal transition at low temperatures in p-HgMnTe [22]. In zero field, such crystals typically have resistivities exceeding 10^6 Ω-cm, whereas a few tesla can reduce their resistivities to 1-10 Ω-cm. Transport experiments [22] and renormalization group calculations suggest that this transition results from a combination of disorder and many-body effects. Magnetic fields reduce disorder by suppressing magnetization fluctuations, thereby driving the system into the metallic state.

In fields above 10 T, where fluctuations are suppressed and the magnetic length becomes comparable to the acceptor radius, one anticipates a second transition — back into the insulating state — caused by magnetic compression of the acceptor orbit. The anisotropic "pancake" shape of the acceptor state in p-HgMnTe [23] should make such effects especially prominent. WONG, FONER, and BECLA [24] have recently extended the magneto-transport experiments of WOJTOWICZ et al. [25] to 45 T to explore this regime. As indicated in Fig. 7, they observe sizable variations of the resistivity ratio, $(\rho_\perp/\rho_{||})$, with magnetic field. The resistivities of their samples reach minima near 20 T, and increase slowly thereafter. However, they are still far below those at B = 0.

Band structure calculations imply that the J = 3/2, J_z = -3/2 state remains uppermost in the valence band well beyond 50 T, in contrast to the situation in HgCdTe. Thus, we believe these experiments are probing the unique orbital state first proposed by MYCIELSKI and MYCIELSKI [23].

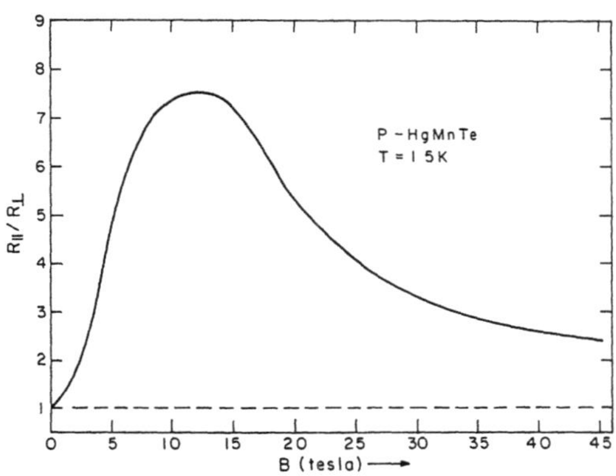

Fig. 7 Resistivity anisotropy vs. field in p-HgMnTe

VI. SFRS in n-CdFeSe

In a tetrahedral environment, the electronic structure of the Fe^{2+} ion is quite different from that of Mn^{2+}. The $(3d)^6$ configuration of Fe^{2+} has a 25-fold orbital and spin degeneracy that is drastically perturbed by crystal field and spin orbit interactions, whereas the $(3d)^5$ configuration of Mn^{2+} is an L=0 state with only spin degeneracy. LOW and WEGER have shown [26] that in zincblende Fe^{2+} has a singlet ground state with a low-lying triplet state; in wurtzite the excited state is further split into a singlet and doublet by the uniaxial crystal field. Thus, Fe-doped II-VI's are van Vleck paramagnets with sizable magnet susceptibilities, but no local moments at low temperature. Susceptibility measurements [27] substantiate these conclusions and yield singlet-triplet splittings in the 1-2 meV range.

HEIMAN et al. [28] have recently compared the SFRS spectra of n-CdFeSe to those of n-CdMnSe, and found them to be strikingly different from one another. Two key features of the SFRS spectrum of n-$Cd_{1-x}Mn_xSe$ — the zero field splitting and large, fluctuation-induced line width — are absent in n-$Cd_{1-x}Fe_xSe$. Figures 8 and 9 compare the Raman splittings and linewidths of similar $Cd_{1-x}Fe_xSe$ and $Cd_{1-x}Mn_xSe$ samples. Note that there is no zero field splitting in the former, and that its linewidth at low field is almost an order of magnitude smaller than that of $Cd_{1-x}Mn_xSe$. These observations imply an absence of magnetic fluctuations in CdFeSe — as would be anticipated in van Vleck systems with only field-induced moments — whereas fluctuations dominate the low field CdMnSe SFRS spectrum. The experiments also imply that there are no donor-BMP in CdFeSe.

The large width of the CdMnSe SFRS line is a result of magnetization fluctuations within the donor orbit; it is quantitatively explained by the theory [5,8]. On the other hand, in CdFeSe which has no local moments, the SFRS linewidth results from compositional fluctuations. A simple theory predicts $\Gamma \sim \Delta E \sim M$, as observed.

Finally, there is a sizable anisotropy (40%) of the susceptibility of CdFeSe. This effect results from the splitting of the excited triplet state of Fe^{2+} by the uniaxial crystal field. No anisotropy has been observed in Mn-based materials.

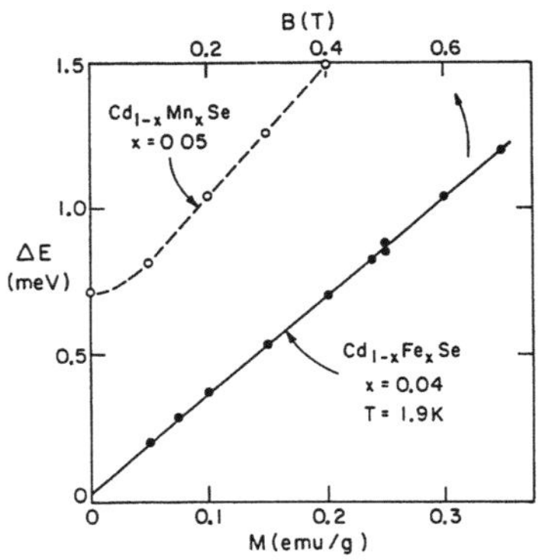

Fig. 8 Spin flip energy vs M. scale Field (B) is Fe sample; divide by 4 for Mn sample

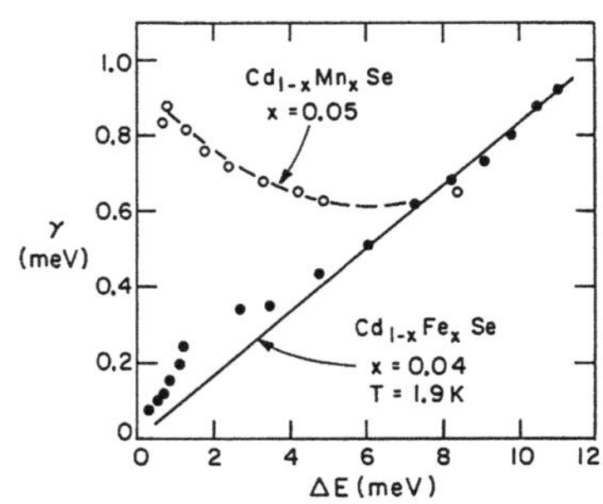

Fig. 9 Full width, half maximum (γ) vs. spin flip energy (ΔE)

This research is supported by NSF grant DMR-8504366 and by DARPA contract N00014-86-K-0760.

VII. Bibliography

1. Y. Shapira, S. Foner, D.H. Ridgely, K. Dwight, and A. Wold, Phys. Rev. B30, 4021 (1984); R.L. Aggarwal, S.N. Jasperson, Y. Shapira, S. Foner, T. Sakakibara, T. Goto, N. Miura, K. Dwight, and A. Wold, Proceedings of the 17th International Conference on the Physics of Semiconductors, San Francisco, 1984 (Springer-Verlag, New York, 1985).
2. Andrzej Mycielski, J Appl. Phys. 63, 3279 (1988).
3. P.A. Wolff, S.Y. Yuen, R.R. Galazka, and A. Mycielski, J. Vac. Sci. Tech. (in press).
4. M. Nawrocki, R. Planel, G. Fishman, and R.R. Galazka, Phys. Rev. Lett. 46, 735 (1981).
5. D. Heiman, P.A. Wolff, and J. Warnock, Phys. Rev. B27, 4848 (1983).
6. D. Alov, S. Gubarev, and V. Timofeev, Sov. Phys. JETP 57, 1052 (1983).
7. D. Peterson, A. Petrou, M. Datta, A. Ramdas, and S. Rodriguez, Solid State Commun. 43, 667 (1983).
8. T. Dietl and J. Spalek, Phys. Rev. Lett. 48, 355 (1982).
9. M. Sawicki, T. Dietl, J. Kossut, J. Igalson, T. Wojtowicz, and W. Plesiewicz, Phys. Rev. Lett. 56, 508 (1986).
10. M.A. Novak, O. G. Symko, D.J. Zheng, and S. Oseroff, Phys. Rev. B33, 6391 (1986).
11. E.D. Isaacs, D. Heiman, M.J. Graf, B.B. Goldberg, R. Kershaw, D. Ridgely, K. Dwight, A. Wold, J. Furdyna, and J.S. Brooks, Phys. Rev. B37, 7108 (1988).
12. E.D. Isaacs and D. Heiman, Rev. Sci. Instrum. 58, 1672 (1987).
13. E.D. Isaacs, D. Heiman, P. Becla, Y. Shapira, R. Kershaw, K. Dwight, and A. Wold, Phys. Rev. B (in press).
14. T. Nhung and R. Planel, Proceedings of the 16th International Conference on the Physics of Semiconductors, Montpellier, 1982, Physica 117B/118B, 488 (1983); T.H. Nhung, R. Planel C. Benoit a la Guillaume, and A. Bhattacharjee, Phys. Rev. B31, 2388 (1985).
15. P.A. Wolff, D. Heiman, E.D. Isaacs, P. Becla, S. Foner, L.R. Ram-Mohan, D.H. Ridgely, K. Dwight, and A. Wold, Proceedings of the Conference on High Magnetic Fields in Semiconductor Physics, Wurzburg, 1986 (Springer-Verlag, Berlin, 1987); P.A. Wolff and L.R. Ram-Mohan, Materials Research Society Symposia Proceedings, Vol. 89, Boston, 1986; L.R. Ram-Mohan and P.A. Wolff, Phys. Rev. B (in press). See also L. Swierkowski and T. Dietl, Acta Physica Polonica A73, 431 (1988).
16. M. Bugajski, P. Becla, P.A. Wolff, D. Heiman, and L.R. Ram-Mohan, Phys. Rev. B (to be published).
17. E.D. Isaacs, D. Heiman, X. Wang, P. Becla, T. Nakao, S. Takayama, and N. Miura (to be published).
18. E.D. Honig, K. Dwight, and A. Wold (to be published).
19. Y. Shapira, E.J. McNiff, Jr., N.F. Oliveira, Jr., E.D. Honig, K. Dwight, and A. Wold, Phys. Rev. B37, 411 (1988).
20. N.F. Oliveira, Jr., S. Foner, Y. Shapira, and T.B. Reed, Phys. Rev. B5, 2634 (1972).
21. D. Heiman, E.D. Isaacs, P. Becla, and S. Foner, Materials Research Society Symposia Proceedings, Vol. 89, Boston, 1986.
22. T. Wojtowicz, T. Dietl, M. Sawicki, W. Plesiewicz, and J. Jarozynski, Phys. Rev. 56, 2419 (1986), and references therein.
23. A. Mycielski and J. Mycielski, J. Phys. Soc. Japan 49, Suppl. A, 807 (1980).

24. S. Wong, S. Foner, and P. Becla (to be published).
25. T. Wojtowicz, T.R. Gawron, J.L. Robert, A. Raymond, C. Bousquet, and A. Mycielski, J. Cryst. Growth 72, 385 (1985).
26. W. Low and M. Weger, Phys. Rev. 118, 1119 (1960).
27. J.P. Mahoney, C.C. Lin, W.H. Brumage, and F. Dorman, J. Chem. Phys. 53, 4286 (1970).
28. D. Heiman, A. Petrou, S.H. Bloom, Y. Shapira, E.D. Isaacs, and W. Giriat, Phys. Rev. Lett. 60, 1876 (1988).

Recent High Magnetic Field Investigations of Semiconductors in Nijmegen

J.A.A.J. Perenboom and J. Singleton

High Field Magnet Laboratory and Research Institute for Materials, University of Nijmegen, Toernooiveld, NL-6525 ED Nijmegen, The Netherlands

Recent work on far-infrared and interband magneto-optical studies done at the High Field Magnet Laboratory of the University of Nijmegen will be reviewed.

1. Introduction

The University of Nijmegen contains one of the world's foremost magnet facilities. Two hybrid magnets give fields of 25 T and 30 T, and a variety of Bitter solenoids are used to provide up to 20 T [1]. The laboratory is equipped with both pulsed and continuous far-infrared lasers, and near-infrared and visible time-resolved (ps and sub-ps) luminescence measurements are undertaken. Cryogenic equipment includes a top-loading 150-mK dilution refrigerator for use in the 20-T magnet, and a variety of ^{3}He-cryostats for work down to 0.3 K: one of the latter can be used for optical measurements at 0.4 K.

In magnetotransport, RIKKEN *et al.* have studied the effects of macroscopic current redistribution in a quasi-two-dimensional electron gas (2DEG) where spatial inhomogeneities had been induced by differences in illumination or applied gate-voltage over parts of the samples [2]; the subKelvin temperatures have been exploited in the study of the influence of long-range impurities on the magnetoresistance in the 2DEG near integer and fractional quantum Hall plateaus reported by KOENRAAD *et al.* at this conference [3], and in a study of the anisotropic magnetoresistance of the 2D layered charge-transfer salt $\beta''-(BEDT-TTF)_2AuBr_2$ [4]: at temperatures below 2 K a number of Shubnikov-de Haas frequencies have been found, corresponding to small pockets in the Fermi surface of only a few percent of the room-temperature Brillouin zone area, and the strong reduction of the magnetoresistance in the high-conductivity a-direction, observed as a magnetic field is applied, points to the presence of linear scattering centers parallel to the a-axis, such as microcracks or domain boundaries. As the preliminary part of a wider study of II-VI and magnetic semiconductors, KUSTERS *et al.* have examined the properties of $Nd_{0.5}Pb_{0.5}MnO_3$ [5], this work will be discussed in Section 4. Our work in semiconductor physics chiefly concerns the effects of reduced dimensionality and the dynamics of carrier energy relaxation: in Sections 2 and 3 of this paper a review will be presented of some of our recent work in magneto-optics.

Springer Series in Solid-State Sciences, Vol. 87
High Magnetic Fields in Semiconductor Physics II Editor: G. Landwehr

2. Far-Infrared Spectroscopy

Application of a magnetic field B will lead to quantisation of the motion of charge carriers in the plane perpendicular to $\boldsymbol{B}$, and the energy levels can then be labeled as $E(N,k_z)=(N+\frac{1}{2})\hbar\omega_c+E_z$. Transitions between these Landau levels can be simply probed and observed in cyclotron resonance (CR). Figure 1 shows the dependence of the CR transition energy on the applied magnetic field, observed in epitaxially-grown bulk GaAs [6]. There is a significant deviation from the straight line representing the relation $\hbar\omega_c=e\hbar B/m_0^*$. This is partly due to band non-parabolicity and partly to coupling with the longitudinal optic phonon, especially so at energies approaching $\hbar\omega_{LO}$= 36.75 meV. In determining the strength of the polaron effect the contribution of the band non-parabolicity must be taken into account. The spin doublet splitting of the CR which can be resolved in ultra-pure GaAs does not depend on polaron coupling [7] and so allows a direct comparison with band-structure calculations to obtain the band non-parabolicity. As can be seen in Fig. 2, the measured anisotropy and magnitude of the spin doublet splitting are very well reproduced with the 5-band $\boldsymbol{k}\cdot\boldsymbol{p}$ model of ZAWADZKI *et al.* [8].

In the case of the 2DEG in GaAs-(Al,Ga)As heterojunctions, a number of new effects reveal themselves: the integer and fractional Quantum Hall Effects, and the formation of electric subbands $E_z=E_0$, E_1, *etc.* due to the confinement by the

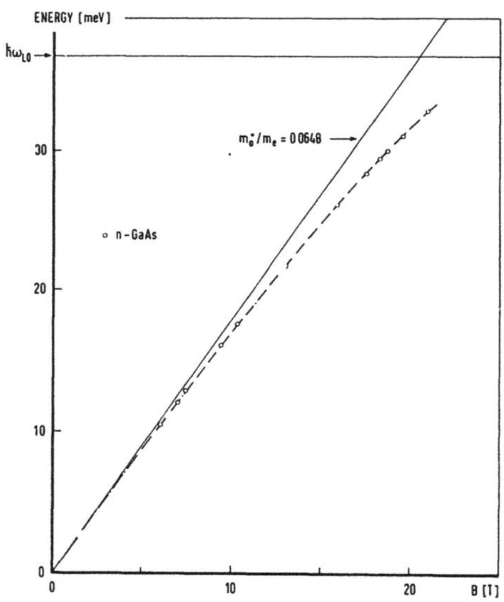

Figure 1. Dependence on magnetic field B of the cyclotron resonance transition energy in an epitaxially-grown n-GaAs sample. The deviation from the zero-order transition energy $\hbar\omega_c=e\hbar B/m_0^*$ is due to band non-parabolicity and polaron effects.

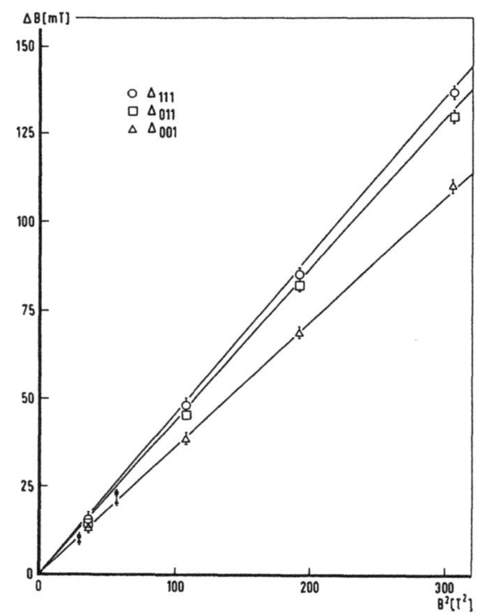

Figure 2. Values of the spin doublet splitting measured in ultra-pure n-GaAs in magnetic fields up to 20 T, and along the (001), (011) and (111) crystal directions. The lines represent results of five-level $\boldsymbol{k}\cdot\boldsymbol{p}$ model calculations.

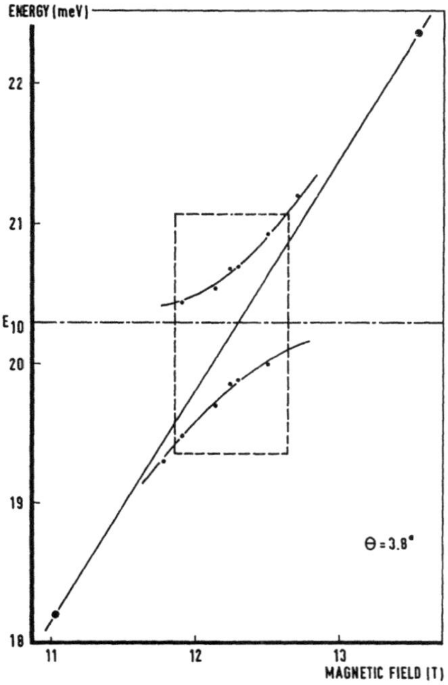

Figure 3. Cyclotron resonance transition field, observed in a high-mobility GaAs-(Al,Ga)As heterojunction, with the axis of the applied magnetic field tilted θ= 3.8° from the normal to the interface. The lifting of the degeneracy of the N=1 Landau level in the lowest electric subband and the N=0 Landau level in the second subband leads to the anticrossing at $E_{10}=E_1-E_0$.

potential at the interface. The confinement and filling factor effects will modify the strength of the polaron coupling, and resonant subband-Landau-level coupling is observed due to hybridisation of Landau levels belonging to different electric subbands when the applied magnetic field has a small component along the plane of the 2DEG [9,10]. This is illustrated in Fig. 3 where we have plotted the position of the CR observed when $\boldsymbol{B}$ is tilted 3.8° from the normal to the interface: the anticrossing allows to determine the subband separation E_{10} and so study for instance its variation upon changing of the electron concentration.

Much experimental and theoretical interest has been generated by the apparently contradictory reports of 'reduced' [6,11] and 'enhanced' [12,13] resonant polaron effects in the 2DEG. There have also been suggestions that coupling may occur to phonons with lower frequency than the LO-phonon involved in the bulk polaron effect [14,15]. In order to resolve some of these contradictions and to check which of the optic phonons is involved in the resonant polaron effect, we have performed cyclotron resonance on the well-characterised GaAs-(Ga,Al)As heterojunction system for 2D carrier densities N_s in the range 0.8×10^{11} cm^{-2} to 5.4×10^{11} cm^{-2}. No transmission signal can be seen in the Reststrahlen band of GaAs, and so reflectivity was used to detect the CR at energies up to 35.6 meV, only 1.1 meV below the LO-phonon energy.

Figure 4 shows the CR observed in these heterojunctions, and it is clear that the shape of the resonance is quite different for energies below or within the Reststrahlen band. Using a model calculation of the multilayer dielectric response

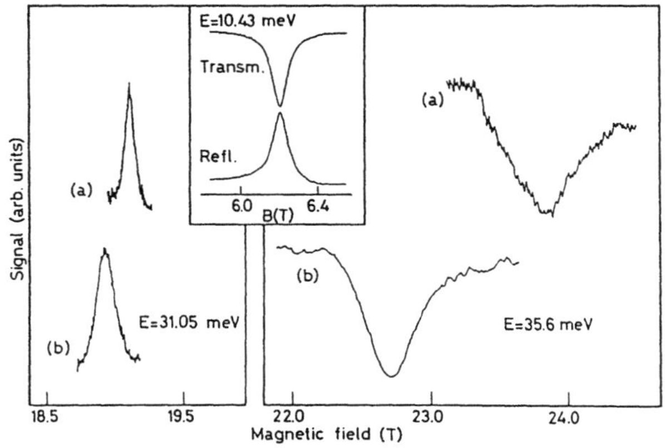

Figure 4. Cyclotron resonance in the reflectivity of a GaAs-(Al,Ga)As heterojunction observed at two different energies, below and within the Reststrahlen band respectively. The upper traces (a) are for a carrier density N_s of 3.4×10^{11} cm^{-2}, and the lower traces (b) for 5.4×10^{11} cm^{-2}. The inset shows simultaneously recorded reflection and transmission traces at low field.

of the heterojunction, we have been able to reproduce the observed waveforms and so accurately determine the effective mass m^* from the reflectivity results [16]. As is reported elsewhere in these proceedings [17], the resonant polaron effect was found to occur at the LO-phonon energy, in agreement with theoretical predictions: no evidence was found for coupling to lower energy phonons.

It is also very apparent from Fig. 4, that the CR observed at the smaller values of N_S is shifted to higher magnetic fields, indicating a strong carrier-density dependence of the strength of the polaron effect. The observed variation of m^* is shown in Fig. 5: It is seen that the effective mass (after subtraction of the

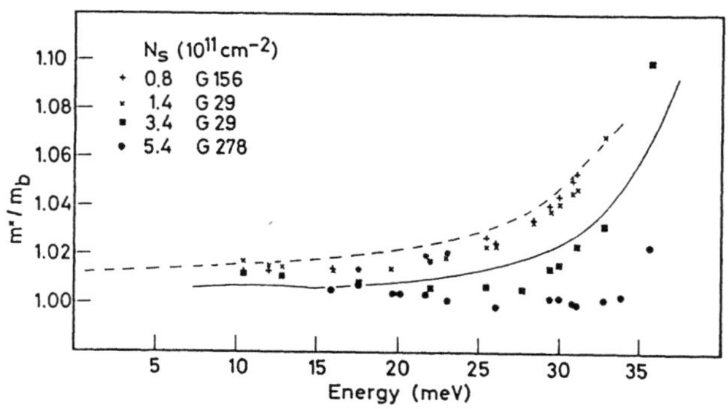

Figure 5. The normalised effective mass m^*/m_b as a function of energy, derived from the position of the cyclotron resonance for four different concentrations N_s of the 2D electron gas. The data for bulk GaAs are shown as a dash-dotted line. The solid line represents model calculations by PEETERS *et al.* (Ref. 18) for $N_s = 4.0\times10^{11}$ cm^{-2}.

non-parabolicity, by normalising on the band structure dependent value m_b) increases as the energy of the LO-optic phonon is approached, but this increase in effective mass becomes progressively less pronounced when we vary the carrier density from 0.8×10^{11} cm^{-2} to 5.4×10^{11} cm^{-2}. This is a manifestation of the reduction in the resonant polaron contribution to the effective mass. In the limit of low N_s, the non-resonant polaron contribution to the effective mass in 2D tends towards that in the bulk (dash-dotted line), but the resonant part is found to be slightly enhanced with respect to the bulk value. Both terms decrease strongly as N_s increases, due to occupation effects, and this has enabled most of the existing apparently contradictory results to be explained [16].

3. Time-Resolved Interband Spectroscopy

Quantum wires and quantum boxes hold great promise as (opto-)electronic devices, but few reliable structures for optical work have been produced [19,20]. However, the dispersion relationships and carrier relaxation rates of wires and boxes can be simulated by the application of magnetic fields to bulk semiconductors and quantum wells respectively: the magnetic field reduces the effective dimensionality of a system by two, so that the bulk density of states becomes one-dimensional in character, whilst that of a 2DEG becomes completely quantised. In the Nijmegen High-Field Magnet Laboratory, time-resolved luminescence measurements of carrier energy relaxation are carried out on such reduced-dimensionality systems.

In our set-up optical excitation is achieved with light pulses of only 2-ps duration from a dye laser which is synchronously pumped by a mode-locked Kr-ion laser with a repetition rate of 82 MHz [21]. The radiation is split into two pulse trains: one is focussed to a 15-μm diameter spot on the sample, the other is sent through a variable delay path. The luminescence from the sample is combined with the delayed probe pulse, and the spectral distribution of sum-frequencies generated by a $LiIO_3$ crystal is determined using a 1-m grating monochromator. The time resolution of the lightgate system is 5 ps. Presently a colliding-pulse ring-laser is under construction, giving 60-fs pulses, also in conjunction with the 25-T hybrid magnet.

Results on bulk GaAs, which showed a decrease in hot-carrier energy relaxation rate with increasing magnetic field [22], have now been extended to 5-nm wide GaAs-(Ga,Al)As quantum wells [23].

In Fig. 6 the luminescence spectra for the quantum well are shown as a function of energy and time for a magnetic field of 20 T, applied perpendicular to the well plane. The carriers were excited non-resonantly by 610-nm radiation from a rhodamine-6G dye laser. Landau-level structure is clearly observed in the spectra at times 75 ps after excitation. With increasing delay time the carriers cool down by phonon emission and recombine, which both lead to depopulation of the higher Landau levels, as shown by the decreasing spectral range of the subsequent spectra. Figure 6 also shows the results of model calculations of the luminescence, including inhomogeneous broadening due to well-width fluctuations, and the time-dependence of the carrier density, carrier effective temperature and homogeneous Landau-level width: such numerical calculations are very sensitive to the parameters chosen, and

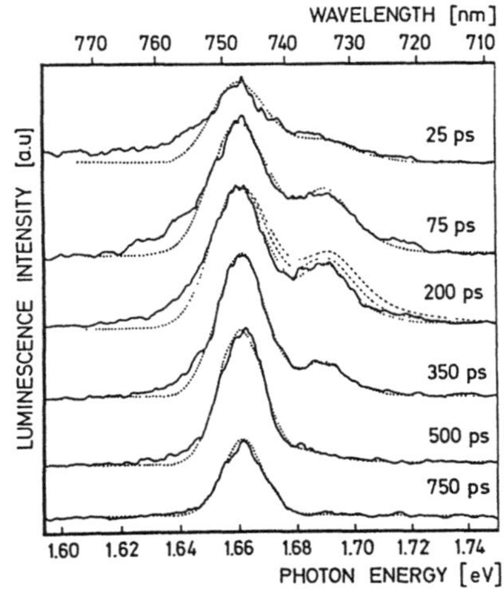

Figure 6. Luminescence spectra from a GaAs-(Al,Ga)As quantum well structure, measured at B=20 T and at different values of the delay time. The dotted lines represent a model calculation based on a thermal distribution over a Gaussian-broadened Landau-level density of states of the excited charge carriers, from which their temperature is deduced. The dashed lines at 200 ps delay time represent the calculated contours obtained with variations of 10 K in charge carrier temperature.

enable the carrier density and their temperature to be accurately deduced (see Fig. 7). The cooling rate of the charge carriers is found to decrease with applied magnetic field, as was observed in bulk GaAs: this effect is due to a reduction of phase-space available for optic-phonon emission associated with the Landau-level quantisation. Surprisingly, for high magnetic fields the cooling rate is enhanced. When the magnetic field is applied along the planes of the quantum well structure however, the confinement of the wavefunctions is due to the quantum well and not to the magnetic field, and no influence of the field is to be expected. This is clearly illustrated in Fig. 8, which shows the times taken for the carriers to cool to 100 K, derived from the fits to the experimental data.

It is evident that a) the carrier cooling rate in the quantum well is strongly reduced with respect to bulk GaAs (*c.f.* Ref. 22); and b) that the carrier cooling rate reduces at fields up to 8 T, but above that field increases again. As the initial cooling rate due to optic-phonon emission is much smaller in the quantum well than in the bulk [24,25], acoustic-phonon emission, usually ignored in the bulk, must be taken into account. REINEN *et al.* have recently shown that all of these results can be described by a model which includes the field dependence of the acoustic-phonon emission rate [26]: their model takes into account electron/polar-optic-phonon scattering electron/acoustic-deformation-potential scattering, and electron/piezo-electric-phonon scattering and allows to calculate the energy relaxation rates as a function of magnetic field, effective carrier temperature and density of photo-excited carriers; the model assumes that the acoustic phonons are in thermal equilibrium with the bath and allows a non-equilibrium distribution of optic phonons. It was found that the energy relaxation rate $<dE/dt>_{ac}$ due to electron/acoustic-deformation-potential scattering principally depends on the density of final states. In the extreme quantum limit, at high magnetic fields, only the lowest Landau level is occupied and the degeneracy of this level increases with B, and it can be seen in Fig. 9 that $<dE/dt>_{ac}$ increases significantly with decreasing filling factor ν [23,26].

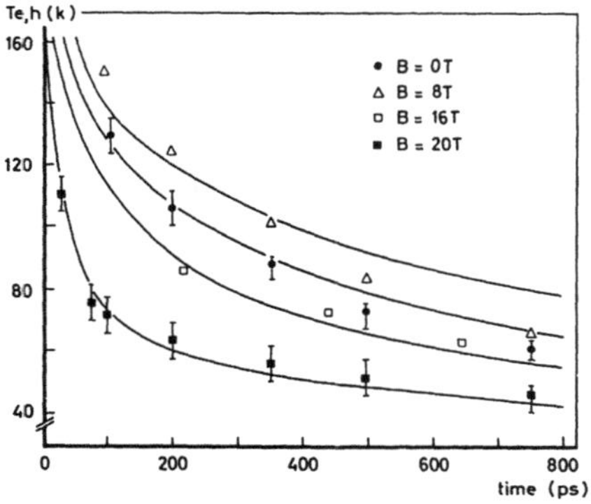

Figure 7. The temperature of the photo-excited charge carriers in a GaAs-(Al,Ga)As quantum well structure, as a function of time and measured for four different values of the applied magnetic field. For low magnetic fields, the cooling rate of the charge carriers is found to decrease (as for bulk GaAs). The strong increase of the cooling rate at higher fields is attributed to enhanced relaxation by acoustic phonon emission.

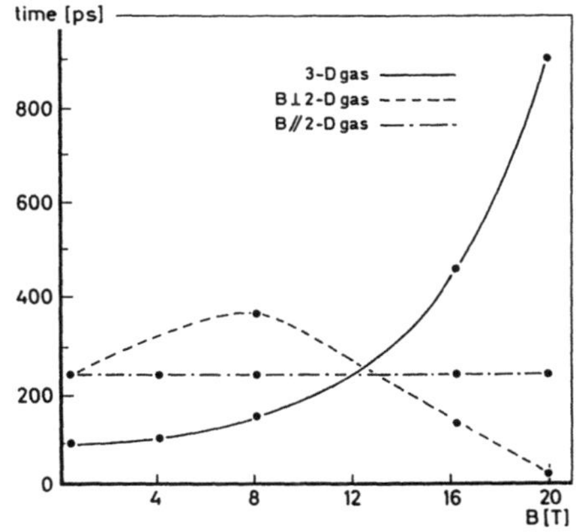

Figure 8. Time needed for the charge carriers to cool down to T=100 K. For bulk GaAs the cooling rate decreases with applied magnetic field. For GaAs-(Al,Ga)As quantum wells the cooling rate is reduced: with the magnetic field applied along the plane of the 2DEG, no field-effect is observed; with the magnetic field applied perpendicular to the 2DEG, the cooling rate initially decreases as in bulk GaAs, but increases strongly at higher fields.

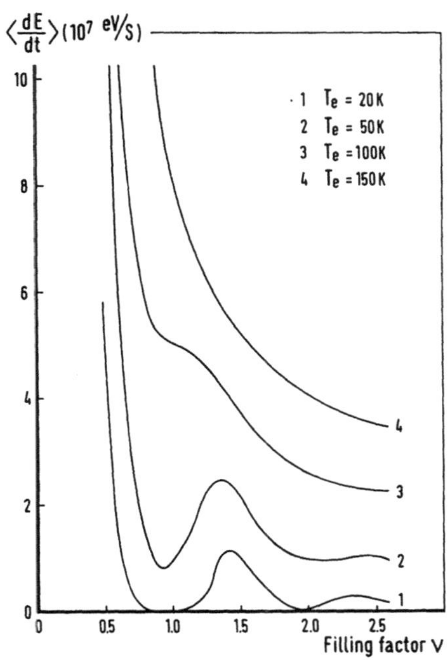

Figure 9. Model calculation of the variation of the energy relaxation rate $<dE/dt>_{ac}$ due to electron-acoustic deformation potential scattering in a 5-nm wide quantum well. Note the strong oscillations of the relaxation rate with filling factor ν and the increase in the extreme quantum limit when the degeneracy of the lowest Landau level increases with B.

Rather different results occurred when the carriers were resonantly excited to within three optic-phonon energies of the lowest heavy-hole to electron transition in a modulation-doped 5-nm wide quantum well. In addition to slow spontaneous emission processes with decay times ~200 ps, an initial, very rapid recombination process was found to take place: at zero magnetic field and a bath temperature of 4.2 K, the carriers within one optic-phonon energy from the 2D band-edge recombine within 15 ps, in accordance with previous results [27], and the carrier lifetime is found to increase with increasing magnetic field until a value of 65 ps is reached close to 12 T, followed by a slight decrease to 50 ps at 20 T. We believe that this effect mirrors the contribution of the different cooling mechanisms discussed above. At a higher bath temperature of 77 K, the carrier lifetime increases by a factor 2-3 at all fields, but the overall field-dependence remains the same. We have observed similar results for luminescence from a wide well surrounded by two superlattices, where the carriers were resonantly pumped into the superlattice mini-bands. Such ultra-short lifetimes have previously been attributed to a large stimulated emission rate [27], and BERENDSCHOT *et al.* show in another paper at this conference [28] that the variation in the carrier lifetime can be explained by the magnetic-field and temperature dependences of the time that the carriers take to cool down to a low-enough carrier temperature for stimulated emission to take place.

4. Magnetic Semiconductors

As the preliminary part of a wider study of II-VI and magnetic semiconductors, the properties of $Nd_{0.5}Pb_{0.5}MnO_3$ have been examined [5]. The substance is a distorted cubic perovskite, and conduction occurs due to the mixed Mn^{3+}/Mn^{4+} valence. $Nd_{0.5}Pb_{0.5}MnO_3$ shows, at zero field, a large peak in the resistivity as a function of temperature at 185 K, just above the temperature at which the Mn sublattice orders ferromagnetically: below this temperature the resistance falls sharply by two orders of magnitude. In the range between 190 K and 300 K, the resistivity varies as $\varrho=\varrho_0 e^{E_a/kT}$, with an activation energy of E_a = 95±5 meV. The application of a magnetic field leads to a rapid decrease in resistivity at all temperatures: the largest relative negative magnetoresistance occurs close to the peak in the zero-field resistivity. As the magnetic field is increased, the peak in the resistivity as a function of temperature becomes smaller and moves to higher temperature. Hall data indicate that the carrier density in the sample below 100 K is roughly constant at about 5×10^{20} cm^{-3}, giving a zero-field mobility at 100 K of ~1 cm^{2}V^{-1}s^{-1}. At temperatures close to and above the peak in the resistivity, the Hall data imply a mobility of 0.1 cm^{2}V^{-1}s^{-1}. The free-carrier density must therefore vary as a function of magnetic field and temperature in order to give the observed variation in resistivity. At the highest fields ($\geqslant$10 T), the low temperature carrier density of 5×10^{20} cm^{-3} is recovered. All of these features are indicative of conduction by magnetic polarons [29,30,31].

Magnetic polarons are localised as a consequence of magnetic interactions or at impurities, so that conduction proceeds via thermal hopping. This explains the activated behaviour of the resistivity between 190 K and 300 K, and the apparent variable number of carriers observed in the Hall data. Magnetic polarons can only exist over a limited range of temperatures and magnetic fields close to the onset of ferromagnetism, and their formation will become impossible if the magnetic ions are ferromagnetically ordered, so that the hopping conduction behaviour will disappear below the onset of order: this explains the rapid fall-off in resistivity observed below this point [29]. Likewise, an external magnetic field will increase the ferromagnetic order in the crystal, so that magnetic polaron formation will be inhibited [30,31]. This is observed as the very prominent negative magnetoresistance in the region 180 K - 300 K. These observations are supported by spin-polarised neutron scattering data on the same set of samples, which have revealed magnetic polaron hopping activation energies and mobilities close to those derived from the magnetotransport data.

Acknowledgement

The authors acknowledge stimulating discussions with T.T.J.M. Berendschot, R.M. Kusters, C.J.G.M. Langerak, F.L. Pratt, H.A.J.M. Reinen, and H. Sigg. Part of this work was supported by the Stichting voor Fundamenteel Onderzoek der Materie (F.O.M.) with financial support from the Nederlandse Organisatie voor Wetenschappelijk Onderzoek (N.W.O.).

References

1. K. van Hulst and J.A.A.J. Perenboom: IEEE Trans. Magn. MAG-**24**, 1397 (1988)
2. G.L.J.A. Rikken, J.A.M.M. van Haaren, A.P. van Gelder, H. van Kempen, P. Wyder, H.-U. Habermeier and K. Ploog: Phys. Rev. B **37**, 10229 (1988)
3. P.M. Koenraad, F.A.P. Blom, J.P. Cuypers, C.T. Foxon, J.A.A.J. Perenboom, S.J.R.M. Spermon and J.H. Wolter: these proceedings: P10
4. F.L. Pratt, W. Hayes, A.J. Fisher, J. Singleton, S.J.R.M Spermon, M. Kurmoo and P. Day: Proc. Conf. Synthetic Metals, Santa Fe, July 1988, to be published
5. R.M. Kusters, J. Singleton, D.A. Keen, R. McGreevy and W. Hayes: Proc. 2nd Int. Symp. on High Field Magnetism, Leuven, July 1988, to be published in Physica B+C
6. H. Sigg, P. Wyder and J.A.A.J. Perenboom: Phys. Rev. B **31**, 5253 (1985)
7. H. Sigg, J.A.A.J. Perenboom, P. Pfeffer and W. Zawadzki: Solid State Commun. **61**, 685 (1987)
8. W. Zawadzki, P. Pfeffer and H. Sigg: Solid State Commun. **53**, 777 (1985)
9. H. Sigg, C.J.G.M. Langerak and J.A.A.J. Perenboom: Springer Series in Solid State Sciences **71**, 248 (1987)
10. G.L.J.A. Rikken, H. Sigg, C.J.G.M. Langerak, H.W. Myron, J.A.A.J. Perenboom and G. Weimann: Phys. Rev. B **34**, 5590 (1986)
11. M. Ziesmann, D. Heitmann and L.L. Chang: Phys. Rev. B **35**, 4541 (1987)
12. M. Horst, U. Merkt and J.P. Kotthaus: Phys. Rev. Lett. **50**, 754 (1983)
13. J. Singleton, R.J. Nicholas and F. Nasir: Solid State Commun. **58**, 833 (1986)
14. M.A. Brummell, R.J. Nicholas, M.A. Hopkins, J.J. Harris and C.T. Foxon: Phys. Rev. Lett. **58**, 77 (1987)
15. R.J. Nicholas, L.C. Brunel, S. Huant, K. Karrai, J.C. Portal, M.A. Brummell, M. Razeghi, K.Y. Cheng and A.Y. Cho: Phys. Rev. Lett. **55**, 883 (1985)
16. C.J.G.M. Langerak, J. Singleton, P.J. van der Wel, J.A.A.J. Perenboom, D.J. Barnes, R.J. Nicholas, M.A. Hopkins and C.T.B. Foxon: to be published
17. C.J.G.M. Langerak, J. Singleton, D.J. Barnes, P.J. van der Wel, R.J. Nicholas, M.A. Hopkins, J.A.A.J. Perenboom and C.T.B. Foxon: these proceedings: P9

18. F.M Peeters, Wu Xiaoguang and J.T. Devreese: Surf. Sci. **196**, 437 (1988)
19. L. Esaki: IEEE J. Quantum Electron. QE-**22**, 1611 (1986)
20. Y. Arakawa and A. Yariv: IEEE J. Quantum Electron. QE-**22**, 1887 (1986)
21. R.W.J. Hollering, T.T.J.M. Berendschot, H.J.A. Bluyssen, P.Wyder, M. Leys and J. Wolter: Solid State Commun. **57**, 527 (1986)
22. R.W.J. Hollering, T.T.J.M. Berendschot, H.J.A. Bluyssen, H.A.J.M. Reinen and P. Wyder: Springer Series in Solid State Sciences **71**, 466 (1987)
23. R.W.J. Hollering, T.T.J.M. Berendschot, H.J.A. Bluyssen, H.A.J.M. Reinen and P. Wyder: Phys. Rev. B (to be published)
24. J. Shah, A. Pinczuck, A.C. Gossard and W. Wiegmann: Phys. Rev. Lett. **54**, 2045 (1985)
25. W.W. Rühle and H.J. Polland: Phys. Rev. B **36**, 1683 (1987)
26. H.A.J.M. Reinen, T.T.J.M. Berendschot, R.J.H. Kappert and H.J.A. Bluyssen: Solid State Commun. **65**, 1495 (1988)
27. J. Dubard, J.L. Oudar, F. Alexandre, D. Hulin and A. Orszag: Appl. Phys. Lett. **50**, 821 (1987)
28. T.T.J.M. Berendschot, H.A.J.M. Reinen, P.C.M. Christianen, H.J.A. Bluyssen and H.P. Meier: these proceedings: P58
29. Y. Shapira, S. Foner, N.F. Oliveira and T.B. Reed: Phys. Rev. B **10**, 4765 (1974)
30. S. Washburn, R.A. Webb, S. von Molnar, F. Holtzberg, G. Flouquet and G. Remenyi: Phys. Rev. B **30**, 6224 (1984)
31. M. Sawicki, T. Dietl, J. Kossut, J. Igalson and W. Plesiewicz: Phys. Rev. Lett. **56**, 508 (1986)

Index of Contributors

Springer Series in Solid-State Sciences

Editors: M. Cardona P. Fulde K. von Klitzing H.-J. Queisser

1 **Principles of Magnetic Resonance** 3nd Edition By C. P. Slichter
2 **Introduction to Solid-State Theory** 2nd Printing. By O. Madelung
3 **Dynamical Scattering of X-Rays in Crystals** By Z. G. Pinsker
4 **Inelastic Electron Tunneling Spectroscopy** Editor: T. Wolfram
5 **Fundamentals of Crystal Growth I** Macroscopic Equilibrium and Transport Concepts. 2nd Printing By F. Rosenberger
6 **Magnetic Flux Structures in Superconductors** By R. P. Huebener
7 **Green's Functions in Quantum Physics** 2nd Edition By E. N. Economou
8 **Solitons and Condensed Matter Physics** 2nd Printing Editors: A. R. Bishop and T. Schneider
9 **Photoferroelectrics** By V. M. Fridkin
10 **Phonon Dispersion Relations in Insulators** By H. Bilz and W. Kress
11 **Electron Transport in Compound Semiconductors** By B. R. Nag
12 **The Physics of Elementary Excitations** By S. Nakajima, Y. Toyozawa, and R. Abe
13 **The Physics of Selenium and Tellurium** Editors: E. Gerlach and P. Grosse
14 **Magnetic Bubble Technology** 2nd Edition By A. H. Eschenfelder
15 **Modern Crystallography I** Symmetry of Crystals. Methods of Structural Crystallography By B. K. Vainshtein
16 **Organic Molecular Crystals** Their Electronic States. By E. A. Silinsh
17 **The Theory of Magnetism I** Statics and Dynamics. 2nd Edition By D. C. Mattis
18 **Relaxation of Elementary Excitations** Editors: R. Kubo and E. Hanamura
19 **Solitons,** Mathematical Methods for Physicists. 2nd Printing By G. Eilenberger
20 **Theory of Nonlinear Lattices** 2nd Edition By M. Toda
21 **Modern Crystallography II** Structure of Crystals. By B. K. Vainshtein, V. M. Fridkin, and V. L. Indenbom
22 **Point Defects in Semiconductors I** Theoretical Aspects By M. Lannoo and J. Bourgoin
23 **Physics in One Dimension** Editors: J. Bernasconi, T. Schneider
24 **Physics in High Magnetic Fields** Editors: S. Chikazumi and N. Miura
25 **Fundamental Physics of Amorphous Semiconductors** Editor: F. Yonezawa
26 **Elastic Media with Microstructure I** One-Dimensional Models. By I. A. Kunin
27 **Superconductivity of Transition Metals** Their Alloys and Compounds. By S. V. Vonsovsky, Yu. A. Izyumov, and E. Z. Kurmaev
28 **The Structure and Properties of Matter** Editor: T. Matsubara
29 **Electron Correlation and Magnetism in Narrow-Band Systems** Editor: T. Moriya
30 **Statistical Physics I** 2nd Edition By M. Toda, R. Kubo, N. Saito
31 **Statistical Physics II** By R. Kubo, M. Toda, N. Hashitsume
32 **Quantum Theory of Magnetism** By R. M. White
33 **Mixed Crystals** By A. I. Kitaigorodsky
34 **Phonons: Theory and Experiments I** Lattice Dynamics and Models of Interatomic Forces. By P. Brüesch
35 **Point Defects in Semiconductors II** Experimental Aspects By J. Bourgoin and M. Lannoo
36 **Modern Crystallography III** Crystal Growth 2nd Edition By A. A. Chernov
37 **Modern Crystallography IV** Physical Properties of Crystals By L. A. Shuvalov
38 **Physics of Intercalation Compounds** Editors: L. Pietronero and E. Tosatti
39 **Anderson Localization** Editors: Y. Nagaoka and H. Fukuyama
40 **Semiconductor Physics** An Introduction 3rd Edition By K. Seeger
41 **The LMTO Method** Muffin-Tin Orbitals and Electronic Structure By H. L. Skriver
42 **Crystal Optics with Spatial Dispersion, and Excitons** By V. M. Agranovich and V. L. Ginzburg
43 **Resonant Nonlinear Interactions of Light with Matter** By V. S. Butylkin, A. E. Kaplan, Yu. G. Khronopulo, and E. I. Yakubovich
44 **Elastic Media with Microstructure II** Three-Dimensional Models By I. A. Kunin
45 **Electronic Properties of Doped Semiconductors** By B. I. Shklovskii and A. L. Efros
46 **Topological Disorder in Condensed Matter** Editors: F. Yonezawa and T. Ninomiya
47 **Statics and Dynamics of Nonlinear Systems** Editors: G. Benedek, H. Bilz, and R. Zeyher
48 **Magnetic Phase Transitions** Editors: M. Ausloos and R. J. Elliott
49 **Organic Molecular Aggregates,** Electronic Excitation and Interaction Processes Editors: P. Reineker, H. Haken, and H. C. Wolf

MIX
Papier aus verantwortungsvollen Quellen
Paper from responsible sources
FSC® C105338

If you have any concerns about our products,
you can contact us on
ProductSafety@springernature.com

In case Publisher is established outside the EU,
the EU authorized representative is:
Springer Nature Customer Service Center GmbH
Europaplatz 3, 69115 Heidelberg, Germany

Printed by Libri Plureos GmbH
in Hamburg, Germany